OPTIMIZATION

OPTIMIZATION

Foundations and Applications

RONALD E. MILLER

A Wiley-Interscience Publication

JOHN WILEY & SONS, INC.

New York • Chichester • Weinheim • Brisbane • Singapore • Toronto

For ordering and customer service, call 1-800-CALL-WILEY.

Library of Congress Cataloging-in-Publication Data:

Miller, Ronald E.
    Optimization : Foundations and applications / Ronald E. Miller.
        p.   cm.
    "A Wiley-Interscience publication."
    Includes bibliographical references and index.
    ISBN 0-471-32242-3 (cloth : alk. paper). — ISBN 0-471-35169-5 (pbk. : alk. paper)
        1. Mathematical optimization.   I. Title.
    QA402.5.M553   2000
    519.3—dc21   99-21921

Printed in the United States of America

10 9 8 7 6 5 4 3 2

For M.G.B.

CONTENTS

PREFACE

The material covered in this book is intended for students from and practitioners in a wide variety of disciplines outside the natural sciences in which quantitative analysis has come to play an increasingly important role. I have in mind many of the social sciences—regional science, sociology, geography, political science, and economics (although there are a number of mathematical foundations texts designed expressly for economists)—as well as areas such as management, information and decision sciences, city and regional planning, urban studies, public policy analysis, and certain kinds of engineering programs (especially those offered in conjunction with arts and sciences curricula in such cross-discipline fields as environmental studies).

Mathematical modeling of human activity has played an increasingly important role in these disciplines. In conjunction with the growth in power and accessibility of computers, analysts find themselves facing problems in which the aim is to find the "best" way to accomplish some objective, usually with the complication of constraints on the available choices. Approaches to these kinds of problems constitute the subject matter of this book.

I have tried whenever possible to accompany algebraic results with geometric interpretations in an effort to emphasize the intuitive appeal of the mathematics, even though this means that the illustrations must be confined to only two or three variables, so as to be able to use the geometry of two or three dimensions. Mathematical details that I believe to be important for a thorough understanding of a topic but that may not be of interest to all readers are given in end-of-chapter appendixes. Many texts cover one or more of the topics in this book in much more mathematical and abstract detail—using theorems, lemmas, and so on, and including proofs. In contrast, I have tried whenever possible to employ more informal arguments.

I have intentionally not directed the materials at any particular discipline, so that, for example, a reader whose field is management science will not be overpowered by examples from economics. Also, in a classroom environment, instructors can introduce illustrations and examples that are appropriate to their own disciplines and to the makeup of the classes that they are teaching.

Many readers will be familiar with some or much of the material in the Foundations sections (Parts I and II), and may find these chapters to be unnecessary or at most to serve as a review to be gone over rather quickly. Part I (Foundations: Linear Models) covers linear algebra and the analysis and solution of linear equation systems. Part II (Foundations: Nonlinear Models) contains classical calculus results for problems requiring unconstrained and equality constrained maximization and minimization of functions of one or of several variables. It also introduces the special problems caused by inequality constraints.

Part III (Applications: Iterative Methods for Nonlinear Problems) is the first of three Applications sections. Here, iterative methods for root finding and for unconstrained maximization and minimization are explored. This material is much less likely to be familiar to many readers. In solving one or more nonlinear equations (root finding, Chapter 5), we find a striking contrast with the straightforward and exact methods that can be used with linear equations (Section 1.3 and Chapter 2). In the case of unconstrained maximization and minimization (Chapter 6), the procedures described are representative of the kinds of techniques that are actually used in most computer programs for finding maxima and minima.

I have included this material in response to student requests, which have increased over the years, for help in understanding the descriptions of iterative methods that are given in computer software manuals (and in advertising for those software packages). With regard to equations, you can read that "the algebraic equation solver (root finder) contains six numerical methods including bisection and Newton–Raphson" or "the *root* function solves for zeros using the secant method." In optimization packages, you might learn that "a BFGS (Broyden–Fletcher–Goldfarb–Shanno) quasi-Newton algorithm is used that updates the Hessian at each step," or the "classical Levenberg–Marquardt method" is embedded in these programs, or the "Polak–Ribiere version of the conjugate gradient descent method" is employed, or "the descent method is Nelder–Mead simplex search." The material in Part III is designed to help you understand references like these.

Part IV (Applications: Constrained Optimization in Linear Models) explores the inequality constrained linear programming problem. The basics are covered in Chapter 7 (simplex method, duality, sensitivity), and some refinements and extensions, including transportation and assignment problems, integer linear programming, hierarchical optimization, and linear goal programming, are explored in Chapter 8. Chapter 9 contains an introduction to the relatively newer subject of interior point methods for linear programs.

Finally, in Part V (Applications: Constrained Optimization in Nonlinear Models) the complexities of inequality constrained maximization and minimization in nonlinear problems (introduced briefly in Section 4.5) are placed squarely in the nonlinear programming arena. The basics (such as Kuhn–Tucker conditions and constraint qualifications) are contained in Chapter 10, and refinements, including nonlinear duality and several alternative kinds of computational methods, are explored in Chapter 11. Much of this material can be viewed as an extension, to the inequality constrained case, of the kinds of iterative methods for unconstrained maximization and minimization that were examined in Chapter 6.

Final answers (without all the steps to solution) to all problems in the book are available as Adobe Acrobat pdf files at the accompanying FTP site:

ftp://ftp.wiley.com/public/sci_tech_med/optimization

Users will need the free Adobe Acrobat Reader to view the files.

This material might serve as the basis for a one-semester or a two-semester course, depending in part on prerequisites. For example, one semester's worth of linear programming might be based on Chapters 1, 2, and 7–9. (If the interior point methods of Chapter 9 were to be stressed, then Chapter 6—and possibly 5—would also be appropriate.) A one-semester course on nonlinear optimization would depend more on Section 1.6; Chapters 3 and 4 (perhaps all presumed prerequisites); then Chapters 5, 6, 10, and 11. I would also hope that the material might serve as an appropriate reference for those who encounter optimization problems in government or consulting positions.

Conspicuously absent in this text is any discussion of what are generally known as dynamic optimization techniques: calculus of variations, the maximum principle, optimal control theory, and dynamic programming. This material, which deals with a rather different set of optimization issues, is inherently more complicated. It addresses problems that are perhaps less generally met in practice in the disciplines to which this book is primarily directed, and its inclusion would have made an already long book much longer.

I have used this material, packaged in various ways, in teaching both undergraduates and graduate students from a variety of disciplines over some three decades at the University of Pennsylvania. In deciding what to include in this book, and how to include it, I have been guided by their questions over the years, both in the classroom and when I have been a member of M.A. thesis and Ph.D. dissertation committees. I am deeply grateful to those students for their contributions.

RONALD E. MILLER

PART I

FOUNDATIONS: LINEAR METHODS

In this part we review a number of elementary but essential definitions and operations in matrix algebra. Matrix representations and manipulations are the basic tools with which we can examine solutions to systems of linear equations. Once solution possibilities for linear equation systems are understood, it is easy to explore the ramifications of systems of linear inequalities. And systems of linear equations or linear inequalities are very often an important part of mathematical models of real-world problems. In addition, particular kinds of nonlinear functions called *quadratic forms* have a straightforward matrix representation. This representation, in turn, serves as a foundation for sets of rules for determining the sign of a quadratic form, and these sign rules on quadratic forms play an important role in distinguishing maxima from minima using classical derivative-based reasoning from differential calculus

1

MATRIX ALGEBRA

This chapter introduces notation that encompasses groups or collections of elements (usually numbers) and presents for these groups of elements a series of definitions and operations that in many ways parallel those for ordinary numbers. The definitions and the rules for operating on such ordinary numbers are collectively referred to as an *algebra.* The groups of numbers that are of interest in this chapter are called *matrices*— a single group is a matrix— and so we have an algebra of matrices. These rules for defining and manipulating matrices are absolutely essential for an understanding and application of a wide variety of *linear models* that are used in the analysis of real-world problems. One of the most popular of these linear models is linear programming, an optimization technique; we will examine how the fundamentals discussed in this and the next chapter relate to that approach in Part IV. Additionally, however, we will find that matrix representations and concepts are basic to an understanding of mathematical principles that underpin a great many nonlinear models as well. For example, some of the rules for distinguishing maximum points from minimum points in nonlinear problems use definitions from matrix algebra. This will become clear in other parts of this book.

Matrices are vitally important to the representation and analysis of systems of linear equations, which are often a basic component of models (both linear and nonlinear) of real-world problems. In fact, matrix algebra is also often known as *linear algebra,* precisely because it offers a convenient way of representing, analyzing, and solving systems of linear equations. We can often learn a great deal about a problem—the kinds of solutions that it has, or, indeed, whether it has any solutions at all—by studying the structure and properties of its matrix representation.

The fundamental definitions and operations of matrix algebra, and their interpretation, are the subject matter of Chapter 1. In Chapter 2 we examine linear equation systems in light of this terminology.

1.1   MATRICES: DEFINITION AND GENERAL REPRESENTATION

A *matrix* is a collection of *elements* arranged in a grid—a pattern (tabular array) of rows and columns. Matrices are defined in this "rectangular" way so that they can be used to represent systems of linear relations among variables. In all cases that will be of interest in this book, the elements will be either known numbers or unknown numbers whose values we want to determine.

The general case is a matrix with m rows and n columns. It is customary to use an uppercase (bold) letter to denote a matrix and the same lowercase letter, with two subscripts, to indicate its elements. If $m = 2$ and $n = 3$, and using double-subscript notation, a_{ij}, to denote the element in row i and column j of the matrix, we have

$$\mathbf{A} = \begin{bmatrix} a_{11} & a_{12} & a_{13} \\ a_{21} & a_{22} & a_{23} \end{bmatrix}$$

This matrix could also be represented as

$$\mathbf{A} = [a_{ij}] \qquad (i = 1,2; j = 1,2,3)$$

A particular example of such a matrix is

$$\mathbf{M} = \begin{bmatrix} 2 & 1 & 3 \\ 4 & 6 & 12 \end{bmatrix}$$

These are said to be 2×3 (read "2 by 3") matrices or matrices of *dimension* 2×3. The number of rows is always indicated first.

When $m = n$, the matrix is *square*. If $m = 1$ (a matrix with only one row), it is called a *row vector;* if $n = 1$ (a matrix with only one column), it is called a *column vector*. As an illustration, a specific 1×4 matrix $\mathbf{R}$ (a row vector) is[1]

$$\mathbf{R} = [20 \quad 16 \quad 3 \quad 62]$$

and a particular 3×1 column vector $\mathbf{C}$ is

$$\mathbf{C} = \begin{bmatrix} 5 \\ 10 \\ 15 \end{bmatrix}$$

If both $m = 1$ and $n = 1$, we have a matrix containing only one element. For the materials in this book, it will not be useful or necessary to make a distinction between a single number and a 1×1 matrix.

1.2   THE ALGEBRA OF MATRICES

1.2.1   Matrix Operations: Addition and Subtraction

Addition   Addition of matrices, represented as $\mathbf{A} + \mathbf{B}$, is accomplished by the simple rule of adding elements *in corresponding positions;* for this reason, it is often referred to as *el-*

---

[1]Sometimes commas are used between the elements in a row vector to avoid confusion, so we could write $\mathbf{R} =$ [20,16,3,62].

ement-by-element addition. This means $a_{ij} + b_{ij}$, for all i and j; and this, in turn, means that only matrices that have exactly the same dimensions can be added. (This is known as the *conformability requirement for addition.*) Given **M**, above, if $\mathbf{N} = \begin{bmatrix} 1 & 2 & 3 \\ 3 & 2 & 1 \end{bmatrix}$, then their sum, $\mathbf{S} = \mathbf{M} + \mathbf{N}$, will be another 2×3 matrix, where

$$\mathbf{S} = \mathbf{M} + \mathbf{N} = \begin{bmatrix} (2+1) & (1+2) & (3+3) \\ (4+3) & (6+2) & (12+1) \end{bmatrix} = \begin{bmatrix} 3 & 3 & 6 \\ 7 & 8 & 13 \end{bmatrix}$$

Subtraction   Subtraction is defined in a completely parallel way, namely, subtraction of elements *in corresponding positions* (element-by-element subtraction), so again only matrices of exactly the same dimensions can be subtracted. For example, the difference matrix, $\mathbf{D} = \mathbf{M} - \mathbf{N}$, will be another 2×3 matrix, where

$$\mathbf{D} = \mathbf{M} - \mathbf{N} = \begin{bmatrix} (2-1) & (1-2) & (3-3) \\ (4-3) & (6-2) & (12-1) \end{bmatrix} = \begin{bmatrix} 1 & -1 & 0 \\ 1 & 4 & 11 \end{bmatrix}$$

1.2.2   Matrix Definitions: Equality and the Null Matrix

Equality   The notion of equality of two matrices is also very straightforward. Two matrices are equal if they have the same dimensions and if elements in corresponding positions are equal (element-by-element equality). So $\mathbf{A} = \mathbf{B}$ when $a_{ij} = b_{ij}$ for all i and j.

The Null Matrix   A *zero* in ordinary algebra is the number that, when added to (or subtracted from) a number, leaves that number unchanged.[2] The parallel notion in matrix algebra is a *null* (or *zero*) *matrix*, defined simply as a matrix containing only zeros. Suppose

$$\mathbf{0} = \begin{bmatrix} 0 & 0 & 0 \\ 0 & 0 & 0 \end{bmatrix}$$

Then it is obvious that $\mathbf{M} + \mathbf{0} = \mathbf{M} - \mathbf{0} = \mathbf{M}$.

1.2.3   Matrix Operations: Multiplication

Multiplication of a Matrix by a Number   If a matrix is multiplied by a number (called a *scalar* in matrix algebra), each element in the matrix is simply multiplied by that number. For example, using the matrices **M**, **R** and **C** from above, we obtain

$$2\mathbf{M} = \begin{bmatrix} 4 & 2 & 6 \\ 8 & 12 & 24 \end{bmatrix}; \quad (-1.5)\mathbf{R} = [-30 \quad -24 \quad -4.5 \quad -93]; \quad (0.2)\mathbf{C} = \begin{bmatrix} 1 \\ 2 \\ 3 \end{bmatrix}$$

Multiplication of a Matrix by Another Matrix   Unlike addition, subtraction, equality, and the notion of a zero element, multiplication of two matrices is not a straightforward and logical extension of multiplication in ordinary algebra. In fact, multiplication of two

---

[2]Officially, this is known as the *identity element* for addition (or subtraction).

matrices is defined in what appears at first to be a completely illogical way. But we will see that the reason for the definition is precisely because of the way in which matrix notation is used for systems of linear relations, especially linear equations.

Using **M**, again, and a 3 ×3 matrix $\mathbf{Q} = \begin{bmatrix} 2 & 0 & 4 \\ 1 & 1 & 2 \\ 3 & 4 & 5 \end{bmatrix}$, the product $\mathbf{P} = \mathbf{MQ}$ is a 3×3 matrix that is found as

$$\mathbf{P} = \mathbf{MQ} = \begin{bmatrix} 2 & 1 & 3 \\ 4 & 6 & 12 \end{bmatrix} \begin{bmatrix} 2 & 0 & 4 \\ 1 & 1 & 2 \\ 3 & 4 & 5 \end{bmatrix} = \begin{bmatrix} 14 & 13 & 25 \\ 50 & 54 & 88 \end{bmatrix} \tag{1.1}$$

which comes from

$$\begin{bmatrix} (4 + 1 + 9) & (0 + 1 + 12) & (8 + 2 + 15) \\ (8 + 6 + 36) & (0 + 6 + 48) & (16 + 12 + 60) \end{bmatrix}$$

The rule is that for each element, p_{ij}, in the product, go *across row i* in the matrix on the left (here **M**) and *down column j* in the matrix on the right (here **Q**), multiplying pairs of elements and summing. So, for p_{23} we take the elements in row 2 of **M**, [4   6   12], and match them with the elements in column 3 of **Q**, $\begin{bmatrix} 4 \\ 2 \\ 5 \end{bmatrix}$, multiplying and adding together; this gives $(4)(4) + (6)(2) + (12)(5) = (16 + 12 + 60) = 88$. For this example in which **M** is 2×3 and **Q** is 3×3, the elements in the product matrix can be represented as

$$p_{ij} = m_{i1}q_{1j} + m_{i2}q_{2j} + m_{i3}q_{3j} \qquad (i = 1,2; j = 1,2,3)$$

or

$$p_{ij} = \sum_{k=1}^{3} m_{ik}q_{kj} \qquad (i = 1,2; j = 1,2,3) \tag{1.2}$$

This definition for multiplication of matrices means that in order to be *conformable for multiplication,* the number of *columns* in the matrix on the left must be the same as the number of *rows* in the matrix on the right. Look again at the p_{ij}; for the three elements in (any) row i of **M**—m_{i1}, m_{i2}, and m_{i3}—there need to be three "corresponding" elements in (any) column j of **Q**—q_{1j}, q_{2j}, and q_{3j}.

The definition of matrix multiplication also means that the product matrix, **P**, will have the same number of rows as **M** and the same number of columns as **Q**:

$$\begin{array}{ccc} \mathbf{P} & = & \mathbf{M} & \mathbf{Q} \\ (m \times n) & & (m \times r) & (r \times n) \end{array} \tag{1.3}$$

In addition, the definition of multiplication means that, in general, the order in which multiplication is carried out makes a difference. In this example, the product the other way around, **QM**, cannot even be found, since there are three *columns* in **Q** but only two *rows* in **M**.[3] For this reason, there is language to describe the order of multiplication in a

---

[3]Try to carry out the multiplication in the order **QM**, and you can easily see where the trouble arises.

matrix product. In $\mathbf{P} = \mathbf{MQ}$, $\mathbf{M}$ is said to *premultiply* $\mathbf{Q}$ (or to multiply $\mathbf{Q}$ on the left) and, equivalently, $\mathbf{Q}$ is said to *postmultiply* $\mathbf{M}$ (or to multiply $\mathbf{M}$ on the right).

Here is a pair of matrices for which multiplication *is* defined in either order. The results, however, are quite different. Suppose $\mathbf{B} = [1 \quad 2 \quad 3]$; then, in conjunction with $\mathbf{C}$ from above, both products $\mathbf{BC}$ and $\mathbf{CB}$ can be found. These are

$$\underset{(1 \times 3)\ (3 \times 1)}{\mathbf{B} \qquad \mathbf{C}} = [1 \quad 2 \quad 3]\begin{bmatrix} 5 \\ 10 \\ 15 \end{bmatrix} = [70]$$

whereas

$$\underset{(3 \times 1)\ (1 \times 3)}{\mathbf{C} \qquad \mathbf{B}} = \begin{bmatrix} 5 \\ 10 \\ 15 \end{bmatrix}[1 \quad 2 \quad 3] \begin{bmatrix} 5 & 10 & 15 \\ 10 & 20 & 30 \\ 15 & 30 & 45 \end{bmatrix}$$

(You should check these multiplications for practice.)

Even when the dimensions of the matrices are such that multiplication can be done in either order *and* both products have the same dimensions, the results will (generally) differ. For example, if $\mathbf{A} = \begin{bmatrix} 1 & 2 \\ 3 & 4 \end{bmatrix}$ and $\mathbf{B} = \begin{bmatrix} 5 & 6 \\ 7 & 8 \end{bmatrix}$, then (you should also work these out for practice)

$$\mathbf{AB} = \begin{bmatrix} 19 & 22 \\ 43 & 50 \end{bmatrix} \quad \text{and} \quad \mathbf{BA} = \begin{bmatrix} 23 & 34 \\ 31 & 46 \end{bmatrix}$$

It is not difficult to create an example in which the order of multiplication will *not* make a difference; suppose $\mathbf{A} = \begin{bmatrix} 3 & 3 \\ 3 & 3 \end{bmatrix}$ and $\mathbf{B} = \begin{bmatrix} 4 & 4 \\ 4 & 4 \end{bmatrix}$. Then, in this particular case, $\mathbf{AB} = \mathbf{BA}$ (again, you should check this for practice)

$$\mathbf{AB} = \mathbf{BA} = \begin{bmatrix} 24 & 24 \\ 24 & 24 \end{bmatrix}$$

But this is an exception; generally order of multiplication makes a difference, in contrast with ordinary algebra, where $ab = ba$. Here is another counterexample; if $\mathbf{A} = \begin{bmatrix} 1 & 2 \\ 3 & 4 \end{bmatrix}$, suppose that $\mathbf{B} = \begin{bmatrix} 1 & 0 \\ 0 & 1 \end{bmatrix}$. Then you can easily establish that

$$\mathbf{AB} = \mathbf{BA} = \begin{bmatrix} 1 & 2 \\ 3 & 4 \end{bmatrix}$$

This brings us to the notion of an *identity matrix*.

1.2.4 Matrix Definitions: The Identity Matrix

In ordinary algebra, 1 is known as the *identity element for multiplication,* which means that a number remains unchanged when multiplied by it. The analogous concept in matrix algebra was illustrated by the matrix $\mathbf{B} = \begin{bmatrix} 1 & 0 \\ 0 & 1 \end{bmatrix}$, immediately above. An *identity matrix* is one that leaves a matrix unchanged when the matrix is multiplied by it.

Consider $\mathbf{M} = \begin{bmatrix} 2 & 1 & 3 \\ 4 & 6 & 12 \end{bmatrix}$; by what matrix could $\mathbf{M}$ be *post*multiplied so that it remained unchanged? Denote the unknown matrix by $\mathbf{I}$ (this is the standard notation for an identity matrix); we want $\mathbf{MI} = \mathbf{M}$. We know from the rule in Equation (1.3) that $\mathbf{I}$ must be a 3 × 3 matrix; it needs three rows to be conformable to postmultiply $\mathbf{M}$ and three columns because the product, which will be $\mathbf{M}$, with dimensions 2 × 3, gets its second dimension from the number of columns in $\mathbf{I}$. You might try letting $\mathbf{I}$ be a 3 × 3 matrix containing all 1's. It may seem logical but it is wrong, because in that case

$$\mathbf{MI} = \begin{bmatrix} 2 & 1 & 3 \\ 4 & 6 & 12 \end{bmatrix} \begin{bmatrix} 1 & 1 & 1 \\ 1 & 1 & 1 \\ 1 & 1 & 1 \end{bmatrix} = \begin{bmatrix} 6 & 6 & 6 \\ 22 & 22 & 22 \end{bmatrix} \neq \mathbf{M}$$

In fact, the only $\mathbf{I}$ for which $\mathbf{MI} = \mathbf{M}$ will be

$$\mathbf{I} = \begin{bmatrix} 1 & 0 & 0 \\ 0 & 1 & 0 \\ 0 & 0 & 1 \end{bmatrix}$$

You should try it, and also try other possibilities, to be convinced that only this matrix will do the job.

An identity matrix is always square and can be of any size to fulfill the conformability requirement for the particular multiplication operation. It has 1's along the *main diagonal,* from upper left to lower right, and 0's everywhere else. We could also find another identity matrix by which to *pre*multiply $\mathbf{M}$ so that it remains unchanged. You can easily check that in this case what we need is the 2 × 2 matrix $\mathbf{I} = \begin{bmatrix} 1 & 0 \\ 0 & 1 \end{bmatrix}$ and then $\mathbf{IM} = \mathbf{M}$.

1.2.5   Matrix Operations: Transposition

Transposition is a matrix operation for which there is no parallel in ordinary algebra. The *transpose* of an $m \times n$ matrix $\mathbf{A}$, usually denoted by a prime ("'") sign, creates an $n \times m$ matrix in which row i of $\mathbf{A}$ becomes column i of $\mathbf{A'}$. Using $\mathbf{M}$ from above, we have

$$\mathbf{M'} = \begin{bmatrix} 2 & 4 \\ 1 & 6 \\ 3 & 12 \end{bmatrix}$$

Notice that the transpose of an n-element column vector (with dimensions $n \times 1$) is an n-element row vector (with dimensions $1 \times n$). The transpose of a transpose is the original matrix itself;

$$(\mathbf{M'})' = \begin{bmatrix} 2 & 1 & 3 \\ 4 & 6 & 12 \end{bmatrix} = \mathbf{M}$$

1.2.6   Matrix Operations: Partitioning

Sometimes is it useful to divide a matrix into *submatrices,* especially if there is some logical reason to distinguish some rows and columns from others. This is known as *partitioning* the matrix; the submatrices are usually separated by dashed or dotted lines. For example, we might create four submatrices from a 4×4 matrix $\mathbf{A}$

$$\mathbf{A} = \left[\begin{array}{cc:cc} a_{11} & a_{12} & a_{13} & a_{14} \\ a_{21} & a_{22} & a_{23} & a_{24} \\ \hdashline a_{31} & a_{32} & a_{33} & a_{34} \\ a_{41} & a_{42} & a_{43} & a_{44} \end{array} \right] = \left[\begin{array}{c:c} \mathbf{A}_{11} & \mathbf{A}_{12} \\ \hdashline \mathbf{A}_{21} & \mathbf{A}_{22} \end{array} \right]$$

If two matrices are partitioned so that submatrices are conformable for multiplication, then products of partitioned matrices can be found as products of their submatrices. For example, suppose that in conjunction with the 4×4 matrix $\mathbf{A}$ above we have a 4×3 matrix $\mathbf{B}$

$$\mathbf{B} = \left[\begin{array}{cc:c} b_{11} & b_{12} & b_{13} \\ b_{21} & b_{22} & b_{23} \\ \hdashline b_{31} & b_{32} & b_{33} \\ b_{41} & b_{42} & b_{43} \end{array} \right] = \left[\begin{array}{c:c} \mathbf{B}_{11} & \mathbf{B}_{12} \\ \hdashline \mathbf{B}_{21} & \mathbf{B}_{22} \end{array} \right]$$

Here $\mathbf{B}_{11}$ and $\mathbf{B}_{21}$ are 2×2 matrices and $\mathbf{B}_{12}$ and $\mathbf{B}_{22}$ are 2×1 matrices (column vectors).

Then the product $\mathbf{AB}$, which will be a 4×3 matrix, can be found as products of the submatrices in $\mathbf{A}$ and $\mathbf{B}$, because they have been formed so that all submatrices are conformable both for the multiplications and also for the summations in which they appear. Here

$$\mathbf{AB} = \left[\begin{array}{c:c} \mathbf{A}_{11} & \mathbf{A}_{12} \\ \hdashline \mathbf{A}_{21} & \mathbf{A}_{22} \end{array} \right] \left[\begin{array}{c:c} \mathbf{B}_{11} & \mathbf{B}_{12} \\ \hdashline \mathbf{B}_{21} & \mathbf{B}_{22} \end{array} \right] = \left[\begin{array}{c:c} \mathbf{A}_{11}\mathbf{B}_{11} + \mathbf{A}_{12}\mathbf{B}_{21} & \mathbf{A}_{11}\mathbf{B}_{12} + \mathbf{A}_{12}\mathbf{B}_{22} \\ \hdashline \mathbf{A}_{21}\mathbf{B}_{11} + \mathbf{A}_{22}\mathbf{B}_{21} & \mathbf{A}_{21}\mathbf{B}_{12} + \mathbf{A}_{22}\mathbf{B}_{22} \end{array} \right] \quad (1.4)$$

and you can check that all conformability requirements are met.

1.2.7   Additional Definitions and Operations

Diagonal Matrices   Identity matrices are examples of *diagonal matrices.* These are always square, with one or more nonzero elements on the diagonal from upper left to lower right and zeros elsewhere. In general, an $n \times n$ diagonal matrix is

$$\mathbf{D} = \begin{bmatrix} d_1 & 0 & \cdots & 0 \\ 0 & d_2 & \cdots & 0 \\ \vdots & \vdots & \ddots & \vdots \\ 0 & 0 & \cdots & d_n \end{bmatrix}$$

A useful notational device is available for creating a diagonal matrix out of a vector. Suppose $\mathbf{C} = \begin{bmatrix} c_1 \\ c_2 \\ c_3 \end{bmatrix}$; then the diagonal matrix with the elements of $\mathbf{C}$ strung out in order along its main diagonal is denoted by putting a "hat" over the **C**. (Sometimes the "less than" and "greater than" signs "<" and ">" are used, if the vector expression is too wide for a hat.) So

$$\hat{\mathbf{C}} = <\mathbf{C}> = \begin{bmatrix} c_1 & 0 & 0 \\ 0 & c_2 & 0 \\ 0 & 0 & c_3 \end{bmatrix}$$

Using a hat over a square $n \times n$ *matrix* **M** (instead of an n-element vector) denotes the $n \times n$ diagonal matrix formed by setting all m_{ij} $(i \neq j)$ equal to zero. For

$$\mathbf{M} = \begin{bmatrix} 1 & 2 & 3 \\ 4 & 5 & 6 \\ 7 & 8 & 9 \end{bmatrix}, \quad \hat{\mathbf{M}} = \begin{bmatrix} 1 & 0 & 0 \\ 0 & 5 & 0 \\ 0 & 0 & 9 \end{bmatrix}$$

In addition, although it is less frequently used,

$$\check{\mathbf{M}} = \begin{bmatrix} 0 & 2 & 3 \\ 4 & 0 & 6 \\ 7 & 8 & 0 \end{bmatrix}$$

Therefore, $\mathbf{M} = \hat{\mathbf{M}} + \check{\mathbf{M}}$. For a nonsquare matrix, **N**, $\hat{\mathbf{N}} = [n_{ii}]$, the diagonal matrix formed by using only n_{11}, n_{22}, and so on from **N**.

When a diagonal matrix, **D**, *post*multiplies another matrix, **M**, the jth element in the diagonal matrix, d_j, multiplies all the elements in the jth *column* of **M**; when a diagonal matrix *pre*multiplies **M**, d_j multiplies all the elements in the jth *row* of **M**. For example

$$\begin{bmatrix} 2 & 1 & 3 \\ 4 & 6 & 12 \end{bmatrix} \begin{bmatrix} d_1 & 0 & 0 \\ 0 & d_2 & 0 \\ 0 & 0 & d_3 \end{bmatrix} = \begin{bmatrix} 2d_1 & d_2 & 3d_3 \\ 4d_1 & 6d_2 & 12d_3 \end{bmatrix}$$

$$= \begin{bmatrix} \begin{bmatrix} 2 \\ 4 \end{bmatrix} d_1 & \begin{bmatrix} 1 \\ 6 \end{bmatrix} d_2 & \begin{bmatrix} 3 \\ 12 \end{bmatrix} d_3 \end{bmatrix}$$

and

$$\begin{bmatrix} d_1 & 0 \\ 0 & d_2 \end{bmatrix} \begin{bmatrix} 2 & 1 & 3 \\ 4 & 6 & 12 \end{bmatrix} = \begin{bmatrix} 2d_1 & d_1 & 3d_1 \\ 4d_2 & 6d_2 & 12d_2 \end{bmatrix} = \begin{bmatrix} d_1[2 & 1 & 3] \\ d_2[4 & 6 & 12] \end{bmatrix}$$

Summation Vectors   If an $m \times n$ matrix, **M**, is postmultiplied by an n-element column vector of 1's, the result will be an m-element column vector containing the *row sums* of **M**. If **M** is premultiplied by an m-element row vector of 1's, the result will be an n-element row vector containing the *column sums* of **M**. For example,

$$\begin{bmatrix} 2 & 1 & 3 \\ 4 & 6 & 12 \end{bmatrix} \begin{bmatrix} 1 \\ 1 \\ 1 \end{bmatrix} = \begin{bmatrix} 6 \\ 22 \end{bmatrix} \quad \text{and} \quad [1 \quad 1] \begin{bmatrix} 2 & 1 & 3 \\ 4 & 6 & 12 \end{bmatrix} = [6 \quad 7 \quad 15]$$

We will denote an n-element column vector of 1's by **1** and so the corresponding row vector is **1'**. (Frequently these are denoted by **e** and **e'**.) These are often called *summation vectors*.

Inner Products and Outer Products   Two special kinds of products are defined for pairs of n-element vectors. These vector products will be illustrated using two general three-element vectors

$$\mathbf{U} = \begin{bmatrix} u_1 \\ u_2 \\ u_3 \end{bmatrix} \quad \text{and} \quad \mathbf{V} = \begin{bmatrix} v_1 \\ v_2 \\ v_3 \end{bmatrix}$$

Inner Products   The inner product, denoted $\mathbf{U} \circ \mathbf{V}$, is defined as the sum of the products of elements in similar positions:

$$\mathbf{U} \circ \mathbf{V} = u_1 v_1 + u_2 v_2 + u_3 v_3$$

Since this result is a scalar, the operation is also known as the *scalar product,* or, because of the standard notation of a large dot between the vectors, as the *dot product.* Note that the same number results from the ordinary matrix product $\mathbf{U'V}$, although officially this produces a 1×1 "matrix," but the distinction between a scalar and a 1×1 matrix is generally not an important one for our purposes. Specifically, if

$$\mathbf{U} = \begin{bmatrix} 1 \\ 2 \\ 3 \end{bmatrix} \quad \text{and} \quad \mathbf{V} = \begin{bmatrix} 5 \\ 4 \\ 8 \end{bmatrix}$$

then $\mathbf{U} \circ \mathbf{V} = 37$.

Outer Products   An alternative kind of product of two n-element column vectors is the outer product, denoted $\mathbf{U} \otimes \mathbf{V}$, which is the same as the *matrix* product $\mathbf{UV'}$. This means that the result is an $n \times n$ matrix. For the general three-element **U** and **V** above,

$$\mathbf{U} \otimes \mathbf{V} = \begin{bmatrix} u_1 v_1 & u_1 v_2 & u_1 v_3 \\ u_2 v_1 & u_2 v_2 & u_2 v_3 \\ u_3 v_1 & u_3 v_2 & u_3 v_3 \end{bmatrix}$$

and for the specific numerical example,

$$\mathbf{U} \otimes \mathbf{V} = \begin{bmatrix} 5 & 4 & 8 \\ 10 & 8 & 16 \\ 15 & 12 & 24 \end{bmatrix}$$

1.3 LINEAR EQUATION SYSTEMS: A PREVIEW

We explore a pair of linear equations that will serve to illustrate the logic of the definition of matrix multiplication and the usefulness of other matrix algebra concepts. The concern of this chapter (and the next) is exclusively with *linear* equation systems. These are systems whose individual equations contain variables to the first power only; no x^2's, no products of variables, $x_i x_j$, no trigonometric, exponential or more complicated functions of variables are allowed. Each equation can be represented geometrically by a straight line or plane or hyperplane;[4] we look at the geometry later in this chapter and in Chapter 2. Solution approaches for *nonlinear* equations and equation systems are taken up in Chapter 5.

Here is a context in which a pair of linear equations might arise.[5] The mayor of a city has implemented a pair of job retraining programs designed to teach currently unemployed workers new skills in order to make them employable in the current marketplace. Each program runs for 6 months, and program costs are to be covered by an allocation of $1,000,000 from the state. The cost per pupil for the 6 months in program 1 is $2500 and $3500 in program 2—reflecting the fact that program 2 is the more ambitious of the two, requiring higher-skilled instructional staff, more sophisticated computers and/or software for each student, and so on. A new session of classes in both programs will start at the beginning of next July.

The mayor wants to know how many students should be admitted to each program. If we let x_1 and x_2 represent the numbers of students admitted for the next sessions of programs 1 and 2, respectively, then the total cost of running the next pair of programs will be $2500\, x_1 + 3500\, x_2$. It might be reasonable to expect that the mayor would prefer to use all the money allocated to the city. This means that the numbers of admissions, x_1 and x_2, should be selected so that

$$2500\, x_1 + 3500\, x_2 = 1{,}000{,}000$$

Additionally, past experience has shown that a student who completes program 1 can be expected to find part-time employment and earn, on average, $10,000 per year. Graduates of program 2, who have higher skill levels, can be expected to earn $20,000 per year. Admission to either program carries with it the obligation to complete the entire program, so there are no dropouts. While a decrease in the numbers of unemployed is itself a prime goal of the job retraining program, the mayor is particularly interested because the city imposes a tax on wages and salaries earned in the city. This is a flat tax of 2.5%, and the mayor has set a target of $115,000 in new annual tax revenues to be collected as a result of this next retraining program. The expected total new wage earnings from the new employees is therefore $10{,}000\, x_1 + 20{,}000\, x_2$, and so the mayor's target requires that

$$10{,}000\, x_1 + 20{,}000\, x_2 = 4{,}600{,}000$$

(since the 2.5% tax on $4,600,000 will generate the $115,000 target).

The problem, then, is to find the values of x_1 and x_2 that satisfy the following pair of linear equations

---

[4]The prefix "hyper" is usually used to extend three-dimensional geometric concepts to higher-dimensional space, such as hyperline and hyperplane.

[5]The object here is to indicate a kind of situation in which linear equations like these might occur; we need not dwell on or worry too much about the credibility or the details of the story.

$$2500\, x_1 + 3500\, x_2 = 1{,}000{,}000$$
$$10{,}000\, x_1 + 20{,}000\, x_2 = 4{,}600{,}000$$

(1.5)

To simplify the numbers, divide everything on both sides of the first equation by 500 and divide both sides of the second equation by 10,000, giving

$$5x_1 + 7x_2 = 2000$$
$$x_1 + 2x_2 = 460$$

(1.6)

If we define

$$\mathbf{A} = \begin{bmatrix} 5 & 7 \\ 1 & 2 \end{bmatrix}, \qquad \mathbf{X} = \begin{bmatrix} x_1 \\ x_2 \end{bmatrix}, \qquad \mathbf{B} = \begin{bmatrix} 2000 \\ 460 \end{bmatrix}$$

(1.7)

then, given the definitions of matrix multiplication and matrix equality, we see that the two linear equations in (1.6) can be represented by one *matrix* equation

$$\mathbf{AX} = \mathbf{B}$$

(1.8)

For this example, we have

$$\begin{bmatrix} 5 & 7 \\ 1 & 2 \end{bmatrix} \begin{bmatrix} x_1 \\ x_2 \end{bmatrix} = \begin{bmatrix} 2000 \\ 460 \end{bmatrix}$$

and carrying out the matrix multiplication on the left gives

$$\begin{bmatrix} (5x_1 + 7x_2) \\ (x_1 + 2x_2) \end{bmatrix} = \begin{bmatrix} 2000 \\ 460 \end{bmatrix}$$

From the definition of element-by-element matrix equality, we see that this represents exactly the two-equation system in (1.6).

If we had a system of three linear equations with unknowns x_1, x_2, and x_3, the matrix representation would be exactly the same as that in (1.8). Now, $\mathbf{A}$ would be a 3×3 matrix, and both $\mathbf{X}$ and $\mathbf{B}$ would be three-element column vectors. Indeed for *any* system of n linear equations in n unknowns—$x_1, \ldots, x_n$—the matrix representation will be that of (1.8). Only the dimensions of the matrices will change.

In ordinary algebra, we "solve" an equation like $3x = 12$ by dividing both sides by 3. This is the same as multiplying both sides by $\frac{1}{3}$ (the reciprocal of 3; often denoted 3^{-1})—the number that, when 3 is multiplied by it, gives us 1 (the identity element for multiplication). So, we go from $3x = 12$ to $x = 4$ in the logical sequence

$$3x = 12 \Rightarrow \tfrac{1}{3}\, 3x = \tfrac{1}{3}\, 12 \ [\text{or } (3^{-1})(3)x = (3^{-1})(12)]$$
$$\Rightarrow 1x = 4 \Rightarrow x = 4$$

Of course, in ordinary algebra the transition from $3x = 12$ to $x = 4$ is virtually immediate. The point here is to set the stage for a parallel approach to systems of linear equations, as in (1.6).

Given the representation in (1.8), it is clear that a way of "solving" this system for the unknowns would be to "divide" both sides by the matrix $\mathbf{A}$, or, alternatively, multiply both sides by the "reciprocal" of $\mathbf{A}$. Parallel to the notation for the reciprocal of a number, this is denoted $\mathbf{A}^{-1}$. If we could find such a matrix, with the matrix property that $(\mathbf{A}^{-1})(\mathbf{A}) = \mathbf{I}$, we would proceed in the same way:

$$\mathbf{AX} = \mathbf{B} \Rightarrow (\mathbf{A}^{-1})\mathbf{AX} = (\mathbf{A}^{-1})\mathbf{B} \Rightarrow \mathbf{IX} = (\mathbf{A}^{-1})\mathbf{B} \Rightarrow \mathbf{X} = (\mathbf{A}^{-1})\mathbf{B} \qquad (1.9)$$

and the values of the unknowns, in $\mathbf{X}$, would be found as a matrix operation in which the vector $\mathbf{B}$ is premultiplied by the matrix $\mathbf{A}^{-1}$.

The sequence in (1.9) is completely general; it is valid whether we have two linear equations in x_1 and x_2; three linear equations in x_1, x_2, and x_3; or n linear equations ($n > 3$) with n unknowns, when $\mathbf{A}$ would be an $n \times n$ matrix and $\mathbf{B}$ and $\mathbf{X}$ would be n-element column vectors. In matrix algebra $\mathbf{A}^{-1}$ is usually called the *inverse* of $\mathbf{A}$, and it is to inverses that we now turn.

1.3.1 Matrix Operations: Division

In matrix algebra, "division" by a matrix is represented by multiplication by the inverse. For the moment, we will explore inverses only for square matrices. This means that if we are dealing with the coefficient matrix from an equation system, as in (1.7) or (1.8), there are the same number of unknowns as equations in the system.[6] The defining characteristic, for a square matrix $\mathbf{A}$, is that

$$\mathbf{A}^{-1}\mathbf{A} = \mathbf{AA}^{-1} = \mathbf{I} \qquad (1.10)$$

(This is another exception to the general rule that order of multiplication makes a difference.)

Just as the number 0 has no reciprocal, a square null matrix of any size will not have an inverse, and we will also find that there are many other matrices for which inverses cannot be found. They are called *singular;* by contrast, a *nonsingular* matrix is one that has an inverse. So the question of the *existence* of an inverse always arises—some matrices have them and some do not. *Uniqueness,* however, is not an issue; if an inverse can be found, then it is the only one.[7]

Finding inverses is a rather tedious mathematical procedure, but modern computers can do it (or find an approximation to an inverse) very quickly. However, it is useful to have at least a rudimentary understanding of how an inverse is found, in order to understand what makes some matrices singular. When the coefficient matrix from a linear equation system turns out to be singular, we may be interested in learning what that means about the underlying equations.

The matrix $\mathbf{A}$ in (1.7) is *nonsingular.* We can find its inverse in a completely brute-force

---

[6]There are more advanced concepts of "pseudo" inverses (or "right" and "left" inverses) for nonsquare matrices, and we will explore and apply these in Chapter 2 when we deal with a variety of different systems of linear equations.

[7]Suppose that $\mathbf{B}$ is an inverse of $\mathbf{A}$, so (i) $\mathbf{AB} = \mathbf{I}$. Suppose that $\mathbf{C}$ is also an inverse of $\mathbf{A}$, and premultiply both sides of (i) by $\mathbf{C}$; this gives (ii) $\mathbf{C}(\mathbf{AB}) = \mathbf{CI} = \mathbf{C}$. Reexpress (ii) as (iii) $(\mathbf{CA})\mathbf{B} = \mathbf{C}$; since $\mathbf{C}$ is also an inverse of $\mathbf{A}$, $\mathbf{CA} = \mathbf{I}$ and (iii) becomes $\mathbf{IB} = \mathbf{C}$, or $\mathbf{B} = \mathbf{C}$, which is what we wanted to show.

way, from the fundamental definition that $\mathbf{AA}^{-1} = \mathbf{I}$. We know that $\mathbf{A}^{-1}$ will be a 2×2 matrix (why?) and that the right-hand side is a 2×2 identity matrix. If we represent the four unknown elements in $\mathbf{A}^{-1}$ as $\begin{bmatrix} s & t \\ u & v \end{bmatrix}$, we have

$$\begin{bmatrix} 5 & 7 \\ 1 & 2 \end{bmatrix} \begin{bmatrix} s & t \\ u & v \end{bmatrix} = \begin{bmatrix} 1 & 0 \\ 0 & 1 \end{bmatrix}$$

Multiplying the two matrices on the left, we have

$$\begin{bmatrix} (5s + 7u) & (5t + 7v) \\ (s + 2u) & (t + 2v) \end{bmatrix} = \begin{bmatrix} 1 & 0 \\ 0 & 1 \end{bmatrix}$$

and from the definition of matrix equality as element-by-element equality, this means

$$\begin{aligned} 5s + 7u = 1 \qquad & 5t + 7v = 0 \\ s + 2u = 0 \qquad & t + 2v = 1 \end{aligned}$$

From the left pair of equations, it is easy to find (bottom equation) that $s = -2u$. Putting that into the top equation on the left, $u = \frac{1}{3}$ and so $s = \frac{2}{3}$. Similarly, from the right pair of equations we find (top equation) $t = -\frac{7}{5}v$. Putting that into the bottom equation gives $v = -\frac{5}{3}$ and so $t = -\frac{7}{3}$. These are the four elements of the inverse, so

$$\mathbf{A}^{-1} = \begin{bmatrix} \frac{2}{3} & -\frac{7}{3} \\ -\frac{1}{3} & \frac{5}{3} \end{bmatrix}$$

which you can easily (and should) check by finding that $\mathbf{AA}^{-1} = \mathbf{I}$ and $\mathbf{A}^{-1}\mathbf{A} = \mathbf{I}$.

Since we have found $\mathbf{A}^{-1}$ for the equations in (1.6), the solution to those equations is exactly as given in (1.9),

$$\mathbf{X} = \mathbf{A}^{-1}\mathbf{B} = \begin{bmatrix} \frac{2}{3} & -\frac{7}{3} \\ -\frac{1}{3} & \frac{5}{3} \end{bmatrix} \begin{bmatrix} 2000 \\ 460 \end{bmatrix} = \begin{bmatrix} 260 \\ 100 \end{bmatrix}$$

You can easily check that $x_1 = 260$ and $x_2 = 100$ are the (only) solutions to the two equations in (1.6). These are the appropriate number of students to admit into each of the two job retraining programs if both the allocated budget is to be completely used up and the mayor's new tax target is to be met.

Here we have found the values for the unknowns in a problem by initially abstracting from the specific parameters, in (1.6), instead manipulating the matrix representation of the problem using the definitions and operations of matrix algebra, as in (1.9), and then reintroducing the particular values after the general solution was found. This is our motivation for calculating inverses, but we need to look at the problem of finding an inverse for a more general situation than this particular numerical example.

Suppose that we try to find the inverse for *any* 2×2 matrix $\mathbf{A} = \begin{bmatrix} a_{11} & a_{12} \\ a_{21} & a_{22} \end{bmatrix}$. Again, to simplify notation, we can represent the inverse as $\begin{bmatrix} s & t \\ u & v \end{bmatrix}$, and the problem is to express these four elements as functions of the four (known) elements in **A**. The defining relationship is still

$$\begin{bmatrix} a_{11} & a_{12} \\ a_{21} & a_{22} \end{bmatrix} \begin{bmatrix} s & t \\ u & v \end{bmatrix} = \begin{bmatrix} 1 & 0 \\ 0 & 1 \end{bmatrix}$$

Carrying out the multiplication on the left, exactly as in the numerical example, we have

$$
\begin{aligned}
a_{11}s + a_{12}u &= 1 & a_{11}t + a_{12}v &= 0 \\
a_{21}s + a_{22}u &= 0 & a_{21}t + a_{22}v &= 1
\end{aligned}
\tag{1.11}
$$

Now, however, it is not quite so straightforward to solve the left-hand pair of equations for s and u as functions of the a's or the right-hand pair for t and v as functions of those same a's. But it is worth doing, because of the way in which the structure of the inverse matrix emerges.

Consider the left-hand pair. Multiply the top equation on both sides by a_{22} and multiply the bottom equation on both sides by a_{12}. This gives

$$a_{22}a_{11}s + a_{22}a_{12}u = a_{22}$$
$$a_{12}a_{21}s + a_{12}a_{22}u = 0$$

Subtracting the bottom from the top equation causes the term with u to drop out, leaving

$$(a_{22}a_{11} - a_{12}a_{21})s = a_{22}$$

from which it is easy to find that

$$s = \frac{a_{22}}{a_{22}a_{11} - a_{12}a_{21}}$$

So s is completely determined as a function of all four of the (known) elements in **A**. Putting this into the bottom equation gives (after some straightforward algebra)

$$u = \frac{-a_{21}}{a_{22}a_{11} - a_{12}a_{21}}$$

Similar algebra on the right-hand pair of equations will establish (you can work this through if you wish) that

$$t = \frac{-a_{12}}{a_{22}a_{11} - a_{12}a_{21}} \quad \text{and} \quad v = \frac{a_{11}}{a_{22}a_{11} - a_{12}a_{21}}$$

Note that all four elements have the same denominator, so we can factor out this scalar and express the inverse as

$$\mathbf{A}^{-1} = \left(\frac{1}{a_{22}a_{11} - a_{12}a_{21}}\right)\begin{bmatrix} a_{22} & -a_{12} \\ -a_{21} & a_{11} \end{bmatrix} \qquad (1.12)$$

This is an important result. It shows that the inverse for *any* 2×2 matrix is made up of a *scalar* part, $\{1/(a_{22}a_{11} - a_{12}a_{21})\}$, and a *matrix* part, $\begin{bmatrix} a_{22} & -a_{12} \\ -a_{21} & a_{11} \end{bmatrix}$. Obviously, no matter what the elements of $\mathbf{A}$, we can always rearrange and display them in the order shown in the matrix part of (1.12). On the other hand, if the denominator of the scalar in curly brackets in (1.12), $(a_{22}a_{11} - a_{12}a_{21})$, should happen to be zero, then $\mathbf{A}^{-1}$ could not be found; $\mathbf{A}$ would in that case be a singular matrix.

This denominator expression is known as the *determinant* of $\mathbf{A}$, usually denoted by $|\mathbf{A}|$, or sometimes by "det $\mathbf{A}$." It is simply the product of the elements on the *main diagonal* (northwest to southeast) of $\mathbf{A}$, $a_{11}a_{22}$, minus the product on the other diagonal (northeast to southwest), $a_{12}a_{21}$. So for a 2×2 matrix to have an inverse, its determinant must not be zero.

These observations on the structure of the inverse for a 2×2 matrix $\mathbf{A}$ also hold for a square matrix of any size:

$$\mathbf{A}^{-1} = \frac{1}{|\mathbf{A}|} \quad \text{(some matrix)} \qquad (1.13)$$

As in the 2×2 case, the matrix part of the inverse is always just some rearrangement of the elements (more complicated for larger matrices, but nonetheless always defined) and the scalar part is the reciprocal of the determinant of $\mathbf{A}$. So the ability to find an inverse for any square matrix always hinges on whether $|\mathbf{A}| = 0$. For the general definition in (1.13) to make sense for the $n \times n$ case ($n > 2$), we need to understand both the determinant and the matrix part for larger matrices. To this end, we need some terminology—tedious at first, but essential in the end.

Minor of an Element a_{ij}

The minor of an element, a_{ij}, in a square matrix $\mathbf{A}$, m_{ij}, is the *determinant* of the matrix remaining when row i and column j are removed from $\mathbf{A}$. $\qquad (1.14)$

Each element in a square matrix has a unique number, its minor, associated with it. For example, from $\mathbf{A} = \begin{bmatrix} 5 & 7 \\ 1 & 2 \end{bmatrix}$, $m_{12} = 1$. When row 1 and column 2 are removed from $\mathbf{A}$, all that is left is $a_{21} = 1$.[8]

Finding minors becomes more complicated as the size of $\mathbf{A}$ increases. For example, you might find some of the minors for

$$\mathbf{A} = \begin{bmatrix} 1 & 1 & 1 \\ 2 & 0 & 6 \\ 3 & 7 & 1 \end{bmatrix}$$

---

[8]Put exactly, the *matrix* that is left is $[1]$, and the determinant of a 1×1 matrix is defined to be the element itself.

Here are two of them:

$$m_{23} = \begin{vmatrix} 1 & 1 \\ 3 & 7 \end{vmatrix} = 4 \quad \text{and} \quad m_{11} = \begin{vmatrix} 0 & 6 \\ 7 & 1 \end{vmatrix} = -42$$

We need the concept of a minor for the next definition.

Cofactor of an Element a_{ij}

The cofactor of an element a_{ij} in a square matrix **A**, denoted $\mathbf{A}_{ij}$,[9] is the minor multiplied by $(-1)^{i+j}$; $\mathbf{A}_{ij} = (-1)^{i+j}(m_{ij})$.   (1.15)

Thus the cofactor of an element differs from the minor at most by sign. When $i+j$ is an even number, $(-1)^{i+j} = 1$ and the cofactor and the minor will be the same. From the same 2 × 2 **A** matrix, $\mathbf{A}_{12} = (-1)^3(m_{12}) = -1$; from the 3 × 3 **A** matrix above,

$$\mathbf{A}_{23} = (-1)^5(4) = -4 \quad \text{and} \quad \mathbf{A}_{11} = (-1)^2(-42) = -42$$

The definition of cofactors allows us to examine a general definition of a determinant, relevant for square matrices of any size, and it also plays a role in defining the matrix part of the inverse.

Determinant of a Square Matrix (General Definition)   The first use of cofactors is in the general definition of $|\mathbf{A}|$. For any $n \times n$ matrix **A**,

$$|\mathbf{A}| = \sum_{j=1}^{n} a_{ij}\mathbf{A}_{ij} \quad \text{(for any } i\text{)} \quad (1.16a)$$

$$|\mathbf{A}| = \sum_{i=1}^{n} a_{ij}\mathbf{A}_{ij} \quad \text{(for any } j\text{)} \quad (1.16b)$$

Equation (1.16a) says that $|\mathbf{A}|$ can be found by selecting any *row* (any i) in **A** and summing the products of each element in that row and its respective cofactor; (1.16b) says that $|\mathbf{A}|$ can just as well be found by selecting any *column* (any j) in **A** and summing the products of each element in that column and its respective cofactor.

Using $\mathbf{A} = \begin{bmatrix} 1 & 1 & 1 \\ 2 & 0 & 6 \\ 3 & 7 & 1 \end{bmatrix}$ again, we evaluate $|\mathbf{A}|$ using (1.16a) across row 3 and (1.16b) down column 1.

Across row 3: $|\mathbf{A}| = a_{31}\mathbf{A}_{31} + a_{32}\mathbf{A}_{32} + a_{33}\mathbf{A}_{33} =$

$$(3)\left[(-1)^4 \begin{vmatrix} 1 & 1 \\ 0 & 6 \end{vmatrix}\right] + (7)\left[(-1)^5 \begin{vmatrix} 1 & 1 \\ 2 & 6 \end{vmatrix}\right] + (1)\left[(-1)^6 \begin{vmatrix} 1 & 1 \\ 2 & 0 \end{vmatrix}\right] =$$

$$(3)(6) + (7)(-4) + (1)(-2) = -12$$

---

[9]Recall that in Section 1.2 we used $\mathbf{A}_{ij}$ to indicate submatrices in a partitioning of a matrix **A**. The context should make clear when $\mathbf{A}_{ij}$ indicates a cofactor and when it represents a submatrix.

Down column 1: $|\mathbf{A}| = a_{11}\mathbf{A}_{11} + a_{21}\mathbf{A}_{21} + a_{31}\mathbf{A}_{31} =$

$$(1)\left[(-1)^2\begin{vmatrix} 0 & 6 \\ 7 & 1 \end{vmatrix}\right] + (2)\left[(-1)^3\begin{vmatrix} 1 & 1 \\ 7 & 1 \end{vmatrix}\right] + (3)\left[(-1)^4\begin{vmatrix} 1 & 1 \\ 0 & 6 \end{vmatrix}\right] =$$

$$(1)(-42) + (2)(6) + (3)(6) = -12$$

(Try some other rows or columns; you will always find $|\mathbf{A}| = -12$.)

There is a fairly simple "diagonals" rule for the (3×3) case. If you take a 3×3 matrix and repeat columns 1 and 2 as columns 4 and 5, then $|\mathbf{A}|$ is the sum of the products of three northwest to southeast diagonals minus the sum of the products of three northeast to southwest diagonals. Specifically, let $\mathbf{A} = \begin{bmatrix} a & b & c \\ d & e & f \\ g & h & i \end{bmatrix}$ (avoiding a_{ij} notation for simplici-

ty) and create an augmented matrix $\mathcal{A} = \begin{bmatrix} a & b & c & a & b \\ a & e & f & d & e \\ g & h & i & g & h \end{bmatrix}$. Then $|\mathbf{A}|$ is found by

adding up the products of the elements on the three "↘" diagonals in $\begin{bmatrix} a & b & c & a & b \\ d & e & f & d & e \\ g & h & i & g & h \end{bmatrix}$

—and subtracting the products on the three "↙" diagonals in $\begin{bmatrix} a & b & c & a & b \\ d & e & f & d & e \\ g & h & i & g & h \end{bmatrix}$. So $|\mathbf{A}|$

$= (aei + bfg + cdh) - (ceg + afh + bdi)$. Using this rule on our 3×3 $\mathbf{A}$, above, we obtain

$$|\mathbf{A}| = (0) + (18) + (14) - (0) - (42) - (2) = -12$$

The trouble is that there is not any easily visualized diagonals rule in the general case of a square matrix of size 4×4 or larger; that is why the two definitions in (1.16) are important.

The Adjoint of a Matrix $\mathbf{A}$ The second use for cofactors is in defining the matrix part of the general definition of $\mathbf{A}^{-1}$. This matrix is known as the *adjoint of* $\mathbf{A}$, often denoted [adj $\mathbf{A}$]. It is compactly defined as

$$\text{adj } \mathbf{A} = [\mathbf{A}'_{ij}] \tag{1.17}$$

In words, the adjoint of an $n \times n$ matrix $\mathbf{A}$ is an $n \times n$ matrix whose elements are the *cofactors of the transpose* of $\mathbf{A}$. If you look back at $\mathbf{A}^{-1}$ for the general 2×2 matrix, in (1.12), you will see that the matrix part is exactly adj $\mathbf{A}$. This is always true; for any $n \times n$ matrix, the matrix part of the definition in (1.13) is [adj $\mathbf{A}$], so

$$\mathbf{A}^{-1} = \frac{1}{|\mathbf{A}|}[\text{adj } \mathbf{A}] \tag{1.18}$$

and now we know, at least in principle, how to calculate both $\mathbf{A}$ and [adj $\mathbf{A}$] for any square matrix $\mathbf{A}$. [We will see later exactly why (1.18) is always true.]

Useful Properties of Determinants   Since the unique solution, $\mathbf{X} = \mathbf{A}^{-1}\mathbf{B}$, provided in (1.9) is completely general, for n linear equations in n unknowns, it would be useful to know under what circumstances an inverse matrix cannot be found for $\mathbf{A}$; this means, under what kinds of conditions will $|\mathbf{A}| = 0$. Several mathematical properties of determinants help to isolate cases in which the matrix is singular.

Property 1.   The determinant of a square matrix $\mathbf{A}$ is equal to the determinant of the transpose; $|\mathbf{A}| = |\mathbf{A}'|$.

For the 2×2 case, this follows directly from the general definition in (1.16), because the elements and their corresponding cofactors from *row i* of $\mathbf{A}$ will be exactly the same as the elements and their corresponding cofactors from *column i* of $\mathbf{A}'$. The demonstration for larger matrices is left as an exercise (Problem 1.15).

Property 2.   If any column or row of a square matrix $\mathbf{A}$ contains all zeros, $|\mathbf{A}| = 0$.

This follows immediately from (1.16) by simply choosing the row or column of zeros along or down which to evaluate $|\mathbf{A}|$.

Property 3.   Multiplication of all of the elements in any column or row of a square matrix $\mathbf{A}$ by a constant, say, c, creates a new matrix, $\mathscr{A}$, for which $|\mathscr{A}| = c|\mathbf{A}|$.

Again, this follows directly from (1.16). Just choose the row or column whose elements have been multiplied by c to evaluate $|\mathscr{A}|$.

Property 4.   If any two rows or columns of a square matrix $\mathbf{A}$ are interchanged, $|\mathbf{A}|$ changes sign.

A general proof of this property does not follow quite as directly from (1.16), but it is easy to explore the 2×2 and 3×3 cases and discover that it is true for them. It leads immediately to two useful observations, both of which identify singular matrices.

Property 4a.   If two (or more) rows or two (or more) columns in a square matrix $\mathbf{A}$ are *equal*, then $|\mathbf{A}| = 0$.

This follows directly from Property 4; if two equal rows (or columns) are interchanged, there can be no effect on the numerical value of the determinant, yet it must change sign. The only n for which $+n = -n$ is $n = 0$.

Property 4b.   If two (or more) rows or two (or more) columns in a square matrix $\mathbf{A}$ are *proportional*, then $|\mathbf{A}| = 0$.

This follows from Properties 3 and 4a, since the constant of proportionality can be factored out and then the matrix has two (or more) rows or columns that are equal.

Specifically, Properties 2, 4a, and 4b describe a series of increasingly subtle cases in which matrices are singular. Here are three matrices that satisfy these three properties, respectively:

$$\mathbf{A}_1 = \begin{bmatrix} 1 & 4 & 8 \\ 0 & 0 & 0 \\ 7 & 6 & 6 \end{bmatrix}, \quad \mathbf{A}_2 = \begin{bmatrix} 1 & 4 & 8 \\ 0 & 3 & 9 \\ 1 & 4 & 8 \end{bmatrix}, \quad \mathbf{A}_3 = \begin{bmatrix} 1 & 2 & 8 \\ 2 & 4 & 1 \\ 3 & 6 & 1 \end{bmatrix}$$

You can easily check that $|\mathbf{A}_i| = 0$ for all three cases, using the diagonals rule for 3×3 matrices or by using the definitions in (1.16). You also know that these matrices are singular because $\mathbf{A}_1$ has a row of zeros (Property 2), $\mathbf{A}_2$ has two equal rows (Property 4a) and $\mathbf{A}_3$ has two proportional columns (Property 4b). Therefore, if any of these $\mathbf{A}_i$ matrices contained the coefficients from a system of three linear equations, we know that the system does not have a unique solution $\mathbf{X} = (\mathbf{A}_i)^{-1}\mathbf{B}$. These properties are *sufficient* to guarantee that $|\mathbf{A}| = 0$; they are not *necessary*. For example, $\mathbf{A}_4 = \begin{bmatrix} 1 & 2 & 3 \\ 4 & 5 & 6 \\ 7 & 8 & 9 \end{bmatrix}$ is also singular

(you might check this). We will see why in Section 1.5.

One further mathematical property of determinants will enable us to see precisely why $\mathbf{A}^{-1} = (1/|\mathbf{A}|)[\text{adjoint } \mathbf{A}]$ for any square matrix $\mathbf{A}$. [We derived this result for the 2×2 case and then stated that it was true for a square matrix of any size, in (1.18).]

Property 5.   Evaluation of a determinant by *alien cofactors* always yields a value of zero. This means

$$\sum_{j=1}^{n} a_{ij}\mathbf{A}_{i'j} = 0 \qquad (\text{where } i \neq i') \tag{a}$$

and

$$\sum_{i=1}^{n} a_{ij}\mathbf{A}_{ij'} = 0 \qquad (\text{where } j \neq j') \tag{a}$$

In (a), the evaluation is carried out by choosing elements from row i and cofactors from some other row, i' (which is what makes them "alien"). In (b) the evaluation is done by selecting elements from down column j and cofactors from some other (alien) column, j'. In either case, the result is always zero. You can check this on $\mathbf{A} = \begin{bmatrix} 1 & 1 & 1 \\ 2 & 0 & 6 \\ 3 & 7 & 1 \end{bmatrix}$, which we

have already seen is nonsingular. (Problem 1.18 asks you to show this result for *any* square matrix $\mathbf{A}$.)

Now we return to the issue of the inverse for a square matrix of any size. Let $\mathbf{A}$ be an $n \times n$ matrix; its adjoint is

$$\text{adj } \mathbf{A} = \begin{bmatrix} \mathbf{A}_{11} & \mathbf{A}_{21} & \cdots & \mathbf{A}_{n1} \\ \mathbf{A}_{12} & \mathbf{A}_{22} & \cdots & \mathbf{A}_{n2} \\ \vdots & \vdots & & \vdots \\ \mathbf{A}_{1n} & \mathbf{A}_{2n} & \cdots & \mathbf{A}_{nn} \end{bmatrix}$$

Form the product of $\mathbf{A}$ and its adjoint (we will soon see why this is interesting to do):

$$\mathbf{A}[\text{adj } \mathbf{A}] = \begin{bmatrix} \sum_{j=1}^{n} a_{1j}\mathbf{A}_{1j} & \sum_{j=1}^{n} a_{1j}\mathbf{A}_{2j} & \cdots & \sum_{j=1}^{n} a_{1j}\mathbf{A}_{nj} \\ \sum_{j=1}^{n} a_{2j}\mathbf{A}_{1j} & \sum_{j=1}^{n} a_{2j}\mathbf{A}_{2j} & \cdots & \sum_{j=1}^{n} a_{2j}\mathbf{A}_{nj} \\ \vdots & \vdots & & \vdots \\ \sum_{j=1}^{n} a_{nj}\mathbf{A}_{1j} & \sum_{j=1}^{n} a_{nj}\mathbf{A}_{2j} & \cdots & \sum_{j=1}^{n} a_{nj}\mathbf{A}_{nj} \end{bmatrix}$$

While this expression may at first appear complicated, it is easily simplified. Notice that all of the elements on the *main diagonal* of this product matrix, $\sum_{j=1}^{n} a_{1j}\mathbf{A}_{1j}$,

$\Sigma_{j=1}^{n} a_{2j} \mathbf{A}_{2j}, \ldots, \Sigma_{j=1}^{n} a_{nj} \mathbf{A}_{nj}$, are just alternative expressions for $\mathbf{A}$, as in (1.13), evaluated along row 1, row 2, $\ldots$, row n. On the other hand, all other elements in this product are of the form $\Sigma_{j=1}^{n} a_{ij} \mathbf{A}_{i'j}$ (elements from one row and cofactors from some other row, making them alien), and hence all of these off-diagonal elements are equal to zero. Simplifying, then

$$\mathbf{A}[\text{adj } \mathbf{A}] = \begin{bmatrix} |\mathbf{A}| & 0 & \cdots & 0 \\ 0 & |\mathbf{A}| & \cdots & 0 \\ \vdots & \vdots & & \vdots \\ 0 & 0 & \cdots & |\mathbf{A}| \end{bmatrix} = |\mathbf{A}| \begin{bmatrix} 1 & 0 & \cdots & 0 \\ 0 & 1 & \cdots & 0 \\ \vdots & \vdots & & \vdots \\ 0 & 0 & \cdots & 1 \end{bmatrix} = |\mathbf{A}|\mathbf{I}$$

Dividing both sides by $|\mathbf{A}|$, a scalar, gives

$$\mathbf{A}\left[\frac{1}{|\mathbf{A}|}[\text{adj } \mathbf{A}]\right] = \mathbf{I}$$

which means that the matrix in large brackets postmultiplying $\mathbf{A}$ is indeed $\mathbf{A}^{-1}$, since the product of the two gives the identity matrix. That is,

$$\mathbf{A}^{-1} = \frac{1}{|\mathbf{A}|}[\text{adj } \mathbf{A}]$$

as was asserted in (1.18).

The Inverse of a Diagonal Matrix   A useful fact about diagonal matrices is that the inverse of a diagonal matrix is another diagonal matrix, each of whose elements is just the reciprocal of the original element. For $\mathbf{C} = [c_1, c_2, c_3]'$, this means

$$(\hat{\mathbf{C}})^{-1} = \begin{bmatrix} \dfrac{1}{c_1} & 0 & 0 \\[2mm] 0 & \dfrac{1}{c_2} & 0 \\[2mm] 0 & 0 & \dfrac{1}{c_3} \end{bmatrix}$$

You can easily check that $(\hat{\mathbf{C}})(\hat{\mathbf{C}})^{-1} = (\hat{\mathbf{C}})^{-1}(\hat{\mathbf{C}}) = \mathbf{I} = \begin{bmatrix} 1 & 0 & 0 \\ 0 & 1 & 0 \\ 0 & 0 & 1 \end{bmatrix}$.

The Inverse of a Partitioned Matrix   The elements of an inverse matrix can also be found in blocks, using partitioning. Consider again a 4×4 matrix $\mathbf{A}$ partitioned into four 2×2 submatrices

$$\mathbf{A} = \begin{bmatrix} \mathbf{A}_{11} & \mathbf{A}_{12} \\ \hline \mathbf{A}_{21} & \mathbf{A}_{22} \end{bmatrix}$$

The partitioning has been carried out so that the two on-diagonal submatrices, $\mathbf{A}_{11}$ and $\mathbf{A}_{22}$, are square (both 2×2 matrices in this example).

Denote $\mathbf{A}^{-1}$ by $\mathbf{R}$ (for *reciprocal*), and assume that it is partitioned in the same way, into four 2×2 submatrices:

$$\mathbf{R} = \left[\begin{array}{c|c} \mathbf{R}_{11} & \mathbf{R}_{12} \\ \hline \mathbf{R}_{21} & \mathbf{R}_{22} \end{array}\right]$$

Then the defining relation for the inverse, $\mathbf{A}\mathbf{A}^{-1} = \mathbf{A}\mathbf{R} = \mathbf{I}$, is

$$\left[\begin{array}{c|c} \mathbf{A}_{11} & \mathbf{A}_{12} \\ \hline \mathbf{A}_{21} & \mathbf{A}_{22} \end{array}\right]\left[\begin{array}{c|c} \mathbf{R}_{11} & \mathbf{R}_{12} \\ \hline \mathbf{R}_{21} & \mathbf{R}_{22} \end{array}\right] = \mathbf{I} \tag{1.19}$$

The multiplication on the left can be carried out following (1.4) for the product of any pair of conformably partitioned matrices; the 4×4 identity matrix can similarly be partitioned into

$$\mathbf{I} = \left[\begin{array}{cc|cc} 1 & 0 & 0 & 0 \\ 0 & 1 & 0 & 0 \\ \hline 0 & 0 & 1 & 0 \\ 0 & 0 & 0 & 1 \end{array}\right] = \left[\begin{array}{c|c} \mathbf{I} & \mathbf{0} \\ \hline \mathbf{0} & \mathbf{I} \end{array}\right]$$

Equation (1.19) represents the following four *matrix* equations

(i) $\mathbf{A}_{11}\mathbf{R}_{11} + \mathbf{A}_{12}\mathbf{R}_{21} = \mathbf{I}$ (ii) $\mathbf{A}_{21}\mathbf{R}_{11} + \mathbf{A}_{22}\mathbf{R}_{21} = \mathbf{0}$ (1.20a)

(i) $\mathbf{A}_{11}\mathbf{R}_{12} + \mathbf{A}_{12}\mathbf{R}_{22} = \mathbf{0}$ (ii) $\mathbf{A}_{21}\mathbf{R}_{12} + \mathbf{A}_{22}\mathbf{R}_{22} = \mathbf{I}$ (1.20b)

just as the 2×2 case was made up of four simple algebraic equations in (1.11). [You should check that all products and terms in (1.20) meet the conformability requirements for multiplication, addition, and equality.]

The elements in the four $\mathbf{A}$ submatrices are known and the elements in the four $\mathbf{R}$ submatrices are what we want to determine. Assume that the inverse for $\mathbf{A}_{22}$ exists and can be found. Then, from (1.20a)(ii), we obtain

$$\mathbf{R}_{21} = (\mathbf{A}_{22})^{-1}(-\mathbf{A}_{21}\mathbf{R}_{11}) \tag{1.21}$$

Substituting this result into (1.20a)(i) gives

$$\mathbf{A}_{11}\mathbf{R}_{11} - \mathbf{A}_{12}(\mathbf{A}_{22})^{-1}\mathbf{A}_{21}\mathbf{R}_{11} = [\mathbf{A}_{11} - \mathbf{A}_{12}(\mathbf{A}_{22})^{-1}\mathbf{A}_{21}]\mathbf{R}_{11} = \mathbf{I}$$

and therefore

$$\mathbf{R}_{11} = [\mathbf{A}_{11} - \mathbf{A}_{12}(\mathbf{A}_{22})^{-1}\mathbf{A}_{21}]^{-1} \tag{1.22}$$

From (1.20b)(ii), multiplying by $(\mathbf{A}_{22})^{-1}$ again

$$(\mathbf{A}_{22})^{-1}\mathbf{A}_{21}\mathbf{R}_{12} + \mathbf{R}_{22} = (\mathbf{A}_{22})^{-1}$$

and so

$$\mathbf{R}_{22} = (\mathbf{A}_{22})^{-1} - (\mathbf{A}_{22})^{-1}\mathbf{A}_{21}\mathbf{R}_{12} = (\mathbf{A}_{22})^{-1}[\mathbf{I} - \mathbf{A}_{21}\mathbf{R}_{12}] \tag{1.23}$$

Substituting into (1.20b)(i), we have

$$\mathbf{A}_{11}\mathbf{R}_{12} + \mathbf{A}_{12}[(\mathbf{A}_{22})^{-1} - (\mathbf{A}_{22})^{-1}\mathbf{A}_{21}\mathbf{R}_{12}] = \mathbf{0}$$

or

$$[\mathbf{A}_{11} - \mathbf{A}_{12}(\mathbf{A}_{22})^{-1}\mathbf{A}_{21}]\mathbf{R}_{12} = -\mathbf{A}_{12}(\mathbf{A}_{22})^{-1}$$

but, using the result in (1.22), this becomes

$$\mathbf{R}_{12} = -\mathbf{R}_{11}\mathbf{A}_{12}(\mathbf{A}_{22})^{-1} \tag{1.24}$$

Putting all the pieces together, we obtain

$$\mathbf{A}^{-1} = \mathbf{R} = \left[\begin{array}{c:c} [\mathbf{A}_{11} - \mathbf{A}_{12}(\mathbf{A}_{22})^{-1}\mathbf{A}_{21}]^{-1} & -\mathbf{R}_{11}\mathbf{A}_{12}(\mathbf{A}_{22})^{-1} \\ \hdashline (\mathbf{A}_{22})^{-1}(-\mathbf{A}_{21}\mathbf{R}_{11}) & (\mathbf{A}_{22})^{-1}[\mathbf{I} - \mathbf{A}_{21}\mathbf{R}_{12}] \end{array} \right]$$

Therefore, from (1.21) through (1.24), we see that it is possible to find all the elements of the inverse matrix $\mathbf{R}$, provided we solve equations in the correct order. First we must find $(\mathbf{A}_{22})^{-1}$. Then (1) find $\mathbf{R}_{11}$ using (1.22); (2) use the result for $\mathbf{R}_{11}$ in (1.21) and (1.24) to find the elements of the "off-diagonal" blocks $\mathbf{R}_{21}$ and $\mathbf{R}_{12}$; (3) finally, use $\mathbf{R}_{12}$ in (1.23) to find $\mathbf{R}_{22}$. The process of solving these four matrix equations parallels exactly the logic used on the four regular equations in (1.11).

Finding $\mathbf{A}^{-1}$ in this way requires finding inverses to two smaller matrices—$\mathbf{A}_{22}$, which is 2×2 in this example, and $[\mathbf{A}_{11} - \mathbf{A}_{12}(\mathbf{A}_{22})^{-1}\mathbf{A}_{21}]$, which is also 2×2—and performing several matrix multiplications. For $\mathbf{A}$ matrices that are large, this approach may be easier than using the general definition in (1.18), which requires the determinant and the adjoint of the entire $\mathbf{A}$ matrix. In addition, for $\mathbf{A}$ matrices with particular structures, the solution via (1.21) through (1.24) may be especially simple. For example, suppose that $\mathbf{A}$ can be partitioned into

$$\left[\begin{array}{c:c} \mathbf{I} & \mathbf{A}_{12} \\ \hdashline \mathbf{0} & \mathbf{A}_{22} \end{array} \right]$$

Then

$$\mathbf{A}^{-1} = \left[\begin{array}{c:c} \mathbf{I} & -\mathbf{A}_{12}(\mathbf{A}_{22})^{-1} \\ \hdashline \mathbf{0} & (\mathbf{A}_{22})^{-1} \end{array} \right]$$

for which only one smaller inverse, $(\mathbf{A}_{22})^{-1}$, needs to be found.

Cramer's Rule (G. Cramer, Swiss, 1704–1752)   This rule says that each of the x_i in an $n \times n$ system of linear equations, $\mathbf{AX} = \mathbf{B}$, can be found as the ratio of two determinants:

$$x_i = \frac{|\mathbf{A}_{Bi}|}{|\mathbf{A}|}$$

The numerator is the determinant of a matrix formed from $\mathbf{A}$ by replacing its ith column (when solving for x_i) by $\mathbf{B}$. This approach to equation solving requires evaluating $n + 1$ determinants of size $n \times n$. For $n > 3$, this is not usually a very efficient procedure. However, in light of our understanding of the general form of the inverse and of the general definition of a determinant, it is easy to see where this rule comes from. From (1.18), $\mathbf{A}^{-1} = (1/|\mathbf{A}|)\,[\mathrm{adj}\ \mathbf{A}]$, so

$$\mathbf{X} = \mathbf{A}^{-1}\mathbf{B} = \frac{1}{|\mathbf{A}|}[\mathrm{adj}\ \mathbf{A}]\mathbf{B} = \frac{1}{|\mathbf{A}|}
\begin{bmatrix}
\mathbf{A}_{11} & \mathbf{A}_{21} & \cdots & \mathbf{A}_{n1} \\
\mathbf{A}_{12} & \mathbf{A}_{22} & \cdots & \mathbf{A}_{n2} \\
\vdots & \vdots & & \vdots \\
\mathbf{A}_{1n} & \mathbf{A}_{2n} & \cdots & \mathbf{A}_{nn}
\end{bmatrix}
\begin{bmatrix}
b_1 \\ b_2 \\ \vdots \\ b_n
\end{bmatrix}$$

(Remember that the subscripts on the elements in the adjoint are reversed from their usual order because we are dealing with cofactors of the *transpose* of $\mathbf{A}$.) The value for a particular x_i is given by $(1/|\mathbf{A}|)$ times the product of the ith row of the adjoint and $\mathbf{B}$:

$$x_i = \frac{1}{|\mathbf{A}|}(\mathbf{A}_{1i}b_1 + \mathbf{A}_{2i}b_2 + \cdots + \mathbf{A}_{ni}b_n) \tag{1.25}$$

Define $\mathbf{A}_{Bi}$ as the original $\mathbf{A}$ matrix with $\mathbf{B}$ replacing its ith column:

$$\overbrace{\phantom{i\text{th column}}}^{i\text{th column}}$$

$$\mathbf{A}_{Bi} =
\begin{bmatrix}
a_{11} & a_{12} & \cdots & b_1 & \cdots & a_{1n} \\
a_{21} & a_{22} & \cdots & b_2 & \cdots & a_{2n} \\
\vdots & \vdots & & \vdots & & \vdots \\
a_{n1} & a_{n2} & \cdots & b_n & \cdots & a_{nn}
\end{bmatrix}$$

Using (1.16), evaluate $|\mathbf{A}_{Bi}|$ down this new ith column, giving

$$|\mathbf{A}_{Bi}| = b_1\mathbf{A}_{1i} + b_2\mathbf{A}_{2i} + \cdots + b_n\mathbf{A}_{ni}$$

But this is just the term in parentheses on the right-hand side of (1.25), so that expression can be condensed to

$$x_i = \frac{1}{|\mathbf{A}|}|\mathbf{A}_{Bi}| = \frac{|\mathbf{A}_{Bi}|}{|\mathbf{A}|}$$

which is just Cramer's rule.

Here is an illustration. For our job retraining example, in (1.6), we had

$$\mathbf{A} = \begin{bmatrix} 5 & 7 \\ 1 & 2 \end{bmatrix} \quad \text{and} \quad \mathbf{B} = \begin{bmatrix} 2000 \\ 460 \end{bmatrix}$$

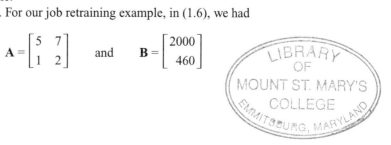

so

$$\mathbf{A}_{B1} = \begin{bmatrix} 2000 & 7 \\ 460 & 2 \end{bmatrix} \quad \text{and} \quad \mathbf{A}_{B2} = \begin{bmatrix} 5 & 2000 \\ 1 & 460 \end{bmatrix}$$

Using Cramer's rule

$$x_1 = \begin{vmatrix} 2000 & 7 \\ 460 & 2 \end{vmatrix} \div \begin{vmatrix} 5 & 7 \\ 1 & 2 \end{vmatrix} = \frac{780}{3} = 260$$

$$x_2 = \begin{vmatrix} 5 & 2000 \\ 1 & 460 \end{vmatrix} \div \begin{vmatrix} 5 & 7 \\ 1 & 2 \end{vmatrix} = \frac{300}{3} = 100$$

which is the same as our earlier results.

1.3.2   The Geometry of Simple Equation Systems: Solution Possibilities

To illustrate the direction in which our study of linear equation systems will move, consider a trivially simple ordinary algebraic equation, $ax = b$. We already looked at the solution procedure for $3x = 12$ earlier in this section. As long as $a \neq 0$, the unique solution is $x = (1/a)b = b/a = a^{-1}b$. When $a = 0$, so that the "equation" is $0x = b$, there are two possibilities (this is not very interesting for such a simple example, but it points the way to similar kinds of situations for much more complex systems of linear equations):

1. If $b = 0$, then x can take on *any* value; there are infinitely many "solutions" to this equation.
2. If $b \neq 0$, then there is *no* value of x that satisfies the equation; there are no "solutions" because the equation is *inconsistent*.

So, when a unique solution does not exist, we can distinguish an inconsistent equation having no solutions from an equation with an infinite number of solutions. In the former case, we might say that too much is required of the unknown (none can satisfy the equation), while in the latter case too little is required (too many unknowns can satisfy the equation). These distinctions hold for larger systems. In a particular situation it may be important to know, when a *unique* solution cannot be found, whether it is because there are no solutions at all or there are too many solutions. It is often preferable to learn that solutions do exist (although they are not unique) rather than to learn that none can possibly be found.

The 2×2 matrix case—two linear equations in two unknowns—illustrates these possibilities. In each case, we supplement the algebra with a geometric picture of the equations in x_1, x_2 space, where values of x_1 are plotted along one axis [usually the horizontal (abscissa)] and x_2 along a second axis [usually the vertical (ordinate)]. This is known as "solution space" for the problems, because if one or more solutions exist, they will appear as a point or points in this same figure. With only two unknowns we (conveniently) need only two axes, so the cases are easily illustrated.

Case 1: Unique Solution   The equations are

$$(1) \quad 3x_1 + 2x_2 = 7$$
$$(2) \quad 2x_1 + 5x_2 = 12$$

(1.26)

In $\mathbf{AX} = \mathbf{B}$ form

$$\mathbf{A} = \begin{bmatrix} 3 & 2 \\ 2 & 5 \end{bmatrix}, \mathbf{B} = \begin{bmatrix} 7 \\ 12 \end{bmatrix}, \text{ and } \mathbf{X} = \begin{bmatrix} x_1 \\ x_2 \end{bmatrix}$$

Since $|\mathbf{A}| \neq 0$, we know that a unique solution can be found algebraically as $\mathbf{X} = \mathbf{A}^{-1}\mathbf{B}$. (Try it; the solution is $\mathbf{X} = \begin{bmatrix} 1 \\ 2 \end{bmatrix}$.)

The geometry for this example is shown in Figure 1.1. The straight lines representing the relations in equations (1) and (2) in (1.26) intersect in a single point. This point simultaneously satisfies both equations; its coordinates represent the unique solution to the problem.

Case 2: No Solution   Suppose that the equations now are

$$(1) \quad 3x_1 + 2x_2 = 7$$
$$(2) \quad 6x_1 + 4x_2 = 11$$

(1.27)

The first equation is the same as in the previous example. Now $\mathbf{A} = \begin{bmatrix} 3 & 2 \\ 6 & 4 \end{bmatrix}$, $\mathbf{B} = \begin{bmatrix} 7 \\ 11 \end{bmatrix}$,

and $\mathbf{X}$ is the same. In this case, $|\mathbf{A}| = 0$, and we know that there can be no unique solution, $\mathbf{X} = \mathbf{A}^{-1}\mathbf{B}$.

The geometry is shown in Figure 1.2. There is no intersection of the lines in solution space; they are parallel. The requirements on x_1 and x_2 are inconsistent; there is no solution.

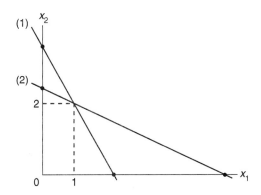

FIGURE 1.1   Solution-space representation of (1.26).

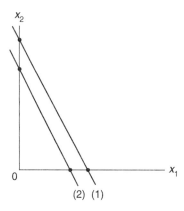

FIGURE 1.2   Solution-space representation of (1.27).

Case 3: Infinite Number of Solutions   We keep the same first equation and change only the right-hand side of the second, to

$$\begin{aligned}(1) \quad & 3x_1 + 2x_2 = 7 \\ (2) \quad & 6x_1 + 4x_2 = 14\end{aligned} \tag{1.28}$$

Now $\mathbf{A} = \begin{bmatrix} 3 & 2 \\ 6 & 4 \end{bmatrix}$ (as in case 2), $\mathbf{B} = \begin{bmatrix} 7 \\ 14 \end{bmatrix}$, and $\mathbf{X}$ is the same. Again, since $|\mathbf{A}| = 0$, we know that there is no unique solution.

The geometry is shown in Figure 1.3. We see that (1) and (2) represent exactly the same line in solution space [equation (2) is just twice equation (1)]. So any point along the single line in Figure 1.3 (which extends infinitely far in both northwesterly and southeasterly directions) satisfies both equations; there are infinitely many solutions in this case.

To sum up, with two linear equations and two unknowns, the absence of a unique solution is indicated by $|\mathbf{A}| = 0$. This is the parallel to $a = 0$ in the $ax = b$ case from ordinary al-

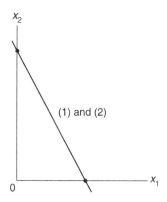

FIGURE 1.3   Solution-space representation of (1.28).

gebra. When $a = 0$ in the $ax = b$ case, the distinction between no solution and an infinite number of solutions hinged on b. The same is true for matrix equations. The only distinction between cases 2 and 3 (where $|\mathbf{A}| = 0$) was in the composition of $\mathbf{B}$. To understand completely the role of $\mathbf{B}$ in those cases, it is useful to turn to the visual aid provided by the geometry of vectors, from which the notions of linear dependence and independence will emerge. Then we will be able to classify all possible solution types for *any* linear equation. This will occupy us in Chapter 2.

1.4   THE GEOMETRY OF VECTORS

In this section we concentrate on a visual representation for two-element column vectors. The geometry would apply equally well to row vectors, and the logic of the representation can be extended to vectors with any number of elements—even though it is impossible to show for vectors of more than three elements, since we will use one axis for each element in vector space figures.

1.4.1   Linear Combinations of Vectors

We are concerned with *column* vectors because we will adopt a column-oriented approach for studying linear equation systems. For this purpose, it is useful to have the definition of a *linear combination of* vectors. Given n vectors, $\mathbf{V}_1, \mathbf{V}_2, \ldots, \mathbf{V}_n$, containing m elements each, and given n scalars, $\alpha_1, \ldots, \alpha_n$, then

$$\mathbf{V}_0 = \alpha_1 \mathbf{V}_1 + \alpha_2 \mathbf{V}_2 + \cdots + \alpha_n \mathbf{V}_n = \sum_{i=1}^{n} \alpha_i \mathbf{V}_i \tag{1.29}$$

is called a *linear combination* of the vectors $\mathbf{V}_1, \mathbf{V}_2, \ldots, \mathbf{V}_n$. When all $\alpha_i \geq 0$, this is called a *nonnegative linear combination* of the $\mathbf{V}_i$, and when, in addition to being nonnegative, $\sum_{i=1}^{n} \alpha_i = 1$, it is called a *convex combination*. (We will meet this last concept again in Chapter 3 when we deal with convex sets.)

The process of forming a linear combination of vectors encompasses only two kinds of operations—multiplication of vectors by scalars and addition of vectors (or subtraction, if some scalars are negative). The notion of linear combinations provides another way of expressing linear equations. Consider the general pair of linear equations in x_1 and x_2:

$$a_{11}x_1 + a_{12}x_2 = b_1$$
$$a_{21}x_1 + a_{22}x_2 = b_2$$

We already know that this can be represented, with appropriate matrix and vector definitions, as $\mathbf{AX} = \mathbf{B}$. Suppose, in addition, that we partition $\mathbf{A}$ into its two columns

$$\mathbf{A} = [\mathbf{A}_1 \,\vdots\, \mathbf{A}_2]$$

where $\mathbf{A}_j$ represents the jth column of $\mathbf{A}$. Then, in conjunction with $\mathbf{X} = \begin{bmatrix} x_1 \\ x_2 \end{bmatrix}$ and $\mathbf{B} = \begin{bmatrix} b_1 \\ b_2 \end{bmatrix}$, we can write this pair of linear equations in linear combination form (1.29) as

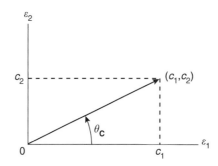

FIGURE 1.4   Vector-space representation of **C**.

$$\mathbf{A}_1 x_1 + \mathbf{A}_2 x_2 = \mathbf{B} \tag{1.30}$$

The structure is the same for n linear equations in n unknowns. Let the $n \times n$ matrix **A** be partitioned by columns:

$$\mathbf{A} = [\mathbf{A}_1 \mathbin{\vdots} \mathbf{A}_2 \mathbin{\vdots} \ \cdots \ \mathbin{\vdots} \mathbf{A}_j \mathbin{\vdots} \ \cdots \ \mathbin{\vdots} \mathbf{A}_n]$$

Then the $n \times n$ system represented in $\mathbf{AX} = \mathbf{B}$ is equivalently shown as

$$\mathbf{A}_1 x_1 + \mathbf{A}_2 x_2 + \cdots + \mathbf{A}_j x_j + \cdots + \mathbf{A}_n x_n = \mathbf{B}$$

We want to develop a geometric understanding of this linear combination form for expressing linear equations.

It is straightforward to associate a picture with any two-element column vector, $\mathbf{C} = \begin{bmatrix} c_1 \\ c_2 \end{bmatrix}$. Using a pair of axes like the horizontal and vertical in solution space, measure a distance equal to c_1 along the horizontal axis and measure c_2 on the vertical axis. A point in this two-dimensional space with coordinates (c_1, c_2) is uniquely associated with the vector **C**. The (algebraic) elements in **C** locate the (geometric) point in a figure whose horizontal and vertical axes carry the labels "first element" (for which we use the label ε_1) and "second element" (ε_2), respectively. Usually the point (c_1, c_2) is connected to the origin in the picture and an arrowhead is put at the point in question; this directed line is called a *ray* from the origin, as in Figure 1.4.

Any two-element vector can be completely described either by its two *coordinates* in this geometric space or by its *length* and also its *direction,* measured by the angle that it makes with the horizontal axis (often denoted θ_C). We examine these two spatial attributes in turn.

Length   The length, or the *norm,* of a vector **C**, usually denoted $\|\mathbf{C}\|$, is $\sqrt{(c_1^2 + c_2^2)}$. This is a straightforward application of the Pythagorean theorem.[10] The analogous concept for an n-element vector is $(\Sigma_{i=1}^n c_i^2)^{1/2}$.

---

[10]There are many possible definitions of the norm of a *matrix.* One that is often used is the maximum of the column sums of the absolute values in the matrix. There are also other definitions of *vector* norms (Section 6.4).

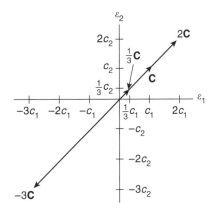

FIGURE 1.5   Scalar multiples of the vector **C**.

Direction   The size of θ_C is measured by either of two trigonometric relations: its sine or its cosine. These are defined as the following ratios: $\sin \theta_C = c_2/\|C\|$ and $\cos \theta_C = c_1/\|C\|$.

With this correspondence between the algebraic and geometric representations for a two-element column vector, several matrix algebra operations, applied to vectors, have immediate geometric interpretations. In particular, we explore the geometry of (1) multiplication by a scalar and (2) addition and subtraction, since these are the fundamental operations used in linear combinations.

1.4.2   The Geometry of Multiplication by a Scalar

Multiplication of a vector by a scalar simply stretches or shrinks the vector.[11] For example, $2C$ doubles the length of the arrow representing C; its norm is changed (doubled) but its direction is not. Specifically,

$$\|2C\| = \sqrt{(2c_1)^2 + (2c_2)^2} = 2\sqrt{(c_1^2 + c_2^2)} = 2\|C\|$$

$$\sin \theta_{2C} = \frac{2c_2}{\|2C\|} = \frac{c_2}{\|C\|} = \sin \theta_C$$

Similarly, $\frac{1}{3}C$ shrinks the arrow to one-third of its previous length, again with no change in direction. However, $-3C$ stretches the arrow to 3 times its original length *and* changes its orientation by $180°$—flips it over.[12] Figure 1.5 illustrates this process.

A vector is stretched or shrunk depending on whether the absolute value of the scalar is greater or less than 1; the orientation in space is reversed or not depending on whether the sign of the scalar is negative or positive.

---

[11]The fact that multiplication of a vector by an ordinary number changes only the *scale* of the vector constitutes the origin of the term *scalar* for an ordinary number in matrix algebra.

[12]Proofs of the changes in θ_C that accompany multiplication by a negative number rely on some fundamental identities from trigonometry that need not concern us here.

1.4.3   The Geometry of Addition (and Subtraction)

Addition of two vectors is also easily represented in vector geometry. Given $\mathbf{C} = \begin{bmatrix} c_1 \\ c_2 \end{bmatrix}$ and $\mathbf{D} = \begin{bmatrix} d_1 \\ d_2 \end{bmatrix}$, we know that the sum will be

$$\mathbf{S} = \mathbf{C} + \mathbf{D} = \begin{bmatrix} c_1 + d_1 \\ c_2 + d_2 \end{bmatrix}$$

Figure 1.6 shows all three vectors.

Since order of addition is immaterial, we can visualize the summation process in either of two ways. In Figure 1.7a we begin with $\mathbf{C}$ located and then add $\mathbf{D}$, using (c_1, c_2) as the point of origin for that addition operation. This adds d_1 to the horizontal coordinate and d_2 to the vertical. In Figure 1.7b we begin with $\mathbf{D}$ in place and then add c_1 and c_2 at that point. By combining Figures 1.7a and 1.7b, we see that vector addition creates the diagonal of the parallelogram that is formed by $\mathbf{C}$ and $\mathbf{D}$ as adjacent sides. Subtraction involves the same principles. The geometry is illustrated in Figures 1.8 and 1.9.

1.4.4   Additional Vector Geometry

It will turn out to be useful to have a definition of the area enclosed by two vectors, since it plays a role in understanding the concepts of linear dependence and independence (Section 1.5). We continue to consider two-element column vectors; the generalization to n-element vectors should be clear in each case.

Distance between Two Vectors   Using $\mathbf{C}$ and $\mathbf{D}$ from above, the distance between their endpoints in vector space is

$$\sqrt{(c_1 - d_1)^2 + (c_2 - d_2)^2}$$

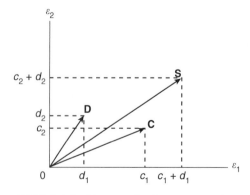

FIGURE 1.6   The sum (**S**) of vectors **C** and **D**.

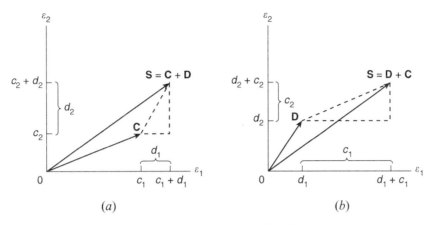

FIGURE 1.7   (*a*) Adding **D** to **C**; (*b*) adding **C** to **D**.

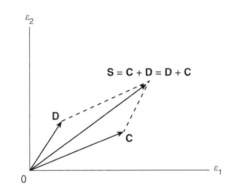

FIGURE 1.8   The sum (**S**) as the parallelogram diagonal.

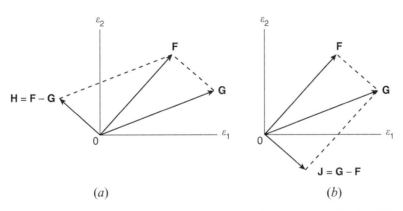

FIGURE 1.9   (*a*) The difference (**H** = **F** − **G**) as a parallelogram side; (*b*) the difference (**J** = **G** − **F**) as a parallelogram side.

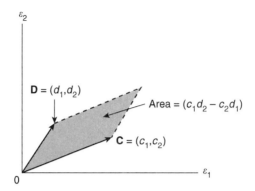

FIGURE 1.10   Parallelogram as the area between vectors **C** and **D**.

(Check Fig. 1.6.[13]) The geometric notion of *distance* between **C** and **D** is equivalent to the norm of the vector representing their *difference,* and so the notation $\|\mathbf{C} - \mathbf{D}\|$ is appropriate.

Angle Included between Two Vectors   The size of the angle included between **C** and **D**, denoted by ϕ, is measured by its cosine, which can be shown to be[14]

$$\cos \phi = \frac{c_1 d_1 + c_2 d_2}{\|\mathbf{C}\| \, \|\mathbf{D}\|} = \frac{\mathbf{C} \circ \mathbf{D}}{\|\mathbf{C}\| \, \|\mathbf{D}\|}$$

Area Enclosed by Two Vectors   Imagine the parallelogram formed by **C** and **D** as adjacent sides, as in Figure 1.10. The *area* of this figure can be shown to be precisely ($c_1 d_2 - c_2 d_1$). (Again, the demonstration requires rules from trigonometry, and we omit it.)

This is an important result. Create a matrix made up of the two column vectors, $[\mathbf{C} \vdots \mathbf{D}]$. Then the area of the parallelogram in Figure 1.10 is just $|\mathbf{C} \vdots \mathbf{D}|$, the *determinant* of the matrix made up of the two columns. Of particular interest is the case where the area is zero. Geometrically, no area means that **C** and **D** are collinear—they lie along the same ray, and hence there is either no angle between them or else the angle is 180°, so they point in exactly opposite directions. Since **C** and **D** both emanate from the origin, they are also said to be parallel. In either case, no parallelogram can be formed, as Figure 1.11 illustrates.

For the 2 × 2 case, we have seen that when $|\mathbf{A}| = 0$ the equation system $\mathbf{AX} = \mathbf{B}$ has no unique solution because an inverse cannot be found. Now the singularity of **A** is seen to have a particular geometric interpretation; its columns are collinear (lie on the same line). We generalize these observations in the next section.

---

[13]For three-element vectors (three-dimensional vector space) the differences between coordinates on all three axes must be used. Hence, a term $(c_3 - d_3)^2$ would be added under the square root. (If you draw well in three dimensions, you can derive this result.) For *n*-dimensional vectors, we have $\left[\sum_{i=1}^{n} (c_i - d_i)^2 \right]^{1/2}$.

[14]The demonstration depends on the *law of cosines* from trigonometry, and we need not bother with it.

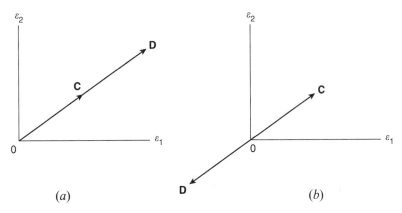

FIGURE 1.11   (*a*) Angle of 0° between **C** and **D**; (*b*) angle of 180° between **C** and **D**.

1.5   LINEAR DEPENDENCE AND INDEPENDENCE

The concepts of linear dependence and independence between vectors and of the rank of a matrix provide a general framework within which to study the kinds of solutions that are possible for linear equation systems of any size.

From the linear combination representation of systems of linear equations, we see that the problem of solving the equation system can be viewed as one of finding the appropriate scalars, $x_1, \ldots, x_n$, to express **B** as a linear combination of $\mathbf{A}_1, \ldots, \mathbf{A}_n$. For a system of two linear equations in two unknowns, $\mathbf{A}_1, \mathbf{A}_2$, and **B** can all be represented in two-dimensional vector space. The linear combination on the left in (1.30) employs two operations: (1) possible stretching or shrinking of each vector (multiplication of each $\mathbf{A}_i$ by its associated x_i) and then (2) adding these new vectors together.

A general illustration is shown in Figure 1.12. From our understanding of the geometry of multiplication of a vector by a scalar (stretching or shrinking the vector) and of vector addition (diagonal of a parallelogram), it should be apparent that *some* stretching and combining of $\mathbf{A}_1$ and $\mathbf{A}_2$ will lead to **B**. In Figure 1.12, draw a line from **B** parallel to $\mathbf{A}_2$ (toward the ε_1 axis) and another from **B** that is parallel to $\mathbf{A}_1$ (toward the ε_2 axis). Then the

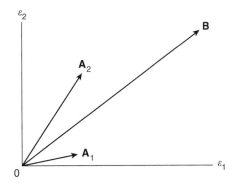

FIGURE 1.12   Vector-space representation of (1.30).

original $\mathbf{A}_1$ and $\mathbf{A}_2$ must be extended until they intersect these lines (at $\mathbf{A}_1^*$ and $\mathbf{A}_2^*$ in the figure), since $\mathbf{B}$ will be the diagonal of that parallelogram.

As long as $\mathbf{A}_1$ and $\mathbf{A}_2$ are not parallel, it will always be possible to find scalars to adjust $\mathbf{A}_1$ and $\mathbf{A}_2$ so that their sum is $\mathbf{B}$. (You should sketch other $\mathbf{A}_1$, $\mathbf{A}_2$, and $\mathbf{B}$ vectors and imagine the pushing and pulling required to express $\mathbf{B}$ as a linear combination of the $\mathbf{A}_i$'s.)

Any pair of nonparallel two-element vectors is said to *span* two-dimensional vector space. These *spanning* vectors are also called *basic* vectors or are said to provide a *basis* for that space. They may be thought of as a (new) pair of axes for the space; the fact that they may not be perpendicular, as we normally think of axes, is not important. What is important is that any other point in two-dimensional vector space, like $\mathbf{B}$, can be located using coordinates along *these* axes. The measures of unit length in the two directions are just the original lengths (*norms*) of $\mathbf{A}_1$ and $\mathbf{A}_2$. Suppose for a particular example in $\mathbf{AX} = \mathbf{B}$ form that we find $x_1 = 1.5$ and $x_2 = 1.3$. In linear combination form, this means

$$(1.5)\mathbf{A}_1 + (1.3)\mathbf{A}_2 = \mathbf{B}$$

The 1.5 measures distance out the $\mathbf{A}_1$ direction (1.5 times the length of $\mathbf{A}_1$) and 1.3 does the same thing for the $\mathbf{A}_2$ direction. This is illustrated (approximately) in Figure 1.13.

In fact, as with "ordinary" axes and units of length along them, any point $\mathbf{B}$ is located *uniquely* relative to the axes formed by $\mathbf{A}_1$ and $\mathbf{A}_2$. And if $\mathbf{A}_1$ and $\mathbf{A}_2$ are not parallel, then $|\mathbf{A}| = |\mathbf{A}_1 \vdots \mathbf{A}_2| \neq 0$, as we saw in Section 1.3. So the requirement of nonsingularity of $\mathbf{A}$, which we needed in order to find $\mathbf{A}^{-1}$, is the same as the requirement that the columns of $\mathbf{A}$ be nonparallel.

All of this is easily visualized in two dimensional vector space, for two linear equations in two unknowns. Again, we need to be able to generalize to larger systems, where the geometry will fail us. For that we need additional terminology. When $|\mathbf{A}| = |\mathbf{A}_1 \vdots \mathbf{A}_2| \neq 0$, the columns of $\mathbf{A}$ are said to be *linearly independent*. The property of linear independence can be described in a more general way:

If the only x_1 and x_2 which satisfy $\mathbf{A}_1 x_1 + \mathbf{A}_2 x_2 = 0$ are $x_1 = 0$ and $x_2 = 0$, then $\mathbf{A}_1$ and $\mathbf{A}_2$ are linearly independent. Conversely, if there exist x_1 and x_2 *not both zero* for which $\mathbf{A}_1 x_1 + \mathbf{A}_2 x_2 = 0$, then $\mathbf{A}_1$ and $\mathbf{A}_2$ are *linearly dependent*.

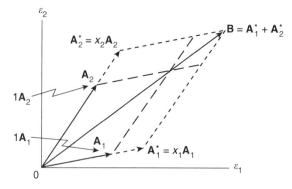

FIGURE 1.13   Solution values to (1.30) as scalars in the linear combination.

This statement may appear cumbersome, but it has a very straightforward geometric interpretation. The right-hand side is the origin in two-dimensional vector space, $\begin{bmatrix} 0 \\ 0 \end{bmatrix}$. If $\mathbf{A}_1$ and $\mathbf{A}_2$ are linearly independent, as in Figure 1.14*a,* then the only way to stretch or shrink them before adding them together to get to the origin is to shrink them both to zero—$(0)\mathbf{A}_1 + (0)\mathbf{A}_2$. By contrast, if $\mathbf{A}_1$ and $\mathbf{A}_2$ are linearly dependent, as in Figure 1.14*b,* then there are many ways (an infinite number, in fact) of stretching, shrinking and combining them to get to the origin. From the figure, it appears that something like twice $\mathbf{A}_2$ ($x_2 = 2$) added to 4 times $\mathbf{A}_1$ *in the opposite direction* ($x_1 = -4$) would lead to $\begin{bmatrix} 0 \\ 0 \end{bmatrix}$. As would $(-2)\mathbf{A}_1 + (1)\mathbf{A}_2$, or $(-10)\mathbf{A}_1 + (5)\mathbf{A}_2$, and so forth. This logic holds for any number of vectors with any number of elements each.

LINEAR DEPENDENCE AND INDEPENDENCE   The vectors $\mathbf{V}_1, \ldots, \mathbf{V}_n$ in *m*-dimensional space are *linearly dependent* if there are *n* scalars $x_1, \ldots, x_n$ *not all zero* for which

$$\mathbf{A}_1 x_1 + \cdots + \mathbf{A}_n x_n = 0$$

If the *only* set of x_i for which $\mathbf{A}_1 x_1 + \cdots + \mathbf{A}_n x_n = 0$ holds is $x_1 = \cdots = x_n = 0$, then $\mathbf{A}_1, \ldots, \mathbf{A}_n$ are *linearly independent.*

An alternative definition of linear dependence is as follows:

If $\mathbf{A}_n$ is a *linear combination* of $\mathbf{A}_1, \ldots, \mathbf{A}_{n-1}$, then $\mathbf{A}_1, \ldots, \mathbf{A}_n$ are *linearly dependent* vectors.

To see why this is true, look at the linear combination

$$\mathbf{A}_n = \mathbf{A}_1 x_1 + \cdots + \mathbf{A}_{n-1} x_{n-1}$$

where not all $x_1, \ldots, x_{n-1}$ are zero. Then a simple rearrangement gives

$$\mathbf{A}_1 x_1 + \cdots + \mathbf{A}_{n-1} x_{n-1} + \mathbf{A}_n(-1) = 0$$

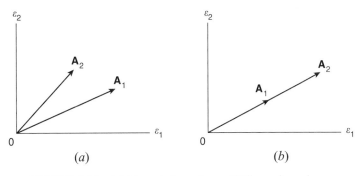

FIGURE 1.14   (*a*) Linear independence; (*b*) linear dependence.

Since not all $x_i = 0$ ($i = 1, \ldots, n-1$) and, moreover, $x_n = -1$, the requirement of linear dependence of $A_1, \ldots, A_n$ is met.[15]

From this it follows that not more than two vectors can be linearly independent in two-dimensional vector space. Once we have two linearly independent vectors (spanning vectors, basis vectors), as in A_1 and A_2 in Figure 1.12, then *any* other two-element vector, B, can be expressed as some linear combination of A_1 and A_2; this means that the three vectors—A_1, A_2, and B—must be linearly dependent. The generalization to any number of dimensions is as follows:

> The maximum number of linearly independent vectors in m-dimensional vector space (vectors of m elements each) is m; more than m vectors in m-dimensional vector space must necessarily be linearly dependent.

The concept of linear dependence and independence provides a convenient framework within which to classify equation systems, and this is still our primary objective. Essential for this purpose is the definition of the *rank* of a matrix:

RANK OF A MATRIX   The rank of A, usually denoted $\rho(A)$, is defined as the number of linearly independent columns in A. It is also defined as the size (dimension) of the largest nonsingular matrix that can be found in A.

We explore these definitions for a few examples. Let

$$A = \begin{bmatrix} 1 & 2 & 3 \\ 2 & 7 & 1 \end{bmatrix}, \qquad B = \begin{bmatrix} 1 & 2 & 2 \\ 2 & 7 & 4 \end{bmatrix}, \qquad C = \begin{bmatrix} 1 & 2 & 3 \\ 2 & 4 & 6 \end{bmatrix},$$

$$D = \begin{bmatrix} 1 & 2 & 3 \\ 2 & 5 & 6 \\ 2 & 4 & 6 \end{bmatrix}, \qquad E = \begin{bmatrix} 1 & 0 & 0 \\ 0 & 1 & 0 \\ 0 & 0 & 1 \end{bmatrix}, \qquad F = \begin{bmatrix} 1 & 2 \\ 2 & 5 \\ 3 & 4 \end{bmatrix}$$

1. $\rho(A) = 2$. A_1 and A_2 are linearly independent (as are A_1 and A_3 or A_2 and A_3). An easy way to see that is to note that $|A_1 \vdots A_2| \neq 0$, $|A_1 \vdots A_3| \neq 0$ and $|A_2 \vdots A_3| \neq 0$. Since each column in A has two elements (A has two rows), once we have identified two columns of A as linearly independent, we know that $\rho(A) = 2$. The maximum number of linearly independent two-element columns is 2. You can add as many new linearly independent columns to A as you wish; they will not increase $\rho(A)$.

This illustrates a general fact—the maximum number of linearly independent columns in an $m \times n$ matrix A is the *smaller* of the two dimensions. This is because the size of the largest square (sub)matrix in A will be the smaller of m and n, and only if at least one such submatrix is nonsingular will $\rho(A)$ take its largest possible value. Since the determinant of a matrix is the same as the determinant of the transpose of the matrix, it also follows that $\rho(A) = \rho(A')$. This, in turn, means that the number of linearly independent columns in A is

---

[15]For m-element vectors, when A_n is the null vector (the origin, in m-dimensional vector space), then the scalars in $A_n = A_1 x_1 + \cdots + A_{n-1} x_{n-1}$ *may* all be zero, but since $x_n = -1$ in $A_1 x_1 + \cdots + A_{n-1} x_{n-1} + A_n(-1) = 0$, the complete set of vectors, $A_1, \ldots, A_n$, still qualifies as linearly dependent. This is the reasoning behind the convention that the null vector is considered a linear combination of any other vectors (of the same dimension), and hence any set of vectors that includes the null vector must necessarily be linearly dependent.

the same as the number of linearly independent rows in **A**. (This is easy to see in the 2×2 case but it is less obvious for larger matrices.) Therefore, a third way to define $\rho(\mathbf{A})$ is as the number of *linearly independent rows* in **A**.

2. $\rho(\mathbf{B}) = 2$. Here $\mathbf{B}_1$ and $\mathbf{B}_2$ are the same as $\mathbf{A}_1$ and $\mathbf{A}_2$ in **A**. The fact that $\mathbf{B}_1$ and $\mathbf{B}_3$ are linearly dependent plays no role, as long as some other pair are linearly independent. The important observation that establishes $\rho(\mathbf{B}) = 2$ is that $|\mathbf{B}_1 \ \vdots \ \mathbf{B}_2| \neq 0$ (or that $|\mathbf{B}_2 \ \vdots \ \mathbf{B}_3| \neq 0$); the fact that $|\mathbf{B}_1 \ \vdots \ \mathbf{B}_3| = 0$ is irrelevant.

3. $\rho(\mathbf{C}) = 1$. Here $|\mathbf{C}_1 \ \vdots \ \mathbf{C}_2| = 0$, $|\mathbf{C}_1 \ \vdots \ \mathbf{C}_3| = 0$ and $|\mathbf{C}_2 \ \vdots \ \mathbf{C}_3| = 0$. There is no 2×2 submatrix in **C** that is nonsingular. There are, however, 1×1 nonsingular submatrices (any individual element; remember that the determinant of a 1×1 matrix is defined to be the element itself). Since $\mathbf{C}_2 = 2\mathbf{C}_1$ and $\mathbf{C}_3 = 3\mathbf{C}_1$, it follows immediately that, for example, $(2)\mathbf{C}_1 + (-1)\mathbf{C}_2 = \mathbf{0}$, $(3)\mathbf{C}_1 + (-1)\mathbf{C}_3 = \mathbf{0}$, or $(12)\mathbf{C}_1 + (-3)\mathbf{C}_2 + (-2)\mathbf{C}_3 = \mathbf{0}$, demonstrating linear dependence of the three columns.[16]

4. $\rho(\mathbf{D}) = 2$. We know that $|\mathbf{D}| = 0$, because $\mathbf{D}_3 = (3)\mathbf{D}_1$ [Property 4(b) for determinants; proportional columns]. So $\rho(\mathbf{D}) < 3$. There are plenty of nonsingular 2×2 submatrices in **D**, so $\rho(\mathbf{D}) = 2$. For example, choosing only rows and columns 1 and 2, $\begin{vmatrix} 1 & 2 \\ 2 & 5 \end{vmatrix} = 1$ ($\neq 0$). Neither columns $\mathbf{D}_1$ and $\mathbf{D}_2$ nor columns $\mathbf{D}_2$ and $\mathbf{D}_3$ are proportional; they are linearly independent in three-dimensional vector space. Instead of spanning this space, the three vectors lie in the same *plane,* so any vector **B** that does not lie on that plane cannot be expressed as a linear combination of the $\mathbf{D}_i$'s.

5. $\rho(\mathbf{E}) = 3$. In this case (the 3×3 identity matrix), $|\mathbf{E}| = 1$ (a fact that is always true for identity matrices), so the three columns are linearly independent.

6. $\rho(\mathbf{F}) = 2$. Again, the 2×2 submatrix formed by choosing the first two rows of **F** is nonsingular. Even though there are three rows, they are not linearly independent. Since these rows are two-element vectors, once a linearly independent *pair* of them has been found, the third (remaining) row can be expressed as a linear combination of those two basic vectors. For example, using $\mathbf{R}_1 = [1 \quad 2]$ and $\mathbf{R}_2 = [2 \quad 5]$, we obtain

$$\mathbf{R}_3 = [3 \quad 4] = (7)\mathbf{R}_1 + (-2)\mathbf{R}_2$$

1.6 QUADRATIC FORMS

Thus far in this chapter we have emphasized the usefulness of matrix representation for linear equation systems and the operations of matrix algebra for finding solutions (where possible) to those systems. Matrices are also very helpful in representing specific kinds of nonlinear functions called *quadratic forms.* Quadratic forms, in turn, play an important role in the theory of maximization and minimization of functions, to which we will turn in later parts of this book. In particular, it will be useful to explore conditions under which the *sign* of the quadratic form can be established unambiguously. We examine the nature of quadratic forms and their signs in this section.

---

[16]When $\rho(\mathbf{M}) = 1$, all the columns of **M** are parallel; each is just a multiple (stretching or shrinking) of any other. So it might seem more correct to say that there are *no* linearly independent columns in **M**. However, given any *one* column, $\mathbf{M}_i$, any other column, $\mathbf{M}_j$, is a linear combination of (proportional to) $\mathbf{M}_i$. In that sense, the number of linearly *independent* columns in **M** may be (and is) considered to be 1.

1.6.1   Basic Structure of Quadratic Forms

A quadratic form of two variables, x_1 and x_2, is any function in which each variable may appear in only two ways: squared and/or multiplied by the other variable. For example,

$$2x_1^2 + 6x_1x_2 + 5x_2^2$$

is a quadratic form in x_1 and x_2. The name simply describes a particular algebraic structure. (A quadratic form may be a function of any number of variables, as we shall see.) Again, precisely because of the way in which matrix multiplication is defined, this function can be written

$$[x_1 \quad x_2]\begin{bmatrix} 2 & 3 \\ 3 & 5 \end{bmatrix}\begin{bmatrix} x_1 \\ x_2 \end{bmatrix}$$

where the two 3's are derived as $\frac{6}{2}$, one-half of the coefficient attached to the x_1x_2 term. Carrying out the multiplication of the second and third terms first, we have

$$[x_1 \quad x_2]\begin{bmatrix} (2x_1 + 3x_2) \\ (3x_1 + 5x_2) \end{bmatrix}$$

and $2x_1^2 + 6x_1x_2 + 5x_2^2$ follows from multiplication of these two vectors.

The general quadratic form for two variables, $Q(x_1, x_2)$, can be represented as

$$Q(x_1, x_2) = a_{11}x_1^2 + (a_{12} + a_{21})x_1x_2 + a_{22}x_2^2 \tag{1.31}$$

Since any coefficient multiplying x_1x_2 can always be represented as the sum of two equal terms, this could as well be written for the general case as

$$Q(x_1, x_2) = a_{11}x_1^2 + 2a_{12}x_1x_2 + a_{22}x_2^2 \tag{1.32}$$

In matrix terms:

$$Q(x_1, x_2) = [x_1 \quad x_2]\begin{bmatrix} a_{11} & a_{12} \\ a_{21} & a_{22} \end{bmatrix}\begin{bmatrix} x_1 \\ x_2 \end{bmatrix} \tag{1.33}$$

or, letting $\mathbf{X} = \begin{bmatrix} x_1 \\ x_2 \end{bmatrix}$ and $\mathbf{A} = \begin{bmatrix} a_{11} & a_{12} \\ a_{21} & a_{22} \end{bmatrix}$ (where $a_{12} = a_{21}$, and $\mathbf{A}$ is said to be *symmetric*), (1.33) is just

$$\mathbf{X'AX} \tag{1.34}$$

The beauty of (1.34) is that it represents a quadratic form of any number of variables; nothing changes except the dimensions of the matrices involved. Thus

$$Q(x_1, x_2, x_3) = \mathbf{X'AX}$$

where

$$\mathbf{X} = \begin{bmatrix} x_1 \\ x_2 \\ x_3 \end{bmatrix} \text{ and } \mathbf{A} \text{ is the } 3 \times 3 \text{ } symmetric \text{ matrix } \begin{bmatrix} a_{11} & a_{12} & a_{13} \\ a_{21} & a_{22} & a_{23} \\ a_{31} & a_{32} & a_{33} \end{bmatrix},$$

where for all off-diagonal elements, $a_{ij} = a_{ji}$. An additional representation is

$$Q(x_1, x_2, x_3) = \sum_{i=1}^{3} \sum_{j=1}^{3} a_{ij} x_i x_j$$

This makes clear the quadratic nature of the function. When $i = j$, a variable is squared; otherwise (when $i \neq j$) *pairs* of variables are multiplied together. (The *number* of variables influences only the range of the subscripts i and j.)

For n variables, then

$$Q(x_1, \ldots, x_n) = \sum_{i=1}^{n} \sum_{j=1}^{n} a_{ij} x_i x_j = \mathbf{X}'\mathbf{A}\mathbf{X} \tag{1.35}$$

where $\mathbf{X}$ is an n-element column vector of the variables and $\mathbf{A}$ is an $n \times n$ symmetric matrix of the coefficients a_{ij}.

1.6.2 Rewritten Structure of Quadratic Forms

Consider $Q(x_1, x_2)$ as written in (1.32):

$$a_{11}x_1^2 + 2a_{12}x_1x_2 + a_{22}x_2^2$$

Note the difficulty that arises if we are trying to determine conditions on the values of the a_{ij} under which this expression would *always* be positive or *always* be negative, regardless of the (nonzero) values taken by x_1 and x_2. In the form of (1.32) the question could not be answered. The middle term, $2a_{12}x_1x_2$, would take on differing signs depending on the signs of x_1 and x_2. (Note that $a_{11}x_1^2$ will always have the same sign as a_{11} and $a_{22}x_2^2$, will always have the same sign as a_{22}—except for the case in which $x_1 = 0$ or $x_2 = 0$, and then these terms are themselves zero.)

What is needed, then, is an expression in which the x's all appear in terms that are squared; this removed any ambiguity about signs. One trick that helps here is to modify (1.32) by the process of "completing the square." Consider the first two terms; if $(a_{12}^2/a_{11})x_2^2$ is added to them, we would have

$$a_{11}x_1^2 + 2a_{12}x_1x_2 + \left(\frac{a_{12}^2}{a_{11}}\right)x_2^2$$

$$= a_{11}\left[x_1^2 + 2\left(\frac{a_{12}}{a_{11}}\right)x_1x_2 + \left(\frac{a_{12}^2}{a_{11}^2}\right)x_2^2\right]$$

$$= a_{11}\left[x_1 + \left(\frac{a_{12}}{a_{11}}\right)x_2\right]^2$$

But then the added term must also be subtracted from (1.32), if the expression is to remain unchanged. Thus from the third term in (1.32) we subtract $(a_{12}^2/a_{11})x_2^2$:

$$a_{22}x_2^2 - \left(\frac{a_{12}^2}{a_{11}}\right)x_2^2$$

$$= \left[a_{22} - \left(\frac{a_{12}^2}{a_{11}}\right)\right]x_2^2 = \left(\frac{a_{11}a_{22} - a_{12}^2}{a_{11}}\right)x_2^2$$

Putting this all together, (1.32) becomes

$$Q(x_1, x_2) = a_{11}\left[x_1 + \left(\frac{a_{12}}{a_{11}}\right)x_2\right]^2 + \left(\frac{a_{11}a_{22} - a_{12}^2}{a_{11}}\right)x_2^2 \tag{1.36}$$

which is a *sum of squares*. The sign of $Q(x_1, x_2)$ for cases in which x_1 and x_2 are both nonzero thus depends entirely on the two terms multiplying the squared expressions, namely on a_{11} and $(a_{11}a_{22} - a_{12}^2)/a_{11}$.

In principle, a quadratic form of three, four, ..., n variables can be written in this "sum of squares" form; in practice, the algebra becomes both tedious and tricky. But we will return to this basic notion—coefficients multiplying squared terms—as a guide to establishing the sign of a quadratic form, after we have explored some more general kinds of minors that can be defined for any square matrix.

The following terminology is in general use for describing quadratic forms. If Q is always positive ($Q > 0$) for *any* values that the x's assume (except when all x's are zero), it is termed *positive definite*.[17] If Q is always negative ($Q < 0$) except when all x's are zero, it is *negative definite*. If Q is either positive or zero ($Q \geq 0$) for all values of x (again, except when all are zero; specifically, this means that $Q = 0$ for some set of x_i not all zero), it is termed *positive semidefinite*.[18] And if the inequality is reversed ($Q \leq 0$), the quadratic form is said to be *negative semidefinite*. For example:

(a) $x_1^2 + 6x_2^2$ is positive definite.

(b) $-x_1^2 - 6x_2^2$ is negative definite.

(c) $x_1^2 - 4x_1x_2 + 4x_2^2 + x_3^2 = (x_1 - 2x_2)^2 + x_3^2$ is positive semidefinite. (It is never negative, but equals zero at $x_1 = 2, x_2 = 1, x_3 = 0$; $x_1 = 10, x_2 = 5, x_3 = 0$; and so on.)

(d) $-(x_1 - 2x_2)^2 - x_3^2$ is therefore negative semidefinite; it is just the negative of the expression in (c).

(e) $x_1^2 - 6x_2^2$ cannot be classed as either definite or semidefinite; it is *indefinite*.

1.6.3 Principal Minors and Permutation Matrices

Our present concern is to examine general rules that establish the signs of quadratic forms of any number of variables. These rules make use of definitions of several expanded classes of minors that can be found for a square matrix. Initially we define *leading principal minors* of a matrix. Then we expand the definition to the more general class of *principal*

---

[17]"Definitely positive" would seem to be more logical, and just "positive" would be in keeping with the "greater than zero" notation in general use in algebra.

[18]Again, "nonnegative" would seem to do as well.

minors (not only *leading*), using *permutation* matrices in the definition. We will find that these two classes of minors appear in the rules for establishing conditions for positive (or negative) definiteness and positive (or negative) semidefiniteness of a quadratic form. These rules, in turn, are used in distinguishing maxima from minima in optimization problems (in many places later in this book).

Leading Principal Minors The minor of an element a_{ij} in a square matrix $\mathbf{A}$, m_{ij}, was defined, in Section 1.3, as the determinant of the matrix remaining after one row (the ith) and one column (the jth) are removed from $\mathbf{A}$. Note that for elements a_{ii}, along the main diagonal of $\mathbf{A}$, the corresponding minors, m_{ii}, will be determinants of matrices derived from $\mathbf{A}$ by removing the ith row and its *corresponding* (ith) column.

The concept of a minor can be extended to include removal of more than one row and corresponding column. *Leading principal minors* constitute one such extension. We will use $|\mathbf{A}_k|$ to denote the kth-*order leading principal minor,* defined as the determinant of the matrix made up of the *first k* rows and columns in $\mathbf{A}$ ($k = 1, \ldots, n$). The vertical bars serve to remind us that we are dealing with a determinant and hence with a square matrix; $\mathbf{A}_k$ alone would still be used for the kth column of $\mathbf{A}$. [This can also be thought of as the determinant of the matrix formed from $\mathbf{A}$ by *dropping* the *last* $(n - k)$ rows and columns.] When $\mathbf{A}$ is of order 4, we have

$$|\mathbf{A}_1| = a_{11}, \qquad |\mathbf{A}_2| = \begin{vmatrix} a_{11} & a_{12} \\ a_{21} & a_{22} \end{vmatrix}, \qquad |\mathbf{A}_3| = \begin{vmatrix} a_{11} & a_{12} & a_{13} \\ a_{21} & a_{22} & a_{23} \\ a_{31} & a_{32} & a_{33} \end{vmatrix}, \qquad |\mathbf{A}_4| = |\mathbf{A}|$$

These n determinants, as a group, are the *leading principal minors* of $\mathbf{A}$. "Leading" serves to remind us that they are found from a particular sequence of matrices of increasing size, beginning in the northwest corner and expanding toward the southeast, simultaneously adding the next row and column at each step. "Principal" is used to indicate that they are related to elements located on the *main diagonal* of $\mathbf{A}$, so that if a given row is added the associated column is added also. (These matrices are also called the *nested* or *successive* or *naturally ordered principal minors* of $\mathbf{A}$.) Certain results on the sign of a quadratic form $\mathbf{X'AX}$ can be established by sign conditions on the leading principal minors of the $\mathbf{A}$ matrix in that quadratic form, as we will see below.

Permutation Matrices and Principal Minors In order to define a more general class of *principal minors* (not just "leading"), it is helpful to be able to use permutation matrices. We will find that this larger class of minors plays a role in establishing other (weaker) results on the sign of a quadratic form.

Permutation Matrices Consider a general 3×3 matrix $\mathbf{A}$ and another matrix, $\mathbf{P}$, which is the 3×3 identity matrix in which columns 2 and 3 (or rows 2 and 3, it amounts to the same thing) have been interchanged:

$$\mathbf{A} = \begin{bmatrix} a_{11} & a_{12} & a_{13} \\ a_{21} & a_{22} & a_{23} \\ a_{31} & a_{32} & a_{33} \end{bmatrix} \quad \text{and} \quad \mathbf{P} = \begin{bmatrix} 1 & 0 & 0 \\ 0 & 0 & 1 \\ 0 & 1 & 0 \end{bmatrix}$$

The product $\mathbf{PAP'}$ produces a new matrix which is the old $\mathbf{A}$ matrix with those same rows and columns interchanged (permuted).

*Pre*multiplication of **A** by **P** interchanges only *rows* 2 and 3 of **A**:

$$
\mathbf{PA} = \begin{bmatrix} 1 & 0 & 0 \\ 0 & 0 & 1 \\ 0 & 1 & 0 \end{bmatrix} \begin{bmatrix} a_{11} & a_{12} & a_{13} \\ a_{21} & a_{22} & a_{23} \\ a_{31} & a_{32} & a_{33} \end{bmatrix} = \begin{bmatrix} a_{11} & a_{12} & a_{13} \\ a_{31} & a_{32} & a_{33} \\ a_{21} & a_{22} & a_{23} \end{bmatrix}
$$

*Post*multiplication of **A** by **P'** interchanges only *columns* 2 and 3 of **A**:

$$
\mathbf{AP'} = \begin{bmatrix} a_{11} & a_{12} & a_{13} \\ a_{21} & a_{22} & a_{23} \\ a_{31} & a_{32} & a_{33} \end{bmatrix} \begin{bmatrix} 1 & 0 & 0 \\ 0 & 0 & 1 \\ 0 & 1 & 0 \end{bmatrix} = \begin{bmatrix} a_{11} & a_{13} & a_{12} \\ a_{21} & a_{23} & a_{22} \\ a_{31} & a_{33} & a_{32} \end{bmatrix}
$$

The combined operation interchanges both rows and columns 2 and 3 of **A** simultaneously:

$$
\mathbf{PAP'} = \begin{bmatrix} 1 & 0 & 0 \\ 0 & 0 & 1 \\ 0 & 1 & 0 \end{bmatrix} \begin{bmatrix} a_{11} & a_{12} & a_{13} \\ a_{21} & a_{22} & a_{23} \\ a_{31} & a_{32} & a_{33} \end{bmatrix} \begin{bmatrix} 1 & 0 & 0 \\ 0 & 0 & 1 \\ 0 & 1 & 0 \end{bmatrix} = \begin{bmatrix} a_{11} & a_{13} & a_{12} \\ a_{31} & a_{33} & a_{32} \\ a_{21} & a_{23} & a_{22} \end{bmatrix}
$$

There are six possible arrangements of the three rows and columns in **A**, and so there are six possible permutation matrices (including the 3×3 identity matrix, which leaves **A** unchanged).[19] For later reference, we list them here, starting with the 3×3 identity matrix:

$$
\mathbf{P}^{(1)} = \begin{bmatrix} 1 & 0 & 0 \\ 0 & 1 & 0 \\ 0 & 0 & 1 \end{bmatrix}, \quad \mathbf{P}^{(2)} = \begin{bmatrix} 1 & 0 & 0 \\ 0 & 0 & 1 \\ 0 & 1 & 0 \end{bmatrix}, \quad \mathbf{P}^{(3)} = \begin{bmatrix} 0 & 1 & 0 \\ 1 & 0 & 0 \\ 0 & 0 & 1 \end{bmatrix},
$$

$$
\mathbf{P}^{(4)} = \begin{bmatrix} 0 & 1 & 0 \\ 0 & 0 & 1 \\ 1 & 0 & 0 \end{bmatrix}, \quad \mathbf{P}^{(5)} = \begin{bmatrix} 0 & 0 & 1 \\ 0 & 1 & 0 \\ 1 & 0 & 0 \end{bmatrix}, \quad \mathbf{P}^{(6)} = \begin{bmatrix} 0 & 0 & 1 \\ 1 & 0 & 0 \\ 0 & 1 & 0 \end{bmatrix}
$$

Identifying exactly what is being done by any particular permutation matrix can be a little tricky at first. You might carry out the multiplications $[\mathbf{P}^{(i)}]\mathbf{A}[\mathbf{P}^{(i)}]'$ for each of the six permutation matrices above to see that they generate matrices in which the original three rows and columns of **A** appear in the following orders:

$$
[1,2,3], \quad [1,3,2], \quad [2,1,3], \quad [2,3,1], \quad [3,2,1], \quad [3,1,2]
$$

We will use permutation matrices below as one way of identifying all of the principal minors of a matrix.

---

[19]If you are familiar with the arithmetic of permutations, recall that the number of distinguishably different *arrangements* of n objects is $n!$ (read "n factorial" and defined only for n equal to a positive integer). This denotes the product of n and all the integers below it, down to 1—$n! = n(n-1) \ldots (n-2)(3)(2)(1)$. Here $n = 3$, so the number of permutations is 6.

Principal Minors   A *principal minor* is the determinant of the matrix that remains after *any particular* set of rows and their corresponding columns have been removed from **A**. The leading principal minors of **A** are a subset of all of the principal minors of **A**; recall that $|\mathbf{A}_k|$ is found as the determinant of the matrix remaining after *all* rows and columns from $(k + 1)$ onward are dropped from **A**.

A *first-order principal minor* is the determinant of the matrix that is left from **A** when *all but* 1 row and corresponding column are removed from **A**. Again, "principal" identifies the fact that we are dealing with minors of main-diagonal elements. Now, however, *any* set of $(n - 1)$ rows and corresponding columns may be deleted. For the general 3×3 matrix, there are three possible first-order principal minors;[20] rows and columns 2 and 3, 1 and 3, or 1 and 2 may be removed, leaving, in turn, a 1×1 matrix with the single element a_{11}, a_{22}, or a_{33}.

As to notation, it is useful to continue to use a variant of $|\mathbf{A}_1|$ (the first-order *leading* principal minor) in order to remember that these are minors derived from a specific matrix **A**. (A number of alternatives presented in the literature use letters unrelated to the one that designates the particular matrix.) We choose to use $\mathscr{A}_1$, but since there are *three* of these first-order principal minors, we need another index to distinguish among them. We will use $\mathscr{A}_1(1) = a_{11}$, $\mathscr{A}_1(2) = a_{22}$, and $\mathscr{A}_1(3) = a_{33}$; the number in parentheses serves simply as a counter for the various possible first-order principal minors. [Note that $\mathscr{A}_1(1)$ is also the one and only first-order *leading* principal minor of **A**.]

Second-order principal minors are found as determinants of the matrix remaining after *any* set of $(n - 2)$ rows and their corresponding columns are removed from **A**. For the $n = 3$ case, there are three possible second-order principal minors, resulting from removal of row and column 3, or 2, or 1.[21] In the more general notation of this section,

$$|\mathscr{A}_2(1)| = \begin{vmatrix} a_{11} & a_{12} \\ a_{21} & a_{22} \end{vmatrix}$$

results when row and column 3 are removed. Similarly

$$|\mathscr{A}_2(2)| = \begin{vmatrix} a_{11} & a_{13} \\ a_{31} & a_{33} \end{vmatrix} \qquad \text{and} \qquad |\mathscr{A}_2(3)| = \begin{vmatrix} a_{22} & a_{23} \\ a_{32} & a_{33} \end{vmatrix}$$

are the second-order principal minors generated by removing row and column 2 or 1, respectively.

Finally, the *nth-order principal minor* is the determinant of **A** itself—officially, it is the determinant of the matrix remaining when $(n - n) = 0$ rows and columns are deleted from **A**. Therefore, for the $n = 3$ example

$$|\mathscr{A}_3(1)| = |\mathbf{A}|$$

---

[20]If you are familiar with the arithmetic of *combinations*, this is immediate from the formula for the number of combinations of three things (here rows and columns of **A**), taken two at a time (the two to be deleted)—$C_2^3 = 3!/2!1! = 3$. (Recall that this is the same as C_1^3, the number of combinations of three things taken one at a time—in this case the one row and column not deleted.)

[21]Referring to the general notation of a (simple) minor, in Section 1.3, these are m_{33}, m_{22}, and m_{11}. This will always be true. The $(n - 1)$st-order principal minors in an $n \times n$ matrix will be the same as the minors of the elements on the main diagonal of that matrix.

A 3×3 matrix has a total of seven principal minors: three first-order, three second-order, and one third-order. The number of principal minors gets large quickly as the size of the matrix increases. A 4×4 matrix has 15, a 5×5 matrix has 31 (in fact, each larger matrix has twice the number of principal minors of the previous matrix plus one), . . . , and a 10×10 matrix has 1023.[22]

It can be tedious, in anything but a rather small matrix, to keep track of all the possible sets of rows and columns that can be removed to generate principal minors. However, there is a fairly automatic way to create all the possible principal minors of a matrix; it uses permutation matrices and the definition of *leading* principal minors. Put compactly, a *k*th-*order principal minor of* **A** is found as the *k*th order *leading* principal minor of **PAP**′, for *any* permutation matrix **P**. By using *all* possible permutation matrices and examining, for each, the first-, second-, . . . , *n*th order *leading* principal minors, one generates all principal minors—in fact, each one more than once!

For the $n = 3$ case, although there are six permutations $[\mathbf{P}^{(i)}]\mathbf{A}[\mathbf{P}^{(i)}]'$, there are only three *distinguishably different* first-order principal minors of **A**. We found them above, as $|\mathcal{A}_1(1)|$, $|\mathcal{A}_1(2)|$, and $|\mathcal{A}_1(3)|$. Two of the six permutations of **A** [using $\mathbf{P}^{(1)}$ or $\mathbf{P}^{(2)}$, from above] leave a_{11} in the upper left corner, another two [using $\mathbf{P}^{(3)}$ or $\mathbf{P}^{(4)}$] put a_{22} there, and the remaining two [$\mathbf{P}^{(5)}$ and $\mathbf{P}^{(6)}$] put a_{33} in that position. We now need additional notation to represent the various rearrangements of **A** caused by pre- and postmultiplication by a particular permutation matrix. For that purpose, we can define

$$\mathbf{A}^{(i)} = [\mathbf{P}^{(i)}]\mathbf{A}[\mathbf{P}^{(i)}].$$

Then, consistent with earlier notation for leading principal minors, we use $|\mathbf{A}_k^{(i)}|$ for the *k*th-order leading principal minor of $[\mathbf{P}^{(i)}]\mathbf{A}[\mathbf{P}^{(i)}]'$ ($i = 1, \ldots, 6$). For the 3×3 case, we have $|\mathbf{A}_1^{(1)}| = a_{11} = |\mathbf{A}_1^{(2)}|$, $|\mathbf{A}_1^{(3)}| = a_{22} = |\mathbf{A}_1^{(4)}|$, and $|\mathbf{A}_1^{(5)}| = a_{33} = |\mathbf{A}_1^{(6)}|$. [Again, $|\mathbf{A}_1^{(1)}| = |\mathbf{A}_1^{(2)}| = |\mathcal{A}^{(1)}|$ (from earlier) is also $|\mathbf{A}_1|$, the *first-order leading principal minor* of **A**.]

There are also six possible second-order principal minors that would be found as the second-order leading principal minors of all the permutations **PAP**′. For the 3×3 case, they are[23]

$$|\mathbf{A}_2^{(1)}| = \begin{vmatrix} a_{11} & a_{12} \\ a_{21} & a_{22} \end{vmatrix}, \qquad |\mathbf{A}_2^{(2)}| = \begin{vmatrix} a_{11} & a_{13} \\ a_{31} & a_{33} \end{vmatrix}, \qquad |\mathbf{A}_2^{(3)}| = \begin{vmatrix} a_{22} & a_{21} \\ a_{12} & a_{11} \end{vmatrix},$$

$$|\mathbf{A}_2^{(4)}| = \begin{vmatrix} a_{22} & a_{23} \\ a_{32} & a_{33} \end{vmatrix}, \qquad |\mathbf{A}_2^{(5)}| = \begin{vmatrix} a_{33} & a_{32} \\ a_{23} & a_{22} \end{vmatrix}, \qquad |\mathbf{A}_2^{(6)}| = \begin{vmatrix} a_{33} & a_{31} \\ a_{13} & a_{11} \end{vmatrix}$$

Again, not all of these determinants are distinguishably different. Recall from Section 1.3 that interchange of any pair of rows *or* columns in a square matrix changes the sign (but not the absolute value) of the determinant of that matrix. Therefore, if each row interchange is accompanied by a corresponding column interchange, the effect is to leave the determinant completely unchanged. Thus it is clear[24] that $|\mathbf{A}_2^{(1)}| = |\mathbf{A}_2^{(3)}|$, $|\mathbf{A}_2^{(2)}| = |\mathbf{A}_2^{(6)}|$ and

---

[22]Again, if you are comfortable with combinatorics, this is easily shown. For example, for $n = 5$, the number of first-, second-, . . . , fifth-order principal minors is $C_4^5 + C_3^5 + C_2^5 + C_1^5 + C_0^5 = 5 + 10 + 10 + 5 + 1 = 31 (= 2^5 - 1)$.

[23]The order in which the permutation matrices are listed is immaterial. Any one of the six could be counted as $i = 1$, any other one as $i = 2$, and so on.

[24]This is obvious in the case of determinants of 2×2 matrices, where the diagonals rule can easily be used—namely, that the determinant is equal to the product of the elements on the main diagonal less the product of the elements along the other diagonal.

$|\mathbf{A}_2^{(4)}| = |\mathbf{A}_2^{(5)}|$. $[|\mathbf{A}_2^{(1)}| (= |\mathbf{A}_2^{(3)}| = |\mathscr{A}_2(1)|$, from earlier) is also $|\mathbf{A}_2|$, the *second-order leading principal minor* of $\mathbf{A}$.]

Finally, one could also find six *third-order* principal minors, $|\mathbf{A}_3^{(1)}|, \ldots, |\mathbf{A}_3^{(6)}|$, from the determinants of the six possible $[\mathbf{P}^{(i)}]\mathbf{A}[\mathbf{P}^{(i)}]'$ permutations of $\mathbf{A}$, when *no* rows or columns are removed. There is no point in exploring these, however, since they will all be equal (again because interchange of any number of rows and the simultaneous interchange of the same columns leaves the determinant of the matrix unchanged). [Each $|\mathbf{A}_3^{(i)}| = |\mathscr{A}_3(1)|$ is also $|\mathbf{A}_3|$, the *third-order leading principal minor* of $\mathbf{A}$.] Therefore, finding principal minors of a matrix $\mathbf{A}$ by looking at the leading principal minors of all possible permutations $\mathbf{PAP}'$ is sure to find them all but it is, in the process, very repetitive.

The General Structure of a Quadratic Form Recall the $\mathbf{A}$ matrix in (1.33) and (1.34). The first critical term in (1.36) is just a_{11}, and the second critical term is the determinant of $\mathbf{A}$ divided by a_{11}. These are precisely the two leading principal minors of $\mathbf{A}$. Using the notation for those minors, (1.36) may be written

$$Q(x_1, x_2) = |\mathbf{A}_1| \left[x_1 + \left(\frac{a_{12}}{a_{11}} \right) x_1 \right]^2 + \left[\frac{|\mathbf{A}_2|}{|\mathbf{A}_1|} \right] x_2^2 \tag{1.37}$$

There is a problem if $|\mathbf{A}_1| (= a_{11}) = 0$. (Suppose that a_{11} is the only zero-valued element in $\mathbf{A}$.) This would mean that the quadratic form was really just

$$2a_{12}x_1x_2 + a_{22}x_2^2$$

There is nothing to prevent us changing x_1 to x_2 and vice versa. Then the quadratic form is

$$a_{11}x_1^2 + 2a_{12}x_1x_2 + 0x_2^2$$

for which $|\mathbf{A}_1| = a_{11} (\neq 0)$ and $|\mathbf{A}_2| = |\mathbf{A}| = a_{11}a_{12} (\neq 0)$ and the representation in (1.37) requires no divisions by zero.

Although the process of completing the squares is much more complicated, the general quadratic form for three variables can be shown to be

$$Q(x_1, x_2, x_3) = |\mathbf{A}_1| \, (\cdot)^2 + \left[\frac{|\mathbf{A}_2|}{|\mathbf{A}_1|} \right] (\cdot)^2 + \cdots + \left[\frac{|\mathbf{A}_3|}{|\mathbf{A}_2|} \right] (\cdot)^2 \tag{1.38}$$

The squared terms contain functions of the x's and a's, which, from the point of view of the *sign* of (1.38), are unimportant. [Here, there would be problems if either $|\mathbf{A}_1|$ or $|\mathbf{A}_2|$ (or both) were zero. They might be avoided through renumbering variables, as in the example for $Q(x_1, x_2)$, but it is less clear that this can always be done effectively because of the more complex nature of $\mathbf{A}_2$.] Finally, for n variables

$$Q(x_1, \ldots, x_n) = |\mathbf{A}_1| \, (\cdot)^2 + \left[\frac{|\mathbf{A}_2|}{|\mathbf{A}_1|} \right] (\cdot)^2 + \cdots + \left[\frac{|\mathbf{A}_n|}{|\mathbf{A}_{n-1}|} \right] (\cdot)^2 \tag{1.39}$$

In each case, the coefficients on the squared terms are ratios of successive pairs of leading principal minors of $\mathbf{A}$. This representation requires that none of the leading principal minors $|\mathbf{A}_1|, \ldots, |\mathbf{A}_{n-1}|$ be zero.

1.6.4 The Sign of a Quadratic Form

For the general quadratic form of two variables, $Q(x_1, x_2)$, to be *positive definite* it is therefore both necessary and sufficient that the coefficients on both squared terms be strictly positive; from (1.37) this means that (1) $|\mathbf{A}_1| > 0$ and (2) $|\mathbf{A}_2|/|\mathbf{A}_1| > 0$. Since the denominator in (2) is covered by the rule in (1), this can be put simply as (1) $|\mathbf{A}_1| > 0$ and (2) $|\mathbf{A}_2| > 0$. In words, the leading principal minors of $\mathbf{A}$ must be positive. Similarly, (1.37) will be *negative definite* if and only if (1) $|\mathbf{A}_1| < 0$ and (2) $|\mathbf{A}_2|/|\mathbf{A}_1| < 0$ or, more simply, (1) $|\mathbf{A}_1| < 0$ and (2) $|\mathbf{A}_2| > 0$.

Because of the generalization of the sum-of-squares form in (1.38) and (1.39), it follows that

> A quadratic form of n variables, $Q(x_1, \ldots, x_n)$, will be (a) *positive definite* if and only if all leading principal minors of $\mathbf{A}$ are positive ($|\mathbf{A}_1| > 0, \ldots,$ $|\mathbf{A}_n| > 0$) and (b) *negative definite* if and only if the leading principal minors *alternate* in sign, beginning negative ($|\mathbf{A}_1| < 0, |\mathbf{A}_2| > 0, \ldots, |\mathbf{A}_n| < 0$ if n is odd and $|\mathbf{A}_n| > 0$ is n is even).   (1.40)

There are similar (but much less obvious and more complicated) rules that establish positive and negative semidefiniteness. In particular, they do not emerge from inspection of the general structure of a quadratic form in (1.39). They are stated in terms of the principal minors of $\mathbf{A}$. Since it is positive and negative *definiteness* that enables us to establish the signs of quadratic forms, rules for semidefiniteness are of less immediate importance. They are included here for primarily for completeness:

> A quadratic form is (a) *positive semidifintie* ($Q \geq 0$) if and only if *all* principal minors are nonnegative (that is, *all* $|\mathscr{A}_1| \geq 0$, *all* $|\mathscr{A}_2| \geq 0, \ldots,$ *all* $|\mathscr{A}_n| \geq 0$) and (b) *negative semidefinite* if and only if all odd-order principal minors are nonpositive (*all* $|\mathscr{A}_1| \leq 0$, *all* $|\mathscr{A}_3| \leq 0$, etc.) and all even-order principal minors are nonnegative (*all* $|\mathscr{A}_2| \geq 0$, *all* $|\mathscr{A}_4| \geq 0$, etc.).   (1.41)

To return to the previous examples, where $\mathbf{X} = [x_1, x_2]'$ or $\mathbf{X} = [x_1, x_2, x_3]'$, as appropriate:

(a) $x_1^2 + 6x_2^2 = \mathbf{X}' \begin{bmatrix} 1 & 0 \\ 0 & 6 \end{bmatrix} \mathbf{X}$; $|\mathbf{A}_1| = 1 > 0$; $|\mathbf{A}_2| = 6 > 0$, thus positive definite, from (1.40).

(b) $-x_1^2 - 6x_2^2 = \mathbf{X}' \begin{bmatrix} -1 & 0 \\ 0 & -6 \end{bmatrix} \mathbf{X}$; $|\mathbf{A}_1| = -1 < 0$; $|\mathbf{A}_2| = 6 > 0$, thus negative definite, again from (1.40), and

(c) $x_1^2 - 4x_1x_2 + 4x_2^2 + x_3^2 = \mathbf{X}' \begin{bmatrix} 1 & -2 & 0 \\ -2 & 4 & 0 \\ 0 & 0 & 1 \end{bmatrix} \mathbf{X}$; $|\mathbf{A}_1| = 1 > 0$; $|\mathbf{A}_2| = 0$; $|\mathbf{A}_3| = 0$; thus

this is *neither* positive definite nor negative definite. If we were to explore principal minors, however, we would find $|\mathscr{A}_1(1)| = |\mathscr{A}_1(2)| = 1$, $|\mathscr{A}_1(3)| = |\mathscr{A}_1(4)| = 4$, and $|\mathscr{A}_1(5)| = |\mathscr{A}_1(6)| = 1$. Furthermore, you can check that $|\mathscr{A}_2(1)| = |\mathscr{A}_2(3)| = 0$, $|\mathscr{A}_2(2)| = |\mathscr{A}_2(6)| = 1$, and $|\mathscr{A}_2(4)| = |\mathscr{A}_2(5)| = 4$; finally all third-order principal minors are the same, $|\mathscr{A}_3(1)| = 0$. Thus, from (1.41), the quadratic form is positive semidefinite.

(d) $-x_1^2 + 4x_1x_2 - 4x_2^2 - x_3^2 = \mathbf{X}'\begin{bmatrix} -1 & 2 & 0 \\ 2 & -4 & 0 \\ 0 & 0 & -1 \end{bmatrix}\mathbf{X}$; $|\mathbf{A}_1| = -1 < 0$; $|\mathbf{A}_2| = 0$; $|\mathbf{A}_3| = 0$;

thus, exactly as in (c), above, this is *neither* positive definite nor negative definite. And also, as in (c), examination of the principal minors of the **A** matrix in this case will establish, using (1.41), that the quadratic form is negative semidefinite.

(e) $x_1^2 - 6x_2^2 = \mathbf{X}'\begin{bmatrix} 1 & 0 \\ 0 & -6 \end{bmatrix}\mathbf{X}$; $|\mathbf{A}_1| = 1 > 0$, $|\mathbf{A}_2| = -6 < 0$, so neither positive definite

nor negative definite. Nor are conditions for semidefiniteness met. The fact that the two first-order principal minors are of opposite sign $[|\mathscr{A}_1(1)| = 1, |\mathscr{A}_1(2)| = -6]$, indicates that no requirements in (1.41) can be satisfied by this quadratic form.

The rules of signs of leading principal minors, (1.40), may allow early judgment that a quadratic form is neither positive definite nor negative definite. For example, if $|\mathbf{A}_1| < 0$ and $|\mathbf{A}_2| < 0$, we can stop; also, if $|\mathbf{A}_1| > 0$ and $|\mathbf{A}_2| < 0$, we can stop. If $|\mathbf{A}_1| > 0$, then *no change* of sign of higher-order leading principal minors is allowed, whereas if $|\mathbf{A}_1| < 0$, then signs *must alternate* continuously from that point onward. Hence, in (c) and (d), above, it would have been sufficient, in examining for positive or negative *definiteness*, to stop at $|\mathbf{A}_2|$ in both cases.

1.7   SUMMARY

In this chapter we have looked at the fundamental definitions and operations of matrix algebra. We have seen that these concepts are helpful for finding solutions or for analyzing systems of linear equations, largely through connections between the algebra and the geometry of some of the fundamentals (including linear combinations, linear dependence and independence, and the notion of the rank of a matrix). These applications for linear equation systems will be extended in the next chapter. Moreover, matrix notation and manipulations will be used throughout the book, determinants and signs of quadratic forms are important in the study of maxima and minima (Part II and later), and the methods of linear systems analysis will be particularly relevant for linear and quadratic programming models (Part IV and Chapter 11, respectively).

PROBLEMS

1.1   Construct **A**, given that $a_{32} = 5$, $a_{11} = 11$, $a_{21} = 0$, $a_{12} = -1$, $a_{31} = -3$ and $a_{22} = 0$.

1.2   Add to **A** from Problem 1.1 a matrix **B** for which $b_{i1} = 10$ (for all i) and $b_{i2} = -2$ (for all i).
  (a)   What dimensions must **B** have?
  (b)   Write out **B**.
  (c)   Find **A** + **B**.

1.3   Subtract from **A** + **B** in Problem 1.2 a matrix **C** for which

$$\mathbf{C}' = \begin{bmatrix} -3 & 3 & 5 \\ 2 & -6 & -4 \end{bmatrix}$$

1.4   Demonstrate the *associative property* of matrix addition using the following 1×4 matrices:

$$\mathbf{D} = [4 \quad -5 \quad 3 \quad 1], \mathbf{E} = [1 \quad 1 \quad 0 \quad 1], \mathbf{F} = [6 \quad 5 \quad -4 \quad 1]$$

That is, find $\mathbf{D} + \mathbf{E} + \mathbf{F}$ as (a) $(\mathbf{D} + \mathbf{E}) + \mathbf{F}$ and (b) $\mathbf{D} + (\mathbf{E} + \mathbf{F})$.

1.5   Given $\mathbf{A} = [a_{ij}]$, $i = 1, 2; j = 1, 2, 3$. In which of the following cases can $\mathbf{B}$ be premultiplied by $\mathbf{A}$? In which can $\mathbf{B}$ be postmultiplied by $\mathbf{A}$? (a) $\mathbf{B} = \begin{bmatrix} 1 & 3 \\ 4 & 6 \end{bmatrix}$; (b) $\mathbf{B}' = \begin{bmatrix} 4 & 6 & 8 \\ 2 & 2 & 3 \end{bmatrix}$; (c) $\mathbf{B} = \begin{bmatrix} 4 \\ 1 \end{bmatrix}$; (d) $\mathbf{B} = [b_{ij}]$, $i = 1,2,3; j = 1,2$; or (e) $\mathbf{B} = [b_{ij}]$, $i = 1,2; j = 1,2,3$.

1.6   For $\mathbf{A} = \begin{bmatrix} 1 & 3 \\ 5 & 7 \end{bmatrix}$, $\mathbf{B} = \begin{bmatrix} 2 & 2 & 4 \\ 5 & -6 & 9 \end{bmatrix}$, and $\mathbf{C} = \begin{bmatrix} 0 & 6 & 6 \\ 4 & 1 & 2 \end{bmatrix}$, find $\mathbf{AB}$, $\mathbf{AC}$ and then $(\mathbf{AB}) + (\mathbf{AC})$. Next find $\mathbf{B} + \mathbf{C}$, then $\mathbf{A}(\mathbf{B} + \mathbf{C})$. This illustrates the *distributive property* $\mathbf{A}(\mathbf{B} + \mathbf{C}) = \mathbf{AB} + \mathbf{AC}$.

1.7   For $\mathbf{A} = [5 \quad -6]$, $\mathbf{B} = \begin{bmatrix} 2 & 4 \\ 6 & 7 \end{bmatrix}$, and $\mathbf{C} = \begin{bmatrix} 10 \\ 20 \end{bmatrix}$, find $\mathbf{ABC}$ by (a) premultiplying $(\mathbf{BC})$ by $\mathbf{A}$ and (b) postmultiplying $(\mathbf{AB})$ by $\mathbf{C}$. This illustrates the *associative property* $\mathbf{A}(\mathbf{BC}) = (\mathbf{AB})\mathbf{C}$.

1.8   Consider

$$\mathbf{A} = \begin{bmatrix} 1 & 2 & 5 \\ 4 & 6 & 7 \\ 1 & 0 & 0 \end{bmatrix} \quad \text{and} \quad \mathbf{B} = \begin{bmatrix} 1 & 0 \\ 2 & 2 \\ 3 & 4 \end{bmatrix}$$

Imagine $\mathbf{A}$ and $\mathbf{B}$ partitioned into four submatrices such that $\mathbf{A}_{11} = \begin{bmatrix} 1 & 2 \\ 4 & 6 \end{bmatrix}$ and $\mathbf{B}_{11} = \begin{bmatrix} 1 \\ 2 \end{bmatrix}$. Find $\mathbf{AB}$ by multiplying and adding submatrices of $\mathbf{A}$ and $\mathbf{B}$ only [as in (1.4)]. Check by finding $\mathbf{AB}$ directly, as if the matrices were not partitioned.

1.9   Find $|\mathbf{A}|$ in each of the following cases: (a) $\mathbf{A} = \begin{bmatrix} 1 & 2 \\ 3 & 4 \end{bmatrix}$, (b) $\mathbf{A} = \begin{bmatrix} 6 & 5 & 4 \\ 1 & 1 & 2 \\ 1 & 1 & 3 \end{bmatrix}$ (at least two different ways), (c) $\mathbf{A} = \begin{bmatrix} 6 & 5 & 4 & 3 \\ 1 & 1 & 2 & 0 \\ 1 & 1 & 3 & 0 \\ 4 & 3 & 1 & -1 \end{bmatrix}$ [evaluate down column 4, so as to make use of the result found in (b), above, and to take advantage of the zeros in that column].

1.10 In each of the following cases, how do you know $|\mathbf{A}| = 0$ without actually evaluating the determinant?

$$(a) \ \mathbf{A} = \begin{bmatrix} 1 & 1 \\ 1 & 1 \end{bmatrix}, \ (b) \ \mathbf{A} = \begin{bmatrix} 2 & 3 & 6 \\ 5 & 6 & 15 \\ 21 & 9 & 63 \end{bmatrix}, \ (c) \ \mathbf{A} = \begin{bmatrix} 1 & 2 & 3 \\ 4 & 5 & 6 \\ 0 & 0 & 0 \end{bmatrix},$$

$$(d) \ \mathbf{A} = \begin{bmatrix} 3 & 5 & 1 \\ 3 & 10 & 2 \\ 0 & 5 & 1 \end{bmatrix}, \ (e) \ \mathbf{A} = \begin{bmatrix} 5 & 4 & 1 \\ 6 & 3 & 2 \\ 5 & 4 & 1 \end{bmatrix}.$$

1.11 Describe the kinds of solutions (unique, none, multiple) that each of the following equation systems has (use both solution-space and vector-space geometry):

$$
\begin{array}{ll}
(a) \quad x_1 + 2x_2 = 3 & (b) \quad x_1 + 2x_2 = 2 \\
\quad\;\; 4x_1 + 7x_2 = 1 & \quad\;\; 4x_1 + 7x_2 = 7 \\[2mm]
(c) \quad x_1 + 2x_2 = 3 & (d) \quad x_1 + 2x_2 = 6 \\
\quad\;\; 4x_1 + 8x_2 = 12 & \quad\;\; 4x_1 + 8x_2 = 25
\end{array}
$$

1.12 Consider the equation system

$$-x_1 + x_2 = 10 \text{ and } -3x_1 + 2x_2 = 16$$

(a) Solve for x_1 and x_2 using ordinary algebra (substitution).
(b) Write the system as $\mathbf{AX} = \mathbf{B}$ and find $\mathbf{X} = \mathbf{A}^{-1}\mathbf{B}$.

1.13 Find the inverse (if it exists) for each of the following matrices:

$$(a) \ \mathbf{A} = \begin{bmatrix} 2 & -4 \\ 3 & 16 \end{bmatrix}, \ (b) \ \mathbf{B} = \begin{bmatrix} 2 & 4 \\ 3 & 6 \end{bmatrix}, \ (c) \ \mathbf{C} = \begin{bmatrix} 1 & 0 & 0 \\ 0 & 1 & 0 \\ 0 & 0 & 2 \end{bmatrix},$$

$$(d) \ \mathbf{D} = \begin{bmatrix} 3 & 1 & 1 \\ 2 & -2 & 2 \\ 1 & 0 & 1 \end{bmatrix}.$$

1.14 Determine the ranks of the following matrices:

$$(a) \ \mathbf{A} = \begin{bmatrix} 1 & 2 & -2 \end{bmatrix}, \ (b) \ \mathbf{B} = \begin{bmatrix} 1 & 2 & -2 \\ 1 & 3 & 5 \end{bmatrix}, \ (c) \ \mathbf{C} = \begin{bmatrix} 1 & 2 & 0 \\ 0 & 0 & 0 \end{bmatrix},$$

$$(d) \ \mathbf{D} = \begin{bmatrix} 1 & 2 \\ 3 & 6 \end{bmatrix}, \ (e) \ \mathbf{E} = \begin{bmatrix} 1 & 2 & 0 \\ 3 & 6 & 1 \end{bmatrix}, \ (f) \ \mathbf{F} = \begin{bmatrix} 1 & 0 & 0 \\ 0 & 1 & 0 \\ 0 & 0 & 1 \end{bmatrix},$$

$$(g) \ \mathbf{G} = \begin{bmatrix} 2 & 4 & 8 \\ 1 & 2 & 4 \\ 1 & 0 & 2 \end{bmatrix}.$$

1.15 From the general definition of a determinant, in (1.16), show that $|\mathbf{A}| = |\mathbf{A}'|$.

1.16 From the general definition of [adj **A**], in (1.17), show that [adj **A**]$'$ = [adj (**A**$'$)].

1.17 From the results in Problems 1.15 and 1.16 and the general definition of $\mathbf{A}^{-1}$, in (1.18), show that $(\mathbf{A}^{-1})' = (\mathbf{A}')^{-1}$.

1.18 Demonstrate that alien cofactor evaluation of a determinant (property 5) will always generate a value of zero, for any square $n \times n$ matrix **A**. [*Hint:* Use the fact that an alien cofactor evaluation is equivalent to finding the determinant of a matrix with two rows (i and i') or two columns (j and j') that are equal; then use property 4a.]

1.19 For any $m \times n$ matrix **A** (where $m \neq n$), show that $\mathbf{AA}'$ is symmetric.

1.20 For any pair of matrices **A** and **B** that are conformable for the multiplication **AB**, show that $(\mathbf{AB})' = \mathbf{B}'\mathbf{A}'$.

1.21 For any pair of matrices **A** and **B** that are conformable for the multiplication **AB** and that are also nonsingular, show that $(\mathbf{AB})^{-1} = \mathbf{B}^{-1}\mathbf{A}^{-1}$.

1.22 Construct a numerical example of a nonsingular 3×3 matrix **A**. Show, for that example, that for any nonzero scalar α, $(\alpha\mathbf{A})^{-1} = (1/\alpha)\mathbf{A}^{-1}$. Prove that this will always be true for any nonsingular $n \times n$ matrix **A**.

1.23 Do the following sets of vectors form a basis in the appropriate vector space?

$$(a)\quad \begin{bmatrix} 2 \\ 0 \end{bmatrix}, \begin{bmatrix} 0 \\ 2 \end{bmatrix}$$

$$(b)\quad \begin{bmatrix} 1 \\ 1 \\ 1 \end{bmatrix}, \begin{bmatrix} 1 \\ 1 \\ 2 \end{bmatrix}, \begin{bmatrix} 1 \\ 2 \\ 2 \end{bmatrix}$$

$$(c)\quad \begin{bmatrix} 1 \\ 2 \\ 3 \\ 1 \end{bmatrix}, \begin{bmatrix} 1 \\ 2 \\ 3 \\ 2 \end{bmatrix}, \begin{bmatrix} 1 \\ 2 \\ 1 \\ 3 \end{bmatrix}, \begin{bmatrix} 1 \\ 2 \\ 2 \\ 3 \end{bmatrix}$$

$$(d)\quad \begin{bmatrix} 1 \\ 0 \\ 3 \end{bmatrix}, \begin{bmatrix} -3 \\ 0 \\ 1 \end{bmatrix}$$

1.24 Decide whether each of the following quadratic forms is (1) positive definite (or semidefinite), (2) negative definite (or semi-definite), or (3) indefinite:

$$(a)\quad -x_1^2 - x_2^2 - 2x_3^2 + x_1x_2 + 2x_2x_3$$

$$(b)\quad 2x_1^2 + 6x_2^2 - 6x_1x_2$$

$$(c)\quad x_1^2 + x_2^2 + x_3^2 - x_1x_2 - x_2x_3 - x_1x_3$$

2

SYSTEMS OF LINEAR EQUATIONS

In this chapter we extend the analysis of systems of linear equations that was begun in Section 1.3 for two equations and two unknowns. Initially we reexamine the 2×2 case using the vector concepts introduced in Sections 1.4 and 1.5. Next, we explore the 3×3 case; certain complications exist only in systems of size 3×3 or larger. Then we cover the general $n \times n$ (square) case ("square" describes the shape of the coefficient matrix **A**). Next, we examine solution possibilities for nonsquare linear equation systems with either fewer or more equations than unknowns. Since it is usual to speak of systems of m linear equations in n unknowns, we initially will be concerned with the $m = n$ case. Finally in this chapter, we explore numerical methods for solving systems of linear equations; in these approaches, calculation of an inverse is replaced by a sequence of relatively simple operations on the equations. The object is to convert the original system into an equivalent set of equations that can be solved in an algebraically straightforward sequential way.

The reason that square systems are so fundamental is clear from both algebraic and geometric considerations. In the 1×1 case, $ax = b$, then if $a \neq 0$, we have precisely enough information to find x uniquely, as $x = a^{-1}b$. Any additional *unknowns* would generally place too great a demand on the information contained in a single equation; $a_1x_1 + a_2x_2 = b$ does not allow a unique determination of values for x_1 and x_2, if a_1, a_2 and $b \neq 0$. (For example, $2x_1 + 3x_2 = 4$.) Similarly, again starting with $ax = b$, additional *equations* would generally mean conflicting requirements on the unknown x. If x must also satisfy $cx = d$ ($c \neq 0$), then for this equation alone $x = c^{-1}d$. If $a^{-1}b \neq c^{-1}d$, then *no* x can simultaneously satisfy both equations, and the system of more equations than unknowns is *inconsistent*. (For example, $3x = 9$ and $5x = 40$.) Should it happen that $a^{-1}b = c^{-1}d$, then indeed both equations can be satisfied by the same value of x, but either equation by itself is sufficient to find that value; one of the equations is unnecessary. (For example, $3x = 9$ and $6x = 18$.)

These examples are elementary; the important point is that the same sort of reasoning holds for systems of any size. Beginning with a 2×2 system, the addition of a third variable increases the dimensionality of the solution-space picture by one, but this space contains only two planes (equations), for which a unique solution for all three unknowns is impossible. The planes will generally intersect in a line; they could also be parallel or be indistinguishable (coplanar). Similarly, adding an equation only (having two variables and three equations) will mean three lines in the two-dimensional solution-space picture; in general these lines will not intersect in a single point. If they do, only two of them are necessary to identify that point; if they do not, then the three-equation two-variable system has no solution; it is inconsistent. We will see in this chapter the circumstances under which square systems do not have a unique solution and also the kinds of solutions that can be salvaged in nonsquare cases.

2.1 THE 2 × 2 CASE AGAIN

The usefulness of the geometry of vectors and the concepts of linear dependence and independence is apparent if we look again at equation systems (1.26) through (1.28). If we view these sets of equations in their linear combination form, $\mathbf{A}_1 x_1 + \mathbf{A}_2 x_2 = \mathbf{B}$, we can represent the systems in vector space.

Case 1: Unique Solution

ALGEBRA

$$\begin{bmatrix} 3 \\ 2 \end{bmatrix} x_1 + \begin{bmatrix} 2 \\ 5 \end{bmatrix} x_2 = \begin{bmatrix} 7 \\ 12 \end{bmatrix} \tag{2.1}$$

GEOMETRY

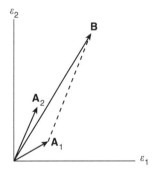

FIGURE 2.1   Vector-space representation of (2.1).

Geometrically, $\mathbf{A}_1$ and $\mathbf{A}_2$ are linearly independent columns, and $\mathbf{B}$ can be expressed as a unique linear combination of them. In (1.26) we found $x_1 = 1$ and $x_2 = 2$. The dashed line in the figure indicates the addition of $2\mathbf{A}_2$ to $1\mathbf{A}_1$, $\begin{bmatrix} 3 \\ 2 \end{bmatrix} (1) + \begin{bmatrix} 2 \\ 5 \end{bmatrix} (2)$, which does indeed give $\begin{bmatrix} 7 \\ 12 \end{bmatrix}$. The vector-space representation in (2.1) does not necessarily lead to a better solution *method* than the solution-space approach of (1.26), provided that a unique solution exists (meaning that **A** is nonsingular). But it provides a different framework for distinguishing the next two cases.

Case 2: No Solution

ALGEBRA

$$\begin{bmatrix} 3 \\ 6 \end{bmatrix} x_1 + \begin{bmatrix} 2 \\ 4 \end{bmatrix} x_2 = \begin{bmatrix} 7 \\ 11 \end{bmatrix} \tag{2.2}$$

GEOMETRY

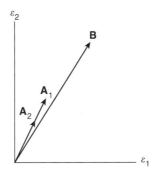

FIGURE 2.2   Vector-space representation of (2.2).

Since **B** is not on the same ray from the origin as the linearly dependent columns **A**₁ and **A**₂, it is impossible to describe its location along the **A**₁**A**₂ ray. Point **B** is linearly *independent* from the linearly *dependent* columns **A**₁ and **A**₂. An efficient way to describe this fact is to note that $\rho(\mathbf{A}) = 1$ whereas $\rho(\mathbf{A} \mid \mathbf{B}) = 2$.

The matrix $(\mathbf{A} \mid \mathbf{B})$, which attaches the right-hand side vector to the matrix of coefficients from the equations, is called the *augmented matrix* of the linear equation system. It obviously has dimensions $m \times (n + 1)$. Where $m = n$, it is clear that $\rho(\mathbf{A} \mid \mathbf{B})$ cannot be greater than this common value representing both the number of equations and the number of unknowns. Thus, case 2 is characterized by two facts: (1) $\rho(\mathbf{A}) = 1$, which is less than n; and (2) $\rho(\mathbf{A} \mid \mathbf{B}) = 2$, which is not equal to $\rho(\mathbf{A})$.

The fact that $\rho(\mathbf{A})$ is less than the number of unknowns (the columns of **A** are linearly dependent) means that the two columns do not define a *pair* of axes in two-dimensional vector space. Either unknown alone is sufficient to contribute a column vector that defines the direction given by both **A**₁ and **A**₂ in Figure 2.2. What we really have is a case of *one* variable and two equations, where the two equations make conflicting demands on the single variable.

Case 3: Infinite Number of Solutions

ALGEBRA

$$\begin{bmatrix} 3 \\ 6 \end{bmatrix} x_1 + \begin{bmatrix} 2 \\ 4 \end{bmatrix} x_2 = \begin{bmatrix} 7 \\ 14 \end{bmatrix} \tag{2.3}$$

GEOMETRY (see Fig. 2.3)

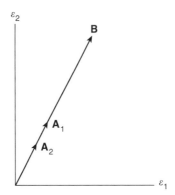

FIGURE 2.3   Vector-space representation of (2.3).

In contrast to case 2, here **B** *is* on the ray defined by both $\mathbf{A}_1$ and $\mathbf{A}_2$; hence **B**'s position can be described in an infinite number of ways, relative to the one direction and the lengths defined by $\mathbf{A}_1$ and $\mathbf{A}_2$. For example, $x_1 = 1$, $x_2 = 2$ will satisfy (2.3), as will $x_1 = -1$, $x_2 = 5$; $x_1 = -2$, $x_2 = 6.5$; $x_1 = -3$, $x_2 = 8$; and so on. $\mathbf{A}_1$, $\mathbf{A}_2$ *and* **B** are linearly dependent; this is reflected in the fact that for this case $\rho(\mathbf{A} \mid \mathbf{B}) = 1$. We already know that $\rho(\mathbf{A}) = 1$, since the coefficient matrices in cases 2 and 3 are identical, so the facts of case 3 are (1) $\rho(\mathbf{A}) = 1$ (< 2) and (2) $\rho(\mathbf{A} \mid \mathbf{B}) = 1$ which is the same as $\rho(\mathbf{A})$. Here we have a case in which the one-variable, two-equation situation of case 2 has been reduced, in effect, to one variable and one equation.

The facts of rank for these three cases are summarized in Table 2.1. This table (see also Fig. 2.4) suggests rules for establishing the *existence* and the *uniqueness* of solutions; these rules turn out to be valid for square linear equation systems of any size and, with slight modifications, for nonsquare systems as well. When $\rho(\mathbf{A}) = \rho(\mathbf{A} \mid \mathbf{B})$, the equations are consistent—they have at least one solution (cases 1 and 3). If the common rank is the maximum possible (which, for the $m = n$ case, is m), a *unique* solution exists (case 1). If the common rank is less than this maximum, *multiple* solutions can be found (case 3). When $\rho(\mathbf{A}) \neq \rho(\mathbf{A} \mid \mathbf{B})$—meaning that $\rho(\mathbf{A}) < \rho(\mathbf{A} \mid \mathbf{B})$—the system is inconsistent and *no* solution can be found that will simultaneously satisfy all the equations (case 2). Through use of the *augmented matrix* for the equation system we have explicitly considered the contents of the right-hand-side vector **B**.

In the job retraining illustration in Equation (1.5) in Chapter 1, the elements of **B** were (1) the state allocation to the city to fund the program and (2) a city tax revenue target set by the mayor. The coefficients attached to the x's (the elements of **A**) were per pupil cost (in the first equation) and expected new income earned (in the second equation); these might have been supplied by a consulting firm. There is no particular reason why the re-

TABLE 2.1   Ranks in Cases 1–3 for $m = n = 2$

	Case 1	Case 2	Case 3
$\rho(\mathbf{A})$	2	1	1
$\rho(\mathbf{A} \mid \mathbf{B})$	2	2	1

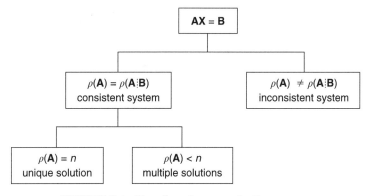

FIGURE 2.4   Tests for existence and uniqueness.

sulting equation system ($\mathbf{AX} = \mathbf{B}$) should necessarily be consistent or indeed even why the estimated figures on education costs and new income earned by graduates of the two programs should lead to a nonsingular $\mathbf{A}$ matrix.

For example, suppose that the estimates of per pupil costs were found to be in error; actually they are $2000 for each participant in program 1 and $4000 for each participant in program 2. With the original right-hand-side vector, $\mathbf{B}$, this leaves us with the following equation system

$$2000x_1 + 4000x_2 = 1{,}000{,}000$$

$$10{,}000x_1 + 20{,}000x_2 = 4{,}600{,}000$$

which reduces, through division of both sides of the first equation by 2000 and of the second equation by 10000, to

$$x_1 + 2x_2 = 500$$

$$x_1 + 2x_2 = 460$$

Using matrix representation, we obtain

$$\mathbf{X} = \begin{bmatrix} x_1 \\ x_2 \end{bmatrix}, \qquad \mathbf{A} = \begin{bmatrix} 1 & 2 \\ 1 & 2 \end{bmatrix}, \qquad \mathbf{B} = \begin{bmatrix} 500 \\ 460 \end{bmatrix}$$

and it is easily seen that $\rho(\mathbf{A}) = 1$ but $\rho(\mathbf{A} \mid \mathbf{B}) = 2$, so the equation system is inconsistent, and there is no way that the mayor can select the numbers of students to admit to the two programs so as to use all the allocated funds from the state and also meet the target for new earned income.

Suppose, however, that the revised cost estimates, given above, were correct, but that the mayor reevaluated the new tax revenue target and decided to increase it to $125,000 (from $115,000). You should check that the equation system representing this problem is now

$$2000x_1 + 4000x_2 = 1{,}000{,}000$$

$$10{,}000x_1 + 20{,}000x_2 = 5{,}000{,}000$$

and that this can be reduced, as in the previous illustration, to

$$x_1 + 2x_2 = 500$$
$$x_1 + 2x_2 = 500$$

This is simply one equation and hence has multiple solutions. [The rules of rank, while unnecessary here, will identify this, since $\rho(\mathbf{A}) = \rho(\mathbf{A} \mid \mathbf{B}) = 1$, indicating a consistent system with multiple solutions.]

2.2 THE 3 × 3 CASE

The examples above, in (2.1)–(2.3), show how the three cases illustrated by equation systems (1.26)–(1.28) can be represented in vector space and classified according to the ranks of the coefficient and augmented matrices, $\mathbf{A}$ and $(\mathbf{A} \mid \mathbf{B})$. The distinctions between equations with a unique solution, with no solution, and with multiple solutions have a clear *geometric* interpretation in vector space that can be understood in terms of the *algebra* of linear dependence and independence. Before we can generalize these ideas completely, it is necessary to recognize that there are additional solution possibilities in cases with more than two variables and equations.

Linear equations with three unknowns are planes in three-dimensional solution space. The natural extensions of the (a) unique, (b) inconsistent, and (c) multisolution examples of Section 2.1 (cases 1–3, respectively) are easy to visualize. The three planes would

(a) intersect in a unique point,
(b) be three distinct and parallel planes,
(c) be *coplanar*—only one plane would be defined by all three equations.

With more than two equations and two unknowns, there can be further variations on the inconsistent and the multiple solution cases. Some examples are:

(d) All three planes could intersect in a common line, allowing a more "restricted" set of multiple solutions—points along a line only, rather than anywhere in the plane, as in (c).

(e) Any pair of equations could be proportional, defining the same plane in solution space and the third plane parallel to them. Then there are multiple solutions that would satisfy the equations associated with the coplanar planes but none that would satisfy all three equations.

(f) Two of the planes could be distinct and parallel while the third "slices" through them, creating two parallel lines in three-dimensional space. In this case, there would be two different sets of multiple solutions to two different pairs of the three equations.

(g) Each pair of planes could intersect in a distinct line, and the three lines formed by such intersections could be parallel. In this case, there would be three different pairs of equations with (different) multiple solutions.

Thus the situation illustrated by (d) adds another set of possible multiple solutions and (e), (f), and (g) add new wrinkles to the "inconsistent" case. While it is true in these last three cases that the entire set of equations is inconsistent and has no common solution(s), it is now possible to find (multiple) solutions to one or more subsets of the equations. In

(e) there is only one such subset of two of the three equations to which common solutions can be found; in (f) there are two sets of (different) pairs of equations to which (multiple) solutions can be found; and in (g) there are three such sets.

The rules of rank that were seen to be useful in sorting out the three possible cases for two equations and two unknowns can be extended to separate out cases like those given in (a)–(g). Whether it is important to make a distinction between, say, complete inconsistency, as in (b), and the kind of inconsistency illustrated in (e)–(g) depends entirely on the situation in which the equation systems arise. At certain times it may be useful to know that, unlike (b), "all is not lost"; then the problem is to identify which subsets of equations *can* be satisfied simultaneously, how many (and, ultimately, which) equations must be ignored to find solutions to the remaining ones, and so on.

2.2.1 Numerical Illustrations

Equations (2.4) provide illustrations for each cases in (a)–(g). They are illustrated in the panels of Figure 2.5.

Case (a): $\rho(\mathbf{A}) = 3, \rho(\mathbf{A} \mid \mathbf{B}) = 3$

$$\begin{aligned} x_1 + 0x_2 + x_3 &= 5 \\ 0x_1 + x_2 + x_3 &= 5 \\ -x_1 - x_2 + x_3 &= 10 \end{aligned} \tag{2.4a}$$

Case (b): $\rho(\mathbf{A}) = 1, \rho(\mathbf{A} \mid \mathbf{B}) = 2$

$$\begin{aligned} -x_1 + x_2 + x_3 &= 10 \\ -x_1 + x_2 + x_3 &= 20 \\ -x_1 + x_2 + x_3 &= 1 \end{aligned} \tag{2.4b}$$

Case (c): $\rho(\mathbf{A}) = 1, \rho(\mathbf{A} \mid \mathbf{B}) = 1$

$$\begin{aligned} -x_1 + x_2 + x_3 &= 10 \\ -2x_1 + 2x_2 + 2x_3 &= 20 \\ -3x_1 + 3x_2 + 3x_3 &= 30 \end{aligned} \tag{2.4c}$$

Case (d): $\rho(\mathbf{A}) = 2, \rho(\mathbf{A} \mid \mathbf{B}) = 2$

$$\begin{aligned} x_1 + 2x_2 - x_3 &= 10 \\ 2x_1 + 4x_2 + x_3 &= 20 \\ 3x_1 + 6x_2 + x_3 &= 30 \end{aligned} \tag{2.4d}$$

Case (e): $\rho(\mathbf{A}) = 1, \rho(\mathbf{A} \mid \mathbf{B}) = 2$

$$\begin{aligned} -x_1 + x_2 + x_3 &= 10 \\ -x_1 + x_2 + x_3 &= 20 \\ -2x_1 + 2x_2 + 2x_3 &= 40 \end{aligned} \tag{2.4e}$$

Case (f): $\rho(\mathbf{A}) = 2, \rho(\mathbf{A} \mid \mathbf{B}) = 3$

$$\begin{aligned} -x_1 + x_2 + x_3 &= 10 \\ -x_1 + x_2 + x_3 &= 20 \\ 2x_1 - 2x_2 + x_3 &= -5 \end{aligned} \tag{2.4f}$$

Case (g): $\rho(\mathbf{A}) = 2, \rho(\mathbf{A} \mid \mathbf{B}) = 3$

$$\begin{aligned} x_1 + 0x_2 + x_3 &= 10 \\ -2x_1 + 0x_2 + x_3 &= 15 \\ -5x_1 + 0x_2 + x_3 &= 5 \end{aligned} \tag{2.4g}$$

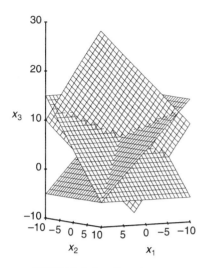

FIGURE 2.5a   Equations (2.4a).

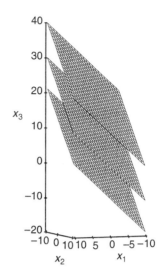

FIGURE 2.5b   Equations (2.4b).

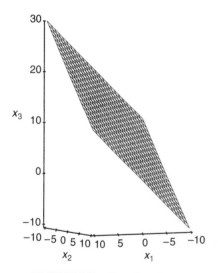

FIGURE 2.5c   Equations (2.4c).

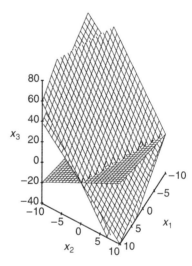

FIGURE 2.5d   Equations (2.4d).

For consistent systems in which $\rho(\mathbf{A}) = \rho(\mathbf{A} \mid \mathbf{B})$, the difference between $\rho(\mathbf{A} \mid \mathbf{B})$ and its maximum possible value [max $\rho(\mathbf{A} \mid \mathbf{B})$] tells us something about the nature of the solution(s). In all the examples above, the maximum possible value for $\rho(\mathbf{A} \mid \mathbf{B})$ is 3. For case (c), with $\rho(\mathbf{A}) = \rho(\mathbf{A} \mid \mathbf{B}) = 1$, [max $\rho(\mathbf{A} \mid \mathbf{B})$] $- \rho(\mathbf{A} \mid \mathbf{B}) = 2$, and these equations have multiple solutions that lie anywhere on a particular *plane* (a *two*-dimensional surface). For case (d), where the common rank of $\mathbf{A}$ and $(\mathbf{A} \mid \mathbf{B})$ is 2, [max $\rho(\mathbf{A} \mid \mathbf{B})$] $- \rho(\mathbf{A} \mid \mathbf{B}) = 1$, and the multiple solutions are those that lie anywhere on a specific *line* (a *one*-dimensional surface). Finally, in case (a), where $\rho(\mathbf{A}) = \rho(\mathbf{A} \mid \mathbf{B}) = 3$, [max $\rho(\mathbf{A} \mid \mathbf{B})$] $- \rho(\mathbf{A} \mid \mathbf{B}) = 0$, and the (unique) solution is a single point (a *zero*-dimensional "surface").

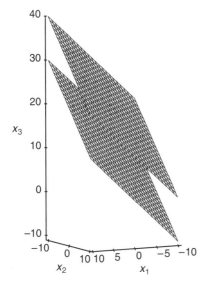

FIGURE 2.5e Equations (2.4e).

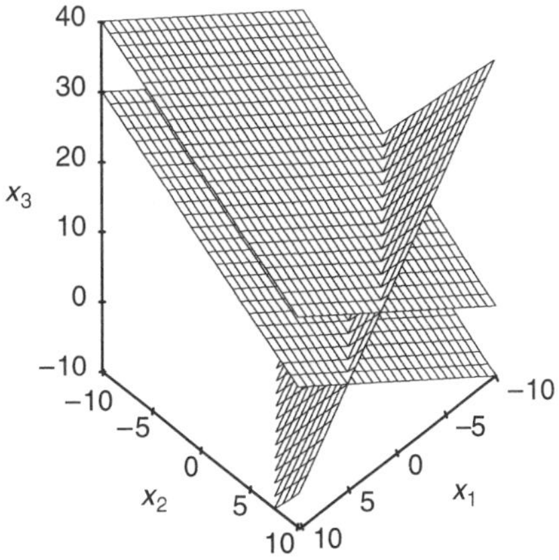

FIGURE 2.5f Equations (2.4f).

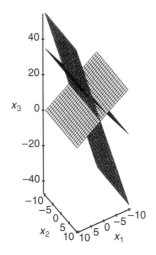

FIGURE 2.5g Equations (2.4g).

For the inconsistent cases, $\rho(\mathbf{A})$ indicates the number of equations that can be simultaneously satisfied by any specific point. In cases (b) and (e), $\rho(\mathbf{A}) = 1$. In case (b), with $\rho(\mathbf{A}) = 1$, we have three parallel planes, so a particular point that lies on one of the planes cannot possibly lie on either of the others; only one equation at a time can be satisfied. In case (e), again with $\rho(\mathbf{A}) = 1$, the first and second equations define the same plane, so the system really consists of only two parallel planes in three-dimensional solution space. Again, only one equation at a time can be satisfied. In case (f), where $\rho(\mathbf{A}) = 2$, we have seen that there are *pairs* of equations that intersect, so in this case it is possible to find points that simultaneously satisfy two of the equations. The same is true in case (g).

TABLE 2.2   Rules of Rank for $m = n = 2$ or 3

	$\rho(\mathbf{A}) = \rho(\mathbf{A} \vdots \mathbf{B})$ (Consistent system)	$\rho(\mathbf{A}) \neq \rho(\mathbf{A} \vdots \mathbf{B})$ (Inconsistent system)
$\rho(\mathbf{A}) = 1$	Multiple solutions; any $(m-1)$ equations are redundant	Complete parallelism; only "partial" solutions if a complete set of $(m-1)$ equations is ignored
$1 < \rho(\mathbf{A}) < n$	Multiple solutions; at least one set of $[m - \rho(\mathbf{A})]$ equations are redundant	"Partial" solutions. At least one set of $[m - \rho(\mathbf{A})]$ equations must be ignored; solutions can then be found for the remaining equations
$\rho(\mathbf{A}) = n$	Unique solution; $\mathbf{X} = \mathbf{A}^{-1}\mathbf{B}$	Impossible, since $\rho(\mathbf{A}) < \rho(\mathbf{A} \vdots \mathbf{B}) \leq n$

2.2.2 Summary of 2 × 2 and 3 × 3 Solution Possibilities

A consistent system is one in which $\rho(\mathbf{A}) = \rho(\mathbf{A} \vdots \mathbf{B})$. Comparison of these two ranks is the test for *existence* of a solution to the entire equation system. For consistent systems, the size of $\rho(\mathbf{A})$ relative to n, the number of unknowns, determines whether the solution is *unique*. When $m = n$, as in the cases examined thus far, the numerical outcome is the same whether one compares $\rho(\mathbf{A})$ to the number of equations (m) or the number of unknowns (n). Logically, however, the question of uniqueness of the solution depends on the number of linearly independent columns in $\mathbf{A}$ relative to the number of unknowns.

If there are multiple solutions in the consistent case, then some of the equations do not provide distinguishably different relations among the variables; they merely repeat information that is contained in one or more of the other equations. The number of unnecessary or *redundant* equations is $[m - \rho(\mathbf{A})]$. For inconsistent systems, this same value indicates how many equations (but not, of course, which ones) must be ignored in the system to remove the inconsistency. These observations are summarized in Table 2.2.

2.3 THE $n \times n$ CASE

The extension to square linear systems of any size, $\mathbf{AX} = \mathbf{B}$, follows directly along the lines explored above; the linear combination form is

$$\mathbf{A}_1 x_1 + \mathbf{A}_2 x_2 + \cdots + \mathbf{A}_n x_n = \mathbf{B}$$

We consider consistent and inconsistent systems in turn. To facilitate comparison with examples later in this chapter, when the number of equations is different from the number of unknowns, we designate the present cases with an "E" (the number of equations is *equal* to the number of unknowns).

2.3.1 Consistent Systems $[\rho(\mathbf{A}) = \rho(\mathbf{A} \vdots \mathbf{B})]$

Case E1: $\rho(\mathbf{A}) = n$.   An $n \times n$ system in which $\rho(\mathbf{A})$ is at its maximum value is always consistent. Any other n-element column $\mathbf{B}$ must necessarily be linearly dependent on the n

columns of **A**, so $\rho(\mathbf{A} \vdots \mathbf{B})$ must be the same as $\rho(\mathbf{A})$. In this case, the unique solution can always be found as $\mathbf{X} = \mathbf{A}^{-1}\mathbf{B}$. This is the "well-behaved" case. Examples for $n = 2$ were shown in Figure 1.1, accompanying Equations (1.26) for solution space and in Figure 2.1 with Equations (2.1) for vector space.

Case E2: $\rho(\mathbf{A}) < n$.   There cannot be a unique solution, since $\mathbf{A}^{-1}$ does not exist. If $1 < \rho(\mathbf{A}) < n$, **B** lies on the line, in the plane, or, more generally, in the vector space of dimension $\rho(\mathbf{A})$ that is spanned by the columns of **A**. Since $\rho(\mathbf{A}) < n$, there is more than one set of columns in **A** that can be used in a linear combination with which to define **B**. If $\rho(\mathbf{A}) = 1$, then all equations define the same plane. The equations in Figures 2.5*c* and 2.5*d* illustrate the $n = 3$ case.

When all elements in **B** are zero, the equations are called *homogeneous*. Homogeneous equations must necessarily be consistent, since adding a column of zeros to any **A** cannot increase its rank, and so $\rho(\mathbf{A}) = \rho(\mathbf{A} \vdots \mathbf{B})$. The only possibilities, then, are that $\rho(\mathbf{A}) = n$ or $\rho(\mathbf{A}) < n$. We consider these in turn.

Case H1: $\rho(\mathbf{A}) = n$.   The equation system must have a unique solution. Since $\mathbf{A}^{-1}$ exists [because $\rho(\mathbf{A}) = n$], the solution is $\mathbf{X} = \mathbf{A}^{-1}\mathbf{B}$; with $\mathbf{B} = \mathbf{0}$, the only solution is $\mathbf{X} = \mathbf{0}$.[1] In the case of homogeneous equations with a unique solution, this is understandably called the trivial solution.

Case H2: $\rho(\mathbf{A}) < n$.   Since homogeneous equations are by definition consistent, we expect multiple solutions. The columns of **A** are now linearly dependent, and we know that there will be solutions in which not all x_i are zero. In fact, we know from the 2×2 and 3×3 examples that an infinite number of solutions can be found, including the one in which all x_i are zero. (The $\mathbf{X} = \mathbf{0}$ solution is *always* possible for homogeneous equations.) Therefore, homogeneous systems of n equations in n unknowns will have a unique but trivial solution if and only if $\rho(\mathbf{A}) = n$, and they will have nontrivial but nonunique solutions if and only if $\rho(\mathbf{A}) < n$.

Examples for the $n = 2$ cases (just to keep the geometry simple) are given below. There is nothing conceptually new in the geometry. The figures parallel those for the 2×2 cases shown earlier except that in solution space the line or lines go through the origin and in vector space **B** is not shown as a distinct arrow since it is the point $[0,0]'$.

ALGEBRA

$$x_1 - 2x_2 = 0$$
$$x_1 + x_2 = 0 \tag{2.5a}$$

GEOMETRY (see Fig. 2.6)

---

[1]This all-zero solution for homogeneous equations also follows directly from the definition of linear independence. The homogeneous equations are

$$\mathbf{A}_1 x_1 + \mathbf{A}_2 x_2 + \cdots + \mathbf{A}_n x_n = \mathbf{0}$$

exactly as in the definition of linear dependence and independence (Section 1.5). Since we are considering the case in which $\rho(\mathbf{A}) = n$, the n columns of **A** are linearly independent (from the definition of the rank of a matrix as the number of linearly independent columns in the matrix); this requires that *all* $x_i = 0$.

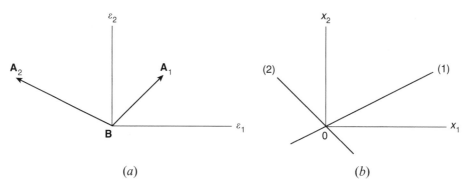

FIGURE 2.6   Representation of (2.5a) in (*a*) vector space; (*b*) solution space.

ALGEBRA

$$x_1 - x_2 = 0$$
$$2x_1 - 2x_2 = 0$$

(2.5b)

GEOMETRY

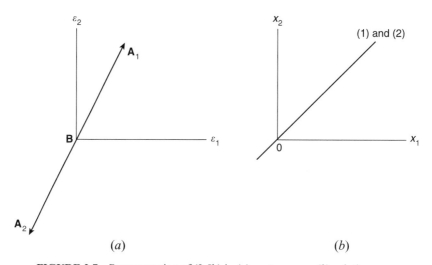

FIGURE 2.7   Representation of (2.5b) in (*a*) vector space; (*b*) solution space.

2.3.2 Inconsistent Systems [$\rho(\mathbf{A}) \neq \rho(\mathbf{A} \vdots \mathbf{B})$]

Case E3: $\rho(\mathbf{A}) = n$.   Since $\rho(\mathbf{A}) \leq \rho(\mathbf{A} \vdots \mathbf{B})$, and since the maximum value for $\rho(\mathbf{A} \vdots \mathbf{B})$ is n, $\rho(\mathbf{A})$ must be strictly less than n and hence this case cannot arise. (It is mentioned only as a logical parallel for case E1, for completeness.)

Case E4: $\rho(\mathbf{A}) < n$.   Since the system is inconsistent, **B** cannot be expressed as a linear combination of the columns of **A**, and no solution to the full set of equations is possible.

"Partial system" solutions can be found, but they require that some of the equations be ignored. The equations in Figures 2.5*b* and 2.5*e–g* provide examples.

2.4 MORE MATRIX ALGEBRA: INVERSES FOR NONSQUARE MATRICES

Linear equation systems in which the number of equations (m) is different from the number of unknowns (n) can still be represented in compact matrix form, precisely because the equations are linear. The coefficient matrix, $\mathbf{A}$, will have dimensions $m \times n$, the column vector of unknowns, $\mathbf{X}$, will contain n elements, $\mathbf{B}$ will be an m-element column vector, and the system can be summarized as $\underset{(m \times n)}{\mathbf{A}} \ \underset{(n \times 1)}{\mathbf{X}} = \underset{(m \times 1)}{\mathbf{B}}$. The concept of an inverse to $\mathbf{A}$, when $m \neq n$, turns out to be useful in the exploration of possible solutions to a nonsquare system of linear equations. In this section we explore such inverses.

Inverses for rectangular matrices have been investigated in a number of references where the topic is treated in far more depth and detail than we need here. The general term for these mathematical constructs is "generalized" inverses. Prominent among them are Moore–Penrose inverses (named after the two mathematicians who proposed and developed them, often denoted, for a matrix $\mathbf{A}$, by $\mathbf{A}^+$, and sometimes called "pseudo-" inverses) and g inverses (also sometimes called "conditional" inverses, and often denoted by $\mathbf{A}^-$). We will be concerned in this section with the relatively simpler concepts of "left" and "right" inverses; they can be used for certain rectangular $\mathbf{A}$ matrices, and the names identify whether they are used to pre- or postmultiply $\mathbf{A}$.

2.4.1 The $m > n$ Case

Given an $m \times n$ matrix $\mathbf{A}$, where $m > n$, suppose that $\mathbf{A}$ has the maximum rank possible— $\rho(\mathbf{A}) = n$. The product $\mathbf{A}'\mathbf{A}$ will be an $n \times n$ matrix, and it can be shown that $\rho(\mathbf{A}'\mathbf{A}) = n$ also, which means that $(\mathbf{A}'\mathbf{A})$ is nonsingular. So, if one were to find the inverse of $(\mathbf{A}'\mathbf{A})$, it would follow that

$$(\mathbf{A}'\mathbf{A})^{-1}(\mathbf{A}'\mathbf{A}) = \mathbf{I}$$

Now suppose that we simply regroup terms in order to visually isolate the final $\mathbf{A}$ in this expression:

$$[(\mathbf{A}'\mathbf{A})^{-1}\mathbf{A}']\mathbf{A} = \mathbf{I}$$

Then it is clear that the matrix in square brackets on the left, $(\mathbf{A}'\mathbf{A})^{-1}\mathbf{A}'$, serves as a kind of inverse to $\mathbf{A}$. If $\mathbf{A}$ is multiplied by it *on the left* the product is an identity matrix, and this is a defining characteristic of an inverse.[2] This is why $(\mathbf{A}'\mathbf{A})^{-1}\mathbf{A}'$ is called the "left inverse" of a rectangular matrix $\mathbf{A}$, when $\mathbf{A}$ has more rows than columns; it is sometimes denoted $\mathbf{A}_L^{-1}$:

$$\mathbf{A}_L^{-1} = (\mathbf{A}'\mathbf{A})^{-1}\mathbf{A}' \tag{2.6}$$

[2]Once $(\mathbf{A}'\mathbf{A})^{-1}$ has been found, it is clear that both $(\mathbf{A}'\mathbf{A})^{-1}(\mathbf{A}'\mathbf{A}) = \mathbf{I}$ and $(\mathbf{A}'\mathbf{A})(\mathbf{A}'\mathbf{A})^{-1} = \mathbf{I}$, but only when $(\mathbf{A}'\mathbf{A})$ is *pre*multiplied by its inverse do we find the matrix $\mathbf{A}$ "alone" at the end of the algebraic expression, allowing us to isolate a left inverse for $\mathbf{A}$. In fact, when $(\mathbf{A}'\mathbf{A})$ is *post*multiplied by its inverse, we see that the expression $[\mathbf{A}(\mathbf{A}'\mathbf{A})^{-1}]$ is the (right) inverse of $\mathbf{A}'$.

For example, let $\mathbf{A} = \begin{bmatrix} 2 & 1 \\ 5 & 3 \\ 8 & 5 \end{bmatrix}$; $\rho(\mathbf{A}) = 2$. Then $\mathbf{A}' = \begin{bmatrix} 2 & 5 & 8 \\ 1 & 3 & 5 \end{bmatrix}$, $\mathbf{A}'\mathbf{A} = \begin{bmatrix} 93 & 57 \\ 57 & 35 \end{bmatrix}$,

and $\mathbf{A}_L^{-1} = \begin{bmatrix} 2.1667 & 0.0667 & -0.8333 \\ -3.5 & -1.0 & 1.5 \end{bmatrix}$. You can easily confirm that $\mathbf{A}_L^{-1}\mathbf{A} = \mathbf{I}$.

To illustrate the importance of the requirement that $\rho(\mathbf{A}) = n$, consider a different matrix; let $\mathbf{A} = \begin{bmatrix} 2 & 1 \\ 6 & 3 \\ 10 & 5 \end{bmatrix}$, where $\rho(\mathbf{A}) = 1$. Then $(\mathbf{A}'\mathbf{A}) = \begin{bmatrix} 70 & 35 \\ 140 & 70 \end{bmatrix}$, which is easily seen

to be singular, so $(\mathbf{A}'\mathbf{A})^{-1}$ does not exist and neither does $\mathbf{A}_L^{-1}$.

2.4.2 The $m < n$ Case

As in the previous case, we continue to assume that $\mathbf{A}$ has the maximum rank possible; here that means $\rho(\mathbf{A}) = m$. Then the product $\mathbf{A}\mathbf{A}'$ will be an $m \times m$ matrix, which, again, can be shown to also have a rank of m. This means that $(\mathbf{A}\mathbf{A}')$ is nonsingular and its inverse could be found, so that $(\mathbf{A}\mathbf{A}')(\mathbf{A}\mathbf{A}')^{-1} = \mathbf{I}$. Regrouping, in the same spirit as for the left-inverse case, we have $\mathbf{A}[\mathbf{A}'(\mathbf{A}\mathbf{A}')^{-1}] = \mathbf{I}$ and so $[\mathbf{A}'(\mathbf{A}\mathbf{A}')^{-1}]$ is termed the *right inverse* of $\mathbf{A}$, when $\mathbf{A}$ has more columns than rows, and it is often denoted $\mathbf{A}_R^{-1}$:

$$\mathbf{A}_R^{-1} = \mathbf{A}'(\mathbf{A}\mathbf{A}')^{-1} \tag{2.7}$$

Note that when $\mathbf{A}$ has $m > n$, a left inverse can be found, provided $\rho(\mathbf{A})$ is maximal (equal to n); when $m < n$, a right inverse is possible, again provided that $\rho(\mathbf{A})$ is maximal (equal to m).

Continuing with the $m < n$ case, consider $\mathbf{A} = \begin{bmatrix} 1 & 3 & 3 \\ 2 & 4 & 6 \end{bmatrix}$ whose rank is 2. Here $\mathbf{A}' = \begin{bmatrix} 1 & 2 \\ 3 & 4 \\ 3 & 6 \end{bmatrix}$, $\mathbf{A}\mathbf{A}' = \begin{bmatrix} 19 & 32 \\ 32 & 54 \end{bmatrix}$, and $\mathbf{A}_R^{-1} = \mathbf{A}'(\mathbf{A}\mathbf{A}')^{-1} = \begin{bmatrix} -0.2 & 0.15 \\ 1 & -0.5 \\ -0.6 & 0.45 \end{bmatrix}$. Again, you can

easily verify that $\mathbf{A}\mathbf{A}_R^{-1} = \mathbf{I}$.

2.4.3 Nonsquare Inverses and Systems of Linear Equations

In a square system of linear equations, $\mathbf{A}\mathbf{X} = \mathbf{B}$ ($m = n$), we multiply both sides *on the left* by $\mathbf{A}^{-1}$ to find $\mathbf{X} = \mathbf{A}^{-1}\mathbf{B}$. In a linear equation system with $m > n$, if $\rho(\mathbf{A}) = n$, we can find a left inverse and hence a "solution" is generated by premultiplying both sides of $\mathbf{A}\mathbf{X} = \mathbf{B}$ by $\mathbf{A}_L^{-1}$:

$$\mathbf{A}_L^{-1}\mathbf{A}\mathbf{X} = \mathbf{A}_L^{-1}\mathbf{B} \Rightarrow \mathbf{X} = \mathbf{A}_L^{-1}\mathbf{B} \quad (\text{since } \mathbf{A}_L^{-1}\mathbf{A} = \mathbf{I})$$

For a *square* nonsingular matrix $\mathbf{A}$, $\mathbf{A}^{-1}$ serves as *both* a left and a right inverse; $\mathbf{A}\mathbf{A}^{-1} = \mathbf{A}^{-1}\mathbf{A} = \mathbf{I}$. An alternative way of viewing the algebraic results for the square case, when $\mathbf{A}$ is nonsingular, is to note that, if $\mathbf{A}\mathbf{X} = \mathbf{B}$, then by multiplying $\mathbf{A}$ *on the right* by $\mathbf{A}^{-1}\mathbf{B}$, it must follow that $\mathbf{A}\mathbf{A}^{-1}\mathbf{B} = \mathbf{B}$, since $\mathbf{A}\mathbf{A}^{-1} = \mathbf{I}$. The conversion from $\mathbf{A}\mathbf{X} = \mathbf{B}$ to $\mathbf{A}\mathbf{A}^{-1}\mathbf{B} = \mathbf{B}$ makes clear that $\mathbf{X} = \mathbf{A}^{-1}\mathbf{B}$ (they occupy corresponding positions on the left-hand sides of the two equations). We knew that already for square linear equation systems because we could carry out the matrix algebra by *premultiplying* both sides of the equations by $\mathbf{A}^{-1}$.

In the $m < n$ case, we would need to multiply $\mathbf{A}$ in $\mathbf{AX} = \mathbf{B}$ *on the right* by $\mathbf{A}_R^{-1}$, to generate an identity matrix. But we cannot get to $\mathbf{A}$ on that side, because it is already postmultiplied by $\mathbf{X}$. However, given a right inverse for $\mathbf{A}$, then, since $\mathbf{AA}_R^{-1} = \mathbf{I}$, $\mathbf{AA}_R^{-1}\mathbf{B} = \mathbf{B}$, and comparison with the original (rectangular) equation system suggests that $\mathbf{A}_R^{-1}\mathbf{B}$ provides a solution for this case of fewer equations than unknowns, since it occupies the same position as $\mathbf{X}$ in $\mathbf{AX} = \mathbf{B}$.

2.5 FEWER EQUATIONS THAN UNKNOWNS ($m < n$)

In the context of the job retraining example, there is no particular reason for the number of retraining programs (two in the example) to be the same as the number of equations (a cost equation and a target revenue equation in the example). In this section we examine the situation of, say, four different job retraining programs and the same two equations ($m < n$). In Section 2.6 we look at the opposite kind of case in which there are, for example, two retraining programs and a third equation relating the variables. As an example, the city may need to provide students in both programs with free weekly transit passes so that they can attend classes. If the city has 8000 such passes to distribute to the students in the next program session, this would add $24x_1 + 24x_2 = 8000$ to the equation system (since each program runs for 24 weeks).

With fewer equations than unknowns ($m < n$), the coefficient matrix in the system $\mathbf{AX} = \mathbf{B}$ is "short and wide;" its greatest possible rank is m, and it necessarily must contain linearly dependent columns (since the maximum number of linearly independent m-element column vectors is m). This means that unique solutions are never possible, because there is more than one set of columns in $\mathbf{A}$ that may be used as the vectors with respect to which $\mathbf{B}$ can be expressed. As usual, the equation system may or may not be consistent.

The smallest example of an $m < n$ system is one linear equation with two unknowns: a single line in two-dimensional solution space. Clearly there are multiple solutions. The next simplest example would be two linear equations in three unknowns, which appear in three-dimensional solution space as two planes; the planes may be parallel (inconsistent system with no solution), be coplanar (consistent system, multiple solutions anywhere on the single plane defined by the two equations) or intersect in a line (again, a consistent system with multiple solutions, but now only those along a single line). We explore the algebra and geometry of some of these possibilities, using the designation "F" for the "fewer equations than unknowns" situation.

2.5.1 Consistent Systems [$\rho(\mathbf{A}) = \rho(\mathbf{A} \vdots \mathbf{B})$]

Case F1: $\rho(\mathbf{A}) = m$.   For the $m < n$ case, when $\mathbf{A}$ is of full rank, then $\rho(\mathbf{A}) = \rho(\mathbf{A} \vdots \mathbf{B})$ and the system *must* be consistent.

Suppose that we choose from among the n columns in $\mathbf{A}$ a set of m of them that are linearly independent [we know that there is at least one such set since $\rho(\mathbf{A}) = m$]. Let $\mathbf{A}^M$ be the $m \times m$ nonsingular matrix composed of these columns, and denote the $m \times (n - m)$ matrix made up of the remaining columns from $\mathbf{A}$ by $\mathbf{A}^N$, so that $\mathbf{A} = [\mathbf{A}^M \vdots \mathbf{A}^N]$. Rearrange the elements in $\mathbf{X}$ similarly, $\mathbf{X} = [\mathbf{X}^M \vdots \mathbf{X}^N]'$, where $\mathbf{X}^M$ contains the x_i's associated with the m columns in $\mathbf{A}^M$ (arranged in the same order as those columns) and $\mathbf{X}^N$ contains the ($n -$

m) remaining unknowns associated with the columns in $\mathbf{A}^N$, also arranged in the same order as those columns. Then $\mathbf{AX} = \mathbf{B}$ can be expressed as

$$[\mathbf{A}^M \mid \mathbf{A}^N] \begin{bmatrix} \mathbf{X}^M \\ \hline \mathbf{X}^N \end{bmatrix} = \mathbf{B} \quad \text{or} \quad \mathbf{A}^M\mathbf{X}^M + \mathbf{A}^N\mathbf{X}^N = \mathbf{B}$$

Using $(\mathbf{A}^M)^{-1}$, which we know exists because $\mathbf{A}^M$ was formed to be nonsingular, we obtain

$$\mathbf{X}^M = (\mathbf{A}^M)^{-1}(\mathbf{B} - \mathbf{A}^N\mathbf{X}^N) \tag{2.8}$$

Given *any* values for the elements in $\mathbf{X}^N$, we can solve for the remaining m unknowns in $\mathbf{X}^M$. One choice would be to set $\mathbf{X}^N = \mathbf{0}$, resulting in

$$\mathbf{X}^M = (\mathbf{A}^M)^{-1}\mathbf{B} \tag{2.9}$$

This is called a *basic* solution to the set of equations, because the m linearly independent vectors in $\mathbf{A}^M$ form a *basis* in m-dimensional vector space, with respect to which *any* other m-element vector can be expressed as a linear combination of those basis vectors. A fundamental characteristic of such basic solutions is that no more than m unknowns (out of the total of n) have a nonzero value. If it turns out that some of the variables in the solution vector $\mathbf{X}^M$ happen also to be zero, that basic solution is said to be *degenerate*.[3]

In general, there will be more than one basic solution to consistent equation systems with $m < n$ and $\rho(\mathbf{A}) = m$, because there will be more than one way to select a set of m columns with which to construct $\mathbf{A}^M$. The *maximum* possible number of basic solutions is the number of ways that m items can be chosen from n, given by the combinatorial formula $C_m^n = n!/[m!(n-m)!]$. However, not all of these choices need necessarily produce a *nonsingular* $\mathbf{A}^M$. For example

$$\begin{aligned} x_1 + 2x_2 + 3x_3 + 3x_4 &= 1 \\ 3x_1 + 2x_2 + 3x_3 + 6x_4 &= 2 \end{aligned} \tag{2.10}$$

where $\rho(\mathbf{A}) = \rho(\mathbf{A} \mid \mathbf{B}) = 2$, $m = 2$, $n = 4$. Basic solutions will involve two of the four columns in $\mathbf{A}$. There are $C_2^4 = 6$ possible ways of choosing the two columns, but one of these ($\mathbf{A}_2$ and $\mathbf{A}_3$) produces a singular $\mathbf{A}^M$. The remaining basic solutions are (you might check at least some of these):

$\mathbf{A}_1$ and $\mathbf{A}_2$: $\mathbf{X}^M = [x_1 \; x_2]' = [\frac{1}{2} \; \frac{1}{4}]'$ so $\mathbf{X} = [\frac{1}{2} \; \frac{1}{4} \; 0 \; 0]'$
$\mathbf{A}_1$ and $\mathbf{A}_3$: $\mathbf{X}^M = [x_1 \; x_3]' = [\frac{1}{2} \; \frac{1}{6}]'$ so $\mathbf{X} = [\frac{1}{2} \; 0 \; \frac{1}{6} \; 0]'$
$\mathbf{A}_1$ and $\mathbf{A}_4$: $\mathbf{X}^M = [x_1 \; x_4]' = [0 \; \frac{1}{3}]'$ so $\mathbf{X} = [0 \; 0 \; 0 \; \frac{1}{3}]'$
$\mathbf{A}_2$ and $\mathbf{A}_4$: $\mathbf{X}^M = [x_2 \; x_4]' = [0 \; \frac{1}{3}]'$ so $\mathbf{X} = [0 \; 0 \; 0 \; \frac{1}{3}]'$
$\mathbf{A}_3$ and $\mathbf{A}_4$: $\mathbf{X}^M = [x_3 \; x_4]' = [0 \; \frac{1}{3}]'$ so $\mathbf{X} = [0 \; 0 \; 0 \; \frac{1}{3}]'$

The last three solutions are, in fact, the same; they are also degenerate, since some (in these cases, one) of the variables in each of the basic solutions has a value of zero.

---

[3]Basic solutions play an important role in developing solutions to linear programming problems, which always contain a linear system that has fewer equations than unknowns, and degeneracy sometimes has important implications when it occurs in solutions to linear programming problems, as we will see in Chapter 7.

Since $\rho(\mathbf{A}) = m$ in (2.10), it is possible to find a right inverse, $\mathbf{A}_R^{-1}$. Here

$$\mathbf{A}_R^{-1} = \mathbf{A}'(\mathbf{A}\mathbf{A}')^{-1} = \begin{bmatrix} -0.2472 & 0.1966 \\ 0.2697 & -0.1236 \\ 0.4045 & -0.1854 \\ -0.1685 & 0.2022 \end{bmatrix}$$

(to four decimal places) and so a solution to the equation system in (2.10) is

$$\mathbf{X} = \mathbf{A}_R^{-1}\mathbf{B} = \begin{bmatrix} -0.2472 & 0.1966 \\ 0.2697 & -0.1236 \\ 0.4045 & -0.1854 \\ -0.1685 & 0.2022 \end{bmatrix}\begin{bmatrix} 1 \\ 2 \end{bmatrix} = \begin{bmatrix} 0.1461 \\ 0.0225 \\ 0.0337 \\ 0.2360 \end{bmatrix}$$

It is easily verified that these values of $x_1, \ldots, x_4$ satisfy the two equations in (2.10).

For a consistent system with fewer linear equations than unknowns, we know that there will always be multiple solutions, and what we have found above, using the right inverse, is simply one of that infinity of solutions. However, this particular solution turns out to have an unusual property. The solution found by using the right inverse to a consistent system with fewer equations than unknowns represents the point satisfying all of the equations that lies *closest to the origin*, using the standard Euclidian measure of distance—the straight-line distance from $\mathbf{X} = \mathbf{A}_R^{-1}\mathbf{B}$ to the origin is less than the distance to the origin from any other point that satisfies the equations.[4]

Case F2: $\rho(\mathbf{A}) = k$, $k < m$.   Since $\rho(\mathbf{A}) \neq m$, a left inverse cannot be found. However, since we are considering a consistent system, we know that a set of $(m - k)$ equations can be removed without any loss of information. Once redundant equations are removed, we have n unknowns related by k $(< m < n)$ equations, and solutions as given in (2.8) and (2.9) can be found.

For example

$$\begin{aligned} x_1 + 2x_2 + \ x_3 + 3x_4 &= 4 \\ 2x_1 + 4x_2 + 2x_3 + 6x_4 &= 8 \\ x_1 + \ x_2 + 3x_3 + 4x_4 &= 2 \end{aligned} \qquad (2.11)$$

for which $\rho(\mathbf{A}) = \rho(\mathbf{A} \,\vdots\, \mathbf{B}) = 2$, $m = 3$, and $n = 4$. The first two equations describe the same plane; either is therefore expendable. Removing the first equation in (2.11) leaves

$$\begin{aligned} 2x_1 + 4x_2 + 2x_3 + 6x_4 &= 8 \\ x_1 + \ x_2 + 3x_3 + 4x_4 &= 2 \end{aligned} \qquad (2.11')$$

---

[4]If you are familiar with the use of calculus for minimization of a multivariable function subject to equality constraints, you can demonstrate this property, with a little patience, for this example. The point $\mathbf{X} = [0.1461 \ \ 0.0225 \ \ 0.0337 \ \ 0.2360]'$ is closer to the origin than any other point that lies on the two hyperplanes defined by the two equations in (2.10).

Details of the minimization problem are as follows: Find a point $\mathbf{X} = [x_1\,x_2\,x_3\,x_4]'$ that minimizes $D = (x_1^2 + x_2^2 + x_3^2 + x_4^2)^{1/2}$ (this is the Euclidian distance, in four-dimensional space from $\mathbf{X}$ to the origin), subject to the two equations in (2.10). The distance from $\mathbf{X} = [0.1461 \ \ 0.0225 \ \ 0.0337 \ \ 0.2360]'$ to the origin is 0.2805. As a curiosity check, you might find another point that satisfies both equations and then calculate D for this new point; it will be larger than 0.2805.

which is a consistent system with $m = 2$ and $n = 4$. If we use $\mathbf{A}^R$ for the coefficient matrix of this reduced system (the "R" here, as a superscript, reminds us that it is a *reduced* version of the original equation system; an "R" as a subscript identifies a right inverse), then here

$$\mathbf{A}^R = \begin{bmatrix} 2 & 4 & 2 & 6 \\ 1 & 1 & 3 & 4 \end{bmatrix}$$

and $\rho(\mathbf{A}^R) = 2 = m$. This falls under the category of case F1, above. You might check that two of the possible *basic* solutions are

$$\begin{bmatrix} x_1 \\ x_2 \end{bmatrix} = \begin{bmatrix} 0 \\ 2 \end{bmatrix} \quad \text{and} \quad \begin{bmatrix} x_3 \\ x_4 \end{bmatrix} = \begin{bmatrix} -2 \\ 2 \end{bmatrix}$$

We could also use the right inverse to the reduced 2×4 system to find a solution. If we also use $\mathbf{X}^R$ and $\mathbf{B}^R$ to indicate that they, too, correspond to the reduced equation system

$$\mathbf{A}^R\mathbf{X}^R = \mathbf{B}^R$$

then

$$(\mathbf{A}^R)_R^{-1} = \begin{bmatrix} 0.0556 & -0.0370 \\ 0.2222 & -0.2593 \\ -0.1667 & 0.3333 \\ 0.0556 & 0.0741 \end{bmatrix}$$

and so

$$\mathbf{X}^R = (\mathbf{A}^R)_R^{-1}\mathbf{B}^R = \begin{bmatrix} 0.3704 \\ 1.2593 \\ -0.6667 \\ 0.5926 \end{bmatrix}$$

is one possible (nonbasic) solution to the reduced equation system (and hence to the original system as well), as can be verified by substitution into the equations.

The Homogeneous Case   The special case of homogeneous equations belongs here in the consistent systems category since, as we have seen, such equations are always consistent. If $\rho(\mathbf{A}) = m$, then solutions of the kind given in (2.8) and (2.9) are possible. A simple geometric example would consist of two distinct planes passing through the origin in three-dimensional solution space and intersecting in a line (also through the origin). Note that the *basic* solution found via (2.9) will in fact be $\mathbf{X} = \mathbf{0}$; the solution is completely degenerate, since the values of all variables included in the basis will be 0. Other (nontrivial) solutions that lie along the line of intersection are possible. These would be found, via (2.8), by specifying any set of values for $\mathbf{X}^N$. Finally, using the right inverse for this $m < n$ case, we would again find the trivial solution, $\mathbf{X} = \mathbf{A}_R^{-1}\mathbf{B} = \mathbf{0}$. Notice that of all points on the line of intersection this is, by definition, "closest" to the origin.

If $\rho(\mathbf{A}) < m$, then in the example the two equations (two planes) would reduce to two equations that describe exactly the same plane, so along with $\mathbf{X} = \mathbf{0}$, there would be multiple solutions, lying anywhere on the plane.

2.5.2 Inconsistent Systems [$\rho(\mathbf{A}) \neq \rho(\mathbf{A} \vdots \mathbf{B})$]

With m equations and n unknowns and $m < n$, when the system is inconsistent it must be true that $\rho(\mathbf{A} \vdots \mathbf{B}) = \rho(\mathbf{A}) + 1$, so $\rho(\mathbf{A}) \leq m - 1$ and a case parallel to F1 cannot occur.

Case F3: $\rho(\mathbf{A}) = m$.   Impossible. (Included only for completeness.)

Case F4: $\rho(\mathbf{A}) = k, k < m$.   As in the $m = n$ case, only "partial" solutions are possible; these can be found only at the cost of *ignoring* equations in the system. There is at least one set of $(m - k)$ equations that, if dropped, leaves a consistent system of k equations in n unknowns for which multiple solutions can be found. For example

$$
\begin{aligned}
x_1 + 2x_2 + x_3 + 3x_4 &= 4 \\
2x_1 + 4x_2 + 2x_3 + 6x_4 &= 5 \\
x_1 + x_2 + 3x_3 + 4x_4 &= 2
\end{aligned}
\tag{2.12}
$$

where $\rho(\mathbf{A}) = 2$, $\rho(\mathbf{A} \vdots \mathbf{B}) = 3$, $m = 3$, $n = 4$. The nature of the inconsistency is revealed by the first two equations, which are "parallel" hyperplanes in four-dimensional solution space. If either of the first two equations is ignored, the remaining system has multiple solutions. For example, ignoring the first equation produces a reduced system for which there are six basic solutions.

In addition, the right inverse for the reduced system could be found and used to generate another solution. Equations (2.12) have the same coefficient matrix as in (2.11) but $\mathbf{B}$ is different and creates, in (2.12), an inconsistent system with $\rho(\mathbf{A}) = 2 < m$, so a right inverse cannot be found for the entire (inconsistent) system. However, dropping the first equation gives $\mathbf{A}^R = \begin{bmatrix} 2 & 4 & 2 & 6 \\ 1 & 1 & 3 & 4 \end{bmatrix}$. This is the same reduced matrix as was used in the previous example, so here, with $\mathbf{B} = \begin{bmatrix} 5 \\ 2 \end{bmatrix}$, we obtain

$$
\mathbf{X}^R = (\mathbf{A}^R)_R^{-1} \mathbf{B}^R = \begin{bmatrix} 0.2037 \\ 0.5926 \\ -0.1667 \\ 0.4259 \end{bmatrix}
$$

These values for $x_1, \ldots, x_4$ satisfy the second and third equations in (2.12); it must be emphasized that $\mathbf{X}^R$ does *not* satisfy the entire system in (2.12).

2.5.3 Summary: Fewer Equations than Unknowns

The essential problem in $m < n$ linear equation systems is one of too few equations or, equivalently, too many unknowns; in any case, the system is "underdetermined." The solution approach is to fix the values of some of the unknowns (*basic* solutions result if those variables are given values of zero).[5] If the system is consistent and $\rho(\mathbf{A}) = m$, then solutions of the sort shown in (2.8) or (2.9) can be found; since the values given to the variables in $\mathbf{X}^N$ are completely arbitrary, the equations necessarily have multiple solutions. Alternatively, the right inverse can be used to find a solution.

If $\rho(\mathbf{A}) < m$, the system may be consistent and it may not. If it is consistent, it has redundant equations; the redundant equations can be removed and one or more of the multiple solutions to the reduced system can be found. If the system is inconsistent, there is no way that a solution can be found to satisfy all of the equations. If we are willing and able to ignore some of the equations, "partial" solutions can be found. In either of these $\rho(\mathbf{A}) < m$ cases, once we have decided which equation to ignore, one or more solutions of the sort shown in (2.8) or (2.9) can be found or, again, the right inverse in (2.7) can be used.

2.6 MORE EQUATIONS THAN UNKNOWNS ($m > n$)

In cases with more equations than unknowns, the coefficient matrix is "tall and narrow"; its greatest possible rank is n. It is easy to visualize the geometric possibilities for the case of three equations and two unknowns ($m = 3$, $n = 2$). These will be three lines in two-dimensional solution space. The lines may all intersect in one point (in which case any one of the equations is unnecessary), they may all be parallel, each pair of equations may intersect in a different point, and so on. We explore the algebra of some of these possibilities in what follows, using the designation "M" to indicate the "more equations than unknowns" scenario.

2.6.1 Consistent Systems [$\rho(\mathbf{A}) = (\mathbf{A} \vdots \mathbf{B})$]

Case M1: $\rho(\mathbf{A}) = n$.   This is the maximum possible rank for $\mathbf{A}$; yet $\mathbf{B}$ is an m-element column vector, and $m > n$. The columns of $\mathbf{A}$ can never constitute a full basis in m-dimensional vector space, since m linearly independent vectors are needed for that purpose. A unique solution could occur only if the consistent equation system had exactly $(m - n)$ *redundant* equations. This is exactly the case when $\rho(\mathbf{A}) = n$. Here is a very simple illustration:

ALGEBRA

$$\begin{aligned} x_1 + 0x_2 &= 2 \\ 0x_1 + x_2 &= 2 \\ x_1 + x_2 &= 4 \end{aligned} \qquad (2.13)$$

GEOMETRY (see Fig. 2.8)

Here $\rho(\mathbf{A}) = 2$, $\rho(\mathbf{A} \vdots \mathbf{B}) = 2$, $m = 3$, $n = 2$. The system is consistent, any $[m - \rho(\mathbf{A})] = 1$ equation can be considered redundant, and the 2×2 reduced system will have a unique solution. You can easily check that no matter which equation is discarded, the remaining two-equation system has the same solution, namely, $\mathbf{X} = \begin{bmatrix} 2 \\ 2 \end{bmatrix}$.

[5]Fixing the value of a particular variable (say, x_i) can be viewed either as *reducing* the number of unknowns by one (since the value of x_i is no longer unknown) or as *increasing* the number of equations by one—namely, by the equation $x_i = c$ (for $c =$ the value chosen for x_i).

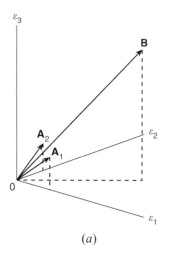

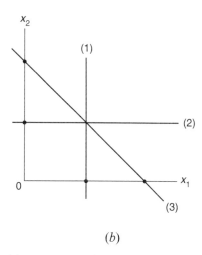

(a)   (b)

FIGURE 2.8   Representation of (2.13) in (a) vector space; (b) solution space.

Suppose we use the left inverse in this case with $m > n$. Here $\mathbf{A} = \begin{bmatrix} 1 & 0 \\ 0 & 1 \\ 1 & 1 \end{bmatrix}$ and $\mathbf{A}_L^{-1} =$

$\begin{bmatrix} 0.6667 & -0.3333 & 0.3333 \\ -0.3333 & 0.6667 & 0.3333 \end{bmatrix}$. (You might go through the calculations to find $\mathbf{A}_L^{-1}$ and also check this by showing that the product $\mathbf{A}_L^{-1}\mathbf{A}$ indeed generates an identity matrix of order 2.) So

$$\mathbf{X} = \mathbf{A}_L^{-1}\mathbf{B} = \begin{bmatrix} 0.6667 & -0.3333 & 0.3333 \\ -0.3333 & 0.6667 & 0.3333 \end{bmatrix} \begin{bmatrix} 2 \\ 2 \\ 4 \end{bmatrix} = \begin{bmatrix} 2 \\ 2 \end{bmatrix}$$

Thus, for a consistent system with more equations than unknowns, using the left inverse of the coefficient matrix provides an automatic way of ignoring redundant equations.

Case M2: $\rho(\mathbf{A}) = k, k < n$.   If $(m - k)$ of the redundant equations are removed, then this is the same situation as in case F2; both reduce to consistent $k \times n$ systems ($k < n$) and can be approached as in (2.8) or (2.9).

The Homogeneous Case   Here, again, the system must always be consistent. If $\rho(\mathbf{A}) = n$, its maximum possible value, then the left inverse could be found and the unique but trivial solution $\mathbf{X} = \mathbf{A}_L^{-1}\mathbf{B} = \mathbf{0}$ would result. The geometry in this case, for $n = 2$, is three or more equations, at least two of which are distinct, that pass through the origin. If $\rho(\mathbf{A}) < n$, there are redundant equations, and there will be multiple solutions. Again, in the case of two unknowns, all equations describe the same line. As before, the solutions could be found as in (2.8) or (2.9), once redundant equations are eliminated from the system.

2.6.2 Inconsistent Systems [$\rho(\mathbf{A}) \neq \rho(\mathbf{A} \mid \mathbf{B})$]

Case M3: $\rho(\mathbf{A}) = n$.   No solutions are possible for the entire equation system. If some ($m - n$) of the equations can be ignored, then a *unique* solution to the reduced $n \times n$ system can be found. For example:

ALGEBRA

$$
\begin{aligned}
x_1 + 0x_2 &= 2 \\
0x_1 + x_2 &= 2 \\
x_1 + x_2 &= 3
\end{aligned}
\tag{2.14}
$$

GEOMETRY

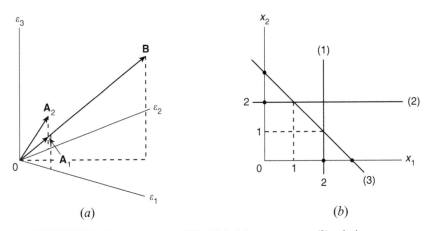

(a) (b)

FIGURE 2.9   Representation of (2.14) in (*a*) vector space; (*b*) solution space.

Here $\rho(\mathbf{A}) = 2$, $\rho(\mathbf{A} \mid \mathbf{B}) = 3$, $m = 3$, $n = 2$. By removing one equation, we obtain a reduced system whose "partial" solution is indicated by one of the three intersections in the solution-space picture.

Note that this equation system differs from that in case M1 only in the third element of **B**. From the rules of rank, we know that the system is inconsistent. Yet, we also know that a left inverse can be found for this coefficient matrix, since $\rho(\mathbf{A})$ is at its maximum value; this left inverse is exactly what we found in exploring case M1. Despite the fact that the equation system is inconsistent, it can still be represented in $\mathbf{AX} = \mathbf{B}$ form and, since $\mathbf{A}_L^{-1}$ exists, we can find $\mathbf{X} = \mathbf{A}_L^{-1}\mathbf{B}$. Here

$$
\mathbf{X} = \begin{bmatrix} 0.6667 & -0.3333 & 0.3333 \\ -0.3333 & 0.6667 & 0.3333 \end{bmatrix} \begin{bmatrix} 2 \\ 2 \\ 3 \end{bmatrix} = \begin{bmatrix} 1.6667 \\ 1.6667 \end{bmatrix}
$$

In fact, this **X** is *not* a solution to the system in (2.14) at all. (The system is inconsistent; we know that there is no solution that satisfies all three equations.) Rather, the point $x_1 = 1.6667$ and $x_2 = 1.6667$, while not even on any one of the equations, is in a particular mathematical sense "closest" to all of them. Think of the equations as $\mathbf{AX} + \mathbf{E} = \mathbf{B}$, so **E** is an "error" vector; $\mathbf{E} = \mathbf{B} - \mathbf{AX}$. An element, e_i, measures the difference between b_i and $\sum_{j=1}^{n} a_{ij}x_j$. Then $\mathbf{X} = \mathbf{A}_L^{-1}\mathbf{B}$ is the vector for which size of **E**, as measured by the sum of the squares of its elements, $\Sigma_{i=1}^{m}\, e_i^2$ or by $\mathbf{E'E}$, is minimized.[6] Here $\mathbf{E} = \begin{bmatrix} 0.3333 \\ 0.3333 \\ -0.3333 \end{bmatrix}$ and

$\mathbf{E'E} = 0.3333$. (If you let $\mathbf{X}$ equal any of the three partial solutions, you will find $\mathbf{E'E} = 1$.) So, while the point does not satisfy any of the equations exactly, it might be said to be "least wrong" relative to all of them. (This is representative of what are known as "least squares" problems in regression and statistics, where a function is sought that best represents a set of observed points—in the sense that the sum of the squares of the distances from the points to the function is minimized.)

Case M4: $\rho(\mathbf{A}) = k$, $k < n$. This situation is identical in principle to that under case M2, except that now when $(m - k)$ equations are omitted the system is changed; requirements on the variables are being ignored. If the former case, $(m - k)$ equations were redundant and could be dropped without loss of information. The approach is the same as under case M2, provided that you can justify ignoring some of the original equations.

2.6.3 Summary: More Equations than Unknowns

In linear systems with more equations than unknowns we have, essentially, too much information about the variables (an "overdetermined" system). For a consistent system (in which one or more solutions exist), at least $(m - n)$ equations must be redundant. If exactly $(m - n)$ may be removed, we have in fact a square system with a unique solution. If more may be removed, then we have an "underdetermined" system with multiple solutions.

If no equations are unnecessary, or if less than $(m - n)$ of them may be removed, then the system is not consistent. The only possibility is for "partial" solutions where $[m - \rho(\mathbf{A})]$ equations are simply ignored. In the particular case when $\rho(\mathbf{A}) = 1$, only one equation at a time can be satisfied; the larger $\rho(\mathbf{A})$, the larger the number of equations that can be satisfied simultaneously.

2.7 NUMERICAL METHODS FOR SOLVING SYSTEMS OF LINEAR EQUATIONS

In this section, we introduce the general ideas behind approaches to solutions of a system of linear equations which do not rely on the inverse of the coefficient matrix. Present-day high-speed computers usually do not find $\mathbf{A}^{-1}$ using the cumbersome calculations required for the determinant and the adjoint matrix, because there are more efficient numerical methods for obtaining the solution values for the unknowns.[7] We examine some of these methods in this section.

---

[6] It is straightforward calculus to show that $\mathbf{X} = (\mathbf{A'A})^{-1}\mathbf{A'B} = \mathbf{A}_L^{-1}\mathbf{B}$ is the minimizer of $\mathbf{E'E}$. The length or *norm* of $\mathbf{E}$ is defined as $\|\mathbf{E}\| = (\Sigma_{i=1}^{m}\, e_i^2)^{0.5}$ (Chapter 1, Section 1.4), so minimization of $\mathbf{E'E}$ is minimization of $\|\mathbf{E}\|^2$. Since the positive root is always used for $\|\mathbf{E}\|$, the minimizer of $\|\mathbf{E}\|^2$ is the same as the minimizer of the norm of $\mathbf{E}$, so we are finding the shortest error vector.

[7] This in no way diminishes the effectiveness of the rules of rank for sorting out the kinds of solutions that one may expect from a particular set of linear equations.

2.7.1 Elementary Matrix Operations and Gaussian Methods (Karl Friedrich Gauss, German, 1777–1855)

Equations (1.26) are repeated here, since we will use them for illustration:

$$3x_1 + 2x_2 = 7$$
$$2x_1 + 5x_2 = 12$$

$$(2.15)$$

We established, in Section 1.3, that the unique solution is $x_1 = 1$, $x_2 = 2$. One straightforward arithmetic approach to finding that solution would proceed as follows: multiply the first equation through by $\frac{2}{3}$, giving $2x_1 + \frac{4}{3}x_2 = \frac{14}{3}$, and then subtract this modified first equation from the second equation. This gives $\frac{11}{3}x_2 = \frac{22}{3}$, from which $x_2 = 2$ follows immediately. Substitution into either of the original equations then gives $x_1 = 1$. With respect to any system of linear equations, it should be clear that the solution will be unchanged by any of the following operations:

1. Interchange of any pair of equations. [In (2.15), it is irrelevant which equation is regarded as first and which as second.]

2. Multiplication of both sides of an equation by a constant (not equal to zero). [The first equation in (2.15) expresses the same information (describes the same line in two-dimensional solution space) as, for example, $6x_1 + 4x_2 = 14$ or as $2x_1 + \frac{4}{3}x_2 = \frac{14}{3}$.]

3. Addition (or subtraction) of a multiple of any one equation to (from) another equation. [In the example, we added $(-\frac{2}{3})$ times the first equation to the second equation (which is equivalent to multiplying equation 1 by $\frac{2}{3}$ and then subtracting it from the second equation.) The resulting equation, $\frac{11}{3}x_1 = \frac{22}{3}$, in conjunction with either of the two original equations, is still a system with the solution $x_1 = 1$, $x_2 = 2$.]

These three operations are known as *elementary transformations* of a system of linear equations; such operations generate *equivalent* systems of equations that have the same solution as the original system. In fact, these transformations are nothing more than formalizations of the most elementary algebraic approach to solving simple and small systems of linear equations, as was used at the outset of this section.

Since all the information that is particular to the equation system is contained in the augmented matrix

$$[\mathbf{A} \vdots \mathbf{B}] = \begin{bmatrix} 3 & 2 & \vdots & 7 \\ 2 & 5 & \vdots & 12 \end{bmatrix}$$

$$(2.16)$$

we can translate the elementary operations into transformations on the matrix $[\mathbf{A} \vdots \mathbf{B}]$. In that case, these are termed *elementary matrix transformations* or *elementary matrix operations*.

(a) Multiplying row 1 by $\frac{2}{3}$ gives

$$[2 \quad \tfrac{4}{3} \vdots \tfrac{14}{3}]$$

$$(2.17)$$

(b) Subtraction of this modified row 1 from row 2 (and replacing row 2 by this result) produces

$$\begin{bmatrix} 3 & 2 & \vdots & 7 \\ 0 & \frac{11}{3} & \vdots & \frac{22}{3} \end{bmatrix} \tag{2.18}$$

Explicitly

$$\begin{bmatrix} 3 & 2 \\ 0 & \frac{11}{3} \end{bmatrix} \begin{bmatrix} x_1 \\ x_2 \end{bmatrix} = \begin{bmatrix} 7 \\ \frac{22}{3} \end{bmatrix}$$

from which, as before, $3x_1 + 2x_2 = 7$ and $\frac{11}{3}x_2 = \frac{22}{3}$. Again, solving the second equation first, $x_2 = 2$, and putting this result into the first equation gives $x_1 = 1$. The coefficient matrix part of the transformed augmented matrix in (2.18), namely, $\begin{bmatrix} 3 & 2 \\ 0 & \frac{11}{3} \end{bmatrix}$, is *upper triangular* (zeros—here only one—below the main diagonal). It is this structure of the coefficient matrix that allows the solution for x_2 first and then x_1.

This solution sequence is known as *backward substitution* or *backward recursion*. Element a_{11} in the original matrix is known as a *pivot* element. After multiplication by the appropriate number (in the example, $\frac{2}{3}$), it is subtracted from the element directly below it in order to create a zero in the a_{21} position. More generally, for $\mathbf{A}$ of order n, the procedure that this example illustrates uses element a_{11} as a pivot to create zeros in *all* remaining locations in column 1 of $\mathbf{A}$—the multipliers are, in turn, $(a_{21}/a_{11}), (a_{31}/a_{11}), \ldots, (a_{n1}/a_{11})$. Denote the modified matrix as $\mathbf{A}^*$. Next, a_{22}^* is the pivot element used to reduce all lower elements in column 2 of $\mathbf{A}^*$ to zero, and so on.

For large systems, a large number of arithmetic operations would be required. This can lead to problems with computational roundoff errors, especially if there are large differences in the sizes of coefficients in different equations in the original system—for example, if $a_{11} = 0.0031$ and $a_{21} = 10,050$.

An alternative upper triangular coefficient matrix is generated by further transforming (2.18) through multiplication of row 1 by $\frac{1}{3}$ and multiplying row 2 by $\frac{3}{11}$. This gives

$$\begin{bmatrix} 1 & \frac{2}{3} & \vdots & \frac{7}{3} \\ 0 & 1 & \vdots & 2 \end{bmatrix} \tag{2.19}$$

and the associated equation system

$$x_1 + \tfrac{2}{3}x_2 = \tfrac{7}{3}$$

$$x_2 = 2$$

with the same solution, $x_2 = 2$ and, putting this into the first 2 equation, $x_1 = \frac{7}{3} - \frac{4}{3} = 1$. The coefficient matrix

$$\begin{bmatrix} 1 & \frac{2}{3} \\ 0 & 1 \end{bmatrix} \tag{2.20}$$

is said to be in *echelon* form, meaning that it is upper triangular and also that the first nonzero element in any row is 1.[8]

---

[8]More generally, an $m \times n$ matrix $\mathbf{Q} = [q_{ij}]$ $(m \le n)$ is said to be in *echelon form* if (a) $q_{ij} = 0$ $(i > j)$, (b) the first nonzero element in each row is 1, and (c) that first nonzero element in any row is further to the right than the first nonzero element in the preceding row.

Clearly, one could go further in simplifying the coefficient matrix, still using only elementary matrix operations. Starting from (2.19), multiply row 2 by $-\frac{2}{3}$, and add the result to row 1:

$$\begin{bmatrix} 1 & 0 & \vdots & 1 \\ 0 & 1 & \vdots & 2 \end{bmatrix} \tag{2.21}$$

In the process, the coefficient matrix has become the identity matrix and the associated equation "system" is now

$$1x_1 + 0x_2 = 1$$

$$0x_1 + 1x_2 = 2$$

In solution space this is a pair of vertical ($x_1 = 1$) and horizontal ($x_2 = 2$) lines that intersect at (1,2).[9]

The procedure that transforms the augmented matrix of the equation system from its original form in (2.16) to the echelon form in (2.19) (after which the values of the unknowns are easily found in reverse order, starting with the last) is known as the method of *Gaussian elimination*. Further transformation to *reduced* echelon form, as in (2.21), is called the *Gauss–Jordan Method* (Camille Jordan, French, 1832–1922). Of the two approaches, Gaussian elimination requires fewer elementary matrix operations and more calculation in the recursive backward solution for the x's; the Gauss–Jordan method requires more elementary matrix operations but then the values of the x's are obvious by inspection, as in the last column in (2.21).

We now come to an extremely important point about the Gaussian and Gauss–Jordan approaches. Writing (2.15) as

$$\mathbf{AX} = \mathbf{IB} \tag{2.22}$$

we recognize that the elementary operations that transformed $\mathbf{A}$ in (2.16) into $\mathbf{I}$ in (2.21) would also transform the $\mathbf{I}$ on the right-hand side in (2.22) into $\mathbf{A}^{-1}$, from the fundamental definition of the inverse, namely, that

$$\mathbf{A}^{-1}\mathbf{AX} = \mathbf{A}^{-1}\mathbf{IB} \quad \text{or} \quad \mathbf{IX} = \mathbf{A}^{-1}\mathbf{B} \tag{2.23}$$

To see how this operates, we create a larger augmentation of the original $\mathbf{A}$ matrix

$$[\mathbf{A} \vdots \mathbf{I}] = \begin{bmatrix} 3 & 2 & \vdots & 1 & 0 \\ 2 & 5 & \vdots & 0 & 1 \end{bmatrix} \tag{2.24}$$

and carry through exactly the same set of elementary transformations as illustrated in (2.17)–(2.19) and (2.21).

---

[9]An $m \times n$ matrix ($m \leq n$) is said to be in *reduced echelon form* if it is in echelon form and if there are also zeros above the first nonzero element in each row. In particular, any transformed augmented matrix from a linear equation system in which the leftmost $m \times m$ submatrix is an identity matrix is in reduced echelon form.

(a) Multiply row 1 by $\frac{2}{3}$:

$$[2 \quad \frac{4}{3} \mid \frac{2}{3} \quad 0]$$

(b) Subtract this from row 2:

$$\begin{bmatrix} 3 & 2 & \vdots & 1 & 0 \\ 0 & \frac{11}{3} & \vdots & -\frac{2}{3} & 1 \end{bmatrix} \tag{2.25}$$

(c) Multiply row 1 by $\frac{1}{3}$ and row 2 by $\frac{3}{11}$:

$$\begin{bmatrix} 1 & \frac{2}{3} & \vdots & \frac{1}{3} & 0 \\ 0 & 1 & \vdots & -\frac{2}{11} & \frac{3}{11} \end{bmatrix}$$

(d) Multiply row 2 by $-\frac{2}{3}$ and add to row 1:

$$\begin{bmatrix} 1 & 0 & \vdots & \frac{15}{33} & -\frac{6}{33} \\ 0 & 1 & \vdots & -\frac{2}{11} & \frac{3}{11} \end{bmatrix} \tag{2.26}$$

You can easily verify that $\mathbf{A}^{-1} = \begin{bmatrix} \frac{15}{33} & -\frac{6}{33} \\ -\frac{2}{11} & \frac{3}{11} \end{bmatrix}$ for our original coefficient matrix $\mathbf{A} = \begin{bmatrix} 3 & 2 \\ 2 & 5 \end{bmatrix}$. In general, it is numerical methods of this sort that computers use to calculate inverses. If a system of equations needs to be solved only once—if there are not several alternative right-hand side **B**'s—then the specific solution can be found using backward substitution from the echelon form (Gaussian elimination), not the reduced echelon form (Gauss–Jordan) that generates the inverse as a byproduct.

2.7.2 Iterative Methods

We illustrate briefly the basic idea behind an *iterative* approach to the solution of the pair of linear equations in (2.15). Iterative methods will play a particularly important role in solution techniques for nonlinear equations and also in optimization techniques, to be discussed in later chapters.

The general idea is simple enough: rewrite the equations so that each one expresses one unknown as a function of all the others, then make an initial guess at the solution, solve for the resulting values, use this result as the next (better) estimate of a solution, and so on. For our specific example, rewrite (2.15) as

$$x_1 = \frac{7 - 2x_2}{3} \quad \text{and} \quad x_2 = \frac{12 - 2x_1}{5} \tag{2.27}$$

Suppose that we begin with an initial guess $\mathbf{X}^0 = \begin{bmatrix} 0 \\ 0 \end{bmatrix}$. Putting these values into the right-hand sides of (2.27) produces a new estimate, denoted $\mathbf{X}^1 = \begin{bmatrix} 2.333 \\ 2.400 \end{bmatrix}$. Continuing in this

TABLE 2.3 Comparison of Two Iterative Approaches to Equations (2.15)

Iteration	Standard Iterative Approach $x_1^k = (7 - 2x_2^{k-1})/3;$ $x_2^k = (12 - 2x_1^{k-1})/5$		Gauss–Seidel Method $x_1^k = (7 - 2x_2^{k-1})/3;$ $x_2^k = (12 - 2x_1^k)/5$	
	x_1	x_2	x_1	x_2
0	0	0	0	0
1	2.333	2.400	2.333	1.467
2	0.733	1.467	1.356	1.858
3	1.356	2.107	1.095	1.962
4	0.929	1.858	1.025	1.990
5	1.095	2.028	1.007	1.997
6	0.981	1.962	1.002	1.999
7	1.025	2.008	1.001	2.000
8	0.995	1.990	1.000	2.000
9	1.007	2.002	—	—
10	0.999	1.997	—	—
11	1.002	2.000	—	—
12	1.000	1.999	—	—
13	1.001	2.000	—	—
14	1.000	2.000	—	—

way[10], finding $\mathbf{X}^k = [x_1^k\, x_2^k]'$ as $x_1^k = (7 - 2x_2^{k-1})/3$ and $x_2^k = (12 - 2x_1^{k-1})/5$, generates the series of results shown in the left part of Table 2.3. It is clear that at about $k = 14$, x_1 and x_2 are correct to three decimal points. Of course, in general the "real" solution will not be known. Therefore some kind of stopping criterion is needed when using iterative methods—for example, when the values of the variables differ from one iteration to the next by less than some small (prespecified) number.

A modest variation on this approach is known as the *Gauss–Seidel* method (Philipp Ludwig von Seidel, German, 1821–1896), the virtue of which is a generally quicker convergence. In our example, we would begin with $\mathbf{X}^0$ and from it generate x_1^1 only. Then, using *this* x_1 (and not x_1^0), we would calculate $x_2^1 = (12 - 2x_1^1)/5$. The superscript 1 on x_2 identifies this as the *first* new estimate (after x_2^0) of x_2; the difference is that it builds on the just-calculated x_1^1, not on x_1^0, as in the previous example. Table 2.3 indicates the faster convergence of this variation on the standard iterative approach. In this example, the Gauss–Seidel method converges on $x_1 = 1$, $x_2 = 2$ in half as many steps, essentially skipping every other value of the generalized iterative approach.

Iterative techniques like this are at the heart of solution methods for systems of nonlinear equations. Chapter 5 covers this material in some detail. At that time we will also explore issues such as alternative kinds of stopping criteria and sensitivity of the iterative results to the choice of an $\mathbf{X}^0$. In addition, it turns out that iterative techniques like the Gauss–Seidel method can be *very* dependent on the *ordering* of the equations.

---

[10]Here and elsewhere in this book, when exploring a sequence of values for $\mathbf{X}$ in an iterative procedure, we will use $\mathbf{X}^k$ for the values of $\mathbf{X} = [x_1, \ldots, x_n]'$ at the kth step. Some texts use $\mathbf{X}^{(k)}$, to avoid confusion with the vector that results when each element x_i ($i = 1, \ldots, n$) is raised to the kth power. It should be clear from the context which meaning is attached to $\mathbf{X}^k$, and omitting parentheses keeps the notation less cluttered.

It is easy to see why this is true in a simple case with two linear equations in two un-knowns. If the equations in (2.15) had been written down in the opposite sequence, namely

$$2x_1 + 5x_2 = 12$$
$$3x_1 + 2x_2 = 7$$

(2.15′)

then the Gauss–Seidel method would use

$$x_1 = -2.5x_2 + 6$$
$$x_2 = -1.5x_1 + 3.5$$

and, starting from the same $\mathbf{X}^0 = \begin{bmatrix} 0 \\ 0 \end{bmatrix}$ that we just used with (2.15), the iterations ex-plode—x_1 toward $+\infty$ and x_2 toward $-\infty$.

Figure 2.10 illustrates exactly what is going on. In Figure 2.10*a* the equations are num-bered in the order in which they appeared in (2.15). Starting at $\mathbf{X}^0 = \begin{bmatrix} 0 \\ 0 \end{bmatrix}$, the fact that $x_2^0 = 0$ identifies $x_1^1 = 2.333$ (where the first equation cuts the x_1 axis). This value of x_1^1, in turn, leads to $x_2^1 = 1.467$ (on the second equation, directly above $x_1^1 = 2.333$). This x_2^1 leads (hor-izontally) to $x_1^2 = 1.356$ (on the first equation), and so on. The iterations follow a "com-pacting staircase" path toward the solution. In Figure 2.10*b* the equations have been renumbered, as they appear in (2.15′). Now the initial $x_2^0 = 0$ leads to $x_1^1 = 6$ (where the present first equation cuts the x_1 axis). This $x_1^1 = 6$ leads (vertically) to $x_2^1 = -5.5$ (on the present second equation), and so on, with the iterations following an "expanding stair-case" path, away from the solution.

It appears that we need to use the equation ordering in which the equation with the lower x_1 axis intercept is labeled (1). But in larger systems, this kind of "preconditioning" of the system of equations can be time consuming and difficult. This ordering problem with the Gauss–Seidel method will return to haunt us in Chapter 5. With more equations and with nonlinear equations it only gets worse.

2.7.3 Factorization of the Matrix A

Many computer approaches to the solution of systems of linear equations rely on various kinds of *factorizations* of the **A** matrix; these factorizations, in turn, are derived from the operations used in one or more Gaussian elimination schemes. For example, using (2.16) again, the multiplier of row 1 that, when subtracted from row 2, produced the zero in the lower left in (2.18), was $\frac{2}{3}$—exactly a_{21}/a_{11}. We construct a 2×2 *lower triangular* matrix, **L**, with ones on the main diagonal and this multiplier as element l_{21}:

$$\mathbf{L} = \begin{bmatrix} 1 & 0 \\ \frac{2}{3} & 1 \end{bmatrix}$$

(2.28)

Denote the transformed **B** vector in (2.18) as **C**:

$$\mathbf{C} = \begin{bmatrix} 7 \\ \frac{22}{3} \end{bmatrix}$$

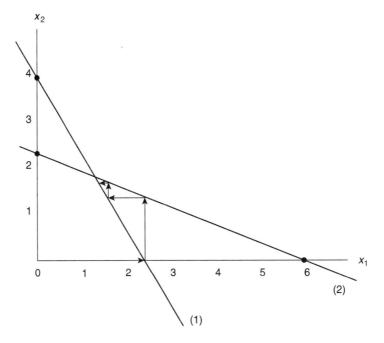

FIGURE 2.10*a*   Equations (2.15).

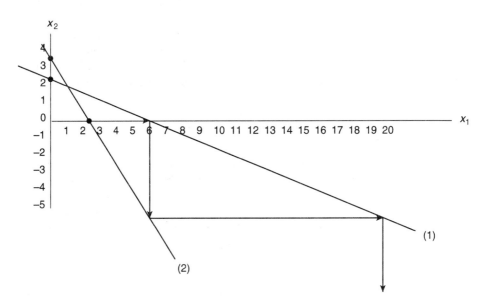

FIGURE 2.10*b*   Equations (2.15′).

Note that $\mathbf{LC} = \mathbf{B}$:

$$\begin{bmatrix} 1 & 0 \\ \frac{2}{3} & 1 \end{bmatrix} \begin{bmatrix} 7 \\ \frac{22}{3} \end{bmatrix} = \begin{bmatrix} 7 \\ 12 \end{bmatrix}$$

Recall the 2×2 *upper triangular* matrix in (2.18); denote it by $\mathbf{U}$:

$$\mathbf{U} = \begin{bmatrix} 3 & 2 \\ 0 & \frac{11}{3} \end{bmatrix} \tag{2.29}$$

Now the (possibly) unexpected result of these definitions of $\mathbf{L}$ and $\mathbf{U}$ is that

$$\mathbf{A} = \mathbf{LU}$$

In words, the Gaussian elimination operations that created $\mathbf{U}$ from $\mathbf{A}$, when recorded in a specific way in a lower triangular matrix $\mathbf{L}$, have created a *decomposition* of $\mathbf{A}$ into the product $\mathbf{LU}$. This means, if $\mathbf{A} = \mathbf{LU}$, that $\mathbf{AX} = \mathbf{B}$ is expressible as

$$\mathbf{LUX} = \mathbf{B}$$

or

$$\mathbf{UX} = \mathbf{L}^{-1}\mathbf{B} \tag{2.30}$$

if $\mathbf{L}$ is nonsingular. You can easily check that $\mathbf{L}^{-1}$ is exactly the matrix on the right in (2.25). We have simply interrupted the conversion of $\mathbf{A}$ in (2.24) into $\mathbf{I}$ and of $\mathbf{I}$ in (2.24) into $\mathbf{A}^{-1}$ at the point where $\mathbf{A}$ has become $\mathbf{U}$ and hence $\mathbf{I}$ has become $\mathbf{L}^{-1}$. Obviously, if $\mathbf{U}$ is nonsingular, then

$$\mathbf{X} = \mathbf{U}^{-1}\mathbf{L}^{-1}\mathbf{B} \tag{2.31}$$

and you can also easily check, for this example, that indeed $\mathbf{A}^{-1}$, which we have in (2.26), is equal to the product $\mathbf{U}^{-1}\mathbf{L}^{-1}$.

In summary, Gaussian elimination converts the original problem $\mathbf{AX} = \mathbf{B}$ into a *pair* of linear systems, in each of which the coefficient matrix is triangular (one upper triangular and one lower triangular):

$$\mathbf{UX} = \mathbf{C} \qquad \text{and} \qquad \mathbf{LC} = \mathbf{B} \tag{2.32}$$

The point is that for any given $\mathbf{B}$, $\mathbf{C}$ can be found by *forward recursion* from $\mathbf{LC} = \mathbf{B}$. For a general 2×2 example

$$\begin{bmatrix} l_{11} & 0 \\ l_{21} & l_{22} \end{bmatrix} \begin{bmatrix} c_1 \\ c_2 \end{bmatrix} = \begin{bmatrix} b_1 \\ b_2 \end{bmatrix}$$

gives (1) $l_{11}c_1 = b_1$ or $c_1 = b_1/l_{11}$ immediately, and then (2) $l_{21}c_1 + l_{22}c_2 = b_2$ gives $c_2 = (1/l_{22})(b_2 - l_{21}c_1)$, which is fully determined since c_1 was found in step 1. Given $\mathbf{C}$, then $\mathbf{X}$ is just as easily found by *backward recursion*. Specifically,

$$\begin{bmatrix} u_{11} & u_{12} \\ 0 & u_{22} \end{bmatrix} \begin{bmatrix} x_1 \\ x_2 \end{bmatrix} = \begin{bmatrix} c_1 \\ c_2 \end{bmatrix}$$

gives (1) $u_{22}x_2 = c_2$ or $x_2 = c_2/u_{22}$ and then (2) $u_{11}x_1 + u_{12}x_2 = c_2$ gives $x_1 = (1/u_{11})(c_2 - u_{12}x_2)$, which is determined because x_2 is known. This general decomposition of $\mathbf{AX} = \mathbf{B}$ into the two triangular linear systems (known as an **LU** *decomposition*) underpins many computer approaches to finding solutions to systems of linear equations. Here is a three-equation example.

2.7.4 Numerical Illustration

Consider the following equation system:

$$x_1 + x_2 + x_3 = 18$$

$$2x_1 + 6x_3 = 6$$

$$3x_1 + 7x_2 + x_3 = 12$$

or

$$\begin{bmatrix} 1 & 1 & 1 \\ 2 & 0 & 6 \\ 3 & 7 & 1 \end{bmatrix} \begin{bmatrix} x_1 \\ x_2 \\ x_3 \end{bmatrix} = \begin{bmatrix} 18 \\ 6 \\ 12 \end{bmatrix} \tag{2.33}$$

You can take advantage of the presence of any zeros in $\mathbf{A}$ by interchanging rows (which is simply reordering the original equations) so as to put the zeros below the main diagonal, since the object of Gaussian elimination is to create an upper triangular coefficient matrix. In this example, interchanging the second and third equations gives, in $\mathbf{AX} = \mathbf{B}$ form:

$$\begin{bmatrix} 1 & 1 & 1 \\ 3 & 7 & 1 \\ 2 & 0 & 6 \end{bmatrix} \begin{bmatrix} x_1 \\ x_2 \\ x_3 \end{bmatrix} = \begin{bmatrix} 18 \\ 12 \\ 6 \end{bmatrix} \tag{2.34}$$

In matrix terms, the interchange of two rows in $\mathbf{A}$ can be accomplished through *pre*-multiplication of $\mathbf{A}$ by an identity matrix in which the same rows have been interchanged. Rearrangement of an identity matrix in this way creates a *permutation* matrix $\mathbf{P}$ (Section 1.6); in the present example $\mathbf{P} = \begin{bmatrix} 1 & 0 & 0 \\ 0 & 0 & 1 \\ 0 & 1 & 0 \end{bmatrix}$, as you can easily verify by calculating the product $\mathbf{PA}$. (Interchange of the second and third *columns* of $\mathbf{A}$ would result from *post*-multiplication by $\mathbf{P}$.)

For illustration, we put the beginning $\mathbf{A}$, $\mathbf{I}$, and $\mathbf{B}$ all together into one 3×7 matrix:

$$\mathbf{M} = [\mathbf{A} \vdots \mathbf{I} \vdots \mathbf{B}] = \begin{bmatrix} 1 & 1 & 1 & 1 & 0 & 0 & 18 \\ 3 & 7 & 1 & 0 & 1 & 0 & 12 \\ 2 & 0 & 6 & 0 & 0 & 1 & 6 \end{bmatrix} \tag{2.35}$$

Element m_{11} is our initial *pivot*. Subtracting 3 times row 1 from row 2 and 2 times row 1 from row 3, so as to create zeros in the remainder of column 1 of **M**, gives

$$\left[\begin{array}{ccc:ccc:c} 1 & 1 & 1 & 1 & 0 & 0 & 18 \\ 0 & 4 & -2 & -3 & 1 & 0 & -42 \\ 0 & -2 & 4 & -2 & 0 & 1 & -30 \end{array}\right] \qquad (2.36)$$

The next pivot is $m_{22} = 4$. Subtracting $-\frac{1}{2}$ times row 2 from row 3 gives

$$\left[\begin{array}{ccc:ccc:c} 1 & 1 & 1 & 1 & 0 & 0 & 18 \\ 0 & 4 & -2 & -3 & 1 & 0 & -42 \\ 0 & 0 & 3 & -\frac{7}{2} & \frac{1}{2} & 1 & -51 \end{array}\right] \qquad (2.37)$$

Since the leftmost 3 × 3 submatrix is now upper triangular, we have $\mathbf{U} = \begin{bmatrix} 1 & 1 & 1 \\ 0 & 4 & -2 \\ 0 & 0 & 3 \end{bmatrix}$

and $\mathbf{C} = \begin{bmatrix} 18 \\ -42 \\ -51 \end{bmatrix}$. From $\mathbf{UX} = \mathbf{C}$, as in (2.32), we find, through backward recursion, that

(a) $3x_3 = -51$, or $x_3 = -17$; (b) $4x_2 - 2(-17) = -42$, or $x_2 = -19$; and (c) $x_1 + (-19) + (-17) = 18$, or $x_1 = 54$. You should verify that for this example

$$\mathbf{A}^{-1} = \begin{bmatrix} \frac{7}{2} & -\frac{1}{2} & -\frac{1}{2} \\ -\frac{4}{3} & \frac{1}{3} & \frac{1}{6} \\ -\frac{7}{6} & \frac{1}{6} & \frac{1}{3} \end{bmatrix}$$

so that the solution found by using the inverse directly is

$$\mathbf{X} = \mathbf{A}^{-1}\mathbf{B} = \begin{bmatrix} 54 \\ -19 \\ -17 \end{bmatrix}$$

It has been asserted that columns 4–6 in (2.37) should contain $\mathbf{L}^{-1}$. Constructing **L** as a lower triangular matrix with ones on the main diagonal and the row multipliers in appropriate locations, we have

$$\mathbf{L} = \begin{bmatrix} 1 & 0 & 0 \\ 3 & 1 & 0 \\ 2 & -\frac{1}{2} & 1 \end{bmatrix}$$

and indeed $\mathbf{L}^{-1} = \begin{bmatrix} 1 & 0 & 0 \\ -3 & 1 & 0 \\ -\frac{7}{2} & \frac{1}{2} & 1 \end{bmatrix}$. Furthermore,

$$\mathbf{LC} = \begin{bmatrix} 1 & 0 & 0 \\ 3 & 1 & 0 \\ 2 & -\frac{1}{2} & 1 \end{bmatrix}\begin{bmatrix} 18 \\ -42 \\ -51 \end{bmatrix} = \begin{bmatrix} 18 \\ 12 \\ 6 \end{bmatrix} = \mathbf{B}$$

as expected from (2.32).

Continuing, for illustration, to create the *reduced* echelon form (Gauss–Jordan method)

(a) Divide row 2 in (2.37) by 4 and row 3 by 3:

$$\left[\begin{array}{ccc:ccc:c} 1 & 1 & 1 & 1 & 0 & 0 & 18 \\ 0 & 1 & -\frac{1}{2} & -\frac{3}{4} & \frac{1}{4} & 0 & -\frac{42}{4} \\ 0 & 0 & 1 & -\frac{7}{6} & \frac{1}{6} & \frac{1}{3} & -17 \end{array}\right]$$

(b) Subtract $-\frac{1}{2}$ times row 3 from row 2 and also subtract 1 times row 3 from row 1:

$$\left[\begin{array}{ccc:ccc:c} 1 & 1 & 0 & \frac{13}{6} & -\frac{1}{6} & -\frac{1}{3} & 35 \\ 0 & 1 & 0 & -\frac{4}{3} & \frac{1}{3} & \frac{1}{6} & -19 \\ 0 & 0 & 1 & -\frac{7}{6} & \frac{1}{6} & \frac{1}{3} & -17 \end{array}\right]$$

(c) Finally, subtract 1 times row 2 from row 1:

$$\left[\begin{array}{ccc:ccc:c} 1 & 0 & 0 & \frac{7}{2} & -\frac{1}{2} & -\frac{1}{2} & 54 \\ 0 & 1 & 0 & -\frac{4}{3} & \frac{1}{3} & \frac{1}{6} & -19 \\ 0 & 0 & 1 & -\frac{7}{6} & \frac{1}{6} & \frac{1}{3} & -17 \end{array}\right]$$

As expected, the final column contains the solution, $\mathbf{X}$, and columns 4–6 contain $\mathbf{A}^{-1}$. [You might check that $\mathbf{A} = \mathbf{LU}$, find $\mathbf{U}^{-1}$, and verify that $\mathbf{A}^{-1} = (\mathbf{LU})^{-1} = \mathbf{U}^{-1}\mathbf{L}^{-1}$.]

Iterative methods can also be applied to this equation system. Rewriting the equations in (2.34) as

$$\begin{aligned} x_1 &= 18 - x_2 - x_3 \\ x_2 &= (12 - 3x_1 - x_3)/7 \\ x_3 &= 1 - \tfrac{1}{3}x_1 \end{aligned}$$

and choosing, arbitrarily, $x_1 = x_2 = x_3 = 0$ as an initial point, the standard iterative procedure converges, not very quickly, to $x_1 = 54.00$, $x_2 = -19.00$, and $x_3 = -17.00$ (all to two-decimal accuracy) after the 57th iteration. By contrast, the Gauss–Seidel method (calculating x_1, x_3, and x_2 in that order) converges to the same solution at the 28th iteration. Table 2.4 indicates something of the convergence.

TABLE 2.4   Two Iterative Approaches to Equations (2.34)

	Standard Iterative Approach			Gauss–Seidel Method		
Iteration	x_1	x_2	x_3	x_1	x_2	x_3
0	0	0	0	0	0	0
5	36.905	−8.277	−7.739	44.629	−15.430	−13.876
10	43.577	−15.050	−14.346	52.258	−18.336	−16.419
20	51.946	−18.413	−16.438	53.940	−18.977	−16.980
28	53.427	−18.890	−16.885	53.996	−18.998	−16.999
40	53.906	−19.000	−16.995	—	—	—
50	53.975	−19.004	−17.002	—	—	—
57	53.998	−19.003	−17.002	—	—	—

TABLE 2.5 Classification of Solutions to m Linear Equations in n Unknowns ($\mathbf{AX} = \mathbf{B}$)

	Consistent System $\rho(\mathbf{A}) = \rho(\mathbf{A} \mathbin{\vdots} \mathbf{B})$		Inconsistent System $\rho(\mathbf{A}) \neq \rho(\mathbf{A} \mathbin{\vdots} \mathbf{B})$
	Unique Solution $\rho(\mathbf{A}) = n$	Multiple Solutions $\rho(\mathbf{A}) < n$	
$m = n$	$\mathbf{X} = \mathbf{A}^{-1}\mathbf{B}$ *Case E1:* 2×2 example Eq. (2.1), Fig. 2.1 3×3 example Eq. (2.4a), Fig. 2.5a	*Case E2:* 2×2 example Eq. (2.3), Fig. 2.3 3×3 examples Eq. (2.4c), Fig. 2.5c Eq. (2.4d), Fig. 2.5d	*Case E4:* 2×2 example Eq. (2.2), Fig. 2.2 3×3 examples Eqs. (2.4b, e, f, g), Figs. 2.5b, e, f, g
	Case H1: 2×2 example Eq. (2.5a), Fig. 2.6	*Case H2:* 2×2 example Eq. (2.5b), Fig. 2.7	
$m < n$	Impossible: $[\rho(\mathbf{A}) \leq m \; (<n)]$	*Case F1:* If $\rho(\mathbf{A}) = m$, solutions as in (2.8) or (2.9) $\mathbf{X}^M = (\mathbf{A}^M)^{-1}(\mathbf{B} - \mathbf{A}^N\mathbf{X}^N)$ or $\mathbf{X}^M = (\mathbf{A}^M)^{-1}\mathbf{B}$ (basic) or, using the right inverse from (2.7), $\mathbf{X} = \mathbf{A}_R^{-1}\mathbf{B}$ 2×4 example Eq. (2.10)	*Case F4:* Partial solutions if $(m - k)$ equations can be ignored; then proceed as in case F1 3×4 example Eq. (2.12)
		Case F2: If $\rho(\mathbf{A}) = k \; (<m)$, there are $(m - k)$ redundant equations; after removing such equations, proceed as above in case F1 3×4 example Eq. (2.11)	

(*continued*)

TABLE 2.5 Classification of Solutions to m Linear Equations in n Unknowns ($AX = B$) *Continued*

	Consistent System $\rho(\mathbf{A}) = \rho(\mathbf{A} \vdots \mathbf{B})$		Inconsistent System $\rho(\mathbf{A}) \neq \rho(\mathbf{A} \vdots \mathbf{B})$
	Unique Solution $\rho(\mathbf{A}) = n$	Multiple Solutions $\rho(\mathbf{A}) < n$	
$m > n$	*Case M1:* If $\rho(\mathbf{A}) = n$, there are redundant equations and the left inverse from (2.6) can be used $\mathbf{X} = \mathbf{A}_L^{-1}\mathbf{B}$ 3×2 example Eq. (2.13), Fig. 2.8	*Case M2:* If $\rho(\mathbf{A}) = k\ (<n)$, there are $(m - k)$ redundant equations; after removing such equations, proceed as above in case F1	*Case M3:* Partial solutions if some equations can be ignored; if $\rho(\mathbf{A}) = n$, a unique solution can be found to the reduced system in which $(m - n)$ equations are ignored If $\rho(\mathbf{A}) = n$ and no equations are ignored, $\mathbf{X} = \mathbf{A}_L^{-1}\mathbf{B}$ finds a solution that is "close" to the equations but satisfies none exactly 3×2 example Eq. (2.14) *Case M4:* If $\rho(\mathbf{A}) = k\ (<n)$, there will be multiple partial solutions, as in case M2

2.8 SUMMARY

In this chapter we have explored the various possibilities that exist for kinds of solutions (unique, multiple, none) for the general linear equation system with m linear equations and n unknowns. Where possible, we have connected a logical geometric interpretation with the characterizations given by the rules of rank, and we have investigated how to find solutions using matrix algebra definitions and operations. These solution methods include both "one shot" approaches, using an inverse to the coefficient matrix, and iterative methods, that approach a solution in a sequence of steps. Table 2.5 summarizes the possibilities via the rules of rank of the equation system.

PROBLEMS

2.1 On the basis of observations on $\rho(\mathbf{A})$ and $\rho(\mathbf{A} \mid \mathbf{B})$, describe the kinds of solutions (unique, none, multiple) that each of the following systems has:

(a) $-x_1 - 3x_2 = 16$
$3x_1 + 9x_2 = -25$

(b) $x_1 + 2x_2 + 3x_3 = 0$
$6x_2 + 9x_3 = 0$

(c) $x_1 + 2x_2 + 2x_3 = 3$
$3x_1 + 4x_2 + 5x_3 = 7$
$4x_1 + 6x_2 + 7x_3 = 10$

(d) $6x_1 + 2x_2 + 8x_3 = 4$
$x_1 + x_2 + 2x_3 = 1$
$3x_1 + 2x_2 + 5x_3 = 1$

2.2 Find all the *basic* solutions to the following equations:

(a) $x_1 + x_2 + 4x_3 = 10$
$2x_1 + 3x_2 + 8x_3 = 10$

(b) $x_1 + 2x_2 + 2x_3 = 3$
$3x_1 + 4x_2 + 5x_3 = 7$
$4x_1 + 6x_2 + 7x_3 = 10$

(c) $x_1 + 2x_2 + 3x_3 + 4x_4 = 10$
$x_1 + 2x_2 + x_3 + 2x_4 = 10$

2.3 What kinds of solutions are possible for the following equations? (If a unique solution exists, find it.)

(a) $x_1 + 2x_2 = 3$
$4x_1 + 7x_2 = 1$
$3x_1 + 6x_2 = 9$

(b) $x_1 + 2x_2 = 3$
$4x_1 + 7x_2 = 1$
$3x_1 + x_2 = 6$

(c) $x_1 + 2x_2 = 3$
$4x_1 + 7x_2 = 1$
$3x_1 + 6x_2 = 6$

(d) $x_1 + 2x_2 = 3$
$4x_1 + 7x_2 = 1$
$x_1 + 3x_2 = 14$

2.4   Do the following equation systems have unique solutions?

(a) $-4x_1 + x_2 + 3x_3 = 10$
$2x_1 + 5x_2 - 7x_3 = 47$
$x_1 - x_2 + 2x_4 = 39$
$-4x_2 + 4x_3 + x_4 = 16$

(b) $x_1 + 3x_3 + 2x_4 = 4$
$-2x_1 + x_2 + 6x_3 + x_4 = 7$
$-x_1 + x_2 + 2x_3 - 4x_4 = 6$
$-3x_3 + 7x_4 = 0$

(c) $x_1 + x_2 = 2; 2x_3 + x_4 = 5; x_1 + x_3 = 9; x_3 + x_4 = 0.$

2.5   By using matrix *algebra*, the rules of rank, and so on, describe the *geometry* in solution space of each of the following:

(a) $3x_1 - x_2 + 2x_3 = 4$
$x_1 + x_2 + x_3 = 10$
$15x_1 - 5x_2 + 10x_3 = 18$

(b) $x_1 + x_2 + x_3 = 1$
$3x_1 + 3x_2 + 3x_3 = 3$
$-5x_1 - 5x_2 - 5x_3 = 5$

(c) $5x_1 - 10x_2 = 20$
$-10x_1 + 5x_2 = 20$
$x_1 - 2x_2 = 4$

(d) $x_1 + x_2 = 4$
$2x_1 + x_2 = 5$
$4x_1 + 3x_2 = 1$

(e) Left-hand side from (a); right-hand side = $\begin{bmatrix} 0 \\ 0 \\ 0 \end{bmatrix}$.

(f) $30x_1 + 5x_2 + 7x_3 = 100; 60x_1 + 10x_2 + 14x_3 = 200.$

(g) Left-hand side from (f); right-hand side = $\begin{bmatrix} 100 \\ 100 \end{bmatrix}$.

2.6   Given the following pair of linear equations:

$$ax_1 + x_2 = 0, \qquad 10x_1 + bx_2 = c$$

(a) Discuss, for the homogeneous case, the conditions on a and b under which multiple solutions will exist.

(b) In the nonhomogeneous case, under what conditions on a and b will the solution value for x_2 be zero?

2.7   Consider the following set of equations:

$$2x_1 + ax_2 - x_3 = 10$$
$$bx_1 - 4x_2 + x_3 = -30$$
$$-x_1 + x_2 \qquad = -20$$

For what values of a and b

(a) Will the system have a unique solution?

(b) Are the equations inconsistent?

(c) Are there infinitely many solutions?

2.8 Given

$$4x_1 + 3x_2 + 6x_3 = 14$$
$$10x_1 + 2x_2 + 4x_3 = 24$$

(a) Find *all* basic solutions.

(b) Find *any one* nonbasic solution.

2.9 Let $\mathbf{A} = \begin{bmatrix} 100 & 30 & 50 \\ 3 & 5 & 2 \\ 206 & 70 & 104 \end{bmatrix}$. Suppose that this is the coefficient matrix from a set of

linear equations.

(a) Can you find a three-element right-hand side column vector $\mathbf{B}_1$ for which the system of equations has a unique solution? Why or why not?

(b) Find a three-element right-hand side column vector $\mathbf{B}_2$ for which the system will have "partial" solutions. Find any *two* of these partial solutions.

2.10 If each of the following were coefficient matrices from a system of linear equations, explain the kind (or kinds) of solutions that the system would have:

(a) $\begin{bmatrix} 1 & 0 & 2 & 0 \\ -7 & 5 & 1 & 0 \\ 3 & -1 & 3 & 1 \\ -6 & 3 & -3 & -1 \end{bmatrix}$ (b) $\begin{bmatrix} 1 & 1 & 1 & 1 \\ -7 & 5 & 1 & 0 \\ 3 & -1 & 3 & 1 \\ -6 & 3 & -3 & -1 \end{bmatrix}$

(c) The same matrix as in (a), but suppose in addition that you are given the right-hand side vector, $\mathbf{B}' = [6 \quad 3 \quad 9 \quad -9]$. Would this change your answer to (a), and if so, how?

(d) The same matrix as in (a), but with the right-hand side vector $\mathbf{B}' = [16 \quad 3 \quad 9 \quad 10]$. Would this change your answer to (a), and if so, how?

2.11 The following results on five different linear equation systems with m equations and n unknowns have been obtained by five different students using their personal computers (PCs). Interpret these results for a person who does not understand the concept of the rank of a matrix; that is, explain the geometry of each situation as indicated by the results on ranks. Write down, in each case, an example of a linear equation system that has the ranks and dimensions given:

(a) $\rho(\mathbf{A}) = 1, \rho(\mathbf{A} \mathrel{|} \mathbf{B}) = 1; m = 2, n = 1$

(b) $\rho(\mathbf{A}) = 2, \rho(\mathbf{A} \mathrel{|} \mathbf{B}) = 3; m = 3, n = 3$

(c) $\rho(\mathbf{A}) = 2, \rho(\mathbf{A} \mathrel{|} \mathbf{B}) = 4; m = 4, n = 4$

(d) $\rho(\mathbf{A}) = 2, \rho(\mathbf{A} \mathrel{|} \mathbf{B}) = 2; m = 3, n = 2$

(e) $\rho(\mathbf{A}) = 0, \rho(\mathbf{A} \mathrel{|} \mathbf{B}) = 1; m = 2, n = 2$

2.12 Investigate the following linear equation system for possible solutions. If there is a unique solution, find it. If not, explain why. If there are multiple solutions, explain why and find any one of them. If there are "partial" solutions, explain why and find all of them:

$$6x_1 - x_2 = 2$$
$$-42x_1 + 7x_2 = -14$$
$$-x_1 + x_2 = 0$$
$$2x_1 = 1$$

2.13 (a) Show that an upper (lower) triangular matrix is nonsingular if and only if all its diagonal elements are nonzero.

(b) Show that the inverse of a nonsingular upper (lower) triangular matrix is also upper (lower) triangular.

(c) Derive $\mathbf{L}$ and $\mathbf{U}$ for an arbitrary 2×2 matrix $\mathbf{A}$. Do the same for a general 3×3 matrix. In each case, verify that $\mathbf{A} = \mathbf{LU}$ and that $\mathbf{A}^{-1} = \mathbf{U}^{-1}\mathbf{L}^{-1}$.

2.14 Three major cities have submitted requests for federal government urban-renewal grants. The cities intend to allocate their grant monies to various needs in their areas in the following way:

City	Low-Cost Housing	Job Programs	Other
A	20%	50%	30%
B	40%	40%	20%
C	50%	10%	40%

The federal fund-granting agency has decided that no matter how much money is appropriated for renewal grants, 40% must go toward low-cost housing and 30% must be used for job programs. How much should the federal government give to each city if

(a) A total of $120 million is appropriated for the grants?

(b) A total of $\$\beta$ million is to be appropriated for the grants, where β has yet to be determined?

2.15 A typical "transportation problem" involves shipping a product from one or more plants where it is produced to one or more markets where it is consumed. Suppose that two factories with 10 and 25 units of product available have to supply three markets needing 15, 15, and 5 units of product. Since total market demand equals total factory supply, it is clear that the total amounts at the two factories will need to be shipped. If x_{ij} represents the number of units shipped from factory i to market j, the amount of product available at each factory generates the following pair of equations showing total shipments from each factory:

$$x_{111} + x_{12} + x_{13} = 10 \qquad \text{and} \qquad x_{21} + x_{22} + x_{23} = 25$$

and the market requirements generate the following three equations showing shipments into each market:

$$x_{11} + x_{21} = 15, \qquad x_{12} + x_{22} = 15, \qquad \text{and} \qquad x_{13} + x_{23} = 5$$

(a) Express this problem as a linear system of five equations in six unknowns.

(b) Express the system in (a) in matrix algebra terms as $\mathbf{AX} = \mathbf{B}$, where $\mathbf{A}$ is a 5×6 matrix (note that it is composed of only zeros and ones).

(c) What is the maximum number of possible basic solutions?

(d) Apply rules of rank (consistency check and uniqueness check) to demonstrate that there are at least two different basic solutions to this problem.

2.16 Construct a set of five linear equations involving two unknowns, x_1 and x_2, that has a unique solution. Explain why there is a unique solution and find that solution. Illustrate using solution-space geometry.

2.17 A government agency oversees public works projects in two regions, A and B. These projects require inputs of human resources and also floor space. The current problem is to plan the number of projects for the next quarter. Let x_A represent the *change* in the number of projects begun in region A—the number next quarter minus the number this quarter. Let x_B be similarly defined for region B. Each project undertaken in A requires 3 workers and 600 m^2 of floor space; in B (whose resources are different and therefore where the character of projects is different from A) the requirements are 6 workers and 2400 m^2 of floor space.

(a) The agency is considering a program in which they would fund an additional 168 workers and also make available an additional total of 6000 m^2 space (through forced vacancies and rehabilitation of unused structures), to be divided between the two regions in whatever manner is needed. To fully use these increased allocations, how would the number of projects undertaken in both regions change from this quarter?

(b) It is later pointed out that the agency's estimate of floor space needs for projects in region B was far too high; in fact, only 1200 m^2 is needed per project. (The labor requirement, however, is not in error.) How would you now answer the question posed in (a), above?

(c) To combat regional unemployment, the agency is considering increasing their funding to an amount that would cover 200 new workers, rather than 168. Given the corrected floor space requirement estimate in part (b), above, what restriction, if any, does this place on the total amount of new floor space that must be made available if all 200 workers are to be employed?

2.18 A municipal government has holdings of two kinds of land: poorly developed and undeveloped. Presently the city owns 1000 parcels of each kind of land. Once a year they buy or sell parcels on the market, at fixed prices. For the next sale, prices have been set at $3000 and $5000 per parcel, for poorly developed and undeveloped land, respectively. (Poorly developed parcels are less expensive because the buyer incurs the additional cost of demolishing the current dilapidated structures on those parcels before the land can be used.) The city's finance director has established that there should be a net *decrease* of $30,000 in city capital tied up in these land holdings. Annual rents realized by the city on the parcels that it holds are $100 and $200 per parcel, respectively, for poorly developed and undeveloped parcels. Can the city meet the finance director's goal while also *increasing* the annual rents that it receives by $4000? If so, how? If not, why not?

2.19 A state commission has responsibility for allocating three kinds of unemployment projects throughout the state. In the current year, 20 projects of each type were funded. Each project can be terminated at the end of the year or renewed for another year; in addition the number of projects of any type can be increased. Projects of types 1, 2, and 3 each generate 10 jobs for unskilled workers, while 20 and 60 skilled jobs are created by projects 1 and 3, respectively (skilled labor is not used in the second type of project). Finally, the projects require $3000, $7000, and $1000 each in yearly state budget support. For the coming year, the governor of the state has determined that (1) there should be an *additional* 180 unskilled jobs created, (2) 60 *more* skilled jobs should be created, and (3) there will be an *increase* of $12,000 in the state budget to fund these increases in the state's job creation goals. How many projects of each type should be planned for the coming year?

In Problems 2.20–2.24, examine what happens when you apply the Gaussian elimination procedure to the equation system.

2.20 $x_1 + x_2 + x_3 = 10$
$2x_1 + 2x_2 + 2x_3 = 20$
$7x_1 + 3x_2 + 3x_3 = 10$

2.21 $x_1 + 2x_2 + 3x_3 = 3$
$2x_1 + 4x_2 + 6x_3 = 3$
$x_1 + x_2 + x_3 = 3$

2.22 See Equations (2.11). This is a consistent equation system with $m < n$; the augmented matrix is

$$(\mathbf{A} \vdots \mathbf{B}) = \begin{bmatrix} 1 & 2 & 1 & 3 & \vdots & 4 \\ 2 & 4 & 2 & 6 & \vdots & 8 \\ 1 & 1 & 3 & 4 & \vdots & 2 \end{bmatrix}$$

2.23 See Equations (2.13). This is a consistent equation system with $m > n$; the augmented matrix is

$$(\mathbf{A} \vdots \mathbf{B}) = \begin{bmatrix} 1 & 0 & \vdots & 2 \\ 0 & 1 & \vdots & 2 \\ 1 & 1 & \vdots & 4 \end{bmatrix}$$

2.24 See Equations (2.14). This is an inconsistent equation system with $m > n$; the augmented matrix is

$$(\mathbf{A} \vdots \mathbf{B}) = \begin{bmatrix} 1 & 0 & \vdots & 2 \\ 0 & 1 & \vdots & 2 \\ 1 & 1 & \vdots & 3 \end{bmatrix}$$

2.25 For Problem 1.12 in Chapter 1:

(a) Use Gaussian elimination (to generate a matrix in echelon form) to solve the equation system.

(b) Use the Gauss–Jordan method (to generate a matrix in reduced echelon form) and thereby solve the equation system.

(c) Identify the **L** and **U** components of the $\mathbf{A} = \mathbf{LU}$ factorization and verify that the solution to $\mathbf{AX} = \mathbf{B}$ can be found as $\mathbf{X} = \mathbf{U}^{-1}\mathbf{L}^{-1}\mathbf{B}$.

2.26 Answer all three parts of Problem 2.25 (above) for Problem 2.4(b) and Problem 2.4(c).

2.27 Try applying the Gauss–Seidel iterative method to solve Problem 1.12 from Chapter 1, using $\mathbf{X}^0 = \begin{bmatrix} 0 \\ 0 \end{bmatrix}$ and employing any stopping criterion that you choose. You will find that the procedure does not converge. Then try reversing the order of the orignal equations, solving the new first equation for x_1 and the new second equation for x_2, and applying Gauss–Seidel from the same $\mathbf{X}^0$. In this case, the procedure converges nicely to the correct solution, $\mathbf{X} = \begin{bmatrix} 4 \\ 14 \end{bmatrix}$. This illustrates the sensitivity of the Gauss–Seidel method to the ordering of the equations.

PART II

FOUNDATIONS: NONLINEAR METHODS

To complement the mathematical foundations provided by linear algebra and systems of linear equations, it is appropriate to review the basics of differential calculus as it applies to problems of maximization and minimization, first without and then with constraints. Many of the modern approaches for realistic constrained optimization problems have their roots in the classical first- and second-order conditions that can be used in relatively simple maximization and minimization problems. However, the classical approach does not deal easily with inequality-constrained problems (or, for that matter, with large problems that have equality constraints), as we will see. That is the motivation for the more modern approaches to constrained optimization, some of which are explored in Part V. Additionally, alternative approaches to unconstrained maximization and minimization that are based on the abilities of modern computers are explored in Chapter 6.

3

UNCONSTRAINED MAXIMIZATION AND MINIMIZATION

In this chapter we assume that you already have some acquaintance with the fundamental concepts of differential calculus. For functions of one variable, $f(x)$, this means essentially the notion of the derivative as the slope of a function (the slope of its tangent) at a point, the application that this has to the location of stationary points of the function (points at which the slope of the tangent is zero), and the classification of such points as maxima, minima, or neither, through the use of the second derivative. An ability to apply the most basic rules of differentiation will also be assumed.

It will also help if you have been exposed to differential calculus with functions of two variables, $f(\mathbf{X})$, where $\mathbf{X} = [x_1 \ x_2]'$, and specifically to the concepts of a partial derivative and of the total differential and to problems of maxima and minima for a function of two variables.

We review these fundamentals and also present the standard results on maximum and minimum points in a second and possibly less familiar form, using differentials. We also examine the connection between second derivative (or differential) conditions and the concepts of concavity and convexity of a function. All of this material provides a foundation on which to construct the extensions of the theory when more (in general, n) variables are involved and when constraints on some or all of the variables are also required. These more complicated issues introduced by constraints are explored in Chapter 4.

Chapters 3 and 4 present the "classical" approaches to maximization and minimization problems in which it is assumed that all necessary derivatives can be found and used to impose first- and second-order conditions. In Part III of this book we will see how the power of modern computers can often be used to implement alternative approaches to the same kinds of maximization and minimization problems.

In Parts IV and V we will extend the investigation of constraints in maximization and minimization problems to the general case in which there may be both equations and inequalities that tie the variables together. This is the subject matter of linear programming [when $f(\mathbf{X})$ and all constraints are linear functions; Chapters 7–9] and of nonlinear programming (when not all functions are linear; Chapters 10 and 11). While the theoretical bases for first- and second-order optimizing conditions bear a resemblance to the materials in this chapter (and Chapter 4), computational procedures for actual maximization and minimization calculations often rely on various kinds of numerical approaches, examples of which are introduced in Part III and continued in Parts IV and V.

3.1   LIMITS AND DERIVATIVES FOR FUNCTIONS OF ONE VARIABLE

3.1.1 Limits

Derivatives are of fundamental importance in maximization and minimization problems. Since the derivative is defined as a limiting process, it is useful to review briefly the notion of a limit. Consider the function $y = f(x) = \frac{1}{2}x^2 - x + 1$; at $x = 2$, $y = f(2) = 1$. For values of x "near to" but larger than 2, $f(x)$ is "near to" 1; for example (all to three-decimal accuracy), at $x = 2.5$, $f(x) = 1.625$; at $x = 2.25$, $f(x) = 1.281$; and so on. At $x = 2.001$, $f(x) = 1.001$. The sequence of values of x—2.5, 2.25, ..., 2.001—is said to approach 2 from the right (from values of x larger than 2). When $f(x)$ approaches a number, α^*, as x approaches a specific value, x_0, from the right, α^* is called the *right-sided limit* of $f(x)$ as x approaches x_0. This is denoted

$$\lim_{x \to x_0^+} f(x) = \alpha^*$$

(The + superscript indicates that the values of x are larger than x_0.)

If we consider values of x "near to" but smaller than 2, $f(x)$, as defined above, is also "near to" 1. When $x = 1.5$, $f(x) = 0.625$; for $x = 1.75$, $f(x) = 0.781$; at $x = 1.999$, $f(x) = 0.999$. The sequence 1.5, 1.75, 1.999 approaches 2 from the left; when $f(x)$ approaches a number α^{**} as x approaches x_0 from the left, α^{**} is called the *left-sided limit* of $f(x)$ as x approaches x_0:

$$\lim_{x \to x_0^-} f(x) = \alpha^{**}$$

When, as in this example, $\alpha^* = \alpha^{**}$, this common value is called *the limit of* $f(x)$ *as* x approaches x_0 (without reference to direction) and is denoted

$$\lim_{x \to x_0} f(x) = \alpha^* \quad (\text{or } \alpha^{**})$$

For $f(x) = \frac{1}{2}x^2 - x + 1$ the statement would be

$$\lim_{x \to 2^+} f(x) = \lim_{x \to 2^-} f(x) = \lim_{x \to 2} f(x) = 1$$

The distinction between right- and left-sided limits is necessary for a number of special kinds of functions. Two illustrations are as follows

Step Functions. Let

$$f(x) = \begin{cases} 1 & \text{when} \quad x \le 5 \\ 3 & \text{when} \quad x > 5 \end{cases}$$

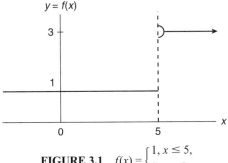

FIGURE 3.1 $f(x) = \begin{cases} 1, x \le 5, \\ 3, x > 5. \end{cases}$

This function is shown in Figure 3.1. At $x_0 = 5$, the right-sided limit is 3, while the left-sided limit is 1. Both $\lim_{x \to 5^+} f(x)$ and $\lim_{x \to 5^-} f(x)$ exist, but since they are unequal, there is no single number that is $\lim_{x \to 5} f(x)$.

Functions that Have Asymptotes. The graph of

$$f(x) = \begin{cases} \dfrac{1}{x-1} & \text{when} \quad x > 1 \\ (x-1)^2 & \text{when} \quad x \le 1 \end{cases}$$

is represented in Figure 3.2. For $x_0 = 1$, $\lim_{x \to 1^-} f(x) = 0$, while as x approaches 1 from the right, $f(x)$ does not approach any finite number; it gets ever larger.

In addition, there may be points at which neither the right- nor the left-sided limit exists. For example, this is true for

$$f(x) = \begin{cases} \dfrac{1}{x} & \text{when} \quad x > 0 \\ \dfrac{-1}{x} & \text{when} \quad x < 0 \end{cases}$$

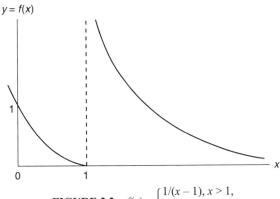

FIGURE 3.2 $f(x) = \begin{cases} 1/(x-1), x > 1, \\ (x-1)^2, x \le 1. \end{cases}$

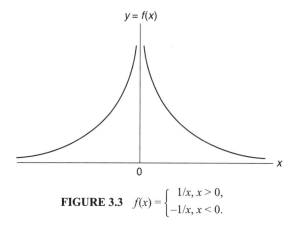

FIGURE 3.3 $f(x) = \begin{cases} 1/x, x > 0, \\ -1/x, x < 0. \end{cases}$

at $x_0 = 0$, as Figure 3.3 suggests. So it may be that at one or more points (a) neither one-sided limit of a function exists, (b) one may exist but not the other, (c) both may exist but their values may differ, or (d) both may exist and have the same value.[1]

With these concepts in mind, the notion of *continuity* of a function is easily defined. If

$$\lim_{x \to x_0} f(x) = f(x_0)$$

the $f(x)$ is said to be continuous at x_0. This says that at x_0 both right- and left-sided limits exist and have the same value, which is $f(x_0)$. If $f(x)$ is not continuous everywhere over the values of x on which it is defined, it is called *discontinuous*. When $f(x)$ is continuous everywhere, its graph has no gaps or spaces in it; it can be drawn without lifting the pencil from the paper. The functions in Figures 3.1–3.3 are seen to be discontinuous.

3.1.2 The Derivative (Algebra)

The derivative of $f(x)$ is defined in terms of limits. It can be described as the "instantaneous rate of change" of $f(x)$ at a specific value of x. This concept is easily shown as the limit of the "average rate of change" of $f(x)$ *over an interval*.

At a specific value of x (say, x_0), the dependent variable has a value of $y_0 = f(x_0)$. Consider some larger value of x (say, x_1), near to x_0. The new value of y will be $y_1 = f(x_1)$. Depending on the specific function, this may be larger than, smaller than, or no different from $f(x_0)$. Then the "average rate of change" of y per unit of x is

$$\frac{f(x_1) - f(x_0)}{x_1 - x_0}$$

which is simply the total change in y, Δy, divided by the total change in x, Δx. This is usually shown as

---

[1]Limits are defined more formally using very small positive quantities δ and ε. The notation $\lim_{x \to x_0^+} f(x) = \alpha^*$ means that $|f(x) - \alpha^*|$ is arbitrarily small for all $x > x_0$ that are "sufficiently close" to x_0. So, if for all $\varepsilon > 0$ we can find a $\delta > 0$ such that for all x in the range $(x_0 < x < x_0 + \delta)$, $|f(x) - \alpha^*| < \varepsilon$, then $\lim_{x \to x_0^+} f(x) = \alpha^*$. And similarly for left-sided limits.

$$\frac{\Delta y}{\Delta x} = \frac{f(x_1) - f(x_0)}{x_1 - x_0} \tag{3.1}$$

If we denote x_1 as $x_0 + \Delta x$, where $\Delta x > 0$, we can also write this average rate of change as

$$\frac{\Delta y}{\Delta x} = \frac{f(x_0 + \Delta x) - f(x_0)}{\Delta x} \tag{3.2}$$

Imagine that Δx, over which this average is calculated, becomes very small. The limit, if it exists, would be denoted (since $\Delta x > 0$)

$$\lim_{\Delta x \to 0^+} \frac{\Delta y}{\Delta x} = \lim_{x_1 \to x_0^+} \frac{f(x_1) - f(x_0)}{x_1 - x_0} = \lim_{\Delta x \to 0^+} \frac{f(x_0 + \Delta x) - f(x_0)}{\Delta x} \tag{3.3}$$

For a value of x, say x_2, near to but smaller than x_0, we could find an average as shown in (3.1) or (3.2), but now where $x_2 = x_0 + \Delta x$ and $\Delta x < 0$. In the limit, when x_2 becomes very close to x_0, we would have (now Δx is negative)

$$\lim_{\Delta x \to 0^-} \frac{\Delta y}{\Delta x} = \lim_{x_2 \to x_0^-} \frac{f(x_2) - f(x_0)}{x_2 - x_0} = \lim_{\Delta x \to 0^-} \frac{f(x_0 + \Delta x) - f(x_0)}{\Delta x} \tag{3.4}$$

When the limits in (3.3) and (3.4) are the same, we have

$$\lim_{\Delta x \to 0^+} \frac{\Delta y}{\Delta x} = \lim_{\Delta x \to 0^-} \frac{\Delta y}{\Delta x} = \lim_{\Delta x \to 0} \frac{\Delta y}{\Delta x}$$

and this limit is called the *derivative of $f(x)$* at the point x_0. It is usually denoted by $(dy/dx)_{x_0}$ or by $f'(x_0)$ (read as "the derivative of y with respect to x, evaluated at $x = x_0$"). From (3.2), we obtain

$$f'(x_0) = \lim_{\Delta x \to 0} \frac{f(x_0 + \Delta x) - f(x_0)}{\Delta x} \tag{3.5}$$

Since the definition of the derivative depends on the existence of a limit, it is clear that functions such as those in Figures 3.1–3.3 will not have derivatives everywhere; we have already seen that they do not have limits everywhere since they are discontinuous. Continuity of a function, however, is not enough to guarantee a derivative at every point. For example, let $y = f(x) = 2 + |x|$; this means that $f(x) = 2 + x$ for $x > 0$ and $f(x) = 2 - x$ for $x < 0$. This function is shown in Figure 3.4.

It is clear from the figure that $\lim_{x \to 0^+} f(x) = 2$, $\lim_{x \to 0^-} f(x) = 2$, and so $\lim_{x \to 0} f(x) = 2$. Moreover, $f(0) = 2$, so the function is continuous at $x = 0$ (as is obvious when we have Fig. 3.4 in front of us). Equally apparent is that, while continuous, the function at $x = 0$ has a kink; it is not "smooth" everywhere. The problem with the derivative becomes clear when we apply (3.3) and (3.4) at $x_0 = 0$. From (3.3), with $\Delta x > 0$, so that $f(x_0 + \Delta x) = 2 + \Delta x$, we have

$$\lim_{\Delta x \to 0^+} \frac{\Delta y}{\Delta x} = \lim_{\Delta x \to 0^+} \frac{2 + \Delta x - 2}{\Delta x} = 1$$

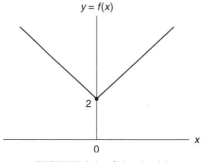

FIGURE 3.4 $f(x) = 2 + |x|$.

and from (3.4), with $\Delta x < 0$ and $f(x_0 + \Delta x) = 2 - \Delta x$, we obtain

$$\lim_{\Delta x \to 0^-} \frac{\Delta y}{\Delta x} = \lim_{\Delta x \to 0^-} \frac{2 - \Delta x - 2}{\Delta x} = -1$$

Since

$$\lim_{\Delta x \to 0^+} \frac{\Delta y}{\Delta x} \neq \lim_{\Delta x \to 0^-} \frac{\Delta y}{\Delta x}$$

the derivative, $\lim_{\Delta x \to 0} (\Delta y / \Delta x)$, does not exist at $x = 0$. (We will soon have a geometric idea of the problem.)

So a function can be continuous at $x = x_0$ and not have a derivative there. However, if there *is* a derivative at $x = x_0$, then we know that $f(x)$ is continuous at that point. A function that has derivatives for all values over which it is defined is said to be everywhere differentiable (or just differentiable, or simply "smooth").

One can in principle calculate the derivative for any differentiable function at any value of x by using this fundamental "delta process." For example, for $f(x) = \frac{1}{2}x^2 - x + 1$, what is its rate of change right at the point where $x = 2$? For *any* x_0 and Δx

$$f(x_0) = \tfrac{1}{2}(x_0)^2 - x_0 + 1$$

$$f(x_0 + \Delta x) = \tfrac{1}{2}(x_0 + \Delta x)^2 - (x_0 + \Delta x) + 1 =$$

$$\tfrac{1}{2}(x_0)^2 + x_0\Delta x + \tfrac{1}{2}(\Delta x)^2 - x_0 - \Delta x + 1$$

so that

$$f(x_0 + \Delta x) - f(x_0) = x_0\Delta x + \tfrac{1}{2}(\Delta x)^2 - \Delta x$$

and, from (3.2)

$$\frac{\Delta y}{\Delta x} = x_0 + \tfrac{1}{2}\Delta x - 1$$

Letting $\Delta x \to 0$, the second term becomes very small and, in the limit, disappears, so

$$\frac{dy}{dx} = \lim_{\Delta x \to 0} \frac{\Delta y}{\Delta x} = x_0 - 1$$

The rate of change of the function $y = f(x) = \frac{1}{2}x^2 - x + 1$ can be found at *any* specific point x_0. To calculate the rate of change at $x_0 = 2$, we see that it is $2 - 1 = 1$; at $x_0 = 3$ the slope is 2; and so on. It is only at the stage represented by dy/dx that the specific value of x is introduced. It is from the fundamental logic of this "delta process" definition of the derivative that the basic rules for differentiation are derived for functions of one variable. The most basic of those rules for differentiation are

(a) If $y = x^n$, $dy/dx = nx^{(n-1)}$
(b) If $y = f(x) + g(x)$, $dy/dx = f'(x) + g'(x)$
(c) If $y = f(x)g(x)$, $dy/dx = f(x)g'(x) + f'(x)g(x)$
(d) If $y = f(x)/g(x)$, $dy/dx = [f'(x)g(x) - f(x)g'(x)]/[g(x)]^2$
(e) If $y = [f(x)]^n$, $dy/dx = n[f(x)]^{(n-1)}f'(x)$

3.1.3 The Derivative (Geometry)

The derivative of $f(x)$ is easily shown geometrically to be the slope of the tangent to the function at x_0, and the tangent is the limiting form of the line through $f(x_0)$ and $f(x_0 + \Delta x)$ on the function, as Δx gets small. For $y = f(x) = \frac{1}{2}x^2 - x + 1$, Figure 3.5 shows the curve representing values taken on by y (vertical axis) over the range of the independent variable, x (horizontal axis), from 1 to 4.

Consider a line measuring the rate of change in y, from an initial point where $x_0 = 2$ and $y_0 = 1$ to the point where $x = 4$ (and $y = 5$). This is shown in Figure 3.6. The vertical displacement (BC) is exactly the numerator of (3.2), and the horizontal displacement (AC) is exactly Δx, the denominator of (3.2). This ratio, $\Delta y/\Delta x$, is $(5 - 1)/(4 - 2) = 2$, the *average* number of units that y changes per unit of change in x over the *interval* from $x = 2$ to $x = 4$; it is precisely the *slope* of the line joining A and B. This line cuts the curve in two points and is called a *secant* to the curve.

Suppose now that Δx had been smaller than 2. If it had been only 1 (x changed from 2 to 3), then this secant line would have sloped up less steeply and, visually, it would have been "closer" to the curve over the range $x = 2$ to $x = 3$. Then $\Delta y/\Delta x$ would be $(2.5 - 1)/(3 - 2) = 1.5$—the average rate of change in y per unit change in x over the interval from $x = 2$ to $x = 3$. As we imagine smaller and smaller changes in x, from a starting point of $x = 2$, the point B rolls down along the curve toward A. If $\Delta x = 0.5$, then $\Delta y/\Delta x = 1.25$; if $\Delta x = 0.001$, $\Delta y/\Delta x$

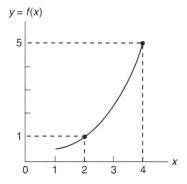

FIGURE 3.5 $f(x) = \frac{1}{2}x^2 - x + 1$, for $1 \le x \le 4$.

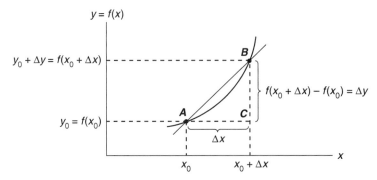

FIGURE 3.6 Geometry of the delta process.

= 1.0005; and so on. Ultimately (in the limit) there is a Δx so close to zero that point B is indistinguishable from A; when this happens, the secant AB becomes the *tangent* to $f(x)$ at A. This shrinkage of Δx toward zero is exactly the limiting process referred to in (3.5), and so the derivative of $y = f(x)$ at a point x_0 is the slope of the tangent to the function at that point.

3.2 MAXIMUM AND MINIMUM CONDITIONS FOR FUNCTIONS OF ONE VARIABLE

Suppose that x is defined over a continuous interval of values, $a \leq x \leq b$, and that $f(x)$ is a smooth function. Then $f(x)$ takes on its *absolute* (or *global*) maximum in that interval at x^o if $f(x) \leq f(x^o)$ for all x in the interval. The function is said to take on a *relative* (or *local*) maximum at x^* if $f(x) \leq f(x^*)$ for all values of x for which $|x^* - x| < \delta$, for some "small" $\delta > 0$. This is often expressed as "for all x in some δ neighborhood of x^*," so that x^* gives the function a larger value than any other "nearby" values of x. Absolute and relative minima are defined similarly—there is a global minimum at x^o if $f(x) \geq f(x^o)$ for all $a \leq x \leq b$ and a local minimum at x^* if $f(x) \geq f(x^*)$ for all x "nearby" to x^*.

3.2.1 The (First) Derivative—a Question of Slope

If we explore the case in which x can vary in an *unconstrained* way over all real numbers, from $-\infty$ to $+\infty$, then geometric logic suggests that if $f(x)$ has a relative maximum or a relative minimum at x^* its tangent at that point must be horizontal, which means that the derivative there must be zero.[2] Since the rate of change is zero, x^* is referred to as a *stationary point* of $f(x)$ (see Figs. 3.7*a* and 3.7*b*). An example of a function like that in Figure 3.7*a* is $f(x) = 2x^2 - 8x + 10$, so that $f'(x) = 4x - 8$, which is zero at $x^* = 2$. An illustration

[2]Geometric intuition is consistent with analytic fact. Consider the definition of a relative maximum given above. For all Δx in the interval $(-\delta < \Delta x < \delta)$, it must be true that $f(x^* + \Delta x) \leq f(x^*)$, or $f(x^* + \Delta x) - f(x^*) \leq 0$. If $\Delta x > 0$ (but still less than δ), this means that $[f(x^* + \Delta x) - f(x^*)]/\Delta x \leq 0$. Since we are examining smooth functions, we know that $f'(x^*)$ exists and is equal to both the right- and left-sided limits, (3.3) and (3.4). Taking the right-sided limit, exactly as in (3.3)—with the reference point now denoted x^*—we have $f'(x^*) \leq 0$. (This relies on the fact that limits of nonpositive or nonnegative expressions are themselves nonpositive or nonnegative, respectively.) Similarly, if $\Delta x < 0$, $[f(x^* + \Delta x) - f(x^*)]/\Delta x \geq 0$. Taking the limit in (3.4), we have $f'(x^*) \geq 0$. Since $f'(x^*) \leq 0$ *and* $f'(x^*) \geq 0$, it follows that $f'(x^*) = 0$. The same holds true for a relative minimum.

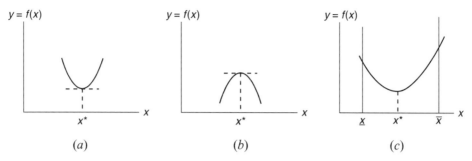

FIGURE 3.7 (*a*) Relative minimum; (*b*) relative maximum; (*c*) relative maxima and minima for $\underline{x}$ $\leq x \leq \bar{x}$.

of a function that has the shape shown in Figure 3.7*b* is $f(x) = -x^2 + 8x + 1$; in that case, $f'(x) = -2x + 8$, so that $f'(x) = 0$ when $x^* = 4$.

A zero-valued derivative is a *necessary* condition for a maximum or a minimum point (either global or local), as long as the independent variable can range completely from $-\infty$ to $+\infty$. However, if x is *constrained* to some interval, $\underline{x} \leq x \leq \bar{x}$, then the derivative need not be zero at a relative maximum or minimum point (see Fig. 3.7*c*). For this case, in the interval over which x is allowed to range, there is an absolute minimum at x^*, a relative maximum at $\underline{x}$ and an absolute maximum at $\bar{x}$.

It seems clear that when x is restricted to less than the entire range of real numbers, it would be necessary to evaluate the function at stationary points as well as at *endpoints*, where x hits boundaries, in order to find global maxima or minima. In this chapter, we concentrate on the unconstrained case. The kinds of modifications that are needed in constrained cases will be explored in Chapter 4 and in Parts IV and V.

The derivative of a function may take on a value of zero at a point that is neither a maximum nor a minimum. If the function is S-shaped, it could have a *point of inflection* where the slope of the tangent is zero. Figure 3.8 illustrates this, using the function $f(x) = (x - 1)^3 + 2$, for which $f'(x) = 3(x - 1)^2 = 0$ occurs at $x^* = 1$.

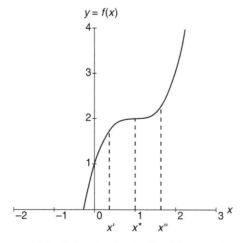

FIGURE 3.8   Point of inflection for $f(x) = (x - 1)^3 + 2$.

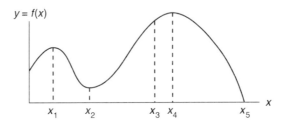

FIGURE 3.9   Function with several stationary points.

By taking the derivative of a function and setting it equal to zero, we search for the one or more values of x for which $f(x)$ *may* have a maximum or a minimum.

1. If no values of x can be found to satisfy $f'(x) = 0$, then $f(x)$ has no stationary point. Straight lines are the illustration. For example, if $f(x) = 3x + 10$—a straight line with a positive slope, so that ever-increasing values of x mean ever-increasing values of $f(x)$— then $f'(x) = 3 \neq 0$ everywhere.[3]

2. If there is one point for which $f'(x) = 0$, you will usually want to identify whether that point represents a maximum or a minimum (or neither). The functions in Figures 3.7*a*, 3.7*b*, and 3.8, which contain x^2, illustrate this possibility.

3. If there is more than one value of x for which $f'(x) = 0$—as with $f(x) = x^3 - 4x^2 + 5x + 2$—it is also usually important to distinguish between maxima, minima, and points of inflection.

Figure 3.9 represents a function that has three stationary points. Assume that to the left of x_1 and to the right of x_5 the curve continues downward infinitely far—there are no additional points at which $f'(x) = 0$. At x_1 the function has a relative maximum, at x_2 it has a relative minimum and at x_4 an absolute maximum. (*If* x were constrained to, say, $x_1 \leq x \leq x_3$, then x_1 and x_2 still represent local maxima and minima, respectively, but x_3 would identify a global maximum.)

There are two straightforward ways to distinguish maxima, minima, and points of inflection for a point x^* where $f'(x) = 0$. They require two additional evaluations of $f(x)$ or of $f'(x)$, respectively. Consider two points, x^*_+ and x^*_- that are slightly to the right and to the left, respectively, of x^*.

1. Find $f(x^*_+)$ and $f(x^*_-)$. If $f(x^*)$ is larger than both, then x^* represents a local maximum; if $f(x^*)$ is smaller than both, then x^* represents a local minimum. Otherwise, if the function is larger on one side of x^* and smaller on the other, x^* represents an inflection point.[4]

2. Find, instead, $f'(x^*_+)$ and $f'(x^*_-)$. If $f'(x^*_+) > 0$ and $f'(x^*_-) < 0$, we have a local minimum at x^*, and if the signs on the two derivatives are reversed, it is a local maximum. Figure 3.10 illustrates. If x^* indicates a point of inflection, then $f'(x^*_+)$ and $f'(x^*_-)$ will have the same sign. (For the case in Fig. 3.8, both derivatives would be positive; they could also both be negative, if the S were reversed, running from upper left to lower right.)

---

[3]Since the independent variable is unconstrained, there cannot be any boundaries into which the straight line might bump.

[4]We neglect for now the possibility that $f(x^*) = f(x^*_+)$ and/or $f(x^*_-)$, which would indicate that $f(x)$ is flat at x^*.

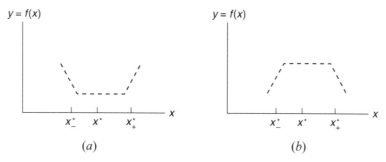

FIGURE 3.10   (*a*) Minimum point at *x**; (*b*) maximum point at *x**.

A third approach for sorting out maxima, minima, and inflection points requires only one additional evaluation, this time of the *second derivative* at *x**.

3.2.2 The Second Derivative—a Question of Shape

In general, the derivative that we have used above in identifying stationary points is itself a function of x (the exception is *linear* functions). It is more exactly known as the *first derivative* of $f(x)$. Therefore, it, too, can be sketched in two-dimensional space, with x (again) on the horizontal axis and $f'(x)$ on the vertical. If it is continuous and smooth (which we will assume unless otherwise noted), it, too, has a tangent at each value of x and hence also a derivative everywhere. Thus, derivatives of derivatives can be found. The derivative of $f'(x)$, called the *second derivative*, is found by differentiating the first derivative, hence the notation $d(dy/dx)/dx = d^2y/dx^2 = f''(x)$. Its relationship to the first derivative is exactly the same as the relationship of the first derivative to the original function; $f''(x)$ measures the rate of change of $f'(x)$ at a point. For example, for $f(x) = x^3 - 4x^2 + 5x + 2$, $f'(x) = 3x^2 - 8x + 5$, which is shown in Figure 3.11, and *its* rate of change at any point x_0 is measured by *its* tangent at x_0.

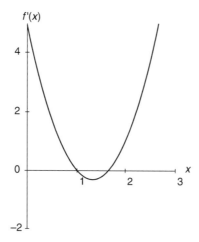

FIGURE 3.11 $f(x) = x^3 - 4x^2 + 5x + 2$; $f'(x) = 3x^2 - 8x + 5$.

3.2.3 Maxima and Minima Using First and Second Derivatives

Consider a point, x^*, at which $f''(x^*) > 0$ and $f'(x^*) = 0$. This means that at x^* the first derivative is getting larger, since *its* rate of change at that particular point, $f''(x^*)$, is positive—$f'(x)$ increases as x increases and decreases as x decreases. Since $f'(x^*) = 0$, this means that just to the left of x^*, $f'(x)$ must be negative, and just to the right of x^*, $f'(x)$ must be positive. This means, in turn, that the function in the neighborhood of x^* looks like Figure 3.10a, and x^* identifies a minimum point for $f(x)$. If $f''(x^*) < 0$, the opposite argument applies, the function looks like Figure 3.10b, and x^* has found a maximum point. These rules on the sign of the second derivative are called *second-order* conditions (because they use the second derivative); they provide information that allows us to distinguish maxima from minima *for points where $f'(x) = 0$.*

If $f''(x^*) = 0$, we are unable to tell whether x^* represents a maximum, a minimum, or neither, and we need further tests. Here are some examples (see also Fig. 3.12):

1. $f(x) = x^4 + 2$, for which $f'(x) = 4x^3$ and $f''(x) = 12x^2$. Setting $f'(x) = 0$ identifies $x^* = 0$; substitution into the second derivative gives $f''(x^*) = 0$. Yet this function, shown in Figure 3.12a, has an unambiguous minimum at $x^* = 0$.

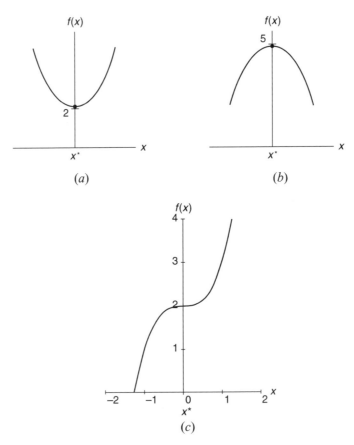

FIGURE 3.12   Three cases in which $f''(x^*) = 0$ at a minimum or a maximum.

TABLE 3.1   Characterization of Maxima and Minima for $f(x)$ at x^*

	Maximum	Minimum
First-order necessary condition	$f'(x^*) = 0$	$f'(x^*) = 0$
Second-order necessary condition	$f''(x^*) \leq 0$	$f''(x^*) \geq 0$
Sufficient conditions	$f'(x^*) = 0$ and $f''(x^*) < 0$	$f'(x^*) = 0$ and $f''(x^*) > 0$

2. For $f(x) = -3x^4 + 5$, we find $x^* = 0$ but $f''(x) = -36x^2$, which is zero when evaluated at x^*. As Figure 3.12b indicates, the function has a maximum at $x^* = 0$.

3. If $f(x) = x^3 + 2$, then, as in the previous two examples, $x^* = 0$ and $f''(x^*) = 0$ also. As seen in Figure 3.12c, this function has a point of inflection at $x^* = 0$. As these examples illustrate, when $f''(x^*) = 0$, we have no idea what kind of a stationary point there is at x^*.

If $f''(x^*) > 0$, we know that x^* cannot be a maximum, so $f''(x^*) \leq 0$ therefore represents another *necessary* condition for a maximum at x^*. Similarly, $f''(x^*) \geq 0$ is a second-order necessary condition for a minimum at x^*. Table 3.1 summarizes the facts thus far regarding maxima and minima for $f(x)$.

Note that the function in Figure 3.12c, with the point of inflection, includes x^3 (x raised to an odd power), whereas those in Figures 3.12a and 3.12b include x^2. This provides a clue to the rule that is required in those cases in which $f''(x^*) = 0$. Since $f''(x)$ is (again, in general) a function of x, we could take *its* derivative, generating a third derivative, d^3y/dx^3 or $f'''(x)$ or $f^{(3)}(x)$. Continuing to take derivatives of derivatives we might find a fourth derivative, $f^{(4)}(x)$, a fifth derivative, $f^{(5)}(x)$, and so on.

Table 3.2 contains these higher-order derivatives for the three functions in Figure 3.12. Since $x^* = 0$ in all three of these cases, not only the second derivatives but also all higher-order ones that involve x will be zero when evaluated at x^*. Finally, at some point for each function, just before the next higher derivative becomes identically zero ("vanishes"), the x will disappear and that derivative will not be zero.

The figures and the table illustrate a general fact. As long as the first nonvanishing derivative is of even order, then x^* will represent a maximum (if the sign of that derivative is negative) or a minimum (if the sign of that derivative is positive). If the first nonvanishing derivative is of odd order, then x^* identifies a point of inflection. These rules are only *illustrated* by the table; they are easily derived from the Taylor series expression for $f(x)$, a topic to which we turn later in this chapter.

TABLE 3.2   Higher-Order Derivatives for Three Functions of x

$f(x)$	$x^4 + 2$	$-3x^4 + 5$	$x^3 + 2$
$f'(x)$	$4x^3$	$-12x^3$	$3x^2$
$f''(x)$	$12x^2$	$-36x^2$	$6x$
$f'''(x)$	$24x$	$-72x$	6
$f^{(4)}(x)$	24	-72	0
$f^{(5)}(x)$	0	0	0

3.2.4 The Differential

The derivative is obtained as the limit of the ratio $\Delta y/\Delta x$, as in (3.5). For very small values of Δx, $\Delta y/\Delta x \cong dy/dx$, or

$$\Delta y \cong \left(\frac{dy}{dx} \right) \Delta x \qquad \text{or} \qquad \Delta y \cong f'(x)\Delta x \tag{3.6}$$

If the change in x is *infinitesimally* small, we term it dx and label the corresponding change in $y = f(x)$ as dy or df. So

$$df \equiv dy = \left(\frac{dy}{dx} \right) dx \qquad \text{or} \qquad df \equiv dy = f'(x)dx \tag{3.7}$$

This expression is the *differential* of the function $y = f(x)$.

There is a straightforward geometric interpretation for (3.7). The differential gives a *linear approximation* to the change in y for a given change in x. This is because the expression in (3.7), when evaluated at a particular point, x_0, is essentially the equation of the tangent to $f(x)$ at x_0.

Since we now want to consider both the *curve*, $y = f(x)$, and a tangent *line* to that curve in the same two-dimensional space, we use $z = g(x)$ for the line.[5] As we saw in Chapter 1, the *general* equation for a line in x, z space is $ax + bz = c$, or $z = g(x) = -(a/b)x + (c/b)$, where $-(a/b)$ is the slope of the line in x, z space and (c/b) is the intercept on the vertical (z) axis. This is the usual "slope intercept" form for a line. For any particular point (x_0, z_0) on the line, $ax_0 + bz_0 = c$. Subtracting the *particular* from the *general* equation, $a(x - x_0) + b(z - z_0) = 0$, or $(z - z_0) = -(a/b)(x - x_0)$.

Considering the tangent to the curve $y = f(x)$ at x_0, we know y_0 $[= f(x_0)] = z_0$ $[= g(x_0)]$[6] *and* we know that the slope of the tangent line, $g(x)$, at x_0 is just the slope of the curve there, namely, $f'(x_0)$. This means that the equation of the line that is tangent to $y = f(x)$ at $x = x_0$ is $(z - z_0) = f'(x_0)(x - x_0)$, where x_0, z_0, and $f'(x_0)$ are known. For a point (x, z) on the tangent line *near to* (x_0, z_0), we can write $x = x_0 + dx$ and $z = z_0 + dz$ so this can be represented as $dz = f'(x_0)dx$, exactly as in (3.7), but using $z = g(x)$ for the line.[7]

As an example, consider $f(x) = x^2 - 2x + 1$, shown in Figure 3.13, for which $f'(x) = 2x - 2$. At $x_0 = 1.5$, $f'(x_0) = 1$ and $y_0 = f(x_0) = 0.25$. The equation of the line that is tangent to $f(x)$ at $x_0 = 1.5$ is then easily found to be $(z - 0.25) = (1)(x - 1.5)$ or $z = x - 1.25$.

If we ask how much y would change if x moved from $x = 1.5$ to $x = 2$ ($dx = 0.5$), we obtain the linear approximation from (3.7). At $x_0 = 1.5$, the estimate of this change is given by $dy = f'(x_0)dx = (1)(0.5) = 0.5$. In Figure 3.13, $AC = dx$ and the true change in y, BC, has been approximated by the tangent line as DC. On this basis we estimate $f(2)$ to be 0.75; in fact, $y = 1.0$ at that point, and the error (in this example, an underestimate of 0.25) is due to the fact that a line is used to approximate a curve.

---

[5]This just means that x is measured along the horizontal axis and both y [values on the curve $f(x)$] and z [values on the line $g(x)$] are measured on the vertical axis.

[6]The curve and its tangent share one point in common, namely, (x_0, y_0) $[= (x_0, z_0)]$.

[7]Since a function $f(x)$ may have the same slope at several points, it is necessary, when writing the equation of the tangent line at a point on $f(x)$, to specify precisely *which* point is involved. This information is lost in $dz = f'(x)dx$ but it is present (x_0 and z_0 appear explicitly) in $(z - z_0) = f'(x_0)(x - x_0)$.

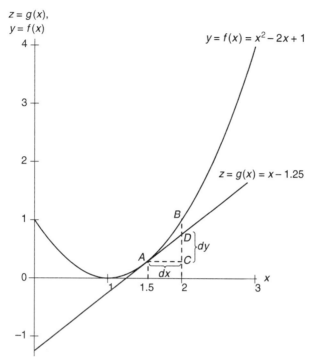

FIGURE 3.13   The differential as a tangent line.

The approximation provided by the differential will be more accurate the smaller the change in x. If $x_0 = 1.5$ changes to 1.6 ($dx = 0.1$), then $dy = f'(x_0)dx = (1)(0.1) = 0.1$, and the new value of y is estimated by the differential to be 0.35, whereas $f(1.6) = 0.36$. The underestimation error is now reduced to 0.01, because we have used the linear approximation over a shorter distance.

3.2.5 Maxima and Minima with Differentials

The conditions for maxima and minima can be stated in terms of differentials just as well as derivatives. For arbitrary small changes dx, $dy = 0$ if and only if $f'(x) = 0$; this is clear from (3.7). Thus the first-order conditions for a maximum *or* a minimum point are that $df = 0$ there. This is denoted, when x^* is the point in question, by $df(x^*) = 0$ or, more simply, when there is no ambiguity, as $df^* = 0$.

Rules parallel to those using the second derivative to distinguish maxima from minima (from neither) can also be formulated using the second differential of y, denoted $d(dy) \equiv d^2y$, or d^2f. The relationship between differentials of higher order is exactly the same as that between derivatives of higher order. The second differential estimates the change in the first differential just as the second derivative approximates the change in the first derivative.

In particular, if $d^2f > 0$ at a point x^* where the first differential is zero [$df^* = 0$ and $d^2f^* > 0$], this means that the first differential is getting larger. Since $df^* = 0$, it must be true that $df(x^*_-) < 0$ for an x^*_- to the left of x^*, and $df(x^*_+) > 0$ at a point x^*_+ to the right of

TABLE 3.3 Alternative Characterization of Maxima and Minima for $f(x)$ at x^*

	Maximum	Minimum
First-order necessary condition	$df(x^*) = 0$	$df(x^*) = 0$
Second-order necessary condition	$d^2f(x^*) \leq 0$	$d^2f(x^*) \geq 0$
Sufficient conditions	$df(x^*) = 0$ and $d^2f(x^*) < 0$	$df(x^*) = 0$ and $d^2f(x^*) > 0$

x^*. Since $df(x^*) = f'(x^*_-)dx$, this means that the tangent to $f(x)$ at x^*_- has a negative slope, since the differential at x^*_- predicts a *decrease* in $f(x)$ for an increase in x from there. Similarly, $df(x^*_+) = f'(x^*_+)dx > 0$ means that the tangent to $f(x)$ at x^*_+ has a positive slope. These observations identify x^* as a minimum point, exactly as in the sketch in Figure 3.10a. The same logic makes clear that $d^2f^* < 0$ is associated with a maximum at x^*. Therefore, Table 3.1 could be rewritten in terms of differentials; this is done in Table 3.3. [Just as with derivatives, the case in which $d^2f^* = 0$ remains unclear, and for the same reasons as we saw when considering $f''(x^*) = 0$.]

The rules in Table 3.3 are equivalent to those in Table 3.1, because techniques for finding derivatives and those for finding differentials are really the same—hence the name *differentiation* covers both. For example, since dy was derived from y by taking the derivative with respect to x and multiplying by dx [$dy = (dy/dx)dx$], the notation $d(dy)$ indicates that this same operation should be carried out on dy to generate the second differential. Using the product rule, and since dx is an arbitrary constant, we have

$$d^2f = d(df) = d[f'(x)dx] = d[f'(x)]dx + [f'(x)]d(dx)$$
$$= [f''(x)dx]dx + f'(x)(0)$$
$$= f''(x)(dx)^2 \tag{3.8}$$

Since dx is squared, the *sign* of d^2f is governed entirely by the sign of $f''(x)$ and the second-order conditions of Table 3.3 are identical to those of Table 3.1.

3.3 TAYLOR'S SERIES, CONCAVITY, AND CONVEXITY OF $f(x)$

Second-order conditions at a point x^* establish the shape of $f(x)$ at that point. The definitions of *concavity* and *convexity* of $f(x)$ are also descriptions of shape, and they are easily generalized to functions of any number of variables. The results in Taylor's series are essential for these definitions.

3.3.1 Taylor's Series for $f(x)$

Taylor's (Brook Taylor, English, 1685–1731) series for $f(x)$ says that an approximation to the value of the function in the neighborhood of a particular point, x_0, can be found to any degree of accuracy, provided the values of $f(x)$ and its derivatives are known at x_0. [Officially, the result below in (3.9) is Taylor's *theorem*, but it has become customary to refer to it as Taylor's series. Some background on Taylor's series is contained in Appendix 3.2.] If we denote a value of x near to x_0 as $x_1 = x_0 + dx$ (so that $dx = x_1 - x_0$), then Taylor's series expresses $f(x_0 + dx)$ as

$$f(x_0 + dx) = f(x_0) + f'(x_0)dx + \left(\frac{1}{2!}\right) f''(x_0)(dx)^2$$

$$+ \left(\frac{1}{3!}\right) f^{(3)}(x_0)(dx)^3 + \cdots + \left(\frac{1}{n!}\right) f^{(n)}(x_0)(dx)^n + R_n \qquad (3.9)$$

where R_n is the "remainder" left unaccounted for after using the first n differentials of $f(x)$. If dx is very small (so that x_1 is very close to x_0), then higher and higher powers of dx will become negligible, especially considering that they are divided by correspondingly higher-order factorials.[8]

Subtracting $f(x_0)$ from both sides of (3.9) produces a series expression for the difference in the values of $f(x)$ at x_0 and at $x_1 = x_0 + dx$. A *first-order* (or *linear*) *approximation* to this difference is generated by eliminating derivatives higher than the first, giving

$$f(x + dx) - f(x_0) \cong f'(x_0)dx \qquad (3.10a)$$

or, using the first differential

$$f(x_0 + dx) - f(x_0) \cong df(x_0) \qquad (3.10b)$$

A *second-order* (or *quadratic*) *approximation* is given by including the second derivative term from (3.9) as well:

$$f(x_0 + dx) - f(x_0) \cong f'(x_0)dx + \frac{1}{2!} f''(x_0)(dx)^2 \qquad (3.11a)$$

or, in terms of differentials,

$$f(x_0 + dx) - f(x_0) \cong df(x_0) + \frac{1}{2!} d^2f(x_0) \qquad (3.11b)$$

We will use these linear and quadratic approximations to $f(x_0 + dx) - f(x_0)$ in exploring the shape of $f(x)$, especially at x^*, a candidate for a maximum or a minimum point.

A relative (local) maximum at x^* was identified (at the beginning of Section 3.2) when $f(x) - f(x^*) \leq 0$ for all x in some neighborhood of x^*. Let $x^* = x_0$ and $x = x_0 + dx = x^* + dx$. Using the second-order Taylor series in (3.11a), the requirement that $f(x) - f(x^*) \leq 0$ translates to

$$f'(x^*)(x - x^*) + \frac{1}{2!} f''(x^*)(x - x^*)^2 \leq 0 \qquad (3.12)$$

Since the "dx neighborhood" includes x's on either side of x^*, the sign of $(x - x^*)$ is indeterminate—positive when x is larger than x^* and negative when x is smaller than x^*.

---

[8]For example, $3! = 6$, $4! = 24$, $5! = 120$, $6! = 720$, ..., $10! = 3,628,800(!)$.

Therefore, the $f'(x^*)(x - x^*)$ term must drop out, and this is only accomplished if $f'(x^*) = 0$. This is, of course, just the first-order necessary condition in Table 3.1. We are left, then, at x^* with

$$f(x) - f(x^*) \cong \frac{1}{2!} f''(x^*)(x - x^*)^2$$

Both $1/2!$ and $(x - x^*)^2$ are unambiguously positive, so $f(x) - f(x^*) \leq 0$ if and only if $f''(x^*) \leq 0$, which is the second-order necessary condition for a maximum in Table 3.1.[9] The same kinds of arguments, using the quadratic approximation provided by Taylor's series, will generate the first- and second-order necessary conditions for a relative minimum that are shown in Tables 3.1 and 3.3.

3.3.2 Concavity and Convexity of $f(x)$

The shape of the function around x^* in Figures 3.7a and 3.7b can be identified in two geometrical ways which serve to establish that the function is always larger or always smaller on either side of x^*.

Location Relative to Secant Lines (*Geometry*) Choose two points, x_-^* and x_+^*, near to x^*, one on each side. Connecting $f(x_-^*)$ and $f(x_+^*)$ produces a secant line that cuts the curve in two places. It is clear that for the function in Figure 3.7a, for any secant produced in this way, $f(x)$ between x_-^* and x_+^* will be *below* its secant; similarly, for any pair of points that produce a secant for $f(x)$ in Figure 3.7b, the function will lie *above* its secant.

Location Relative to Tangent Lines (*Geometry*) For any x chosen near to x^* (on either side) in Figure 3.7a, if the tangent to $f(x)$ is drawn at that point, it is clear that $f(x)$ lies *above* the tangent. The opposite is true for the function in Figure 3.7b. For any x near to x^* the curve lies *below* its tangent.

A function that lies always below its secants or always above its tangents at points near to x^* is said to be *strictly convex at* x^*. If $f(x)$ is always below its secants (always above its tangents) everywhere in the range over which the function is defined, it is said to be *everywhere strictly convex*, or, just *strictly convex*—not only at x^* (Fig. 3.14a). On the other hand, if $f(x)$ is strictly convex over some values of x but linear over some other values of x, so that it *coincides* with secants and tangents over those other values, it is *convex*, but not strictly so (Fig. 3.14b). Such a function is also described as never above its secants or never below its tangents.

Intuitively, a *concave* $f(x)$ is a convex $f(x)$ turned upside down. More precisely, $f(x)$ is *strictly concave at* x^* if it is always above its secants or always below its tangents near x^*. If a function is everywhere above secants (everywhere below tangents) it is *strictly concave*—again, not just at x^*. If a concave function has a linear part, then it is concave but not strictly so. (Turn Figure 3.14 upside down for the general idea.) Table 3.4 summarizes.

---

[9]Exactly the same kind of argument, using (3.11b) instead of (3.11a), will produce the first- and second-order necessary conditions for a maximum in Table 3.3, expressed in terms of differentials at x^* rather than derivatives.

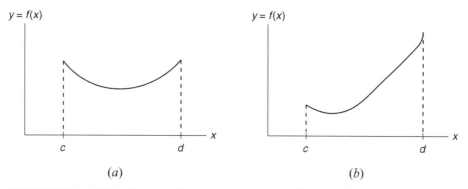

FIGURE 3.14   (a) Strictly convex function over $c \leq x \leq d$; (b) convex function over $c \leq x \leq d$.

TABLE 3.4   Convexity and Concavity of $f(x)$ Classified by Location Relative to Secants and Tangents

Location of $f(x)$ Relative to	Strict Convexity	Convexity	Strict Concavity	Concavity
Secants	Below	Never above (below or coincident)	Above	Never below (above or coincident)
Tangents	Above	Never below (above or coincident)	Below	Never above (below or coincident)

Convex Sets   These geometric characterizations of convexity and concavity have precise algebraic representations. However, in describing the shape characteristics of $f(x)$, it is necessary to be clear about the range of values of x over which $f(x)$ is defined and hence over which the description of shape is valid. Thus, we need to explore the notion of a convex set of points (later in this chapter) when we examine maxima and minima for a function of more than one variable.

For the present simplest case of a function of one variable, it is adequate to note that a convex set of points (values of x) is defined by the requirement that all points on the *line* connecting any two points in the set must also be in the set. So, if x ranges continuously from $-\infty$ to $+\infty$, $f(x)$ is said to be defined over a convex set of points (the line defining the entire x axis). If x can take on all the values only within a restricted range, $c \leq x \leq d$, then $f(x)$ is defined over that part of the x axis that is the line connecting $x = c$ and $x = d$, which is again convex set, as in Figure 3.14.[10] By contrast, if $-\infty \leq x \leq 10$ and $15 \leq x \leq \infty$, these values of x do not constitute a convex set, since the line (portion of the x axis) connecting, say, 2 (in the left interval) to 17 (in the right interval) includes numbers between 10 and 15, which are not in the set of points defined by the two inequalities.

If x ranges continuously over $-\infty \leq x \leq +\infty$ or over $c \leq x \leq d$ and if $f''(x) > 0$ over that range, then the slope of the function is ever increasing, so $f(x)$ must lie everywhere above

[10]A continuous interval that includes its endpoints, like $c \leq x \leq d$, is called a *closed* interval and is sometimes denoted $[c,d]$. (The context should eliminate any possible confusion with a two-element row vector.) Inequalities denoted by "$\leq$" and "$\geq$" are called "weak" because they allow inclusion of the boundaries. An inequality that uses only "$<$" and/or "$>$" is called "strong" because it disallows the boundaries.

its tangent; it is strictly convex. If $f''(x) = 0$ over some part or parts of an interval and $f''(x) > 0$ over the remainder of the interval, this means that the tangent slope remains constant over some parts of the interval; hence $f(x)$ is linear and coincides with its tangent over that range, and so the function is *convex* (but not strictly so). A function is *concave* or *strictly concave* if $f''(x) \leq 0$ or $f''(x) < 0$, respectively. A linear function is *both* concave and convex, although neither *strictly*.

Location Relative to Secant Lines (*Algebra*) Here is a standard definition of convexity for $f(x)$. It does not require derivatives of the function, and its geometric interpretation turns out to be exactly a description of the location of $f(x)$ relative to its secants.

CONVEX FUNCTION $f(x)$.   A function $f(x)$ is *convex* over a convex set of points $c \leq x \leq d$ if for every α in the closed interval $[0,1]$ and for every pair of points (x_1, x_2) in the closed interval $[c,d]$,

$$f[\alpha x_2 + (1 - \alpha)x_1] \leq \alpha f(x_2) + (1 - \alpha)f(x_1) \tag{3.13}$$

The function is strictly convex if $\leq$ is replaced by $<$ and α is in the open interval $0 < \alpha < 1$.

The geometry of this definition is indicated in Figure 3.15. Along the horizontal axis there are three points of interest in the interval $[c,d]$; these are x_1, x_2 and the point represented by the combination $\alpha x_2 + (1 - \alpha)x_1$. Since $0 \leq \alpha \leq 1$, this point must lie in the closed interval $[x_1, x_2]$; if $\alpha = 0$, the point is at x_1, if $\alpha = 1$, it is at x_2, and any other α will identify a point between x_1 and x_2. The length A is simply the value of $f(x)$ for that particular position on the x axis; $A = f[\alpha x_2 + (1 - \alpha)x_1]$. This is the left-hand side of the inequality in (3.13).

At x_1 and x_2 the function takes on the values $f(x_1)$ and $f(x_2)$, respectively. The line joining these two points is just the convex combination $\alpha f(x_2) + (1 - \alpha)f(x_1)$, for all α in $[0,1]$. This is nothing more than the secant to the function through the points $f(x_1)$ and $f(x_2)$. When the weights α and $(1 - \alpha)$ in the combination are the same as those in $\alpha x_2 + (1 - \alpha)x_1$, then the location defined on the line is directly above $\alpha x_2 + (1 - \alpha)x_1$, which is

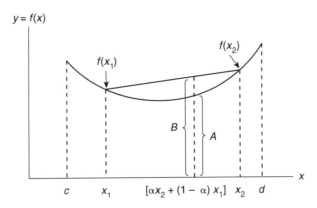

FIGURE 3.15   A convex function over the interval $[c,d]$.

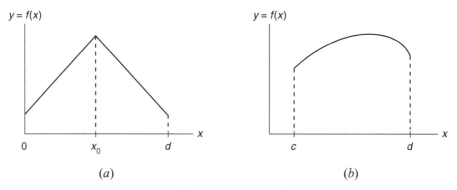

FIGURE 3.16   (*a*) Concave function over [0,*d*]; (*b*) concave function over [*c*,*d*].

the length of B.[11] This is the right-hand side of the inequality in (3.13). Thus $f(x)$ is convex in the interval [*c*, *d*] if $A \leq B$—the function is never above its secant (as in Fig. 3.14*b*); it is strictly convex if $A < B$—always below its secant (as in Fig. 3.14*a*).

A similar definition for a concave function requires only that the inequality be reversed.

CONCAVE FUNCTION $f(x)$.   A function $f(x)$ is *concave* over a convex set of points $c \leq x \leq d$ if for every α in the closed interval [0,1] and for every pair of points (x_1, x_2) in the closed interval [*c*,*d*],

$$f[\alpha x_2 + (1 - \alpha)x_1] \geq \alpha f(x_2) + (1 - \alpha)f(x_1) \tag{3.14}$$

The function is strictly concave if $\geq$ is replaced by $>$ and α is in the open interval $0 < \alpha < 1$. The geometry is that of never below ($\geq$) or always above ($>$) a secant.

The definitions in (3.13) and (3.14) have the advantage that they are valid for functions regardless of whether they have derivatives throughout the range in question. The function in Figure 3.16*a* is seen to be concave, from (3.14), even though a derivative does not exist at the point x_0 in the interval [0,*d*].

Location Relative to Tangent Lines (*Algebra*) Rewrite (3.13) as

$$\alpha f(x_1) + f[\alpha x_2 + (1 - \alpha)x_1] - f(x_1) \leq \alpha f(x_2)$$

Assume that $\alpha \neq 0$ (meaning that $0 < \alpha \leq 1$), divide both sides by α and rearrange the algebra; this gives

$$f(x_1) + \frac{1}{\alpha}\{f[x_1 + \alpha(x_2 - x_1)] - f(x_1)\} \leq f(x_2) \tag{3.15}$$

Now use a first-order Taylor series approximation to rewrite the term in curly brackets in (3.15) as $f'(x_1)[\alpha(x_2 - x_1)]$.[12] Finally, then, (3.15) can be written as

---

[11]This can be shown by straightforward application of the proportional-side property of similar triangles, drawing a line parallel to the x axis through $f(x_1)$.

[12]This requires differentiability of $f(x)$ at x_1. Here x_1 plays the role of x_0 and $\alpha(x_2 - x_1)$ is dx in (3.10).

$$f(x_1) + f'(x_1)(x_2 - x_1) \leq f(x_2) \tag{3.16a}$$

or

$$f(x_2) - f(x_1) \geq f'(x_1)(x_2 - x_1) \tag{3.16b}$$

Figure 3.17 indicates the geometry of the expression in (3.16b); a convex function is never below ($\geq$) its tangents, and a strictly convex function is always above ($>$) its tangents.

In the same way, the expression in (3.14) for concave functions can be rewritten using a first-order Taylor series approximation; it will look like (3.16) except that the inequality will be reversed. That is, for a concave function,

$$f(x_1) + f'(x_1)(x_2 - x_1) \geq f(x_2) \tag{3.17a}$$

or

$$f(x_2) - f(x_1) \leq f'(x_1)(x_2 - x_1) \tag{3.17b}$$

The geometry is that a concave function is never above its tangents; when the inequality is strong, the strictly concave function is always below its tangents.

Insights from the Second-Order Taylor Series Approximation From the second-order Taylor series expression in (3.11a),

$$f(x_0 + dx) - f(x_0) - f'(x_0)dx \cong \frac{1}{2!} f''(x_0)(dx)^2 \tag{3.18}$$

For a convex function, from (3.16b), the left-hand side of (3.18) must be nonnegative and so, for a convex function, the right-hand side must also be nonnegative:

$$\frac{1}{2!} f''(x_0)(dx)^2 \geq 0$$

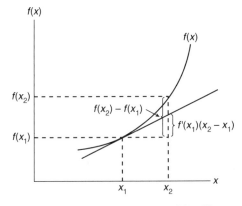

FIGURE 3.17 The geometry of (3.16b).

This means $f''(x_0) \geq 0$; this was the second-order necessary condition for a minimum at x_0, as in Tables 3.1 and 3.3. For a concave function, from (3.17b), the left-hand side of (3.18) will be nonpositive, implying $f''(x_0) \leq 0$. This was the second-order necessary condition for a maximum at x_0, again as in Tables 3.1 and 3.3.

Taylor's series, in (3.9), also provides information to cover the case in which both $f'(x_0)$ and $f''(x_0)$ are zero. The value of $f(x_0 + dx) - f(x_0)$ then depends on the sign of $f^{(3)}(x_0)$ or, should that also be zero, on the sign of the first nonzero higher-order derivative. However, the higher-powered dx terms, $(dx)^3$, $(dx)^4$, $(dx)^5$, and so on, add a complication, since in order to allow variation away from x_0 in *either direction* along the x axis we must allow both $dx < 0$ and $dx > 0$. If the first nonzero derivative is of *even* power [$f^{(4)}(x_0), f^{(6)}(x_0)$, ...] then, since $(dx)^4$, $(dx)^6$, ... and all the factorials are positive, the sign of $f(x_0 + dx) - f(x_0)$ is the same as the sign of that derivative. If the first nonzero derivative is of *odd* power, however [$f^{(3)}(x_0), f^{(5)}(x_0)$, ...] then the sign of the dx term accompanying it [$(dx)^3$, $(dx)^5$, ...] will depend on whether dx is positive or negative and cannot be established unambiguously in advance. Hence the following general rule, which was illustrated in Table 3.2:

If $f'(x_0) = f''(x_0) = \ldots = f^{(n-1)}(x_0) = 0$ and $f^{(n)}(x_0) \neq 0$, then $f(x)$ has a stationary point at x_0 that is a *point of inflection* if n is odd, a *maximum* if n is   (3.19) even and $f^{(n)}(x_0) < 0$ or a *minimum* if n is even and $f^{(n)}(x_0) > 0$.

3.3.3 Local, Global, and Unique Maxima and Minima

If we know, or can deduce, that $f(x)$ is *strictly concave* over $[c,d]$, then we know that there is *at most* one maximum point in that interval, and if a *local* maximum is found, it is the *unique global* maximum of $f(x)$ over $[c,d]$. This is the case in Figure 3.16b. In the same way, a *strictly convex* function over $[c,d]$ has at most one minimum point; the *local* minimum, if found, is a *unique global* minimum in the interval $c \leq x \leq d$. If a function is just concave or convex, not *strictly* so, then a maximum or a minimum may not be unique. Figures 3.18a and 3.18b illustrate these possibilities for a (not strictly) concave function.

If the second derivative (or second differential) of $f(x)$ is *independent* of x^*—that is, if it is a constant, k—then $f(x)$ is concave (if $k \leq 0$) or convex (if $k \geq 0$) *over its entire range*. For maximum problems with concave functions over the entire range of x and for minimum problems with convex functions over the entire range, the first-order conditions on $f'(x)$ are therefore both *necessary and sufficient*. If a function is strictly concave ($k < 0$) or strictly convex ($k > 0$), then the maximum or minimum is unique.

3.3.4 Additional Kinds of Convexity and Concavity

Note that the "bell" shaped function in Figure 3.18c has an unambiguous and unique maximum in the $[c,d]$ interval, yet it is clearly not concave—it is sometimes above and sometimes below its tangent over that interval. This function is called *quasiconcave* in the interval. The definitions of *quasiconcavity* and *quasiconvexity* are less demanding than those of concavity and convexity, and so they admit a larger class of functions. These (and associated) concepts that define alternative kinds of function shape are examined in Ap-

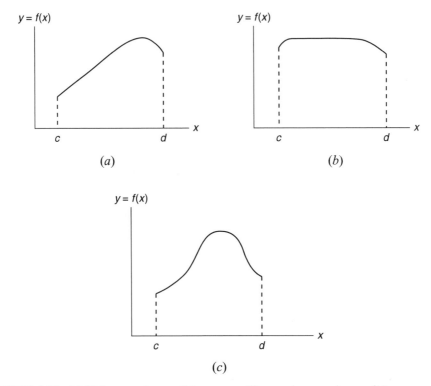

FIGURE 3.18   (*a*) Unique maximum, $f(x)$ concave; (*b*) nonunique maximum, $f(x)$ concave; (*c*) unique maximum, $f(x)$ not concave but quasiconcave.

pendix 3.1. They have assumed particular importance in inequality constrained nonlinear optimization (nonlinear programming) problems for functions of several variables, which are explored in Part V of this book.[13]

3.3.5 Numerical Examples

Here are some concrete illustrations identifying and classifying stationary points for $f(x)$.

(a) $f(x) = -(x - 3)^2 + 5$ (Fig. 3.19*a*). Here $f'(x) = -2(x - 3)$ and $f''(x) = -2$. This is a standard text book kind of illustration—in this case of a strictly concave function $[f''(x) < 0$ everywhere] with a unique global maximum. Setting $f'(x) = 0$, we have a very simple linear equation whose only solution is $x^* = 3$.

(b) $f(x) = (x - 2)^3 + 2$ (Fig. 3.19*b*). The first-order necessary condition is $3(x - 2)^2 = 0$, which leads to the unique solution $x^* = 2$. Since $f''(x) = 6x - 12$, $f''(x^*) = 0$, and we cannot identify what kind of stationary point this represents. Continuing with higher-order derivatives, $f^{(3)}(x) = 6$; since this is the third derivative (odd), we have found neither a

---

[13]For that reason, we will not dwell on these concepts here, but if you read the first part of Appendix 3.1, which deals with functions of a single variable, you will then be able to extend some of the maximum and minimum conditions in this section to cases in which $f(x)$ is only quasiconcave or quasiconvex.

maximum nor a minimum but rather a point of inflection at x^*, from the general rule in (3.14).

(c) $f(x) = e^{-(x-2)^2}$ (Fig. 3.19c). The picture suggests a two-dimensional trace of a three-dimensional bell. Solving $f'(x) = 0$ leads to the unique solution $x^* = 2$, and since $f''(x^*) < 0$, this identifies an unambiguous maximum at x^*: $f(x)$ is strictly concave *only in the neighborhood* of x^*.

It is clear from Figure 3.19c that the function changes from concave to convex as one moves in either the positive or the negative direction from $x^* = 2$; however, it is also clear that the maximum at x^* is a global one. This is a *strictly quasiconcave* function, and local maxima are also global maxima for such functions, just as in the case of strictly concave functions. And the same kind of logic holds for global minima of strictly quasiconvex functions, which would look like Figure 3.19c turned upside down.[14]

(d) $f(x) = (x + 1)^3 + x^2$ (Fig. 3.19d). Now $f'(x) = 3x^2 + 8x + 3$, and $f'(x) = 0$ has two solutions, $x^* = -0.4514$ or $x^{**} = -2.2153$. (Here and throughout these examples, we will work with four figures to the right of the decimal point.) Since $f''(x) = 6x + 8$, $f''(x^*) > 0$ and $f''(x^{**}) < 0$, identifying x^* as a relative minimum [$f(x)$ is convex in the neighborhood of x^*] and x^{**} as a relative maximum [$f(x)$ is concave in the neighborhood of x^{**}].

In unconstrained problems, when $f(x)$ has exactly two solutions to $f'(x) = 0$, where one is a maximum and one a minimum, then neither extremum can be a global one, since $f(x)$ will fall toward $-\infty$ on one side of the maximum and $f(x)$ will rise toward $+\infty$ on the other side of the minimum. However, if x^* identifies a maximum (or a minimum) and x^{**} represents an inflection point, then the extremum at x^* is a global one.

(e) $f(x) = (0.5)[(x + 1)^3 + x^2]^2 - 3$ (Fig. 3.19e). This is an example in which you will find *three* (real) roots to the first-order condition $f'(x) = 0$.[15] (Problem 3.5 asks you to explore two other examples of this sort.)

In this case, $f'(x) = 3x^5 + 20x^4 + 44x^3 + 39x^2 + 17x + 3$, so $f'(x) = 0$ is a fifth-degree equation; it has three real and two imaginary roots.[16] The real roots are $x^* = -3.1479$, $x^{**} = -2.2153$ and $x^{***} = -0.4514$. For this function, $f''(x^*) > 0$, $f''(x^{**}) < 0$ and $f''(x^{***}) > 0$. This algebra is consistent with what we know from the figure—namely that x^* and x^{***} identify local minima [$f(x)$ is convex in the neighborhoods of both x^* and x^{***}] while x^{**} is a local maximum [$f(x)$ is concave in the neighborhood of x^{**}]. Moreover, x^* is a *global* minimum. We can see that from the figure; we could also identify it by substitution into $f(x)$, since $f(x^*) = -3$ and $f(x^{***}) = -2.9320$.[17]

---

[14]More details on quasiconcave and quasiconvex functions of one variable are contained in the first part of Appendix 3.1.

[15]You might convince yourself of the fact that when there are three roots you have either (1) two maxima and one minimum or two minima and one maximum; (2) two inflection points and one extreme point; (3) one maximum, one minimum, and one inflection point; or (d) three inflection points—in which case the function looks like a kind of "staircase," with no maxima or minima.

[16]Solving an equation like this can in itself be a difficult task. Computer software for solving nonlinear equations, such as $f'(x) = 0$ here, may be successful most of the time; however, it may fail to find all the solutions all the time. One way to get some idea of the solution to a nonlinear equation is to use a good computer graphics program to graph $f'(x)$ and simply look at the places where the curve crosses the x axis [which is where the function $f'(x) = 0$]. We explore some approaches to solving nonlinear equations in Chapter 5.

[17]Even without the geometry, we know that the function cannot drop any lower than $f(x^*) = -3$ because to do so would require another local maximum to the left of x^* or to the right of x^{***}, and we have established that there is only one local maximum, at x^{**}.

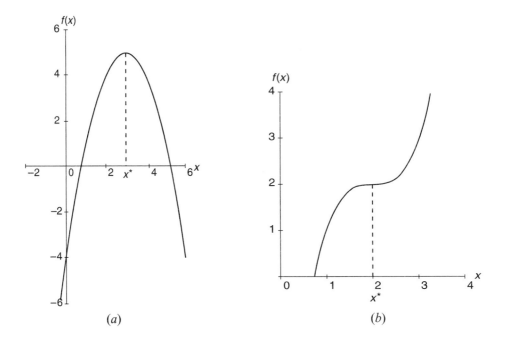

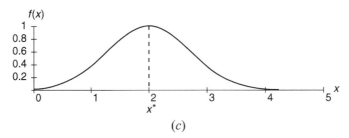

FIGURE 3.19 $(a) f(x) = -(x-3)^2 + 5;$ $(b) f(x) = (x-2)^3 + 2;$ $(c) f(x) = e^{-(x-2)^2}.$

3.4 MAXIMA AND MINIMA FOR FUNCTIONS OF TWO INDEPENDENT VARIABLES

We now consider the case in which y depends on two independent variables, x_1 and x_2; $y = f(x_1, x_2)$. Since only three dimensions are necessary, the conditions for maximum and minimum points are easily visualized. They then easily generalize to functions of any number of variables, which will be the topic of Section 3.6. Our concern throughout will be with functions that are smooth and continuous, so that differentiation is possible everywhere.[18] In general, we will require derivatives at least up to the second order; hence we are interested in functions in which variables are at least squared or appear multiplied to-

---

[18]Also, we continue to assume that the independent variables can range over the entire set of real numbers. In Chapter 4 we will deal explicitly with constraints.

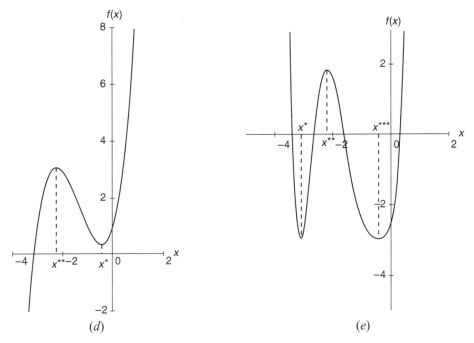

FIGURE 3.19   (d) $f(x) = (x + 1)^3 + x^2$; (e) $f(x) = (0.5)[(x + 1)^3 + x^2]^2 - 3$.

gether. This is logical, since otherwise the "curves" would have no curvature; they would be planes or lines, and the search for maxima and minima via first-order conditions (where the tangent has a slope of zero) would generally be meaningless.

3.4.1 Partial Derivatives, Gradient Vectors, and Hessian Matrices

Partial Derivatives With functions of two (or more) variables, the concept of a *partial derivative* is necessary. If $y = f(x)$ then dy/dx unambiguously denotes the derivative of $f(x)$ with respect to the only independent variable, x. When $y = f(x_1, x_2)$, a change in y may occur when both variables change or when either one of them changes while the other remains constant. Thus we need to specify which variable is changing and which is not; the notation often used is $\partial y/\partial x_1$ and $\partial y/\partial x_2$ (or $\partial f/\partial x_1$ and $\partial f/\partial x_2$), read as the *partial derivative* of the function with respect to x_1 or x_2. The use of ∂f in place of df indicates the similarity of the concept; f_{x_1} and f_{x_2} are also sometimes used, but simpler yet are just f_1 and f_2.

Partial derivatives are designed to measure a change in $f(x_1, x_2)$ when all variables except one remain constant, the technique of partial differentiation is an extremely simple extension of ordinary differentiation. It requires only that the other x (or x's, if there are three or more variables) be regarded as constants. For example, if $\mathbf{X} = [x_1, x_2]'$ and $y = f(\mathbf{X}) = 3x_1^2 + 4x_1 x_2 + x_2^3$, the two possible partial derivatives are

$$\frac{\partial f}{\partial x_1} \equiv f_1 = 6x_1 + 4x_2 \qquad \text{and} \qquad \frac{\partial f}{\partial x_2} \equiv f_2 = 4x_1 + 3x_2^2 \qquad (3.20)$$

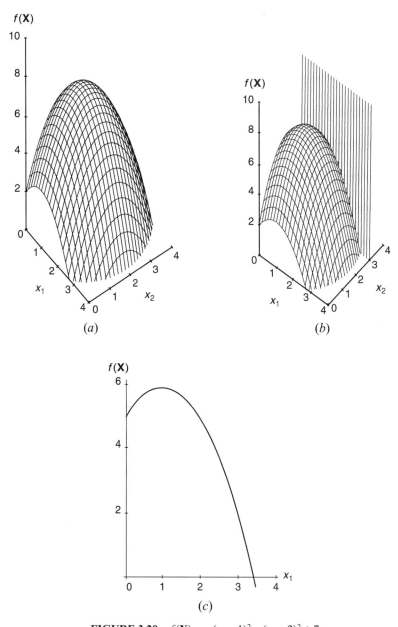

FIGURE 3.20 $f(\mathbf{X}) = -(x_1 - 1)^2 - (x_2 - 2)^2 + 7$.

Note that in general both variables appear in both partial derivatives. In f_1, if a particular value for x_2, say x_2^0, is assumed, then f_1 (evaluated at x_2^0) is a function of only x_1, the variable actually assumed to change. Similarly, f_2 is a function of x_2 relative to some specific x_1^0. Since this is the case, it is easy to visualize the partial derivatives as curves in particular planes in three-dimensional x_1, x_2, y space.

This is illustrated for $f(\mathbf{X}) = -(x_1 - 1)^2 - (x_2 - 2)^2 + 7$, an overturned bowl in three-dimensional space (Figure 3.20a). A plane cutting through the surface perpendicular to the

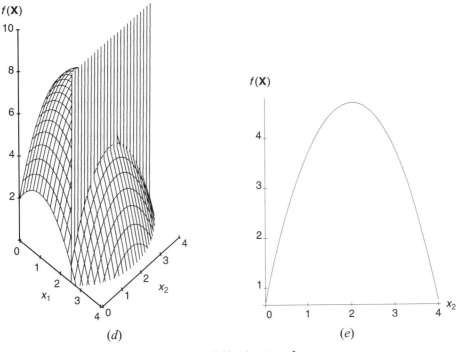

FIGURE 3.20 (*continued*)

x_1, x_2 plane and parallel to the x_1, y plane at a particular value of x_2 is shown in Figure 3.20*b*, where $x_2 = 3$. The contour of $f(\mathbf{X})$ on that $x_2 = 3$ plane is shown in Figure 3.20*c*. Similarly, the plane $x_1 = 2.5$, cutting $f(\mathbf{X})$ and parallel to the x_2, y plane, is shown in Figure 3.20*d*, and the contour of $f(\mathbf{X})$ on that plane is shown in Figure 3.20*e*. Once y is reduced to a function of either x_1 or x_2 alone, then the (partial) derivative has exactly the same interpretation as did dy/dx in the one-variable case; specifically, f_1 and f_2 are the slopes of the tangents to the contour of $f(\mathbf{X})$ in the $x_2 = x_2^0$ or $x_1 = x_1^0$ plane, respectively.[19]

Higher-order partial derivatives can also be found. They are similar to higher-order derivatives in exactly the same way that $\partial y/\partial x_1$ was related to dy/dx. Thus $\partial^2 y/\partial x_1^2 = \partial(\partial y/\partial x_1)/\partial x_1$ is the *second partial derivative* of $f(\mathbf{X})$ with respect to x_1; the more convenient notation is simply f_{11}. From the first partials in (3.20), $f_{11} = 6$ and $f_{22} = 6x_2$. With functions of two or more variables, it is also possible to measure, for example, how the first partial derivative with respect to x_1 is itself affected by changes in x_2. In the example in (3.20) we saw that x_2 entered into f_1; it must be set to some particular value, x_2^0, in order to examine the relationship between y and changes in x_1 alone (Figs. 3.20*b* and 3.20 *d*).The value x_2^0 simply determines the location of the x_1, y plane that is cutting the surface $f(\mathbf{X})$. The notation for this *cross-partial* is $\partial(\partial y/\partial x_1)/\partial x_2$, or $\partial^2 y/(\partial x_1 \partial x_2)$, or simply f_{12}. Note that the partial differentiation operations are carried out in the order of the subscripts.[20] From the example in (3.20), $f_{12} = 4 = f_{21}$.

---

[19]You can be convinced of this by examining the geometry of the delta process to generate a change in either x_1 or x_2, forming $\Delta y/\Delta x_1$ or $\Delta y/\Delta x_2$ [as in (3.2)], and taking limits.

[20]In fact, all the smooth, continuous functions with which we will be dealing have the property that order makes no difference in cross-partials, so that $f_{12} = f_{21}$.

Gradient Vectors The *gradient* of a function is a vector containing the first partial derivatives of the function. (We choose to define it always as a column vector; some texts prefer a row vector. Whether a row or a column is meant should always be clear from the context.) The usual notation is $\nabla f(x_1, x_2)$, or simply ∇f:

$$\nabla f(x_1, x_2) \equiv \nabla f = \begin{bmatrix} f_1 \\ f_2 \end{bmatrix} \tag{3.21}$$

Specifically for $f(\mathbf{X}) = 3x_1^2 + 4x_1x_2 + x_2^3$,

$$\nabla f = \begin{bmatrix} 6x_1 + 4x_2 \\ 4x_1 + 3x_2^2 \end{bmatrix}$$

If the first partial derivatives are evaluated at a particular point $\mathbf{X}^0 = [x_1^0, x_2^0]'$, the gradient at that point is denoted $\nabla f(\mathbf{X}^0)$ or simply ∇f^0, or, when there is absolutely no ambiguity about the function in question, simply ∇^0. For the function in this example, at $\mathbf{X}^0 = [2,5]'$, $\nabla f^0 = \begin{bmatrix} 32 \\ 83 \end{bmatrix}$.

Hessian Matrices A function of two independent variables will have two second partial derivatives (f_{11} and f_{22}) and two (equal) cross-partial derivatives (f_{12} and f_{21}). These are often represented as elements of a *matrix* (just as f_1 and f_2 are elements of the gradient *vector*). If we take the gradient of f_1, the first element in ∇f, we have $\nabla f_1 = [f_{11} \; f_{12}]'$ and, similarly, $\nabla f_2 = [f_{21} \; f_{22}]'$. Building a matrix by stacking the transposes of these two gradients (so that the gradients are rows), we have

$$\begin{bmatrix} [\nabla f_1]' \\ [\nabla f_2]' \end{bmatrix} = \begin{bmatrix} f_{11} & f_{12} \\ f_{21} & f_{22} \end{bmatrix}$$

which is often denoted by $\nabla(\nabla f)$ or $\nabla^2 f$ or, when the function is unambiguous, simply ∇^2. Frequently, this matrix is also denoted by $\mathbf{H}$; it is called the *Hessian* matrix of $f(\mathbf{X})$ (Ludwig Otto Hesse, German, 1811–1874). Thus, for $f(x_1, x_2)$

$$\nabla^2 \equiv \mathbf{H} = \begin{bmatrix} f_{11} & f_{12} \\ f_{21} & f_{22} \end{bmatrix} \tag{3.22}$$

We will see that gradient (∇) and Hessian ($\mathbf{H}$) representations provide compact notation for examining maximum and minimum conditions for $f(x_1, x_2)$—and ultimately, with appropriate extensions, for $f(x_1, \ldots, x_n)$, where $n > 2$.

3.4.2 Maxima and Minima for $f(x_1, x_2)$

The geometry in Figure 3.20 (specifically 3.20*c* and *e*) makes clear that a set of *necessary* conditions for $\mathbf{X}^* = [x_1^* \; x_2^*]'$ to be either a maximum or a minimum point of $f(x_1, x_2)$ is that both f_1 and f_2 be zero there (identifying hilltops on the contours in those two figures). We denote these derivatives evaluated at $\mathbf{X}^*$ by f_1^* and f_2^*, so the necessary conditions are $f_1^* = 0$ and $f_2^* = 0$. These two requirements are compactly expressed as $\nabla f^* = \mathbf{0}$ (where the "$\mathbf{0}$" on the right-hand side now represents a two-element column vector of 0's—as it must to conform to the element-by-element definition of equality in matrix algebra).

In introductory discussions of maxima and minima for $f(x_1, x_2)$ it is customary to note that maxima are distinguished from minima by the sign of the second partial derivative with respect to *either* variable, evaluated at $\mathbf{X}^*$, *provided a certain inequality is satisfied*. And the second partial sign rule is the same as the second derivative sign rule for $f(x)$, namely, $f_{11}^* < 0$ (*or* $f_{22}^* < 0$) identifies a maximum at $\mathbf{X}^*$ and $f_{11}^* > 0$ (*or* $f_{22}^* > 0$) identifies a minimum there, provided in either case that

$$[f_{11}^* f_{22}^* - (f_{12}^*)^2] > 0 \tag{3.23}$$

When (3.23) is not met, the sign tests on the second partial derivatives are inadequate; we cannot identify what kind of point $\mathbf{X}^*$ represents. Differentials provide a more general approach to conditions for maxima and minima for a function of two or more independent variables, and it is these that we now explore. In the process we will see where (3.23) comes from.

3.4.3 The Total Differential for Functions of Two Variables

Since y now depends on the values taken by two independent variables, the differential must reflect the possibility of arbitrary small changes in both of these variables; it must include both dx_1 and dx_2. Parallel to (3.7), the product $f_1 dx_1$ would serve to transmit the small change in x_1 to the function $f(x_1, x_2)$. Similarly, for independent changes in x_2, $f_2 dx_2$ measures their impact on f. Since *both* x_1 and x_2 may change independently, the total effect on $f(\mathbf{X})$ is the sum of the two individual effects:[21]

$$dy \equiv df = \frac{\partial y}{\partial x_1} \, dx_1 + \frac{\partial y}{\partial x_2} \, dx_2 = f_1 dx_1 + f_2 dx_2 \tag{3.24a}$$

Defining a two-element column vector of the small changes dx_1 and dx_2 as

$$\mathbf{dX} \equiv \begin{bmatrix} dx_1 \\ dx_2 \end{bmatrix}$$

and using the gradient of $f(\mathbf{X})$ again, the total differential in (3.24a) can be represented as

$$dy \equiv df = [\nabla f]'[\mathbf{dX}] \tag{3.24b}$$

or, with inner product notation

$$dy \equiv df = [\nabla f] \circ [\mathbf{dX}] \tag{3.24c}$$

As in the case of the differential for a function of one variable, $df(\mathbf{X}^0)$ is essentially the equation of the tangent (now a plane) to the surface $f(\mathbf{X})$ at $\mathbf{X}^0$. The reasoning is parallel to that for $f(x)$, although somewhat more involved. The general equation of a plane in x_1, x_2, z space is $ax_1 + bx_2 + cz = d$ or $z = g(\mathbf{X}) = (-a/c)x_1 + (-b/c)x_2 + (d/c)$. For a particular point

---

[12]This result can be derived by using the fundamental delta process for Δy with both Δx_1 and Δx_2 and taking appropriate limits.

(x_1^0, x_2^0, z^0) on the plane, $ax_1^0 + bx_2^0 + z^0 = d$, and subtracting the particular from the general, we obtain

$$a(x_1 - x_1^0) + b(x_2 - x_2^0) + c(z - z^0) = 0$$

or

$$z - z^0 = \frac{-a}{c}(x_1 - x_1^0) + \frac{-b}{c}(x_2 - x_2^0) \tag{3.25}$$

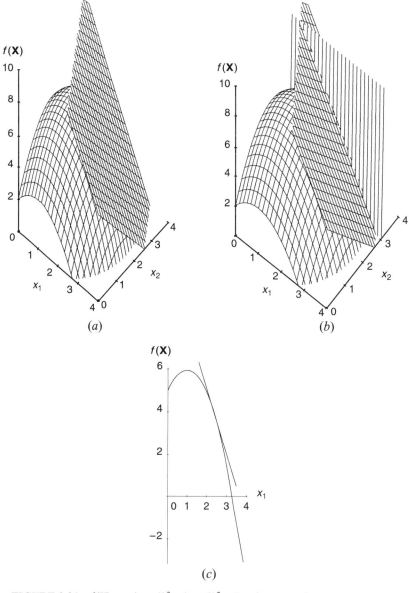

FIGURE 3.21 $f(\mathbf{X}) = -(x_1 - 1)^2 - (x_2 - 2)^2 + 7$ and tangent plane at $x_1 = 2.5$, $x_2 = 3$.

For a tangent plane to $f(\mathbf{X})$ at $\mathbf{X}^0$, we know z^0 $[=f(\mathbf{X}^0)]$—the plane and the surface have $\mathbf{X}^0$ in common. We also know the slopes of two lines that lie in that tangent plane—$\partial f/\partial x_1^0$ [the slope at $x_1 = x_1^0$ of the contour of $f(\mathbf{X})$ in the vertical plane $x_2 = x_2^0$, as in Fig. 3.20c] and $\partial f/\partial x_2^0$ [the slope at $x_2 = x_2^0$ of $f(\mathbf{X})$ in the vertical plane $x_1 = x_1^0$, as in Fig. 3.20e]. In the former case, since $x_2 = x_2^0$, (3.25) reduces to the equation for the tangent *line* to $f(\mathbf{X})$ at x_1^0 in the $x_2 = x_2^0$ plane, namely, $z - z^0 = (-a/c)(x_1 - x_1^0)$, and so $\partial f/\partial x_1^0 \equiv f_1^0 = (-a/c)$. By the same kind of argument, $f_2^0 = (-b/c)$. Letting $dz = z - z^0$, $\mathbf{dX} = \begin{bmatrix} (x_1 - x_1^0) \\ (x_2 - x_2^0) \end{bmatrix}$ and $\nabla f = \begin{bmatrix} f_1 \\ f_2 \end{bmatrix} = \begin{bmatrix} (-a/c) \\ (-b/c) \end{bmatrix}$, (3.25) can be rewritten as $dz = [\nabla f]'(\mathbf{dX})$ as in (3.24b), but now using z for the dependent variable.

Consider $f(\mathbf{X}) = -(x_1 - 1)^2 - (x_2 - 2)^2 + 7$ in Figure 3.20a. Let $\mathbf{X}^0 = \begin{bmatrix} 2.5 \\ 3 \end{bmatrix}$, so $f(\mathbf{X}^0) = 3.75$ and $\nabla f(\mathbf{X}^0) = \begin{bmatrix} -3 \\ -2 \end{bmatrix}$. The equation of the plane that is tangent to $f(\mathbf{X})$ at $\mathbf{X}^0$ is $(z - z^0)$ $[\nabla f(\mathbf{X}^0)]'[\mathbf{X} - \mathbf{X}^0]$, which in this illustration works out to $z = -3x_1 - 2x_2 + 17.25$.

The surface $f(\mathbf{X})$ and this tangent plane at $\mathbf{X}^0 = \begin{bmatrix} 2.5 \\ 3 \end{bmatrix}$ are shown in Figure 3.21a. The vertical plane $x_2 = 3$ has been added in Figure 3.21b (it already appeared in Fig. 3.20b but without the tangent plane). The two planes intersect in a line that is tangent to the $f(\mathbf{X})$ contour in the $x_2 = 3$ plane (Fig. 3.21c). Similarly, Figure 3.21d shows the vertical plane $x_1 = 2.5$ added to Figure 3.21a (this plane was already seen in Fig. 3.20d, without the tangent plane). The intersection of these two planes is also tangent to the contour of $f(\mathbf{X})$ in the $x_1 = 2.5$ plane (Fig. 3.21e).

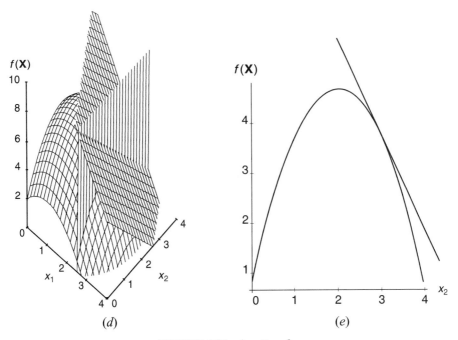

(d)   (e)

FIGURE 3.21   (*continued*)

Suppose $\mathbf{dX} = \begin{bmatrix} 1 \\ 2 \end{bmatrix}$, so $f(\mathbf{X}^0 + \mathbf{dX}) = -8.25$. The linear approximation to the change in $f(\mathbf{X})$

is $df(\mathbf{X}^0) = \begin{bmatrix} -3 & -2 \end{bmatrix} \begin{bmatrix} 1 \\ 2 \end{bmatrix} = -7$. The actual change is $f(\mathbf{X}^0 + \mathbf{dX}) - f(\mathbf{X}^0) = -12$, and the differential underestimates the change by 5 units. As with the $f(x)$ case, as the elements in $\mathbf{dX}$ become smaller, the error in the linear estimate will become smaller. For example, if $\mathbf{dX} = \begin{bmatrix} 0.5 \\ 1 \end{bmatrix}$, the amount of underestimation provided by the differential is reduced to 1.25 units.

An illustration that requires only two dimensions is that of the area (y) of a rectangle as a function of length (x_1) and width (x_2). The functional relationship is $y = f(\mathbf{X}) = x_1 x_2$. While $f(\mathbf{X})$ is a surface in three-dimensional space, the rectangle can be shown in two dimensions, as in Figure 3.22. Suppose $x_1 = 10$ and $x_2 = 5$, so that $y = 50$. Let x_1 change to 10.5 and x_2 change to 5.2, and use (3.24) to approximate the *change* in area. Since $f_1 = x_2$ and $f_2 = x_1$, we have

$$df = x_2 dx_1 + x_1 dx_2 = 5(0.5) + 10(0.2) = 4.5$$

This is the estimate of the increase in area provided by the differential.

We would approximate the *new* area as $50 + 4.5 = 54.5$; in fact, it is $(10.5)(5.2) = 54.6$. Figure 3.22 illustrates the source of the error. The two terms in the differential are the shaded areas; the error is thus caused by omitting the upper right area—$(dx_1)(dx_2)$. In this case, the error is $(0.5)(0.2) = 0.1$, as we saw. Clearly, the smaller the changes in x_1 and x_2, the more accurate is the linear approximation given by the differential.

The second total differential of a function of two variables is found by taking the differential of the first differential; $d(df) \equiv d^2 f$. The exact expression for $d^2 f$ can be derived using partial derivatives, exactly as derivatives were used for (3.8). The rules for differentials parallel those for derivatives, and they can also easily be applied to the first differential in (3.24a):

$$
\begin{aligned}
d^2 f &= d(f_1 dx_1 + f_2 dx_2) \\
&= d(f_1 dx) + d(f_2 dx_2) && \text{(parallel to the rule for the derivative of a sum)} \\
&= d(f_1) dx_1 + d(dx_1) f_1 + \\
&\quad d(f_2) dx_2 + d(dx_2) f_2 && \text{(parallel to the rule for the derivative of a product)}
\end{aligned}
$$

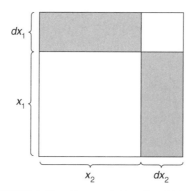

FIGURE 3.22   Changes in length, width and area.

In general, f_1 and f_2 are functions of both variables, so

$$d(f_1) = f_{11}dx_1 + f_{12}dx_2 \qquad \text{and} \qquad d(f_2) = f_{21}dx_1 + f_{22}dx_2$$

[This is a straightforward application of the rule in (3.24a) to two new functions of x_1 and x_2.] Using $d(dx_1) \equiv d^2x_1$ and $d(dx_2) \equiv d^2x_2$, we derive

$$d^2f = (f_{11}dx_1 + f_{12}dx_2)dx_1 + f_1d^2x_1 + (f_{21}dx_1 + f_{22}dx_2)dx_2 + f_2d^2x_2$$

Since we continue to assume that the functions are well behaved, with smooth and continuous second partials and cross-partials, the cross-partial derivatives will be identical and

$$d^2f = f_{11}(dx_1)^2 + 2f_{12}dx_1dx_2 + f_{22}(dx_2)^2 + f_1d^2x_1 + f_2d^2x_2$$

When x_1 and x_2 are *independent* variables, then dx_1 and dx_2 are arbitrary and unrelated small changes (i.e., constants). This means that $d(dx_1) \equiv d^2x_1 = 0$ and $d^2x_2 = 0$ also, and the second total differential for a function of two *independent* variables reduces to

$$d^2f = f_{11}(dx_1)^2 + 2f_{12}dx_1dx_2 + f_{22}(dx_2)^2 \qquad (3.26)$$

Recalling the definition in Section 1.6 of a quadratic form of two variables, $Q(x_1,x_2)$ [as in (1.32)], we recognize that (3.26) is a quadratic form in the variables dx_1 and dx_2. This means that we can write (3.26) as

$$d^2f = [dx_1 \quad dx_2]\begin{bmatrix} f_{11} & f_{12} \\ f_{21} & f_{22} \end{bmatrix}\begin{bmatrix} dx_1 \\ dx_2 \end{bmatrix} \qquad (3.27)$$

or, using the Hessian matrix as defined in (3.22) (where $\mathbf{H}$ is symmetric, since we are dealing with functions for which $f_{ij} = f_{ji}$), this is

$$d^2f = [\mathbf{dX}]'\mathbf{H}[\mathbf{dX}] \qquad (3.28)$$

For $d^2f > 0$ (a positive definite quadratic form) we know from (1.40) in Section 1.6 that the leading principal minors of $\mathbf{H}$ (the $|\mathbf{H}_i|$'s, using the notation of that section) must all be positive. In particular, when the second total differential is evaluated at a candidate $\mathbf{X}^*$ for a maximum or minimum point at which $\nabla f(\mathbf{X}^*) = 0$, the conditions for $d^2f^* > 0$ are $|\mathbf{H}_1^*|$ > 0 and $|\mathbf{H}_2^*|$ $(= |\mathbf{H}^*|) > 0$; that is, $f_{11}^* > 0$ and $[f_{11}^*f_{22}^* - (f_{12}^*)^2] > 0$. For (3.28) to be negative definite at $\mathbf{X}^*$ $(d^2f^* < 0)$, the requirements are $|\mathbf{H}_1^*| < 0$ and $|\mathbf{H}_2^*|$ $(= |\mathbf{H}^*|) > 0$, meaning $f_{11}^* < 0$ and, again, $[f_{11}^*f_{22}^* - (f_{12}^*)^2] > 0$. Note that this last requirement, needed for either positive or negative definiteness of d^2f^*, is exactly (3.23).

For the quadratic form to be positive or negative semidefinite—$d^2f^* \geq 0$ or $d^2f^* \leq 0$—the requirements are given in (1.41) in Section 1.6 in terms of all the principal minors (not just the leading ones) of the symmetric matrix in the quadratic form. (These principal minors are the $|\mathcal{H}_i|$'s in the notation of Section 1.6.) Specifically, for d^2f^* in (3.28) to be positive semidefinite, $|\mathcal{H}^*(1)|$, $|\mathcal{H}^*(2)|$, and $|\mathcal{H}_2^*|$ must all be nonnegative. Here this means $f_{11}^* \geq 0$, $f_{22}^* \geq 0$ and $[f_{11}^*f_{22}^* - (f_{12}^*)^2] \geq 0$. For d^2f^* in (3.28) to be negative semidefinite, the conditions are $|\mathcal{H}^*(1)|$ and $|\mathcal{H}^*(2)|$ nonpositive and $|\mathcal{H}_2^*|$ nonnegative, which here means $f_{11}^* \leq 0$, $f_{22}^* \leq 0$ and $[f_{11}^*f_{22}^* - (f_{12}^*)^2] \geq 0$.

TABLE 3.5   Characterization of Maxima and Minima for $f(x_1, x_2)$ at X*

	Maximum	Minimum
First-order necessary condition	$df(\mathbf{X}^*) = 0$	
	or, using gradients	
	$\nabla f(\mathbf{X}^*) = \mathbf{0}$	
Second-order necessary condition	$d^2f(\mathbf{X}^*) \leq 0$	$d^2f(\mathbf{X}^*) \geq 0$
	or, using *principal minors* of the Hessian	
	$\|\mathcal{H}_1^*(1)\| = f_{11}^* \leq 0,$	$\|\mathcal{H}_1^*(1)\| = f_{11}^* \geq 0,$
	$\|\mathcal{H}_1^*(2)\| = f_{22}^* \leq 0,$	$\|\mathcal{H}_1^*(2)\| = f_{22}^* \geq 0,$
	and $\|\mathcal{H}_2^*\| = \|\mathbf{H}^*\| \geq 0$	
Sufficient conditions	$df(\mathbf{X}^*) = 0$	
	or, using gradients	
	$\nabla f(\mathbf{X}^*) = \mathbf{0}$	
	and	
	$d^2f(\mathbf{X}^*) < 0$	$d^2f(\mathbf{X}^*) > 0$
	or, using *leading principal minors* of the Hessian	
	$\|\mathbf{H}_1^*\| < 0,$	$\|\mathbf{H}_1^*\| > 0,$
	and $\|\mathbf{H}_2^*\| = \|\mathbf{H}^*\| > 0$	

Note: Principal minors and leading principal minors were defined and discussed in Section 1.6 of Chapter 1.

Thus the problem of establishing the sign of the second total differential is just a particular case of the general problem of determining the sign of a quadratic form. This is one of the primary reasons for examining the structure of quadratic forms and their signs in Chapter 1. Table 3.5 displays the results for maxima and minima of $f(x_1, x_2)$.

As with $f(x)$, maxima and minima of $f(x_1, x_2)$ are distinguished by the shape properties of the function—that is, by the concavity or convexity of $f(\mathbf{X})$—either near to a candidate point $\mathbf{X}^*$ or everywhere. We explore the shapes of $f(x_1, x_2)$ in the next section.

3.5 TAYLOR'S SERIES, CONCAVITY, AND CONVEXITY OF $f(\mathbf{X})$

3.5.1 Taylor's Series for $f(\mathbf{X})$

Taylor's series for a function of n variables is [parallel to (3.9), but in terms of differentials] and where $\mathbf{dX} = [dx_1, \ldots, dx_n]'$:

$$f(\mathbf{X}_0 + \mathbf{dX}) = f(\mathbf{X}_0) + df(\mathbf{X}_0) + \left(\frac{1}{2!}\right)d^2f(\mathbf{X}_0) +$$

$$\left(\frac{1}{3!}\right)d^3(\mathbf{X}_0) + \cdots + \left(\frac{1}{n!}\right)d^nf(\mathbf{X}_0) + R_n \tag{3.29}$$

(Details are in Appendix 3.2.) The first-order (linear) approximation to the difference in the values of $f(\mathbf{X})$ at $\mathbf{X}_0$ and at $\mathbf{X}_0 + \mathbf{dX}$ results from elimination of all differentials above the first

$$f(\mathbf{X}_0 + \mathbf{dX}) - f(\mathbf{X}_0) \cong df(\mathbf{X}_0)$$

or, in terms of the gradient,

$$f(\mathbf{X}_0 + \mathbf{dX}) - f(\mathbf{X}_0) \cong [\nabla f(\mathbf{X}_0)]' \mathbf{dX} \qquad (3.30)$$

[cf. (3.10a)]. The second-order (quadratic) approximation to $f(\mathbf{X}_0 + \mathbf{dX}) - f(\mathbf{X}_0)$ results from eliminating all differentials above the second, and recalling that $d^2 f(\mathbf{X}_0) = (\mathbf{dX})' \mathbf{H}(\mathbf{X}_0)(\mathbf{dX})$, where $\mathbf{H}(\mathbf{X}_0)$ is the Hessian matrix evaluated at $\mathbf{X}_0$:

$$f(\mathbf{X}_0 + \mathbf{dX}) - f(\mathbf{X}_0) \cong [\nabla f(\mathbf{X}_0)]' \mathbf{dX} + \tfrac{1}{2}(\mathbf{dX})' \mathbf{H}(\mathbf{X}_0)(\mathbf{dX}) \qquad (3.31)$$

[cf. (3.11a)].

As with $f(x)$, a relative maximum of $f(\mathbf{X})$ at $\mathbf{X}^*$ is defined by the observation that $f(\mathbf{X}) \le f(\mathbf{X}^*)$ for all $\mathbf{X}$ in some neighborhood of $\mathbf{X}^*$. In terms of the second-order approximation in (3.31), this maximum condition can be expressed as

$$[\nabla f(\mathbf{X}^*)]'(\mathbf{X} - \mathbf{X}^*) + \tfrac{1}{2}(\mathbf{X} - \mathbf{X}^*)' \mathbf{H}(\mathbf{X}^*)(\mathbf{X} - \mathbf{X}^*) \le 0 \qquad (3.32)$$

(where $\mathbf{X}^* = \mathbf{X}_0$ and $\mathbf{X} = \mathbf{X}_0 + \mathbf{dX}$). This is the many-variable analog to (3.12).

Again, as with $f(x)$, $(\mathbf{X} - \mathbf{X}^*)$ can be either positive or negative, so to achieve nonpositivity in (3.32), $\nabla f(\mathbf{X}^*)$ must be zero—exactly the first-order necessary condition in Table 3.5. This means, in (3.32),

$$\tfrac{1}{2}(\mathbf{X} - \mathbf{X}^*)' \mathbf{H}(\mathbf{X}^*)(\mathbf{X} - \mathbf{X}^*) \le 0$$

The left-hand side is a quadratic form in $(\mathbf{X} - \mathbf{X}^*)$, and so to satisfy the inequality $\mathbf{H}(\mathbf{X}^*)$ must be negative semidefinite—the second-order necessary condition in Table 3.5. The story for relative minima at $\mathbf{X}^*$ is the same, except that the inequality in (3.32) is reversed, and as a result $\mathbf{H}(\mathbf{X}^*)$ must be positive semidefinite.

3.5.2 Concavity and Convexity of $f(\mathbf{X})$

The geometry for $f(x)$ was straightforward. A convex function was never above its secant lines or below its tangent lines; if the function was *strictly* convex, then it was always below its secants or above its tangents. And similarly for a concave $f(x)$. For $f(x_1, x_2)$ the same observations are true, only now the function is located relative to secant lines or tangent planes. For $f(\mathbf{X}) = f(x_1, \ldots, x_n)$ (when $n > 2$) the terminology is secant hyperlines and tangent hyperplanes, but the geometry eludes us.

In the definition of convexity and concavity of $f(x)$ over an interval $c \le x \le d$, it was assumed that x could range continuously over all possible values beginning with c and ending with d. This assured that $f(x)$ was defined over a convex set of points. A similar requirement is necessary when we consider functions of many variables, and we will soon examine convex sets when two or more variables are involved.

Location Relative to Secants   The algebraic definitions of convexity and concavity for $f(\mathbf{X})$ parallel those for $f(x)$ in (3.13) and (3.14).

CONVEX FUNCTION $f(x)$. A function $f(x_1, \ldots, x_n)$ is *convex* over a convex set of points $\mathbf{X}$ if for every α in [0, 1] and every pair of points $\mathbf{X}' = [x_1', \ldots, x_n']'$ and $\mathbf{X}'' = [x'', \ldots, x'']'$, in the set $\mathbf{X}$:

$$f[\alpha \mathbf{X}'' + (1 - \alpha)\mathbf{X}'] \le \alpha f(\mathbf{X}'') + (1 - \alpha)f(\mathbf{X}') \qquad (3.33)$$

As before, for *strict convexity* the $\leq$ is replaced by $<$, and the range of α is in the *open* interval $0 < \alpha < 1$.

CONCAVE FUNCTION $f(x)$. A function $f(x_1, \ldots, x_n)$ is *concave* over a convex set of points **X** if for every α in [0, 1] and every pair of points $\mathbf{X}' = [x_1', \ldots, x_n']'$ and $\mathbf{X}'' = [x_1'', \ldots, x_n'']'$, in the set **X**:

$$f[\alpha \mathbf{X}'' + (1 - \alpha)\mathbf{X}'] \geq \alpha f(\mathbf{X}'') + (1 - \alpha)f(\mathbf{X}') \qquad (3.34)$$

As before, for *strict concavity* the $\geq$ is replaced by $>$, and the range of α is in the *open* interval $0 < \alpha < 1$.

The fact that convex and concave functions are defined over convex sets serves, as we are about to see, to ensure that if **X**' and **X**'' are in the set, so is $\alpha \mathbf{X}'' + (1 - \alpha)\mathbf{X}'$, for $0 \leq \alpha \leq 1$. The interpretation of convexity and concavity is similar to that for $f(x)$. If $f(\mathbf{X})$ is convex, then all points along the line that is the convex combination joining any two points on the surface $f(\mathbf{X})$ never underestimate the value of the function; if $f(\mathbf{X})$ is *strictly* convex, all these points overestimate $f(\mathbf{X})$. If the function is concave or strictly concave, the terms overestimate and underestimate are interchanged.

When $\mathbf{X} = [x_1, x_2]'$, $f(\mathbf{X})$ is a surface in three-dimensional space. A strictly convex $f(\mathbf{X})$ looks like a bowl, and if you imagine a line drawn between any two points on the bowl the line will always be above the bowl, except at its endpoints. A strictly concave $f(\mathbf{X})$ resembles an overturned bowl, and a line connecting any two points on the surface will always be below the bowl, except at its endpoints.

Location Relative to Tangents Reasoning exactly parallel to that in the $f(x)$ case will make clear that for a convex function $f(\mathbf{X})$ of two or more variables

$$f(\mathbf{X}'') - f(\mathbf{X}') \geq [\nabla f(\mathbf{X}')]'(\mathbf{X}'' - \mathbf{X}') \qquad (3.35)$$

and, for a concave function

$$f(\mathbf{X}'') - f(\mathbf{X}') \leq [\nabla f(\mathbf{X}')]'(\mathbf{X}'' - \mathbf{X}') \qquad (3.36)$$

[cf. (3.16b) and (3.17b)].

As with $f(x)$, the right-hand sides of (3.35) and (3.36) are the first differentials for $f(\mathbf{X})$, with $\mathbf{X}'' = \mathbf{X}' + d\mathbf{X}$—that is, they represent the tangent planes to $f(\mathbf{X})$ at $\mathbf{X}'$. For a convex function, (3.35) indicates that $f(\mathbf{X})$ lies everywhere on or above its tangent planes. If $f(\mathbf{X})$ is concave, then it lies everywhere on or below its tangent planes. Figure 3.21a provides an illustration.

Insights from the Second-Order Taylor Series Approximation When $d^2 f^* \leq 0$, this means that the second-order Taylor's series approximation [as in (3.31)] to $f(\mathbf{X})$ around $\mathbf{X}^*$ is nonpositive; the function is no smaller at $\mathbf{X}^*$ than at nearby points, and $\mathbf{X}^*$ represents a local maximum. Alternatively put, $f(\mathbf{X})$ is *concave* in the neighborhood of $\mathbf{X}^*$. If $d^2 f^* < 0$, then $f(\mathbf{X})$ is larger at $\mathbf{X}^*$ than at nearby points, and $\mathbf{X}^*$ identifies a unique local maximum; $f(\mathbf{X})$ is *strictly concave* in the neighborhood of $\mathbf{X}^*$.

Similarly, when $d^2 f^* \geq 0$, the quadratic approximation to $f(\mathbf{X}^* + d\mathbf{X}) - f(\mathbf{X}^*)$ is nonnegative and the function is *convex* in the neighborhood of $\mathbf{X}^*$, which represents a local minimum. If $d^2f^* > 0$, then $f(\mathbf{X}^* + d\mathbf{X}) - f(\mathbf{X}^*) > 0$, and the function is *strictly convex* near to $\mathbf{X}^*$, which represents a unique local minimum.

If $d^2f < 0$ *everywhere*, then $f(\mathbf{X})$ is everywhere concave, and a point satisfying $\nabla f(\mathbf{X}) = \mathbf{0}$ represents a global maximum. If $d^2f > 0$ everywhere, then $f(\mathbf{X})$ is everywhere convex, and a point satisfying $\nabla f(\mathbf{X}) = \mathbf{0}$ constitutes a global minimum.

As we will see in some examples below, there are functions with completely normal maxima or minima that will not be identifiable by the rules in Table 3.5. Some of these represent the higher-dimensional analog to the two-dimensional case in which both $f'(x_0) = 0$ and $f''(x_0) = 0$. The multivariable Taylor's series in (3.29) indicates that higher-order differentials would be needed in the case where both $df^* = 0$ and $d^2f^* = 0$. However, the structures of $d^3f, d^4f, \ldots, d^nf$ are complex, and we will not pursue this algebra any further.

3.5.3 Convex Sets

The adjective *convex*, applied to a set of points instead of a function, is intuitively very simple—a set is convex if and only if the straight line joining any two points in the set lies entirely in the set also. Put formally, a set of points, S, is convex if and only if, for any two points $\mathbf{X}'$ and $\mathbf{X}''$ that are in the set S, the convex combination

$$\alpha\mathbf{X}' + (1 - \alpha)\mathbf{X}'' \tag{3.37}$$

is also in the set S, for all α in the interval $0 \le \alpha \le 1$. Figure 3.23 illustrates several convex and several nonconvex sets of two variables (in the x_1, x_2 plane).

Intuitively, a nonconvex set has dents or holes in it. The *intersection* of two (or more) convex sets—points simultaneously included in both (or all) sets—is also a convex set. The intersection of two or more *nonconvex* sets may be a convex set and it may be a nonconvex set (Figure 3.24).

We will not consider the influence of constraints on maximization and minimization problems until Chapter 4, but it is useful at this point, when dealing with definitions of convexity and concavity for $f(\mathbf{X})$, to note some ways in which convex sets of points are created from conditions imposed on $\mathbf{X} = [x_1, \ldots, x_n]'$ by equations and inequalities.

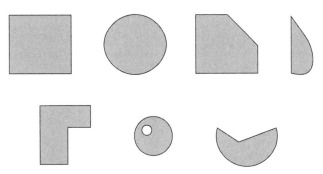

FIGURE 3.23   Convex sets (top); nonconvex sets (bottom).

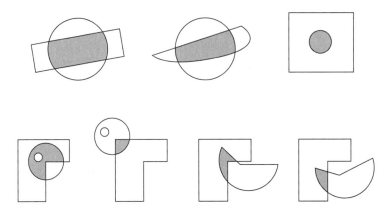

FIGURE 3.24 Intersections of convex sets (top); intersections of nonconvex sets (bottom).

Linear Equations Defining Convex Sets In two-dimensional x_1x_2 space the general representation of a linear equation is $a_1x_1 + a_2x_2 = b$; a linear equation of three variables, $a_1x_1 + a_2x_2 + a_3x_3 = b$, describes a plane in three-dimensional space. In more dimensions we have a hyperplane: $a_1x_1 + a_2x_2 + \cdots + a_nx_n = b$. Let $\mathbf{A}$ be the n-element row vector $[a_1, \ldots, a_n]$ and $\mathbf{X} = [x_1, \ldots, x_n]'$; then a hyperplane in n-dimensional space is defined[22] as $\mathbf{AX} = b$. When $n = 2$ or 3, we have a line or a plane. It is easy to show, using the fundamental definition of convexity, above, that a hyperplane is a convex set of points.

Let $\mathbf{X}' = [x_1', \ldots, x_n']'$ and $\mathbf{X}'' = [x_1'', \ldots, x_n'']'$ be *any* two (distinct) points that lie on the hyperplane, so that $\mathbf{AX}' = b$ and $\mathbf{AX}'' = b$. Define $\mathbf{X}'''$ as a convex combination of these two points; $\mathbf{X}''' = \alpha\mathbf{X}'' + (1-\alpha)\mathbf{X}'$. We want to show that $\mathbf{X}'''$ is also a point on the hyperplane, for any α $(0 \le \alpha \le 1)$. Geometrically, we want to prove that for *any* pair of points, $\mathbf{X}'$ and $\mathbf{X}''$, on the plane, all points on the line connecting $\mathbf{X}'$ and $\mathbf{X}''$ lie on the plane also. Since α is a scalar,

$$\mathbf{AX}''' = \mathbf{A}[\alpha\mathbf{X}'' + (1 - \alpha)\mathbf{X}']$$
$$= \alpha\mathbf{AX}'' + (1 - \alpha)\mathbf{AX}' = \alpha b + (1 - \alpha)b = b$$

For the $n = 2$ case, the interpretation is very simple. The line connecting any two points on a specific line is also part of that line. For the $n = 3$ case, the interpretation is equally simple. All points on the line joining any two points on a particular plane must also lie on that plane.

Thus, a hyperplane defines a convex set of points; also, points that simultaneously satisfy several hyperplanes (their intersection) form a convex set. As examples, two planes in three dimensions will, in general, intersect in a line (a convex set). If the planes are coplanar, their "intersection" is just the plane itself (also a convex set). If they are parallel and not coplanar, there is no intersection at all (no values of the x's can simultaneously satisfy both planes).

---

[22]Note that $\mathbf{A}$ and $\mathbf{X}$ are vectors, hence their product is the scalar b. This is to be distinguished from the matrix representation of a *set* of linear equations—that is, a *set* of hyperplanes—$\mathbf{AX} = \mathbf{B}$, where $\mathbf{A}$ is a *matrix* and $\mathbf{B}$ a vector. In Chapter 2 we saw that such a set of equations has a unique solution if and only if the hyperplanes have one common point.

It is also easy to recognize that *nonlinear* equations do not define convex sets of points. If $g(x_1, x_2) = b$ is a *curve* in x_1x_2 space, then the convex combination of any two points on the curve (the straight line connecting the points) will certainly not, generally, be on the curve.

Inequalities Defining Convex Sets

Linear Inequalities   In two dimensions $a_1x_1 + a_2x_2 \leq b$ defines a line and all points on one side of it. For example, in Figure 3.25, the shaded area in x_1x_2 space represents part of the area defined by $2x_1 + x_2 \leq 10$. This area defined by the inequality is called a *half-space*. When the inequality is weak, the half-space is termed *closed*; this simply means that the boundary (here, the line $a_1x_1 + a_2x_2 = b$) is included. If the inequality is strong, the corresponding half-space is *open*. Note that *open* and *closed* refer to whether the bounding function is included; the terms say nothing about the *extent* of the region. The closed half-space shown in Figure 3.25 is said to be *unbounded* because it extends infinitely far downward and to the left.

Half-spaces (either open or closed) are also convex sets. Again, using $\mathbf{A} = [a_1, \ldots, a_n]$ and $\mathbf{X} = [x_1, \ldots, x_n]'$, a linear inequality can be represented as $\mathbf{AX} \leq b$, or, equivalently, as $\mathbf{AX} - b = 0$. Also using $\mathbf{X}'$ and $\mathbf{X}''$ as before, let $\mathbf{AX}' \leq b$, $\mathbf{AX}'' \leq b$, and $\mathbf{AX}'' \leq \mathbf{AX}'$. (This is perfectly general; if $\mathbf{AX}' \neq \mathbf{AX}''$, the points can always be labelled so that $\mathbf{AX}' < \mathbf{AX}''$.) Again form the convex combination $\mathbf{X}''' = \alpha\mathbf{X}'' + (1 - \alpha)\mathbf{X}'$ ($0 \leq \alpha \leq 1$). We want to show that $\mathbf{AX}''' \leq b$. The approach is similar to that above, for linear equations:

$$\mathbf{AX}''' = \mathbf{A}[\alpha\mathbf{X}'' + (1 - \alpha)\mathbf{X}'] = \alpha\mathbf{AX}'' + (1 - \alpha)\mathbf{AX}'$$

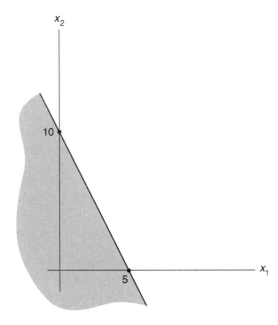

FIGURE 3.25 $2x_1 + x_2 \leq 10$.

Since $\mathbf{AX}'' \leq \mathbf{AX}' \leq b$, replace $\mathbf{AX}''$ with $\mathbf{AX}'$, giving

$$\mathbf{AX}''' \leq \alpha\mathbf{AX}' + (1-\alpha)\mathbf{AX}' = \mathbf{AX}'$$

so $\mathbf{AX}''' \leq \mathbf{AX}'$ and since $\mathbf{AX}' \leq b$, then

$$\mathbf{AX}''' \leq b$$

This will be true for any convex combination of any two $\mathbf{X}'$ and $\mathbf{X}''$ that satisfy the linear inequality. Therefore, also, points that satisfy several linear inequalities simultaneously (their intersection) form a convex set.

Nonlinear Inequalities   Consider an inequality $g(x_1, \ldots, x_n) \leq b$. It is generally useful to write this as

$$h(x_1, \ldots, x_n) \leq 0$$

where

$$h(x_1, \ldots, x_n) \equiv g(x_1, \ldots, x_n) - b$$

just as in the linear case where $h(\mathbf{X}) = \mathbf{AX} - b \leq 0$. We now examine the set of points defined by the inequality $h(\mathbf{X}) \leq 0$ for the particular example in which $h(\mathbf{X})$ is a nonlinear and *convex* function.

As in the case of *linear* inequalities, let $\mathbf{X}' = [x_1', \ldots, x_n']'$ and $\mathbf{X}'' = [x_1'', \ldots, x_n'']'$ be two points that satisfy the inequality; that is, $h(\mathbf{X}') \leq 0$ and $h(\mathbf{X}'') \leq 0$. Furthermore, let $h(\mathbf{X}'') \leq h(\mathbf{X}')$. (This also is completely general; the two points can be labeled so that it is always true.) Consider the convex combination of $\mathbf{X}'$ and $\mathbf{X}''$, $\mathbf{X}''' = \alpha\mathbf{X}'' + (1-\alpha)\mathbf{X}'$, for $0 \leq \alpha \leq 1$. We now want to show that $\mathbf{X}'''$ also satisfies the inequality, namely that $h(\mathbf{X}''') \leq 0$. Here is where convexity of $h(\mathbf{X})$ comes in. Since $h(\mathbf{X})$ is a convex function, we know [from the basic definition of convexity in (3.33)] that

$$h(\mathbf{X}''') = h[\alpha(\mathbf{X}'') + (1-\alpha)\mathbf{X}'] \leq \alpha h(\mathbf{X}'') + (1-\alpha)h(\mathbf{X}')$$

However, since $h(\mathbf{X}'') \leq h(\mathbf{X}') \leq 0$,

$$\alpha h(\mathbf{X}'') + (1-\alpha)h(\mathbf{X}') \leq \alpha h(\mathbf{X}') + (1-\alpha)h(\mathbf{X}') = h(\mathbf{X}') \leq 0$$

and hence $h(\mathbf{X}''') \leq h(\mathbf{X}') \leq 0$, so $h(\mathbf{X}''') \leq 0$, which is what we wanted to show.

Therefore, points satisfying the inequality defined by the *convex* function $h(\mathbf{X}) \leq 0$ constitute a convex set. The same would be true of points that simultaneously satisfy several such constraints, $h^1(\mathbf{X}) \leq 0, \ldots, h^m(\mathbf{X}) \leq 0$, provided all $h^i(\mathbf{X})$ are convex (again, because the intersection of convex sets is itself a convex set).[23]

If $h(\mathbf{X})$ is a *concave* function, the area defined by $h(\mathbf{X}) \geq 0$ can be shown also to define a convex set, by an argument parallel to the one above. [Alternatively, the negative of a

---

[23]We use $h^i(\mathbf{X})$ rather than $h_i(\mathbf{X})$ to designate the ith constraint because the latter is used to indicate the partial derivative of a function $h(x_1, \ldots, x_n)$ with respect to x_i.

convex function is concave; so if $h(\mathbf{X})$ is convex and $h(\mathbf{X}) \leq 0$ defines a convex set, then (multiplying by -1) $-h(\mathbf{X}) \geq 0$ also defines a convex set, and $-h(\mathbf{X})$ is concave.] The same would be true for the set of points that simultaneously satisfy several constraints $h^i(\mathbf{X}) \geq 0$, provided all $h^i(\mathbf{X})$ are concave.

In summary:

1. If $h(\mathbf{X})$ is convex, $h(\mathbf{X}) \leq 0$ defines a convex set.

2. If all of several $h^i(\mathbf{X})$ are convex ($i = 1, \ldots, m$), and if there are points that satisfy all $h^i(\mathbf{X}) \leq 0$ simultaneously, these points are a convex set.

3. If $h(\mathbf{X})$ is concave, $h(\mathbf{X}) \geq 0$ defines a convex set.

4. If all of several $h^i(\mathbf{X})$ are concave ($i = 1, \ldots, m$), and if there are points that satisfy all $h^i(\mathbf{X}) \geq 0$ simultaneously, these points are a convex set.

3.5.4 Numerical Illustrations for the $f(x_1,x_2)$ Case

We first examine two "well behaved" examples that satisfy the sufficient conditions in Table 3.5. These are cases for which $|\mathbf{H_2}^*| \, (= |\mathbf{H}^*| = [f^*_{11}f^*_{22} - (f^*_{12})^2]) > 0$. Then we will explore the kinds of shapes that $f(\mathbf{X})$ may have when $|\mathbf{H}^*| = 0$ or when $|\mathbf{H}^*| < 0$, which means for cases in which the sufficient conditions in that table are not met.

Two Well-Behaved Cases

(a) $f(\mathbf{X}) = (x_1 - 1)^2 + (x_2 - 2)^2$ (Fig. 3.26). This is a bowl with an unambiguous minimum at $x_1^* = 1$, $x_2^* = 2$. The stationary point is easily identified from $\nabla f(\mathbf{X}) = \begin{bmatrix} 2(x_1 - 1) \\ 2(x_2 - 2) \end{bmatrix}$ $= \mathbf{0}$. The fact that it is a minimum follows from examination of the leading principal minors of the Hessian there. For this function, $\mathbf{H} = \begin{bmatrix} 2 & 0 \\ 0 & 2 \end{bmatrix}$, so $|\mathbf{H_1}| = 2$ and $|\mathbf{H_2}| \, (= |\mathbf{H}|) = 4$, and the *sufficient* conditions for a minimum point (Table 3.5) are met. In fact, $f(\mathbf{X})$ is con-

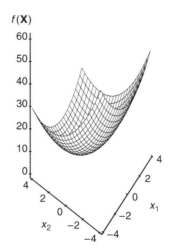

$f(\mathbf{X})$

FIGURE 3.26 $f(\mathbf{X}) = (x_1 - 1)^2 + (x_2 - 2)^2$.

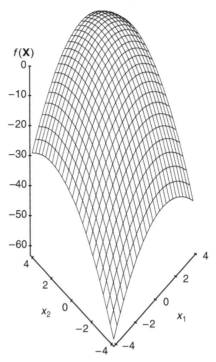

FIGURE 3.27 $f(\mathbf{X}) = -(x_1 - 1)^2 - (x_2 - 2)^2$.

vex everywhere, since the elements in **H** are independent of **X** (which is why stars are not needed on $|\mathbf{H}_1|$ and $|\mathbf{H}_2|$).

(b) $f(\mathbf{X}) = -(x_1 - 1)^2 - (x_2 - 2)^2$ (Fig. 3.27). Since this is just the negative of the previous function, it is not surprising to find that it looks like an overturned bowl. From $\nabla f = \mathbf{0}$ conditions, we find again that $\mathbf{X}^* = \begin{bmatrix} 1 \\ 2 \end{bmatrix}$ is a stationary point. The Hessian is now **H** $\begin{bmatrix} -2 & 0 \\ 0 & -2 \end{bmatrix}$, and it is easily established that the leading principal minors alternate, beginning negative. This identifies $\mathbf{X}^*$ as a maximizing point.

Some Cases that Are Not So Well-Behaved: $|\mathbf{H}^*| = 0$

(c) $f(\mathbf{X}) = (x_1 - 1)^4 + (x_2 - 2)^4$ (Fig. 3.28). This is again a bowl, although it has more steeply sloped sides than the one in Figure 3.26; also, $f(\mathbf{X})$ is very flat near to the minimum point.[24] Here $\nabla f = \begin{bmatrix} 4(x_1 - 1)^3 \\ 4(x_2 - 2)^3 \end{bmatrix}$ and $\nabla f = \mathbf{0}$ leads again to $\mathbf{X}^* = \begin{bmatrix} 1 \\ 2 \end{bmatrix}$. Now, however

$$\mathbf{H} = \begin{bmatrix} 12(x_1 - 1)^2 & 0 \\ 0 & 12(x_2 - 2)^2 \end{bmatrix}$$

[24]For this and the figures that follow, sometimes the scale and/or the orientation has been altered to emphasize the particular shape of the function.

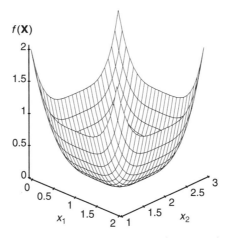

FIGURE 3.28 $f(\mathbf{X}) = (x_1 - 4)^4 + (x_2 - 2)^4$.

so that $\mathbf{H}^* = \begin{bmatrix} 0 & 0 \\ 0 & 0 \end{bmatrix}$, leading to $|\mathcal{H}_1^*(1)| (= |\mathbf{H}_1^*|) = 0$, $|\mathcal{H}_1^*(2)| = 0$ and $|\mathcal{H}_2^*| (= |\mathbf{H}_2^*| = |\mathbf{H}^*|) = 0$. Thus this function at $\mathbf{X}^*$ satisfies both first- and second-order *necessary* conditions for *either* a maximum or a minimum point (Table 3.5), even though the geometry shows that $f(\mathbf{X})$ is everywhere convex and that $\mathbf{X}^*$ is an unambiguous minimum point.

(d) $f(\mathbf{X}) = -(x_1 - 1)^4 - (x_2 - 2)^4$ (Fig. 3.29). This is just the negative of the previous function. Again, as you can easily establish, $\nabla f = \mathbf{0}$ leads to $\mathbf{X}^* = \begin{bmatrix} 1 \\ 2 \end{bmatrix}$, and the principal minors of the Hessian matrix are all equal to zero, so the fact that the function is everywhere concave and has a minimum at $\mathbf{X}^*$ cannot be established by the rules in Table 3.5.

(e) $f(\mathbf{X}) = x_1^3 + x_2^3$ (Fig. 3.30). Here $\nabla f = \begin{bmatrix} 3x_1^2 \\ 3x_2^2 \end{bmatrix}$, so $\mathbf{X}^* = \begin{bmatrix} 0 \\ 0 \end{bmatrix}$. The Hessian in this case is $\mathbf{H} = \begin{bmatrix} 6x_1 & 0 \\ 0 & 6x_2 \end{bmatrix}$ and so, again, all principal minors evaluated at $\mathbf{X}^*$ are zero. As seen in

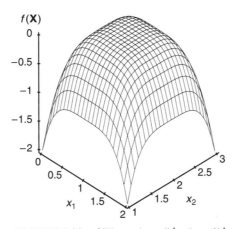

FIGURE 3.29 $f(\mathbf{X}) = -(x_1 - 4)^4 - (x_2 - 2)^4$.

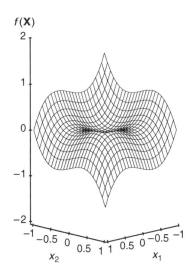

FIGURE 3.30 $f(\mathbf{X}) = x_1^3 + x_2^3$.

the figure, this function has a three-dimensional analog to a *point of inflection* at the origin.

(f) $f(\mathbf{X}) = -x_1^2 + 2x_1x_2 - x_2^2$. Here $\nabla f = \begin{bmatrix} -2x_1 + 2x_2 \\ 2x_1 - 2x_2 \end{bmatrix}$, and imposing the $\nabla f = \mathbf{0}$ requirement generates a pair of linear homogeneous equations with a singular coefficient matrix; hence there are multiple solutions—any values of x_1 and x_2 along the line $x_1 = x_2$. The Hessian matrix here is $\mathbf{H} = \begin{bmatrix} -2 & 2 \\ 2 & -2 \end{bmatrix}$, so, while $|\mathbf{H}_1| < 0$ (the first *sufficient* condition for a maximum), $|\mathbf{H}_2| = 0$. The function everywhere satisfies only the first- and second-order *necessary* conditions for either a maximum or a minimum. This is illustrated in Fig. 3.31. The function has a "ridge line" along $x_1 = x_2$ in x_1, x_2 space. In this case, that line represents the top of the function (a canopy).

(g) $f(\mathbf{X}) = x_1^2 + 2x_1x_2 + x_2^2$. This is similar to the function in (f), above, only the signs on the two squared terms have been changed. The gradient is similar to that in the previous example, namely, $\nabla f = \begin{bmatrix} 2x_1 + 2x_2 \\ 2x_1 + 2x_2 \end{bmatrix}$, and the $\nabla f = \mathbf{0}$ condition is again a pair of linear homogeneous equations with a singular coefficient matrix. Again there are multiple solutions, this time any values for which $x_1 = -x_2$. In this case, $\mathbf{H} = \begin{bmatrix} 2 & 2 \\ 2 & 2 \end{bmatrix}$, so $|\mathbf{H}_1| > 0$ but, again, $|\mathbf{H}_2| = 0$. Now there is a ridge line along the line $x_1 = -x_2$ in x_1, x_2 space. In this case it represents the bottom of a trough (see Fig. 3.32).

A Case that Is Not So Well-Behaved: $|\mathbf{H}^*| < 0$ When $|\mathbf{H}^*| < 0$, this means that $f^*_{11} f^*_{22} < (f^*_{12})^2$, and the two second partial derivatives need not be of the same sign. Suppose that they are not; for example, $f^*_{11} > 0$ and $f^*_{22} < 0$. Then if we examine the curvature of $f(\mathbf{X})$ in a plane perpendicular to the x_2 axis, at x_2^*, we find that the slope of the tangent plane is increasing ($f^*_{11} > 0$). We conclude that $f(\mathbf{X})$ was at a *minimum* in the x_1, y plane at $\mathbf{X}^*$. Similarly, $f^*_{22} < 0$ suggests a *maximum* of $f(\mathbf{X})$ in the x_2, y plane at $x_1 = x_1^*$. Thus $f(\mathbf{X})$ has a

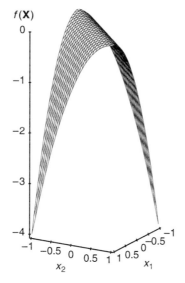

FIGURE 3.31 $f(\mathbf{X}) = -x_1^2 + 2x_1x_2 - x_2^2$.

maximum from one point of view and a minimum from another point of view. This is what is known as a *saddle point* at $\mathbf{X}^*$. Here is an example.

(h) $f(\mathbf{X}) = 1 + x_1^2 - x_2^2$. In this case, $\nabla f = \begin{bmatrix} 2x_1 \\ -2x_2 \end{bmatrix}$, leading to $\mathbf{X}^* = \begin{bmatrix} 0 \\ 0 \end{bmatrix}$. Here $\mathbf{H} = \begin{bmatrix} 2 & 0 \\ 0 & -2 \end{bmatrix}$, so $|\mathbf{H}_1| = 2$ and $|\mathbf{H}_2| = -4$, satisfying neither the second-order *necessary* con-

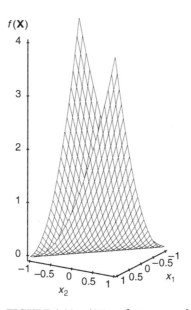

FIGURE 3.32 $f(\mathbf{X}) = x_1^2 + 2x_1x_2 + x_2^2$.

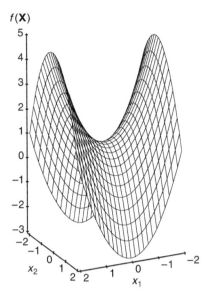

FIGURE 3.33 $f(\mathbf{X}) = 1 + x_1^2 - x_2^2$.

ditions nor the *sufficient* conditions for either a maximum or for a minimum. Figure 3.33 illustrates the saddle shape for this particular function.

3.5.5 Quasiconcave and Quasiconvex Functions $f(x_1, x_2)$

For the case of $f(x)$, we saw in Figure 3.18c that a function with a bell shape (concave over a part of its range and convex over other parts) could also have a global maximum at a point where $f'(x) = 0$. That represented a relaxation of the more strict requirements of concavity (for a global maximum); it was an illustration of a (strictly) quasiconcave function of one variable. The analog to that picture for $f(x_1, x_2)$ is a bell in three dimensions. Figure 3.34a illustrates a strictly quasiconcave function (a bell) with a global maximum, and Figure 3.34b shows a strictly quasiconvex function (an upside-down bell) with a global minimum. Mathematical descriptions of quasiconcavity and quasiconvexity for functions of more than one variable are explored in the second part of Appendix 3.1.

Identification of quasiconcavity and quasiconvexity turns out to be particularly useful, and sometimes easy (from the context of the problem), for functions in nonlinear programming problems. This broadens the classes of functions for which relative maxima or minima will be absolute and unique maxima or minima. We explore this further in Part V.

3.6 MAXIMA AND MINIMA FOR FUNCTIONS OF n INDEPENDENT VARIABLES

We now use the discussion of quadratic forms and their signs to generalize results on maxima and minima for functions of any number of independent variables, $f(\mathbf{X})$, where $\mathbf{X} = [x_1, \ldots, x_n]'$. A point $\mathbf{X}^*$ will be a relative maximum or a relative minimum for $f(\mathbf{X})$ according to the rules given in Table 3.6. This is virtually identical to Table 3.5 for $f(x_1, x_2)$;

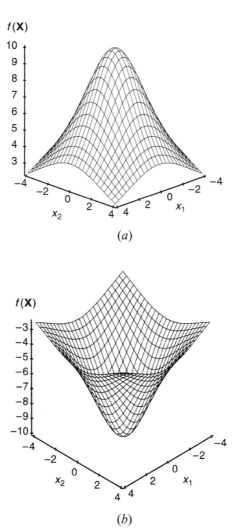

FIGURE 3.34   (a) Strict quasiconcavity $[f(\mathbf{X}) = 100/(x_1^2 + x_2^2 + 10)]$; (b) strict quasiconvexity $[f(\mathbf{X}) = -100/(x_1^2 + x_2^2 + 10)]$.

all that is required is a thorough understanding of the structure of df and d^2f for a function of n variables.

3.6.1 First-Order Conditions

The n-variable analog to df for $f(x_1,x_2)$ in (3.24a) is

$$df = f_1 dx_1 + f_2 dx_2 + \cdots + f_n dx_n \tag{3.38a}$$

Exactly as before, $df = 0$ for independent small changes in $dx_1, dx_2, \ldots, dx_n$ (not all zero) if and only if $f_1 = f_2 = \cdots = f_n = 0$. This is then the first-order condition that any stationary point for a function of n variables must satisfy.

TABLE 3.6   Characterization of Maxima and Minima for $f(x_1, \ldots, x_n)$ at X*

	Maximum	Minimum
First-order necessary condition	$df(\mathbf{X}^*) = 0$ or, using gradients $\nabla f(\mathbf{X}^*) = \mathbf{0}$	
Second-order necessary condition	$d^2f(\mathbf{X}^*) \leq 0$	$d^2f(\mathbf{X}^*) \geq 0$
	or, using *principal minors* of the Hessian	
	All $\|\mathcal{H}_i^*\| \leq 0$ (for i odd) or all $\|\mathcal{H}_i^*\| \geq 0$ (for i even)	All $\|\mathcal{H}_i^*\| \geq 0$
Sufficient conditions	$df(\mathbf{X}^*) = 0$ or, using gradients $\nabla f(\mathbf{X}^*) = \mathbf{0}$ and	
	$d^2f(\mathbf{X}^*) < 0$	$d^2f(\mathbf{X}^*) > 0$
	or, using *leading principal minors* of the Hessian	
	$\|\mathbf{H}_i^*\|$ alternate in sign, beginning negative	All $\|\mathbf{H}_i^*\| > 0$

Using $\nabla f(\mathbf{X}) = [f_1, f_2, \ldots, f_n]'$ and $\mathbf{dX} = [dx_1, dx_2, \ldots, dx_n]'$, (3.38a) can be expressed as

$$df = [\nabla f]'[\mathbf{dX}] \tag{3.38b}$$

or

$$df = [\nabla f] \circ [\mathbf{dX}] \tag{3.38c}$$

exactly as in (3.24b) and (3.24c) for the two-variable case. The first-order condition for a maximum or a minimum is again $\nabla f(\mathbf{X}) = \mathbf{0}$.

3.6.2 Second-Order Conditions

The second total differential for $f(\mathbf{X})$ is exactly as in (3.28) for $f(x_1, x_2)$:

$$d^2f = [\mathbf{dX}]'\mathbf{H}[\mathbf{dX}] \tag{3.39}$$

The only difference is that now $\mathbf{dX}$ is an n-element vector and the Hessian is the $n \times n$ matrix

$$\mathbf{H} = \begin{bmatrix} f_{11} & f_{12} & \cdots & f_{1n} \\ f_{21} & f_{22} & \cdots & f_{2n} \\ \vdots & \vdots & & \vdots \\ f_{n1} & f_{n2} & \cdots & f_{nn} \end{bmatrix} \tag{3.40}$$

Rules for establishing positive or negative definiteness—or positive or negative semidefiniteness—of d^2f in (3.39) at a candidate point $\mathbf{X}^* = [x_1^*, \ldots, x_n^*]'$ depend again on the rules laid out in (1.62a) (Chapter 1, Section 1.6). For $d^2f^* < 0$ ($\mathbf{X}^*$ is a maximum point), we need $|\mathbf{H}_1^*| < 0$, $|\mathbf{H}_2^*| > 0$, $|\mathbf{H}_3^*| < 0$, ..., $|\mathbf{H}_n^*| < 0$ if n is odd and $|\mathbf{H}_n^*| > 0$ if n is even. And for $d^2f^* > 0$ ($\mathbf{X}^*$ identifies a minimum) we need $|\mathbf{H}_1^*| > 0$, $|\mathbf{H}_2^*| > 0$, $|\mathbf{H}_3^*| > 0$, ..., $|\mathbf{H}_n^*| > 0$. In either case, it may be necessary to evaluate all n determinants—the first is simply the element f_{11}^*, but the last is the determinant of the entire $n \times n$ matrix $\mathbf{H}^*$.

Positive or negative *semi*definiteness of the quadratic form in (3.39)—the second-order necessary conditions for either a maximum or a minimum—are established via sign requirements on all principal minors of $\mathbf{H}^*$ [as in (1.62b) in Chapter 1]. The total number of principal minors to be evaluated increases quickly with the size of n. When $n = 2$, there are only three principal minors (as indicated in Table 3.5). For $n = 5$, the number is increased to 31; for $n = 8$ it is 255. At some point, these conditions become relatively difficult if not impractical to apply.[25]

The rules for maxima and minima of functions of one or of two variables can now be seen to be special cases of the general results of this section. And again, if concavity or convexity of the function is known (e.g., from the context of the problem), then the first-order necessary condition, $\nabla f = \mathbf{0}$, is also sufficient to identify a maximum point (concave function) or a minimum point (convex function), which will be an absolute maximum or minimum if $f(\mathbf{X})$ is strictly concave or strictly convex. If the function is known to be quasiconcave or quasiconvex, local maxima or minima are also global maxima or minima, respectively. Definitions of quasiconcavity and quasiconvexity in the multivariable case are in the second part of Appendix 3.1.

3.6.3 Example

We examine the stationary points of

$$f(x_1, x_2, x_3) = 5x_1^2 + 2x_2^2 + x_3^4 - 32x_3 + 6x_1x_2 + 5x_2$$

for maxima and minima. From $\nabla f = \mathbf{0}$ we have

$$10x_1 + 6x_2 = 0$$
$$6x_1 + 4x_2 = -5 \tag{3.41}$$
$$4x_3^3 - 32 = 0$$

The first two (linear) equations are easily solved for x_1 and x_2, and the third equation contains only information about x_3; $\mathbf{X}^* = \begin{bmatrix} 7.5 \\ -12.5 \\ 2 \end{bmatrix}$ satisfies the first-order conditions in (3.41).

The Hessian matrix for this example is

$$\mathbf{H} = \begin{bmatrix} 10 & 6 & 0 \\ 6 & 4 & 0 \\ 0 & 0 & 12x_3^2 \end{bmatrix}$$

---

[25]The number of principal minors is easily derived from the combinatorial formula, $C_r^n = n!/r!(n-r)!$, for counting the number of ways of selecting r items (rows and columns to be dropped) from n (total number of rows and columns). For any n, the sum of all of these possibilities, from $r = 1$ to $r = n$, is $2^n - 1$.

The relevant leading principal minors are $|\mathbf{H}_1|$ $(= |\mathbf{H}_1^*|) = 10 > 0$, $|\mathbf{H}_2|$ $(= |\mathbf{H}_2^*|) = 4 > 0$, and $|\mathbf{H}_3| = 4(12x_3^2)$. Since $x_3^* = 2$, $|\mathbf{H}_3^*| = 192 > 0$, so the leading principal minors are all positive; therefore $d^2 f^* > 0$ and $\mathbf{X}^*$ represents a minimum point.

Since $d^2 f^* > 0$, it is clear that the second-order necessary conditions for a minimum are met. It is also clear that there is no need to examine those conditions in this case, since positive definiteness has been easily established using the *leading* principal minors of $\mathbf{H}$. For practice, however, you might find (1) the three first-order principal minors, $|\mathcal{H}_1^*(1)|$, $|\mathcal{H}_1^*(2)|$, $|\mathcal{H}_1^*(3)|$, (2) the three second-order principal minors, $|\mathcal{H}_2^*(1)|$, $|\mathcal{H}_2^*(2)|$, $|\mathcal{H}_2^*(3)|$; and (3) $|\mathcal{H}_3^*|$ $(= |\mathbf{H}^*|)$—they are all positive.

3.7 SUMMARY

This chapter has provided a review of the way in which differential calculus is used in maximization or minimization problems for functions of first one, then two, and finally n (independent) variables. The approach via differentials was introduced and shown to parallel that using derivatives. This offers no particular advantage in one- and two-variable cases; however, it becomes essential for a discussion of maxima and minima for functions of three or more variables. It also provides the framework within which to study the more realistic cases in which not only are several variables involved, but constraints or restrictions apply to those variables so that they no longer can be regarded as independent of one another; for example, it might be that x_1 and x_2 must sum to a given constant, so if x_1 increases, then x_2 must decrease. We explore the ramifications of such restrictions in some detail in the next chapter.

Using the calculus-based (classical) approach, we are able to investigate functions of any number of independent variables for their maximum and minimum points. The steps are as follows: (1) find the (one or more) $\mathbf{X}^*$ that satisfy $\nabla f = \mathbf{0}$, and (2) evaluate selected minors of $\mathbf{H}$ at (the one or more) $\mathbf{X}^*$, unless concavity or convexity of $f(\mathbf{X})$ can be established by other means—for example, if a cost function (to be minimized) is known to be "U-shaped" (convex).

For complicated functions of many variables, neither of these steps will likely be easy. The set of equations generated by $\nabla f = \mathbf{0}$ may be highly nonlinear. The second partials and cross-partials that make up the Hessian may themselves be complicated functions of $\mathbf{X}$, and evaluation of each of these elements for a large Hessian (large n) and for several candidate points $\mathbf{X}^*$ can be tedious. Once the elements of $\mathbf{H}^*$ are in place, evaluation of a number of principal minors can also be a burdensome task. For reasons such as these, actual computational work for solving maximization and minimization problems often takes the form of *numerical* (or *iterative*) methods, either for solving the system of equations $\nabla f = \mathbf{0}$ when the equations are nonlinear or for searching for maxima or minima directly, bypassing the classical first- and second-order conditions. These are the topics of Chapters 5 and 6, respectively.

APPENDIX 3.1 QUASICONCAVE AND QUASICONVEX FUNCTIONS

Both curves in Figure 3.1.1 have a unique maximum at x^*, where $f'(x) = 0$, and if the functions are defined only over the set $S = \{x | c \le x \le d\}$, then this is a unique global maximum for each of the functions. Function I satisfies either of the standard geometric representations of *strict concavity* over the set; it is everywhere above its secants or below

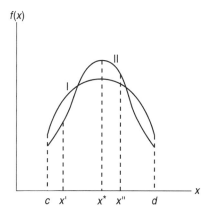

FIGURE 3.1.1   Two functions with a maximum at x^*.

its tangents. Function II clearly does not, since, for example, it is above its tangent at x' and below its tangent at x''. It is an example of a function that is *strictly quasiconcave*.

It is therefore not surprising that variations have been developed on the notions of convexity and concavity that are more relaxed and embrace a wider class of functions. We explore the most prevalent of these relaxations, those of quasiconcavity and quasiconvexity, in this appendix.

For each of the two functions in Figure 3.1.1, the necessary condition $f'(x) = 0$ is also sufficient to identify the global and unique maximum at x^* because of the shapes of the functions (at that point and elsewhere in the interval).[26] In numerical methods for finding maxima and minima (Part III), it is very important to know not only whether a local extreme point found via such methods represents a maximum or a minimum point but also whether the extremum is also a global one and if it is unique. Knowing the shape of $f(x)$ in advance helps with answers to these kinds of questions.

These relaxed types of convexity and concavity have found particular application in economic analysis, where it is often realistic to assume for example that production functions of firms, utility functions of individuals, and so on, satisfy shape requirements that are less exacting than ordinary convexity or concavity. [Examples can be found in, among others, Arrow and Enthoven (1961), Takayama (1985), and Beavis and Dobbs (1990).] In most of these cases, the functions in question depend on more than one variable. However, to initiate the ideas, we look at $f(x)$, where the geometry is particularly simple.

3A.1.1   Quasiconcavity and Quasiconvexity of $f(x)$

Definitions   The first definitions do not require derivatives and are rather similar to (3.9) and (3.10) in the text.[27]

---

[26]Remember that "global" means over the examined interval—in this case $[c,d]$. If a function is known to be strictly concave or strictly quasiconcave *everywhere*, then "global" means over the entire range of x, $-\infty \le x \le +\infty$.

[27]These definitions can be found in Takayama (1985). There are several slight variations in definitions of these and other kinds of alternative concavities and convexities in the literature. For example, see Martos (1965, which includes "quasimonotonic" functions), Bazaraa and Shetty (1979, "strict" and "strong" quasiconcavity or quasiconvexity), Ponstein (1967, "pseudo concavity or convexity"), Zoutendijk (1976, "monotonic strict quasiconcavity/convexity"). Mangasarian (1969) gives a very complete account.

Definition 1   Let $f(x)$ be defined over a convex set of points, S, and consider any *distinct* pair of points, x' and x'', in S. Suppose $f(x'') \geq f(x')$; then $f(x)$ is defined to be *quasiconcave* if, for α in the interval ($0 \leq \alpha \leq 1$):

$$f(x'') \geq f(x') \quad \text{implies} \quad f[\alpha x'' + (1 - \alpha)x'] \geq f(x') \tag{3.1.1}$$

This is sometimes written as

$$f[\alpha x'' + (1 - \alpha)x'] \geq \min[f(x'), f(x'')]$$

where $\min[f(x'), f(x'')]$ indicates the smaller of the two values inside brackets.

Further, $f(x)$ is defined to be *strictly quasiconcave* if the right-hand side inequality in (3.1.1) is strong and for ($0 < \alpha < 1$); $f(x)$ is strictly quasiconcave if

$$f(x'') \geq f(x') \quad \text{implies} \quad f[\alpha x'' + (1 - \alpha)x'] > f(x') \tag{3.1.2}$$

Geometrically, the definition in (3.1.1) says that for a quasiconcave function the value of $f(x)$ evaluated at any point on the line connecting x' and x'' is *no smaller* than $f(x')$—the function never dips below the lower of its two endpoints, for *any* choice of x' and x'' in S. Strict quasiconcavity, as defined in (3.1.2), requires that $f(x)$ evaluated along the line between any x' and x'' in S be *larger* than $f(x')$. This rules out any horizontal segments, which are allowed by (3.1.1).[28]

Figure 3.1.2 illustrates some possibilities for quasiconcave and strictly quasiconcave functions. The function in Figure 3.1.2a is strictly quasiconcave. At a point like x^*, where $f'(x^*) = 0$, the function has a unique global maximum. The function in Figure 3.1.2b is also strictly quasiconcave, but there is an inflection point with $f'(x^*) = 0$ at x^*. For a function like this, setting $f'(x) = 0$ will identify both x^* (not a maximum point) and x^{**} (the unique global maximum). Some of the definitions of additional kinds of quasiconvexity have been designed to rule out the inflection point problem that is illustrated in Figure 3.1.2b).

In Figure 3.1.2c, $f(x)$ is quasiconcave (not strictly), since it is horizontal over $[x', x'']$. A maximum, as at x^*, will be unique and global, but $f(x)$ will also be unchanging in the range $[x', x'']$. Finally, $f(x)$ in Figure 3.1.2d is also only quasiconcave (not strictly), and in this case the function has a maximum for any value of x in the interval $[x', x'']$.[29] So if any maximum point is found, it is global but not unique. Note that for both quasiconcave and strictly quasiconcave functions, the *minimum* over the interval of definition will be at one end or the other.

---

[28]Choose x' and x'' so that $f(x') = f(x'')$. Then (3.1.1) can be satisfied even if $f[\alpha x'' + (1 - \alpha)x'] = f(x') = f(x'')$ for all ($0 \leq \alpha \leq 1$), which would mean that $f(x)$ is horizontal from x' to x''. In addition, $f(x)$ is said to be *explicitly quasiconcave* if the left-hand side inequality in (3.1.2) is also strong, again for $x' \neq x''$, and now for ($0 < \alpha \leq 1$); $f(x)$ is explicitly quasiconcave if

$$f(x'') > f(x') \quad \text{implies} \quad f[\alpha x'' + (1 - \alpha)x'] > f(x')$$

This requires that $f(x)$ evaluated along the line between any x' and x'' be *larger* than $f(x')$, as long as $f(x'') > f(x')$; this again rules out horizontal segments, *except* if they are at the top.

[29]While this function is continuous, it is not differentiable everywhere, since it has corners at x' and x''.

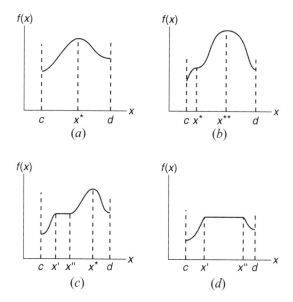

FIGURE 3.1.2   Quasiconcave functions over ($c \leq x \leq d$).

A function $f(x)$ is (strictly) *quasiconvex* if $-f(x)$ is (strictly) quasiconcave. Parallel algebraic statements require only that the final inequalities in (3.1.1) and (3.1.2) be reversed.[30] In particular, let $f(x)$ be defined over a convex set of points, S, and consider any *distinct* pair of points, x' and x'', in S. Suppose $f(x'') \geq f(x')$; then $f(x)$ is defined to be *quasiconvex* if, for α in the interval ($0 \leq \alpha \leq 1$)

$$f(x'') \geq f(x') \quad \text{implies} \quad f[\alpha x'' + (1-\alpha)x'] \leq f(x'') \qquad (3.1.3)$$

This is sometimes written as

$$f[\alpha x'' + (1-\alpha)x'] \leq \max[f(x'), f(x'')]$$

where $\max[f(x'), f(x'')]$ indicates the larger of the two values inside brackets.

Further, $f(x)$ is defined to be *strictly quasiconvex* if the right-hand side inequality in (3.1.3) is strong and for ($0 < \alpha < 1$); $f(x)$ is strictly quasiconvex if

$$f(x'') \geq f(x') \quad \text{implies} \quad f[\alpha x'' + (1-\alpha)x'] < f(x'') \qquad (3.1.4)$$

Geometrically, a quasiconvex function $f(x)$ evaluated at any point on the line connecting x' and x'' is *no larger* than $f(x'')$. The function never rises above the higher of the two endpoints over any interval defined by an x' and an x'' in S. It the function is strictly quasiconvex, it is *always below* that higher endpoint.

---

[30]As in an earlier footnote, a function is *explicitly quasiconvex* if the last inequality in that footnote is changed to "$< f(x'')$."

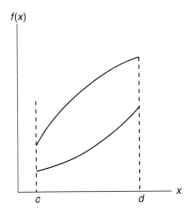

FIGURE 3.1.3   Two functions that are strictly quasiconcave and strictly quasiconvex over ($c \leq x \leq d$).

Concave functions always satisfy Definition 1 for quasiconcavity, and all strictly con-cave functions are strictly quasiconcave; furthermore, any (strictly) convex function is also (strictly) quasiconvex.[31]

Recall that a linear function is both concave and convex, but neither strictly so. Both functions in Figure 3.1.3 are strictly quasiconcave over $c \leq x \leq d$; they are everywhere above $f(x')$ in any $[x', x'']$ interval for which $f(x') < f(x'')$. In addition (and this is where the definitions of quasiconcavity and quasiconvexity take some getting used to), both are also strictly quasiconvex [everywhere below $f(x'')$ in that same interval, again for $f(x') < f(x'')$].

With (ordinary) concave and convex functions, any nonnegative linear combination of concave (convex) functions is itself concave (convex). This sometimes useful property does not carry over to quasiconcave and quasiconvex functions; it is not very difficult (you might want to try it) to come up with two quasiconvex functions whose sum (linear combination with weights of one on each function) is not quasiconvex.

A second definition of quasiconcavity and quasiconvexity is available when $f(x)$ is dif-ferentiable.

Definition 2   The function $f(x)$, defined over a convex set S, is quasiconcave if, for *any* distinct points, x' and x''

---

[31]From the definition in (3.10), if $f(x)$ is concave, then

$$f[\alpha x'' + (1 - \alpha)x'] \geq \alpha f(x'') + (1 - \alpha)f(x')$$

but the right-hand side is just

$$\{\alpha[f(x'') - f(x')]\} + f(x')$$

Assuming that $f(x'') \geq f(x')$, and since $\alpha \geq 0$, the expression in curly brackets is nonnegative, meaning that

$$f[\alpha x'' + (1 - \alpha)x'] \geq f(x')$$

which is the definition of quasiconcavity for $f(x)$ in (3.1.1). Similarly, all strictly concave functions are also strictly quasiconcave, as defined in (3.1.2). Any (strictly) convex function can be shown to be (strictly) quasi-convex by the same kind of argument.

$$f(x'') \geq f(x') \quad \text{implies} \quad f'(x')(x'' - x') \geq 0 \qquad (3.1.5)$$

For $f(x)$ to be quasiconvex, (3.1.5) becomes

$$f(x'') \geq f(x') \quad \text{implies} \quad f'(x'')(x'' - x') \geq 0 \qquad (3.1.6)$$

Note that the derivative is evaluated at x'' rather than at x'.

The geometry of Definition 2 is straightforward. Consider (3.1.5) and assume for concreteness that $f(x'') > f(x')$ and $f'(x) \neq 0$ [we know that $(x'' - x') \neq 0$ since the points are distinct]. The requirement for quasiconcavity is that the signs of $f'(x')$ and $(x'' - x')$ be the same. If x'' is to the right of x' [so that $(x'' - x') > 0$], this means that $f(x)$ must be increasing at x'—$f'(x') > 0$. This ensures that $f(x)$ will not fall below $f(x')$ as x increases toward x''. Similarly, if x'' is to the left of x' [so that $(x'' - x') < 0$], then $f(x)$ must be decreasing at x'—$f'(x') < 0$, and $f(x)$ cannot fall below $f(x')$ as x *decreases* toward x''. Illustrations are given in Figure 3.1.4.

Here is one more definition of quasiconcavity and quasiconvexity for $f(x)$:

Definition 3 Let $f(x)$ be defined over a convex set of points S; the function $f(x)$ is quasiconcave if and only if

$$\text{For any real number } k, \text{ the set } \{x | f(x) \geq k\} \text{ is convex} \qquad (3.1.7)$$

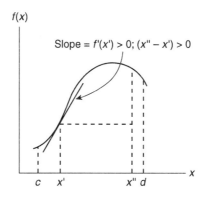

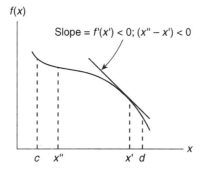

FIGURE 3.1.4   Illustrations of Definition 2 for quasiconcavity.

Similarly, a function $f(x)$ is quasiconvex if and only if

$$\text{For any real number } k, \text{ the set } \{x \mid f(x) \leq k\} \text{ is convex} \tag{3.1.8}$$

The geometry is illustrated in Figure 3.1.5. These properties of convex sets of points as defined by quasiconcave and quasiconvex functions will turn out to be useful when we explore nonlinear programming problems in Part V of this book.

Numerical Illustrations   Here are some specific numerical illustrations of functions that do not satisfy the definitions of traditional convexity or concavity.

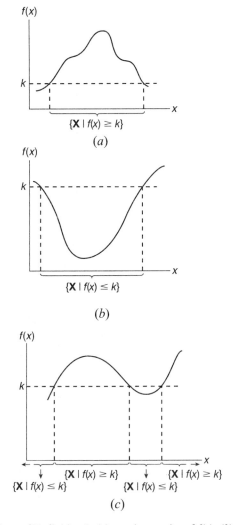

FIGURE 3.1.5   Illustrations of Definition 3: (*a*) quasiconcavity of $f(x)$; (*b*) quasiconvexity of $f(x)$; (*c*) $f(x)$ fails Definition 3 for either quasiconcavity or quasiconvexity.

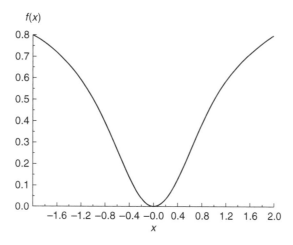

FIGURE 3.1.6 $f(x) = x^2/(x^2 + 1)$.

1. $f(x) = x^2/(x^2 + 1)$ (Figure 3.1.6). It is easy to establish, visually, that $f(x)$ satisfies any of the three definitions for quasiconvexity—for example, as in (3.1.3), $f(x)$ over any pair of points $[x', x'']$ never rises above the higher of its two endpoints. In fact, this function is *strictly* quasiconvex; it is certainly not strictly *convex*.

Since $f'(x) = 2x/(x^2 + 1)^2$ and $f''(x) = -2(3x^2 - 1)/(x^2 + 1)^3$, an unambiguous minimum at $x^* = 0$ is easily identified in the standard way, solving $f'(x) = 0$ for x^* and then examining the sign of $f''(x^*)$, which is positive in this example. For strictly convex functions, we know that first-order conditions are both necessary *and sufficient* to identify a unique minimum. This example illustrates that the same can be true for strictly quasiconvex functions, but only if they do not have inflection points at which $f'(x) = 0$. The distinction for the quasiconvex case is illustrated by turning Figures 3.1.2a and 3.1.2b upside down.

It is clear from the figure that the function changes from convex to concave as x either increases or decreases from the origin. At a point of inflection, where the curvature of the function changes, $f''(x) = 0$. It can also happen that $f''(x) = 0$ at a true maximum or minimum point, as in the examples in Figures 3.13a and 3.13b in the text; but at such maxima or minima, $f'(x) = 0$ also. If a function has a point of inflection that happens to be "horizontal" [as in Fig. 3.1.2b], then $f'(x) = 0$ also there. So, if there are one or more points, x^c, at which $f''(x^c) = 0$, but where $f'(x^c) \neq 0$, then $f(x)$ changes shape at x^c, and so $f(x)$ will be quasiconcave and quasiconvex in a neighborhood of x^c.[32]

2. $f(x) = e^{-(x-2)^2}$. This was Figure 3.19c in the text; it is repeated here in Figure 3.1.7. First- and second-order conditions establish that there is an unambiguous maximum at $x^* = 2$.

This picture suggests a two-dimensional trace of a three-dimensional bell. It is easily seen that this function satisfies any of the three definitions for quasiconcavity. For example, over any $[x', x'']$, $f(x)$ never falls below the lower of its two endpoints. In fact, this $f(x)$

---

[32]For this example, $f''(x) = 0$ when $3x^2 = 1$, which means that $x^c = \sqrt{\frac{1}{3}} = \pm 0.57735$. Let $x^{c(1)} = -0.57735$ and $x^{c(2)} = +0.57735$, and note that $f'(x^{c(1)}) = -0.6495$ ($\neq 0$) and $f'(x^{c(2)}) = +0.6495$ ($\neq 0$). Everywhere to the left of $x^{c(1)}$, $f''(x) < 0$, so the function is concave there; between $x^{c(1)}$ and $x^{c(2)}$, $f''(x) > 0$ and the function is convex, and everywhere to the right of $x^{c(2)}$, $f''(x)$ is again negative so the function is again concave.

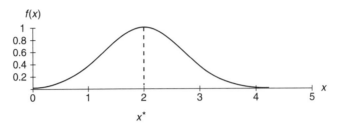

FIGURE 3.1.7 $f(x) = e^{-(x-2)^2}$.

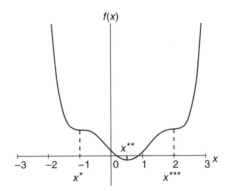

FIGURE 3.1.8 $f(x) = (0.1)(x + 3)^3 (x - 2)^3 + 1$.

is *strictly* quasiconcave, and the point that satisfies the first-order condition $f'(x) = 0$ is the unique global maximum for this function.[33]

3. $f(x) = (0.1)(x + 1)^3 (x - 2)^3 + 1$. This is shown in Figure 3.1.8 and is seen to be strictly quasiconvex.

Here $f'(x) = 0$ is a fifth-degree equation with three real roots, $x^* = -1$, $x^{**} = 0.5$ and $x^{***} = 2$. Examining $f''(x)$ at each of these points, we find $f''(x^*) = 0$ (a point of inflection), $f''(x^{**}) > 0$ (a minimum point) and $f''(x^{***}) = 0$ (another inflection point). In fact, x^{**} represents a unique global minimum to $f(x)$.

3A.1.2 Quasiconcavity and Quasiconvexity of $f(\mathbf{X})$

Definitions   Extensions to a function of n variables are fairly straightforward. The usefulness of the concepts parallels the discussion of the various cases for $f(x)$ in Figure 3.1.2—namely, in helping to identify local, global, and unique maxima and minima, only now for $f(\mathbf{X})$, where $\mathbf{X} = [x_1, \ldots, x_n]'$. Let $\mathbf{X}' = [x', \ldots, x']'$ and $\mathbf{X}'' = [x'', \ldots, x'']'$.

Definition 1'   Let $f(\mathbf{X})$ be defined over a convex set of points, S, and consider any distinct pair of points, $\mathbf{X}'$ and $\mathbf{X}''$, in S. Suppose $f(\mathbf{X}'') \geq f(\mathbf{X}')$; then $f(\mathbf{X})$ is quasiconcave if

$$f(\mathbf{X}'') \geq f(\mathbf{X}') \quad \text{implies} \quad f[\alpha\mathbf{X}'' + (1 - \alpha)\mathbf{X}'] \geq f(\mathbf{X}') \qquad (3.1.1')$$

---

[33]For this function, $f''(x) = 0$ at $x^{c(1)} = 1.29289$ or $x^{c(2)} = 2.70710$, at which points $f'(x) = +0.8578$ and -0.8578, respectively. You can visualize, in Figure 3.1.7, that to the left of $x^{c(1)}$, $f(x)$ is convex $[f''(x) > 0]$, between $x^{c(1)}$ and $x^{c(2)}$, $f(x)$ is concave $[f''(x) < 0]$, and to the right of $x^{c(2)}$, $f(x)$ is again convex $[f''(x) > 0]$.

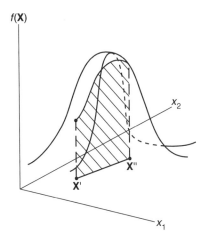

FIGURE 3.1.9   Quasiconcavity of $f(\mathbf{X})$ over $\mathbf{X} \geq \mathbf{0}$.

for $(0 \leq \alpha \leq 1)$. Additionally, $f(\mathbf{X})$ is *strictly* quasiconcave if

$$f(\mathbf{X}'') \geq f(\mathbf{X}') \quad \text{implies} \quad f[\alpha\mathbf{X}'' + (1-\alpha)\mathbf{X}'] > f(\mathbf{X}') \tag{3.1.2'}$$

for $(0 < \alpha < 1)$.[34]

Exactly as with $f(x)$, a function $f(\mathbf{X})$ is (strictly) quasiconvex if $-f(\mathbf{X})$ is (strictly) quasiconcave. In particular, let $f(\mathbf{X})$ be defined over a convex set of points, S, and consider any distinct pair of points, $\mathbf{X}'$ and $\mathbf{X}''$, in S. Suppose $f(\mathbf{X}'') \geq f(\mathbf{X}')$; then $f(\mathbf{X})$ is quasiconvex if

$$f(\mathbf{X}'') \geq f(\mathbf{X}') \quad \text{implies} \quad f[\alpha\mathbf{X}'' + (1-\alpha)\mathbf{X}'] \leq f(\mathbf{X}'') \tag{3.1.3'}$$

for $(0 \leq \alpha \leq 1)$, and $f(\mathbf{X})$ is *strictly* quasiconvex if

$$f(\mathbf{X}'') \geq f(\mathbf{X}') \quad \text{implies} \quad f[\alpha\mathbf{X}'' + (1-\alpha)\mathbf{X}'] < f(\mathbf{X}'') \tag{3.1.4'}$$

for $(0 < \alpha < 1)$.[35]

Figure 3.1.9 suggests the geometry in the $n = 2$ case. For a quasiconcave function $f(\mathbf{X})$, the outline of $f(\mathbf{X})$ on the plane defined by the line connecting *any* pair of points $\mathbf{X}'$ and $\mathbf{X}''$ in S will never fall below $f(\mathbf{X}')$.

When $f(\mathbf{X})$ is differentiable, the parallels to (3.1.5) and (3.1.6), above, make use of the gradient (column vector) of partial derivatives of the function.

Definition 2' $f(\mathbf{X})$, defined over a convex set S, is quasiconcave if, for $\mathbf{X}'' \neq \mathbf{X}'$ in S,

$$f(\mathbf{X}'') \geq f(\mathbf{X}') \quad \text{implies} \quad [\nabla f(\mathbf{X}')]'(\mathbf{X}'' - \mathbf{X}') \geq 0 \tag{3.1.5'}$$

---

[34]In addition, $f(\mathbf{X})$ is *explicitly* quasiconcave if, for $(0 < \alpha \leq 1)$,

$$f(\mathbf{X}'') > f(\mathbf{X}') \quad \text{implies} \quad f[\alpha\mathbf{X}'' + (1-\alpha)\mathbf{X}'] > f(\mathbf{X}')$$

[35]Also, $f(\mathbf{X})$ is *explicitly* quasiconvex if, for $(0 < \alpha \leq 1)$,

$$f(\mathbf{X}'') > f(\mathbf{X}') \quad \text{implies} \quad f[\alpha\mathbf{X}'' + (1-\alpha)\mathbf{X}'] < f(\mathbf{X}'')$$

For quasiconvexity, (3.1.6) becomes

$$f(\mathbf{X}'') \geq f(\mathbf{X}') \quad \text{implies} \quad [\nabla f(\mathbf{X}'')]'(\mathbf{X}'' - \mathbf{X}') \geq 0 \qquad (3.1.6')$$

The parallels to Definition 3 are also straightforward.

Definition 3'   The function $f(\mathbf{X})$ is quasiconcave if and only if

$$\text{For any real number } k, \text{ the set } \{\mathbf{X} | f(\mathbf{X}) \geq k\} \text{ is convex} \qquad (3.1.7')$$

Similarly, a function $f(\mathbf{X})$ is quasiconvex if and only if

$$\text{For any real number } k, \text{ the set } \{\mathbf{X} | f(\mathbf{X}) \leq k\} \text{ is convex} \qquad (3.1.8')$$

The geometry is illustrated in Figure 3.1.10.

In addition, for functions that are twice differentiable, there is a set of necessary conditions and a set of sufficient conditions for quasiconcavity or quasiconvexity of $f(\mathbf{X})$ that depend on signs of leading principal minors of a Hessian-like matrix.

Using first partial derivatives and second partials and cross-partials of $f(\mathbf{X})$, create a matrix $\overline{\mathbf{H}}$, in which the Hessian matrix is bordered by the first partial derivatives of $f(\mathbf{X})$:

$$\overline{\mathbf{H}} = \begin{bmatrix} 0 & f_1 & f_2 & \cdots & f_n \\ f_1 & f_{11} & f_{12} & \cdots & f_{1n} \\ f_2 & f_{21} & f_{22} & \cdots & f_{2n} \\ \vdots & \vdots & \vdots & & \vdots \\ f_n & f_{n1} & f_{n2} & \cdots & f_{nn} \end{bmatrix} = \begin{bmatrix} 0 & \nabla' \\ \nabla & \mathbf{H} \end{bmatrix} \qquad (3.1.9)$$

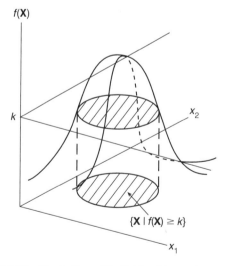

FIGURE 3.1.10   Illustration of Definition 3'; quasiconcavity of $f(\mathbf{X})$.

where ∇ and $\mathbf{H}$ are, as usual, the gradient and the Hessian matrix of $f(\mathbf{X})$. Then the rules for quasiconcavity and quasiconvexity take the form of requirements on the signs of leading principal minors of $\overline{\mathbf{H}}$.

Let the leading principal minors of $\overline{\mathbf{H}}$ be

$$|\overline{\mathbf{H}}_k| = \begin{vmatrix} 0 & \cdots & f_k \\ \vdots & & \vdots \\ f_k & \cdots & f_{kk} \end{vmatrix}$$

for $1 \leq k \leq n$ (this leaves out the zero in the upper left, which, technically speaking, is the first leading principal minor in $\overline{\mathbf{H}}$). Then, for *any* $\mathbf{X}$ satisfying $\mathbf{X} \geq \mathbf{0}$:

Definition 4a   A *necessary* condition for $f(\mathbf{X})$ to be quasiconcave is that the series of leading principal minors must alternate between nonpositive and nonnegative, beginning nonpositive:

$$|\overline{\mathbf{H}}_1| \leq 0, |\overline{\mathbf{H}}_2| \geq 0, \ldots, |\overline{\mathbf{H}}_n| \begin{Bmatrix} \leq 0 \\ \geq 0 \end{Bmatrix} \quad \text{if } n \text{ is } \begin{Bmatrix} \text{odd} \\ \text{even} \end{Bmatrix} \tag{3.1.10}$$

[Note that the condition on $|\overline{\mathbf{H}}_1|$ is always satisfied (why?); it is included here only for completeness.] Put more compactly,

$$(-1)^r |\overline{\mathbf{H}}_r| \geq 0 \quad \text{for all} \quad r = 1, \ldots, n \tag{3.1.10'}$$

Similarly, a necessary condition for $f(\mathbf{X})$ ($\mathbf{X} \geq \mathbf{0}$) to be quasiconvex is that all of these leading principal minors be nonpositive:

$$|\overline{\mathbf{H}}_1| \leq 0, |\overline{\mathbf{H}}_2| \leq 0, \ldots, |\overline{\mathbf{H}}_n| \leq 0 \tag{3.1.11}$$

or simply

$$|\overline{\mathbf{H}}_r| \leq 0 \quad \text{for all} \quad r = 1, \ldots, n \tag{3.1.11'}$$

(As before, the conditions on $|\overline{\mathbf{H}}_1|$ will automatically be met.)

Finally, a stronger condition on the leading principal minors of $\overline{\mathbf{H}}$ produces *sufficient* conditions for quasiconcavity or quasiconvexity over $\mathbf{X} \geq \mathbf{0}$.

Definition 4b   A *sufficient* condition for $f(\mathbf{X})$ to be quasiconcave is that these same leading principal minors alternate between strictly negative and strictly positive, beginning negative:

$$|\overline{\mathbf{H}}_1| < 0, |\overline{\mathbf{H}}_2| > 0, \ldots, |\overline{\mathbf{H}}_n| \begin{Bmatrix} < 0 \\ > 0 \end{Bmatrix} \quad \text{if } n \text{ is } \begin{Bmatrix} \text{odd} \\ \text{even} \end{Bmatrix} \tag{3.1.12}$$

or, compactly,

$$(-1)^r |\overline{\mathbf{H}}_r| > 0 \quad \text{for all} \quad r = 1, \ldots, n \tag{3.1.12'}$$

(Note that now the condition on $|\overline{\mathbf{H}}_1|$ needs to be included explicitly.)

The parallel sufficient condition for quasiconvexity is that all of these leading principal minors be negative:

$$|\overline{\mathbf{H}}_1| < 0, |\overline{\mathbf{H}}_2| < 0, \ldots, |\overline{\mathbf{H}}_n| < 0 \tag{3.1.13}$$

or simply

$$|\overline{\mathbf{H}}_r| < 0 \quad \text{for all} \quad r = 1, \ldots, n \tag{3.1.13'}$$

Numerical Illustrations

1. Figure 3.1.11 illustrates $f(\mathbf{X}) = 100/(x_1^2 + x_2^2 + 10)$. The function looks like a bell in three-dimensional space, centered at the origin. The function is easily seen to satisfy (3.1.29) in Definition 1′ for strict quasiconcavity—on the plane above the line connecting any $\mathbf{X}'$ and $\mathbf{X}''$, the trace of $f(\mathbf{X})$ is always above its lower endpoint. It is also easy to visualize the fact that for any k, the set $\{\mathbf{X} | f(\mathbf{X}) \geq \mathbf{k}\}$ is convex, which establishes quasiconcavity as characterized by Definition 3′.

Solving the two equations in $\nabla f = \mathbf{0}$ leads easily to $\mathbf{X}^* = \begin{bmatrix} 0 \\ 0 \end{bmatrix}$, where $|\mathbf{H}_1^*| < 0$ and $|\mathbf{H}_2^*|$

$(= |\mathbf{H}^*|) > 0$, so $f(\mathbf{X})$ is concave at the origin, identifying $\mathbf{X}^*$ as a local maximum point. For this example,

$$\mathbf{H} = \begin{bmatrix} \dfrac{200(3x_1^2 - x_2^2 - 10)}{D^3} & \dfrac{800x_1x_2}{D^3} \\ \dfrac{800x_1x_2}{D^3} & \dfrac{-200(x_1^2 - 3x_2^2 + 10)}{D^3} \end{bmatrix}$$

where $D = x_1^2 + x_2^2 + 10$, and so the signs of the principal minors and hence the concavity or convexity of the function depend very much on the values of x_1 and x_2 at which they are evaluated. For example, at $\mathbf{X} = \begin{bmatrix} 2 \\ 2 \end{bmatrix}$, $f(\mathbf{X})$ has changed to convex (from concave at the origin). We cannot know from the leading principal minor test at $\mathbf{X}^*$ that this is in fact the global maximum point for this function.

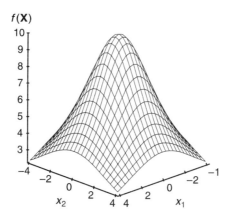

FIGURE 3.1.11 $f(\mathbf{X}) = 100/(x_1^2 + x_2^2 + 10)$.

The observations on quasiconcavity depend on our ability to look at the three-dimensional representation of the function and then be convinced that the requirements in the definitions must hold. Alternatively, we might notice in this case that $f(\mathbf{X})$ satisfies the necessary leading principal minor conditions (Definition 4a) for a quasiconcave function, but these characterize the function only over $\mathbf{X} \geq \mathbf{0}$. Here $|\overline{\mathbf{H}}_1| = -(f_1)^2 = -[-200x_1/(x_1^2 + x_2^2 + 10)^2]^2$, which is always nonpositive (it is equal to zero when $x_1 = 0$, because the top of the bell is at $x_1 = 0$ and $x_2 = 0$), and (you can work through the algebra, which is not much of a task with appropriate mathematical software on a computer) $|\overline{\mathbf{H}}_2| = 8(10^6)(x_1^2 + x_2^2)/(x_1^2 + x_2^2 + 10)^6$, which is always nonnegative. As with $|\overline{\mathbf{H}}_1|$, $|\overline{\mathbf{H}}_2| = 0$ at the top of the bell, when $x_1 = x_2 = 0$.

2. Figure 3.1.12 illustrates the following function:

$$f(\mathbf{X}) = -[e^{-(x_1-2)^2}e^{-(x_2-2)^2}]$$

which represents a quasiconvex function that looks like a hole with a bottom at $\mathbf{X} = \begin{bmatrix} 2 \\ 2 \end{bmatrix}$, where $f(\mathbf{X}) = -1$. Satisfaction of the requirements in Definitions 1' or 3' for quasiconvexity is, again, easily to establish visually, once we have the picture of $f(\mathbf{X})$ in front of us.

What about the algebra? Here

$$|\overline{\mathbf{H}}_1| = -[2e^{(-x_1^2+4x_1-x_2^2+4x_2-8)}(x_1 - 2)]^2$$

which is zero at $x_1 = 2$ (the value of x_1 at the bottom of the hole) and negative otherwise. For this function, we obtain

$$|\overline{\mathbf{H}}_2| = -8e^{(-3x_1^2+12x_1-3x_2^2+12x_2-24)}(x_1^2 - 4x_1 + x_2^2 - 4x_2 + 8)$$

which is negative except when $x_1 = x_2 = 2$, at the bottom of the hole, and then $|\overline{\mathbf{H}}_2| = 0$. Thus $f(\mathbf{X})$ satisfies the two leading principal minor necessary conditions—(3.3.11) of Definition 4a—for quasiconvexity, but not the sufficient conditions. In fact, it is clear that any function with the top of a hill or the bottom of a hole in the nonnegative quadrant will fail the sufficient conditions, since $|\overline{\mathbf{H}}_1| = -(f_1^2) = 0$ there, and

$$|\overline{\mathbf{H}}_2| = -(f_1^2)f_{22} - (f_2^2)f_{11} + 2f_1f_2f_{12} = 0$$

there also. Following are two examples (paragraphs 3 and 4) that meet the sufficiency conditions over $\mathbf{X} \geq \mathbf{0}$.

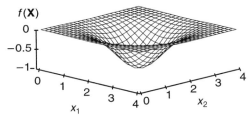

FIGURE 3.1.12 $f(\mathbf{X}) = -[e^{-(x_1-2)^2}e^{-(x_2-2)^2}]$.

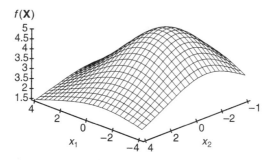

FIGURE 3.1.13 $f(\mathbf{X}) = 100/[(x_1 + 1)^2 + (x_2 + 1)^2 + 20]$.

3. Bell with a top at $\mathbf{X} = \begin{bmatrix} -1 \\ -1 \end{bmatrix}$, where $f(\mathbf{X}) = 5$ (Figure 3.1.13):

$$f(\mathbf{X}) = \frac{100}{(x_1 + 1)^2 + (x_2 + 1)^2 + 20}$$

The requisite algebraic details are

$$f_1 = \frac{-200(x_1 + 1)}{(x_1^2 + 2x_1 + x_2^2 + 2x_2 + 22)^2}$$

which could be zero only if the numerator were zero, namely, if $x = -1$, so $|\overline{\mathbf{H}}_1| = -(f_1^2) < 0$ for all $\mathbf{X} \geq \mathbf{0}$ and

$$|\overline{\mathbf{H}}_2| = 8(10^6) \frac{x_1^2 + 2x_1 + x_2^2 + 2x_2 + 2}{(x_1^2 + 2x_1 + x_2^2 + 2x_2 + 22)^6}$$

Neither the numerator nor denominator can be zero for $\mathbf{X} \geq \mathbf{0}$; moreover, both are strictly positive for $\mathbf{X} \geq \mathbf{0}$. This means that $|\overline{\mathbf{H}}_2| > 0$ for all $\mathbf{X} \geq \mathbf{0}$ and the sufficient conditions for quasiconcavity in (3.1.12) are met.

4. Upside-down bell with a bottom at $\mathbf{X} = \begin{bmatrix} -1 \\ -1 \end{bmatrix}$, where $f(\mathbf{X}) = -2$ (Fig. 3.1.14):

$$f(\mathbf{X}) = \frac{-100}{(x_1 + 1)^2 + (x_2 + 1)^2 + 20} + 3$$

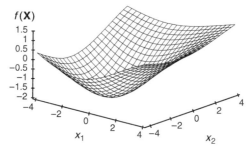

FIGURE 3.1.14 $f(\mathbf{X}) = -100/[(x_1 + 1)^2 + (x_2 + 1)^2 + 20] + 3$.

Except for the constant term, this is just the negative of $f(\mathbf{X})$ in paragraph 3 above, so the sign of f_1 will be opposite to what it was in paragraph 3, but since $|\overline{\mathbf{H}}_1| = -(f_1^2)$, it will remain negative. By contrast, $|\overline{\mathbf{H}}_2|$ will have just the opposite sign from the previous example (you might work out the details to be convinced that this is the case), so for this function $|\overline{\mathbf{H}}_1| < 0$ and $|\overline{\mathbf{H}}_2| < 0$ for all $\mathbf{X} \geq 0$, satisfying the sufficient conditions in (3.1.13) for quasiconvexity.

APPENDIX 3.2   MACLAURIN'S AND TAYLOR'S SERIES

3A.2.1   Maclaurin's Series (Colin Maclaurin, Scotch, 1698–1746)

Consider an nth degree *polynomial* function of x (meaning that x appears squared, cubed, and so on up to the nth power):

$$f(x) = a_0 + a_1x + a_2x^2 + a_3x^3 + \cdots + a_nx^n \tag{3.2.1}$$

or, compactly and since $x^0 \equiv 1$ and $x^1 \equiv x$:

$$f(x) = \sum_{i=0}^{n} a_ix^i$$

Its derivatives are easily found. For example

$$f'(x) = a_1 + 2a_2x + 3a_3x^2 + 4a_4x^3 + \cdots + na_nx^{(n-1)}$$
$$f''(x) = 2a_2 + 3(2)a_3x + 4(3)a_4x^2 + \cdots + n(n-1)a_nx^{(n-2)} \tag{3.2.2}$$
$$f^{(3)}(x) = 3(2)a_3 + 4(3)(2)a_4x + \cdots + n(n-1)(n-2)a_nx^{(n-3)}$$

and, finally

$$f^{(n)} = n(n-1)(n-2) \cdots (2)(1)a_n$$

Evaluate these derivatives at $x = 0$, which means that all x terms in the derivatives will disappear. Using the notation of *factorials*, where $n! = n(n-1)(n-2) \cdots (3)(2)(1)$ (defined for any n that is a positive integer), we obtain

$$f'(0) = a_1 = (1!)a_1$$
$$f''(0) = 2a_2 = (2!)a_2$$
$$f^{(3)}(0) = 3(2)a_3 = (3!)a_3$$

and so forth until

$$f^{(n)}(0) = (n!)a_n$$

It is easy to solve for $a_1, \ldots, a_n$ from these relationships:

$$a_1 = \frac{f'(0)}{1!}, \qquad a_2 = \frac{f''(0)}{2!}, \qquad a_3 = \frac{f^{(3)}(0)}{3!}, \ldots, a_n = \frac{f^{(n)}(0)}{n!}$$

Inserting these values of the a's into (3.2.1) and using the fact that $f(0) = a_0$ and the definition that $0! \equiv 1$, we obtain

$$f(x) = \frac{f(0)}{0!} + \left[\frac{f'(0)}{1!}\right]x + \left[\frac{f''(0)}{2!}\right]x^2 + \left[\frac{f^{(3)}(0)}{3!}\right]x^3 + \cdots + \left[\frac{f^{(n)}(0)}{n!}\right]x^n \quad (3.2.3)$$

or, compactly

$$f(x) = \sum_{i=0}^{n} \left[\frac{f(i)}{i!}\right]x^i$$

This is Maclaurin's series, in which an nth-order polynomial function is expressed *exactly* as a function of its n derivatives.

3A.2.2 Taylor's Series

Consider the polynomial in (3.2.1), above, and let $x = x_0 + dx$. For a *given* value, x, we have a function that depends on dx

$$g(dx) = a_0 + a_1(x_0 + dx) + a_2(x_0 + dx)^2 + a_3(x_0 + dx)^3 + \cdots + a_n(x_0 + dx)^n$$

Derivatives will have exactly the same structure of those for $f(x)$, in (3.2.2), except that $(x_0 + dx)$ will replace x throughout. Using $g'(dx)$ to indicate the first derivative of $g(dx)$ with respect to dx, $g''(dx)$ for the second derivative, and so on:

$$g'(dx) = a_1 + 2a_2(x_0 + dx) + 3a_3(x_0 + dx)^2 + \cdots + na_n(x_0 + dx)^{(n-1)}$$

$$g''(dx) = 2a_2 + 3(2)a_3(x_0 + dx) + 4(3)a_4(x_0 + dx)^2 + \cdots + n(n-1)a_n(x_0 + dx)^{(n-2)}$$

$$g^{(3)}(dx) = 3(2)a_3 + 4(3)(2)a_4(x_0 + dx) + \ldots + n(n-1)(n-2)a_n(x + dx)^{(n-3)}$$

and, finally

$$g^{(n)}(dx) = n(n-1)(n-2) \cdots (2)(1)a_n$$

Expand $g(dx)$ around $dx = 0$, using Maclaurin's series in (3.2.3):

$$g(dx) = \frac{g(0)}{0!} + \left[\frac{g'(0)}{1!}\right]dx + \left[\frac{g''(0)}{2!}\right](dx)^2 + \left[\frac{g^{(3)}(0)}{3!}\right](dx)^3 + \cdots + \left[\frac{g^{(n)}(0)}{n!}\right](dx)^n \quad (3.2.4)$$

But if $dx = 0$, then $x = x_0$, and since $g(dx) = f(x)$, $g(0) = f(x_0)$, $g'(0) = f'(x_0)$, and so on. So (3.2.4) can be reexpressed, replacing each derivative of $g(dx)$, evaluated at $dx = 0$, with the corresponding derivative of $f(x)$, evaluated at $x = x_0$, and replacing the dx terms with the equivalent $(x - x_0)$ expression:

$$f(x) = \frac{f(x_0)}{0!} + \left[\frac{f'(x_0)}{1!}\right](x - x_0) + \left[\frac{f''(x)}{2!}\right](x - x_0)^2 + \left[\frac{f^{(3)}(x_0)}{3!}\right](x - x_0)^3$$

$$+ \cdots + \left[\frac{f^{(n)}(x_0)}{n!}\right](x - x_0)^n \quad (3.2.5)$$

This is an *exact* expression for the *n*th-order polynomial function $f(x)$, in terms of its *n* derivatives, all evaluated at some specific point x_0. For Maclaurin's series, x_0 had to be zero; Taylor showed that a similar result, as in (3.2.4), can be derived for *any* fixed value, x_0.

3A.2.3 Taylor's Theorem

Taylor's *theorem* extends the kind of result in Taylor's *series*, (3.2.5), to *any* function $\phi(x)$, that has derivatives up through the *n*th. It says that the result in (3.2.5) is not necessarily exact, and that after the *n*th derivative term, there will be a *remainder*, which represents the *difference* between the exact value of $\phi(x)$ at x and the value given by the sum of the ($n + 1$) terms, from $(x - x_0)^0$ through $(x - x_0)^n$, in (3.2.5):

$$\phi(x) = \frac{\phi(x_0)}{0!} + \left[\frac{\phi'(x)}{1!}\right](x - x_0) + \left[\frac{\phi''(x_0)}{2!}\right](x - x_0)^2 + \left[\frac{\phi^{(3)}(x_0)}{3!}\right](x_0 - x)^3$$
$$+ \cdots + \left[\frac{\phi^{(n)}(x_0)}{n!}\right](x_0 - x)^n + R_n$$

(3.2.6)

The only difference from (3.2.5)—except that now the function is $\phi(x)$ instead of $f(x)$—is the final term, R_n. A specific form of the R_n term was proposed by Lagrange:

$$R_n = \left[\frac{\phi^{(n+1)}(k)}{(n + 1)!}\right](x - x_0)^{(n+1)}$$

where k is some value *between* x and x_0. Note that this looks just like what one would expect for the "next" term in (3.2.6), if one were needed, except for the fact that the ($n + 1$)st derivative is not evaluated at x_0. That next term is not necessary in (3.2.5), because the expression is exactly correct for a *polynomial* function $f(x)$; it appears in (3.2.6) because $\phi(x)$ is *not* required to be a polynomial function. To the extent that $R_n \rightarrow$

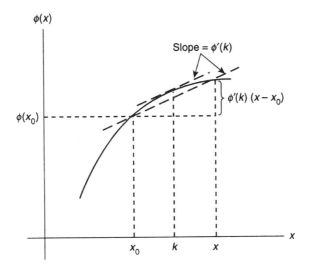

FIGURE 3.2.1   The geometry of R_n.

0 as $n \to \infty$, then the truncated series in (3.2.5) will provide a "good" approximation, for sufficiently large n.

The Lagrange form for R_n has a straightforward geometric interpretation for the case in which $n = 0$. Here

$$R_0 = \phi'(k)(x - x_0)$$

and the approximation provided by (3.2.6) is

$$\phi(x) = \phi(x_0) + \phi'(k)(x - x_0) \tag{3.2.7}$$

Figure 3.2.1 illustrates this relationship. There is a point, $x = k$, between x and x_0, for which the slope of the function, $\phi'(k)$, is equal to the slope of the secant cutting the function at the two points $[x_0, \phi(x_0)]$ and $[x, \phi(x)]$, so that the difference between $\phi(x_0)$ [the first term on the right in (3.2.7)] and $\phi(x)$ [the left-hand side of (3.2.7)] is just $\phi'(k)(x - x_0)$.

REFERENCES

Arrow, Kenneth J. and Alain C. Enthoven, "Quasi-Concave Programming," *Econometrica* **29**, 779–800 (1961).

Bazaraa, Mokhtar S. and C. M. Shetty, *Nonlinear Programming: Theory and Algorithms,* Wiley, New York, 1979.

Beavis, Brian and Ian Dobbs, *Optimization and Stability Theory for Economic Analysis,* Cambridge University Press, New York, 1990.

Mangasarian, Olvi L., *Nonlinear Programming,* McGraw-Hill, New York, 1969.

Martos, B., "The Direct Power of Adjacent Vertex Programming Methods," *Management Science* **12**, 241–252 (1965).

Ponstein, J., "Seven Kinds of Convexity," *SIAM Review* **9**, 115–119 (1967).

Takayama, Akira, *Mathematical Economics* (2nd Edition), Cambridge University Press, New York, 1985.

Zoutendijk, Guus, *Mathematical Programming Methods,* North-Holland, Amsterdam, 1976.

PROBLEMS

3.1 Find the derivative of each of the following functions using the "delta process." Check your answer by finding the derivative exactly:

(a) $f(x) = 4x^2 + 2x + 1$

(b) $f(x) = 12x + 15$

(c) $f(x) = 1/x^2$

3.2 Examine the following functions for maxima, minima, and points of inflection, without using the second derivative:

(a) $f(x) = 3x^2 + 12x + 10$

(b) $f(x) = 3x^2 - 12x + 10$

(c) $f(x) = -3x^2 + 12x + 10$

3.3 Write the equation for the tangent line to the function in each of the following cases:

(a) Problem 3.2(a), above, at $x = 10$
(b) Problem 3.2(b), above, at $x = 2$
(c) Problem 3.2(c), above, at $x = 4$

3.4 Find maxima and/or minima for each of the following functions, using both first and second derivatives. Indicate in each case whether the points are relative or absolute maxima or minima.

(a) $f(x) = 2x^3 + 3x^2 - 12x - 15$
(b) $f(x) = 3x^4 - x^3 + 2$
(c) $f(x) = x^3$
(d) $f(x) = 3x^2 + 12x + 10$
(e) $f(x) = 8x^3 - 9x^2 + 1$
(f) $f(x) = x^4 - 18x^2 + 15$

3.5 Explore the following functions for maxima, minima, and inflection points:

(a) $f(x) = (0.1)(x + 1)^3(x - 2)^3 + 1$. [Here $f'(x) = 0$ is a fifth-degree equation with three real roots. It is shown in Figure 3.1.8 in Appendix 3.1.]
(b) $f(x) = (x + 3)^3(x - 2)^2 + 1$. [Here again there are three real roots to $f'(x) = 0$, which is now a fourth-degree equation.]

3.6 (a) Find the number that, when diminished by its square, is a maximum; (b) find two numbers whose sum is 40 and whose product is as large as possible.

3.7 Find maxima and/or minima for each of the following functions:

(a) $f(\mathbf{X}) = x_1^2 + x_1x_2 + x_2^2 - x_2$
(b) $f(\mathbf{X}) = x_1^2 + x_2^2 + (3x_1 + 4x_2 - 26)^2$
(c) $f(\mathbf{X}) = 9x_1^2 - 18x_1 - 16x_2^2 - 64x_2 - 55$
(d) $f(\mathbf{X}) = 16 - 2(x_1 - 3)^2 - (x_2 - 7)^2$
(e) $f(\mathbf{X}) = -2x_1^3 + 6x_1x_2 - x_2^2 - 4x_2 + 100$

3.8 If $z = xy/(x + y)$, show that $x(\partial z/\partial x) + y(\partial z/\partial y) = z$.

3.9 An excursion boat can be rented for 100 people at $10 per person. For every passenger over 100, the boat owners agree to reduce the fare (for all passengers) by 5¢. How many passengers would maximize the owners' total rental receipts? How much is that maximum rental?

3.10 Costs, y, have been shown to depend in the following way on x_1 and x_2 (total outputs of goods 1 and 2, respectively):

$$y = x_1^2 + 4x_2^2 - 40x_1 + 500$$

(a) What amount of goods 1 and 2 should be produced in order to minimize costs?
(b) What amount of goods 1 and 2 should be produced in order to minimize costs, if there is an upper limit of 12 on x_1?

3.11 Suppose that the cost function in Problem 3.10 has been incorrectly estimated; it actually should be

$$y = x_1^2 + 4x_2^2 - 40x_1 + 500 - 2x_1x_2$$

How would you now answer the question in Problem 3.10, above?

3.12 Suppose that, in the function in Problem 3.11, y represented *profits*, not costs, as a function of x_1 and x_2.

(a) What amount of goods 1 and 2 should be produced to maximize profits?

(b) How much of good 2 should be produced if it is decided that x_1 must be exactly 12?

(c) How would you answer (b) if x_2 could not exceed 5?

3.13 Suppose $f(x) = -2x^2 + 90x + 600$, where $0 \le x \le 50$. What value of x will minimize $f(x)$? How (if at all) would your answer differ if $0 \le x \le 40$?

3.14 For the function $f(x) = ax^3 + bx^2 + cx + d$, find a set of values for a, b, c, and d so that $f(x)$ has a relative maximum at $x = -1$ and a relative minimum at $x = 2$.

The next three problems are best done on a computer with software that finds derivatives and also, for Problem 3.15, has decent three-dimensional graphics, to help you visualize the function.

3.15 Explore the following function for extreme points:

$$f(\mathbf{X}) = 2(x_1^2 + x_2^2)e^{-(x_1^2+x_2^2)}$$

3.16 Here is a version of what is known as the wine importer's problem. (Computer software for finding and evaluating derivatives is helpful for this problem because the second derivative is algebraically fairly complicated.) The value, V, of a particular wine increases as time passes in the following way: $V = 6(2.5)^{t^{0.5}}$, where t is time, in years. The present value of the wine, P, given a discount rate r and continuous appreciation, is $P = e^{-rt}V$. How long should the importer hold the wine before selling it in order to maximize the present value (and, presumably, also the importer's profit) if the discount rate is 8 percent?

3.17 (a) For $f(\mathbf{X}) = 100/(x_1^2 + x_2^2 + 10)$, shown in Figure 3.1.11 in Appendix 3.1, verify that the locus of points along which the shape of the function switches from concave to convex is given by the circle $x_1^2 + x_2^2 = 3.3334$—$f(\mathbf{X})$ is concave for values of x_1 and x_2 inside that circle and convex for values outside the circle. To do this: (1) set $x_2 = 0$, and find where $f_{11} = 0$ (and verify that $f_1 \ne 0$ there, so that the point is not a maximum or minimum), and (2) set $x_1 = 0$, find where $f_{22} = 0$ (and verify that $f_2 \ne 0$ there, to rule out the same maximum or minimum possibilities). (b) Verify that the locus of inflection points for $f(\mathbf{X}) = -[e^{-(x_1-2)2} e^{-(x_2-2)2}]$—shown in Figure 3.1.12 in Appendix 3.1—is a circle entered on (2,2), with radius 1.41421; that is, $(x_1 - 2)^2 + (x_2 - 2)^2 = 2$. For values of x_1 and x_2 that are inside the circle, $f(\mathbf{X})$ is convex, and for values outside the circle, $f(\mathbf{X})$ is concave.

4

CONSTRAINED MAXIMIZATION AND MINIMIZATION

The material in this chapter extends the classical analysis of unconstrained maximum and minimum problems via differential calculus to more realistic situations where the variables are required to satisfy one or several side conditions. Initially these will take the form of equations that express necessary relations among the variables. These constraints may or may not affect the location (and the size) of the maximum or minimum for the function, but a systematic way of dealing with them is needed. Next, we expand the problem to allow for one or more inequality constraints that set upper or lower limits for functions of the variables, rather than imposing strict equalities. In many problems this kind of constraint is more realistic than an equation, and we need a procedure for taking inequalities into account.

The material of this chapter, therefore, comprises classical methods for constrained maximization or minimization problems with functions of many variables; these are constrained optimization problems. A large set of such problems forms the material of mathematical programming, of which linear programming is one special, very important, extremely useful branch. This chapter puts mathematical programming in its general, classical setting and provides a background for more modern approaches that are explored in Chapters 7–9, for the linear case, and in Chapters 10 and 11, for the nonlinear case.

The problem of establishing the sign of a quadratic form when the variables are subject to the complication of a side condition is explored in Section 4.1. This is an extension of the discussion in Section 1.6, for independent variables. The results are of immediate relevance to rules for maxima and minima of functions whose variables must satisfy equality constraints. These applications are examined in Sections 4.2–4.4. In Section 4.5 we investigate the complications that arise in maximum and minimum problems when the variables are constrained by one or more inequalities. One fact that emerges from the material in this chapter is that the classical rules for maxima and minima, especially second-order conditions, are totally impractical for "large" problems with many unknowns and/or constraints. If any information on the shapes of the functions is available from the context of the problem, it is essential that it be used.

4.1   QUADRATIC FORMS WITH SIDE CONDITIONS

Consider again $Q(x_1,x_2) = a_{11}x_1^2 + 2a_{12}x_1x_2 + a_{22}x_2^2$, as in Section 1.6, only now for the case in which x_1 and x_2 are not independent variables but rather must conform to the added and very specific linear homogeneous equation $c_1x_1 + c_2x_2 = 0$. (We will soon see exactly why this particular equation form is of interest.) It is easy to solve for either variable; for example $x_2 = -(c_1/c_2)x_1$, and thus Q can be written as a function of x_1 alone:

$$Q(x_1) = (a_{11}c_2^2 - 2a_{12}c_1c_2 + a_{22}c_1^2)\frac{x_1^2}{c_2^2} \tag{4.1}$$

The sign of $Q(x_1,x_2)$ with the side condition taken explicitly into account, as in (4.1), hinges on the sign of $(a_{11}c_2^2 - 2a_{12}c_1c_2 + a_{22}c_1^2)$. It may not be obvious at first glance, but this expression is precisely the negative of the determinant of a matrix, $\overline{\mathbf{A}}$, whose straightforward structure is made more apparent by partitioning:

$$\overline{\mathbf{A}} = \begin{bmatrix} 0 & c_1 & c_2 \\ \hline c_1 & a_{11} & a_{12} \\ c_2 & a_{21} & a_{22} \end{bmatrix}$$

Then the expression multiplying (x_1^2/c_2^2) in (4.1) is just $-|\overline{\mathbf{A}}|$; $Q(x_1) = -|\overline{\mathbf{A}}|(x_1^2/c_2^2)$. Therefore, $Q > 0$ (for $x_1 \neq 0$) if and only if $-|\overline{\mathbf{A}}| > 0$ or $|\overline{\mathbf{A}}| < 0$, and $Q < 0$ if and only if $-|\overline{\mathbf{A}}| < 0$ or $|\overline{\mathbf{A}}| > 0$. The structure of $\overline{\mathbf{A}}$ should be clear; it consists of the original (symmetric) matrix of coefficients, $\mathbf{A}$, from the quadratic form, bordered by the coefficients of the linear relation between the variables, $c_1x_1 + c_2x_2 = 0$. Letting $\mathbf{C} = [c_1, c_2]$ and denoting by $\mathbf{A}$ the symmetric matrix $\begin{bmatrix} a_{11} & a_{12} \\ a_{21} & a_{22} \end{bmatrix}$, we see that $\overline{\mathbf{A}} = \begin{bmatrix} 0 & \mathbf{C} \\ \hline \mathbf{C}' & \mathbf{A} \end{bmatrix}$.

It is much more complicated to show the case for $Q(x_1, x_2, x_3)$ subject to $c_1x_1 + c_2x_2 + c_3x_3 = 0$. By virtue of the side relation, one of the variables can be eliminated from $Q(x_1, x_2, x_3)$. It then becomes a quadratic form in the two remaining *independent* variables, similar in structure to (1.32) in Chapter 1 except that a_{11}, a_{12}, and a_{22} are replaced by much more complex expressions. It, too, can then be written as a sum of squares as in (1.36). The result is that for positive definiteness of $Q(x_1, x_2, x_3)$ when $c_1x_1 + c_2x_2 + c_3x_3 = 0$, the *last two*[1] leading principal minors of

$$\overline{\mathbf{A}} = \begin{bmatrix} 0 & \mathbf{C} \\ \hline \mathbf{C}' & \mathbf{A} \end{bmatrix}$$

must be *negative,* where, in the three-variable case, $\mathbf{C} = [c_1, c_2, c_3]$ and $\mathbf{A}$ is the symmetric matrix $\mathbf{A} = \begin{bmatrix} a_{11} & a_{12} & a_{13} \\ a_{21} & a_{22} & a_{23} \\ a_{31} & a_{32} & a_{33} \end{bmatrix}$. This means

$$|\overline{\mathbf{A}}_3| = \begin{vmatrix} 0 & c_1 & c_2 \\ \hline c_1 & a_{11} & a_{12} \\ c_2 & a_{21} & a_{22} \end{vmatrix} < 0, \qquad |\overline{\mathbf{A}}_4| = |\overline{\mathbf{A}}| = \begin{vmatrix} 0 & c_1 & c_2 & c_3 \\ \hline c_1 & a_{11} & a_{12} & a_{13} \\ c_2 & a_{21} & a_{22} & a_{23} \\ c_3 & a_{31} & a_{32} & a_{33} \end{vmatrix} < 0$$

---

[1]The *number* of leading principal minors is derived from the number of variables (n) and the number of equations to which they must conform (m); it is $n - m$, which for this case is $3 - 1 = 2$.

For negative definiteness, $Q(x_1,x_2,x_3) < 0$, the signs of these two determinants must alternate, beginning *positive*.

For the general case, $Q(\mathbf{X})$, subject to $\mathbf{CX} = 0$, where $\mathbf{X} = [x_1, \ldots, x_n]'$ and $\mathbf{C} = [c_1, \ldots, c_n]$, the results are a direct extension of the three-variable case. Again, $\overline{\mathbf{A}} = \begin{bmatrix} 0 & \vdots & \mathbf{C} \\ \cdots & \vdots & \cdots \\ \mathbf{C}' & \vdots & \mathbf{A} \end{bmatrix}$, where

$\mathbf{A}$ is the $n \times n$ symmetric matrix associated with the quadratic form.

For $Q > 0$ (positive definiteness) subject to $\mathbf{CX} = 0$, it is both necessary and sufficient that the last $(n - 1)$ leading principal minors (i.e., starting with the third) of $\overline{\mathbf{A}}$ be *negative*:

$$|\overline{\mathbf{A}}_3| < 0, |\overline{\mathbf{A}}_4| < 0, \ldots, |\overline{\mathbf{A}}_{n+1}| < 0 \tag{4.2}$$

$Q < 0$ (negative definiteness), subject to the same side relation, if and only if these same leading principal minors alternate in sign, beginning *positive*.[2]

The generalization, finally, to more than one linear relation of the sort $\Sigma_{i=1}^{n} c_i x_i = 0$ is an extension of the rules in (4.2), but it is much more difficult to derive. Let

$\mathbf{C} = \begin{bmatrix} c_{11} & \cdots & c_{1n} \\ \vdots & & \vdots \\ c_{m1} & \cdots & c_{mn} \end{bmatrix}$, so that $\mathbf{CX} = \mathbf{0}$ represents a system of m linear homogeneous

equations in n variables ($m < n$) with $\rho(\mathbf{C}) = m$. Then

For $Q(\mathbf{X})$ to be positive definite ($Q > 0$) subject to $\mathbf{CX} = \mathbf{0}$, it is necessary and sufficient that the last $n - m$ leading principal minors of the $(m + n) \times (m + n)$ matrix $\overline{\mathbf{A}} = \begin{bmatrix} \mathbf{0} & \vdots & \mathbf{C} \\ \cdots & \vdots & \cdots \\ \mathbf{C}' & \vdots & \mathbf{A} \end{bmatrix}$ have the sign of $(-1)^m$. Q is negative definite ($Q < 0$) if and only if these same leading principal minors alternate in sign, beginning with the sign of $(-1)^{m+1}$. $\tag{4.3}$

The rules in (4.2) for $m = 1$ and the $(n + 1) \times (n + 1)$ matrix $\overline{\mathbf{A}}$ used there are seen to be special cases of the rules in (4.3).

It is worth noticing that the results in (4.3) will be tedious and complicated to apply for many realistic problems that are likely to have many variables and not so many constraints, so that $n - m$ is large. An example is the standard economics problem of maximizing a consumer's utility from consumption of a large number of different goods subject to a (single) budget constraint. For example

---

[2]These rules differ from those for the unconstrained quadratic form (Section 1.6) in two respects. First, the *signs* of the rules are just reversed. This is because the rewritten sums-of-squares expressions for the quadratic forms, taking into account the constraints, involve squared terms multiplied by the *negative* of leading principal minors of $\overline{\mathbf{A}}$. Second, not all of the leading principal minors of $\overline{\mathbf{A}}$ are involved. Since, in the n-variable case, any one variable can be removed because of the linear side relation, the unconstrained quadratic form has $n - 1$ independent variables. It can be written as the sum of $n - 1$ squared terms, each multiplied by a coefficient that, except for the first term, is the ratio of two successive leading principal minors. Thus $n - 1$ leading principal minors are needed. Since $\overline{\mathbf{A}}$ is $(n + 1) \times (n + 1)$, it has a total of $n + 1$ leading principal minors; if two are ignored, exactly the right number will remain. From the algebra of the sum-of-squares expression, we can deduce that $|\overline{\mathbf{A}}_1|$ and $|\overline{\mathbf{A}}_2|$ are the leading principal minors that should be eliminated.

1. With 10 variables and one constraint, there would be nine minors to investigate—determinants of 3×3, 4×4, ..., 11×11 matrices.

2. With 25 variables and one constraint, the number of minors is increased to 24, again starting with the determinant of a 3×3 matrix, but now working up to, finally, a 26×26 matrix.

With many variables and also many constraints, the *number* of leading principal minors decreases but their size increases. For example, with 25 variables and 10 constraints, there would be (only) 15 leading principal minors to evaluate, but they range in size from 21×21 up to 35×35. This is tiresome work, even with modern computers, and whenever there is information, from the context of the problem, about the shapes of the functions, it should be exploited.

4.2   MAXIMA AND MINIMA FOR FUNCTIONS OF TWO DEPENDENT VARIABLES

We begin with the simplest constrained optimization problem in which there is a function of two variables subject to one equality constraint. Then we explore the case of more variables in Section 4.3 and of more constraints in Section 4.4. Unless otherwise noted, the rules will isolate relative (local) maxima and minima; this may be troublesome, as we will see.

4.2.1   First-Order Conditions: Differentials

Consider $f(x_1, x_2)$ where x_1 and x_2 must also satisfy the completely general relation $h(x_1, x_2) = 0$. If the constraint is $g(x_1, x_2) = c$, then $h(x_1, x_2) \equiv g(x_1, x_2) - c = 0$. For example, if $x_1^2 + 3x_2 = 10$, this can be written

$$h(x_1, x_2) = x_1^2 + 3x_2 - 10 = 0$$

We assume, throughout this section, that both variables appear in the constraint. If they did not—for example, if the constraint were $3x_1 = 9$, or $x_1 = 3$—then we would actually have an unconstrained problem with only one variable, x_2.

Geometrically, we are now looking for a maximum or minimum on only that part of the surface $y = f(x_1, x_2)$ that lies above [or below, depending on $f(x_1, x_2)$] $h(x_1, x_2) = 0$ in x_1, x_2 space. Figure 4.1 illustrates a maximum problem in which $f(x_1, x_2)$ has the general shape of an inverted bowl and the constraint is a straight line in x_1, x_2 space. A linear constraint has the effect of building a wall on the x_1, x_2 plane. The wall passes through the surface $f(x_1, x_2)$, and the constrained maximum point must be found on the trace of the $f(x_1, x_2)$ surface on this wall. When $h(x_1, x_2)$ is not a linear function, the $f(x_1, x_2)$ surface will be cut by a curved wall.

The effect of the constraint is to remove the independence of x_1 and x_2. In a simple example, such as $h(x_1, x_2) = x_1 + 3x_2 - 10 = 0$, the required *dependence* between x_1 and x_2 is easily made explicit by solving for one variable in terms of the other; for example, $x_1 = 10 - 3x_2$. In general, however, it may not be easy to solve $h(x_1, x_2) = 0$ for either of the variables as a function of the other, so we need a more generalized approach. This is provided by the fact that the dependence between x_1 and x_2 carries over to dx_1 and dx_2. From any

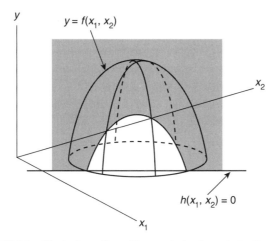

FIGURE 4.1 Illustration of equality-constrained maximization problem.

point $\overline{\mathbf{X}}$ that lies on the constraint, a decision to change $\overline{x}_1$ by an arbitrary amount $d\overline{x}_1$ *must* be accompanied by a corresponding change in $\overline{x}_2, d\overline{x}_2$, of exactly the right size so that the new point $[\overline{x}_1 + d\overline{x}_1, \overline{x}_2 + d\overline{x}_2]$, is also located on the constraint. The nature of this functional dependence between dx_1 and dx_2 is easily established for *any* (differentiable) constraint by taking the total differential of both sides

$$h_1 dx_1 + h_2\, dx_2 = 0 \tag{4.4a}$$

or

$$[\nabla h(\mathbf{X})]'[\mathbf{dX}] = 0 \tag{4.4b}$$

where $\nabla h(\mathbf{X}) \equiv [h_1 \quad h_2]'$ and $\mathbf{dX} = [dx_1\ dx_2]'$.

Notice that if $h(\mathbf{X}) = 0$ is a *linear* function of x_1 and x_2, then the assumption that both variables appear in the constraint assures that $h_1 \neq 0$ and $h_2 \neq 0$. This will be important in what follows, in order to avoid division by zero. If $h(\mathbf{X})$ is a *nonlinear* function, it is not true that $\nabla h(\mathbf{X})$ will always contain only nonzero elements (we will see examples of this later). This, in turn, gives rise to the need for what is generally called a "rank condition" or "constraint qualification"—to eliminate "ill-behaved" problems with one or more zero-valued elements in $\nabla h(\mathbf{X})$. We ignore these subtleties for the moment, in order to proceed with the general logic of conditions for optimization with equality constraints. Later we will explore these rank conditions and constraint qualifications for both equality- and inequality-constrained optimization problems.

It is worth noting that the left-hand side in either (4.4a) or (4.4b) is the first differential of the function $h(\mathbf{X})$ that defines the left-hand side of the constraint, and the right-hand side of (4.4a) or (4.4b) is the differential of a constant (viz., the zero that is the right-hand side of the constraint). From (4.4a), for example

$$dx_2 = -\left(\frac{h_1}{h_2}\right)dx_1 \tag{4.5}$$

gives the explicit functional dependence between dx_1 and dx_2. (Here we require $h_2 \neq 0$, as anticipated above.)

The *logic* of the first differential condition for an unconstrained maximum or minimum at $\mathbf{X}^*$—that the slope of the tangent at $\mathbf{X}^*$ be zero—still holds, only now the tangent is not a *plane* touching $f(x_1, x_2)$ in three-dimensional space but rather a *line* touching the contour of $f(\mathbf{X})$ that is traced on the wall in two-dimensional space above $h(\mathbf{X}) = 0$. To impose the requirement that $\mathbf{X}^*$ be on the constraint, we need simply substitute (4.5) into the first differential for $f(\mathbf{X})$, $f_1 dx_1 + f_2 dx_2$, giving

$$df = f_1 dx_1 + f_2 \left[-\left(\frac{h_1}{h_2} \right) dx_1 \right]$$

This takes into account the precise dependence between dx_1 and dx_2, and so now it *is* appropriate to require that this rewritten expression for df be zero for either a maximum or a minimum point

$$df = f_1 dx_1 + f_2 \left[-\left(\frac{h_1}{h_2} \right) dx_1 \right] = 0 \tag{4.6}$$

from which it follows that

$$\frac{f_1}{f_2} = \frac{h_1}{h_2} \qquad \text{or} \qquad \frac{f_1}{h_1} = \frac{f_2}{h_2} \tag{4.7}$$

In taking the first differential of $h(\mathbf{X}) = 0$ in order to establish the relationship between dx_1 and dx_2, we have lost information on the size of the constant term, if there was one, in the constraint. For example, both

$$x_1^2 + 3x_1 x_2 + 4x_2^2 = 2 \qquad \text{and} \qquad x_1^2 + 3x_1 x_2 + 4x_2^2 = 700$$

when expressed as $h(x_1, x_2) = 0$, have $dh = (2x_1 + 3x_2)dx_1 + (3x_1 + 8x_2)dx_2$. Therefore, to the necessary conditions in (4.7) we need to add the original statement of the constraint, to recapture the lost constant term.[3] The conditions that must hold at a stationary point of $f(x_1, x_2)$ subject to $h(x_1, x_2) = 0$ are therefore

$$\frac{f_1}{h_1} = \frac{f_2}{h_2} \tag{4.8a}$$

and

$$h(x_1, x_2) = 0 \tag{4.8b}$$

These results can also be derived in a way that lends itself to straightforward generalization to both more variables and also more constraints. Consider the first differential nec-

---

[3]Since information loss occurs between $h(x_1, x_2)$ and dh, you might ask: why use the differential form? Why not use the information in $h(x_1, x_2)$ directly? The reason is that the first-order condition is on df, which is a function of dx_1 and dx_2; by using dh, which is *always* of the form $h_1 dx_1 + h_2 dx_2$, we have immediately the necessary relationship between the changes in x_1 and x_2. In addition, as already noted, in general it may not be easy at all to solve $h(x_1, x_2)$ for one of the variables in order to substitute into $f(x_1, x_2)$.

essary condition for a maximum or minimum of $f(x_1,x_2)$ in the *independent variables* case, in conjunction with (4.4a) or (4.4b):

$$f_1 dx_1 + f_2 dx_2 = 0$$

$$h_1 dx_1 + h_2 dx_2 = 0$$

This is a pair of linear homogeneous equations; the first equation is what we are accustomed to imposing in an *unconstrained* maximization or minimization problem, and the second is the differential of the constraint in a *constrained* problem. By considering the equations simultaneously, the precise dependence between dx_1 and dx_2 (from the second equation) is imposed on the first equation as well, as in (4.6).

In matrix form

$$\begin{bmatrix} f_1 & f_2 \\ h_1 & h_2 \end{bmatrix} \begin{bmatrix} dx_1 \\ dx_2 \end{bmatrix} = \begin{bmatrix} 0 \\ 0 \end{bmatrix}$$

or, using gradients,

$$\begin{bmatrix} [\nabla f(\mathbf{X})]' \\ [\nabla h(\mathbf{X})]' \end{bmatrix} [\mathbf{dX}] = \mathbf{0}$$

For nontrivial solutions (allowing values other than $dx_1 = 0$ and $dx_2 = 0$), the coefficient matrix must be singular (Chapter 2); here this means

$$\begin{vmatrix} f_1 & f_2 \\ h_1 & h_2 \end{vmatrix} = 0$$

For this small example with two unknowns and one constraint, this determinental condition is easily seen to be $f_1 h_2 - f_2 h_1 = 0$ or $f_1/h_1 = f_2/h_2$, exactly as in (4.7) and (4.8a).

The matrix singularity condition can also be expressed (again, from Chapter 2) by requiring that the rows of $\begin{bmatrix} f_1 & f_2 \\ h_1 & h_2 \end{bmatrix}$ be linearly dependent: $[f_1 \quad f_2] = \alpha[h_1 \quad h_2]$. If we consider the transpose of these row vectors, the relationship remains the same:

$$\begin{bmatrix} f_1 \\ f_2 \end{bmatrix} = \alpha \begin{bmatrix} h_2 \\ h_2 \end{bmatrix}$$

or, compactly

$$\nabla f = \alpha[\nabla h]$$

In two-dimensional vector space, α is the scalar by which ∇h must be stretched or shrunk to reach ∇f; the two gradients must be collinear, pointing either in the same or in exactly opposite directions.

The following are alternative ways that this determinental condition is sometimes expressed

$$\begin{cases} f_1 = \alpha h_1 \\ f_2 = \alpha h_2 \end{cases} \quad \text{or} \quad \begin{cases} f_1 - \alpha h_1 = 0 \\ f_2 - \alpha h_2 = 0 \end{cases} \quad \text{or} \quad \begin{cases} \dfrac{f_1}{h_1} = \alpha \\ \dfrac{f_2}{h_2} = \alpha \end{cases} \tag{4.9}$$

The factor of proportionality, α, could have appeared explicitly in (4.8a) as the common value for the two ratios that are required to be equal. However, its interpretation as the scalar showing the relationship between the two rows of the coefficient matrix for the pair of linear homogeneous equations in dx_1 and dx_2 would not have been clear, and this proportionality requirement between rows will be of interest when the problem is expanded to more variables and more constraints. As before, the original constraint $h(x_1,x_2) = 0$ must be added to (4.9) because of the information lost in taking differentials. Thus the necessary conditions for a stationary point can also be written as

$$
\text{(a)} \begin{cases} f_1 - \alpha h_1 = 0, f_2 - \alpha h_2 = 0 \\ \quad\quad \text{or} \\ \nabla f - \alpha[\nabla h] = \mathbf{0} \end{cases} \quad\quad \text{and} \quad\quad \text{(b)} \;\; h(x_1,x_2) = 0 \quad\quad (4.10)
$$

4.2.2   First-Order Conditions: Lagrange Multipliers (Joseph Louis Lagrange, French, 1736–1813)

The method of Lagrange multipliers is a systematic way of generating the conditions in (4.10). Create what is called the Lagrangian function, L, by subtracting the constraint, multiplied by an unknown new variable, λ, from $f(\mathbf{X})$:

$$
L(x_1,x_2,\lambda) = f(x_1,x_2) - \lambda[h(x_1,x_2)] \quad\quad (4.11)
$$

Note that for those x_1 and x_2 that satisfy the constraint, $h(x_1,x_2) = 0$, the Lagrangian function L is exactly the same as $f(x_1,x_2)$. Then the trick is to take the three first partial derivatives of L *as if* all of the variables were *independent*, and set these derivatives equal to zero. This sets up the necessary conditions for a stationary point as if L were a function of three independent variables. Here this first-order condition is $dL = 0$, which means

$$
\nabla L \equiv \begin{bmatrix} \dfrac{\partial L}{\partial x_1} \\[6pt] \dfrac{\partial L}{\partial x_2} \\[6pt] \dfrac{\partial L}{\partial \lambda} \end{bmatrix} = \begin{bmatrix} f_1 - \lambda h_1 \\ f_2 - \lambda h_2 \\ -h(x_1,x_2) \end{bmatrix} = \begin{bmatrix} 0 \\ 0 \\ 0 \end{bmatrix} \quad\quad (4.12)
$$

The first two conditions in (4.12) are just $\nabla f - \lambda[\nabla h] = \mathbf{0}$; exactly as in (4.10a), with λ playing the role of α, the factor of proportionality. For generalization to larger problems, it will be useful to define $\nabla L_{\mathbf{X}}$ as the vector of first partial derivatives of L with respect to each of the x's. Using that notation, the first two conditions in (4.12) could also be represented as $\nabla L_{\mathbf{X}} = \mathbf{0}$. The last condition in (4.12), from the partial derivative of L with respect to λ, serves to reproduce the constraint—after both sides are multiplied by (-1); this is just (4.10b). Again, for later generalization to problems with more than one constraint, we can denote $\partial L/\partial\lambda$ as ∇L_{Λ}—in which case $\nabla L \equiv \begin{bmatrix} \nabla L_{\mathbf{X}} \\ \hline \nabla L_{\Lambda} \end{bmatrix}$—and so the third condition in (4.12) can be compactly represented as $\nabla L_{\Lambda} = \mathbf{0}$.

Therefore, constructing the Lagrangian function, taking partial derivatives as if all variables were independent, and setting these partials equal to zero is just a simple and automatic way of generating the first-differential (necessary) conditions for a maximum or minimum of $f(x_1, x_2)$ subject to $h(x_1, x_1) = 0$.[4] How difficult these conditions are to solve depends entirely on how complex the first derivative functions are. For a function $f(\mathbf{X})$ that includes one or more x's cubed (or raised to a power higher than three), these $\nabla L = \mathbf{0}$ conditions will contain one or more nonlinear equations.

4.2.3 Second-Order Conditions

Once the point or points satisfying (4.8)—or (4.9) or (4.12)—have been found, we need to distinguish maxima from minima (from neither). For this we used the sign of the second total differential in the case of independent variables (Tables 3.5 and 3.6). The issue is, as always, the *shape* of $f(\mathbf{X})$ at the stationary point(s).

Recall that the definitions of convexity and concavity of $f(\mathbf{X})$ (Chapter 3) depended on the function being defined over a convex set of points. For $f(x_1, x_2)$ with no constraints, the set over which $f(\mathbf{X})$ is defined is the entire $x_1 x_2$ plane (a convex set). In a problem with a constraint, $h(x_1, x_2) = 0$, the issue turns on the nature of that constraint. If $h(x_1, x_2)$ happens to be linear, then indeed $f(x_1, x_2)$ is defined over a convex set (a line), and convexity/concavity information about $f(x_1, x_2)$—if available—could be used to distinguish maxima from minima. However, if $h(x_1, x_2)$ is a nonlinear function, then the set of points satisfying the constraint is *not* a convex set. In that case, the sign of the second total differential must again be established, but account must be taken of the nature of the *dependence* between x_1 and x_2.

The general expression for d^2f was given in Chapter 3, as

$$d^2f = f_{11}(dx_1)^2 + 2f_{12}(dx_1)(dx_2) + f_{22}(dx_2)^2 + f_1 d^2x_1 + f_2 d^2x_2$$

When x_1 and x_2 (and hence dx_1 and dx_2) were independent, we saw that d^2x_1 and d^2x_2 were both zero, and the last two terms were dropped, resulting in (3.26). Now the variables are dependent; either one may be considered the (single) independent variable. For example, consider x_1 as the independent variable; dx_1 can be chosen arbitrarily but dx_2 is then determined through the functional relationship in (4.5)—$dx_2 = -(h_1/h_2)dx_1$. Therefore, while $d^2x_1 \equiv d(dx_1)$ is still zero, $d^2x_2 \equiv d(dx_2)$ is not, since it depends on dx_1. The second total differential now has one more term than previously in (3.26); it is

$$d^2f = f_{11}(dx_1)^2 + 2f_{12}(dx_1)(dx_2) + f_{22}(dx_2)^2 + f_2 d^2x_2 \tag{4.13}$$

Now d^2f is no longer a quadratic form in dx_1 and dx_2, because of the new last term in (4.13). The rules of sign for quadratic forms from Section 1.6 cannot be applied, and yet it is the sign of d^2f that distinguishes maxima from minima.

The d^2x_2 term spoils the symmetry. If it could be replaced by terms with only $(dx_1)^2$, $dx_1 dx_2$, and $(dx_2)^2$ (and d^2x_1, which is equal to zero), (4.13) could be rewritten as a qua-

---

[4]From the point of view of the necessary conditions, $\lambda[h(\mathbf{X})]$ could just as well be *added* to $f(\mathbf{X})$ to create the Lagrangian. This would simply define $\partial L/\partial x_1 = f_1 + \lambda h_1$ and $\partial L/\partial x_2 = f_2 + \lambda h_2$ in (4.12), and it would generate $h(x_1, x_2) = 0$ directly, without multiplication by (-1). It is subtracted here to maintain consistency with the discussion of the determinental condition and because of later interpretations to be made of λ.

dratic form. Here is how this can be accomplished. Take the total differential of $h(x_1, x_2) = 0$ twice

$$h_{11}(dx_1)^2 + 2h_{12}(dx_1)(dx_2) + h_{22}(dx_2)^2 + h_1 d^2x_1 + h_2 d^2x_2 = 0 \qquad (4.14)$$

[The left-hand side of (4.14), for $h(\mathbf{X})$, is identical to d^2f for $f(\mathbf{X})$.] Since $d^2x_1 = 0$, re-arrangement of (4.14) gives

$$d^2x_2 = -\left(\frac{1}{h_2}\right)[h_{11}(dx_1)^2 + 2h_{12}(dx_1)(dx_2) + h_{22}(dx_2)^2]$$

Putting this into (4.13)—note that it is multiplied by f_2—and collecting terms on $(dx_1)^2$, $dx_1 dx_2$ and $(dx_2)^2$, we have

$$d^2f = \left[f_{11} - \left(\frac{f_2}{h_2}\right)h_{11}\right](dx_1)^2 + 2\left[f_{12} - \left(\frac{f_2}{h_2}\right)h_{12}\right](dx_1)(dx_2)$$

$$+ \left[f_{22} - \left(\frac{f_2}{h_2}\right)h_{22}\right](dx_2)^2 \qquad (4.15)$$

or

$$d^2f = [\mathbf{dX}]'\begin{bmatrix} v_{11} & v_{12} \\ v_{21} & v_{22} \end{bmatrix}[\mathbf{dX}]$$

where $v_{jk} = f_{jk} - (f_2/h_2)h_{jk}$, $v_{jk} = v_{kj}$, the f_{ij} are elements of the Hessian of $f(\mathbf{X})$ and the h_{ij} are elements of the Hessian of $h(\mathbf{X})$.

The expression in (4.15) is considerably more complicated than (3.26), but it is now a *quadratic form* in dx_1 and dx_2. It is, however, a quadratic form of two *dependent* variables, and the exact nature of the dependence between dx_1 and dx_2 is given by $h_1 dx_1 + h_2 dx_2 = 0$, as in (4.4a). We are interested, for second-order conditions, in the sign of the quadratic form in (4.15) when the variables are subject to

$$[\nabla h] \circ [\mathbf{dX}] = h_1 dx_1 + h_2 dx_2 = 0$$

This was exactly the subject matter in Section 4.1, and the results in (4.2) are of interest here. It is because of the particularly simple structure of the first total differential of $h(x_1, x_2) = 0$ that a side condition of the form $c_1 x_1 + c_2 x_2 = 0$ was studied in that section.

For d^2f^* to be positive definite, so that $\mathbf{X}^*$ identifies a minimum point, the determinant of

$$\overline{\mathbf{V}}^* = \begin{bmatrix} 0 & h_1^* & h_2^* \\ h_1^* & v_{11}^* & v_{12}^* \\ h_2^* & v_{21}^* & v_{22}^* \end{bmatrix} \qquad (4.16)$$

must be negative. For $d^2f^* < 0$, so $\mathbf{X}^*$ is a maximum, $|\mathbf{V}^*|$ must be positive.

Since the determinant is evaluated at $\mathbf{X}^*$ (where $df = 0$), the f_k^*/h_k^* terms in each v_{jk}^* could be replaced by f_1^*/h_1^* or by λ^* throughout because of (4.12)—for example, $v_{ii}^* = f_{ii}^* - \lambda^* h_{ii}^*$ and $v_{jk}^* = f_{jk}^* - \lambda^* h_{jk}^*$. Note that these are just the second partials and cross-partials of the Lagrangian function in (4.11) with respect to x_1 and x_2, evaluated at $\mathbf{X}^*$ (and the associated λ^*); that is, they are the elements of the Hessian matrix for L (not f). The matrix in (4.16) is an example of a "bordered Hessian."[5] It is a structure that occurs repeatedly in conditions for constrained maxima and minima. An alternative and more compact representation for (4.16) that makes clear the components of which it is composed is

$$\overline{\mathbf{V}}^* = \left[\begin{array}{c|c} 0 & [\nabla h^*]' \\ \hline [\nabla h^*] & \nabla^2 L^* \end{array} \right]$$

(Remember that $\nabla^2 f$ is used to denote the Hessian matrix of $f(\mathbf{X})$ in cases when it is necessary to identify the function explicitly.)

4.2.4   Example

Examine $f(\mathbf{X}) = 6x_1 x_2$ for maxima and minima if the variables are also required to satisfy $h(\mathbf{X}) = 2x_1 + x_2 - 10 = 0$.[6] By straightforward application of (4.10) we see that we need, initially, $\nabla f = \begin{bmatrix} 6x_2 \\ 6x_1 \end{bmatrix}$ and $\nabla h = \begin{bmatrix} 2 \\ 1 \end{bmatrix}$. Using (4.10a), part of the first differential condition is that $6x_2/2 = 6x_1/1$, or $x_2 = 2x_1$. Putting this relationship into the constraint in (4.10b) gives $x_1^* = 2.5$ and therefore $x_2^* = 5$.

To identify what kind of point $\mathbf{X}^* = \begin{bmatrix} 2.5 \\ 5 \end{bmatrix}$ is, we turn to the sign of $d^2 f^*$. Here

$$\nabla^2 f = \begin{bmatrix} 0 & 6 \\ 6 & 0 \end{bmatrix} \quad \text{and} \quad \nabla^2 h = \begin{bmatrix} 0 & 0 \\ 0 & 0 \end{bmatrix}$$

Since all second partials and cross-partials of the constraint are zero, it happens in this example that $v_{jk} = f_{jk}$. The bordered Hessian matrix in question is

$$\overline{\mathbf{V}}^* = \mathbf{V} = \left[\begin{array}{c|cc} 0 & 2 & 1 \\ \hline 2 & 0 & 6 \\ 1 & 6 & 0 \end{array} \right] \quad \text{and} \quad |\overline{\mathbf{V}}^*| = |\mathbf{V}| = 12 + 12 = 24 > 0$$

so $d^2 f < 0$, and $\mathbf{X}^*$ represents a *maximum* of the function subject to the constraint.

Using the Lagrange method on this problem, we construct[7]

$$L = 6x_1 x_2 - \lambda(2x_1 + x_2 - 10)$$

and imposing (4.12), we obtain

$$L_1 = 6x_2 - 2\lambda = 0$$
$$L_2 = 6x_1 - \lambda = 0$$
$$L_\lambda = -(2x_1 + x_2 - 10) = 0$$

---

[5]If you think of λ as the "first" unknown in the Lagrangian function in (4.11), then instead of (4.12), we would

have $\nabla L = \begin{bmatrix} -h(x_1, x_2) \\ f_1 - \lambda h_1 \\ f_2 - \lambda h_2 \end{bmatrix}$, and finding the Hessian matrix for $L(\lambda, x_1, x_2)$ *with respect to all three unknowns* would

generate exactly the matrix in (4.16), including the first row and column "borders."

[6]Obviously, in this simple case we could solve $2x_1 + x_2 = 10$ for either x_1 or x_2 and substitute into $6x_1 x_2$, making it a function of one variable only. We are interested here in more general procedures for the cases in which it is not easy—or, indeed, it is impossible—to eliminate one of the variables in this way.

[7]On the form of the constraint in L, as well as the use of $-\lambda$, see the end of this section.

The first two equations lead to $6x_2 = 2\lambda$ and $6x_1 = \lambda$. Therefore $3x_2 = 6x_1$ or $x_2 = 2x_1$; this result, put into L_λ gives, as before, $2x_1 + 2x_1 = 10$, so that $x_1^* = 2.5$ and $x_2^* = 5$.

This second approach adds one piece of information that is absent from direct application of the first-order conditions in (4.8): the value of λ^*. Although λ is not one of the variables whose optimum values are of *direct* interest in the problem, it turns out that λ^* provides possibly useful information about the constraint. In this example, $\lambda^* = (6x_2^*)/2 = 15 = (6x_1^*/1)$. Note that $f(\mathbf{X}^*) = 75$.

Suppose now that the constraint were $2x_1 + x_2 = 11$, rather than 10—the right-hand side of the constraint has increased by one unit.[8] Clearly the relationships from (4.8a) will not change, since the constant term in the constraint is in no way involved in h_1 or h_2. Substituting $x_2 = 2x_1$ into the new constraint gives the new values $x_1^{**} = 2.75$, $x_2^{**} = 5.5$, and $f(\mathbf{X}^{**}) = 90.75$. The change in the value of $f(\mathbf{X})$ is 15.75, and λ^* is 15. This is no accident; the value of λ^* gives an *approximation* to the amount that $f(\mathbf{X}^*)$ will change for a *unit* change in the constant term, c, in the constraint. Equally importantly (for later discussion of inequality constraints), the *sign* of λ^* indicates the *direction* of the change. A positive λ^* means that if c increases, so does f^*; $\lambda^* < 0$ means that an increase in c is accompanied by a decrease in f^*.

This *marginal valuation* property of λ^* can be demonstrated in the following way, which requires the concept of a *total derivative*. From

$$df = f_1 dx_1 + \cdots + f_i dx_i + \cdots + f_n dx_n$$

as in (3.38a), a total derivative of $f(\mathbf{X})$ with respect to x_i is defined by dividing both sides of df by dx_i, so that

$$\frac{df}{dx_i} = f_1\left(\frac{dx_1}{dx_i}\right) + \cdots + f_i + \cdots + f_n\left(\frac{dx_n}{dx_i}\right)$$

This provides a measure of the change in $f(\mathbf{X})$ with respect to x_i when all other variables, $x_j (j = 1, \ldots, n; j \neq i)$ are also functions of x_i and are allowed to vary when x_i varies. (Compare with the *partial derivative* $\partial f/\partial x_i$, in which all other variables, x_j, are held constant.)

The optimal value of the Lagrangian function for this equality-constrained problem is

$$L^* = f(x_1^*, x_2^*) - \lambda^*[h(x_1^*, x_2^*)] = f(x_1^*, x_2^*) - \lambda^*[g(x_1^*, x_2^*) - c]$$

The values of x_1^*, x_2^*, and λ^* depend on c, because of the first-order conditions in (4.12). Specifically

$$\nabla L_{\mathbf{X}^*} = \begin{bmatrix} L_1^* \\ L_2^* \end{bmatrix} = \begin{bmatrix} f_1^* - \lambda^* h_1^* \\ f_2^* - \lambda^* h_2^* \end{bmatrix} = \mathbf{0}$$

and

$$\nabla L_{\mathbf{\Lambda}^*} = L_\lambda^* = -h(x_1^*, x_2^*) = -[g(x_1^*, x_2^*) - c] = 0$$

---

[8]Since $\mathbf{X}^*$ represents a maximum, it is clear that the larger the value of $2x_1 + x_2$ allowed, the larger the value of f^*. Hence we could also say that the constraint was *relaxed* by one unit.

Then the effect of a change in c on L^*, transmitted via changes in the values of x_1^*, x_2^*, and λ^*, is given by the total derivative of L^* with respect to c;

$$\frac{dL^*}{dc} = L_1^*\left(\frac{dx_1^*}{dc}\right) + L_2^*\left(\frac{dx_2^*}{dc}\right) + L_\lambda^*\left(\frac{d\lambda^*}{dc}\right) + L_c^*$$

From first-order conditions, $L_1^* = L_2^* = L_\lambda^* = 0$; at an optimum point $L_c^* = \lambda^*$ and $L^* = f^*$, and we are left with $dL^*/dc = df^*/dc = \lambda^*$, indicating that the optimum value of the Lagrange multiplier measures the effect on $f(\mathbf{X}^*)$ of a "small" change in the right-hand side of the constraint.

4.2.5 Geometry of the First-Order Conditions

The first-order conditions $[\nabla f] = \alpha[\nabla h]$ or $[\nabla f] = \lambda[\nabla h]$, along with $h(\mathbf{X}) = 0$, have a straightforward geometric interpretation. Recall (Chapter 1) that a vector in vector space is uniquely defined by its length (norm) and the angle that it makes with any other vector used as a fixed reference (in two-dimensional space this is usually the horizontal axis). Also in Chapter 1, a measure of the size of the angle ϕ between two n-element vectors, $\mathbf{V}$ and $\mathbf{W}$ (both drawn as emanating from the origin), was given by the cosine of that angle, which was

$$\cos \phi = \frac{\mathbf{V} \circ \mathbf{W}}{\|\mathbf{V}\| \, \|\mathbf{W}\|}$$

where $\|\mathbf{V}\| = (\Sigma_{i=1}^n v_i^2)^{\frac{1}{2}}$, the length of $\mathbf{V}$. At the points where ϕ is equal to 0°, 90°, 180°, 270°, and 360°, $\cos \phi$ takes on values +1, 0, –1, 0, and +1, respectively. Since it is conventional to measure the angle between two vectors as the *included* angle, it cannot exceed 180°. Therefore, two vectors are *orthogonal* (meet at 90°) if and only if $\mathbf{V} \circ \mathbf{W} = 0$.

Normals Consider any *homogeneous* linear equation of two variables, $a_1 x_1 + a_2 x_2 = 0$. Letting $\mathbf{A} = [a_1, a_2]'$ and $\mathbf{X} = [x_1, x_2]'$, this equation can be written as $\mathbf{A}'\mathbf{X} = 0$ or, in inner product form, as $\mathbf{A} \circ \mathbf{X} = 0$. Any point $\mathbf{X}^* = [x_1^*, x_2^*]'$ that satisfies the equation defines a unique *vector* from the origin to that point; this vector is simply a part of the line defined by the equation, and $\mathbf{A}$ is orthogonal to it. This is illustrated in Figure 4.2, in which one set

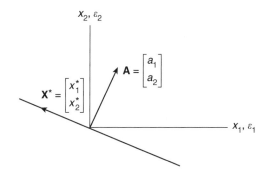

FIGURE 4.2 Schematic representation of the equation $\mathbf{A} \circ \mathbf{X} = 0$.

of axes can accommodate both solution space and vector space, since two dimensions are needed in each case. Because **A** will be orthogonal to *all* vectors defined by *all* points **X** = $[x_1, x_1]'$ that satisfy the equation, it can be described simply as orthogonal to the line; **A** is also said to be *normal* to the line.

This observation on normality is easily extended to *any* line in two-dimensional solution space. In particular, it applies as well to any *nonhomogeneous* equation $a_1 x_1 + a_1 x_1 = $ **A**∘**X** = $b (b \neq 0)$, since the value of the right-hand side influences the *intercept* of the line on the vertical axis, but not its slope. [The slope of both **A**∘**X** = 0 and **A**∘**X** = b is $(-a_1/a_2)$; the intercepts of the two equations on the vertical axis are 0 and (b/a_2), respectively.] Consider any two lines, **A**∘**X** = b and **C**∘**X** = d. If these *lines* are orthogonal (**A**∘**C** = 0) then their slopes, $(-a_1/a_1)$ and $(-c_1/c_2)$, must be negative reciprocals.

This normality characteristic is easily generalized to more than two dimensions. In three dimensions, with appropriate definitions, a single homogeneous linear equation $a_1 x_1 + a_2 x_2 + a_3 x_3 = 0$ has the same representation, **A**∘**X** = 0 (both vectors now contain three elements). Geometrically, the equation defines a plane through the origin of three-dimensional solution space, and **A** is orthogonal to every line through the origin that lies on that plane; **A** is said to be orthogonal to the plane. Similar extensions hold for a nonhomogeneous linear equation of three variables and ultimately for any linear equation of any number of variables. For n variables $(n > 3)$, the n-element vector **A** is orthogonal to the hyperplane defined by the equation.

Normality of Gradients   For all *linear* functions, the *gradient* of the function is normal to the *contour lines* of the function. For $f(\mathbf{X}) = a_1 x_1 + \cdots + a_n x_n = $ **A**∘**X** = b, $\nabla f = \mathbf{A}$, and we have just seen that this is orthogonal to $f(\mathbf{X}) = b$, for any value of b, including zero. When $n = 2$, $f(\mathbf{X})$ defines a plane in three-dimensional space and ∇f is a two-element vector. The contours of the $f(\mathbf{X})$ surface take the form of parallel lines in two-dimensional x_1, x_2 space, one line for each value of b, and the two-element gradient is orthogonal to all of these lines. The same ideas hold for $n \geq 2$ but are difficult to visualize.

For nonlinear functions, there is a slightly more subtle orthogonality relationship—the gradient at any point is normal to the *tangent* to the *contour curve* of the function at that point. Figure 4.3 illustrates for $f(\mathbf{X}) = x_1^2 + x_2^2$, which is a bowl in three-dimensional space. At $\mathbf{X}^0 = \begin{bmatrix} 1 \\ 1 \end{bmatrix}$, $f(\mathbf{X}^0) = 2$, $\nabla f(\mathbf{X}^0) = \begin{bmatrix} 2 \\ 2 \end{bmatrix}$, and the equation of the tangent at that point [in x_1, x_2 space, where $f(\mathbf{X}) = 0$] is

$$[\nabla f(\mathbf{X}^0)] \circ \begin{bmatrix} (x_1 - x_1^0) \\ (x_2 - x_2^0) \end{bmatrix} = 0 \qquad \text{or} \qquad x_1 + x_2 = 2$$

This defines a *line* in x_1, x_2 space (dashed in the figure), to which $\nabla f(\mathbf{X}^0)$ is orthogonal.

An important feature of gradients is that ∇f, evaluated at a particular point, indicates the direction of *maximum increase* of $f(\mathbf{X})$ from that point (and the negative of the gradient indicates the direction of *maximum decrease* of the function). This property will be very important when we explore numerical methods for finding maxima and minima (Chapter 6); for the present, it will help us with a geometric interpretation of the first-order conditions, as in (4.10).

It is not very difficult to see why this is true. Recall that each partial derivative of a function of n variables translates a (small) change in one of the variables into a change in

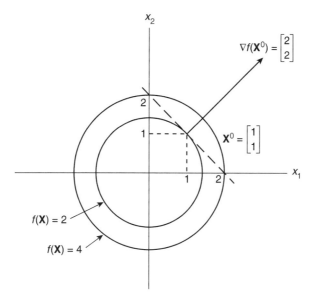

FIGURE 4.3 ∇f for $f(\mathbf{X}) = x_1^2 + x_2^2$ at $\mathbf{X}^0 = \begin{bmatrix} 1 \\ 1 \end{bmatrix}$.

$f(\mathbf{X})$. From a particular point, $\mathbf{X}^0$, and for any *given* small changes in $x_1, \ldots, x_n$, $d\mathbf{X}$ is a vector in n-dimensional space that indicates a *direction* in that space. For the sake of notational simplicity, let the small change dx_j be denoted c_j, so that $d\mathbf{X} = \mathbf{C}$ and $df(\mathbf{X}^0) = [\nabla f(\mathbf{X}^0)] \circ [\mathbf{C}]$. Suppose that these changes in the x_j are chosen so that the *length* of the vector of changes is exactly one unit ($\|\mathbf{C}\| = 1$). Our concern is with a set of directions, as indicated by the elements in the vector $\mathbf{C}$ and not with the length of that vector.

That established, a reasonable question might be: For what direction is the change in $f(\mathbf{X})$ maximized? The problem is straightforward—find the elements of $\mathbf{C}$ so as to

$$\text{Maximize}\quad df(\mathbf{X}^0) = [\nabla f(\mathbf{X}^0)] \circ [\mathbf{C}]$$
$$\text{subject to}\quad \|\mathbf{C}\| = 1 \tag{4.17}$$

Since the coefficients in $\nabla f(\mathbf{X}^0)$ are known, the function to be maximized in this equality-constrained problem is linear in the c's, but the constraint is not. Hence classical Lagrange multiplier techniques could be applied. (This is left as an exercise if you are interested.)

A bit of reflection, however, ought to suggest a maximizing solution almost by inspection. (Or retrospection, which is why we recalled facts from Chapter 1 about the cosine of the angle between two vectors earlier in this section.) We want to maximize the inner product of two vectors—$\nabla f(\mathbf{X}^0)$, which is known, and $\mathbf{C}$, which is unknown. We saw that the inner product appears as the numerator in the expression for the cosine of the angle between two vectors. For this problem, we have

$$\cos \phi = \frac{[\nabla f(\mathbf{X}^0)] \circ [\mathbf{C}]}{\|\nabla f(\mathbf{X}^0)\|}$$

since $\|\mathbf{C}\| = 1$. Thus our function to be maximized in (4.17) can alternatively be expressed as $(\cos \phi)\|\nabla f(\mathbf{X}^0)\|$. In fact, it is even simpler, since the norm of the gradient at $\mathbf{X}^0$ is just a (positive) constant, call it k. The problem therefore reduces to one of maximizing $(k)\cos \phi$, which occurs when $\cos \phi$ takes on its maximum possible value of $+1$, when $\phi = 0^0$. The implication: $[\nabla f(\mathbf{X}^0)]\circ[\mathbf{C}]$ is maximized at $\mathbf{X}^0$ when there is *no* angle between the two vectors, which means when the vector of changes, $\mathbf{C}$, points in the *same direction* as the gradient there, $\nabla f(\mathbf{X}^0)$.

This approach to the problem also shows the implication of $\phi = 180^\circ$, for which $\cos \phi = -1$. Since -1 is the minimum possible value for $\cos \phi$, the direction of *maximum decrease* in $f(\mathbf{X})$ at a particular point is in the direction exactly opposite to the gradient at that point, in the direction $[-\nabla f(\mathbf{X}^0)]$.

Vector Geometry of the First-Order Conditions   The first-order requirement for a maximum or a minimum point, $\nabla f = \alpha[\nabla h]$ or, using Lagrange multipliers, $\nabla f = \lambda[\nabla h]$, says that the two gradients, ∇f and ∇h, must be proportional at a candidate point (since one is just a scalar multiple of the other). Additionally, of course, the point must lie on the constraint.

As an illustration, the linear constraint in the preceding example is

$$h(\mathbf{X}) = 2x_1 + x_2 - 10 = \begin{bmatrix} 2 \\ 1 \end{bmatrix} \circ \begin{bmatrix} x_1 \\ x_2 \end{bmatrix} - 10 = 0$$

and $\nabla h = \begin{bmatrix} 2 \\ 1 \end{bmatrix}$ is orthogonal to that line in x_1, x_2 space. Here $f(\mathbf{X}) = 6x_1x_2 = 0$ when x_1 and/or $x_2 = 0$. The function grows ever-larger for increasing values of x_1 and/or x_2 in the first quadrant, where $x_1 > 0$ and $x_2 > 0$ (and for decreasing values of x_1 and/or x_2 in the third quadrant where $x_1 < 0$ and $x_2 < 0$; but since no points in this quadrant satisfy the constraint, we need not consider it further). In the second and fourth quadrants, where x_1 and x_2 are of opposite sign, and where parts of the constraint are located, $f(\mathbf{X})$ decreases as x_1 and x_2 move away from the origin. Since this is a maximization problem, we concentrate on the first (positive) quadrant only.

Gradients to the objective function contours are shown at five points on the constraint in Figure 4.4. At $\mathbf{X}^1 = \begin{bmatrix} 0.5 \\ 9 \end{bmatrix}$ and $\mathbf{X}^4 = \begin{bmatrix} 4.5 \\ 1 \end{bmatrix}$, on the contour along which $f(\mathbf{X}) = 27$, the respective gradients are $\nabla f(\mathbf{X}^1) = \begin{bmatrix} 54 \\ 3 \end{bmatrix}$ and $\nabla f(\mathbf{X}^4) = \begin{bmatrix} 6 \\ 27 \end{bmatrix}$. At $\mathbf{X}^2 = \begin{bmatrix} 1 \\ 8 \end{bmatrix}$ and $\mathbf{X}^3 = \begin{bmatrix} 4 \\ 2 \end{bmatrix}$, on a higher contour along which $f(\mathbf{X}) = 48$, $\nabla f(\mathbf{X}^2) = \begin{bmatrix} 48 \\ 6 \end{bmatrix}$ and $\nabla f(\mathbf{X}^3) = \begin{bmatrix} 12 \\ 24 \end{bmatrix}$. All of these

gradient *directions* are shown in Figure 4.4. (There is not enough room to show the true *length* in each case, but direction is all that matters at this point.) Notice that ∇f at each of these points is *not* orthogonal to the line $h(\mathbf{X}) = 0$. Finally, at $\mathbf{X}^* = \begin{bmatrix} 2.5 \\ 5 \end{bmatrix}$, which is on the constraint and also on the contour along which $f(\mathbf{X}) = 75$, $\nabla f(\mathbf{X}^*) = \begin{bmatrix} 30 \\ 15 \end{bmatrix}$, and it is clear that ∇f^* and ∇h^* are, indeed, collinear—specifically, $\nabla f^* = (15)[\nabla h^*]$.

Since ∇f and ∇h are orthogonal to the contours of $f(\mathbf{X})$ and to $h(\mathbf{X})$, respectively, the two requirements at $\mathbf{X}^*$, $\nabla f(\mathbf{X}^*) = \lambda[\nabla h(\mathbf{X}^*)]$ and $h(\mathbf{X}^*) = 0$, mean that at $\mathbf{X}^*$ the two gra-

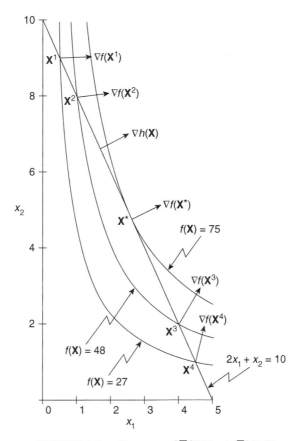

FIGURE 4.4.   Geometry of $\nabla f(\mathbf{X}^*) = \lambda[\nabla h(\mathbf{X}^*)]$.

dients either point in exactly the same direction (in which case $\lambda > 0$) or else in exactly opposite directions (in which case $\lambda < 0$). We explore the $\lambda = 0$ possibility in what follows.

4.2.6   Two Examples with Gradients that Are Null Vectors

The next examples illustrate that a problem in which $\nabla f(\mathbf{X}^*) = \mathbf{0}$ can be easily handled by the conditions in (4.10) or (4.12), while one in which $\nabla h(\mathbf{X}^*) = \mathbf{0}$ [or worse, where $\nabla h(\mathbf{X}) = \mathbf{0}$ everywhere] will fail to satisfy those conditions.

An Illustration with $\boldsymbol{\nabla}f(\mathbf{X}^*) = \mathbf{0}$ In this example the unconstrained maximum point happens to satisfy the constraint exactly:

$$\text{Maximize}\quad f(\mathbf{X}) = -(x_1 - 5)^2 - (x_2 - 5)^2 + 20$$

$$\text{subject to}\quad h(\mathbf{X}) = x_1 + x_2 - 10 = 0$$

The geometry in x_1, x_2 space is straightforward and is indicated in Figure 4.5. The constraint is a straight line with a slope of -1 that intersects the vertical (x_2) axis at 10, and $f(\mathbf{X})$ is a concave function (the signs of the leading principal minors of its Hessian matrix alternate,

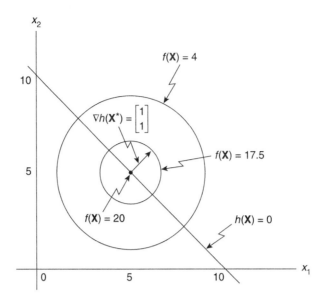

FIGURE 4.5 Contours of $f(\mathbf{X}) = -(x_1 - 5)^2 - (x_2 - 5)^2 + 20$ and $h(\mathbf{X}) = x_1 + x_2 - 10 = 0$.

beginning negative). It has the shape of an overturned bowl, with circular contours in x_1, x_2 space and an unconstrained maximum at the point where $\nabla f(\mathbf{X}) = \begin{bmatrix} -2(x_1 - 5) \\ -2(x_2 - 5) \end{bmatrix} = \mathbf{0}$; this means at $\mathbf{X}^* = \begin{bmatrix} 5 \\ 5 \end{bmatrix}$. It is clear that $\mathbf{X}^*$ also satisfies the constraint.

For this problem, $\nabla h(\mathbf{X}) = \nabla h(\mathbf{X}^*) = \begin{bmatrix} 1 \\ 1 \end{bmatrix}$, and the collinearity requirements

$$\nabla f(\mathbf{X}^*) = \alpha[\nabla h(\mathbf{X}^*)] \qquad \text{or} \qquad \nabla f(\mathbf{X}^*) = \lambda[\nabla h(\mathbf{X}^*)]$$

are easily met by setting $\alpha = 0$ (or $\lambda = 0$), so that $\begin{bmatrix} 0 \\ 0 \end{bmatrix} = (0)\begin{bmatrix} 1 \\ 1 \end{bmatrix}$.

This illustrates that there is no particular trouble in applying the collinearity conditions in (4.10) or (4.12) in an equality-constrained problem in which the extreme point happens to lie exactly on the constraint, so that $\nabla f(\mathbf{X}^*) = \mathbf{0}$. The same cannot be said for problems in which $\nabla h(\mathbf{X}^*) = \mathbf{0}$.

An Illustration with $\nabla h(\mathbf{X}^*) = \mathbf{0}$ Consider the problem of maximizing the same $f(\mathbf{X})$ but subject to $h(\mathbf{X}) = 0.5x_1^2 + x_2 - 5 = 0$. The geometry is shown in Figure 4.6. The constraint is represented by the curve $x_2 = -0.5x_1^2 + 5$, and the constrained maximum is at the point where this curve is tangent to the highest $f(\mathbf{X})$ contour.

In order to use (4.10) or (4.12), we need the usual two gradients. Here $\nabla f(\mathbf{X}) = \begin{bmatrix} -2(x_1 - 5) \\ -2(x_2 - 5) \end{bmatrix}$, and now $\nabla h(\mathbf{X}) = \begin{bmatrix} x_1 \\ 1 \end{bmatrix}$. From (4.10a), $-2(x_1 - 5) = \alpha x_1$ and $-2(x_2 - 5) = \alpha$, so that $x_2 = (6x_1 - 5)/x_1$. Putting this into the constraint leads to a nonlinear equation

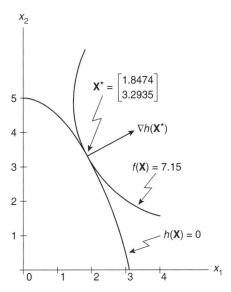

FIGURE 4.6   Contours of $f(\mathbf{X}) = -(x_1 - 5)^2 - (x_2 - 5)^2 + 20$ and $h(\mathbf{X}) = 0.5x_1^2 + x_2 - 5 = 0$ [and also $h(\mathbf{X}) = (0.5x_1^2 + x_2 - 5)^3 = 0$].

whose solution is $\mathbf{X}^* = \begin{bmatrix} 1.8474 \\ 3.2935 \end{bmatrix}$. This means that $\nabla f(\mathbf{X}^*) = \begin{bmatrix} 6.3052 \\ 3.4130 \end{bmatrix}$ and, along with

$\nabla h(\mathbf{X}^*) = \begin{bmatrix} 1.8474 \\ 1 \end{bmatrix}$, we find $\alpha^* = 3.4130$. so that the two gradients are, indeed,

collinear at $\mathbf{X}^*$— $\begin{bmatrix} 6.3052 \\ 3.4130 \end{bmatrix} = (3.4130) \begin{bmatrix} 1.8474 \\ 1 \end{bmatrix}$.

There was nothing remarkable about this illustration, but we use it as a background against which to view the next example. Suppose, instead, that the constraint had been

$$h(\mathbf{X}) = (0.5x_1^2 + x_2 - 5)^3 = 0$$

so that

$$\nabla h(\mathbf{X}) = \begin{bmatrix} 3x_1(0.5x_1^2 + x_2 - 5)^2 \\ 3(0.5x_1^2 + x_2 - 5)^2 \end{bmatrix}$$

The important point is that the constraint curve in Figure 4.6 represents this *new constraint* exactly also; the locus of points in x_1, x_2 space along which $(0.5x_1^2 + x_2 - 5)^3 = 0$ is $0.5x_1^2 + x_2 - 5 = 0$ or $x_2 = -0.5x_1^2 + 5$, exactly as before. The trouble is that the multiplicative term $(0.5x_1^2 + x_2 - 5)^2$ in both elements of $\nabla h(\mathbf{X})$ is *also* equal to zero at any point on this curve, which means that $\nabla h(\mathbf{X}) = \mathbf{0}$ *everywhere along this constraint*—despite the fact that the solution-space geometry of the two problems is exactly the same. In conjunction with $\nabla f(\mathbf{X}) = \begin{bmatrix} -2(x_1 - 5) \\ -2(x_2 - 5) \end{bmatrix}$, the requirements in (4.10a) are then

$$\begin{bmatrix} -2(x_1 - 5) \\ -2(x_2 - 5) \end{bmatrix} = \alpha \begin{bmatrix} 0 \\ 0 \end{bmatrix}$$

which is only possible if $\nabla f(\mathbf{X}) = \mathbf{0}$, meaning $\mathbf{X} = \begin{bmatrix} 5 \\ 5 \end{bmatrix}$, and this is impossible because of the constraint.

In this particular example, we can find $\mathbf{X}^*$ by solving $h(\mathbf{X}) = 0$ for, say, x_2, substituting into $f(\mathbf{X})$, and then finding the unconstrained maximum of this function of (now) only x_1. (This is not an effective *general* procedure.) This leads to $\mathbf{X}^* = \begin{bmatrix} 1.8474 \\ 3.2935 \end{bmatrix}$, just as in the previous example. Note that the first-order conditions in (4.10) or (4.12) will not hold at this constrained maximum point. Geometrically, since $\nabla h(\mathbf{X})$ is everywhere the null vector, no amount of stretching or shrinking will ever produce $\nabla f(\mathbf{X}^*) = \begin{bmatrix} 6.3052 \\ 3.4130 \end{bmatrix}$.

This illustrates the kind of impasse that can occur when a problem has *constraints* with null gradients. There may be a unique constrained maximum or minimum point that the conditions in (4.10) or (4.12) will fail to identify, and so for such problems these cannot be necessary conditions. We will explore such constraint peculiarities further in what follows in this chapter and also later, in Chapter 10, for problems with inequality constraints. In essence, most first-order necessary conditions for constrained maximization and minimization problems are designed for problems with well-behaved constraints that do not exhibit the kinds of peculiarities illustrated in the two-variable case by $\nabla h(\mathbf{X}) = \mathbf{0}$. Also, for problems with more variables and more constraints (whether equations or inequalities), the constraint peculiarities can be more subtle than in this illustration.

4.2.7   A Note on the Form of the Lagrangian Function

The Lagrangian function in (4.11) is in the format $L = f(\mathbf{X}) - \lambda[g(\mathbf{X}) - c]$. This reflects two choices: (1) the operation attaching the second term—$\lambda[g(\mathbf{X}) - c]$—was subtraction; and (2) in rearranging the constraint equation $g(\mathbf{X}) = c$, the right-hand side, c, was moved to the left.

If the Lagrangian were constructed by *adding* the second term, so that $L = f(\mathbf{X}) + \lambda[g(\mathbf{X}) - c]$, the $\partial L/\partial x_j = 0$ conditions would produce exactly the same necessary relationships between the x_j's; only the definition of λ in terms of the x_j would be opposite in sign to that derived from (4.11). The $\partial L/\partial \lambda$ expression would reproduce the original constraint exactly, instead of its negative. Therefore the same $\mathbf{X}^*$ would be found. However, since λ, and in particular λ^*, would be of *opposite* sign, then $\lambda^* = -(df^*/dc)$ or $(df^*/dc = -\lambda^*)$. This distinction will turn out to be particularly important when we consider inequality constraints in Section 4.5. We prefer the form in which the constraint term is subtracted, because then λ^* and df^*/dc are of the *same* sign, but this is just a matter of taste.

In attaching the constraint, it would also have been possible to use $[c - g(\mathbf{X})]$, since for $\mathbf{X}$ satisfying $g(\mathbf{X}) = c$, this would also be zero. A Lagrangian function in the form $L = f(\mathbf{X}) - \lambda[c - g(\mathbf{X})]$ will generate the same necessary relationships from $\partial L/\partial x_j$, except that, again, λ will have a sign opposite to that from (4.11). Therefore, in this form, $\lambda^* = -(df^*/dc)$, as is also true with $L = f(\mathbf{X}) + \lambda[g(\mathbf{X}) - c]$.

Finally, if L is formed as $f(\mathbf{X}) + \lambda[c - g(\mathbf{X})]$, then the $\partial L/\partial x_j$ will be identical to those derived from (4.11), and the original constraint will be exactly reproduced by $\partial L/\partial \lambda$; multiplication by (-1), as with (4.11), will not be necessary.

Therefore, for either

$$L = f(\mathbf{X}) - \lambda[g(\mathbf{X}) - c] \qquad \text{or} \qquad L = f(\mathbf{X}) + \lambda[c - g(\mathbf{X})]$$

the property $\lambda^* = df^*/dc$ holds. In this text we will use $L = f(\mathbf{X}) - \lambda[g(\mathbf{X}) - c]$, throughout. Other texts sometimes use other versions. The second-order conditions are unchanged, whatever form the Lagrangian function takes.

4.3   EXTENSION TO MORE DEPENDENT VARIABLES

4.3.1   First-Order Conditions

Let $\mathbf{X} = [x_1, \ldots, x_n]'$ (where $n > 2$) and consider maximization or minimization of $f(\mathbf{X})$ subject to one equality constraint $h(\mathbf{X}) = 0$. The necessary conditions are easily found as a direct extension of the Lagrange multiplier approach in (4.10) or (4.12), as

$$(a) \begin{cases} f_1 - \lambda h_1 = \cdots = f_n - \lambda h_n = 0 \\ \text{or} \quad \nabla f - \lambda[\nabla h] = \mathbf{0} \\ \text{or} \quad \nabla f = \lambda[\nabla h] \end{cases} \qquad \text{and} \qquad (b) \ h(\mathbf{X}) = 0 \qquad (4.18)$$

These conditions are derived from the fact that, as before, we require $df = 0$ provided that we take into account the relationships between the dx_j that are found by taking the total differential of both sides of the constraint. We have two homogeneous linear equations involving now the n unknowns dx_j

$$f_1 dx_1 + f_2 dx_2 + \cdots + f_n dx_n = 0$$
$$h_1 dx_1 + h_2 dx_2 + \cdots + h_n dx_n = 0$$

or, in matrix form

$$\begin{bmatrix} f_1 & f_2 & \cdots & f_n \\ h_1 & h_2 & \cdots & h_n \end{bmatrix} \begin{bmatrix} dx_1 \\ dx_2 \\ \vdots \\ dx_n \end{bmatrix} = \mathbf{0} \qquad (4.19)$$

and letting $\mathbf{dX} \equiv [dx_1, dx_2, \ldots, dx_n]'$, this can be expressed as

$$\begin{bmatrix} [\nabla f]' \\ [\nabla h]' \end{bmatrix} [\mathbf{dX}] = \mathbf{0}$$

For these equations to have nontrivial basic solutions—to allow variation in all the x_j's through nonzero values of the associated dx_j's—requires that the rank of the $2 \times n$ coefficient matrix in (4.19) be less than 2 (Section 2.5). This can be achieved by *proportionality* between the two rows:

$$[f_1 \quad f_2 \cdots f_n] = \lambda[h_1 \quad h_2 \cdots h_n] \qquad \text{or} \qquad \nabla f = \lambda[\nabla h]$$

which is exactly the requirement in (4.18a).

These conditions can also be generated directly by partial differentiation of the the Lagrangian function $L = f(\mathbf{X}) - \lambda h(\mathbf{X})$ and setting $\nabla L = \mathbf{0}$. Extending earlier notation, $\nabla L_{\mathbf{X}} \equiv [\partial L/\partial x_1, \ldots, \partial L/\partial x_n]'$ and $\nabla L_\Lambda \equiv \partial L/\partial \lambda$, the $\nabla L = \mathbf{0}$ condition means $\nabla L_{\mathbf{X}} = \mathbf{0}$ and $\nabla L_\Lambda = 0$, which are exactly the requirements in (4.18a) and (4.18b), respectively.

The problem that would be caused by a constraint for which $\nabla h = \mathbf{0}$ is just an n-dimensional generalization of the illustration above for $f(x_1, x_2)$. Again, if $\nabla h(\mathbf{X}^*) = \mathbf{0}$, then the only way to satisfy the requirements in (4.18a) is if $\nabla f(\mathbf{X}^*) = \mathbf{0}$ also, meaning that the unconstrained optimum point happens to satisfy the constraint—in which case the problem is not one of constrained optimization at all. Obviously, this is not generally the case; usually $\nabla f(\mathbf{X}^*) \neq \mathbf{0}$, and so the conditions in (4.18a), $\nabla f(\mathbf{X}^*) = \lambda[\nabla h(\mathbf{X}^*)]$, cannot be met when $\nabla h(\mathbf{X}^*) = \mathbf{0}$. Therefore, as in the case of a function of two variables, the requirements of (4.18a) are necessary conditions for a constrained optimum only for problems in which $\nabla h(\mathbf{X}^*) \neq \mathbf{0}$.

4.3.2   Second-Order Conditions

Conditions for distinguishing maxima from minima for a function of n variables subject to one constraint parallel those for determining the sign of the general n-variable quadratic form $Q(x_1, \ldots, x_n)$ subject to $c_1 x_1 + \cdots + c_n x_n = 0$, which was explored in Section 4.1. The second differential elements in $d\mathbf{X}$ are not independent, because of the linear side condition. The revised expression for $d^2 f^*$ is exactly parallel to that in the case of $f(x_1, x_2)$ in Section 4.2 [as in (4.15)], namely, $d^2 f^* = [d\mathbf{X}]\mathbf{V}^*[d\mathbf{X}]$, where $\mathbf{V}^*$ is the $n \times n$ matrix of elements $v_{jk}^* = f_{jk}^* - \lambda^* h_{jk}^*$ (the Hessian matrix of the Lagrangian function), evaluated at $\mathbf{X}^*$. Then border this matrix with the elements of ∇h^*, giving

$$\overline{\mathbf{V}}^* = \begin{bmatrix} 0 & [\nabla h^*]' \\ \hline [\nabla h^*] & \mathbf{V}^* \end{bmatrix} \tag{4.20}$$

[This is exactly (4.16), only now ∇h^* is an n-element vector and $\mathbf{V}^*$ is an $n \times n$ matrix.]

For a minimum at $\mathbf{X}^*$, the results in (4.2) require that all leading principal minors of $\overline{\mathbf{V}}^*$, beginning with the third, must be *negative*. For $d^2 f^* < 0$, and hence a maximum at $\mathbf{X}^*$, these same leading principal minors must *alternate*, beginning *positive*.

If the constraint is linear, so that it defines a plane (or hyperplane)—which is a convex set of points—then information about the convexity or concavity of $f(\mathbf{X})$ would serve to distinguish maxima and minima without recourse to second differential conditions.

4.3.3   Example

We add a third variable to the example of the last section. Let $f(x_1, x_2, x_3) = 6x_1 x_2 - 10x_3^2$, where the variables are required to satisfy $h(\mathbf{X}) = 2x_1 + x_2 + 3x_3 - 10 = 0$. Here $\nabla f = [6x_2, 6x_1, -20x_3]'$ and $\nabla h = [2, 1, 3]'$ and the requirements in (4.18a) lead to $x_2 = 2x_1$ and $x_3 = -0.9x_1$. Substitution into the constraint gives $x_1 = 7.6923$; consequently $x_2 = 15.3846$ and $x_3 = -6.9231$.

For the second differential test we also need second partials and cross partials of $f(\mathbf{X})$ and $h(\mathbf{X})$. Here

$$\nabla^2 f = \begin{bmatrix} 0 & 6 & 0 \\ 6 & 0 & 0 \\ 0 & 0 & -20 \end{bmatrix} \quad \text{and} \quad \nabla^2 h = \begin{bmatrix} 0 & 0 & 0 \\ 0 & 0 & 0 \\ 0 & 0 & 0 \end{bmatrix}$$

Since all $h_{ij} (= h^*_{ij}) = 0$, $v^*_{ij} = f^*_{ij}$, and since all second partials and cross-partials of $f(\mathbf{X})$ are constants, $f^*_{ij} = f_{ij}$ throughout, so we can omit all the asterisks and we have

$$\overline{\mathbf{V}} = \begin{bmatrix} 0 & 2 & 1 & 3 \\ 2 & 0 & 6 & 0 \\ 1 & 6 & 0 & 0 \\ 3 & 0 & 0 & -20 \end{bmatrix}$$

from which

$$|\overline{\mathbf{V}}_3| = \begin{vmatrix} 0 & 2 & 1 \\ 2 & 0 & 6 \\ 1 & 6 & 0 \end{vmatrix} = 12 + 12 = 24 > 0$$

and

$$|\overline{\mathbf{V}}_4| = |\overline{\mathbf{V}}| = -156 < 0$$

The correct leading principal minors alternate in sign, beginning positive, so $\mathbf{X}^* = [7.6923, 15.3846, -6.9231]'$ represents a maximum for $f(\mathbf{X}) = 6x_1x_2 - 10x_3^2$ subject to $2x_1 + x_2 + 3x_3 = 10$.

4.4 EXTENSION TO MORE CONSTRAINTS

The nature of the dependence among the n variables may be expressed by more than one constraint. Consider $f(\mathbf{X})$ with m constraining equations $h^1(\mathbf{X}) = 0, \ldots, h^m(\mathbf{X}) = 0$.[9] It is appropriate to assume that there are fewer constraints than variables ($m < n$) because of the structure of the first differential conditions, shown below. For the case of *linear* equations, we saw in Section 2.6 that the existence of more equations than variables always implies either redundant equations (which can be dropped) or an inconsistent system in which not all equations can be met simultaneously. We denote $\partial h^i / \partial x_j$ by h^i_j, the partial derivative of the ith constraint with respect to the jth variable, and we use ∇h^i to represent the n-element gradient vector of partial derivatives of the ith constraint with respect to $x_1, \ldots, x_n$.

4.4.1 First-Order Conditions

The first differential conditions are the same as always; we impose the requirement that $df = 0$, but subject to the connections between the dx_j ($j = 1, \ldots, n$) that are implicit in each of the m constraints $h^i(\mathbf{X}) = 0$ ($i = 1, \ldots, m$).

$$\begin{aligned} df &= f_1 dx_1 + f_2 dx_2 + \cdots + f_n dx_n = 0 \\ dh^1 &= h^1_1 dx_1 + h^1_2 dx_2 + \cdots + h^1_n dx_n = 0 \\ &\vdots \qquad \vdots \qquad \vdots \qquad \qquad \vdots \qquad \vdots \\ dh^m &= h^m_1 dx_1 + h^m_2 dx_2 + \cdots + h^m_n dx_n = 0 \end{aligned} \qquad (4.21a)$$

---

[9]Recall from Section 3.3 that we use $h^i(\mathbf{X})$ rather than $h_i(\mathbf{X})$ to designate the ith constraint because the latter is generally used to indicate the partial derivative of a function $h(x_1, \ldots, x_n)$ with respect to x_i.

In gradient notation, the first equation in (4.21a) is simply $[\nabla f]\circ[\mathbf{dX}] = 0$ and the remaining set of m equations is

$$[\nabla h^i]\circ[\mathbf{dX}] = 0 \qquad (i = 1, \ldots, m)$$

Finally, letting

$$\mathbf{M} = \begin{bmatrix} f_1 & f_2 & \cdots & f_n \\ h_1^1 & h_2^1 & \cdots & h_n^1 \\ \vdots & \vdots & & \vdots \\ h_1^m & h_2^m & \cdots & h_n^m \end{bmatrix} = \begin{bmatrix} [\nabla f]' \\ [\nabla h^1]' \\ \vdots \\ [\nabla h^m]' \end{bmatrix}$$

we can represent the equations in (4.21a) as

$$\mathbf{M}[\mathbf{dX}] = \mathbf{0} \tag{4.21b}$$

This set of $m + 1$ homogeneous linear equations in the n variables $dx_1, \ldots, dx_n$ is the n-variable, m-constraint extension of the two-variable, one-constraint case in Section 4.2. The system of equations is always consistent, as we saw in Chapter 2. For nontrivial solutions to all square $(m + 1) \times (m + 1)$ subsystems of (4.21)—that is, to all *basic* systems—the rank of the coefficient matrix must be less than $m + 1$. In fact, if redundant constraints have been eliminated (and if the problem is well behaved, so that $\nabla h^i(\mathbf{X}) \neq 0$, for any i), the rank will be m. Then, for example, the first row of $\mathbf{M}$ can be expressed as a linear combination of the remaining m (linearly independent) rows:

$$[f_1, f_2, \ldots, f_n] = \sum_{i=1}^{m} \lambda_i[h_1^i, \ldots, h_n^i] \qquad \text{or} \qquad \nabla f = \sum_{i=1}^{m} \lambda_i[\nabla h^i]$$

These, plus the constraints, make up the first-order conditions:

$$\nabla f = \sum_{i=1}^{m} \lambda_i[\nabla h^i] \qquad \text{or} \qquad \nabla f - \sum_{i=1}^{m} \lambda_i[\nabla h^i] = \mathbf{0} \tag{4.22a}$$

$$h^i(\mathbf{X}) = 0 \qquad (i = 1, \ldots, m) \tag{4.22b}$$

As before, the Lagrange approach, setting first partials equal to zero, generates exactly these conditions. Here

$$L(x_j, \lambda_i) = f(\mathbf{X}) - \sum_{i=1}^{m} \lambda_i[h^i(\mathbf{X})] \tag{4.23}$$

and it is easily seen that the extension of first-order conditions from earlier sections is again simply stated as

$$\nabla L_{\mathbf{X}} = \mathbf{0} \qquad \text{and} \qquad \nabla L_{\Lambda} = \mathbf{0} \tag{4.24}$$

where, as before, $\nabla L_{\mathbf{X}} = [\partial L/\partial x_1, \ldots, \partial L/\partial x_n]'$ and now, with m constraints, $\nabla L_{\Lambda} = [\partial L/\partial \lambda_1, \ldots, \partial L/\partial \lambda_m]'$.

The *Jacobian* (Carl Gustav Jacob Jacobi, German, 1804–1851) matrix of the system of constraining equations $h^i(\mathbf{X}) = 0$ $(i = 1, \ldots, m)$ is defined as

$$\mathbf{J} = \begin{bmatrix} \dfrac{\partial h^1}{\partial x_1} & \cdots & \dfrac{\partial h^1}{\partial x_n} \\[2mm] \dfrac{\partial h^2}{\partial x_1} & \cdots & \dfrac{\partial h^2}{\partial x_n} \\[1mm] \vdots & & \vdots \\[1mm] \dfrac{\partial h^m}{\partial x_1} & \cdots & \dfrac{\partial h^m}{\partial x_n} \end{bmatrix} = \begin{bmatrix} [\nabla h^1]' \\[2mm] [\nabla h^2]' \\[1mm] \vdots \\[1mm] [\nabla h^m]' \end{bmatrix}$$

This is sometimes written as $\mathbf{J} = [\partial(h^1, \ldots, h^m)/\partial(x_1, \ldots, x_n)]$. Using Jacobian notation, the matrix in (4.21b) is $\mathbf{M} = \begin{bmatrix} [\nabla f]' \\ \mathbf{J} \end{bmatrix}$. Letting $\mathbf{h}(\mathbf{X}) = [h^1(\mathbf{X}), \ldots, h^m(\mathbf{X})]'$, and $\boldsymbol{\Lambda} = [\lambda_1, \ldots, \lambda_m]'$, the first-order conditions in (4.22) are

(a) $\nabla f = \mathbf{J}' \boldsymbol{\Lambda}$ and (b) $\mathbf{h}(\mathbf{X}) = \mathbf{0}$ (4.22′)

(You might write out a small example—say, three variables and two constraints—to see that this is correct.)

The vector geometry of these conditions is a logical extension of that for problems with only one constraint (Sections 4.2 and 4.3), but it is less easy to visualize because of the dimensions required. The trouble with dimensions arises because the smallest number of constraints (more than one) is when $m = 2$, and since $m < n$, this must be associated with $n \geq 3$. For $n = 3$ (the smallest possible), $f(x_1, x_2, x_3)$ is a surface in four-dimensional space (!), although the *contours* of $f(\mathbf{X})$ will be surfaces in only three dimensions. If you visualize well in three-dimensional space, suppose that there are two linear constraints, so that they will be planes. They will intersect in a line [the assumption that the rank of the coefficient matrix in (4.21) is m ensures that these planes are neither parallel nor coplanar], and their gradients at any point along that line will be distinguishably different vectors in three-dimensional space. The conditions in (4.22a) require that at a candidate point, $\mathbf{X}^*$ [which, by (4.22b), must be on that line of intersection of the two planes], $\nabla f(\mathbf{X}^*)$ must be able to be expressed as a linear combination of those two constraint gradients [again, assuming that $\nabla h^i(\mathbf{X}^*) \neq \mathbf{0}$ for $i = 1,2$, so that $\nabla h^1(\mathbf{X}^*)$ and $\nabla h^2(\mathbf{X}^*)$ span the three-dimensional space in which $\mathbf{X}^*$ lies].

4.4.2 Second-Order Conditions

The second differential conditions require positive definiteness or negative definiteness of a more widely bordered Hessian matrix. These conditions parallel the results in Section 4.1, specifically in (4.3), for a quadratic form in $x_1, \ldots, x_n$ subject to m linear homogeneous equations $\mathbf{CX} = \mathbf{0}$ $(m < n)$. Obviously, for problems in which m and n are large, the work required to evaluate $(n - m)$ leading principal minors becomes immense, both because each determinant is large and because there may be many of them. The *size* of the determinants depends on the sizes of m and n (the last leading principal minor is always a determinant of size $m + n$), whereas the *number* that must be evaluated depends on the difference between the two. These conditions are presented here primarily to indicate the symmetry with all that has gone before and also to serve as a background against which to compare the simplicity of some of the methods that will be explored later in this book to deal with both linear and nonlinear programming problems with large numbers of variables and constraints, although the latter are generally inequalities.

Consider the partitioned and more widely bordered Hessian matrix

$$
\overline{\mathbf{V}} = \left[
\begin{array}{c:c}
\begin{matrix} \mathbf{0} \\ (m \times m) \end{matrix} & \begin{matrix} \mathbf{J} \\ (m \times n) \end{matrix} \\
\hdashline
\begin{matrix} \mathbf{J}' \\ (n \times m) \end{matrix} & \begin{matrix} \mathbf{V} \\ (n \times n) \end{matrix}
\end{array}
\right]
\tag{4.25}
$$

The upper left is a null matrix. The off-diagonal blocks are the Jacobian and its transpose, a logical extension of the single row and column border in (4.20). The elements of $\mathbf{V}$ are extensions of the bracketed terms in expression (4.15). Specifically, $\mathbf{V} = [v_{jk}]$ where

$$
v_{jk} = f_{jk} - \sum_{i=1}^{m} \lambda_i h_{jk}^i
$$

Following (4.3), the rank of the Jacobian matrix must be m. Then for $d^2 f^* > 0$ (for $\mathbf{X}^*$ to represent a constrained minimum point), the last $n - m$ leading principal minors of $\overline{\mathbf{V}}^*$ must have the sign $(-1)^m$. For $d^2 f^* < 0$ (for a maximum at $\mathbf{X}^*$), these same leading principal minors must alternate in sign, starting with $(-1)^{m+n}$.

4.4.3   Example

Consider the preceding example to which a second constraint is added:

$$
\text{Maximize } f(\mathbf{X}) = 6x_1 x_2 - 10x_3^2
$$

where the variables must satisfy

$$
h^1(\mathbf{X}) = 2x_1 + x_2 + 3x_3 - 10 = 0
$$

(as in Section 4.3) and also

$$
h^2(\mathbf{X}) = x_1 + x_2 + 2x_3 - 8 = 0
$$

For this problem

$$
\nabla f = \begin{bmatrix} 6x_2 \\ 6x_1 \\ -20x_3 \end{bmatrix}, \quad
\nabla h^1 = \begin{bmatrix} 2 \\ 1 \\ 3 \end{bmatrix}, \quad
\nabla h^2 = \begin{bmatrix} 1 \\ 1 \\ 2 \end{bmatrix}
$$

and, using (4.22), we have

$$
\text{(a)} \quad
\begin{cases}
6x_2 - 2\lambda_1 - \lambda_2 = 0 \\
6x_1 - \lambda_1 - \lambda_2 = 0 \\
-20x_3 - 3\lambda_1 - 2\lambda_2 = 0
\end{cases}
$$

$$(b) \begin{cases} 2x_1 + x_2 + 3x_3 = 10 \\ x_1 + x_2 + 2x_3 = 8 \end{cases}$$

From the first equation in (a), $\lambda_2 = 6x_2 - 2\lambda_1$; putting this result into the other two equations in (a) and rearranging gives $6x_1 + 6x_2 + 20x_3 = 0$. When this is combined with the other two equations in x_1, x_2, and x_3 in (b), we have a system of three (linear) equations in three unknowns whose coefficient matrix is nonsingular. Hence a unique solution exists; it is $\mathbf{X}^* = \begin{bmatrix} 8 \\ 12 \\ -6 \end{bmatrix}$.

For the second differential test we also need the elements for $\overline{\mathbf{V}}$ in (4.25). Here

$$\nabla^2 f = \begin{bmatrix} 0 & 6 & 0 \\ 6 & 0 & 0 \\ 0 & 0 & -20 \end{bmatrix}, \quad \nabla^2 h^1 = \begin{bmatrix} 0 & 0 & 0 \\ 0 & 0 & 0 \\ 0 & 0 & 0 \end{bmatrix}, \quad \nabla^2 h^2 = \begin{bmatrix} 0 & 0 & 0 \\ 0 & 0 & 0 \\ 0 & 0 & 0 \end{bmatrix}$$

Since all second partials and cross-partials are constants, and since all elements of ∇h^1 and ∇h^2 are also constants, all elements of $\overline{\mathbf{V}}^*$ are constants and so we do not need asterisks. Moreover, since all elements in $\nabla^2 h^1$ and $\nabla^2 h^2$ are zero (because both constraints are linear), $v_{jk} = f_{jk}$ for this example. Thus

$$\overline{\mathbf{V}} = \overline{\mathbf{V}}^* = \begin{bmatrix} 0 & 0 & 2 & 1 & 3 \\ 0 & 0 & 1 & 1 & 2 \\ 2 & 1 & 0 & 6 & 0 \\ 1 & 1 & 6 & 0 & 0 \\ 3 & 2 & 0 & 0 & -20 \end{bmatrix}$$

and the sign of the last $n - m = 1$ leading principal minor, which is just the sign of $|\overline{\mathbf{V}}|$, is of interest. For a minimum, the sign must be positive [the same as the sign of $(-1)^2$] and for a maximum it must be negative [the same as that of $(-1)^3$]. Straightforward application of the general definition of the determinant (along row 2 is a possible choice because of the small size of the nonzero elements) gives $|\overline{\mathbf{V}}^*| = -8$. Thus $\mathbf{X}^*$ represents a constrained maximum point for this problem.

In this particular illustration, with three variables and two constraints, the five equations in (a) and (b) are all linear. That makes them very easy to solve. In general, some or all of these equations will be nonlinear, and they may not be so easy to solve. Also, in general, the elements in $\overline{\mathbf{V}}$ will be functions of the x's, and if there are several solutions to the equations in (a) and (b), the last $n - m$ leading principal minors of $\overline{\mathbf{V}}$ will have to be evaluated for each of those solutions. Note, too, that if there are many more variables than constraints, there may be a great many principal minors to evaluate. For these kinds of reasons, approaches to nonlinear equations and alternatives to the second differential conditions are of great practical importance, as we will find in the remaining chapters of this book.

4.5   MAXIMA AND MINIMA WITH INEQUALITY CONSTRAINTS

In many real-world situations, the constraints in a problem take the form not of equations but rather of inequalities, setting upper and/or lower limits on combinations of the variables. For example, consumers are often faced with a number of alternative ways in which to spend their incomes; unless the set of choices includes every imaginable good, the budget constraint takes the form of an inequality, setting an upper limit (total disposable income) on the amount that can be spent. In this section we consider the implications of inequality constraints on the approach to locating maxima and minima. Geometrically, for $f(x_1,x_2)$, we are restricted not to just the part of a surface above a *line* or *curve* (as in Fig. 4.1) but rather to that part of $f(\mathbf{X})$ above an entire *area*, as in Figure 4.7. The set of points defined by the inequality constraint(s) is called the *feasible region* for the problem, and $f(\mathbf{X})$ is the *objective function*.

In general, inequality constrained problems are most effectively confronted using the techniques of mathematical programming—either linear programming, for the case in which $f(\mathbf{X})$ and all constraints are linear functions of the x's, or nonlinear programming, for the case in which $f(\mathbf{X})$ and/or one or more of the constraints are nonlinear functions. These topics are covered in Parts IV and V of this book. In the present section, we conclude our investigation of constrained optimization problems in this chapter by taking an initial and exploratory look at the kinds of approaches that might be used with inequality constraints.

4.5.1   Standard Forms for Inequality-Constrained Problems

We will consider the *standard form* for a maximization problem with one inequality constraint to be

$$\text{Maximize} \quad f(\mathbf{X}) \quad \text{subject to} \quad h(\mathbf{X}) \leq 0$$

where $\mathbf{X} = [x_1, \ldots, x_n]'$ and where $h(\mathbf{X}) = g(\mathbf{X}) - c$, so that the original form of the constraint was $g(\mathbf{X}) \leq c$. This is perfectly general in the sense that the direction of any inequality can be reversed by multiplying both sides by (-1); $g(\mathbf{X}) \leq c \; [h(\mathbf{X}) \leq 0]$ is the

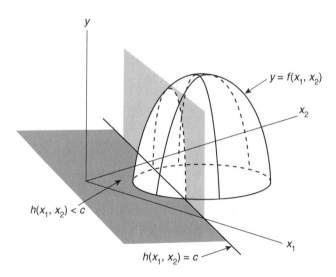

FIGURE 4.7   Inequality-constrained maximization problem.

same requirement as $-g(\mathbf{X}) \geq -c$ $[-h(\mathbf{X}) \geq 0]$. Logically, however, we generally expect that the constraint on a maximization problem will set an upper limit—$g(\mathbf{X}) \leq c$—keeping the variables in that problem from becoming extremely large. [A consumer trying to maximize utility faces an upper (budget) limit on the amount that can be spent.]

If there are several constraints, $g^i(\mathbf{X}) \leq c_i$, the maximization problem in standard form will be

$$\text{Maximize} \quad f(\mathbf{X}) \quad \text{subject to} \quad h^i(\mathbf{X}) \leq 0 \quad (i = 1, \ldots, m)$$

where $h^i(\mathbf{X}) = g^i(\mathbf{X}) - c_i$. In contrast to equality-constrained problems, it is not necessary that $m < n$.

Similarly, the standard form for a single inequality-constrained minimization problem will be

$$\text{Minimize} \quad f(\mathbf{X}) \quad \text{subject to} \quad h(\mathbf{X}) \geq 0$$

The variables in a minimization problem will be bounded from below—$g(\mathbf{X}) \geq c$—keeping them from becoming extremely small. If there are several constraints, $g^i(\mathbf{X}) \geq c_i$, the minimization problem in standard form will be

$$\text{Minimize} \quad f(\mathbf{X}) \quad \text{subject to} \quad h^i(\mathbf{X}) \geq 0 \quad (i = 1, \ldots, m)$$

4.5.2   More on Inequalities and Convex Sets

Linear Inequality Constraints   In Chapter 3 we explored conditions under which inequality constraints defined convex sets of points. With linear inequalities, the set of points satisfying all of them (the feasible region for the problem) is always convex. Here is an illustration. Let $h^1(\mathbf{X}) = x_1 + 2x_2 - 8 \leq 0$, $h^2(\mathbf{X}) = x_1 + x_2 - 5 \leq 0$, and $h^3(\mathbf{X}) = x_1 - 4 \leq 0$. These are illustrated in Figure 4.8. The shaded area, including the boundary sides and the corners, is the convex feasible region defined by these three linear inequalities.

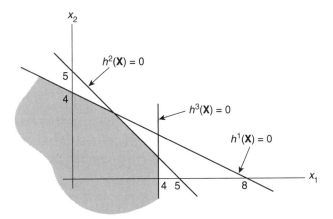

FIGURE 4.8   Convex feasible region.

Additional constraints would either cut away at the feasible region by adding a new side and reducing the feasible region [as would happen if $h^4(\mathbf{X}) = x_1 + 10x_2 - 38 \leq 0$ were added to the figure (you might try it)] or else they would be redundant because their boundary lies outside the already defined region [as would be the case if $h^5(\mathbf{X}) = x_1 + x_2 - 10 \leq 0$ were added to Fig. 4.8; again, you might try it]. This is why there is no reason to require fewer constraints than unknowns when the constraints are inequalities.

Nonlinear Inequality Constraints   With nonlinear constraints, $h(\mathbf{X}) \leq 0$ defines a convex set of points when $h(\mathbf{X})$ is itself a *convex* (or *quasiconvex*) function; also, several constraints $h^i(\mathbf{X}) \leq 0$ $(i = 1, \ldots, m)$ define a convex set of points when each of the $h^i(\mathbf{X})$ is a convex (or quasiconvex) function of the x's .[10]

Suppose $h^1(\mathbf{X}) = x_1^2 + 4x_2^2 - 16$ and $h^2(\mathbf{X}) = (x_1 - 3)^2 + (x_2 - 3)^2 - 9$. Both of these functions are easily found to be *convex* by the leading principal minors test on their Hessian matrices. Figure 4.9a shows the functions in three-dimensional $x_1,x_2,h(\mathbf{X})$ space. Contours on the x_1x_2 plane and the *convex* feasible region (shaded) defined by $h^1(\mathbf{X}) \leq 0$ and $h^2(\mathbf{X}) \leq 0$ are shown in Figure 4.9b.

By contrast, suppose that the two constraints are $h^1(\mathbf{X})$, as above, and $h^3(\mathbf{X}) = -(x_1 - 3)^2 - (x_2 - 3)^2 + 9$, which is the negative of $h^2(\mathbf{X})$, above. Logic suggests, and the leading principal minors test confirms, that $h^3(\mathbf{X})$ is a *concave* function. Figure 4.10a shows these two functions in three-dimensional space, and the *nonconvex* feasible region in x_1,x_2 space defined by $h^1(\mathbf{X}) \leq 0$ and $h^3(\mathbf{X}) \leq 0$ is indicated (shaded) in Figure 4.10b.

Conversely, an inequality constraint of the form $h(\mathbf{X}) \geq 0$ defines a convex set of points when $h(\mathbf{X})$ is a *concave* (or *quasiconcave*) function of the variables; if several $h^i(\mathbf{X})$ are each concave (or quasiconcave) functions, then the set of points satisfying all the constraints $h^i(\mathbf{X}) \geq 0$ is also a convex set. We will see that if convexity/concavity information is available for the functions in a problem [$f(\mathbf{X})$ and the $h^i(\mathbf{X})$], it can play a significant role in distinguishing maxima from minima in inequality constrained problems.

4.5.3   One Inequality Constraint

One Approach: Ignore It   If a problem has just one inequality constraint, then a straightforward approach is simply to ignore the constraint and see what happens. Find the one or more values of $\mathbf{X}^*$ that satisfy $\nabla f(\mathbf{X}) = \mathbf{0}$ for an unconstrained maximum or minimum, examine second-order conditions to sort out maxima, minima, and points of inflection, and check whether $\mathbf{X}^*$ satisfies the inequality constraint.

For a maximization problem with the constraint $h(\mathbf{X}) \leq 0$, if $\mathbf{X}^*$ represents a maximum point and $h(\mathbf{X}^*) \leq 0$, then the problem is solved; the unconstrained maximum point lies in the feasible region. If $h(\mathbf{X}^*) > 0$, then the hilltop lies outside the boundary (as would be the case in the illustration in Fig. 4.7), and so the problem should be redone, this time with the constraint imposed as an equality; in this case, Lagrangian techniques would be appropriate. The logic is that in moving toward the absolute maximum

---

[10]Subtraction of a constant term (like c) from a concave (convex) function leaves a concave (convex) function; only position, not shape, has been changed. Thus, since $h(\mathbf{X}) \equiv g(\mathbf{X}) - c$, the concavity (convexity) of either $h(\mathbf{X})$ or $g(\mathbf{X})$ implies the concavity (convexity) of the other; they have the *same* shape. If we had defined $h(\mathbf{X})$ as $c - g(\mathbf{X})$, then $h(\mathbf{X})$ and $g(\mathbf{X})$ would be of *opposite* shape.

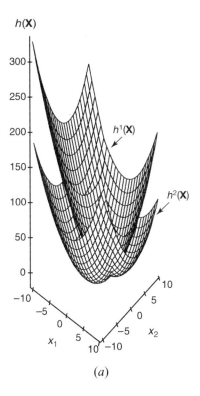

(a)

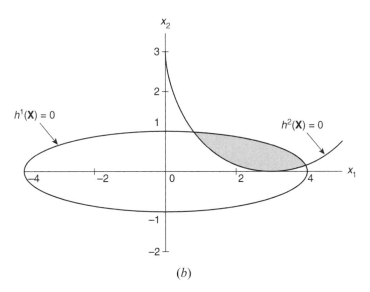

(b)

FIGURE 4.9   (a) $h^1(\mathbf{X}) = x_1^2 + x_2^2 - 16$; $h^2(\mathbf{X}) = (x_1 - 3)^2 + (x_2 - 3)^2 - 9$; (b) feasible region defined by $h^1(\mathbf{X}) \le 0$ and $h^2(\mathbf{X}) \le 0$.

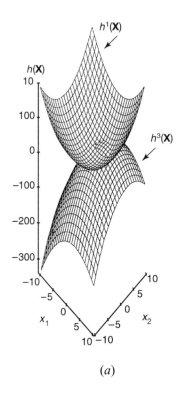

(a)

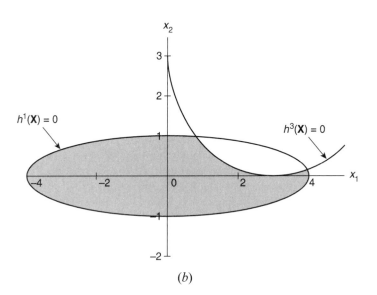

(b)

FIGURE 4.10 (a) $h^1(\mathbf{X}) = x_1^2 + x_2^2 - 16$, $h^3(\mathbf{X}) = -(x_1 - 3)^2 - (x_1 - 3)^2 + 9$; (b) feasible region defined by $h^1(\mathbf{X}) \le 0$ and $h^3(\mathbf{X}) \le 0$.

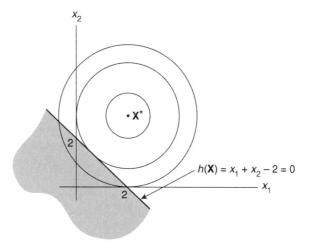

FIGURE 4.11   Contours of $f(\mathbf{X})$ and feasible region for example (a).

(which cannot be attained, because of the constraint), we would be willing to stop only when the boundary is hit.[11]

Similarly, for a minimization problem in standard form, an unconstrained minimum at $\mathbf{X}^*$ must be checked in the constraint $h(\mathbf{X}) \geq 0$, and logic is similar. If $h(\mathbf{X}^*) \geq 0$, the problem is solved; if $h(\mathbf{X}^*) < 0$, the unconstrained minimum point is outside the boundary set by the constraint. Again, the problem should be redone with the constraint imposed as an equality, using Lagrangian techniques.

Numerical Examples and the $\lambda^* \geq 0$ Rule   Here are some numerical examples that illustrate the link between the algebra and the geometry for a maximization problem with one inequality constraint. These provide an introduction (without any proofs) of what will turn out to be a general approach to problems with any number of inequality constraints. (You can easily construct minimization problems that illustrate these same principles.)

(a)   Maximize $f(\mathbf{X}) = -x_1^2 - x_2^2 + 4x_1 + 6x_2$ subject to $h(\mathbf{X}) = x_1 + x_2 - 2 \leq 0$. The geometry in x_1, x_2 space is shown in Figure 4.11. Contours of $f(\mathbf{X})$ are circles; the smaller the circle, the larger the value of $f(\mathbf{X})$. Ignoring the constraint, it is easily established that first-order conditions are met only at $\mathbf{X}^* = \begin{bmatrix} 2 \\ 3 \end{bmatrix}$. The Hessian matrix in this case is made up of constants—$\mathbf{H}^* = \mathbf{H} = \begin{bmatrix} -2 & 0 \\ 0 & -2 \end{bmatrix}$. The leading principal minors alternate, beginning

---

[11]To simplify the discussion, we will generally assume in this section that there is only one value of $\mathbf{X}^*$ that satisfies first-order conditions for an unconstrained maximum. There could be none [if $f(\mathbf{X})$ were a tilted plane, for example]. If there are several relative maxima in a maximization problem, then each must be checked in the constraint. If *all* those relative maxima satisfy the constraint (more than one hilltop in the feasible region), then $f(\mathbf{X})$ must be evaluated for each in order to determine the best. If at least one violates the constraint (hilltops both inside and outside the feasible region), then the problem needs to be done again with the constraint as an equation, and $f(\mathbf{X})$ evaluated with the solution to that problem must then be compared to the best of the unconstrained solutions in the feasible region.

negative, and so $f(\mathbf{X})$ is strictly concave (looks like an overturned bowl) and $\mathbf{X}^*$ represents a maximum point. However, since $h(\mathbf{X}^*) = 3 \ (> 0)$, the constraint is violated. Therefore we impose the constraint as an equation.

ALGEBRA. The Lagrangian function is

$$L = f(\mathbf{X}) - \lambda[h(\mathbf{X})] = -x_1^2 - x_2^2 + 4x_1 + 6x_2 - \lambda(x_1 + x_2 - 2)$$

and the associated first-order conditions are:

$$L_1 = -2x_1 + 4 - \lambda = 0$$
$$L_2 = -2x_2 + 6 - \lambda = 0$$
$$L_\lambda = -(x_1 + x_2 - 2) = 0$$

This leads to a new candidate point, $\mathbf{X}^* = \begin{bmatrix} 0.5 \\ 1.5 \end{bmatrix}$.

From examination of the second-order conditions in the unconstrained case, we know that $f(\mathbf{X})$ is a strictly concave function whose hilltop is outside the boundary set by $x_1 + x_2 = 2$. Therefore, we know that imposing the constraint as an equation is appropriate. This is corroborated by the fact that $\lambda^* = 3 \ (> 0)$, found by substituting $\mathbf{X}^*$ into either of the first two first-order conditions. From Section 4.2 we know that $\lambda^* = df^*/dc$; so $\lambda^* > 0$ indicates that an outward movement of the boundary line (a relaxation of the constraint) would generate a larger value for $f(\mathbf{X}^*)$. This is an algebraic signal, from the Lagrange multiplier, that the hilltop is outside the boundary and therefore that imposing the constraint as an equation is the correct procedure.

GEOMETRY. We saw, also in Section 4.2, that the first-order conditions for $\mathbf{X}^*$ to either maximize or minimize $f(\mathbf{X})$, subject to one *equality* constraint, $h(\mathbf{X}) = 0$, required $\nabla f^* = \lambda^*[\nabla h^*]$. For this example, we have (Fig. 4.12a)

$$\mathbf{X}^* = \begin{bmatrix} 0.5 \\ 1.5 \end{bmatrix}, \qquad \nabla f^* = \begin{bmatrix} 3 \\ 3 \end{bmatrix}, \qquad \lambda^* = 3, \qquad \nabla h = \begin{bmatrix} 1 \\ 1 \end{bmatrix} \text{ (everywhere)}$$

so indeed at $\mathbf{X}^*$ the collinearity condition is met:

$$\begin{bmatrix} 3 \\ 3 \end{bmatrix} = (3) \begin{bmatrix} 1 \\ 1 \end{bmatrix}$$

The unconstrained hilltop is now marked by a flag.

For an upper-bound inequality constraint, $h(\mathbf{X}) \leq 0$, ∇h evaluated anywhere on the boundary points in a direction away from the feasible region. For a lower bound constraint, $h(\mathbf{X}) \geq 0$, as would occur in a minimization problem in standard form, ∇h evaluated anywhere on the boundary $h(\mathbf{X}) = 0$ points *into* the feasible region. You can easily see this by exploring a few simple numerical examples.

Therefore, when we impose the constraint in this problem as an equation, because the hilltop is outside the feasible region, we recognize that λ^* must be positive in $\nabla f^* = \lambda^*[\nabla h^*]$. Not only must ∇f^* and ∇h^* be collinear, but both must point in the same direction, outward from the boundary and toward the hilltop, as in the illustration in Figure 4.12a.

(b) If it were the (unusual) case that $\mathbf{X}^*$ happened to be exactly on the boundary set by $h(\mathbf{X}) = 0$, then $\nabla f^* = 0$. This would happen with $f(\mathbf{X}) = -x_1^2 - x_2^2 + 4x_1 + 6x_2$ if the constraint were $h(\mathbf{X}) = x_1 + x_2 - 5 \leq 0$ (Fig. 4.12b). Since it will not *generally* be the case that $\nabla h^* = 0$ also, the $\nabla f^* = \lambda^*[\nabla h^*]$ requirement can always be met in this situation with $\lambda^* = 0$. Therefore, the requirement that $\lambda^* \geq 0$ covers both cases of the hilltop on the boundary as well as strictly outside it.

(c) Maximize $f(\mathbf{X}) = -x_1 - x_2 + 4x_1 + 6x_2$ [as in (a) and (b)], subject to $x_1 + x_2 \leq 10$. The unconstrained solution for this objective function has already been found in (a) at $\mathbf{X}^* = \begin{bmatrix} 2 \\ 3 \end{bmatrix}$. Since $f(\mathbf{X})$ is strictly concave, $\mathbf{X}^*$ represents a maximum; it also satisfies the constraint, and so the problem is solved.

ALGEBRA. As an illustration of the effectiveness of the $\lambda^* \geq 0$ rule, suppose that we (incorrectly) went ahead and imposed the constraint in this problem as an equality. The condition in $\nabla L_{\mathbf{X}} = \mathbf{0}$ is the same as in (a); only L_λ has changed and now we have $L_\lambda = -(x_1 + x_2 - 10) = 0$, which leads to $x_1^* = 4.5$ and $x_2^* = 5.5$. The second-order conditions on $\overline{\mathbf{V}}^*$ indicate that $\mathbf{X}^*$, indeed, represents a maximum point for $f(\mathbf{X})$ along the boundary provided by the constraint as an equation. However, substituting $\mathbf{X}^*$ into either of the $\nabla L_{\mathbf{X}} = \mathbf{0}$ equations leads to $\lambda^* = -5$ (< 0), signaling trouble.

GEOMETRY. The constraint as an equation is $x_1 + x_2 = 10$, and the unconstrained maximum of $f(\mathbf{X})$ lies *inside* that boundary. The issue is to establish that fact when we do not have the picture in front of us. This is what λ^* does. Here $\mathbf{X}^* = \begin{bmatrix} 4.5 \\ 5.5 \end{bmatrix}$, $\nabla f^* = \begin{bmatrix} -5 \\ -5 \end{bmatrix}$, while $\nabla h = \begin{bmatrix} 1 \\ 1 \end{bmatrix}$, as in (a). The negative elements in ∇f^* mean that *decreases* in x_1 and x_2 from their current values at $\mathbf{X}^*$ would *increase* $f(\mathbf{X})$; the hilltop is *inside* the boundary set by the constraint. As usual, ∇h^* points *outward* from the feasible region. Therefore, to satisfy $\nabla f^* = \lambda^*[\nabla h^*]$ requires $\lambda^* = -5$, thus violating the $\lambda^* \geq 0$ requirement when the constraint is imposed as an equality (see Fig. 4.12c).

(d) Maximize $f(\mathbf{X}) = x_1^2 + x_2^2 - 4x_1 - 6x_2$, subject to $h(\mathbf{X}) = x_1 + x_2 - 2 \leq 0$, as in (a). Here $f(\mathbf{X})$ is the negative of the objective function in (a), so while the contours look the same in two-dimensional x_1, x_2 space, we are now looking down into the bottom of a bowl rather than onto a hilltop in Figure 4.9; smaller circles now identify lower values of $f(\mathbf{X})$.

If we go ahead and impose the usual first-order conditions for an unconstrained extremum, we again find $\mathbf{X}^* = \begin{bmatrix} 2 \\ 3 \end{bmatrix}$, which is bad in all respects; it violates the constraint, as before, and the Hessian matrix test on $f(\mathbf{X})$ identifies it as a strictly convex function so that $\mathbf{X}^*$ represents a minimum instead of a maximum.

In this case, the fact that the constraint is violated does *not* mean that we should redo the problem with the constraint as an equation, because we have found the wrong kind of extreme point. In fact, $f(\mathbf{X})$ has no finite maximum (for example, it becomes increasingly large for larger and larger negative values of x_1 and x_2).

Suppose that we (again, incorrectly) had gone ahead and imposed the constraint as an equality.

ALGEBRA. As in (a), we would find $\mathbf{X}^* = \begin{bmatrix} 0.5 \\ 1.5 \end{bmatrix}$. We would also find $\lambda^* = -3$ (< 0), telling us that moving the equality constraint outward would lead to *lower* values of

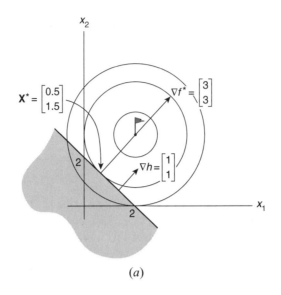

(a)

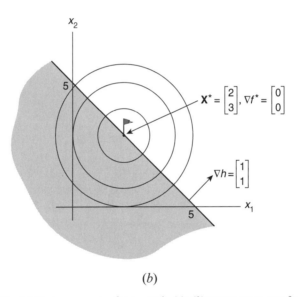

(b)

FIGURE 4.12   (a) Vector geometry for example (a); (b) vector geometry for example (b).

$f(\mathbf{X}^*)$. We also find that now $\nabla f^* = \begin{bmatrix} -3 \\ -3 \end{bmatrix}$. These facts are letting us know that the maximum (which occurs when x_1 and x_2 are infinitely small) is indeed inside the boundary set by the constraint.

GEOMETRY.   Figure 4.12d illustrates this problem. It is the same as in Figure 4.12a, except that now the flag indicates the bottom of a bowl and ∇f^* points in the opposite direction. It is impossible to satisfy $\nabla f^* = \lambda^* \nabla h^*$ with $\lambda^* \geq 0$, so the equality constrained approach is wrong.

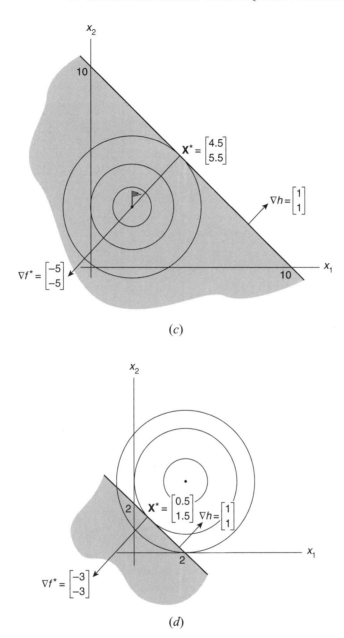

FIGURE 4.12   (*c*) vector geometry for example (c); (*d*) vector geometry for example (d).

4.5.4 More than One Inequality Constraint

With more than one inequality constraint, if the unconstrained hilltop lies outside the feasible region, it need not violate all the constraints. Even if all constraints *are* violated, you cannot conclude that the problem should be reworked with all constraints as equations. That simple kind of "either/or" approach that worked well with only one constraint is now inappropriate, and a more systematic technique is needed.

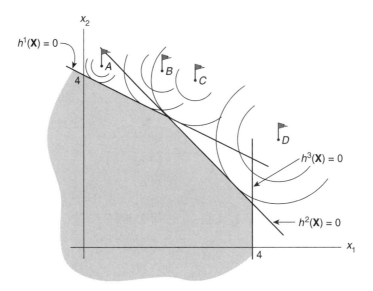

FIGURE 4.13   Alternative hilltops outside feasible region.

The kinds of issues raised by several inequality constraints are easily illustrated using the example of three linear inequality constraints in x_1, x_2 space that was shown in Figure 4.10. Four alternative possible locations (A through D) for a hilltop that is outside the feasible region are shown in Figure 4.13; the constraints are those from Figure 4.10. Suppose that the objective function to be maximized is an overturned bowl so that it has circular contours. For circles centered on any particular hilltop, a larger radius means lower values of $f(\mathbf{X})$. Table 4.1 contains some pertinent facts about points A through D.

At A, where only $h^1(\mathbf{X}) \leq 0$ is violated, it is indeed correct to do the problem with that constraint imposed as an equation. But at B, where both $h^1(\mathbf{X}) > 0$ and $h^2(\mathbf{X}) > 0$, the correct constrained maximum is found along only $h^1(\mathbf{X}) = 0$, because that constraint turns out to be the more restrictive when hill climbing toward B. With a hilltop at C, it is appropriate to impose both of the first two constraints as equations. But for a point like D, where all three constraints are violated, it is clearly not correct to assume that all three constraints should be imposed as equalities. The constrained maximum point is at a point on $h^2(\mathbf{X}) = 0$ and inside the boundaries imposed by the first and third constraints.

The exception would be if, say, three equations with two unknowns each intersected in a unique point. We saw in Chapter 2 that this would be a case of $m > n$ with a unique solution $[\rho(\mathbf{A}) = \rho(\mathbf{A} \vdots \mathbf{B})]$, so any one of the three equations could be eliminated without changing the solution. With three *inequality* constraints of two unknowns each, if the three intersect in a common point, then one of them (but not *any* one) can be eliminated without changing the feasible region; you can easily verify this with a sketch for the $m = 3$, $n = 2$ case.

You can easily imagine alternative illustrations that would put the constrained maximum point on only $h^3(\mathbf{X}) = 0$. You can also imagine alternative locations for the hilltop outside feasible region that would put the constrained maximum point at the intersection of $h^2(\mathbf{X}) = 0$ and $h^3(\mathbf{X}) = 0$ (but not at the intersection of the first and third constraints; why?).

TABLE 4.1   Alternative Hilltops Outside the Feasible Region

Location of Unconstrained Hilltop	Constraints Violated at Unconstrained Hilltop	Location of Constrained Maximum Point
A	$h^1(\mathbf{X}) \leq 0$	On $h^1(\mathbf{X}) = 0$
B	$h^1(\mathbf{X}) \leq 0$ and $h^2(\mathbf{X}) \leq 0$	On $h^1(\mathbf{X}) = 0$
C	$h^1(\mathbf{X}) \leq 0$ and $h^2(\mathbf{X}) \leq 0$	On $h^1(\mathbf{X}) = 0$ and $h^2(\mathbf{X}) = 0$
D	$h^1(\mathbf{X}) \leq 0$, $h^2(\mathbf{X}) \leq 0$ and $h^3(\mathbf{X}) \leq 0$	On $h^2(\mathbf{X}) = 0$

These geometric considerations point the way to a systematic approach to inequality constrained problems. It is reasonable to expect that the maximum point will be located either inside the feasible region or else somewhere on the boundary, either on an edge or at a corner.[12] This logic also applies to a problem in which some or all the constraints are nonlinear functions. A consistent sequence of steps for a problem with m inequality constraints would be:

(i)   Ignore all constraints.

(ii) Impose constraints as equalities, one at a time.

(iii) Impose constraints as equalities, two at a time.

(iv) Impose constraints as equalities, three at a time (if possible), ..., then $(m-1)$ at a time (if possible), then m at a time (again, if possible).

In each case, examine the other constraints to see whether the candidate $\mathbf{X}^*$ satisfies them. The trouble is, satisfaction of all constraints may not be enough. For example, consider the hilltop at D in Figure 4.13.

(i) *Ignore Constraints.* This generates D, which is rejected because it violates all three constraints.

(ii) *Equalities, One at a Time.* With $h^1(\mathbf{X}) = 0$, a point of tangency between a contour of $f(\mathbf{X})$ and that constraint would be outside the boundary set by the second constraint. Along the constraint $h^2(\mathbf{X}) = 0$, the correct constrained maximum will be found. Along $h^3(\mathbf{X}) = 0$, the local maximum would violate both the first and second constraints.

(iii) *Equalities, Two at a Time.* With both $h^1(\mathbf{X}) = 0$ and $h^2(\mathbf{X}) = 0$, their intersection at $\mathbf{X} = \begin{bmatrix} 2 \\ 3 \end{bmatrix}$ is found. While this clearly satisfies all three constraints, we can see from the figure that this is not the best constrained maximum point; it is on a lower objective function contour than the (true) maximum along $h^2(\mathbf{X}) = 0$. What is needed is an algebraic signal to reject a point like $\mathbf{X} = \begin{bmatrix} 2 \\ 3 \end{bmatrix}$, for cases when the geometry of the problem is not available to us.

---

[12]The notions of "edge" and "corner" become increasingly complicated with increases in the number of variables, n, and hence in the number of dimensions for the feasible region geometry. We ignore these complications for the present.

This signal will come from the signs of the λ's for the equality constraints at that point. In this case (and as we will see in a numerical illustration later), λ_1 would be negative because relaxation of that constraint (moving it outward a little), in conjunction with the stationary $h^2(\mathbf{X}) = 0$ boundary, would lead to a new intersection point on a *lower* contour of $f(\mathbf{X})$.

You can work through the details for the case of both $h^2(\mathbf{X}) = 0$ and $h^3(\mathbf{X}) = 0$, which leads to $\mathbf{X} = \begin{bmatrix} 4 \\ 1 \end{bmatrix}$, where λ_3 would be negative. Notice that the point identified by the final *pair* of equations, $h^1(\mathbf{X}) = 0$ and $h^3(\mathbf{X}) = 0$, is $\mathbf{X} = \begin{bmatrix} 4 \\ 2 \end{bmatrix}$, which would immediately be rejected because it violates the second constraint.

(iv) *Equalities, Three at a Time.* This is impossible for this example that has three linear equalities with two unknowns.

What we learn from all this is that a *systematic* approach is needed—one that sorts out constraints that will be *binding* (met as equations) from those that can be ignored (*nonbinding*, met as strict inequalities) at the optimum solution to a constrained maximization or minimization problem. It is just such a procedure that we now explore.

4.5.5   A Systematic Approach: The Kuhn–Tucker Method (Harold W. Kuhn, USA, 1925– and Albert W. Tucker, USA, 1905–1995)[13]

One Inequality Constraint   We illustrate the basic ideas of the Kuhn–Tucker approach by looking initially at a problem with one inequality constraint. We continue to assume that all constraints in any problem satisfy Kuhn and Tucker's constraint qualification so that the $\nabla h^i(\mathbf{X}^*) = 0$ impasse mentioned in Sections 4.2–4.4 does not arise. Kuhn and Tucker suggest that we face the inequality constraint $h(\mathbf{X}) \le 0$ squarely and convert it to an equation by adding a variable to the left-hand side to make up the difference between $h(\mathbf{X})$ and 0, for any set of values of the x's . Since the left-hand side of the inequality must be less than or equal to the right-hand side, this added variable must be nonnegative. It is often called a "slack" variable, since it serves to take up the slack between the left- and right-hand sides of the constraint. If we add the slack variable, s, squared, then we are assured that the new term on the left will be nonnegative; so $h(\mathbf{X}) \le 0$ becomes $h(\mathbf{X}) + s^2 = 0$. The maximization problem is now

$$\text{Maximize} \quad f(\mathbf{X})$$
$$\text{subject to} \quad h(\mathbf{X}) + s^2 = 0$$

This is a straightforward equality constrained maximization problem with $n + 1$ variables (the n original x's and one slack variable, s) and one constraint.[14] Lagrange multiplier techniques are now completely appropriate. Forming the Lagrange function

---

[13]Kuhn and Tucker presented this general approach to inequality-constrained optimization problems (Kuhn and Tucker, 1950). There is some question regarding proper attribution of these ideas. For example, Takayama (1985) cites reference to an apparently little-known M.A. thesis at the University of Chicago (W. Karush, *Minimization of Functions of Several Variables with Inequalities as Side Conditions*). Some writers now refer to the *Karush–Kuhn–Tucker* (KKT) conditions. Also, since there is a Lagrange multiplier flavor to the Kuhn–Tucker approach, the term *Kuhn–Tucker–Lagrange* is sometimes used. Additionally, Fritz John (1948) published similar work. We continue with the label *Kuhn–Tucker* conditions, whether it is historically perfectly accurate or not, because it is firmly extablished in much of the literature in this field.

[14]Note that, had only s been added, it would then have been necessary to append a new constraint requiring that $s \ge 0$, and we are back to a problem with an inequality constraint, which is just what we are trying to avoid.

$$L = f(\mathbf{X}) - \lambda[h(\mathbf{X}) + s^2] \tag{4.26}$$

the usual kinds of first-order conditions, $\nabla L = \mathbf{0}$, are generated. This is now a set of $(n + 2)$ equations:

$$\frac{\partial L}{\partial x_j} = f_j - \lambda h_j = 0 \qquad (j = 1, \ldots, n) \tag{4.27a}$$

$$\frac{\partial L}{\partial \lambda} = -[h(\mathbf{X}) + s^2] = 0 \tag{4.27b}$$

$$\frac{\partial L}{\partial s} = -2\lambda s = 0 \tag{4.27c}$$

Expressed in terms of the partial gradients of L, the conditions in (4.27a) are found from $\nabla L_{\mathbf{X}} = \mathbf{0}$, which is a vector of n equations; those in (4.27b) and (4.27c) are from $\nabla L_{\Lambda} = 0$ and $\nabla L_{\mathbf{S}} = 0$, respectively.

Consider (4.27c). The minus sign and the 2 can be ignored (multiply both sides of the equation by $-\frac{1}{2}$). The implication of $\lambda s = 0$ is that either $\lambda = 0$ or $s = 0$, or possibly both. But if $s = 0$, then from (4.27b), again after eliminating the redundant negative sign, $h(\mathbf{X}) = 0$. Therefore s, which was added to the problem only to convert it to equality-constraint form, can be eliminated from the first-order conditions in (4.27), and the results in (4.27b) and (4.27c) can be replaced by a single equation involving λ and $h(\mathbf{X})$ only:

$$\lambda[h(\mathbf{X})] = 0 \tag{4.27d}$$

Thus the n equations in (4.27a) and the one in (4.27d) contain the remaining first-order conditions in terms of the original variables, x_j, and the Lagrange multiplier, λ. In this process, however, the *direction* of the original inequality has been lost, and so $h(\mathbf{X}) \leq 0$ must be introduced as an explicit requirement. The complete first-order conditions are therefore

$$f_j - \lambda h_j = 0 \, (j = 1, \ldots, n) \qquad \text{or} \qquad \nabla f = \lambda[\nabla h] \tag{4.28a}$$

$$\lambda[h(\mathbf{X})] = 0 \tag{4.28b}$$

$$h(\mathbf{X}) \leq 0 \tag{4.28c}$$

and we know from the illustrations earlier in this section that we must have $\lambda^* \geq 0$, since we employed the Lagrange multiplier approach on an equality constraint. These results, the Kuhn–Tucker conditions, are often stated as follows:

The necessary conditions for a point $\mathbf{X}^*$ to be a relative maximum of $f(\mathbf{X})$ subject to $h(\mathbf{X}) \leq 0$ are that a nonnegative λ exists such that it and $\mathbf{X}^*$ sat- (4.29)
isfy (4.28)

Alternatively, the nonnegativity of λ is simply added to (4.28), giving

$$f_j - \lambda h_j = 0 \, (j = 1, \ldots, n) \qquad \text{or} \qquad \nabla f = \lambda[\nabla h] \tag{4.29'a}$$

$$\lambda[h(\mathbf{X})] = 0 \tag{4.29'b}$$

$$h(\mathbf{X}) \leq 0 \qquad\qquad (4.29'\text{c})$$

$$\lambda \geq 0 \qquad\qquad (4.29'\text{d})$$

Necessary conditions for a single *equality* constrained problem were given in (4.18); they are repeated here, for reference:

$$\nabla f = \lambda[\nabla h] \qquad\qquad (4.18\text{a})$$

$$h(\mathbf{X}) = 0 \qquad\qquad (4.18\text{b})$$

Conditions (4.18a) are required for either an equality or an inequality constrained problem. However, (4.29'd) alters the geometric interpretation of conditions (4.29'a) when the constraint is an inequality. For an equality constraint, $\nabla f = \lambda[\nabla h]$ meant that at a maximum point the two gradients had to point either in the same direction $\lambda > 0$) or in precisely opposite directions ($\lambda < 0$).[15] Now, for an inequality constrained problem, since $\lambda < 0$ is not allowed, the two gradients in conditions (4.29'a) must point in the same direction (again, ignoring the $\lambda = 0$ possibility).

In addition, (4.18b), which reproduces the equality constraint, is replaced in (4.29') by three conditions: two inequalities [the original constraint, $h(\mathbf{X}) \leq 0$, and the nonnegativity requirement on λ] and a nonlinear equation, $\lambda[h(\mathbf{X})] = 0$.

This mix of inequalities and nonlinear equations, in (4.29'b–4.29'd), changes the approach to a solution. One technique is to find the value (or values) of the variables that satisfy the *equations* in (4.29'a) and (4.29'b) and then check it (or them) in the *inequalities* (4.29'c) and (4.29'd). In particular, the nonlinear equation $\lambda[h(\mathbf{X})] = 0$ requires that one find solutions by assuming (i) $\lambda = 0$ and (ii) $h(\mathbf{X}) = 0$. In general, then, there will be more than one $\mathbf{X}^*$ to check in the inequalities.

If the shape characteristics of the functions are known exactly, we can make a more powerful statement:

> If $f(\mathbf{X})$ is *concave* and the constraint function, $h(\mathbf{X})$, is *convex* (or *quasi-convex*), then conditions (4.29) are also *sufficient* for a local maximum.   (4.30)

Under these concavity/convexity assumptions, the Kuhn–Tucker conditions become *both* necessary and sufficient for a maximum; a point satisfying them is a maximum, and if a point is a maximum, it must satisfy them. Finally, if $f(\mathbf{X})$ is *strictly* concave, the maximum will be an *absolute* one.[16]

Geometry of the Kuhn–Tucker Conditions   Consider the implications of (4.29'b). If $\lambda = 0$, the conditions in (4.29'a) reduce to $\nabla f = \mathbf{0}$, precisely as in an *unconstrained* problem. Geometrically, we are finding the unconstrained hilltop. If it turns out to be inside the boundary, as required by (4.29'c), it will be the solution to the problem; if it violates the constraint, it will be rejected.

On the other hand, if $h(\mathbf{X}) = 0$ in (4.29'b), then the constraint is being treated as an

---

[15]We explored the geometric characteristics of the special case in which $\lambda = 0$ in Section 4.2.
[16]We examine possible relaxations of the shape conditions on both $f(\mathbf{X})$ (such as *quasiconcavity*) and on constraints in more detail in Chapter 10.

equation, and we find the maximizing point along the boundary. If $\lambda > 0$ at that point, as required by (4.29′d), it is the correct optimum.

So the algebra of the requirements in (4.29′) forces us to explore what we know to be the two logical possibilities for a maximization problem with one inequality constraint—either the maximum is inside the boundary, in which case the constraint can be ignored ($\lambda = 0$), or else the maximum is outside of the boundary, in which case the constrained maximum will be on the boundary, where the constraint is an equation [$h(\mathbf{X}) = 0$].

Example   In the example in Section 4.2 we looked at the problem of maximizing $f(\mathbf{X}) = 6x_1x_2$, subject to $2x_1 + x_2 = 10$. There we found a maximum at $x_1^* = 2.5$ and $x_2^* = 5$. Suppose now that the constraint had been an inequality: $2x_1 + x_2 \leq 10$ (or $2x_1 + x_2 - 10 \leq 0$).

Using the Kuhn–Tucker conditions in (4.29′), we have

$$\begin{bmatrix} 6x_2 \\ 6x_1 \end{bmatrix} = \lambda \begin{bmatrix} 2 \\ 1 \end{bmatrix} \tag{4.31a}$$

$$\lambda(2x_1 + x_2 - 10) = 0 \tag{4.31b}$$

$$(2x_1 + x_2 - 10) \leq 0 \tag{4.31c}$$

$$\lambda \geq 0 \tag{4.31d}$$

From (4.31a) it is easily established that

$$6x_2 = 2\lambda, \qquad 6x_1 = \lambda, \qquad \text{and so} \qquad x_2^* = 2x_1^* \tag{4.32}$$

From (4.31b), we need to examine the implications of $\lambda = 0$ and of $2x_1 + x_2 = 10$. In particular, if $\lambda = 0$, does (4.31c) hold, and if $2x_1 + x_2 = 10$, does (4.31d) hold? If there is more than one alternative that does meet all conditions, evaluate $f(\mathbf{X}^*)$ and then select that point for which $f(\mathbf{X}^*)$ is largest. Only if the shape (convexity and concavity) conditions are also met can we be sure that these conditions are sufficient also.

Case 1   Let $\lambda^* = 0$. Then, from (4.32), $\mathbf{X}^* = \begin{bmatrix} 0 \\ 0 \end{bmatrix}$. The constraints in (4.31c) and (4.31d) are clearly met, and so this $\mathbf{X}^*$ (the origin) satisfies all requirements of (4.29), the Kuhn–Tucker *necessary* conditions for a maximum. Here $f^* = 0$. However, for this particular function, where $f(\mathbf{X}) = 6x_1x_2$, $\mathbf{H} = \mathbf{H}^* = \begin{bmatrix} 0 & 6 \\ 6 & 0 \end{bmatrix}$, so that $|\mathbf{H}_1^*| = 0$ and $|\mathbf{H}_2^*| = -36$, and $f(\mathbf{X})$ is neither strictly concave nor strictly convex.

In this case, the origin, found when the constraint is ignored ($\lambda^* = 0$) represents a saddle point for $f(\mathbf{X})$. In the first and third quadrants, where both variables are positive or both negative, $f(\mathbf{X}) > 0$. In the second and fourth quadrants, where one variable is positive and the other negative, $f(\mathbf{X}) < 0$. There is clearly no finite absolute maximum, since increasingly large negative values for both x_1 and x_2 generate increasingly large values for $f(\mathbf{X})$, while still satisfying the inequality constraint that $2x_1 + x_2 \leq 10$.

Case 2   Let $2x_1 + x_2 = 10$ in (4.31b). Putting the results in (4.32) into this equation, we find that $\mathbf{X}^* = \begin{bmatrix} 2.5 \\ 5 \end{bmatrix}$. From (4.31a), this establishes that $\lambda^* = 15$, which satisfies (4.31d).

Hence, again, all conditions in (4.29′) are satisfied; here $f(\mathbf{X}^*) = 75$. From this we can conclude that $\mathbf{X}^* = \begin{bmatrix} 2.5 \\ 5 \end{bmatrix}$ represents a *local* (relative) maximum to $6x_1x_2$ subject to $2x_1 + x_2 \leq 10$. This is certainly true; small deviations (while still satisfying the constraint) around this point give smaller values to $f(\mathbf{X})$.

Several Inequality Constraints   The problem now is

$$\text{Maximize } f(\mathbf{X})$$
$$\text{subject to} \qquad h^1(\mathbf{X}) \leq 0, \ldots, h^m(\mathbf{X}) \leq 0 \tag{4.33}$$

Generalization to more than one inequality constraint is similar to the extension for equality-constrained cases. The Kuhn–Tucker approach is applicable no matter how many variables (n) and how many constraints (m). (Remember that it is not necessary to assume that $m < n$, as in equality-constrained problems.)

After adding a new squared slack variable to the left-hand side of *each* inequality in order to convert it to an equation, the appropriate Lagrange function is a direct m-constraint analog to the equality-constrained case in (4.23):

$$L = f(\mathbf{X}) - \sum_{i=1}^{m} \lambda_i[h^i(\mathbf{X}) + s_i^2]$$

Using $\mathbf{\Lambda} = [\lambda_1, \ldots, \lambda_m]'$, and $\mathbf{S} = [s_1, \ldots, s_m]'$, the first-order conditions require as usual that $\nabla L = \mathbf{0}$, which means $\nabla L_{\mathbf{X}} = \mathbf{0}$ (n equations), $\nabla L_{\mathbf{\Lambda}} = \mathbf{0}$ (m equations), and $\nabla L_{\mathbf{S}} = \mathbf{0}$ (another m equations). These $n + 2m$ equations are the m-constraint extension of the requirements in (4.27).

In particular, the $\nabla L_{\mathbf{\Lambda}} = \mathbf{0}$ conditions are

$$\frac{\partial L}{\partial \lambda_i} = -[h^i(\mathbf{X}) + s_i^2] = 0 \qquad (i = 1, \ldots, m)$$

and the requirements in the $\nabla L_{\mathbf{S}} = \mathbf{0}$ equations are

$$\frac{\partial L}{\partial s_i} = -2\lambda_i s_i = 0 \qquad (i = 1, \ldots, m)$$

These are the m equation extensions of (4.27b) and (4.27c), and they can be reexpressed, eliminating the slack variables s_i through the same kind of argument as in the case of only one constraint. Eliminating the -2 from the second set of equations, we have $\lambda_i s_i = 0$; so either $\lambda_i = 0$ or $s_i = 0$, or both. When $s_i = 0$, then $h^i(\mathbf{X}) = 0$ (from the first set of equations). Therefore these $2m$ equations can be expressed as

$$\lambda_i[h^i(\mathbf{X})] = 0 \qquad (i = 1, \ldots, m)$$

As before, the constraints must be reintroduced explicitly, and we know that the λ_i must be nonnegative. Therefore, the Kuhn–Tucker necessary conditions for the problem in (4.33) are

$$f_j - \sum_{i=1}^{m} \lambda_i \frac{\partial h_i}{\partial x_j} = 0 \quad (j = 1, \ldots, n) \qquad \text{or} \qquad \nabla f = \sum_{i=1}^{m} \lambda_i [\nabla h^i] \qquad (4.34\text{a})$$

$$\lambda_i [h^i(\mathbf{X})] = 0 \quad (i = 1, \ldots, m) \qquad\qquad\qquad (4.34\text{b})$$

$$h^i(\mathbf{X}) \leq 0 \quad (i = 1, \ldots, m) \qquad\qquad\qquad (4.34\text{c})$$

$$\lambda_i \geq 0 \quad (i = 1, \ldots, m) \qquad\qquad\qquad (4.34\text{d})$$

Again, these are also sufficient conditions for a maximum if $f(\mathbf{X})$ is a *concave* function and all the constraint functions $h^i(\mathbf{X})$ $(i = 1, \ldots, m)$ are *convex* or *quasiconvex*. If $f(\mathbf{X})$ is *strictly* concave, the maximum is an *absolute* one.[24]

As in Section 4.4, using $\mathbf{h}(\mathbf{X}) = [h^1(\mathbf{X}), \ldots, h^m(\mathbf{X})]'$ and recalling the notation for the Jacobian to a set of functional relations, $\mathbf{J} = [\partial h^i / \partial x_j]$, the problem can be expressed as

$$\begin{aligned} \text{Maximize} \quad & f(\mathbf{X}) \\ \text{subject to} \quad & \mathbf{h}(\mathbf{X}) \leq \mathbf{0} \end{aligned} \qquad (4.33')$$

and the Kuhn–Tucker necessary conditions are

$$\nabla f = \mathbf{J}'\boldsymbol{\Lambda} \qquad\qquad\qquad (4.34'\text{a})$$

$$\boldsymbol{\Lambda} \circ [\mathbf{h}(\mathbf{X})] = \mathbf{0} \qquad\qquad\qquad (4.34'\text{b})$$

$$\mathbf{h}(\mathbf{X}) \leq \mathbf{0} \qquad\qquad\qquad (4.34'\text{c})$$

$$\boldsymbol{\Lambda} \geq \mathbf{0} \qquad\qquad\qquad (4.34'\text{d})$$

The first-order conditions for a problem with m *equality* constraints were given in (4.22); they are repeated here:

$$\nabla f = \mathbf{J}'\boldsymbol{\Lambda} \qquad\qquad\qquad (4.22\text{a})$$

$$\mathbf{h}(\mathbf{X}) = \mathbf{0} \qquad\qquad\qquad (4.22\text{b})$$

The relationships between these two sets of conditions are readily apparent. They follow exactly the same pattern as that already noted in comparing (4.29') and (4.18) for the cases of only one constraint.

Geometry Again   The conditions in (4.34) or (4.34') have, once again, intuitively appealing geometric interpretations. For simplicity only, we consider the case of two linear constraints and assume that $f(\mathbf{X})$ is strictly concave, so that there is just one hilltop. That hilltop may be inside the boundaries set by the two constraints, it may be situated so that at the highest feasible point only one constraint is binding, or it may be that both constraints are effective at the constrained maximum point.

---

[24]Again, the kinds of conclusions that are possible when $f(\mathbf{X})$ is only *quasiconcave* will be examined later, in Chapter 10.

This is illustrated in Figure 4.14, along with the appropriate gradients. The important point is to recognize that we are assured of exploring all of these alternatives when we exhaust all the options presented by the first-order conditions in (4.34b) or (4.34′b). We examine these in turn:

(i) Let $\lambda_1 = 0$ and $\lambda_2 = 0$. Then, from either (4.34a) or (4.34′a), $\nabla f = \mathbf{0}$, and the $\mathbf{X}^*$ (or $\mathbf{X}^*$'s) found are candidates for the *unconstrained* maximum point. If, indeed, the maximum is located inside both boundaries, then this exploration will identify it, and the point will satisfy both of the constraints in (4.34c) and the two nonnegativities in (4.34d). This is illustrated by the hilltop at A in Figure 4.14.

(ii) Let $h^1(\mathbf{X}) = 0$ and $\lambda_2 = 0$, so we are exploring along the boundary set by the first constraint. Condition (4.34a) then requires that $\nabla f = \lambda_1[\nabla h^1]$ only (since $\lambda_2 = 0$), and nonnegativity of λ_1 means that we are looking for a point at which ∇f and ∇h^1 are collinear and point in the same direction. A hilltop at B, as in Figure 4.14, would be detected by this boundary exploration and the $\mathbf{X}^*$ found along this boundary would be the correct maximizing point—it satisfies the second inequality in (4.34c) and, since the hilltop is outside the $h^1(\mathbf{X}) = 0$ boundary, λ_1^* will be positive [satisfying (4.34d)].

(iii) Let $\lambda_1 = 0$ and $h^2(\mathbf{X}) = 0$ [this is just the reverse of the case in (ii)]. Now we are searching for the maximum point along the second constraint boundary; also $\nabla f = \lambda_2[\nabla h^2]$ and the same interpretation as in (ii) is appropriate. The situation in Figure 4.14 with a hilltop at D illustrates a problem for which this exploration would identify the correct maximizing point.

(iv) Let $h^1(\mathbf{X}) = 0$ and $h^2(\mathbf{X}) = 0$. In this case, we are examining the point at which both constraints are equations. Then, from (4.34a), we obtain

$$\nabla f = \lambda_1[\nabla h^1] + \lambda_2[\nabla h^2] \tag{4.35}$$

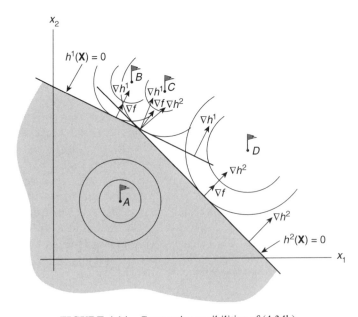

FIGURE 4.14   Geometric possibilities of (4.34b).

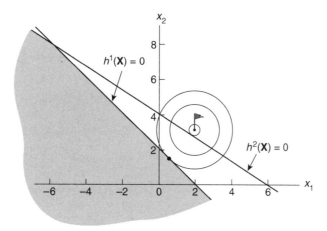

FIGURE 4.15   Geometry of example with two constraints.

Since, from (4.34d), $\Lambda \geq \mathbf{0}$, we are searching for a point at which ∇f is expressible as a nonnegative linear combination of the *two* constraint gradients. Recalling the geometry of linear combinations (multiplication of a vector by a scalar and addition of two vectors, from Section 1.4), the requirement in (4.35) means that ∇f must lie in the *cone* defined by ∇h^1 and ∇h^2 at $\mathbf{X}^*$.[25] For a problem with a hilltop at C, as in Figure 4.14, we would find the right $\mathbf{X}^*$ in this exploration.

Thus, imposition of the requirements in (4.34b) in the Kuhn–Tucker first-order conditions is just an automatic way of forcing a systematic exploration of all possible locations for the constrained maximizing point. We ignore the rare possibility in which the unconstrained maximum lies exactly on one of the boundaries or, even less probable, at the intersection of the two. In such atypical cases, simple unconstrained exploration will find the hilltop (as would the more tedious systematic exploration along the boundaries).

Example with Two Constraints   Consider maximizing $f(\mathbf{X}) = -x_1^2 - x_2^2 + 4x_1 + 6x_2$, subject to $h^1(\mathbf{X}) = x_1 + x_2 - 2 \leq 0$ and $h^2(\mathbf{X}) = 2x_1 + 3x_2 - 12 \leq 0$. The geometry in x_1, x_2 space is shown in Figure 4.15.

The Kuhn–Tucker conditions for the problem, directly from (4.34), are

$$\begin{cases} -2x_1 + 4 - \lambda_1 - 2\lambda_2 = 0 \\ -2x_2 + 6 - \lambda_1 - 3\lambda_2 = 0 \end{cases} \tag{4.36a}$$

$$\begin{cases} \lambda_1(x_1 + x_2 - 2) = 0 \\ \lambda_2(2x_1 + 3x_2 - 12) = 0 \end{cases} \tag{4.36b}$$

---

[25]In two-dimensional vector space, a cone is generated by all nonnegative linear combinations of two linearly independent vectors. Multiplication of a vector by a scalar simply stretches it (if the scalar is greater than 1) or shrinks it (if the scalar is less than 1). And the vector that is the sum of two others is the diagonal of the parallelogram for which the two original vectors for adjacent sides. In three dimensions, a cross section of a cone defined by three linearly independent vectors will be a triangle, and so forth. In general, the term is a *convex polyhedral cone*.

$$\begin{cases} x_1 + x_2 - 2 \le 0 \\ 2x_1 + 3x_2 - 12 \le 0 \end{cases} \tag{4.36c}$$

$$\lambda_1 \ge 0, \lambda_2 \ge 0 \tag{4.36d}$$

As usual, we begin with the implications of the conditions in (4.36b).

(i) Let $\lambda_1 = 0$ and $\lambda_2 = 0$, so we are searching for the unconstrained hilltop. From (4.36a), we find $\mathbf{X}^* = \begin{bmatrix} 2 \\ 3 \end{bmatrix}$, which violates both inequality constraints in (4.36c).

(ii) Let $x_1 + x_2 = 2$ and $\lambda_2 = 0$. This is exploration along the boundary set by the first constraint. With $\lambda_2 = 0$, the conditions in (4.36a) determine that $-x_1 + x_2 = 1$. These two linear equations in x_1 and x_2 have a unique solution $\mathbf{X}^* = \begin{bmatrix} 0.5 \\ 1.5 \end{bmatrix}$. You should check that all other conditions in (4.36) are met—in the process you will find $\lambda_1^* = 3$. Therefore $x_1^* = 0.5$ and $x_2^* = 1.5$ satisfy all the Kuhn–Tucker necessary conditions for a constrained maximum to this problem. The constraints are linear and hence convex, and therefore the feasible region is convex. We have already seen that this particular $f(\mathbf{X})$ is strictly concave, so the Kuhn–Tucker necessary conditions in (4.36) are also sufficient for a maximum.

(iii) Let $\lambda_1 = 0$ and $2x_1 + 3x_2 = 12$; now we are exploring along the boundary set by the second constraint. The conditions in (4.36a), along with $\lambda_1 = 0$, establish that $-3x_1 + 2x_2 = 0$. This, in conjunction with the equality constraint, defines a pair of linear equations in x_1 and x_2 with a unique solution $\mathbf{X}^* = \begin{bmatrix} 1.846 \\ 2.769 \end{bmatrix}$. This point violates the first inequality constraint [the first requirement in (4.36c)], and so the solution is invalidated.

(iv) Let $x_1 + x_2 = 2$ and $2x_1 + 3x_2 = 12$ Now both constraints are equations; their unique solution is $\mathbf{X}^* = \begin{bmatrix} -6 \\ 8 \end{bmatrix}$. However, putting this information into the requirements in (4.36a) generates $\lambda_2^* = -26$, which violates (4.36d) and disqualifies this solution.

This leaves $\mathbf{X}^* = \begin{bmatrix} 0.5 \\ 1.5 \end{bmatrix}$, found in (ii), as the constrained maximum point for this problem. Since we already established that $f(\mathbf{X})$ is everywhere concave (the elements of the Hessian are all constants), once $\mathbf{X}^*$ in (ii) was identified as both necessary and also sufficient, we could have stopped looking further. The calculations in (iii) and (iv) were carried out simply to illustrate how the Kuhn–Tucker conditions allow one to explore *all* potential locations for the constrained maximizing point.

As usual, the ease of solution of each of the possible cases depends entirely on the structure of the first partial derivative equations. There is no guarantee that any of these equations will be linear. If you have access to a good mathematical software computer program that solves systems of nonlinear equations effectively, then the four equations in four unknowns [from (4.36a) and (4.36b), for this particular example] could be solved simultaneously. The same four solutions as we found in (i) through (iv) would emerge; each would then have to be checked against the inequality requirements in (4.36c) and (4.36d), and only $\mathbf{X}^* = \begin{bmatrix} 0.5 \\ 1.5 \end{bmatrix}$ would pass all of these tests.

The number of options that must be considered in an inequality constrained problem grows quickly with the number of constraints. Each constraint contributes a nonlinear equation in the form of (4.34b); and for each such equation *two* options—$\lambda_i = 0$ and $h^i(\mathbf{X})$ = 0—must be investigated. With m constraints there are 2^m such options to be examined, one at a time. [For 8 constraints, this means 256 individual possibilities; for 10 constraints, the number quadruples to 1024(!).]

Some combinations of options may be impossible. This was illustrated with the three linear constraints and two unknowns in Figure 4.8 [and (4.13)]. In that example, there are eight (2^3) possibilities. One is the unconstrained case; three are the explorations along one constraint as an equation; another three come from imposing two of the three constraints as equations. The last case would be impossible, since it requires that all three constraints be imposed as equations.

Minimization Problems Using the Kuhn–Tucker Approach Minimization problems with inequality constraints can be approached in exactly the same Kuhn–Tucker manner. The only (but important) qualification is that the problems must be in *standard form*. For a problem with one constraint, this means

$$\text{Minimize } f(\mathbf{X})$$
$$\text{subject to } h(\mathbf{X}) \geq 0 \tag{4.37}$$

The Lagrangian function is formed by *subtracting* a nonnegative variable from the left-hand side of the constraint to make it an equality:

$$L = f(\mathbf{X}) - \lambda[h(\mathbf{X}) - s^2]$$

[cf. (4.26)]. Parallel to (4.27), we have the $\nabla L = \mathbf{0}$ conditions

(a) $\dfrac{\partial L}{\partial x_j} = f_j - \lambda h_j = 0$ (where $j = 1, \ldots, n$)

(b) $\dfrac{\partial L}{\partial \lambda} = -[h(\mathbf{X}) - s^2] = 0$

(c) $\dfrac{\partial L}{\partial s} = -2\lambda s = 0$

from which conditions exactly analogous to (4.28) can be deduced (except, of course, for the direction of the inequality):

$$f_j - \lambda h_j = 0 \ (j = 1, \ldots, n) \quad \text{or} \quad \nabla f = \lambda[\nabla h] \tag{4.38a}$$

$$\lambda[h(\mathbf{X})] = 0 \tag{4.38b}$$

$$h(\mathbf{X}) \geq 0 \tag{4.38c}$$

and the the Kuhn–Tucker necessary conditions are as follows:

The necessary conditions for a point $\mathbf{X}^*$ to be a relative minimum of $f(\mathbf{X})$ subject to $h(\mathbf{X}) \geq 0$ are that a nonnegative λ exist such that it and $\mathbf{X}^*$ satisfy (4.38).   (4.39)

The generalization to m constraints is completely parallel to the statement in (4.34):

$$f_j - \sum_{i=1}^{m} \lambda_i \frac{\partial h_i}{\partial x_j} = 0 \qquad (j = 1, \ldots, n) \tag{4.40a}$$

$$\lambda_i [h^i(\mathbf{X})] = 0 \qquad (i = 1, \ldots, m) \tag{4.40b}$$

$$h^i(\mathbf{X}) \geq 0 \qquad (i = 1, \ldots, m) \tag{4.40c}$$

$$\lambda_i \geq 0 \qquad (i = 1, \ldots, m) \tag{4.40d}$$

or, in compact form, as in (4.34'):

$$\nabla f = \mathbf{J}' \mathbf{\Lambda} \tag{4.40'a}$$

$$\mathbf{\Lambda} \circ [\mathbf{h}(\mathbf{X})] = 0 \tag{4.40'b}$$

$$\mathbf{h}(\mathbf{X}) \geq \mathbf{0} \tag{4.40'c}$$

$$\mathbf{\Lambda} \geq \mathbf{0} \tag{4.40'd}$$

For the minimization problem, these Kuhn–Tucker necessary conditions are also sufficient if $f(\mathbf{X})$ is a *convex* function and the constraint $h(\mathbf{X})$ [or constraints, $h^i(\mathbf{X})$] are *concave* or *quasiconcave*, so that the region defined by them is a convex set. And the minimum will be an *absolute* one if $f(\mathbf{X})$ is *strictly* convex.

For example, minimize $f(\mathbf{X}) = x_1^2 + x_2^2 + x_3^2$ subject to (a) $x_1 + x_2 + x_3 \geq 0$ and (b) $2x_1 + 2x_3 \geq 2$. Following (4.40), the first-order conditions are

$$\text{(a)} \quad \begin{cases} 2x_1 - \lambda_1 - 2\lambda_2 = 0 \\ 2x_2 - \lambda_1 = 0 \\ 2x_3 - \lambda_1 - 2\lambda_2 = 0 \end{cases}$$

$$\text{(b)} \quad \begin{cases} \lambda_1(x_1 + x_2 + x_3) = 0 \\ \lambda_2(2x_1 + 2x_3 - 2) = 0 \end{cases}$$

$$\text{(c)} \quad \begin{cases} x_1 + x_2 + x_3 \geq 0 \\ 2x_1 + 2x_3 - 2 \geq 0 \end{cases}$$

$$\text{(d)} \quad \lambda_1 \geq 0, \lambda_2 \geq 0$$

Starting, as usual, with the requirements in (b), you can easily explore the four possibilities. The only point that satisfies all of these first-order conditions is $\mathbf{X}^* = \begin{bmatrix} 0.5 \\ 0 \\ 0.5 \end{bmatrix}$.

Since $\mathbf{H} = \begin{bmatrix} 2 & 0 & 0 \\ 0 & 2 & 0 \\ 0 & 0 & 2 \end{bmatrix}$, the leading principal minors are all positive, and $f(\mathbf{X})$ is identified as a strictly convex function. Both constraints are linear and hence also concave, and the Kuhn–Tucker conditions for this problem are also sufficient, so $\mathbf{X}^*$ is the unique absolute minimum point.

4.5.6   Further Complications

Nonnegative Variables   A completely general *mathematical programming problem* consists of either of the following

$$\text{Maximize}\quad f(\mathbf{X}) \qquad\qquad \text{Minimize}\quad f(\mathbf{X})$$

$$\text{subject to}\quad h^i(\mathbf{X}) \le 0 \qquad \text{subject to}\quad h^i(\mathbf{X}) \ge 0$$

($i = 1, \ldots, m$), which are the kinds of problems that we have explored thus far in this chapter, with the added stipulation that all of the x's be nonnegative; $\mathbf{X} \ge \mathbf{0}$.

In real-world problems, it is often appropriate to disallow negative values of the unknowns—you cannot build fewer than *no* Cadillacs at a particular automobile assembly plant next quarter, you cannot rehabilitate a negative number of dilapidated houses in an urban area, and so on. These added constraints further complicate the already difficult Kuhn–Tucker conditions, and we defer consideration of them until later chapters. If it happens that *all* functions in a problem are linear—that is, $f(\mathbf{X})$ and all of the $h^i(\mathbf{X})$ constraints are linear functions—then, with the nonnegativity requirements included, we have a *linear programming problem*. The special features of such problems, and the accompanying (simpler) solution procedures that work for them, are explored in Part IV (Chapters 7–9).

When not all functions are linear, which means when *at least one* is not, and when $\mathbf{X} \ge \mathbf{0}$ is also required, then we have a *nonlinear programming problem*, and this is the subject matter of Part V (Chapters 10 and 11). We will find that the increased complexity of the Kuhn–Tucker conditions for such problems has motivated alternative kinds of solution procedures.

Linearly Dependent Constraint Gradients   Here are two kinds of problems for which the Kuhn–Tucker necessary conditions are invalid. Both fail the Kuhn–Tucker constraint qualification or regularity condition; they are not "well behaved" problems.

Example 1   In Section 4.2 we saw that the necessary condition for $\mathbf{X}^*$ to be a constrained maximum for a well-behaved problem with one equality constraint was $\nabla f^* = \lambda^*[\nabla h^*]$. We also saw that a constraint like $h(\mathbf{X}) = (0.5x_1^2 + x_2 - 5)^3 = 0$ rendered a problem not well behaved in the sense that $\nabla h(\mathbf{X}) = \mathbf{0}$ everywhere, even though there was a true constrained maximum at a point for which $\nabla f(\mathbf{X}^*) \ne \mathbf{0}$ (Fig. 4.10). Since $\nabla f^* = \lambda^*[\nabla h^*]$ is also part of the first-order conditions for an inequality constrained problem, the "null gradient" situation creates difficulties here, too.

Figure 4.16 contains the two-dimensional geometry for the following problem:

$$\text{Maximize } f(\mathbf{X}) = -(x_1 - 5)^2 - (x_2 - 4)^2$$

$$\text{subject to } h^1(\mathbf{X}) = x_1 \le 2 \qquad \text{and} \qquad h^2(\mathbf{X}) = (0.5x_1^2 + x_2 - 5)^3 \le 0$$

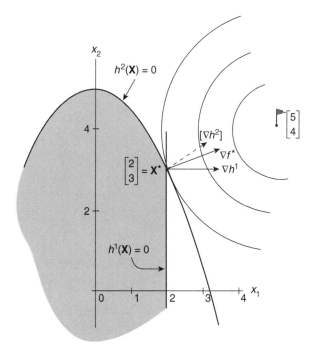

FIGURE 4.16 $f(\mathbf{X}) = -(x_1 - 5)^2 - (x_2 - 4)^2$, $h^1(\mathbf{X}) = x_1 - 2$, $h^2(\mathbf{X}) = (0.5x_1^2 + x_2 - 5)^3$.

The objective function is strictly concave with an unconstrained hilltop at $\mathbf{X} = \begin{bmatrix} 5 \\ 4 \end{bmatrix}$. From the figure it is easily seen that the constrained maximum point is at $\mathbf{X}^* = \begin{bmatrix} 2 \\ 3 \end{bmatrix}$, where vectors representing what would normally be the gradients (normals) for each of the two constraints have been drawn. The algebraic problem is that while $\nabla h^1(\mathbf{X}) = \begin{bmatrix} 1 \\ 0 \end{bmatrix}$ everywhere, $\nabla h^2(\mathbf{X}) = \begin{bmatrix} 0 \\ 0 \end{bmatrix}$ along the boundary defined by $h^2(\mathbf{X}) = 0$.

Any time the unconstrained hilltop for $f(\mathbf{X})$ subject to $h^i(\mathbf{X}) \leq 0$ is located outside of the feasible region in such a way that *two* constraints will be binding at the constrained optimum point, the gradients of *both* of those active constraints are needed for the nonnegative linear combination $\lambda_1^* \nabla h^1(\mathbf{X}^*) + \lambda_2^* \nabla h^2(\mathbf{X}^*) = \nabla f(\mathbf{X}^*)$. In two-dimensional vector space it takes two linearly independent vectors—here $\nabla h^1(\mathbf{X}^*)$ and $\nabla h^2(\mathbf{X}^*)$—to span that space. Then $\nabla f(\mathbf{X}^*)$ must be expressible as a nonnegative linear combination of those two spanning vectors [which means, as we have seen, that $\nabla f(\mathbf{X}^*)$ must lie in the cone defined by the two gradient vectors]. When one (or worse, both) of the constraint gradients are null vectors, they fail to span the space.

Example 2   Figure 4.17 illustrates a problem with a feasible region that is nonconvex in a very specific way. At $\mathbf{X}^*$, which is clearly the constrained maximum point, the two nonlinear constraint boundaries are tangent, and therefore their gradients at that point are linearly dependent (they point in exactly opposite directions). Again, they fail to span the

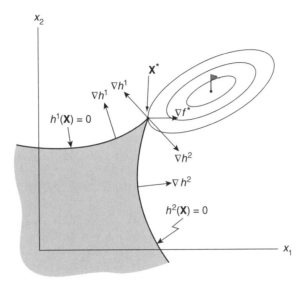

FIGURE 4.17   Nonconvex feasible region with outward pointing "cusp."

space, even though neither is a null vector. Here $\lambda_1^* \nabla h^1(\mathbf{X}^*) + \lambda_2^* \nabla h^2(\mathbf{X}^*)$ (with $\lambda_1^*, \lambda_2^* \geq$ 0) does not define a cone, but only a single line. And since $\nabla f(\mathbf{X}^*)$ does not lie on that line, it cannot be expressed as a nonnegative linear combination of the two gradients.

We examine these troublesome kinds of possibilities in some detail in Chapter 10, after developing Kuhn–Tucker conditions for the complete nonlinear programming problem.

4.6   SUMMARY

In this chapter we have seen how the fundamental logic of the calculus rules for finding stationary points of a function of several variables and distinguishing maxima from minima (from neither) can be extended to more realistic problems in which independence among the variables in $f(\mathbf{X})$ is removed by first one and then several equality constraints. The first-order conditions, necessary for any kind of stationary point, involve a proportionality requirement between $\nabla f(\mathbf{X})$ and $\nabla h(\mathbf{X})$ [or $\nabla h^i(\mathbf{X})$], plus the constraints $h(\mathbf{X}) = 0$ [or $h^i(\mathbf{X}) = 0$] considered explicitly. The method of Lagrange multipliers is simply an easily remembered procedure for generating the first-order requirements. Analysis of second-order conditions—to determine what kinds of stationary points were found—required that we establish the sign of a quadratic form whose variables are subject to a particular linear relation (or set of relations); this was the second differential test.

Next we considered the implications of inequalities rather than equations as constraints on the variables in $f(\mathbf{X})$. We found it useful to be able to employ some fundamentals from convex set theory. With only one inequality constraint, it is reasonable to begin by finding unconstrained maxima or minima (checking second-order conditions to be certain that the right kind of extreme point has been found). If the constraint is not violated, then the problem is solved. If it *is* violated, then the problem should be treated as if the constraint were an equation, using Lagrangian techniques.

In more general cases, with several inequality constraints, the Kuhn–Tucker approach is appropriate. It specifies a set of first-order conditions for those problems that do not exhibit the anomaly of linearly dependent constraint gradients. The question of maxima versus minima is often handled by appealing to the convexity/concavity properties of $f(\mathbf{X})$ and each constraint function $h^i(\mathbf{X})$.

It should be emphasized that necessary conditions by themselves *can* be very useful. They allow us to eliminate points as possible maxima or minima if it can be demonstrated that those points cannot possibility fulfill the conditions. Even though a point that meets the necessary conditions *may not* be a maximum (for example), a point that does not satisfy the conditions *cannot* be a maximum, except for problems that fail the constraint qualification. Moreover, the necessary conditions sometimes have an interesting interpretation in the context of the problem being dealt with; they may serve to characterize the stationary points in a way that provides useful insight into the nature of the optimum, even if it cannot easily be found.

As an example, consider $\lambda_i[h^i(\mathbf{X})] = 0$, in (4.34b), where $h^i(\mathbf{X}) \leq 0$ is a rewrite of $g^i(\mathbf{X}) \leq c_i$. Suppose that i represents a resource that is in scarce supply (only c_i units of it are available) and $g^i(\mathbf{X})$ is a function representing how that resource is used up when $x_1, \ldots, x_n$ are different from zero. When $h_i(\mathbf{X}^*) \neq 0$, that means $g^i(\mathbf{X}^*) < c_i$; this says that $\mathbf{X}^*$ does not use up all the available c_i. Then we know that $\lambda_i^* = 0$; in words, a value of zero is assigned to a resource that is in excess supply (not completely used up) in an optimal solution. Conversely, when $\lambda_k^* \neq 0$ (which means $\lambda_k^* > 0$), (4.34b) requires that $h_k(\mathbf{X}^*) = 0$ [meaning $g^k(\mathbf{X}^*) = c_k$]. In words, if the kth resource *is* completely used up in an optimal solution, it is given a *positive* valuation—we would benefit from having more units of k than the c_k that are available.

REFERENCES

John, Fritz, "Extremum Problems with Inequalities as Subsidiary Conditions," in *Studies and Essays*, Courant Anniversary Volume, Interscience, New York, 1948.

Kuhn, Harold, and Albert W. Tucker, "Nonlinear Programming," in Jerzy Neyman, ed., *Proceedings of the Second Berkeley Symposium on Mathematical Statistics and Probability*, University of California Press, Berkeley, 1950, pp. 481–492. [Reprinted in Peter Newman, ed., *Readings in Mathematical Economics*, Vol. I, *Value Theory*, Johns Hopkins Press, Baltimore, 1968.]

Takayama, Akira, *Mathematical Economics*, 2nd ed., Cambridge University Press, New York, 1985.

PROBLEMS

4.1 Maximize $f(\mathbf{X}) = 6x_1^2 + 5x_2^2$ subject to the constraint that $x_1 + 5x_2 = 3$.

4.2 Minimize the function in Problem 4.1 subject to the same constraint.

4.3 Find the maximum or the minimum of

$$f(\mathbf{X}) = -x_1^2 - x_2^2 - x_3^2 + 4x_1 + 6x_2 + 2x_3$$

$$\text{subject to } 2x_1 + 3x_2 + x_3 = 10.$$

4.4 Maximize or minimize $f(\mathbf{X}) = 5x_1^2 + 6x_2^2 - 3x_1x_2$ subject to $2x_1 + 3x_2 = 58$.

4.5 Maximize or minimize $f(\mathbf{X}) = 4x_1^2 + 2x_2^2 + x_3^2 - 4x_1x_2$ subject to $x_1 + x_2 + x_3 = 15$ and $2x_1 - x_2 + 2x_3 = 20$.

4.6 Find the maximum or minimum of $f(\mathbf{X})$ in Problem 4.3, subject to both $2x_1 + 3x_2 + x_3 = 10$ and $x_1 + 2x_2 + 4x_3 = 7$.

4.7 Minimize $f(\mathbf{X}) = 6x_1^2 + 5x_2^2$
  (a) Subject to $x_1 + 5x_2 \leq 3$.
  (b) Subject to $x_1 + 5x_2 \geq 3$.

4.8 Find the maximum of $f(\mathbf{X}) = 2x_1^2 + 12x_1x_2 - 7x_2^2$
  (a) Subject to $2x_1 + 5x_2 = 98$.
  (b) Subject to $2x_1 + 5x_2 \leq 98$.

4.9 Do Problem 4.3 as if the constraint were $2x_1 + 3x_2 + x_3 \leq 10$.

4.10 Do Problem 4.6 as if both constraints were "less than or equal to" inequalities.

4.11 (a) Maximize $f(x) = x^4 - 18x^2 + 15$, subject to $x \leq 2$.
  (b) Maximize the same $f(x)$, subject to $x \geq 2$.
  (c) Maximize the same $f(x)$, subject to $x \geq 4$.
  (d) Minimize the same $f(x)$, subject to $x \leq 2$.
  (e) Minimize the same $f(x)$, subject to $x \geq 2$.
  (f) Minimize the same $f(x)$, subject to $x \geq 4$.

4.12 Find the dimensions of a rectangular field with the maximum possible area, given a roll of 100 ft (linear) of wire fencing that can be used to enclose the field.

4.13 Here are two alternative extensions of Problem 4.12 to three dimensions.
  (a) Find the dimensions of a rectangular volume with maximum (cubic) area that has a total surface area of 600 ft^2.
  (b) Find the dimensions of a rectangular volume with maximum (cubic) area that has a total "edge length" of 120 ft.

4.14 Show that of all rectangular solids than enclose a given volume, a cube is the one that has *minimum* surface area.

4.15 A Norman window consists of a rectangle surmounted by a semicircle. Let h be the height of the rectangular part and r be the radius of the semicircular part. For a given window perimeter, what values for h and r will maximize the amount of light that can come through the window?

4.16 Find the area of the largest rectangle that can be inscribed in a circle of radius k.

PART III

APPLICATIONS: ITERATIVE METHODS FOR NONLINEAR PROBLEMS

The systems of equations examined in Part I were linear, and the representations and operations from matrix algebra could therefore be applied. It is obvious that connections between variables in real-world problems need not necessarily be expressible as relations that are always linear. In this part, we will see how some of the basic notions from Part II can be brought to bear on the problem of finding a solution to a single nonlinear equation or to a system of nonlinear equations. These kinds of iterative, or numerical, methods are at the heart of most computer software for solving such problems. They also provide an alternative approach to solving maximization and minimization problems. The ideas are firmly based on aspects of classical calculus techniques, but the approaches take advantage of the speed and power of modern computers to solve such problems in a different and effective way.

5

SOLVING NONLINEAR EQUATIONS

In this chapter we explore methods for finding solutions to either one nonlinear equation with one unknown or several nonlinear equations with several unknowns. Equations may arise as a part of the process of solving a problem—for example, in first-order conditions for maxima and minima, as we saw in Chapters 3 and 4. Despite textbook examples to the contrary, these will in general not all be *linear* equations. Systems of equations may also turn out to be an essential part of the mathematical description, or model, of a real-world situation, such as, in the social sciences, an input–output model (linear) or an econometric or computable general equilibrium model (virtually always nonlinear) of an economy.

In the case of *linear* equations, we know how to proceed. One linear equation in one unknown is trivial; $ax = b$ becomes $x = b/a$, unless $a = 0$, in which case there is no equation, anyway. Systems of several linear equations in several unknowns were examined in Chapters 1 and (especially) 2, using concepts from matrix algebra.

For *nonlinear* equations, solution procedures are not nearly so straightforward. Here are two fairly simple examples of one nonlinear equation in one unknown: (i) $x^2 = 4$ and (ii) $x^3 + x^2 - x = 2$. [The second can also be expressed, through factoring, as $(x - 1)(x + 1)^2 = 1$]. It is generally helpful to think of such equations rewritten in $f(x) = 0$ form. Here these would be (i) $f(x) = x^2 - 4 = 0$ and (ii) $f(x) = x^3 + x^2 - x - 2 = (x - 1)(x + 1)^2 - 1 = 0$.

The one or more x's that satisfy the equations are the *roots* or *zeros* of that equation. We show these two *functions* of x in Figures 5.1a and 5.1b. The *equation* is solved by finding the one or more x's for which $f(x) = 0$; geometrically, this is where the function crosses (or is tangent to) the x axis. In Figure 5.1a, we see that the x-axis crossings are at $x = 2$ and $x = -2$, as we expect; the equation is very simple and we recognize immediately that its solutions are $x = +2$ and $x = -2$. Note that *any function* $f(x) = x^2 + c$ will have the same shape as in Figure 5.1a. If $c = 0$, so $f(x) = x^2$, the function coincides with the x axis at just one point (there would be just one root); this would be at $x = 0$, where the curve is *tangent* to the axis (Fig. 5.1c). And if the function were $f(x) = x^2 + 1$, then the bottom of the function would be further up the vertical axis, with its lowest point at $f(x) = 1$, when $x = 0$. In that case, $f(x)$ does not cross the horizontal axis at all, since the solutions to the *equation* $x^2 + 1 = 0$ are $x = \pm\sqrt{-1}$, and these are not representable along the real number axis for x (Fig. 5.1d).

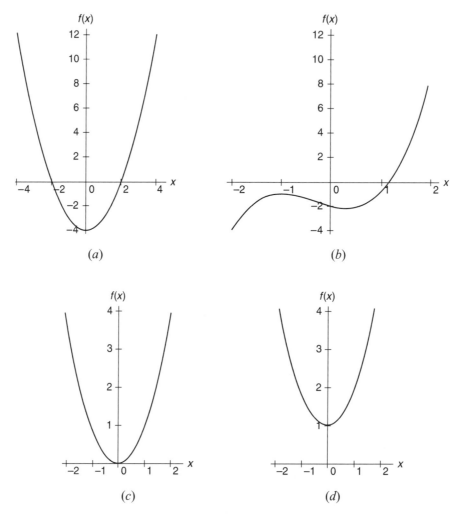

FIGURE 5.1   Nonlinear function examples. (a) $f(x) = x^2 - 4$; (b) $f(x) = (x-1)(x+1)^2 - 1$; (c) $f(x)$ $= x^2$; (d) $f(x) = x^2 + 1$.

Similarly, you can imagine alternative locations for the second function, depending on what constant is added to or subtracted from $(x-1)(x+1)^2$. For various values of c, $f(x) =$ $(x-1)(x+1)^2 + c$ could be located so that there is one (negative) root (Fig. 5.1e, where c $= 2$), two roots (Figs. 5.1f and 5.1g, where $c = 0$ and 1.2, respectively) or three roots (Fig. 5.1h, where $c = 0.5$).

For *several* nonlinear equations in several unknowns—$f^1(\mathbf{X}) = 0, \ldots, f^n(\mathbf{X}) = 0$—the matrix algebra results for linear systems are completely inappropriate; the nonlinear equations cannot be expressed in $\mathbf{AX} = \mathbf{B}$ (or $\mathbf{AX} - \mathbf{B} = \mathbf{0}$) form, and consequently inverses cannot be used in exploring for solutions.

However, with systems of linear equations, we saw in Section 2.6 that numerical or iterative procedures like the Gauss–Jordan or Gauss–Seidel techniques could also be used. The approaches investigated in this chapter for one or more nonlinear equations are also

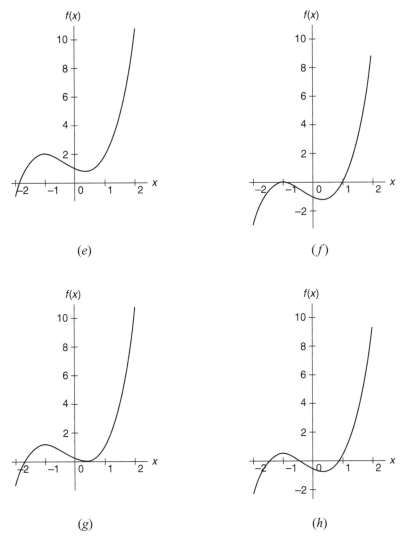

FIGURE 5.1   (e) $f(x) = (x - 1)(x + 1)^2 + 2$; (f) $f(x) = (x - 1)(x + 1)^2$; (g) $f(x) = (x - 1)(x + 1)^2 + 1.2$; (h) $f(x) = (x - 1)(x + 1)^2 + 0.5$.

classifiable as iterative methods. They describe a series of repeated steps, starting from some specified initial point, whose object is to *approach* the solution, or, importantly, to approach *one of the solutions*, when there are several (as is the case in various panels of Fig. 5.1).

Techniques such as these are used in computer programs for solving nonlinear equations, as might be expected, since one of the (many) virtues of computers is their ability to perform an enormous number of repeated arithmetic operations very very quickly.[1] A

---

[1] Iterative methods are also at the heart of most computer programs for maximization and minimization problems, as we will see in Chapter 6.

small sample of some advertising brochures for mathematical software for root finding (and of the manuals that accompany that computer software) turned up references to many specific iterative methods. For example:

> Gradients are calculated automatically using an adaptive finite-difference method.

> . . . derivatives . . . are replaced by differences (numerical differentiation).

> . . . the *root* function solves for zeros using the secant method.

> If you specify one starting value of x, . . . Newton's method is used. If you specify two starting values . . . a variant of the secant method is used.

> The algebraic equation solver (root finder) contains six numerical methods including bisection and Newton–Raphson.

It is exactly these kinds of numerical methods that we explore in this chapter. At least a minimal introduction to these approaches is necessary if you are to be an intelligent and informed user of any of these modern mathematical packages. For example, how can you choose among the "six numerical methods including bisection and Newton–Raphson" for root finding without some understanding of how these techniques work and of the kinds of situations in which they get into trouble? Also, at least a general idea of how the methods work helps you understand what is going on when, for example, different programs give different solutions to the same problem or when the software responds with the message that it can find *no* solution at all.

In general terms, an iterative method consists of three basic components:

1. A set of initial values for the variables, $\mathbf{X}^0$
2. An algorithm for getting to the next (and, generally, "better") values of the variables, $\mathbf{X}^{k+1}$, from the current values, $\mathbf{X}^k$, where $k = 0, \ldots, n$
3. A stopping criterion; that is, a rule for deciding when n is large enough, which means, when $\mathbf{X}^n$ is "close enough" to an exact solution.[2]

In any thorough mathematical discussion of iterative methods, concepts such as "precision" and "error" of machine calculations are considered.[3] These computer characteristics are important in identifying the seriousness of roundoff error in machine calculations; different iterative techniques make different demands on computer accuracy and hence are subject to differing effects of roundoff error. These kinds of concerns are beyond the scope of this text; you can find them discussed in many texts on numerical methods (e.g., Dennis and Schnabel, 1983; Press et al., 1986). Another important characteristic by which various iterative methods are distinguished is their speed of convergence. We will examine this idea for the numerical examples in the sections that follow.

---

[2]To be unambiguously clear, $\mathbf{X}^{(k)}$ should be used to denote the value of $\mathbf{X}$ at the kth step, to distinguish it from $\mathbf{X}$ raised to the kth power. However, it should always be clear from the context how $\mathbf{X}^k$ is to be interpreted. See footnote 10, Chapter 2.

[3]For example, one measure of computer precision is given by *macheps* (machine epsilon), which is the smallest positive number τ for which $1 + \tau > 1$. The smaller this measure, the more "precise" is the computer. The effectiveness of stopping rules for iterative methods is dependent on the precision of the computer being used.

5.1   SOLUTIONS TO $f(x) = 0$

In this and following sections, we will divide the exposition into two groups of methods: those that do not rely on derivatives and those that do use derivatives, whether exact or approximate.

5.1.1   Nonderivative Methods

Bracketing and Bisection   An important notion for some nonderivative methods is that of *bracketing* a root. Since a root of $f(x)$ is a value of x for which $f(x) = 0$, it is clear that a root is contained (bracketed) in an interval $a \leq x \leq b$ if $f(a)$ and $f(b)$ have opposite signs. If this is the case, then, for continuous functions, at least at one value of x between $x = a$ and $x = b$, $f(x) = 0$, and so $f(x)$ has a root at that value (or those values). In what follows, we will assume that a and b are sufficiently close so that, when $f(a)$ and $f(b)$ have opposite signs, there is just *one* root between $x = a$ and $x = b$. [An alternative way of identifying a bracketing interval (a,b) is to note that $f(a)f(b) < 0$ when $f(a)$ and $f(b)$ are of opposite signs.] If we assume that $f(x)$ is continuous and that we can compute values of $f(x)$ wherever necessary, then as soon as we have identified values of $x = a$ and $x = b$ for which $f(a)f(b) < 0$, we can begin to "close in" on the root. One common and simple method for such closing in is the *bisection* method.

Given an $x = a$ and $x = b$ for which $f(a)f(b) < 0$, find the value of $f(x)$ at the *midpoint* between a and b, call it c, and use this value of x to replace that endpoint of the current interval ($a \leq x \leq b$) for which $f(c)$ has the same sign. Deciding on the initial values of a and b is another matter. As with all iterative techniques, the better the starting point(s), the better the results. Continue in this manner until it is time to stop. There are many kinds of possible stopping rules; we explore two. One identifies a minimal interval of uncertainty when two successive values of x, x^k, and x^{k+1}, are "close enough" to one another. For example, when $|x^k - x^{k-1}| < \bar{\varepsilon}$, for some small predetermined $\bar{\varepsilon} > 0$. A second rule is to stop when $f(x)$ for the most recently found point is "sufficiently close" to zero; for example, when $|f(x^k)| < \delta$, where δ is a prespecified small positive number. The logic of the bisection method is indicated in Figure 5.2.

Speed of convergence is a measure of relative attractiveness of alternative techniques that is often used to distinguish among them. We illustrate how the speed of convergence of a method can be examined, using the iterative procedure in the bisection method as a concrete example. In the numerical illustrations that follow, one indication of the relative speed of convergence of alternative techniques is provided by the number of iterations that are required to achieve a predetermined level of accuracy. This is only one of many possible ways of examining relative efficiency of alternative iterative methods. Computational time is another. For example, if one method takes only five iterations, but each iteration requires 3 s of computing time, then it is not necessarily "better" than an alternative iterative method that requires 80 iterations where each takes only 0.1 s each. Computing time is a function of (among others) the computer used (what kind of chip, operating at what megahertz, etc.?) and of the sophistication of the particular program(s) and software used to implement the iterative method(s). These same considerations will be relevant in Chapter 6, when we explore alternative iterative methods for unconstrained maximization and minimization problems.

Given an initial choice of $x = a$ and $x = b$, the first "interval of uncertainty" is $|a - b|$ units long. Define that interval to be ε^0. Suppose that we define a desired level of accuracy in

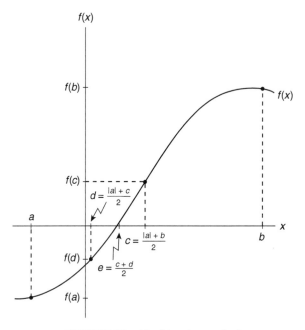

FIGURE 5.2   The bisection method.

terms of the size of this interval of uncertainty; we will be satisfied when we have found a root for $f(x)$ if x lies in an interval of size $\bar{\varepsilon}$. Consider the progression of these intervals of uncertainty, noting that the bisection method cuts the interval exactly in half at each step—if the interval is ε^k at the kth step, at the $(k+1)$st step it will be of length $\varepsilon^{k+1} = \varepsilon^k/2$. Therefore

$$\varepsilon^1 = \frac{\varepsilon^0}{2}$$

$$\varepsilon^2 = \frac{\varepsilon^1}{2} = \frac{\varepsilon^0}{4} = \frac{\varepsilon^0}{(2)^2}$$

$$\vdots$$

$$\varepsilon^k = \frac{\varepsilon^{k-1}}{2} = \frac{\varepsilon^0}{(2)^k}$$

The question, then, is for what value of k will $\varepsilon^k = \bar{\varepsilon}$? Straightforward algebra, setting the expression above for ε^k equal to $\bar{\varepsilon}$, leads to $k = \log_2(\varepsilon^0/\bar{\varepsilon})$. For example, if $\varepsilon^0 = 5$ and $\bar{\varepsilon} = 0.001$, we find that $k = 12.2877$, which means effectively that after the 13th iteration, the interval of uncertainty will be 0.001 or less.

From our observation that $\varepsilon^{k+1} = \varepsilon^k/2 = (1/2)(\varepsilon^k)^1$, the speed of convergence is described as *linear*, because the exponent to which ε^k is raised is one. If this exponent is greater than one, and the constant multiplying ε^k is less than one, then convergence is described as *superlinear*—superlinear convergence occurs if $\varepsilon^{k+1} = c(\varepsilon^k)^m$ when $c < 1$ and $m > 1$. In comparing alternative numerical methods, analysts are often interested in the rate of convergence of each technique; other things being equal, the faster the rate of convergence, the more preferred is the technique.

The Secant Method   Instead of simply bisecting the interval on the x axis at each step, and evaluating $f(x)$ for each of those bisection points, the secant method builds a sequence of candidate values of x that take into account the *characteristics* of $f(x)$. Specifically, assume that (as in the bisection method) we have an initial pair of values for x, a and b. In fact, for the secant method (and others to follow in this section), these initial points need not bracket a root; for the bisection method, initial bracketing is essential. Then the secant method "builds" a secant through the points $f(a)$ and $f(b)$; this is tantamount to approximating $f(x)$ in the interval $a \leq x \leq b$ by the *linear* function that is the secant through $f(a)$ and $f(b)$—hence the name. Figure 5.3 illustrates the general idea.

Given this linear approximation to $f(x)$, find where it crosses the x axis [this is a root of this *linear approximation* to $f(x)$] and call the point $x = c$. Then build another secant through the points $f(b)$ and $f(c)$, generating another linear approximation which crosses the x axis at $x = d$. Using that point, construct yet another secant through $f(c)$ and $f(d)$, and so forth; at each step, the secant (linear approximation to the function) is constructed through the values of $f(x)$ for the two *most recently found* values of x. Notice that in this example, the point $x = d$, which is found as the intersection of the second secant with the x axis, is to the left of the original left-hand point, $x = a$, suggesting that the estimates are getting worse, not better. However, as soon as the next secant is constructed, through $f(c)$ and $f(d)$ (secant 3 in the figure), we find an intersection with the x axis ($x = e$) which is

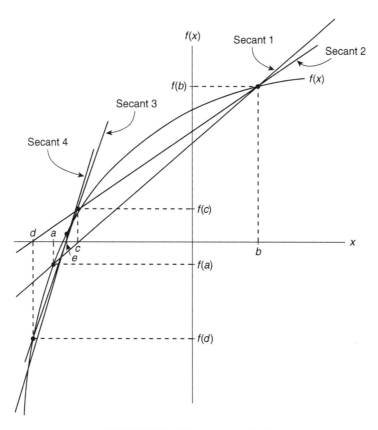

FIGURE 5.3   The secant method.

closer to the root of the function. And the intersection of secant 4 [connecting $f(d)$ and $f(e)$] with the x axis ($x = f$) is even better. To see that initial bracketing is not needed for the secant method, suppose in Figure 5.3 that point a did not exist and that the two *initial* points were b and c. The same sequence—$(b, c) \rightarrow d$, $(c, d) \rightarrow e$—would still occur.

This is the logic and the *geometry* of the secant method; what about the *algebra*? We look at the initial secant, through $f(a)$ and $f(b)$; its slope is

$$\frac{f(b) - f(a)}{b - a}$$

But from the geometry in Figure 5.3, it is clear that this slope is also equal to

$$\frac{f(b) - 0}{b - c}$$

The points a and b are known, and we assume that is it possible to evaluate $f(x)$ at those points, so $f(a)$ and $f(b)$ can be found. Then setting the two slopes equal to one another, we find c, the "new" value on the x axis, from straightforward algebra:

$$c = b - f(b) \left[\frac{b - a}{f(b) - f(a)} \right]$$

This is an unambiguous rule for finding c when $a, b, f(a)$, and $f(b)$ are known. Then b and c are used in the next step in exactly the same way as a and b in the first step, leading to a new point d from which another iteration proceeds, using c and d, and so on.

In general, at the kth iteration,

$$x^{k+1} = x^k - f(x^k) \left[\frac{x^k - x^{k-1}}{f(x^k) - f(x^{k-1})} \right] \tag{5.1}$$

The method fails if, at any step k, $[f(x^k) - f(x^{k-1})] = 0$, because (5.1) would require division by zero. But $[f(x^k) - f(x^{k-1})]$ is also the numerator in the expression for the slope of the secant connecting points $f(x^k)$ and $f(x^{k-1})$—$[f(x^k) - f(x^{k-1})]/(x^k - x^{k-1})$. Geometrically, when the numerator is zero, the secant approximation to the tangent never cuts the x axis (it is parallel to that axis) and no "next" value, x^{k+1}, can be found.

The Regula Falsi Method   *Regula falsi* means "false position." This procedure follows the general outline of the secant method, with the difference that, once $x = c$ is found (exactly as in the secant method), then *either* $x = a$ or $x = b$ is used in conjunction with $x = c$ to build the next secant, depending on whether $f(a)$ or $f(b)$ is of *opposite* sign from $f(c)$. Suppose that it is $f(a)$, so that the next secant is built through $f(a)$ and $f(c)$. [Relative to the secant method, this is building the secant in the wrong (false) position; perhaps that is the origin of the name.] We are assured by this convention that the true root has been *bracketed* in the interval $a \leq x \leq c$, and so the rate of convergence might be expected to be faster. We will see in the examples to follow that this expectation, appealing as it may be, is often wrong.

The method is illustrated in Figure 5.4. Note that since x^{k+1} is always found using x^k and some earlier value of x, call it x^e, for which $f(x^k)f(x^e) < 0$, $f(x^k)$ cannot equal $f(x^e)$ and

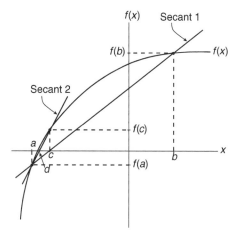

FIGURE 5.4   The regula falsi method.

so the zero-slope problem of the secant method cannot arise. In this illustration, using the same function as in Figure 5.3, we see that after two iterations, at the point $x = d$, we have come quite close to the true root. In Figure 5.3, the secant method gave us an $x = d$ that was worse than the original $x = a$.

Brent's Method   The secant or false position methods are usually faster than bisection. However, many variants of nonderivative methods use a combination approach; they intersperse bisection with either secant or false position searches. Another successful method, associated with the work of Brent, replaces the assumption that the nonlinear function $f(x)$ can be approximated *linearly* in some small neighborhood with the assumption that it can be (better) approximated by a *quadratic* function [Brent (1973); also attributed to van Wijngaarden and Dekker, cited in Press et al. (1986), Section 9.3].

More precisely, an *inverse* quadratic function is "fitted"—meaning that x is assumed to be a quadratic function of y, instead of the usual format $y = f(x)$, where if f were assumed quadratic, then y would be a quadratic function of x. The algebra is a little complicated. Assume that the most recent three points used in the approximation are a, b, and c with associated values $f(a)$, $f(b)$, and $f(c)$. The problem is then to find the particular function

$$x = q(y) = k_1 y^2 + k_2 y + k_3$$

that goes through the three (x, y) points $[a, f(a)]$, $[b, f(b)]$ and $[c, f(c)]$. Inserting these values generates three *linear* equations in the three unknowns k_1, k_2, and k_3. The algebra is tedious by hand but quickly done with computer software for algebraic manipulations.[4] After a good deal of rearrangement of results, the inverse polynomial through these points is

$$x = \frac{[y - f(a)][y - f(b)]c}{[f(c) - f(a)][f(c) - f(b)]} + \frac{[y - f(b)][y - f(c)]a}{[f(a) - f(b)][f(a) - f(c)]} + \frac{[y - f(c)][y - f(a)]b}{[f(b) - f(c)][f(b) - f(a)]}$$

---

[4]The results here are particularizations of a general result originally due to Lagrange (who did *not* have access to a computer), for fitting a polynomial of degree $N - 1$ through N points. In the present case, this means a second-degree polynomial (quadratic) function through three points.

To find the "root" associated with this nonlinear approximation, set $y = 0$ and solve for x (the algebra, again, is tedious if done by hand): $x = b + (P/Q)$, where

$$P = S[T(R - T)(c - b) - (1 - R)(b - a)]$$
$$Q = (T - 1)(R - 1)(S - 1)$$

and $R = f(b)/f(c)$, $S = f(b)/f(a)$, and $T = f(a)/f(c)$.

This identifies the root to the quadratic approximation to $f(x)$ through the three most recent points $[a, f(a)]$, $[b, f(b)]$, and $[c, f(c)]$. Clearly, there is a good deal of calculation required in each Brent approximation of this sort; the tradeoff between speed of convergence and time to calculate each step depends on how much better a quadratic approximation to $f(x)$ is than a linear one—it depends on the mathematical characteristics of the particular $f(x)$ whose roots we want to find.

There are essentially two potential algebraic problems with the Brent method. Since $x = b + (P/Q)$, the technique is in trouble if either $P = 0$ or $Q = 0$ (or, indeed, both). When $P = 0$, values of x are repeated and eventually explode. When $Q = 0$, the division required by the (P/Q) term is impossible. In either case, a practical technique is simply to restart the procedure with a different set of initial values.[5]

Numerical Illustrations Tables 5.1–5.4 contain numerical illustrations for each of these approximation techniques. The first test function (Tables 5.1 and 5.2) is $f(x) = x_2 - 4$; the second (Tables 5.3 and 5.4) is $f(x) = (x - 1)(x + 2)^2 - 1$. (These functions were shown in Fig. 5.1.) In all cases, the iterations were stopped after the kth step if $|f(x^k)| < 0.0001$ (= 1×10^{-4}).[6]

In Table 5.1, in the early iterations for each technique the results have been rounded to four decimals; in later steps, six decimals have been used in order to illustrate the precise way in which each technique achieves the level of accuracy given in $\delta = 10^{-4}$. This is such a simple problem that no sophisticated nonlinear equation solution procedure is needed at all; it was chosen simply to make it easier to relate successive solutions in any of the iterative methods to the *true* solution.[7]

For the bisection, secant, and regula falsi methods, the initial values, $x^0 = a$ and $x^1 = b$, were chosen to be $x^0 = 1$ and $x^1 = 4$. These certainly bracket a root—$f(x^0) = -3$ and $f(x^1) = 12$, so that $f(1)f(4) < 0$. [Of course, one of the roots is known in this case to be 2. The point is that the $f(x^0)f(x^1) < 0$ test does not require knowledge of the root, which is what

---

[5]It is clear that $Q = 0$ if T and/or R and/or S equals one; from the definitions of R, S, and T, this means if at least two of the values $f(a)$, $f(b)$ and $f(c)$ are equal. These values are the three points (distances above and/or below the horizontal axis) through which an inverse quadratic function is to be fitted. If at least two of these points lie on the same horizontal line, there is not enough "curvature" for the curve. (Remember that we are looking for x as a quadratic function of y, which means that in x, y space the approximation will look like a "U" but on its side—rotated 90°.) Also from the definitions of R, S, and T, it is clear that trouble emerges if $f(a)$ or $f(c) = 0$. But of course if $f(a) = 0$, that means that a root has been found at $x = a$, and if $f(c) = 0$, a root has already been located at $x = c$.

[6]There is nothing sacred about this particular value for the stopping criterion. In any given situation, the person solving the problem must decide on the appropriate amount of accuracy in the result.

[7]For example, if the problem had been $x^2 - 3 = 0$, the solutions are $\pm\sqrt{3} = \pm 1.732050807. \ldots$, and it is not as easy to recognize (e.g., visually, in a table) how close $x^k = 1.732157841$ is to the true solution as it is to relate an $x^k = 2.000122$ to the actual value of $x = 2$ for the $x^2 - 4 = 0$ problem.

TABLE 5.1   Iterative Methods for Finding Roots to $f(x) = x^2 - 4$

	Bisection		Secant		Regula Falsi		Brent's Method	
k	x^k	$f(x^k)$	x^k	$f(x^k)$	x^k	$f(x^k)$	x^k	$f(x^k)$
0	1.0	−3.0	1.0	−3.0	1.0	−3.0	1.0	−3.0
1	4.0	12.0	4.0	12.0	4.0	12.0	4.0	12.0
2	2.5	2.25	1.6	−1.44	1.6	−1.44	1.6	−1.44
3	1.75	−0.9375	1.8571	−0.5510	1.8571	−0.5510	2.0945	0.3870
4	2.125	0.5156	2.0165	0.0664	1.9512	−0.1927	1.9942	−0.0232
5	1.9375	−0.2461	1.9994	-2.44×10^{-3}	1.9836	−0.0653	2.000031	1.24×10^{-4}
6	2.0313	0.1260	1.999997	-1.00×10^{-5}	1.9945	−0.0219	2.0	-8.27×10^{-9}
7	1.9844	−0.0623			1.9982	-7.31×10^{-3}		
8	2.0078	0.0313			1.9994	-2.44×10^{-3}		
9	1.9961	−0.0156			1.999796	-8.13×10^{-3}		
10	2.001953	7.82×10^{-3}			1.999932	-2.71×10^{-4}		
11	1.999023	-3.91×10^{-3}			1.999977	-9.03×10^{-5}		
12	2.000488	1.95×10^{-3}						
13	1.999756	-9.77×10^{-4}						
14	2.000122	4.88×10^{-4}						
15	1.999939	-2.44×10^{-4}						
16	2.000031	1.22×10^{-4}						
17	1.999985	-6.10×10^{-5}						

TABLE 5.2   Iterations Required from Alternative Starting Points for $f(x) = x^2 - 4$

Starting Values (x^0, x^1)	Bisection	Secant	Regula Falsi
(1, 4)	16	5	10
(0, 5)	17	7	15
(−6, 0)	17	8	9
(0, 10)	18	8	30
(0, 100)	21	13	300
(0, 1000)	25	18	2,996
(0, 10,000)	28	23	29,958
(0, 100,000)	30	27	—[a]

Starting Values (x^0, x^1, x^2)	Brent
(1, 1.6, 4)	4
(0, 0.8, 5)	6
(−6, −0.6667, 0)	7
(0, 0.4, 10)	9
(0, 0.04, 100)	18
(0, 0.004, 1000)	18
(0, 4×10^{-4}, 10,000)	23
(0, 4×10^{-5}, 100,000)	28

[a]Patience ran out before this starting point was tried with regula falsi. *Conjecture:* The number of necessary iterations would be above 299,000.

TABLE 5.3   Iterative Methods for Finding Roots to $f(x) = (x - 1)(x + 1)^2 - 1$

	Bisection		Secant		Regula Falsi		Brent	
k	x^k	$f(x^k)$	x^k	$f(x^k)$	x^k	$f(x^k)$	x^k	$f(x^k)$
0	0	−2.0	0	−2.0	0	−2.0	0	−2.0
1	3.0	31.0	3.0	31.0	3.0	31.0	3.0	31.0
2	1.5	2.125	0.1818	−2.1427	0.1818	−2.1427	0.1818	−2.1427
3	0.75	−1.7656	0.3640	−2.1833	0.3640	−2.1833	−2.3709	−7.3354
4	1.125	−0.4355	−9.4525	−747.7706	0.5375	−2.0934	1.0685	−0.7069
5	1.3125	0.6711	0.3928	−2.1779	0.6932	−1.8795	1.2386	0.1954
6	1.2185	0.0769	0.4215	−2.1689	0.8251	−1.5826	1.2042	−0.0079
7	1.1719	−0.1893	7.3779	446.6517	0.9307	−1.2582	1.205557	−7.03×10⁻⁵
8	1.1953	−0.0587	0.4551	−2.1537	1.0114	−0.9537		
9	1.2070	0.0084	0.4884	−2.1334	1.0708	−0.6964		
10	1.2012	−0.0253	3.9776	72.7742	1.1132	−0.4946		
11	1.2041	−0.0085	0.5877	−2.0393	1.1428	−0.3443		
12	1.205566	−1.75×10⁻⁵	0.6801	−1.9029	1.1632	−0.2363		
13			1.9697	7.5520	1.1771	−0.1606		
14			0.9397	−1.2269	1.1865	−0.1084		
15			1.0836	−0.6369	1.1928	−0.0729		
16			1.2390	0.1938	1.1971	−0.0488		
17			1.2021	−0.0198	1.1999	−0.0326		
18			1.205479	−5.21×10⁻⁴	1.2018	−0.0218		
19			1.205570	1.44×10⁻⁶	1.2030	−0.0145		
20					1.2039	−0.0097		
21					1.2044	−0.0065		
22					1.2048	−0.0043		
23					1.2051	−0.0029		
24					1.2052	−0.0019		
25					1.2053	−0.0013		
26					1.2054	−8.47×10⁻⁴		
27					1.205472	−5.64×10⁻⁴		
28					1.205504	−3.76×10⁻⁴		
29					1.205526	−2.50×10⁻⁴		
30					1.205541	−1.67×10⁻⁴		
31					1.205550	−1.11×10⁻⁴		
32					1.205557	−7.39×10⁻⁵		

we are looking for.] So you just keep trying initial candidates for x^0 and x^1 until a pair is found for which $f(x^0)f(x^1) < 0$.

In solving $x^2 = 4$ [finding roots for $f(x) = x^2 - 4$], we realize that there are two solutions, $x = \pm 2$. Alternative starting points to those used in Table 5.1 will lead to *either* $x = 2$ or $x = -2$. For bisection, a pair of starting points that bracket $x = -2$ would lead to that solution. The other methods do not *require* bracketing; pairs of negative starting points would generally lead to $x = -2$. Note that no matter where you start, you will (at best) find only *one* root. If you have reason to believe that there is more than one root, and if you would like to find all of them, then you must use numerous alternative starting points, however (intelligently or randomly) selected.

For Brent's method, *three* initial values are required, so $x^0 = 1$, $x^1 = 4$, and $x^2 = 1.6$ were

**TABLE 5.4   Iterations Required from Alternative Starting Points for $f(x) =$
$(x - 1)(x + 1)^2 - 1$**

Starting Values (x^0, x^1)	Bisection	Secant	Regula Falsi
(0, 3)	11	18	31
(−2, 0)	—[a]	10	16
(0, 1)	—[a]	6	6
(−2, −3)	—[a]	13	18
(0, 10)	18	12	262
(0, 100)	22	40	22421
(0, 1000)	23	19	—[b]
(0, 10,000)	27	19	
(0, 100,000)	32	19	

Starting Values (x^0, x^1, x^2)	Brent
(0, 0.181818, 3)	6
(−2, 0, 2)	6
(0, 1, 2)	5
(−1.6923, −2, −3)	20
(0, 0.0183, 10)	18
(0, 1.98×10^{-4}, 100)	28
(0, 2.00×10^{-6}, 1000)	27
(0, 2.00×10^{-8}, 10,000)	27
(0, 2.00×10^{-10}, 100,000)	27

[a]This set of starting values does not bracket a root and therefore cannot be used for bisection.
[b]In view of the results in Table 5.2, further starting points were not explored for regula falsi. *Conjecture:* Each succeeding starting point would require approximately 100 times the iteration count for the immediately preceding starting point.

used (where x^2 is the first point *derived* from $x^0 = 1$ and $x^1 = 4$ in the secant and regula falsi methods). This might seem to bias the example illustrations in favor of Brent, since the
third of the "initial" values has already been found by the intelligent workings of the secant or regula falsi method. In fact, as later figures will show, Brent's method performs
about as well for a wide variety of initial values.

Bisection stopped at $k = 17$, which means 16 iterations were required after having chosen $x^0 = 1$ and $x^1 = 4$. The secant method needed only five iterations (again, after 1 and 4
were chosen), regula falsi needed 10 (so in this example it was *not* better than the secant
method) and Brent's technique was best, as might have been expected; it needed four iterations (after the initial three points were supplied).

Iteration counts for several sets of alternative starting points are shown in Table 5.2. In
terms of the way in which the information is presented in Table 5.1, this means that $k - 1$
is reported for the bisection, secant, and regula falsi techniques while $k - 2$ is reported for
Brent's method. For techniques requiring two initial values (all but Brent), the (x^0, x^1) values that were used are:

(a) (0, 5) which brackets the positive root but provides a wider initial interval of un-
    certainty than did (1, 4).

(b) (–6, 0) which is wider still and brackets the negative root.

(c) A series of increasingly outrageous (wider) initial intervals that bracket the positive root, starting with (0, 10) and going up to (0, 100,000). They were examined in order to explore the sensitivity of the techniques to very wide variations in the size of the initial interval of uncertainty.

Comparable starting values for x^0, x^1, and x^2 are employed for Brent's method. In each case, the third "initial" point for Brent's method is the first point derived by the secant method from its initial two points. The first entry in each of the columns in Table 5.2 summarizes the iteration count from the starting points in Table 5.1, for comparison.

We see that the secant and Brent approaches are comparably fast and similarly sensitive to initial points. Bisection generally requires more iterations, although it is *relatively* less sensitive to wide variations in the initial interval of uncertainty. Performance of regula falsi is erratic. From some starting points it performs somewhere between the bisection and secant approaches (first three rows in Table 5.2); from the other points, starting with (0,10), it is inferior and gets rapidly much worse as the initial interval of uncertainty increases. For this function, all methods except regula falsi are seen to be relatively unfazed by very wide initial intervals.

The problem with regula falsi is that it insists on building a series of very steep secants, the first of which—when the starting point is (0, 1000), for example—goes through $f(0) =$ –4 and $f(1000) = 999996$. This secant cuts the x axis at $x^2 = 0.004$, and successive secants are very close together; their intersections move very very slowly from 0.004 in the direction of 2. The secants are even more vertical and consequently closer together for $x^1 =$ 10,000 and worse yet for $x^1 = 100,000$. This is a situation in which the apparent virtue of regula falsi (always using bracketing points for its next iteration) is in fact a vice. All of these observations are relevant, of course, only for $f(x) = x^2 - 4$ and for the sets of starting points and the stopping criterion chosen, but they provide some illustration of how the four techniques behave relative to one another.

The same kinds of illustrative results are presented in Table 5.3, for the more complicated function $f(x) = (x - 1)(x + 1)^2 - 1$. The initial values used are $x^0 = 0$ and $x^1 = 3$, which bracket the root. (Again, for Brent's method, these two values were used along with $x^2 = 0.1818$, which is the first point derived via the secant or regula falsi approaches.) Comparisons can easily be made with the results in Table 5.1. For example, for this function, bisection actually needs fewer iterations than for $f(x) = x^2 - 4$, Brent stays the same at four while the secant method now needs 18 and regula falsi uses 31 (again, all after the selected initial values).

Note how the secant method jumps around and takes chances. For example, at $k = 4$ it has identified an x^4 for which $f(x^4) = -748$, and three steps later, at $k = 7$, $f(x^7) = 447$; but it corrects for these lapses of judgment effectively. Regula falsi, by contrast, plods along, cutting down the interval of uncertainty (very) slowly but surely. It does not get into $x =$ 1.2 ... territory until $k = 18$ and then takes an agonizing additional 14 steps.

Table 5.4 contains results similar to those in Table 5.2, for several alternative starting points. For the approaches that require two initial points, these (x^0, x^1) values are:

(a) (–2,0) which does not contain the root, so bisection cannot be used.

(b) (0,1) which also does not contain the root, but is "closer" to it.

(c) (–2,–3), which does not contain the root and is further from it than the non-bracketing intervals defined in either (a) or (b).

(d) The series of increasingly wide initial intervals of uncertainty that was used in the previous example.

As in Table 5.2, comparable starting values for x^0, x^1, and x^2 are employed for Brent's method, and the first element in each column of the table summarizes the iteration count for the initial points from Table 5.3. (Again, the iteration counts reported are values of $k - 1$ for all methods except Brent's, where it is $k - 2$.)

You can make the same kinds of comparisons as were done with the results in Tables 5.1 and 5.2. For this function and these starting points, the secant method is usually best, followed fairly closely sometimes by bisection and sometimes by Brent; and regula falsi comes in (again) a poor fourth.

5.1.2   Derivative Methods

Newton's name is associated with a large number of the root finding methods that employ derivatives or approximations to them. One of the most widely used of these is usually known as the *Newton–Raphson* method (Isaac Newton, English, 1642–1727; Joseph Raphson, English, 1648–1715). Two variants that we will also look at briefly are the *modified Newton* and *quasi-Newton* methods. In all cases it is necessary to evaluate both the function and its derivative (or else to *approximate* the derivative) at various values of x.

The Newton–Raphson Method   This approach is still one of creating a linear approximation of $f(x)$ near to a particular value x^k. With derivative methods, this linear approximation is obtained by using the first two terms of a Taylor series expansion of $f(x)$ in the neighborhood of x^k. In general, this means using

$$f(x + dx) \cong f(x) + f'(x)(dx)$$

In particular, letting $x = x^k$ (the "current" value of x) and $dx = x^{k+1} - x^k$, we have

$$f(x^{k+1}) = f(x^k) + f'(x^k)(x^{k+1} - x^k) \tag{5.2}$$

To find where this line crosses the x axis [i.e., to find a root to this linear approximation to $f(x)$ in the neighborhood of x^k], set $f(x^{k+1}) = 0$; this gives $x^{k+1} - x^k = -f(x^k)/f'(x^k)$, or

$$x^{k+1} = x^k - \frac{f(x^k)}{f'(x^k)} \tag{5.3}$$

Figure 5.5*a* illustrates several steps in a Newton–Raphson approach. Note that the procedure breaks down if $f'(x^k) = 0$, reflecting exactly the same kind of geometric problem as was identified earlier for the secant approach.

The Modified Newton Method   While evaluation of the derivative at each step may not be difficult for $f(x)$, it can become computationally burdensome for a function of several variables, $f(\mathbf{X})$. For that reason, the modified Newton method replaces $f'(x^k)$ at each step with $f'(x^0)$, so (5.3) is simplified to

$$x^{k+1} = x^k - \frac{f(x^k)}{f'(x^0)} \tag{5.4}$$

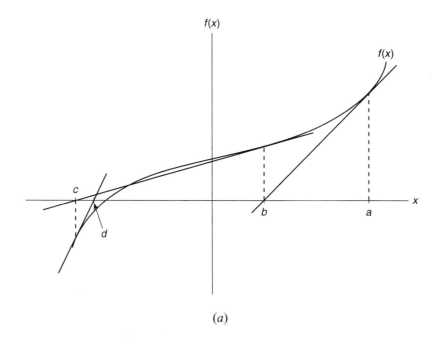

(*a*)

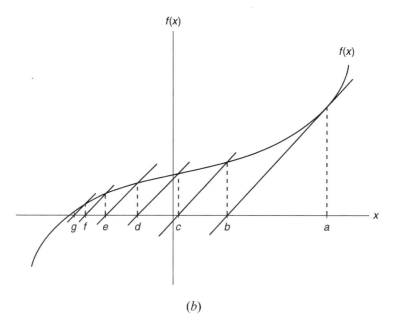

(*b*)

FIGURE 5.5   (*a*) The Newton–Raphson method; (*b*) the modified Newton method.

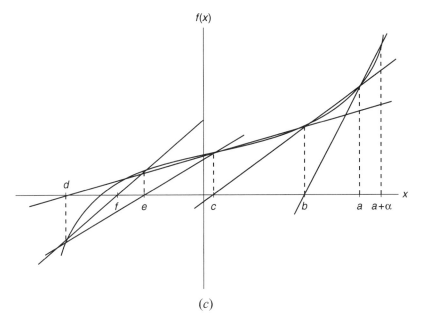

(c)

FIGURE 5.5 (c) the quasi-Newton method.

This may not seem intuitively so appealing, but Newton methods are most effectively employed near to a root, when $dx = x^{k+1} - x^k$ is "small," so that the first-order Taylor series approximation is relatively accurate. In that case, it is quite possible that the derivative does not change greatly for successive steps. This method is illustrated in Figure 5.5b. [For reasons made clear by the figure, this is also known as the "parallel chord" method.]

The Quasi-Newton Method   This variation on Newton–Raphson can be described as building on the first-order Taylor series approximation, replacing the derivative in (5.3) with a forward-difference approximation, but evaluated at x^{k-1}. (This and other approaches to estimating derivatives without using differentiation—that is, by finite differences— are explored in Appendix 5.1.) This means that

$$f'(x^k) \cong \frac{f(x^k) - f(x^{k-1})}{x^k - x^{k-1}}$$

Putting this into (5.3) gives

$$x^{k+1} = x^k - f(x^k)\left[\frac{x^k - x^{k-1}}{f(x^k) - f(x^{k-1})}\right] \tag{5.5}$$

Note that this expression for x^{k+1} is algebraically identical to the secant approach in (5.1), where we saw that *two* initial points were needed from which to generate the first secant approximation to the derivative. In the quasi-Newton technique, we specify only *one* initial point, x^0, so an initial estimate of $f'(x^0)$ is created using a forward-difference approximation

$$f'(x^0) \cong \frac{f(x^0 + \alpha) - f(x^0)}{\alpha}$$

for an arbitrarily chosen small $\alpha > 0$. This amounts to approximation of a tangent by a secant that cuts $f(x)$ at $f(x^0)$ and at $f(x^0 + \alpha)$. The smaller α, the better the approximation. Once you have (an approximation to) $f'(x^0)$, you can determine x^1, which, in conjunction with the secant through $f(x^0)$ and $f(x^1)$, will identify x^2, as in (5.5), and the procedure is off and running, as in Figure 5.5c.

The quasi-Newton approach is grounded in the derivative-based notion of a first-order Taylor series, including an $f'(x)$ term. But substitution of the finite-difference approximation to that derivative reveals this approach to be a masquerading secant technique (with a slight variant for its initial step), so it could equally well be classified as a nonderivative method.

Since derivative approximations are used in iterative methods, the procedures are often described as *updating* schemes. In general, this turns out to be of interest generally in multidimensional problems. It is included here in the $f(x)$ case primarily for illustration and to set the stage for its appearance later for multidimensional problems. The new derivative approximation at x^k can be expressed as the old derivative approximation at x^{k-1} plus a change to that old approximation:

$$f'(x^k) \equiv f'(x^{k-1}) + \Delta[f'(x^{k-1})]$$

This is accomplished in a way that seems needlessly complicated at this point, but it will be a useful format later for the multidimensional case.

At x^{k-1}, the approximation to $f'(x^{k-1})$ is related to the *previous* point, x^{k-2}, by

$$f'(x^{k-1}) \cong \frac{f(x^{k-1}) - f(x^{k-2})}{x^{k-1} - x^{k-2}} = \frac{A}{B}$$

as in Figure 5.6. At x^k, this slope, A/B, is also seen to be equal to C/D, where $C = f'(x^{k-1})(x^k - x^{k-1})$ and $D = x^k - x^{k-1}$. Then the derivative approximation at x^k will be

$$f'(x^k) \cong \frac{F}{D} = \frac{C}{D} - \frac{E}{D} = f'(x^{k-1}) - \left(\frac{E}{D}\right)$$

So the derivative adjustment (or updating) term, $\Delta[f'(x^{k-1})]$, is $-E/D$. Since $F/D - C/D = -E/D$, and $F = f(x^k) - f(x^{k-1})$, we see that this adjustment term is

$$\Delta[f'(x^{k-1})] = \frac{f(x^k) - f(x^{k-1})}{x^k - x^{k-1}} - \frac{f'(x^{k-1})(x^k - x^{k-1})}{x^k - x^{k-1}}$$

and so

$$f'(x^k) \cong f'(x^{k-1}) + \frac{f(x^k) - f(x^{k-1}) - f'(x^{k-1})(x^k - x^{k-1})}{x^k - x^{k-1}} \tag{5.6}$$

Numerical Illustrations Tables 5.5 through 5.9 contain results for these derivative-based approximation techniques, for the same two functions as in the previous section. As

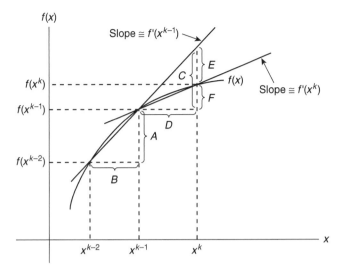

FIGURE 5.6   Updating $f'(x^{k-1})$.

before, iterations were stopped after the kth step if $|f(x^k)| < 10^{-4}$. Also as before, displays to six decimal accuracy are used near the root in order to indicate more precisely how the approximation technique approaches the solution.

The results in Tables 5.5 and 5.6 are for $f(x) = x^2 - 4$. Note that we cannot begin either the Newton–Raphson or modified Newton iterations at $x^0 = 0$ because $f'(0) = 0$ for this function, and both of these methods require division by $f'(x^0)$—at the first iteration only

TABLE 5.5   Derivative-Based Iterative Methods for Finding Roots to $f(x) = x^2 - 4$ (with $x^0 = 0.01$)

	Newton–Raphson		Modified Newton		Quasi-Newton	
k	x^k	$f(x^k)$	x^k	$f(x^k)$	x^k	$f(x^k)$
0	0.01	−3.9999	0.01	−3.9999	0.01	−3.9999
1	200.005	39998.00	200.05	39998.00	190.481	36279.17
2	100.01	9998.50	-2×10^6	4×10^{12}	0.031	−3.999
3	50.03	2498.63	-2×10^{-14}	4×10^{28}	0.052	−3.997
4	25.05	623.66	(Explosion)		48.220	2321.168
5	12.61	154.92			0.135	−3.982
6	6.46	37.76			0.217	−3.953
7	3.54	8.54			11.449	127.074
8	2.34	1.45			0.556	−3.691
9	2.02	0.10			0.863	−3.255
10	2.000143	5.71×10^{-4}			3.156	5.962
11	2.000000	2.04×10^{-8}			1.673	−1.201
12					1.922	−0.307
13					2.007	0.029
14					1.999858	-5.67×10^{-4}
15					2.000000	-1.01×10^{-6}

TABLE 5.6   Derivative-Based Iterative Methods for Finding Roots to $f(x) = x^2 - 4$ (with $x^0 = 4$)

	Newton–Raphson		Modified Newton		Quasi-Newton	
k	x^k	$f(x^k)$	x^k	$f(x^k)$	x^k	$f(x^k)$
0	4	12	4	12	4	12
1	2.5	2.25	2.5	2.25	2.5002	2.2509
2	2.05	0.2025	2.2188	0.9229	2.1539	0.6393
3	2.000610	2.44×10^{-3}	2.1034	0.4243	2.0165	0.0664
4	2.000000	3.72×10^{-7}	2.0504	0.2040	2.000610	2.44×10^{-3}
5			2.0249	0.1001	2.000003	1.01×10^{-5}
6			2.0124	0.0496		
7			2.0062	0.0247		
8			2.0031	0.0123		
9			2.001536	6.15×10^{-3}		
10			2.000768	3.07×10^{-3}		
11			2.000384	1.54×10^{-3}		
12			2.000192	7.67×10^{-4}		
13			2.000096	3.84×10^{-4}		
14			2.000048	1.92×10^{-4}		
15			2.000024	9.59×10^{-5}		

with Newton–Raphson [as in (5.3)] and for all iterations with modified Newton [as in (5.4)]. The quasi-Newton approach, since it finds the derivative at x^0 with a finite-difference approximation, *could* be started at $x^0 = 0$. (The result is not shown in any of the tables, but the quasi-Newton technique reaches $x^k = 2.000000$ from $x^0 = 0$ at $k = 22$.)

A straightforward way out of the dilemma when $f'(x^0) = 0$ is to perturb x^0 slightly away from zero, say, to $x^0 = 0.01$. Then the division-by-zero problem disappears. Results for this starting point are shown in Table 5.5. Notice that all three approaches make some bad calls at the outset. The reason for that should be clear. Both Newton–Raphson and Modified

TABLE 5.7   Iterations Needed from Alternative Starting Points for $f(x) = x^2 - 4$

x^0	Newton–Raphson	Modified Newton	Quasi-Newton
–4	4	15	5
3	3	9	5
4	4	15	5
5	4	21	6
–7	5	33	6
–8	5	39	7
–9	5	45	7
10	5	51	7
100	9	588	12
1000	12	5978	17
10,000	15	59899	22
100,000	19	—[a]	26

[a]*Conjecture:* This starting point for modified Newton would require approximately 10 times as many iterations as were required for the starting point at 10,000.

TABLE 5.8 **Derivative-Based Iterative Methods for Finding Roots to $f(x) =$ $(x - 1)(x + 1)^2 - 1$**

	Newton–Raphson		Modified Newton		Quasi-Newton	
k	x^k	$f(x^k)$	x^k	$f(x^k)$	x^k	$f(x^k)$
0	0	−2.0	0	−2	0	−2.0
1	−0.67	−1.19	−2	−4	−2.00	−4.01
2	−0.88	−1.03	−6	−176	1.98	7.82
3	−1.05	−1.00	−182	−5,995,264	−0.64	−1.20
4	−1.22	−1.10	−5,995,446	$−2.16 \times 10^{20}$	−0.30	−1.64
5	−1.40	−1.39	(Explodes)		−1.62	−1.99
6	−1.66	−2.15			5.84	225.43
7	−2.11	−4.87			−1.55	−1.77
8	−4.26	−56.86			−1.49	−1.66
9	−2.06	−4.44			−0.94	1.08
10	−3.75	−37.05			−0.01	−1.99
11	−1.54	−1.75			−1.89	−3.31
12	−1.88	−3.25			2.82	25.51
13	−2.77	−12.81			−1.35	−1.29
14	0.999	−1.005			−1.15	−1.05
15	1.167	−0.218			−0.28	−1.66
16	1.194	−0.068			−2.64	−10.83
17	1.202	−0.022			0.15	−2.12
18	1.204	−0.007			0.82	−1.59
19	1.205	−0.002			2.84	26.02
20	1.2054	$−7.83 \times 10^{-4}$			0.94	−1.23
21	1.205525	$−2.57 \times 10^{-4}$			1.02	−0.90
22	1.205548	$−8.44 \times 10^{-5}$			1.26	0.31
23					1.20	−0.05
24					1.2052	1.89×10^{-3}
25					1.205572	1.23×10^{-4}

Newton use $f'(x^0)$ as a divisor of $f(x^0)$ at their initial iteration, thereby creating a relatively large number for this example, since $f'(0.01) = 0.02$. The modified Newton technique perpetuates this initial bad luck by always using $f'(x^0)$ [in place of $f'(x^k)$ at each step] and for that reason, as shown by the results in the table, it explodes. (Can you see why the explosion is toward $-\infty$ in this example?) The quasi-Newton method is almost as bad initially (sending us out to $x^1 = 190.48$ instead of $x^1 = 200$, as in Newton–Raphson and modified Newton), but it, like Newton–Raphson also, is able to correct for this lapse of judgment, which is temporary for these two methods and permanent for modified Newton.

Comparison with the results for the nonderivative methods in Table 5.1 for the same function and almost the same starting point suggests no great advantage at all for the derivative-based methods. Both secant and Brent nonderivative methods are superior to Newton–Raphson and quasi-Newton in terms of number of iterations required.[8] And any-

[8]Quasi-Newton results will always be replicated by the secant method if you use the x^0 and x^1 found in the initial quasi-Newton step as the two opening points. For example, from the results in Table 5.5, using $x^0 = 0.01$ and $x^1 = 190.4814286$ (which was rounded to 190.481 in the table), the secant approach will generate exactly the same sequence as shown for quasi-Newton in Table 5.5.

TABLE 5.9   Iterations Needed from Alternative Starting Points for $f(x) = (x - 1)(x + 1)^2 - 1$

x^0	Newton–Raphson	Modified Newton	Quasi-Newton
3.0	13	52	7
0.181818	46	(Explodes)	11
−2.0	23	19	11
10	16	591	11
100	22	56,845	19
1,000	38	—[a]	27
10,000	33		36
100,000	39		44

[a]Further experiments with modified Newton were foregone. The iteration count appears to increase by a factor of 100 from one starting point to the next.

thing is superior to modified Newton, which just does not work. The reason for the poor showing of the derivative methods is because they are put off by an initial derivative that is close to zero. This puts the initial estimates of x [and, as a consequence, $f(x)$] badly out of kilter.[9]

In Table 5.6 we see the benefits of a different starting point at which the derivative is not close to zero. Here $x^0 = 4$ has been used; this corresponds to x^1 for the approximation methods in Table 5.1. Note that now the derivative-based methods do very well, indeed, rewarding the user for the trouble of bothering with derivatives (or their approximations)—they get closer to the root faster, even though $x^0 = 4$ is further away from the root than was $x^0 = 1$ in Table 5.5.

Results in Tables 5.1, 5.5, and 5.6 suggest, for one specific function only, that bisection is by far the most pedestrian approach (a fact that might have been expected, considering its simplicity), while all others are very roughly twice as fast. We have also seen that this depends on choosing an initial starting point that does not generate $f'(x^0) = 0$ for the derivative-based methods, and that even $f'(x^0)$ near to zero should be avoided.

Table 5.7 contains iteration counts for a number of alternative starting points for $f(x) = x^2 - 4$. For these examples it is clear that Newton–Raphson and quasi-Newton are just about equally efficient and relatively insensitive to selection of the starting point. Modified Newton is far worse.

Results from using derivative-based methods on the second test function, $f(x) = (x - 1)(x + 1)^2 - 1$, are shown in Table 5.8. For this function it is possible to start at $x^0 = 0$, since $f'(0) \neq 0$. Newton–Raphson required 22 iterations, quasi-Newton took 25, and modified Newton failed again, sending values for x [and $f(x)$] explosively toward $-\infty$. Looking at the results in Table 5.3 for this function with nonderivative methods, we see that the two successful derivative methods out performed only regula falsi. Otherwise, nonderivative methods were better.

Finally, Table 5.9 includes iteration counts for several alternative starting points for $f(x) = (x - 1)(x + 1)^2 - 1$. You might try to see why (1) Newton–Raphson does so much better for $x^0 = 0$ (shown in Table 5.8) than for $x^0 = 0.181818$ (which is closer to the root); (2) why $x^0 = -2$ is just about as good a starting point for Newton–Raphson as $x^0 = 0$, even

[9]Note in Table 5.5 that Newton–Raphson works back from $x^1 = 200$ by essentially cutting the bad guesses in half through about $k = 7$, whereas quasi-Newton goes through a sequence of bad over- and underestimates until it settles down at about $k = 11$.

though it is farther from the root; (3) why, in the cases shown in Table 5.9, quasi-Newton nearly always out performs Newton–Raphson; or (4) why modified Newton is better than Newton–Raphson for $x^0 = -2$, worse for $x^0 = 3$, and impossible for $x^0 = 0.181818$.

5.2 SOLUTIONS TO F(X) = 0

We now consider *systems* of equations, $f^1(x_1, \ldots, x_n) = 0, \ldots, f^n(x_1, \ldots, x_n) = 0$. As usual, letting $\mathbf{X} = [x_1, \ldots, x_n]'$ and $\mathbf{F}(\mathbf{X}) = [f^1(\mathbf{X}), \ldots, f^n(\mathbf{X})]'$, the system can be described compactly as $\mathbf{F}(\mathbf{X}) = \mathbf{0}$. In the case of an $n \times n$ *linear* equation system, $\mathbf{F}(\mathbf{X}) \equiv \mathbf{AX} - \mathbf{B}$. In general, finding solutions to a system of *nonlinear* equations is enormously difficult; the difficulty depends, of course, on the nature and extent of the nonlinearities in the equations. There is no approach that will *guarantee* a solution.

For linear equations (Chapters 1 and 2), when there were two equations in two unknowns (x_1 and x_2), we could easily sketch the geometry in solution space, with x_1 on the horizontal axis and x_2 on the vertical axis. Geometrically, the unique solution (if one existed) was the point of intersection of the two lines. Algebraically, the solution was easily found using the inverse of the matrix of coefficients of the equation system. (Other possibilities, in addition to unique solutions, were parallel lines, with no solution, and a pair of equations whose solution space images were collinear, with multiple solutions.) Iterative approaches for systems of linear equations are also available (Section 2.7). These included the Gauss–Seidel technique, which turns out to be useful for nonlinear equation systems, also.

As an example, the equations in (1-26) were

$$3x_1 + 2x_2 = 7$$
$$2x_1 + 5x_2 = 12$$

The solution-space representation was shown in Figure 1.1 in Chapter 1; it is repeated in Figure 5.7a. In Section 5.1 it was useful to convert an equation like $x^2 = 4$ to $f(x) = x^2 - 4 = 0$. This allowed a two-dimensional representation of an essentially one-dimensional fact (namely, $x = \pm 2$) in $x, f(x)$ space. Similarly, the pair of linear equations from (1.26) can be visualized as linear functions (planes) in three-dimensional $x_1, x_2, f(x_1, x_2)$ space. From that point of view, the two linear equations are

$$f^1(x_1, x_2) = 3x_1 + 2x_2 - 7 = 0$$
$$f^2(x_1, x_2) = 2x_1 + 5x_2 - 12 = 0$$

and the two lines in Figure 5.7a indicate where the three-dimensional functions $f^1(\mathbf{X})$ and $f^2(\mathbf{X})$ (planes in this case) cut the x_1, x_2 plane. This three-dimensional view is suggested in Figure 5.7b.

For two *nonlinear* equations in x_1 and x_2, instead of *planes* in three dimensions, we have curved surfaces. The problem of solving the pair of nonlinear equations is equivalent to

1. Finding the *curves* (not *lines*, as in the case of linear equations) that indicate where the functions $f^1(\mathbf{X})$ and $f^2(\mathbf{X})$ cut the x_1, x_2 plane.
2. Then finding where those curves intersect, if at all, in x_1, x_2 space.

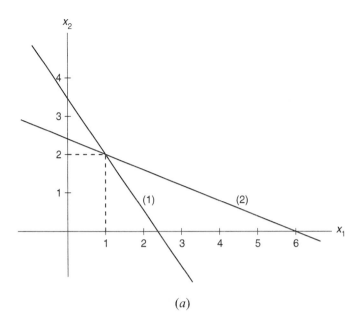

(a)

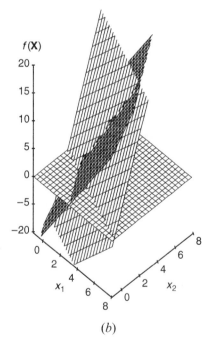

(b)

FIGURE 5.7   (a) Solution-space representation of (1) $3x_1 + 2x_2 = 7$ and (2) $2x_1 + 5x_2 = 12$; (b) $f^1(\mathbf{X})$ and $f^2(\mathbf{X})$ in three-dimensional space.

Figure 5.8 suggests some options. In particular, in Figures 5.8a and 5.8b, $\mathbf{X}^*$ represents a solution. Whether it is unique or not depends entirely on the behavior of the two functions $f^1(\mathbf{X}) = 0$ and $f^2(\mathbf{X}) = 0$ over the rest of the x_1,x_2 plane. In (a), it is a unique solution; in Figure 5.8b it is not. In Figures 5.8c and 5.8d we find no solutions. These illustrations highlight the fact that, with nonlinear functions, there are no simple and general rules (like the rules of rank for linear equation systems) for distinguishing unique, multiple, and inconsistent (no solution) cases. In particular, there is no simple relationship describing the connection between solutions in cases where there is more than one. Recall that in the linear case the solutions were proportional. This will return to haunt us.

5.2.1 Nonderivative Methods

Selective Solution and Substitution   In principle, you can always try to solve one or more equations so as to express one unknown as a function of the remaining unknowns, then substitute these results into the remaining equations, forming a smaller system. Eventually, in some cases, the original system can be reduced in this way to one equation in one unknown. That one equation can then be solved either directly or by methods that were discussed in Section 5.1 (depending on its complexity), and values of the remaining variables can then be found in some sequence by substitution.

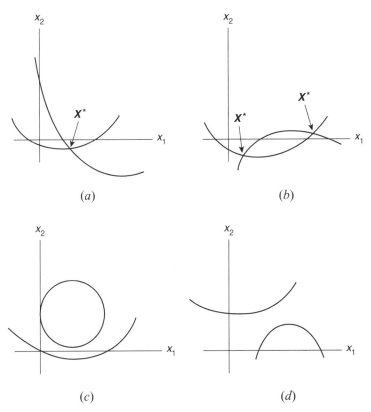

FIGURE 5.8   Solution-space representations of $f^1(\mathbf{X})$ and $f^2(\mathbf{X})$: (a) unique solution; (b) nonunique solution; (c) no solution; (d) no solution.

We use a specific example of four equations in four unknowns to illustrate this technique. This example will be explored in more detail below, as the third (and largest) of three numerical illustrations. It is introduced here at this time because it conveys more of the spirit of sequential solution and substitution approaches than do smaller systems of nonlinear equations. The equation system is

$$2x_1 + x_2 = 3 \tag{5.7}$$

$$x_1 x_3 + x_2 - x_4 = 0 \tag{5.8}$$

$$4x_3 - x_4^2 = 20 \tag{5.9}$$

$$x_1 + (0.5)x_4 = 0 \tag{5.10}$$

This system is not extremely nonlinear, but it is adequate as an illustration. Here it is possible to solve (5.7) easily for either x_1 or x_2; for example

$$x_2 = 3 - 2x_1 \tag{5.7'}$$

Similarly, (5.10) is easily solved for either x_1 or x_4; again, for example

$$x_1 = -(0.5)x_4 \tag{5.10'}$$

From (5.9),

$$x_3 = \frac{20 + x_4^2}{4} \tag{5.9'}$$

and all of this information can be put into (5.8), giving, after some simplification

$$-\tfrac{1}{8}x_4^3 - \tfrac{20}{8}x_4 = -3 \tag{5.8'}$$

The value(s) of x_4 that are consistent with this one nonlinear equation can be found by the techniques discussed in earlier parts of this chapter. To three decimals, the only real number solution is found to be $x_4 = 1.128$. (The other two solutions to this cubic equation are $x_4 = -0.5641 \pm 4.5776 \sqrt{-1}$.) Then, from (5.9'), $x_3 = 5.318$; from (5.10'), $x_1 = -0.564$; and, from (5.7'), $x_2 = 4.128$. This approach is useful only when the number of equations is relatively small and at least some of the equations lend themselves to fairly straightforward algebraic solution for one of the unknowns as a function of the others. This qualification holds for the next variation on this technique as well.

Complete Algebraic Solution and Substitution   If you have available any of the mathematical toolbox computer software that does straightforward algebra—for example, solving equations for specified variables and also performing algebraic substitutions efficiently—then a *systematic* sequential approach can be tried.[10] This is nothing more than the approach just discussed, but executed in a rigorous and systematic way.

---

[10]There are many kinds of such software, including (but by no means limited to) Derive, Mathcad, Matlab, Mathematica, and Maple.

Solve the first equation for x_1 as a function of the remaining variables, $x_2, \ldots, x_n$. Put this algebraic result into *each* of the remaining equations that contains x_1. Then solve the second equation for x_2 as a function of $x_3, \ldots, x_n$, and put that result into *each* of the remaining equations that contains x_2. Continuing in this way, the last equation will be an expression involving x_n only. Then solve this equation for x_n. (Here, again, the methods discussed in Section 5.1 may be appropriate.) There may be more than one solution value for x_n. If there are several, the following set of steps must be followed for *each* x_n in turn. Substitute back into the equation for x_{n-1}, giving one or more numerical values for x_{n-1}, and so forth until the final step, when one or more sets of values for $x_n, \ldots, x_2$ are substituted into the first equation, thus determining x_1. The forward substitution part is *algebraic*; the backward substitution is *numeric*.

In any sequential procedure like this, the *ordering* of the equations can make a big difference in the ease of solution—through rearrangement it may be possible to order the equations so that the first is easily solved for x_1, the next for x_2, and so forth. But in large and complicated systems, the best ordering, if there is one, may not be at all obvious. The approach will not work if, at step k, it is not possible to solve for x_k as a function of x_{k+1}, $\ldots, x_n$.

Using Equations (5.7)–(5.10) as an illustration, without any reordering, we proceed as follows. Solve (5.7) for x_1, giving

$$x_1 = \frac{3 - x_2}{2} \tag{5.7''}$$

Substitution of this fact about x_1 into Equations (5.8) and (5.10) [Eq. (5.9) does not contain x_1] leaves three equations that depend on x_2 through x_4:

$$\frac{(3 - x_2)x_3}{2} + x_2 - x_4 = 0$$

$$4x_3 - x_4^2 = 20$$

$$\frac{3 - x_2}{2} + (0.5)x_4 = 0$$

Solving the first of these for x_2 (now as a function of x_3 and x_4 only) gives

$$x_2 = \frac{3x_3 - 2x_4}{x_3 - 2} \tag{5.8''}$$

From the second equation above,

$$x_3 = \frac{20 + x_4^2}{4} \tag{5.9''}$$

Finally, substituting (5.8'') and then (5.9'') into the third equation gives a single equation in x_1 only:

$$\frac{8(x_4 - 3)}{x_4^2 + 12} + x_4 = 0 \tag{5.10''}$$

This completes the *forward substitution* of purely algebraic expressions. Specific numerical results come from successive *backward substitutions* in the order x_4, x_3, x_2, x_1. The solutions to Equation (5.10''), found using algebraic software or any of the methods in Section 5.1, are $x_4 = 1.128$, along with $-0.5641 \pm 4.5776 \sqrt{-1}$. Using the real number solution in Equation (5.9''), $x_3 = 5.318$. Continuing, using the known values for x_3 and x_4 in Equation (5.8''), $x_2 = 4.128$. Finally, using x_2 in Equation (5.7''), $x_1 = -0.564$.

Gaussian Methods   A basic and frequently used technique for systems of nonlinear equations, when derivatives are not used, is essentially the Gauss–Seidel iterative scheme that was already examined in the context of a system of *linear* equations (Section 2.7). When the equations are nonlinear, the approach is identical in philosophy but more difficult in implementation, and unfortunately it can be *very* sensitive to the choice of a starting point.

The basic idea is as follows:

(a) Begin with an initial guess for the values of the elements in the solution vector $\mathbf{X}$: $\mathbf{X}^0 = [x_1^0, x_2^0, \dots, x_n^0]'$.

(b) Start with $f^1(\mathbf{X}) = 0$, and *solve* for x_1 as a function of $x_2, \dots, x_n$. This is the same as the initial step in the sequential procedure that was just discussed, above. This gives $x_1 = g^1(x_2, \dots, x_n)$. Given the initial values for $x_2, \dots, x_n$ from $\mathbf{X}^0$, *evaluate* $g^1(x_2^0, \dots, x_n^0)$ for a "better" value for x_1, and denote it x_1^1.

(c) In the next equation, $f^2(\mathbf{X}) = 0$, consider x_2 as the dependent variable. That is, solve $f^2(\mathbf{X}) = 0$ for x_2 as a function of *all* the other x's , including x_1. Given the value of x_1 calculated in the previous step, x_1^1, and given the initial values of the other variables, $x_3^0, \dots, x_n^0$, evaluate $g^2(x_1^1, x_3^0, \dots, x_n^0)$ for a value for x_2 that is "better" than x_2^0; denote this x_2^1.

(d) Continue in this fashion, solving in turn, for $i = 3, \dots, n$:

$$f^i(\mathbf{X}) = 0 \qquad \text{for} \qquad x_i = g^i(\mathbf{X}_i)$$

where $\mathbf{X}_i$ denotes the vector containing all unknowns except for x_i. In each case, find x_i by evaluating $g^i(\mathbf{X}_i)$ using the values of $x_1, \dots, x_{i-1}$ found in the *immediately preceding* steps and using $x_{i+1}^0, \dots, x_n^0$ from the *initial values* in $\mathbf{X}^0$.

(e) After one iteration of n steps through this sequence, there will be a new set of values for all n of the variables. Denote this vector $\mathbf{X}^1$, use it in place of $\mathbf{X}^0$, and repeat the entire procedure to produce another set of n new values, $\mathbf{X}^2$. Once the *solutions* $x_i = g^i(\mathbf{X}_i)$ have been found, each subsequent step requires *evaluations* only. [Of course, solutions $g^i(\mathbf{X}_i)$ may be impossible or one or more of the expressions $g^i(\mathbf{X}_i)$ may be (algebraically) very complicated and cumbersome.]

(f) Carry out the previous step, in (e), where now $\mathbf{X}^1$ plays the role of $\mathbf{X}^0$ and $\mathbf{X}^2$ plays the role of $\mathbf{X}^1$. This will generate yet another set of n new values, $\mathbf{X}^3$.

(g) Continue until some appropriate stopping criterion has been met. A stopping rule parallel to the one used in Section 5.1 when searching for a root to $f(x) = 0$ could be used, although it would be more complicated to implement, because we are dealing now with n equations, $f^i(\mathbf{X}) = 0$ $(i = 1, \dots, n)$ and not just one. At the kth step, after finding $\mathbf{X}^k$, *each* of the functions $f^1(\mathbf{X}), \dots, f^n(\mathbf{X})$ could be evaluated for that value, $\mathbf{X}^k$, and then a criterion could be constructed on the basis of the differences of $|f^1(\mathbf{X}^k)|, \dots, |f^n(\mathbf{X}^k)|$ from 0. The problem is that there are n such differences. The criterion could be based on the *largest* differences among the $f^i(\mathbf{X})$ at the kth step, $\max_i |f^i(\mathbf{X}^k)|$, or it could be some aggregate measure, such as the sum of all absolute values of these differences (so that positive and negative deviations from zero do not cancel one another), $\sum_{i=1}^{n} |f^i(\mathbf{X}^k)|$. In general, for large n and complicated functions, these n functional evaluations at each step may take a lot of time and/or computer memory.

An alternative kind of stopping criterion examines the sizes of the differences in successive values of the elements in $\mathbf{X}^{k+1}$ and $\mathbf{X}^k$; the logic is that if the values of the x's are not changing very much from one iteration to the next, we must be "near to" a root. [This simpler kind of criterion could also be used in the case of root finding for $f(x)$, but the use of $|f(x^k)|$ itself is not difficult with just one function.] Again, it would be possible to construct a criterion based on the largest difference at any step, $\max_i |x_i^{k+1} - x_i^k|$ or on some aggregate measure. For the numerical illustrations that follow in this section, the $\max_i |f^i(\mathbf{X}^k)| < \delta$ criterion is used.

We use the set of equations in (5.7)–(5.10) to illustrate a number of the characteristics of the Gauss–Seidel method. In particular, we have the following solutions $x_i = g^i(\mathbf{X}_i)$:

$$x_1 = g^1(x_2,x_3,x_4) = \frac{3 - x_2}{2} \tag{5.11}$$

$$x_2 = g^2(x_1,x_3,x_4) = x_4 - x_1 x_3 \tag{5.12}$$

$$x_3 = g^3(x_1,x_2,x_4) = \frac{20 + x_4^2}{4} \tag{5.13}$$

$$x_4 = g^4(x_1,x_2,x_3) = -2x_1 \tag{5.14}$$

[In this relatively simple example, not all x's appear in each of the equations, and both g^1 and g^4 are very simple functions; in general, one has to be prepared for the eventuality that *all* $n-1$ unknowns, except for x_i, will appear in each g^i ($i = 1, \ldots, n$) and that some of these g^i will be very messy functions.]

Numerical Illustration: Two Equations in Two Unknowns   For both the Gauss–Seidel approach and several derivative-based methods, to be discussed below, we use three different nonlinear equation systems containing, in turn, two, three, and four equations with two, three, and four unknowns, respectively. The two-equation system is

$$x_1 + x_2 = 2 \tag{5.15a}$$

$$x_1^2 - 2x_1 + x_2^2 - 4x_2 = 4 \tag{5.15b}$$

or, in $f^i(\mathbf{X}) = 0$ form:

$$f^1(\mathbf{X}) = x_1 + x_2 - 2 = 0 \tag{5.16a}$$

$$f^2(\mathbf{X}) = x_1^2 - 2x_1 + x_2^2 - 4x_2 - 4 = 0 \tag{5.16b}$$

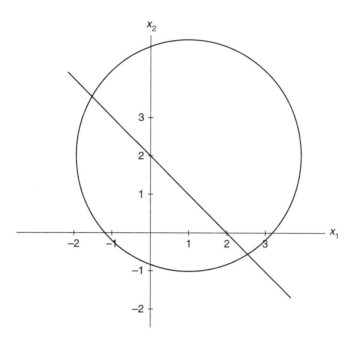

FIGURE 5.9   Solution-space representation of (1) $x_1 + x_2 = 2$ and (2) $x_1^2 - 2x_1 + x_2^2 - 4x_2 = 4$.

The x_1, x_2-space geometry is sketched in Figure 5.9. The two functions intersect at two points, which means that this pair of equations has two distinct solutions.[11] From the figure it appears that the solutions are at approximately $x_1 = -1.5$ and $x_2 = 3.6$ and at $x_1 = 2.5$, $x_2 = -0.5$. In fact, to three decimal points, the two solutions are $\mathbf{X}^* = \begin{bmatrix} -1.562 \\ 3.562 \end{bmatrix}$ and $\mathbf{X}^{**} = \begin{bmatrix} 2.562 \\ -0.562 \end{bmatrix}$, as we are about to see. As with any system of nonlinear equations, the objective is to be able to find these points of geometric intersection algebraically, using the numerical methods discussed in this section.

Algebraic Solution and Substitution   Here, it is straightforward to solve (5.15a) for either x_1 or x_2. The most systematic approach (unnecessary for such a small system as this) is to solve $f^i(\mathbf{X}) = 0$ for x_i. Thus, from (5.15a), we have $x_1 = 2 - x_2$. Putting this result into (5.15b) gives $x_2^2 - 3x_2 - 2 = 0$, from which (using the general formula for solution of a quadratic equation, computer software that does algebra, or one of the approaches in Section 5.1) $x_2 = (3 \pm \sqrt{17})/2 = 3.562$ or -0.562, and thus, going back to (5.15a), the two solutions are

$$\mathbf{X}^* = \begin{bmatrix} -1.562 \\ 3.562 \end{bmatrix} \quad \text{and} \quad \mathbf{X}^{**} = \begin{bmatrix} 2.562 \\ -0.562 \end{bmatrix} \tag{5.17}$$

[11]Note the difference here from the linear equations case, where if there is more than one solution, then there are infinitely many (proportional) solutions.

Gauss–Seidel Results from using the Gauss–Seidel approach (which will often be called simply "Gauss" in what follows) are shown in Table 5.10. The stopping criterion is parallel to the rule used in the single-equation cases in Section 5.1; the iterations continue until *both* functions are within 10^{-4}, at least, of zero. Again, for one example only, the steps to solution are shown explicitly, simply to give, for one example, a complete account of how the solution is approached by this method.

Note that in this two-equation case, x_2 is found in the second step in each iteration using the just-derived new value of x_1 in $g^2(x_1)$. Thus the solution derived in this step *must* satisfy the second equation exactly (or as closely as possible with the precision used for the calculations). That explains the 0's as the second element in $\mathbf{F}(\mathbf{X}^k) = [f^1(\mathbf{X}^k), f^2(\mathbf{X}^k)]$ in the upper panel of the table. The Gauss–Seidel sequence for this problem generates a set of estimates for $f^1(\mathbf{X}^k)$ that alternatively overshoot and undershoot 0, and each successive estimate is something like roughly half as bad as in the preceding step. From the $\mathbf{F}(\mathbf{X}^k)$ column in Table 5.10 we can also see when and where the iterations would have stopped under less exacting values of δ. For example, with $\delta = 10^{-3}$ we would have been satisfied with $\mathbf{X} = \begin{bmatrix} 2.561 \\ -0.562 \end{bmatrix}$ after 15 iterations.

This relatively simple pair of equations illustrates a problem that can arise with Gauss. In this case, where there are only two equations and two unknowns, we start by finding

$$x_1 = g^1(x_2) = 2 - x_2$$

TABLE 5.10 **Results for the Gauss–Seidel Iterative Method for Two Equations in Two Unknowns**

k	$[\mathbf{X}^k]'$	$[\mathbf{F}(\mathbf{X}^k)]'$
0	[0, 0]	[−2, −4]
1	[2, −0.8284]	[−0.8284, 0]
2	[2.8284, −0.3784]	[0.4500, 0]
3	[2.3784, −0.6646]	[−0.2861, 0]
4	[2.6646, −0.4958]	[0.1687, 0]
5	[2.4958, −0.6005]	[−0.1046, 0]
6	[2.6005, −0.5374]	[0.0631, 0]
7	[2.5374, −0.5761]	[−0.0387, 0]
8	[2.5761, −0.5526]	[0.0235, 0]
9	[2.5526, −0.5670]	[−0.0144, 0]
10	[2.5670, −0.5582]	[0.0087, 0]
11	[2.5582, −0.5636]	[−0.0053, 0]
12	[2.5636, −0.5603]	[0.0032, 0]
13	[2.5603, −0.5623]	[−0.0020, 0]
14	[2.5623, −0.5611]	[0.0012, 0]
15	[2.561095, −0.561832]	[−7.36×10⁻⁴, 0]
16	[2.561832, −0.561383]	[4.49×10⁻⁴, 0]
17	[2.561383, −0.561656]	[−2.74×10⁻⁴, 0]
18	[2.561656, −0.561490]	[1.67×10⁻⁴, 0]
19	[2.561490, −0.561591]	[−1.02×10⁻⁴, 0]
20	[2.561591, −0.561529]	[6.20×10⁻⁵, 0]

Note: Stopping criterion: max $[|f^1(\mathbf{X}^k)|, |f^2(\mathbf{X}^k)|] \leq 10^{-4}$. From alternative starting points (0, 1), (0, 2), (0, 3), (0, 4), (−1, 3) and (A, 0), where "A" is any real number, either 20 or 21 iterations were also required.

Then

$$x_2 = g^2(x_1) = 2 \pm \sqrt{-(x_1^2) + 2x_1 + 8}$$

This means, first, that the two possibilities represented by the "+" and the "−" on the right-hand side of x_2 have to be examined separately. Secondly, when there is a square root term, as in x_2, an iterative sequence may well lead to imaginary numbers (if a negative number is generated under the square root sign). Programs that expect to work with real numbers only will stop when faced with the need to find the square root of a negative number as a step in a sequence of repeated operations. This happens in this example. When the square root term is *added* in $g^2(x_1)$, x_2 grows larger and x_1 becomes smaller; when $x_1 < -2$, the quantity under the square root sign is negative. Therefore, the results for Gauss reported for this example are all for the case in which

$$x_2 = g^2(x_1) = 2 - \sqrt{-(x_1^2) + 2x_1 + 8}$$

and this is why the results show the technique always approaching the **X**** solution in (5.17) and never **X***. The same problem (square root of a negative number) can also arise when using the x_2 above (with the second term subtracted) if $x_2^0 > 4$.

Numerical Illustration: Three Equations in Three Unknowns   For this illustration, we use

$$x_1^2 + x_2^2 + x_3^2 = 125 \tag{5.18a}$$

$$x_1 + x_2 + x_3 = 15 \tag{5.18b}$$

$$x_1 x_2 + x_2 x_3 + x_1 x_3 + x_1 x_2 x_3 = 50 \tag{5.18c}$$

In $f^i(\mathbf{X})$ notation, these are

$$f^1(\mathbf{X}) = x_1^2 + x_2^2 + x_3^2 - 125 = 0 \tag{5.19a}$$

$$f^2(\mathbf{X}) = x_1 + x_2 + x_3 - 15 = 0 \tag{5.19b}$$

$$f^3(\mathbf{X}) = x_1 x_2 + x_2 x_3 + x_1 x_3 + x_1 x_2 x_3 - 50 = 0 \tag{5.19c}$$

The geometry of the $f(x)$ representation would require four dimensions—one each for x_1, x_2 and x_3, and one along which to measure $f(\mathbf{X})$. Instead of that, then, consider the "solution space" geometry of the functions in (5.18). Just as the equations in (5.15) were two-dimensional solution-space representations of the three-dimensional functions in (5.16), so the equations in (5.18) can be thought of as three-dimensional solution-space representations for the four-dimensional equations in (5.19).

From that viewpoint, (5.18a) is a sphere with a center at the origin and a radius of $\sqrt{125}$, and (5.18b) is a plane oriented so that it intersects each of the three axes at 15. Values of x_1, x_2, and x_3 that simultaneously satisfy these two equations would lie along the intersection of the plane with the sphere—a set of points in a (tilted) circle. Finally, then, the solutions to all three equations will be found where that circle intersects the three-dimensional solution-space representation of (5.18c). The geometry of function (5.18c) is quite complicated; it is basically a rather flat plane with valleys, folds, and spikes near the origin. The various panels in Figure 5.10 show pairs of equations together and also all three, although by then the geometry is rather cluttered. For example, Figure 5.10(a) shows the sphere [(5.18a)] and the plane [(5.18b)].[12]

This geometry can be generated with any of the mathematical toolbox computer software that incorporates three-dimensional graphics. To solve the problem algebraically it is necessary (using math toolbox algebraic capabilities, if needed) to solve (5.18c) only for, say, x_3 as a function of x_1 and x_2.

Algebraic Solution and Substitution It turns out to be very difficult to solve these equations sequentially in the order in which they are presented, arriving finally at $x_3 = g^3(x_3)$, and then solving that equation for x_3. This is another indication of the fact that *ordering* of equations can make a huge difference in the success of a technique. Imagine that (5.18a) is moved to below (5.18c)—so the ordering of the equations is (b), (c), (a). Then, if (5.18b) is solved first, for, say, x_1, we have

$$x_1 = 15 - x_2 - x_3 \tag{5.18'b}$$

Putting this result into (5.18c) and solving for x_2 gives (at this point the algebra gets really messy and a computer package is essential)

$$x_2 = \frac{\sqrt{x_3^3 - 33x_3^2 + 255x_3 + 25} + (15 - x_3)\sqrt{x_3 + 1}}{2\sqrt{x_3 + 1}} \tag{5.18'c}$$

and

$$x_2 = \frac{-\sqrt{x_3^3 - 33x_3^2 + 255x_3 + 25} + (x_3 - 15)\sqrt{x_3 + 1}}{2\sqrt{x_3 + 1}} \tag{5.18'c'}$$

Finally (and now the algebra is even worse), putting (5.18'b) into (5.18a) and then either (5.18'c) or (5.18'c') into that revised expression for x_1 generates in (5.18a) an expression involving x_3 only. It boils down to

$$x_3 = \frac{\left[\sqrt{x_3^3 - 33x_3^2 + 255x_3 + 25} + (15 - x_3)\sqrt{x_3 + 1}\right]^2}{4(x_3 + 1)}$$

$$\pm \frac{\left[\sqrt{x_3^3 - 33x_3^2 + 255x_3 + 25} + (x_3 - 15)\sqrt{x_3 + 1}\right]^2}{4(x_3 + 1)}$$

$$+ x_3^2 - 125 \tag{5.18'a}$$

[12]Note that there could be *no* intersections of the two equations—for example, if the sphere had a smaller radius or the plane were located further out from the origin. Also, there could be one unique solution if the plane were located so that it happened to be just tangent to the sphere.

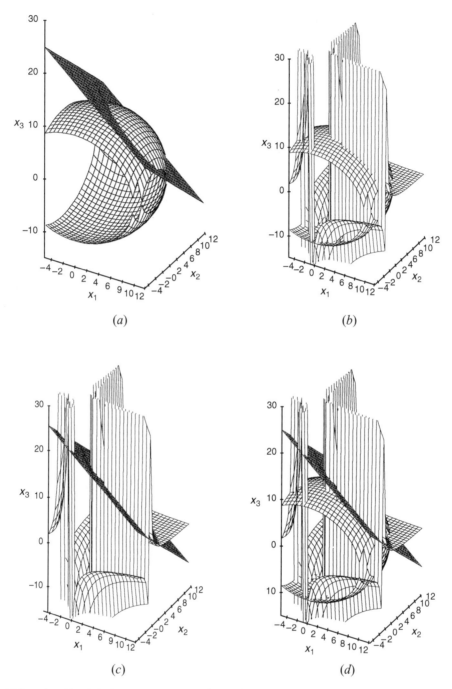

FIGURE 5.10   Solution-space representations of (5.18): (*a*) functions (5.18a) and (5.18b); (*b*) functions (5.18a) and (5.18c); (*c*) functions (5.18b) and (5.18c); (*d*) all three functions in (5.18).

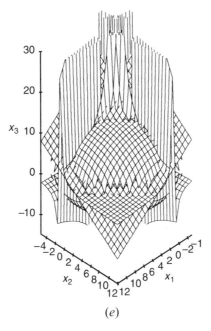

(e)

FIGURE 5.10 (e) alternative view of all three functions in (5.18).

Under various combinations of "+" and "−" for the square root terms, this equation has the three solutions $x_3 = 0, 5, 10$. In conjunction with the earlier equations, it is easily established that there are six solutions to the entire system of equations:

$$\begin{bmatrix} 0 \\ 5 \\ 10 \end{bmatrix}, \quad \begin{bmatrix} 0 \\ 10 \\ 5 \end{bmatrix}, \quad \begin{bmatrix} 5 \\ 0 \\ 10 \end{bmatrix}, \quad \begin{bmatrix} 5 \\ 10 \\ 0 \end{bmatrix}, \quad \begin{bmatrix} 10 \\ 5 \\ 0 \end{bmatrix}, \quad \begin{bmatrix} 10 \\ 0 \\ 5 \end{bmatrix}$$

Gauss–Seidel   For these equations we generate

$$\text{(a)} \quad x_1 = g^1(x_2, x_3) = \sqrt{125 - x_2^2 - x_3^2}$$

$$\text{(b)} \quad x_2 = g^2(x_1, x_3) = 15 - x_1 - x_3$$

$$\text{(c)} \quad x_3 = g^3(x_1, x_2) = (50 - x_1 x_2)/[x_1(x_2 + 1) + x_2]$$

Table 5.11 contains results from using Gauss for this set of nonlinear equations. The starting points are:

(a) The origin, reflecting absolutely no information about possible solutions.
(b) (1,6,11), which is relatively near to one of the solutions to this set of equations, namely, (0,5,10).
(c) (1,2,3), which, again, reflects no particular insight about the possible location of a solution.

TABLE 5.11   Results for the Gauss–Seidel Iterative Method for Three Equations in Three Unknowns

	A. Iteration Counts from Alternative Starting Points[a]		
Starting Point $(\mathbf{X}^0)$	Solution	Starting Point $(\mathbf{X}^0)$	Solution
0	16	0	IM[b]
0		5	
0		9	
1	IM	0.1	16
6		0	
11		0.01	
1	18	1.1	15
2		1	
3		1.01	
4	21		
5			
6			

	B. Iteration Counts for Alternative Values of δ			
Starting Point	$\delta = 10^{-3}$	$\delta = 10^{-4}$	$\delta = 10^{-6}$	$\delta = 10^{-10}$
(0, 0, 0)	14	16	22	33
(1, 2, 3)	15	18	23	34
(4, 5, 6)	18	21	27	38
(0.1, 0, 0.01)	14	16	22	33
(1.1, 1, 1.01)	12	15	21	32

[a]Solution reached was always, to at least four-decimal accuracy, $\mathbf{X} = [10, 5, 0]'$. Stopping criterion: $\max_i [|f^i(\mathbf{X}^k)|] \le \delta$ (where $\delta = 10^{-4}$).
[b]"IM" indicates lack of success for Gauss because imaginary numbers occurred during the iterations.

    (d) (4,5,6), which was chosen because it satisfies Equation (5.18b) exactly.

    (e) (0,5,9), which, like (b), is quite close to the solution (0,5,10).

In addition, two "perturbations" of the origin were also used as starting points (these are the last two sets of entries in the $\mathbf{X}^0$ column on the right in Table 5.11, part A). These new starting points were not at all necessary for Gauss, but they turn out to have meaning when we investigate derivative-based methods, below. We see that in two cases Gauss was unsuccessful because an imaginary number arose in the course of the iterations.

To at least four decimal accuracy, all the solutions found from the successful starting points in Table 5.11, part A are the same, and the variation in number of iterations required (between 15 and 21) is not particularly large. Note that only one of the six possible solutions to this set of equations is found by Gauss–Seidel from these starting points. If the equations were renumbered in the Gauss–Seidel approach (so that they would be solved in some different order), other solutions might be identified. In part B of Table 5.11 we find no great sensitivity to starting points in any column (for a given value of δ) and not surprising kinds of variations across stopping criteria—for example, approximately twice as many iterations needed for the very exacting requirement that $\delta = 10^{-10}$ as for $\delta = 10^{-4}$.

Numerical Illustration: Four Equations in Four Unknowns   For this example we continue to use Equations (5.7)–(5.10). We do not attempt any geometric interpretation, since even solution space would require four dimensions.

Algebraic Solution and Substitution   We have already examined how this approach works on these equations, in the initial description of sequential approaches.

Gauss–Seidel   For this set of equations, it turns out that Gauss is *extremely* sensitive to starting point. In fact, using the stopping criterion of $\delta = 0.01$, it exploded (failed to converge) from each of the following starting points (which will be used again in illustrating derivative-based methods):

(a) The origin, as usual representing no information at all regarding the equations.
(b) $(1,1,1,1)$, which is just one of many possibilities chosen because it is easily seen to satisfy Equation (5.7).
(c) $(0,0,6,2)$, which satisfies Equation (5.9).
(d) $(-2,0,1,4)$, which satisfies Equation (5.10).
(e) $(-1,1,2,2)$, which also satisfies Equation (5.10).

Gauss was totally useless from these starting points. To salvage the procedure, Table 5.12 contains results from starting points that are (much) nearer to the actual solution. Recall that substitution methods, above, led to $\mathbf{X}^* = [-0.564, 4.128, 5.318, 1.128]'$ (to three

TABLE 5.12   Results for the Gauss–Seidel Iterative Method for Four Equations in Four Unknowns[a]

Starting Point ($\mathbf{X}^0$)	Original	Equation Ordering			
		$\{2,3,4,1\}$	$\{3,4,1,2\}$	$\{4,1,2,3\}$	$\{4,2,3,1\}$
−0.5	Exp[b]	Exp	Exp	Exp	Exp
4.0					
5.0					
1.0					
−0.55	−0.550	Exp	−0.550	−0.550	Exp
4.1	4.015		4.016	4.015	
5.3	5.303		5.303	5.303	
1.1	1.100		1.100	1.100	
−0.56	Same	Exp	−0.550	−0.550	−0.544
4.1			4.036	4.035	4.088
5.3			5.303	5.314	5.314
1.1			1.120	1.120	1.120
−0.57	Same	−0.561	−0.550	−0.550	−0.570
4.1		4.121	4.056	4.055	4.161
5.3		5.303	5.303	5.325	5.325
1.1		1.140	1.140	1.140	1.140

[a]In each case, the number of iterations was 1. Stopping criterion: $\max_i [|f^i(\mathbf{X}^k)|] \leq 0.09$.
[b]"Exp" indicates that the procedure exploded in the sense that one or more of the x's $\to \infty$ or $-\infty$.

decimal accuracy). To two decimal accuracy, the smallest δ that did not lead to explosion from a starting point of $(-0.55, 4.1, 5.3, 1.1)$—"near to" $\mathbf{X}^*$—was $\delta = 0.09$. This was the stopping criterion that was used in this table.

As already mentioned (Section 2.7 and earlier in this section), iterative procedures like Gauss–Seidel can be sensitive to the *order* in which the equations are presented and solved. Using Equations (5.7)–(5.10) in that order, we would find $x_1 = (3 - x_2)/2$ from (5.7), $x_2 = x_4 - x_1 x_3$ from (5.8), and so forth. The natural Gauss–Seidel iterative sequence would then be: (a) substitute from $\mathbf{X}^0$ into the equation for x_1, giving x_1^1, (b) using x_1^1 along with x_3^0 and x_4^0, find x_2^1 [as in (5.8)], (c) then find x_3^1 and finally x_4^1.

In general, there is no reason to care which of four equations is listed first, which second, etc. For example, we could interchange Equations (5.7) and (5.8). Then solving (5.8) for x_1 gives $x_1 = (x_4 - x_2)/x_3$ (note that there will be a problem in this case if $x_3 = 0$), and solving (5.7) for x_2 gives $x_2 = 3 - 2x_1$ (the expressions for x_3 and x_4 would be unchanged). Note that there are 24 potential arrangements of the four equations, although not all are meaningful, since not all equations contain all unknowns. For example, the first equation could not be interchanged with the third or fourth equation, since it does not conatin x_3 or x_4. We have not examined this kind of sensitivity in Table 5.12, but Problem 5.10 invites you to do so (with the help of a computer). (This was also explored in Problem 2.27, but for only two linear eqations in two unknowns.)

In addition, the iterative solution sequence for the x's found from the original equation ordering could be rearranged, for example by first using the equation for x_2 to find x_2^1 from the elements of $\mathbf{X}^0$, next using the equation for x_3 to find x_3^1 given x_2^1, x_4^0 and x_1^0, then finding x_4^1 and finally x_1^1, etc. It is this kind of ordering sensitivity that is explored in Table 5.12; this particular rearranged solution sequence for the four unknowns is identified in that table as $\{2, 3, 4, 1\}$.

Perhaps most notable from Table 5.12 is the fact that even the solution sequence for the unknowns can mean the difference between whether or not the procedure converges at all, from a given $\mathbf{X}^0$. And when there is convergence for more than one ordering, the solutions may differ. In addition, Gauss can be very sensitive to choice of an $\mathbf{X}^0$. Thus, in the absence of some quite clear indications about the probable location of the solution to the set of nonlinear equations, the Gauss–Seidel method can easily fail. This is just a (possibly dramatic) reflection of the general difficulty of the problem of finding solutions to such sets of equations.

5.2.2 Derivative Methods

For *systems* of equations, derivative methods are essentially multidimensional analogs of the Newton–Raphson and associated techniques for a single equation. Starting again with the first two terms in a Taylor series expansion, we have the following for each function $f^i(\mathbf{X})$:

$$f^i(\mathbf{X}^{k+1}) \cong f^i(\mathbf{X}^k) + [\nabla f^i(\mathbf{X}^k)]'[\mathbf{X}^{k+1} - \mathbf{X}^k] \qquad (5.20)$$

Considering one of these equations for each i ($i = 1, \ldots, n$), let $\mathbf{F}(\mathbf{X}^{k+1}) = [f^1(\mathbf{X}^{k+1}), \ldots, f^n(\mathbf{X}^{k+1})]'$, and $\mathbf{F}(\mathbf{X}^k) = [f^1(\mathbf{X}^k), \ldots, f^n(\mathbf{X}^k)]'$. Finally, arrange each of the gradients, transposed into row vectors, as in (5.20), into an $n \times n$ matrix:

$$\begin{bmatrix} [\nabla f^1(\mathbf{X}^k)]' \\ \vdots \\ [\nabla f^n(\mathbf{X}^k)]' \end{bmatrix} = \begin{bmatrix} \dfrac{\partial f^1}{\partial x_1^k} & \cdots & \dfrac{\partial f^1}{\partial x_n^k} \\ \vdots & & \vdots \\ \dfrac{\partial f^n}{\partial x_1^k} & \cdots & \dfrac{\partial f^n}{\partial x_n^k} \end{bmatrix} = \begin{bmatrix} \dfrac{\partial f^i}{\partial x_j^k} \end{bmatrix}$$

This is the *Jacobian* matrix of the equation system (Section 4.4), evaluated at $\mathbf{X}^k$, and denoted $\mathbf{J}(\mathbf{X}^k)$. Then the entire set of equations given by (5.20) for $i = 1, \ldots, n$ is

$$\mathbf{F}(\mathbf{X}^{k+1}) \cong \mathbf{F}(\mathbf{X}^k) + \mathbf{J}(\mathbf{X}^k)(\mathbf{X}^{k+1} - \mathbf{X}^k) \tag{5.21}$$

The Newton–Raphson Method   As in the Newton–Raphson case for a single equation, the expression in (5.21) provides a set of *linear approximations* to $f^1(\mathbf{X}), \ldots, f^n(\mathbf{X})$ in the neighborhood of $\mathbf{X}^k$. These planes (hyperplanes) intersect in a line (hyperline), and by setting $\mathbf{F}(\mathbf{X}^{k+1}) = \mathbf{0}$ we find the point at which the line intersects the horizontal plane; that is, we find a root of the linear approximations to the system of equations. Therefore

$$\mathbf{0} = \mathbf{F}(\mathbf{X}^k) + \mathbf{J}(\mathbf{X}^k)(\mathbf{X}^{k+1} - \mathbf{X}^k)$$

and so

$$\mathbf{X}^{k+1} = \mathbf{X}^k - [\mathbf{J}(\mathbf{X}^k)]^{-1}\mathbf{F}(\mathbf{X}^k) \tag{5.22}$$

This is the multidimensional analog to the single equation Newton–Raphson procedure in (5.3).

   This approach requires that the $n \times n$ Jacobian matrix and then its inverse be found at each iteration. For sufficiently large n, this requirement can be computationally burdensome for two reasons: (a) in general, each element of the Jacobian, $\partial f^i / \partial x_j$, will be a function of all of the x's , and hence finding $\mathbf{J}(\mathbf{X}^k)$ requires n^2 functional evaluations; and (b) finding the inverse itself, once the elements of $\mathbf{J}(\mathbf{X}^k)$ have been filled in, can be a huge computational task. And, of course, the matrix $\mathbf{J}$ must be nonsingular. These are multidimensional extensions of the problems associated with Newton–Raphson for a single equation.

The Modified Newton Approach   Parallel to the single-equation case, the modified Newton approach replaces $\mathbf{J}(\mathbf{X}^k)$—and hence $[\mathbf{J}(\mathbf{X}^k)]^{-1}$—at each step with $\mathbf{J}(\mathbf{X}^0)$—and therefore an inverse for the Jacobian needs to be found only once, at the outset, for $\mathbf{J}(\mathbf{X}^0)$. The success of this approach depends on how "close" successive values of $\mathbf{X}^k$ are to one another, how accurate the linear approximations are as representatives of the nonlinear functions $f^i(\mathbf{X})$, and hence how much the curvature (slope) changes from one iteration to the next.

   A variant of this approach goes one step further and replaces $\mathbf{J}(\mathbf{X}^0)$ with an identity matrix, $\mathbf{I}$, or with a scalar multiple of an identity matrix, $\alpha\mathbf{I}$, for some selected value of α. Clearly, this represents a kind of minimal-information approach and, as might be expected, is generally not very successful.

Quasi-Newton Methods   In the single-equation problem, the derivative, $f'(x^k)$, is needed at each step in order to be able to divide by it (multiply by its reciprocal), as in (5.3). The parallel in the case of a system of nonlinear equations is that the $n \times n$ Jacobian matrix, $\mathbf{J}(\mathbf{X}^k)$, is needed at each iteration to permit multiplication by its reciprocal (inverse), as in (5.22). Because of this, *updating* schemes play a much larger role in iterative methods for solving *systems* of nonlinear equations than they do in the single-equation case.

Consider the initial step. First an $\mathbf{X}^0$ is chosen. Then $\mathbf{F}(\mathbf{X}^0)$ and $\mathbf{J}(\mathbf{X}^0)$ are found. Then, using the multidimensional Newton–Raphson approach, as in (5.22), $\mathbf{X}^1$ is found as

$$\mathbf{X}^1 = \mathbf{X}^0 - [\mathbf{J}(\mathbf{X}^0)]^{-1}\mathbf{F}(\mathbf{X}^0) \tag{5.23}$$

The next iteration would be

$$\mathbf{X}^2 = \mathbf{X}^1 - [\mathbf{J}(\mathbf{X}^1)]^{-1}\mathbf{F}(\mathbf{X}^1) \tag{5.24}$$

This requires that the set of nonlinear functions, $\mathbf{F}(\mathbf{X})$, be evaluated at $\mathbf{X}^1$ and also that the Jacobian, or some approximation to it, be found at $\mathbf{X}^1$—the latter so that its inverse can be used in moving to $\mathbf{X}^2$.

The important results for derivative methods for a function of one variable are repeated here. From the Taylor series approximation, we have $x^{k+1} = x^k - f(x^k)/f'(x^k)$. [This was (5.3).] A secant approximation to the tangent at x^k was given as

$$f'(x^k) \cong \frac{f(x^k) - f(x^{k-1})}{x^k - x^{k-1}}$$

Alternatively expressed, this is

$$f'(x^k)(x^k - x^{k-1}) = f(x^k) - f(x^{k-1}) \tag{5.25}$$

and so (5.3) became, in (5.5)

$$x^{k+1} = x^k - f(x^k)\left[\frac{x^k - x^{k-1}}{f(x^k) - f(x^{k-1})}\right]$$

which is the quasi-Newton procedure for solving $f(x) = 0$. Note that identification of the next point, x^{k+1}, requires the current and previous points, x^k and x^{k-1}, and the value of $f(x)$ at those two points. For systems of nonlinear equations, a similar approach is needed.

One very successful method, due to Broyden (1965), can be thought of as the multidimensional extension of this secant approximation to the tangent. In forming an estimate of $\mathbf{J}(\mathbf{X}^1)$, which is just the multivariable, multifunction analog of $f'(x^1)$, Broyden's method also makes use of two sets of information that are at hand after the move from $\mathbf{X}^0$ to $\mathbf{X}^1$. Specifically, $\mathbf{J}(\mathbf{X}^1)$ is approximated by $\mathcal{J}^1 \equiv \mathcal{J}(\mathbf{X}^1)$, where

$$\mathcal{J}^1(\mathbf{X}^1 - \mathbf{X}^0) = \mathbf{F}(\mathbf{X}^1) - \mathbf{F}(\mathbf{X}^0)$$

[note the parallel to (5.25) for the single-variable case] or, with $\Delta\mathbf{X} = \mathbf{X}^1 - \mathbf{X}^0$ and $\Delta\mathbf{F} = \mathbf{F}(\mathbf{X}^1) - \mathbf{F}(\mathbf{X}^0)$:

$$\mathcal{J}^1(\Delta\mathbf{X}) = \Delta\mathbf{F} \tag{5.26}$$

[Dennis and Schnabel (1983, Chapter 8) give a much more elaborate and detailed mathematical derivation of Broyden's method, if you are interested.] In fact, since the Broyden method is an iterative scheme, the idea is to find, at each iteration as represented in (5.22), the next Jacobian approximation expressed as the previous Jacobian approximation plus some change in (updating of) that previous approximation. This means, find $\mathcal{J}^{k+1}$ as

$$\mathcal{J}^{k+1} = \mathcal{J}^k + \Delta\mathcal{J}^k \tag{5.27}$$

Setting $k = 0$ and postmultiplying both sides by $\Delta\mathbf{X}$ yields

$$\mathcal{J}^1(\Delta\mathbf{X}) = \mathcal{J}^0(\Delta\mathbf{X}) + \Delta\mathcal{J}^0(\Delta\mathbf{X}) \tag{5.28}$$

and, using (5.26),

$$\mathcal{J}^0(\Delta\mathbf{X}) + \Delta\mathcal{J}^0(\Delta\mathbf{X}) = \Delta\mathbf{F}$$

or

$$\Delta\mathcal{J}^0(\Delta\mathbf{X}) = \Delta\mathbf{F} - \mathcal{J}^0(\Delta\mathbf{X}) \tag{5.29}$$

Putting this into (5.28), we obtain

$$\mathcal{J}^1(\Delta\mathbf{X}) = \mathcal{J}^0(\Delta\mathbf{X}) + [\Delta\mathbf{F} - \mathcal{J}^0(\Delta\mathbf{X})] \tag{5.30}$$

which relates the next Jacobian estimate to the current one and to the known values of $\Delta\mathbf{F}$ $[= \mathbf{F}(\mathbf{X}^1) - \mathbf{F}(\mathbf{X}^0)]$ and $\Delta\mathbf{X}$ $[= \mathbf{X}^1 - \mathbf{X}^0]$ as well. [Mathematically, this is just (5.26) with the term $\mathcal{J}^0(\Delta\mathbf{X})$ both added to and subtracted from the right-hand side; using (5.28) just makes clear exactly what that term should be.]

However, we really want an expression for $\Delta\mathcal{J}^0$ alone, separated from $\Delta\mathbf{X}$, as it appears in (5.29). Some matrix algebra manipulations help in that regard. If the bracketed term on the right in (5.30) is both multiplied and divided by $(\Delta\mathbf{X})'(\Delta\mathbf{X})$ (a scalar), we have

$$\mathcal{J}^1(\Delta\mathbf{X}) = \mathcal{J}^0(\Delta\mathbf{X}) + \frac{[\Delta\mathbf{F} - \mathcal{J}^0(\Delta\mathbf{X})](\Delta\mathbf{X})'(\Delta\mathbf{X})}{(\Delta\mathbf{X})'(\Delta\mathbf{X})} \tag{5.31}$$

or, factoring out $\Delta\mathbf{X}$ from the second term,

$$\mathcal{J}^1(\Delta\mathbf{X}) = \mathcal{J}^0(\Delta\mathbf{X}) + \left(\frac{[\Delta\mathbf{F} - \mathcal{J}^0(\Delta\mathbf{X})](\Delta\mathbf{X})'}{(\Delta\mathbf{X})'(\Delta\mathbf{X})} \right)(\Delta\mathbf{X})$$

From (5.28), which has exactly the same structure, we see that the term in large parentheses plays the role of $\Delta\mathcal{J}^0$:

$$\Delta\mathcal{J}^0 = \frac{[\Delta\mathbf{F} - \mathcal{J}^0(\Delta\mathbf{X})](\Delta\mathbf{X})'}{(\Delta\mathbf{X})'(\Delta\mathbf{X})} \tag{5.32}$$

The important point is that the algebra has produced an updating scheme, to transform $\mathcal{J}^0$ into $\mathcal{J}^1$, that depends only on $\mathcal{J}^0$ and the two n-element column vectors $\Delta\mathbf{F}$ and $\Delta\mathbf{X}$.

Note that $[\Delta\mathbf{F} - \mathcal{J}^0(\Delta\mathbf{X})]$ is also an n-element column vector, so the $n \times n$ matrix $\Delta\mathcal{J}^0$ is found as the product of an n-element column vector and an n-element row vector, with each term in the product matrix divided by the scalar $(\Delta\mathbf{X})'(\Delta\mathbf{X})$. For any n-element column vector $\mathbf{V}$ and n-element row vector $\mathbf{W}$ it is easily shown that $\mathbf{VW}$ is a matrix whose rank is one. (This is illustrated in Appendix 5.2.) Thus the transformation of $\mathcal{J}^0$ into $\mathcal{J}^1$, using (5.31), is known as a "rank 1 update" of $\mathcal{J}^0$. For the $(k+1)$st iteration, then, we have

$$\mathbf{X}^{k+1} = \mathbf{X}^k - [\mathcal{J}^k]^{-1}\mathbf{F}(\mathbf{X}^k) \tag{5.33}$$

where $\mathbf{J}(\mathbf{X}^k)$ has been approximated by $\mathcal{J}^k$ and

$$\mathcal{J}^k = \mathcal{J}^{k-1} + \left(\frac{[\Delta\mathbf{F} - \mathcal{J}^{k-1}(\Delta\mathbf{X})](\Delta\mathbf{X})'}{(\Delta\mathbf{X})'(\Delta\mathbf{X})}\right) \tag{5.34}$$

where $\Delta\mathbf{F} = \mathbf{F}(\mathbf{X}^k) - \mathbf{F}(\mathbf{X}^{k-1})$ and $\Delta\mathbf{X} = (\mathbf{X}^k - \mathbf{X}^{k-1})$. This is Broyden's Jacobian update at the kth step. Note the resemblance of (5.34) to the expression for the $f(x^k)$ update for a single function of one variable in (5.6).

Since it is actually the *inverse* of the Jacobian, $[\mathbf{J}(\mathbf{X}^k)]^{-1}$, that is needed for the $(k+1)$st iteration, an alternative approach would be to try to find the approximation $(\mathcal{J}^k)^{-1}$ *directly*, through an updating scheme that builds on the previous *inverse*, $(\mathcal{J}^{k-1})^{-1}$ as

$$(\mathcal{J}^k)^{-1} = (\mathcal{J}^{k-1})^{-1} + \Delta[(\mathcal{J}^{k-1})^{-1}] \tag{5.35}$$

This would eliminate the need to invert the $n \times n$ Jacobian matrix approximation at each step.

A result originally presented by Sherman and Morrison (1950) and then extended by Woodbury (1950) provides a connection between changes in one or more elements of a matrix and the consequent changes in the inverse to that matrix. The Sherman–Morrison–Woodbury (SMW) formula is discussed in Appendix 5.2. When this formula is applied to the problem of estimating $(\mathcal{J}^k)^{-1}$ through an update of $(\mathcal{J}^{k-1})^{-1}$, we have

$$(\mathcal{J}^k)^{-1} = (\mathcal{J}^{k-1})^{-1} + \frac{[\Delta\mathbf{X} - (\mathcal{J}^{k-1})^{-1}\,\Delta\mathbf{F}](\Delta\mathbf{X})'(\mathcal{J}^{k-1})^{-1}}{(\Delta\mathbf{X})'(\mathcal{J}^{k-1})^{-1}\,\Delta\mathbf{F}} \tag{5.36}$$

and this is used in the iterative procedure in (5.33) to approximate $(\mathcal{J}^k)^{-1}$ directly at each step.

Numerical Illustration: Two Equations in Two Unknowns   The equations were given above, in (5.15a) and (5.15b). Similar starting points are used, and the results for Gauss are also repeated here, to facilitate comparison, in Table 5.13.

We see from the table that

1. Gauss was successful from all starting points [except for (0,5), which we already knew would not work], always converging to $\mathbf{X}^{**}$ and relatively insensitive to variations in starting point.

2. Newton often required slightly more than half as many iterations. However, from three of the starting points, with which Gauss was not troubled at all, Newton needed

TABLE 5.13   Results for Alternative Iterative Methods for Two Equations in Two Unknowns[a]

Starting Point $(\mathbf{X}^0)'$	Gauss	Newton	Modified Newton Using αI $\alpha = 1$	$\alpha = 5$	Using $\mathbf{J}(\mathbf{X}^0)$	Broyden
			$\delta = 10^{-4}$			
(0,0)	X** (20)	X** (13)	Exp	X* (33)	Exp	X** (8)
(2,0)	X** (20)	X** (11)	Exp	X* (34)	X** (12)	X** (5)
(−1,−1)	X** (20)	X** (14)	Exp	Exp	Exp	X** (9)
(−1,3)	X** (20)	X* (162)	Exp	X* (29)	X* (12)	X* (5)
(0,1)	X** (21)	X** (12)	Exp	X* (32)	S	S
(0,2)	X** (20)	X** (13)	Exp	X* (31)	Exp	X* (8)
(0,3)	X** (20)	S	Exp	X* (30)	C	X* (6)
(0,4)	X** (21)	X* (166)	Exp	X* (30)	X* (13)	X* (6)
(0,5)	IM	X* (167)	Exp	X* (29)	X* (6)	X* (5)
			$\delta = 10^{-5}$			
(0,0)	25	15	Exp	43	Exp	9
(−1,3)	25	200	Exp	39	14	6
			$\delta = 10^{-10}$			
(0,0)	48	25	Exp	75	Exp	10
(−1,3)	49	393	Exp	71	26	7

[a]Iteration counts in parentheses. Stopping criterion: $\max_i \ [|f^i(\mathbf{X}^k)|] \leq \delta$. *Key:* $\mathbf{X^{**}} = \begin{bmatrix} 2.562 \\ -0.562 \end{bmatrix}$; $\mathbf{X^*} = \begin{bmatrix} -1.562 \\ 3.562 \end{bmatrix}$; "Exp" means that the value of at least one variable exploded; "C" means that the procedure failed to converge; "IM" means that an imaginary number was generated during the iterations; "S" indicates that the inverse of a singular matrix was required.

around 160 iterations to propel itself to $\mathbf{X^*}$ (Gauss always chose $\mathbf{X^{**}}$). This illustrates the kind of sensitivity to starting point that these kinds of iterative procedures can exhibit.

3. Modified Newton with $\mathbf{J}^k = \mathbf{I}$ was totally unsuccessful, always exploding. However, with $\mathbf{J}^k = \alpha\mathbf{I}$ and $\alpha = 5$, it exploded only once, although it was consistently relatively slow. In these examples, it always converged to $\mathbf{X^*}$.

4. Modified Newton using $\mathbf{J}^k = \mathbf{J}^0$ exhibits a great deal of sensitivity to starting point, and it encountered a number of problems. In three cases, it exploded. Once it was stopped

when trying to invert a singular matrix, and once it simply failed to converge. When it did converge, it once found **X**** and three times ended up at **X***. Note that this wide variety of behavior occurred with relatively slight changes in the starting point—from (0,1) through (0,4).

5. Broyden, like Gauss and Newton, was unsuccessful only once, when inversion of a singular matrix was asked for. It was generally about twice as fast (in terms of number of iterations) as Newton; on occasion, as from (−1,3), (0,4) and (0,5), it was vastly superior to Newton.

The bottom of Table 5.13 contains results for one smaller and one larger value of δ in the stopping criterion. The same relative successes of alternative methods occur. Broyden is always the winner, and Newton is sometimes second best, sometimes terrible. Gauss is reliable but not spectacular in terms of iteration count.

Numerical Illustration: Three Equations in Three Unknowns Again we use the three-equation system in (5.18a)–(5.18c), above. The outcomes are shown in Table 5.14. Again, results for Gauss are repeated for comparison. The origin, as usual representing no information at all regarding the equations, was totally unsuccessful as a starting point for the derivative-based techniques because the initial Jacobian is singular, so none of those techniques can get started. For that reason, two permutations of the origin were used; they are shown at the bottom of the table.

Some of the highlights from Table 5.14 are:

1. Gauss works fairly well from many of the starting points, and the number of iterations required is relatively insensitive to X^0. It always found the same solution.

2. Newton is much more affected by starting point choice, in terms of both number of iterations and solution found. In this illustration, three of the possible six solutions were found by Newton, depending on the starting point used.

3. Modified Newton is, as usual, relatively unreliable. When it works, it works very well; but it very often fails.

4. The Broyden methods generally work quickly and usually give the same result, although three starting points [(1,2,3) as well as the two perturbations of the origin] give them unexpected trouble. This is unfortunate, in light of the fact that Gauss reaches a solution from those starting points without any apparent trouble at all. The Broydens, when they work, always find the same solution—which is a different one from that found by Gauss.

Differences in the effectiveness of Broyden versus its Sherman–Morrison–Woodbury version would become more pronounced in much larger and more complex problems. In such problems, having to find the inverse of the (approximated) Jacobian at each step (as in Broyden) might be expensive and time consuming, relative to the SMW variant in which inverses are not required.

Numerical Illustration: Four Equations in Four Unknowns We again use the four equations in (5.7)–(5.10). Results are shown in Table 5.15. Here it was again necessary to perturb the origin as an initial starting point since the initial Jacobian in all the derivative-based methods happens to be singular there.

TABLE 5.14   Results for Alternative Iterative Methods for Three Equations in Three Unknowns[a]

Starting Point ($\mathbf{X}^0$)	Gauss	Newton	Modified Newton Using $\mathbf{J}(\mathbf{X}^0)$	Broyden	Broyden–SMW
0	10	S	S	S	S
0	5				
0	0				
	(16)				
1	IM	0	0	0	0
6		5	5	5	5
11		10	10	10	10
		(5)	(7)	(7)	(8)
1	10	0	Exp	0	C
2	5	10		5	
3	0	5		10	
	(18)	(16)		(68)	
4	10	0	Exp	0	0
5	5	5		5	5
6	0	10		10	10
	(21)	(7)		(9)	(10)
0	IM	0	0	0	0
5		5	5	5	5
9		10	10	10	10
		(4)	(9)	(5)	(6)
0.1	10	10	Exp	C	C
0	5	5			
0.01	0	0			
	(16)	(30)			
1.1	10	10	Exp	C	C
1	5	5			
1.01	0	0			
	(15)	(25)			

[a]Iteration counts in parentheses. Solution values to at least four-decimal accuracy. Stopping criterion: $\max_i [|f^i(\mathbf{X}^k)|] \leq 10^{-4}$. *Key:* "S" indicates that the initial Jacobian matrix was singular and hence this technique, which requires the inverse of the Jacobian, cannot get started. "Exp" indicates that values for one or more of the x's exploded. "IM" indicates that an imaginary number was generated in the iterations. "C" indicates lack of convergence; the procedure was stopped after 5000 iterations with no evidence of convergence.

Some observations from Table 5.15 are that (1) Gauss is totally unsuccessful, (2) Newton never fails, (3) modified Newton is generally useless, and (4) Broyden and Broyden with the SMW variation also are always successful. They are a bit faster from most of the starting points but somewhat slower from $\mathbf{X}^0$ near to the origin.

Table 5.16 contains results for Newton and the two Broydens from a series of increasingly outrageous and uninformed starting points. They illustrate that Newton performs excellently in all situations. Broyden uses more and more iterations as the starting point moves further from the solution. Broyden with SMW behaves similarly from the first two

TABLE 5.15   Iterations for Alternative Iterative Methods for Four Equations in Four Unknowns[a]

Starting Point (X^0)	Gauss	Newton	Modified Newton Using J(X^0)	Broyden	Broyden–SMW
1 1 1 1	Exp	15	Exp	9	9
0 0 6 2	Exp	7	13	6	6
–2 0 1 4	Exp	13	Exp	10	10
–1 1 2 2	Exp	19	Exp	7	7
0 0.01 0.1 0	Exp	10	Exp	18	18
0 0 0.001 0	Exp	10	Exp	67	68

[a]In all cases, the solution reached was (to three decimals) $X^* = [-.564, 4.128, 5.318, 1.128]'$. Stopping criterion: $\max_i [|f^i(X^k)|] \leq 10^{-4}$.

starting points. Beyond that, it is sometimes very much better than Broyden alone [from (100, 100, 100)] and sometimes very much worse [from (1000, 1000, 1000)]. Finally, only Newton gets it right from the starting point where all x's equal 10,000—and reaches the correct solution from there in the same number of iterations as from (100, 100, 100)!

5.3   SUMMARY

In this chapter we have examined iterative methods that move toward solutions to $f(x) = 0$ and $F(X) = 0$. In the case of $f(x) = 0$, algebraic procedures are useful for certain well-specified functions; for example, $ax^2 + bx + c = 0$ is known to have solutions $x = (-b \pm \sqrt{b^2 - 4ac})/2a$. General algebraic solutions of this sort for more complex functions of x are often very cumbersome or impossible to find. Nonderivative iterative methods are successful in root finding, although they may be quite slow in some cases. The deriva-

TABLE 5.16 Iterations for Alternative Iterative Methods for Four Equations in Four Unknowns from More Extreme Starting Points[a]

Starting Point (X^0)	Newton	Broyden	Broyden–SMW
10 10 10 10	10	12	10
20 20 20 20	9	15	15
100 100 100 100	9	440	131
1000 1000 1000 1000	9	465	740
10000 10000 10000 10000	9	C[b]	C

[a]In all cases, the solution reached was (to three decimals) **X*** = [−.564, 4.128, 5.318, 1.128]′. Stopping criterion: $\max_i [|f^i(\mathbf{X}^k)|] \leq 10^{-4}$.
[b]"C" indicates lack of convergence; the procedure was stopped after 4000 iterations with no clear evidence of convergence.

tive-based methods, particularly Newton–Raphson, can be very efficient, provided one does not begin from a "bad" starting point (but how can you know?) and if derivatives are easily available.

For **F(X)** = **0**, the iterative procedures are much more complex because finding roots to a system of nonlinear equations is, in general, a very difficult task. Sequential solution techniques, in which equations are solved one at a time for one variable as a function of the others, are feasible only when the equations have relatively simple structures, so that such equation-by-equation solutions can be found. Not only must such solutions be possible, but they must also be relatively "simple"—if, at any step i, there are multiple solutions $x_i = g^i(\mathbf{X}_i)$ and some or all of them are algebraically complicated, then substitution into other equations or evaluation for specific values of the $\mathbf{X}_i$ can become tedious. Derivative methods, which are extensions of the Newton–Raphson approach for $f(x) = 0$, can be very effective. For large problems in which evaluation and/or inversion of the Jacobian at each step is intractable, an estimation routine as in Broyden's updating method, with or without the SMW extension to direct estimation of the Jacobian inverse, can be impressively successful. The speed and capacity of present-day computers make possible the application of these kinds of iterative techniques to an ever-increasing set of practical, real-world problems.

APPENDIX 5.1 FINITE-DIFFERENCE APPROXIMATIONS TO DERIVATIVES, GRADIENTS, AND HESSIAN MATRICES

Iterative methods are generally classified according to whether they require derivatives for the operation of their algorithm. In some cases, when the function or functions are easily differentiated, the problem solver can simply enter and label the functional relationships that represent each of the derivatives—first-order, higher-order, partial, cross-partial—that the algorithm requires. On the other hand, the derivatives may be generated when and as needed in the algorithm by approximation techniques, which we now examine briefly.

5A.1.1 Functions of One Variable

Here the problem is to find $f'(x)$ given $f(x)$. There are essentially two approaches (with one variant).

The Forward-Difference Method For some (however appropriately defined) small parameter α, where $\alpha > 0$, the derivative of $f(x)$ at x^k is approximated by

$$f'(x^k) \cong \frac{f(x^k + \alpha) - f(x^k)}{\alpha} \tag{5.1.1}$$

The geometry is straightforward; the slope of the tangent to $f(x)$ at x^k is approximated by the slope of the secant that cuts $f(x)$ at $f(x^k)$ and at $f(x^k + \alpha)$, as illustrated by the line FD in Figure 5.1.1. [Since α is positive, the second cut of the function by the secant is to the right of $f(x^k)$.] If the function is evaluated at $x^k - \alpha$ [which is the same as if α were negative in (5.1.1)], the secant cuts the function to the left of $f(x^k)$—which might logically be called a "backward difference method".

The Central-Difference Method Using the same notion of a small parameter $\alpha > 0$, this approach finds the derivative of $f(x)$ at x^k as

$$f'(x^k) \cong \frac{f(x^k + \alpha) - f(x^k - \alpha)}{2\alpha} \tag{5.1.2}$$

Now the slope of the tangent is approximated by the slope of the secant that cuts the function at $f(x^k - \alpha)$ and at $f(x^k + \alpha)$; this is illustrated in Figure 5.1.1 by the line CD. Note that, in general, the slope of the secant used in the central difference method is likely to be closer to that of the tangent (the dashed line in Fig. 5.1.1) than is the slope of the approximating secant line in the forward difference method. But this comes at a price. Assuming that one already has $f(x^k)$, the central-difference method requires *two* additional functional evaluations (at $x^k + \alpha$ and $x^k - \alpha$); the forward-difference method requires only *one* (at $x^k + \alpha$).

Table 5.1.1 illustrates these rather straightforward procedures for the two test functions from the text: (i) $f(x) = x^2 - 4$ and (ii) $f(x) = (x - 1)(x + 1)^2 - 1$. In both cases, approximations to the derivatives are found at the point $x^k = 2$. From simple differentiation, we know that for the function in (a) (of Table 5.1.1), $f'(x) = 2x$, and so $f'(2) = 4$; and for the function in (b), $f'(x) = 3x^2 + 2x - 1$, and so $f'(2) = 15$. The object, of course, is to approximate these two tangent slopes without differentiating.

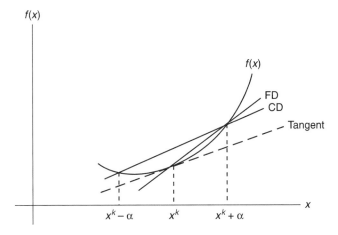

FIGURE 5.1.1   Forward- and central-difference methods of approximating a tangent.

Note, in Table 5.1.1, that for a given value of α, forward and backward approximations for the derivative of $f(x) = x^2 - 4$ are "symmetric" in the sense that they over- and underestimate by the same amount; hence, the central difference approximation is exactly correct. (You should appreciate the reason for this from the character of the derivative for this function.) In the results for $f(x) = (x - 1)(x + 1)^2 - 1$ in Table 5.1.1, extra decimal places have been introduced in order to indicate the increasing precision for smaller and smaller α. In addition, two relatively large values of α have been included at the bottom of the table; even though estimates from either the forward-difference or the backward-difference method are very poor, note that the central difference approximation provides an average over each of the two very bad estimates that is quite effective.

TABLE 5.1.1   Approximation of Derivatives (at $x^k = 2$)

α	Forward Difference $[f(x^k+\alpha)-f(x^k)]/\alpha$	Backward Difference $[f(x^k-\alpha)-f(x^k)]/\alpha$	Central Difference $[f(x^k+\alpha)-f(x^k-\alpha)]/2\alpha$
	(a) $f(x) = x^2 - 3$		
0.5	4.5	3.5	4.0
0.25	4.25	3.75	4.0
0.1	4.1	3.9	4.0
0.01	4.01	3.99	4.0
	(b) $f(x) = (x - 1)(x + 1)^2 - 1$		
0.5	18.75	11.75	15.25
0.25	16.81	13.31	15.06
0.1	15.71	14.31	15.01
0.01	15.0701	14.9301	15.0001
0.001	15.0070	14.9930	15.0000
1.0	23.0	9.0	16.0
2.0	33.0	-5.0	19.0

5A.1.2 Functions of Several Variables

For $f(\mathbf{X})$, where $\mathbf{X} = [x_1, \ldots, x_n]'$, there are again two fundamental approaches to the approximation of a partial derivative, now at a point $\mathbf{X}^k = [x_1^k, \ldots, x_n^k]'$. These are logically parallel to the two methods for $f(x)$.

The Forward-Difference Method Here, at $\mathbf{X}^k$ and for a specified $\alpha > 0$, we obtain

$$\frac{\partial f}{\partial x_j} = f_j \cong \frac{f(x_1^k, \ldots, x_j^k + \alpha, \ldots, x_n^k) - f(\mathbf{X}^k)}{\alpha} \tag{5.1.3}$$

[Compare with (5.1.1).] This is sometimes written more compactly, using $\mathbf{I}_j$ for the jth column of an $n \times n$ identity matrix

$$\frac{\partial f}{\partial x_j} = f_j \cong \frac{f(\mathbf{X}^k + \alpha \mathbf{I}_j) - f(\mathbf{X}^k)}{\alpha} \tag{5.1.4}$$

For $\alpha < 0$, (5.1.4) provides an analog to the backward-difference method.

The Central-Difference Method In a manner that is completely parallel to the approach for the one-variable case, for $\alpha > 0$

$$\partial f/\partial x_j = f_j \cong \frac{f(\mathbf{X}^k + \alpha \mathbf{I}_j) - f(\mathbf{X}^k - \alpha \mathbf{I}_j)}{2\alpha} \tag{5.1.5}$$

[Compare with 5.1.2).] Both of these approaches for approximating the partial derivatives of $f(\mathbf{X})$ at $\mathbf{X}^k$ have geometric parallels to those shown in Figure 5.1.1. [For $f(\mathbf{X})$, there will be n such pictures; in each case, the horizontal axis measures a particular x_j and $f(\mathbf{X})$ is measured on the vertical axis.]

5A.1.3 Systems of Equations

For a system of equations, $\mathbf{F}(\mathbf{X}) = \begin{bmatrix} f^1(\mathbf{X}) \\ f^2(\mathbf{X}) \\ \vdots \\ f^n(\mathbf{X}) \end{bmatrix} = \mathbf{0}$, the parallel to a gradient is the *Jacobian*

matrix (Section 4.4)

$$\mathbf{J}(\mathbf{X}) = \frac{\partial f^i}{\partial x_j} = [f_j^i] = \begin{bmatrix} f_1^1 & f_2^1 & \cdots & f_n^1 \\ f_1^2 & f_2^2 & \cdots & f_n^2 \\ \vdots & \vdots & & \vdots \\ f_1^n & f_2^n & \cdots & f_n^n \end{bmatrix}$$

Since the kth *row* of $\mathbf{J}(\mathbf{X})$ is just $[\boldsymbol{\nabla} f^k]'$, the gradient of the kth function, an alternative representation is

$$\mathbf{J}(\mathbf{X}) = \begin{bmatrix} [\boldsymbol{\nabla} f^1]' \\ \vdots \\ [\boldsymbol{\nabla} f^n]' \end{bmatrix}$$

and the methods for partial derivative approximations for $f(\mathbf{X})$, above, can be used for each of the elements in each of the rows in the Jacobian. In compact terms, under the *forward-difference method*, the jth column of $\mathbf{J}(\mathbf{X})$ at $\mathbf{X}^k$ is approximated by $\mathscr{J}_j^k$, where

$$\mathscr{J}_j^k = \frac{\mathbf{F}(\mathbf{X}^k + \alpha \mathbf{I}_j) - \mathbf{F}(\mathbf{X}^k)}{\alpha} \tag{5.1.6}$$

Remember that $\mathbf{F}(\mathbf{X})$ is the column vector containing the n *functions* $f^1(\mathbf{X})$ through $f^n(\mathbf{X})$, so that

$$\mathbf{F}(\mathbf{X} + \alpha \mathbf{I}_j) = [f^1(\mathbf{X} + \alpha \mathbf{I}_j), \ldots, f^n(\mathbf{X} + \alpha \mathbf{I}_j)]'$$

If the *central-difference method* is used, the approximation for the jth column of the Jacobian is

$$\mathscr{J}_j^k = \frac{\mathbf{F}(\mathbf{X}^k + \alpha \mathbf{I}_j) - \mathbf{F}(\mathbf{X}^k - \alpha \mathbf{I}_j)}{2\alpha} \tag{5.1.7}$$

For a system of n equations in n unknowns, the central-difference method will require $2n^2$ functional evaluations—two evaluations for each of n variables in each of n functions—and the forward-difference method will require n^2 such evaluations.

5A.1.4 Hessian Matrices

Since the elements in Hessian matrices are second partials and cross-partial derivatives, f_{ii} and f_{ij}, similar approximation methods should be (and are) available. Consider the case in which $\nabla f(\mathbf{X})$ is available (either having been entered into the computer or approximated as indicated above). Then, since $\mathbf{H}(\mathbf{X}) \equiv \nabla[\nabla f(\mathbf{X})] \equiv \nabla^2 f(\mathbf{X})$, the approach for systems of equations, above, can be applied, with $\mathbf{F}(\mathbf{X}) \equiv \nabla f(\mathbf{X})$. Except for the most elementary functions, the elements of $\nabla f(\mathbf{X})$ will themselves be functions of $\mathbf{X}$. When evaluated at a particular $\mathbf{X}^k$, the gradient is a vector of constants.

Specifically, for the Hessian approximation, $\mathbf{F}(\mathbf{X}) = [f_1, f_2, \ldots, f_n]'$, so a typical element in the Jacobian for this particular $\mathbf{F}(\mathbf{X})$ will be $\partial f_i / \partial x_j$ or f_{ij}. For well-behaved functions, the Hessian matrix is symmetric; $f_{ij} = f_{ji}$. However, when using approximation methods, there is no guarantee that the estimate for $\partial f_i / \partial x_j$ will turn out to be the same as the estimate for $\partial f_j / \partial x_i$. A simple way to force the required symmetry is to replace both f_{ij} and f_{ji}, for each specific i–j pair, with their average value, $(f_{ij} + f_{ji})/2$. Thus an approximation at $\mathbf{X}^k$ to the symmetric Hessian can be found as $\mathscr{H}^k$, where

$$\mathscr{H}^k = \tfrac{1}{2}[\mathscr{J}^k + (\mathscr{J}^k)'] \tag{5.1.8}$$

where $\mathscr{J}^k = [\mathscr{J}_1^k \mid \mathscr{J}_2^k \mid \cdots \mid \mathscr{J}_n^k]$ and the $\mathscr{J}_j^k$ are estimated by either (5.1.6) or (5.1.7).

If $\nabla f(\mathbf{X})$ is not available, and you want a direct approximation of the Hessian elements at $\mathbf{X}^k$, a logical extension of the *forward-difference* approximation technique (with $\alpha > 0$) for $\partial f / \partial x_j$ is constructed as follows. A forward-difference estimate of the slope of $f(\mathbf{X})$ in the direction x_j at $\mathbf{X}^k$, namely, $\partial f / \partial x_j^k$, is given [as in (5.1.4)] by

$$S_1 = \frac{f(\mathbf{X}^k + \alpha \mathbf{I}_j) - f(\mathbf{X}^k)}{\alpha} = \frac{N_1}{\alpha}$$

A forward-difference estimate of the same partial derivative (slope), now at the point $\mathbf{X}^k$ + $\alpha\mathbf{I}_i$ (indicating an x_i different from that at $\mathbf{X}^k$) can be represented as $\partial f/\partial x_j^{k'}$ (where k' represents the perturbation of $\mathbf{X}^k$ by an amount α in the x_i direction). This estimate is

$$S_2 = \frac{f(\mathbf{X}^k + \alpha\mathbf{I}_i + \alpha\mathbf{I}_j) - f(\mathbf{X}^k + \alpha\mathbf{I}_i)}{\alpha} = \frac{N_2}{\alpha}$$

Therefore, the *change* in the value of $\partial f/\partial x_j^k$ for an α-sized change in x_i, which is $\partial/\partial x_i^k(\partial f/\partial x_j^k) = \partial f/\partial x_i^k \partial x_j^k = h_{ij}^k$, is given by $(S_2 - S_1)/\alpha$:

$$h_{ij}^k = \frac{N_2 - N_1}{\alpha^2} = \frac{f(\mathbf{X}^k + \alpha\mathbf{I}_i + \alpha\mathbf{I}_j) - f(\mathbf{X}^k + \alpha\mathbf{I}_i) - [f(\mathbf{X}^k + \alpha\mathbf{I}_j) - f(\mathbf{X}^k)]}{\alpha^2}$$

The geometry for $f(x_1, x_2)$ is illustrated in Figure 5.1.2.

It should be clear that the number of functional evaluations increases dramatically when an $n \times n$ Hessian matrix is approximated in this way. (You might speculate on what the analog to the *central-difference* approximation would look like for h_{ij}^k; as usual, the number of function evaluations will be doubled, as compared to the number needed for the forward difference approach.)

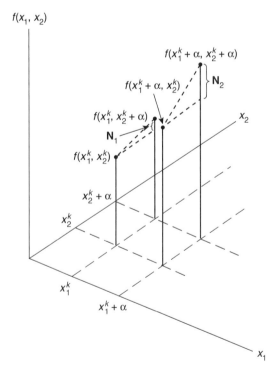

FIGURE 5.1.2   Forward-difference approximation to $h_{12}^k = \partial^2 f/\partial x_1^k \partial x_2^k$.

APPENDIX 5.2 THE SHERMAN–MORRISON–WOODBURY FORMULA

A formula that has turned out to be very useful in a number of iterative techniques for both nonlinear equations and optimization problems was initially presented by Sherman and Morrison (1950) and extended by Woodbury (1950). The work was inspired by the fact that in the early days of computers finding the inverse of anything but a small matrix was a very large task, in terms of memory requirements and computational time. Thus there were many schemes in which an attempt was made to exploit the particular structure or characteristics (e.g., "sparseness") of a matrix whose inverse was required. As computer technology has developed, many such concerns either disappear or else do not arise except for extremely large matrices—but of course increasingly large problems *are* being fed into computers precisely because of their increased speed and capacity.

In any event, Sherman and Morrison developed a simple formula for the following situation: Suppose that you have a matrix, $\mathbf{A}$, and also its inverse, $\mathbf{A}^{-1}$. Then suppose that only one element of $\mathbf{A}$ is changed, producing a matrix $\mathbf{A}^*$. Is it possible to find $(\mathbf{A}^*)^{-1}$ by "adjusting" $\mathbf{A}^{-1}$, which is already known? The answer is "yes," and the adjustment is quite simple.

Example 1 To begin with a concrete example, suppose $\mathbf{A} = \begin{bmatrix} 1 & 1 & 1 \\ 2 & 0 & 6 \\ 3 & 7 & 1 \end{bmatrix}$, and also that $\mathbf{A}^{-1}$ has been found:

$$\mathbf{A}^{-1} = \begin{bmatrix} 3.5 & -0.5 & -0.5 \\ -1.333 & 0.167 & 0.333 \\ -1.167 & 0.333 & 0.167 \end{bmatrix}$$

Consider an $\mathbf{A}^*$ that differs from $\mathbf{A}$ only in that 3 has been added to element a_{23}, changing it from a 6 to a 9. First, note that $\mathbf{A}^*$ can be very simply expressed as $\mathbf{A} + \Delta\mathbf{A}$, where in this case $\Delta\mathbf{A} = \begin{bmatrix} 0 & 0 & 0 \\ 0 & 0 & 3 \\ 0 & 0 & 0 \end{bmatrix}$. In this small illustration, the inverse of $\mathbf{A}$ can easily be found to be (to three decimals)

$$(\mathbf{A}^*)^{-1} = \begin{bmatrix} 2.625 & -0.25 & -0.375 \\ -1.042 & 0.083 & 0.292 \\ -0.583 & 0.167 & 0.083 \end{bmatrix}$$

The idea is to find an alternative to the direct computation of $(\mathbf{A}^*)^{-1}$, making use of $\mathbf{A}^{-1}$. This is what Sherman and Morrison accomplished.

The heart of the procedure is captured in two matrices (for the moment these are vectors). Let $\mathbf{U} = \begin{bmatrix} 0 \\ 1 \\ 0 \end{bmatrix}$ and $\mathbf{V} = \begin{bmatrix} 0 & 0 & 3 \end{bmatrix}$; then $\Delta\mathbf{A}$ can be represented in terms of $\mathbf{U}$ and $\mathbf{V}$ as

$$\Delta\mathbf{A} = \mathbf{UV} \qquad \text{or} \qquad \Delta\mathbf{A} = \mathbf{U} \otimes \mathbf{V}$$

(Recall that $\otimes$ denotes the *outer product* of two n-element vectors, whether column vectors or row vectors or one of each.) The trick here is to let $\mathbf{U}$ be the ith column from an identity matrix that is the same size as **A** (here 3 × 3), where i serves to indicate the *row* in **A** in which the change occurs (here $i = 2$). Similarly, **V** is constructed by (a) forming a three-element null row vector and then (b) replacing the jth element of this vector, where j represents the *column* in **A** in which the change occurs (here $j = 3$), by the amount of the change, Δa_{ij} (here $\Delta a_{23} = +3$).[13]

Note that for *any* pair of n-element vectors **U** and **V** (where **U** is a row and **V** is a column vector), the product **UV** $(= \mathbf{U} \otimes \mathbf{V})$ will be an $n \times n$ matrix whose rank is 1. This is apparent in the current example, since $\Delta \mathbf{A} = \mathbf{UV}$ is a 3 × 3 matrix with only one nonzero element. For the general three-element case, with $\mathbf{U} = \begin{bmatrix} u_1 \\ u_2 \\ u_3 \end{bmatrix}$ and $\mathbf{V} = [v_1 \quad v_2 \quad v_3]$, we obtain

$$\mathbf{UV} = \begin{bmatrix} u_1 v_1 & u_1 v_2 & u_1 v_3 \\ u_2 v_1 & u_2 v_2 & u_2 v_3 \\ u_3 v_1 & u_3 v_2 & u_3 v_3 \end{bmatrix}$$

This clearly has linearly dependent rows (and columns), from the way in which it was formed. Rows are proportional to the row vector **V** and columns are proportional to the column vector **U**. For example, row 2 is (u_2/u_1) times row 1, column 2 is (v_2/v_1) times column 1, and so on. Results are the same in the n-element case.

In many iterative procedures, a given matrix is corrected or "updated" at each iteration. In those cases, the relationship between the matrix at iteration k and at iteration $k+1$ is given as

$$\mathbf{A}^{k+1} = \mathbf{A}^k + \Delta \mathbf{A}^k$$

and, if $\Delta \mathbf{A}^k$ can be expressed as a product **UV**, it is often referred to as a "rank 1 correction" or a "rank 1 update" of **A**.

Using $\mathbf{A}^{-1}$, **U** and **V**, Sherman and Morrison demonstrated that

$$(\mathbf{A}^*)^{-1} = \mathbf{A}^{-1} - \Delta(\mathbf{A}^{-1}) = \mathbf{A}^{-1} - \frac{(\mathbf{A}^{-1}\mathbf{U})(\mathbf{V}\mathbf{A}^{-1})}{(1 + \mathbf{V}\mathbf{A}^{-1}\mathbf{U})} \tag{5.2.1}$$

Note that the numerator of $\Delta(\mathbf{A}^{-1})$ is the product of an n-element column vector $(\mathbf{A}^{-1}\mathbf{U})$ and an n-element row vector $(\mathbf{V}\mathbf{A}^{-1})$. The denominator is simply a scalar.[14] Thus, given a rank 1 correction of **A**, (5.2.1) describes a "rank 1 correction" of the *inverse* of **A**.

---

[13]If an amount is subtacted from a_{ij}, then Δa_{ij} is negative. If a_{ij} is multiplied by some factor, the modification must be translated into a term to be added or subtacted. In this example, if a_{23} is multiplied by 5, $\Delta a_{23} = +24$, if it is multiplied by (1/3), $\Delta a_{23} = -4$, and so forth.

[14]This is often presented using inner and outer product notation:

$$(\mathbf{A}^*)^{-1} = \mathbf{A}^{-1} - (\mathbf{A}^{-1}\mathbf{U}) \otimes (\mathbf{V}\mathbf{A}^{-1})/[1 + (\mathbf{V}') \circ (\mathbf{A}^{-1}\mathbf{U})]$$

Some authors define *both* **U** and **V** as column vectors, in which case **V** must be transposed before premultiplying **A**. Also, some authors define both inner and outer products for pairs of n-element *column* vectors only, in which case further transpositions are required. Since the inner and outer product representation is not very useful in more general cases, in which there are changes in several rows and columns of a matrix, we ignore these subtleties and use the notation of regular matrix multiplication in what follows.

In this example

$$\mathbf{A}^{-1}\mathbf{U} = \begin{bmatrix} -0.5 \\ 0.167 \\ 0.333 \end{bmatrix}, \ \mathbf{VA}^{-1} = [-3.5 \quad 1 \quad 0.5] \text{ and } \mathbf{VA}^{-1}\mathbf{U} = 1$$

so that

$$\Delta(\mathbf{A}^{-1}) = (0.5)\begin{bmatrix} 1.75 & -0.5 & -0.25 \\ -0.583 & 0.167 & 0.083 \\ -1.167 & 0.333 & 0.167 \end{bmatrix}$$

The rank of this matrix is also 1, for example since in the numerator of $\Delta(\mathbf{A}^{-1})$ the column vector $\mathbf{A}^{-1}\mathbf{U}$ premultiplies the row vector $\mathbf{VA}^{-1}$, so each of the rows of the product $(\mathbf{A}^{-1}\mathbf{U})(\mathbf{VA}^{-1})$ is proportional to $(\mathbf{VA}^{-1})$ and each column is proportional to $(\mathbf{A}^{-1}\mathbf{U})$—in exactly the same way that each row in $\mathbf{UV}$ was seen to be proportional to $\mathbf{V}$. The scalar $\mathbf{VA}^{-1}\mathbf{U}$ does not destroy this proportionality. [Rounding may make this less apparent, but it is also clear that column 2 is twice column 3 and column 1 is (-7) times column 3]. Using (5.2.1)

$$(\mathbf{A}^*)^{-1} = \begin{bmatrix} 2.625 & -0.25 & -0.375 \\ -1.042 & 0.083 & 0.292 \\ -0.583 & 0.167 & 0.083 \end{bmatrix}$$

which corresponds to the earlier result when $(\mathbf{A}^*)^{-1}$ was found directly. When n is large, the number of arithmetic operations required for the new inverse using (5.2.1) is much smaller than would be needed to find the inverse directly.

The Sherman-Morrison procedure works in exactly the same way if the roles of $\mathbf{U}$ and $\mathbf{V}$ are "reversed" in the following sense. Let $\mathbf{V}$ be the jth row from an identity matrix that is the same size as $\mathbf{A}$, where j indicates the *column* in $\mathbf{A}$ in which the change occurs (in the example, $j = 3$). Similarly, $\mathbf{U}$ is constructed from a three-element null column vector in which the ith element (where i is the *row* in $\mathbf{A}$ in which the change occurs; in the example, $i = 2$) is replaced by Δa_{ij} ($\Delta a_{23} = +3$ in the example). You should use (5.2.1) in this case—that is, now with $\mathbf{U} = \begin{bmatrix} 0 \\ 3 \\ 0 \end{bmatrix}$ and $\mathbf{V} = [0 \quad 0 \quad 1]$—and note that the same values are found for $(\mathbf{A}^*)^{-1}$.

Of course, the procedure breaks down if the denominator in $\Delta(\mathbf{A}^{-1})$ in (5.2.1) is zero, which means if $\mathbf{VA}^{-1}\mathbf{U} = -1$. The next example illustrates (but does not prove) that if this is the case, then $\mathbf{A}^*$ is singular and so of course no inverse could be found for it.

Example 2   Let $\mathbf{A} = \begin{bmatrix} 1 & 1 & 1 \\ 2 & 2 & 6 \\ 3 & 7 & 1 \end{bmatrix}$ and suppose that 4 is subtracted from element a_{23};

$\Delta a_{23} = -4$. (It should be clear that the new matrix, $\mathbf{A}^*$, will be singular, since row 2 will be just twice row 1.) As in Example 1, $\mathbf{U} = \begin{bmatrix} 0 \\ 1 \\ 0 \end{bmatrix}$ but now $\mathbf{V} = [0 \quad 0 \quad -4]$ and $\mathbf{A}^{-1} =$

$$\begin{bmatrix} 2.5 & -0.375 & -0.25 \\ -1.0 & 0.125 & 0.25 \\ -0.5 & 0.25 & 0 \end{bmatrix}$$, so that $\mathbf{VA}^{-1}\mathbf{U}$ is found indeed to be equal to -1 (you might calculate it, for practice); hence the denominator in (5.2.1) is zero.

While the original Sherman and Morrison results concerned the inverse of a matrix in which only one element was changed, Woodbury (1950) generalized the formula to allow changes in more than one element. If all the changes are in one row (or column), the procedure is an extremely straightforward extension of the results above. This is illustrated in Example 3, below.

Example 3   Given the same original $\mathbf{A}$ matrix as in Example 1, suppose that 2, 5, and 3, respectively, are to be added to the elements of row 2. Since the changes are in row 2 of $\mathbf{A}$, $\mathbf{U}$ is the same as in the previous two examples. Now, however, all three zeros in $[0 \quad 0 \quad 0]$ are replaced; since $\Delta a_{21} = 2$, $\Delta a_{22} = 5$ and $\Delta a_{23} = 3$, $\mathbf{V} = [2 \quad 5 \quad 3]$. Thus

$$\Delta\mathbf{A} = \mathbf{UV} = \begin{bmatrix} 0 & 0 & 0 \\ 2 & 5 & 3 \\ 0 & 0 & 0 \end{bmatrix} \quad \text{and} \quad \mathbf{A}^* = \begin{bmatrix} 1 & 1 & 1 \\ 4 & 5 & 9 \\ 3 & 7 & 1 \end{bmatrix}$$

Using this $\mathbf{U}$ and $\mathbf{V}$ and the same $\mathbf{A}^{-1}$ as in Example 1 in the definition of $\Delta(\mathbf{A}^{-1})$ in (5.2.1), we have

$$\Delta(\mathbf{A}^{-1}) = \frac{1}{1.833}\begin{bmatrix} 1.583 & -0.417 & -0.583 \\ -0.528 & 0.139 & 0.194 \\ -1.056 & 0.278 & 0.389 \end{bmatrix}$$

As in Example 1, both $\Delta\mathbf{A}$ $(= \mathbf{UV})$ and $\Delta(\mathbf{A}^{-1})$ are rank 1 matrices. Using (5.2.1) again, $(\mathbf{A}^*)^{-1}$ is found to be

$$(\mathbf{A}^*)^{-1} = \begin{bmatrix} 3.5 & -0.5 & -0.5 \\ -1.333 & 0.167 & 0.333 \\ -1.167 & 0.333 & 0.167 \end{bmatrix} - \begin{bmatrix} 0.864 & -0.227 & -0.318 \\ -0.288 & 0.076 & 0.106 \\ -0.576 & 0.152 & 0.212 \end{bmatrix} =$$

$$\begin{bmatrix} 2.636 & -0.273 & -0.182 \\ -1.045 & 0.091 & 0.227 \\ -0.591 & 0.181 & -0.045 \end{bmatrix}$$

Except for rounding, this is the same as $(\mathbf{A}^*)^{-1}$ found directly. (Try it.)

If the changes had been concentrated in one *column* of $\mathbf{A}$ instead of one row, the roles of $\mathbf{U}$ and $\mathbf{V}$ would be reversed. Suppose $\Delta a_{11} = 10$, $\Delta a_{21} = 5$ and $\Delta a_{31} = 4$, so that $\Delta\mathbf{A} = \begin{bmatrix} 10 & 0 & 0 \\ 5 & 0 & 0 \\ 4 & 0 & 0 \end{bmatrix}$. You should work through the application of (5.2.1) when $\mathbf{U} = \begin{bmatrix} 10 \\ 5 \\ 4 \end{bmatrix}$ and $\mathbf{V} = [1 \quad 0 \quad 0]$ and then check that this indeed generates the correct elements for $(\mathbf{A}^*)^{-1}$.

Consider next the case in which there are changes in several elements located in several rows and columns. The Woodbury extension of (5.2.1) for this case is

$$(\mathbf{A}^*)^{-1} = \mathbf{A}^{-1} - \Delta(\mathbf{A}^{-1}) = \mathbf{A}^{-1} - [\mathbf{A}^{-1}\mathbf{U}(\mathbf{I} + \mathbf{VA}^{-1}\mathbf{U})^{-1}\mathbf{VA}^{-1}] \tag{5.2.2}$$

where $\mathbf{I}$ is an identity matrix whose order is the same as the smaller of the dimensions of both $\mathbf{U}$ and $\mathbf{V}$. We illustrate again with the $\mathbf{A}$ matrix from Example 1.

Example 4 Suppose now that elements a_{22}, a_{23}, a_{32} and a_{33} are changed; specifically Δa_{22}

$= 6$, $\Delta a_{23} = 7$, $\Delta a_{32} = 9$ and $\Delta a_{33} = 2$, so that $\mathbf{A}^* = \mathbf{A} + \Delta\mathbf{A}$, where $\Delta\mathbf{A} = \begin{bmatrix} 0 & 0 & 0 \\ 0 & 6 & 7 \\ 0 & 9 & 2 \end{bmatrix}$. In

this case, $\mathbf{U}$ is composed of *two* columns—the second and the third—from a 3×3 identity matrix, representing the fact that there are changes in elements in both rows 2 and 3 of $\mathbf{A}$. Similarly, $\mathbf{V}$ will be a 2×3 matrix whose first row incorporates the changes for the first of the changing rows in $\mathbf{A}$ (in this example, row 2 of $\mathbf{A}$) and whose second row displays the changes for the second of the changing rows in $\mathbf{A}$ (here this is row 3 of $\mathbf{A}$). For this example, then

$$\mathbf{U} = \begin{bmatrix} 0 & 0 \\ 1 & 0 \\ 0 & 1 \end{bmatrix} \quad \text{and} \quad \mathbf{V} = \begin{bmatrix} 0 & 6 & 7 \\ 0 & 9 & 2 \end{bmatrix}$$

so that $\Delta\mathbf{A} = \mathbf{UV}$. (Try it.)

Note that the product $\mathbf{UV}$ is now a matrix of rank 2. This is generally true for any 3×2 matrix $\mathbf{U}$ and 2×3 matrix $\mathbf{V}$ (with no null columns in $\mathbf{U}$ and no null rows in $\mathbf{V}$); $\mathbf{UV}$ will be a rank 2 matrix and hence $\Delta\mathbf{A}$ is a "rank 2 correction" or "rank 2 update" of $\mathbf{A}$. In fact, for any $\mathbf{U}$ of size $n \times 2$ and $\mathbf{V}$ of size $2 \times n$ (provided there are no null columns in $\mathbf{U}$ and no null rows in $\mathbf{V}$) the product $\mathbf{UV}$ will produce a matrix of rank 2.

The requisite pieces for (5.2.2) for this example are

$$\mathbf{A}^{-1}\mathbf{U} = \begin{bmatrix} -0.5 & -0.5 \\ 0.167 & 0.333 \\ 0.333 & 0.167 \end{bmatrix}, \quad \mathbf{VA}^{-1} = \begin{bmatrix} -16.167 & 3.333 & 3.167 \\ -14.333 & 2.167 & 3.333 \end{bmatrix}$$

and

$$\mathbf{VA}^{-1}\mathbf{U} = \begin{bmatrix} 3.333 & 3.167 \\ 2.167 & 3.333 \end{bmatrix}$$

Therefore, to three decimals, $(\mathbf{I} + \mathbf{VA}^{-1}\mathbf{U}) = \begin{bmatrix} 0.364 & -0.266 \\ -0.182 & 0.364 \end{bmatrix}$, and $\Delta(\mathbf{A}^{-1}) =$

$\begin{bmatrix} 2.171 & -0.409 & -0.451 \\ -1.103 & 0.167 & 0.256 \\ -1.069 & 0.242 & 0.195 \end{bmatrix}$ and so, using (5.2.2) and $\mathbf{A}^{-1}$ from Example 1:

$$(\mathbf{A}^*)^{-1} = \begin{bmatrix} 1.329 & -0.091 & -0.049 \\ -0.231 & 0 & 0.077 \\ -0.098 & 0.091 & -0.028 \end{bmatrix}$$

[You should find $(\mathbf{A}^*)^{-1}$ directly as a check.] The rank of $\Delta(\mathbf{A}^{-1})$ is 2, and this is also true in general for any $n \times 2$ matrix $\mathbf{U}$ and $2 \times n$ matrix $\mathbf{V}$. In those cases, (5.2.2) describes a "rank 2 correction" to $\mathbf{A}^{-1}$ induced by a "rank 2 correction" to $\mathbf{A}$.

Note that when $\mathbf{U}$ has only one column and $\mathbf{V}$ has only one row, (5.2.2) reduces directly to (5.2.1), since in that case $(\mathbf{I} + \mathbf{VA}^{-1}\mathbf{U})^{-1}$ is just the reciprocal of a scalar. Note also that for (5.2.2) to be useful, $(\mathbf{I} + \mathbf{VA}^{-1}\mathbf{U})$ must be nonsingular; this is just the matrix analog to $1 + \mathbf{VA}^{-1}\mathbf{U} \neq 0$ in the illustration of Example 1. Also as in the case of Example 1, if $(\mathbf{I} + \mathbf{VA}^{-1}\mathbf{U})$ is singular, the new matrix, $\mathbf{A}^*$, is singular. The following variation on this example illustrates this point. Let $\Delta\mathbf{A} = \begin{bmatrix} 0 & 0 & 0 \\ 0 & 5 & -4 \\ 4 & 2 & 6 \end{bmatrix}$, so that $\mathbf{U} = \begin{bmatrix} 0 & 0 \\ 1 & 0 \\ 0 & 1 \end{bmatrix}$ and $\mathbf{V} =$

$\begin{bmatrix} 0 & 5 & -4 \\ 4 & 2 & 6 \end{bmatrix}$. You should check that in this case $\mathbf{VA}^{-1}\mathbf{U} = \begin{bmatrix} -0.5 & 1.0 \\ 0.333 & -0.333 \end{bmatrix}$ and there-

fore $(\mathbf{I} + \mathbf{VA}^{-1}\mathbf{U}) = \begin{bmatrix} 0.5 & 1.0 \\ 0.333 & 0.667 \end{bmatrix}$, which is clearly singular.

One final example illustrates that the $\mathbf{U}$ and $\mathbf{V}$ representations remain valid when the changes in $\mathbf{A}$ are not in adjacent rows and columns.

Example 5   Once again using $\mathbf{A}$ from Example 1, let $\Delta\mathbf{A} = \begin{bmatrix} 1 & 3 & 0 \\ 0 & 0 & 0 \\ 0 & 0 & 5 \end{bmatrix}$, so that $\mathbf{A}^* =$

$\begin{bmatrix} 2 & 4 & 1 \\ 2 & 0 & 6 \\ 3 & 7 & 6 \end{bmatrix}$. For this case, let $\mathbf{U} = \begin{bmatrix} 1 & 0 \\ 0 & 0 \\ 0 & 1 \end{bmatrix}$ (identifying rows 1 and 3 of $\mathbf{A}$ as those in

which the changes occur) and $\mathbf{V} = \begin{bmatrix} 1 & 3 & 0 \\ 0 & 0 & 5 \end{bmatrix}$ (representing the magnitudes of the

changes in rows 1 and 3 of $\mathbf{A}$). It is clear that $\mathbf{UV} = \Delta\mathbf{A}$ and that $\Delta\mathbf{A}$ is a rank 2 matrix. You should work through (5.2.2) for this example to find that (to three decimal accuracy)

$$(\mathbf{A}^*)^{-1} = \begin{bmatrix} 0.913 & 0.370 & -0.522 \\ -0.130 & -0.196 & 0.217 \\ -0.304 & 0.043 & 0.174 \end{bmatrix}$$

The Woodbury approach in (5.2.2) is useful only if the number of rows (or columns) in which changes in $\mathbf{A}$ occur is small relative to n, the size of $\mathbf{A}$. If $\mathbf{U}$ is a matrix of dimension $n \times q$, meaning that there are changes in q of the rows of $\mathbf{A}$, then $\mathbf{V}$ will be of dimension $q \times n$. Using (5.2.2) requires the inverse of the matrix $(\mathbf{I} + \mathbf{VA}^{-1}\mathbf{U})$, which is a square matrix of dimension q. Therefore, if q is nearly as large as n, the advantage of using (5.2.2) disappears, since finding the inverse of $\mathbf{A}^*$ directly would not require much more computation. This fact is usually recognized by noting that the approach of (5.2.2) is useful only if $q \ll n$ (read as "q is much smaller than n").

REFERENCES

Brent, Richard P., *Algorithms for Minimization without Derivatives*, Prentice-Hall, Englewood Cliffs, NJ, 1973.

Broyden, C. G., "A Class of Methods for Solving Nonlinear Simultaneous Equations," *Math. Comput.* **19**, 577–593 (1965).

Dennis, J. E., Jr., and Robert B. Schnabel, *Numerical Methods for Unconstrained Optimization and Nonlinear Equations*, Prentice-Hall, Englewood Cliffs, NJ, 1983.

Press, William H., Brian P. Flannery, Saul A. Teukolsky, and William T. Vetterling, *Numerical Recipes. The Art of Scientific Computing*, Cambridge University Press, Cambridge, UK, 1986.

Sherman, Jack, and Winifred J. Morrison, "Adjustment of an Inverse Matrix Corresponding to a Change in One Element of a Given Matrix," *Ann. Math. Stat.*, **21**, 124–127 (1950).

Woodbury, Max A., *Inverting Modified Matrices,* Memorandum 42, Statistical Research Group, Princeton University, Princeton, NJ, 1950.

PROBLEMS

5.1 If you have computer software that includes one or more of the numerical methods for root finding that are explored in this chapter, then you can examine the relative success of alternative methods on various test functions. You can also use alternative starting points as well as stopping criteria to see how the outcomes are affected for a given test function. Even without a computer, you could use the two test functions for $f(x)$ in Section 5.1 with alternative starting points and stopping criteria (e.g., as in Tables 5.2, 5.4, 5.7, and 5.9). By calculating the first few steps in each case, you can get an idea of how the method moves away from the starting point that you have chosen and how quickly or slowly the iterations appear to converge. Since both of these test functions have more than one solution, you can also see how the particular solution you reach is influenced by the starting point you choose.

5.2 Solve the nonlinear equation $f(x) = x^2 - 1 = 0$ by using

(a) The Newton–Raphson approach, starting at $x^0 = 2$, carrying five places to the right of the decimal point, and stopping when $|x^{k+1} - x^k| < 0.00001$.

(b) The secant method, with the same starting point and stopping criterion, and in which x^1 (since you need two "initial" points for the secant method) is the same as the x^1 generated in the first iteration of Newton–Raphson, in (a).

(c) The modified Newton approach, with everything else as in (a).

(d) Bisection, again with conditions as in (a).

5.3 (a) Since $x^0 = 0$ is just as close to a solution for Problem 5.2 as is $x^0 = 2$ (the starting point suggested for the problem), it would seem that $x^0 = 0$ should be just as good a starting point for the approaches in (a) through (d) of that problem. Is this true? Why or why not?

(b) What results will you get if you choose $x^0 = -0.01$, instead of $x^0 = 2$, in (a)–(d) of Problem 5.2?

5.4 For $f(x) = x^2 + 2$, try the secant, bisection, Newton–Raphson, modified Newton, and quasi-Newton approaches. (You could also use Brent, if you have software that includes it as an option.) Sketch the curve and notice that there is no (real number) solution to the equation $f(x) = 0$. How is this indicated in each of the procedures?

5.5 If $f(x) = x^2$, then there is just one solution to $f(x) = 0$. You can test this (but not "prove" it) by trying a large number of alternative starting points and finding that you are always led to the same (unique) solution, namely, $x = 0$.

5.6 Consider $f(x)$ defined over $0 \le x \le 1$ (the *unit interval*). If one can "trap" a root to $f(x) = 0$ within an interval of length 0.1 (e.g, between $x = 0.3$ and $x = 0.4$), then the point is located within an *interval of uncertainty* that is 10% of the length of the unit interval. Suppose that $f(x_1, x_2)$ is defined over $0 \leq x_1, x_2 \leq 1$—that is, over the *unit square*. Now, if a root to $f(x_1, x_2) = 0$ is located in a *square of* uncertainty that is 10% of the *area* of the unit square [the two-dimensional analog to 10% of the *length* of the unit interval in the case of $f(x)$], what is the interval of uncertainty on *each of the x's*? Extend this reasoning to $f(\mathbf{X})$ for three additional cases, where $\mathbf{X}$ is a 3-, 10- and 20-element vector, respectively, with explicit numerical results for the range of uncertainty on each of the roots to $f(x) = 0$ in each case. (In the case of $\mathbf{X} = [x_1, x_2, x_3]'$, the idea is that the root is located within a *cube* of uncertainty that is 10% of the *volume* of the *unit cube*, etc.)

5.7 Consider the following pair of equations:

$$x_1 + x_2 = 3$$
$$(x_1 - 1)^2 + (x_2 - 3)^2 = 10$$

Examine the first few iterations that would be made by each of the following methods, always starting at the origin:

(a) Gauss–Seidel

(b) Newton–Raphson

(c) Modified Newton

(d) Broyden, using the Jacobian inverse updating formula in (5.36)

If you have computer software that employs these methods, you can print out each of the first few iterations and watch what is happening. If not, you can perform the operations with the help of any matrix algebra computer package (especially for manipulating the algebra needed in Gauss–Seidel and to find the Jacobian inverses in the derivative-based methods). For (c) and (d), you can also compare the *estimates* of $(\mathbf{J}^k)^{-1}$ that are provided at each step by modified Newton and Broyden with the *actual* inverse, computed in Newton–Raphson. Note that this pair of equations has two solutions:

$$\mathbf{X}^1 = \begin{bmatrix} -1.6794 \\ 4.6794 \end{bmatrix} \quad \text{and} \quad \mathbf{X}^2 = \begin{bmatrix} 2.6794 \\ 0.3206 \end{bmatrix}$$

If you used iterative methods software for (b) and (d), you will always have found $\mathbf{X}^2$ (or a solution close to it). Note that there is no signal that there might be more than one solution. This is in sharp contrast to the linear $n \times n$ case, $\mathbf{AX} = \mathbf{B}$, where the clue to multiple solutions is that $\rho(\mathbf{A}) = \rho(\mathbf{A} \vdots \mathbf{B}) < n$. With these nonlinear equations, only if you try enough different starting points will you (perhaps) eventually be led to $\mathbf{X}^1$. This is true in general of systems of nonlinear equations, and it is why you should always try several different starting points.

5.8 Suppose that the equations in the previous question were

$$x_1 + x_2 = 8.47212$$
$$(x_1 - 1)^2 + (x_2 - 3)^2 = 10$$

Use each of the approaches in (a) through (d) in that question. Sketch the geometry in $x_1 x_2$ space. You will find that there is now only *one* solution.

5.9 Finally, let the pair of equations be:

$$x_1 + x_2 = 10$$

$$(x_1 - 1)^2 + (x_2 - 3)^2 = 10$$

Try using the approaches of (a) through (d) on these two equations. If you use software with one or more of the iterative methods, you will find that it never converges (at least with any reasonably "accurate" stopping rule, such as $\delta = 10^{-4}$.) If you sketch the solution space geometry, you will see what is wrong; the two equations do not intersect at all in the x_1x_2 plane. This is the case of no solutions, which for a square linear system is indicated by $\rho(\mathbf{A}) \neq \rho(\mathbf{A} \mid \mathbf{B})$. There is no comparable convenient signal for nonlinear equations.

5.10 Here are two examples in which you are asked to use the Gauss–Seidel approach on larger systems of linear equations—four equations with four unkowns. Linearity simplifies the operation of solving the individual equations for one variable as a function of the others; however, having four equations expands the number of portential orderings of the equations (to 24). To do these effectively you will need to write a small computer program or use software for Gauss–Seidel that allows you to easily change the order of the equations and alter the starting point.

(a) Problem 2.4(c) from Chapter 2. Because not all variables appear in all equations, there are many orderings that are not meaningful—for example, interchange of the first two equations, because equation (2) cannot be solved for x_1. In fact, of the 24 possible orderings of these four equations, only two—{3, 1, 2, 4} and {3, 1, 4, 2}—allow you to solve for the unknowns in the order in which the equations are presented (the first equation for x_1, the second for x_2, and so on). Using Gauss–Seidel with the equation ordering {3, 1, 2, 4} leads easily to the solution in around 18 iterations [using a stopping criterion of $\max_i [|f^i(\mathbf{X}^k)|]$ $\leq 10^{-4}$], from a very wide variety of starting points. On the other hand, the equation ordering {3, 1, 4, 2} converges only from those $\mathbf{X}^0$ that are very close to the actual solution, which is $\mathbf{X} = [4 \quad -2 \quad 5 \quad -5]'$.

(b) Problem 2.4(b) from Chapter 2. In this case, 16 of the 24 possible orderings are meaningless [because of the absence of x_2 from equation (1) and both x_1 and x_2 from equation (4)]. In seven of the remaining eight possible orderings, Gauss–Seidel explodes—even from $\mathbf{X}^0 = [2 \quad 8 \quad 0.5 \quad 0.2]'$. (The solution, to three decimals, is $\mathbf{X} = [2.071 \quad 7.929 \quad 0.500 \quad 0.214]'$.) Only when the equations are ordered {1, 3, 2, 4} and solved in that order does the procedure converge—quickly and from a very wide variety of starting points.

5.11 For $f(\mathbf{X}) = 16 - 2(x_1 - 3)^2 - (x_2 - 7)^2$, find approximations to the gradient at the point $\mathbf{X} = \begin{bmatrix} 1 \\ 1 \end{bmatrix}$ using

(a) A forward-difference approximation.

(b) A backward-difference approximation.

(c) A central-difference approximation.

Use several values of α in each case (e.g., $\alpha = 0.2, 0.1, 0.01, 0.001$). Compare your results, in each case, with the actual value of the gradient at that point.

5.12 (a) Find the forward difference approximation to the Hessian matrix for the function in Problem 5.11, again at $\mathbf{X} = \begin{bmatrix} 1 \\ 1 \end{bmatrix}$, and compare your results with the actual value of the Hessian at that point.

(b) Derive the *central-difference* formulation for the approximation to the Hessian for a typical on-diagonal element h_{ii} and also for a typical off-diagonal element h_{ij}. Evaluate this Hessian approximation also at $\mathbf{X} = \begin{bmatrix} 1 \\ 1 \end{bmatrix}$, and compare with the actual value of the Hessian at that point and with your results in part (a) of this question. What does this suggest (and why is it always true) about the relative accuracy of forward- versus central-difference approximations to the Hessian of a function when the elements of the gradient vector, ∇f, are linear functions of $\mathbf{X}$ (so that the elements of the Hessian matrix are constants)?

5.13 Answer questions 5.11 and 5.12 for $f(\mathbf{X}) = x_1^3 + 5x_1x_2^2 - 6x_1^2x_2 + 3x_2^3$, using several values of α and any specific value of $\mathbf{X}$ that you wish.

5.14 Explore the forward-difference approach to approximating the *second derivative* of $f(x) = x^2 - 2$ at $x^k = 2$. The forward-difference approximation to the first derivative at x^k is

$$f'(x^k) \cong \frac{f(x^k + \alpha) - f(x^k)}{\alpha}$$

as in (5.1.1). Following this logic, we have

$$f''(x^k) \cong \frac{f'(x^k + \beta) - f'(x^k)}{\beta}$$

Show that this expands to

$$f''(x^k) \cong \frac{f(x^k + \beta + \alpha) - f(x^k + \beta) - f(x^k + \alpha) + f(x^k)]}{\alpha\beta}$$

and that, at $x^k = 2$, with $\alpha = 0.1$ and $\alpha = 0.2$, the approximation is exactly equal to the second derivative.

6

SOLVING UNCONSTRAINED MAXIMIZATION AND MINIMIZATION PROBLEMS

In this chapter we examine iterative methods for finding maxima or minima of functions. These methods *approach* the solution through a series of guesses—where each successive guess is (usually; not always) better than the previous one—closer to the "true" solution. These techniques, also known as numerical methods or search methods, are to be distinguished from those in which the first- and second-order conditions from classical calculus are used directly. The classical first-order condition requires root finding—solution of the single equation $f'(x) = 0$ for a function of one variable and solution of the system of equations $\nabla f(\mathbf{X}) = \mathbf{0}$ for a function of more than one variable. In general the equations will not be linear. Classical second-order conditions use the sign of the second derivative [for $f(x)$] or second differential [for $f(\mathbf{X})$], evaluated at the root(s), to distinguish maximum points or minimum points from those that are neither, or they depend on other evidence of concavity or convexity of the function.

As noted in Chapter 5, an iterative method consists of essentially three components:

1. A set of initial values for the unknowns, $\mathbf{X}^0$. [For $f(x)$ there is only one unknown and hence only one initial value, x^0.]
2. An algorithm for getting to the next values, $\mathbf{X}^{k+1}$, from the current values, $\mathbf{X}^k$, where $k = 0, 1, \ldots, n$.
3. A rule for deciding when the current estimate, $\mathbf{X}^k$, is "close enough" to an (unknown) exact solution.

In common with the iterative methods for root finding in Chapter 5, the methods that are examined in this chapter are also built from variants of these three components. However, as we will see, since the objective is different—finding a maximum and/or minimum of a function, not finding where the function takes on the specific value of zero—the algorithms will turn out to be quite different from those in Chapter 5 for root finding.

It is again useful to distinguish methods for a function of one variable, $f(x)$, from those for a function of several variables, $f(\mathbf{X})$. It is also again helpful, especially with methods for $f(\mathbf{X})$, to distinguish those that rely on derivatives (actual or estimated) from those that do not. The principles underlying searches for maxima and for minima are identical—take steps that increase $f(\mathbf{X})$ for a maximum ("hill climbing" or "ascent" methods) or take steps that decrease $f(\mathbf{X})$ for a minimum ("descent" methods). For that reason, it is adequate to concentrate attention on only one of the two; in general in this chapter, we will use minimization problems as the prototype

As was true of iterative methods for finding roots to nonlinear equations in the last chapter, it is numerical methods such as we are about to explore in this chapter that are at the heart of computer software packages for maximization and minimization problems. Again, here is a sampling from manuals accompanying some of this software:

> . . . employs a quasi-Newton algorithm that generally leads to superlinear convergence . . . uses quadratic penalty functions . . . updates the Hessian at each step.

> [and] step length methods. The primary descent method for the single dependent variable [case] is the classical Levenberg–Marquardt method . . . . Also, the Polak–Ribiere version of the conjugate gradient descent method is provided.

> . . . includes three minimization techniques: (1) Powell's method . . . a hybrid of Gauss-Newton and steepest descent methods . . . (2) The Marquardt–Levenberg method . . . varying between the inverse Hessian method and the steepest descent method . . . (3) The simplex (search) method . . . due to Nelder and Mead.

> . . . uses the iterative Levenberg-Marquardt method . . . [or] a quasi-Newton method (a variation of the gradient method) . . . creates a Jacobian matrix at each step . . . computes the Gauss-Newton step for each variable.

> . . . works by following the path of steepest descent from each point that it reaches. The principal algorithms for unconstrained minimization are the Nelder–Mead simplex search method and the BFGS quasi-Newton method . . . [other problems] are solved using the Gauss–Newton and Levenberg–Marquardt methods. The routines offer a choice of algorithms and line search strategies.

Again, the intelligent user of these programs should have some understanding of what such descriptions mean. That is why these methods are explored in this chapter.

6.1 MINIMIZATION OF $f(x)$

Derivatives play a minor role in *iterative* methods for a function of one variable. If $f'(x)$ is easily derived, and if the roots of $f'(x) = 0$ are easily found [and, ideally, if $f''(x)$ is easily evaluated at the root or roots], then the minimization problem has been solved. In this section, then, we concentrate on iterative methods that do not make use of derivatives.

Search methods are labeled "simultaneous" when, at the outset, an entire set of points is selected at which to evaluate the function in question. In the second major group of search techniques, the choice of the evaluation points is made sequentially; each new choice of a value for x is based on information gathered from the preceding choices. We look at these two approaches in turn.

6.1.1   Simultaneous Methods

One of the most general of the simultaneous approaches is *exhaustive search*. This means just what it says. Choose an interval over which to search (perhaps on the basis of some knowledge or even just a guess about the shape of the function) and also select a number of points (values of x) within that interval. How many points there should be, whether equally spaced and so on, are all choices to be made by the analyst. Then find the value of $f(x)$ at each of the selected points and see what you learn about the shape of the function. On the basis of information from the first search, select a second interval (smaller than the previous interval, if the initial search appears to have trapped an extreme point) and an accompanying number of points in that interval (usually closer together than in the previous step, in order to focus in on the maximum or minimum point with more accuracy) and find $f(x)$ at those points. On the basis of the information at this step, either continue on with another step or stop.

Suppose that the left endpoint of the chosen interval is denoted by x_0, the right endpoint is x_n and that intervening points are numbered from left to right.[1] Then if $f(x_0) < f(x_1) < \cdots < f(x_n)$, it might be inferred that a minimum (if one exists at all) is to the left of x_0 or a maximum (if it exists) is to the right of x_n. If the space between the successive evaluation is "too large," then there could be *local* maxima or minima along the way (Fig. 6.1a). On the other hand, if the value of $f(x)$ increases over part of the interval (say, from x_0 to x_i) and then decreases thereafter (from x_j to x_n), one would infer that a local maximum must be located around the x_i, x_j area, and a more thorough search (points of evaluation closer together) could be undertaken in the (x_i, x_j) interval (Fig. 6.1b). Similarly, if $f(x)$ decreases from x_0 to x_g and then increases from x_h to x_n, the presence of a minimum around (x_g, x_h) has been uncovered (Fig. 6.1c).

For $f(x_1, x_2)$, the search would be over a two-dimensional checkerboard, rather than along a line, as for $f(x)$. For $f(x_1, x_2, x_3)$, the search would be over a three-dimensional grid. Clearly, the number of points within the chosen line interval [for $f(x)$], or rectangular area [for $f(x_1, x_2)$], or rectangular volume [for $f(x_1, x_2, x_3)$], and so on, grows astronomically with the number of variables, and this approach is really only sensible for searching over one or at most two dimensions. Also, it is not very practical for those cases in which the functional evaluations themselves are expensive. Alternatives to equally spaced intervals are to choose evaluation points randomly or in some specific, nonrandom manner.

Equal-interval search is illustrated in Tables 6.1 and 6.2. The functions whose minima are sought are $f(x) = x^2 - 3$ and $f(x) = (x-1)(x+1)^2 - 1$, respectively. The second of these was used in Chapter 5 for root finding; the first is similar to $f(x) = x^2 - 4$ from that chapter. They are sketched in Figure 6.2. Remember that search methods are used when we do *not* have an exact picture of $f(x)$ (like those in Fig. 6.2) and/or when $f'(x)$ is not easily found and/or when $f'(x) = 0$ is not easily solved.

We use these relatively simple example functions only so that the geometry of search methods can be easily understood and represented. (For equal-interval search there is nothing subtle about either the algebra or the geometry; for more complex iterative meth-

---

[1]For a function of one variable, $f(x)$, we will use $x_0, x_1, \ldots, x_k$ to indicate a series of $k+1$ values for that variable; so, for example, x^2 or x^3 will still denote the variable squared or cubed, and not particular values of x. For a function of n variables, $f(x_1, \ldots, x_n)$, a series of $k+1$ values for any one of them, x_i, will be denoted by $x_i^0, x_i^1, \ldots, x_i^k$.

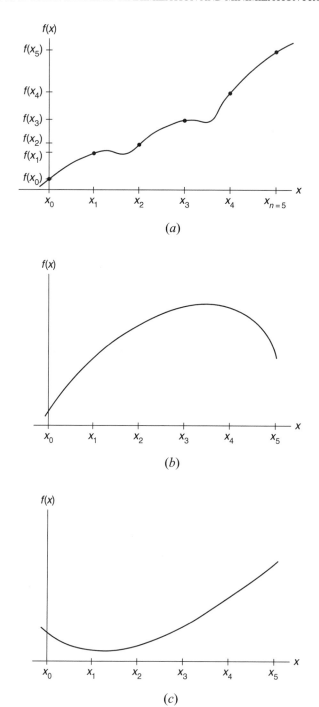

FIGURE 6.1   (*a*) Local maxima and minima between x_0 and x_5; (*b*) local maximum between x_3 and x_4; (*c*) local minimum between x_1 and x_2.

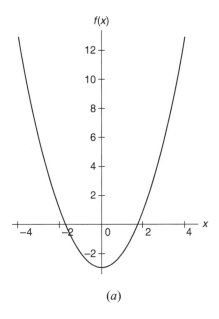

(a)

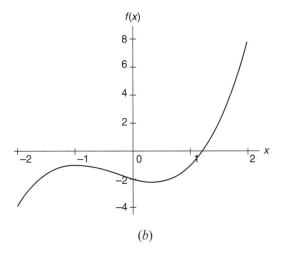

(b)

FIGURE 6.2   Test functions: (a) $f(x) = x^2 - 3$; (b) $f(x) = (x - 1)(x + 1)^2 - 1$.

ods a geometric explanation of the algebra will be more helpful.) In each table, the choice of search interval and spacing between evaluation points for the next step is based on information that is learned from the current step about the shape of the function. So the next step benefits from the current one, but within each step a decision on evaluation points is made at the beginning.

In Table 6.1, there are two sequences of steps toward a minimum. Both sequences start with the search interval (–5, 4). In the first (part A of the table), the initial spacing between evaluation points is 1. Thus $x = 0$, the point at which the minimum occurs, is one of the evaluation points in the first step. Of course, the fact that $x = 0$ is indeed the minimum

TABLE 6.1   Equal-Interval Search for Minimum of $f(x) = x^2 - 3$

x_i	$f(x_i)$

A. Initial Interval (−5, 4); Initial Spacing between Points 1.0

Step 1. Search Interval: (−5, 4); Spacing between Evaluation Points: 1.0

x_i	$f(x_i)$
−5	22
−4	13
−3	6
−2	1
−1	−2
0	−3
1	−2
2	1
3	6
4	13

Step 2. Search Interval: (−1, 1); Spacing between Evaluation Points: 0.25

x_i	$f(x_i)$
−1	−2
−0.75	−2.44
−0.5	−2.75
−0.25	−2.94
0	−3
0.25	−2.94
0.5	−2.75
0.75	−2.44
1	−2

Step 3. Search Interval: (−0.25, 0.25); Spacing between Evaluation Points: 0.05

x_i	$f(x_i)$
−0.25	−2.9375
−0.20	−2.9600
−0.15	−2.9775
−0.10	−2.9900
−0.05	−2.9975
0	−3
0.05	−2.9975
0.10	−2.9900
0.15	−2.9775
0.20	−2.9600
0.25	−2.9375

B. Initial Interval (−5, 4); Initial Spacing between Points 0.75

Step 1. Search Interval: (−5, 4); Spacing between Evaluation Points: 0.75

x_i	$f(x_i)$
−5	22
−4.75	15.063
−3.5	9.250
−2.75	4.563
−2	1
−1.25	−1.438
−0.5	−2.750

TABLE 6.1   *continued*

x_i	$f(x_i)$

B. Initial Interval $(-5, 4)$; Initial Spacing between Points 0.75 (*continued*)

Step 1. Search Interval: $(-5, 4)$; Spacing between Evaluation Points: 0.75
(cont.)

x_i	$f(x_i)$
0.25	−2.940
1	−2
1.75	0.063
2.5	3.250
3.25	7.563
4	13

Step 2. Search Interval $(-0.5, 1.00)$; Spacing between Evaluation Points: 0.15

x_i	$f(x_i)$
−0.5	−2.7500
−0.35	−2.8775
−0.20	−2.9600
−0.05	−2.9975
0.10	−2.9900
0.25	−2.9375
0.40	−2.8400
0.55	−2.6975
0.70	−2.5100
0.85	−2.2775
1.00	−2.0000

Step 3. Search Interval: $(-0.20, 0.10)$; Spacing between Evaluation Points: 0.03

x_i	$f(x_i)$
−0.20	−2.9600
−0.17	−2.9711
−0.14	−2.9804
−0.11	−2.9879
−0.08	−2.9936
−0.05	−2.9975
−0.02	−2.9996
0.01	−2.9999
0.04	−2.9984
0.07	−2.9951
0.10	−2.9900

would not be known; if it were, there would be no need to search. However, it does appear from the 10 values of $f(x_i)$ in step 1 that a minimum to $f(x) = x^2 - 3$ is probably in the neighborhood of $x = 0$. It is not obvious, from this initial step, whether $f(x)$ takes on a smaller value at some point between the current evaluation points in the neighborhood of $x = 0$. Step 2 exploits these observations (narrowing both the search interval and the distance between evaluation points) and step 3 provides even more support for the conclusion that the function is symmetric around a minimum at $x = 0$.

The first step in part B of Table 6.1 differs from that in part A only in that the spacing between evaluation points is now 0.75. This is done intentionally so as to "miss" the exact minimum at $x = 0$. Again, details of each subsequent step are chosen on the basis of information gained from the current step.

How do you know when to stop? The question is not as easy to answer as in iterative methods for root finding. There you know that $f(x) = 0$ for a root; hence for a current estimate, x_k, $|f(x_k)| < \varepsilon$ is a logical stopping requirement (for some small $\varepsilon > 0$) in root-finding procedures. For maximization and minimization problems, the value of $f(x)$ at the maximum or minimum point is not known, so there is no unambiguous value (like zero in the root-finding case) with which to compare $f(x_k)$ at each step. Therefore, many stopping rules for simultaneous iterative methods for maximization and minimization are based on the amount of *change* that occurs in some value from one step to the next. The general idea is that once these improvements between steps become very small, there is little point in continuing.

For simultaneous methods, stopping rules can be based on absolute values of differences between consecutive values of x^* (the best x at each step), or between consecutive values of $f(x^*)$, or on some variant of either of these. When the measured difference becomes less than some small positive ε, the search can be stopped. Let x_{k-1}^* and x_k^* be the best values of x at the $(k-1)$st and kth steps. Then the first rule is based on $|x_{k-1}^* - x_k^*| < \varepsilon$ and the second on $|f(x_{k-1}^*) - f(x_k^*)| < \varepsilon$; these can be represented compactly by $|\Delta x^*| < \varepsilon$ and $|\Delta f(x^*)| < \varepsilon$, respectively. Notice that the first criterion is a measure of the improvement, at each step, in locating the best x and the second criterion is a measure of improvement in the minimized value of the function.

The significance of the size of a difference in $|\Delta x^*|$ or $|\Delta f(x^*)|$ might appropriately be viewed relative to the size of the x_k^*'s or $f(x_k^*)$'s. For example, a $|\Delta f(x^*)|$ of 0.001 is more impressive if $f(x_{k-1}^*)$ and $f(x_k^*)$ are in the neighborhood of 10,000 than if they are both near to 0.005. For this reason, *relative* difference measures are often used. One example is

$$|\Delta f(x^*)| \div (0.5) \left[|f(x_{k-1}^*)| + |f(x_k^*)| \right] \leq \varepsilon$$

which expresses the change relative to the average value of the previous and current f^* (absolute values are used to avoid negative and positive f^*'s from offsetting each other). Multiplied by 100, this represents a kind of "average percentage difference" in $f(x^*)$ from the $(k-1)$st to the kth steps. We will use this criterion in many of the numerical illustrations later in this chapter.

One example of yet another criterion is the following: Given that x^* is the best estimate of the minimizing x at the current step, let x^- and x^+ be the two *adjacent* values on either side of x at this step.[2] Then another stopping rule is $|f(x^*) - f(x^m)| < \varepsilon$, where $f(x^m) = $ min $[f(x^-), f(x^+)]$, the smaller of $f(x^-)$ and $f(x^+)$.

Some of these criteria are illustrated for the results in Table 6.1. For part A of the table, we find

1. Consecutive values of x^*; $|\Delta x^*| < \varepsilon$. From step 1 to 2, $|0 - 0| = 0$; from step 2 to 3, again, $|0 - 0| = 0$.
2. Consecutive values of $f(x^*)$; $|\Delta f(x^*)| < \varepsilon$. From step 1 to 2, $|(-3) - (-3)| = 0$; from step 2 to 3, $|(-3) - (-3)| = 0$. Either of these two criteria would stop at step 2, for *any* $\varepsilon > 0$, and that would be appropriate, since the best x has indeed been found.

---

[2]In two-dimensional search over a grid, note that there will be more than two "adjacent" points. In three dimensions, the number of points adjacent to any given point is even larger.

3. Values adjacent to the current best; $|f(x^*) - f(x^m)| < \varepsilon$. [Because of the symmetry of $f(x)$ in this illustration, and because of the particular (equal) spacing of evaluation points around x^*, it happens that $f(x^-) = f(x^+) = f(x^m)$ at each step.] At step 1, $|f(0) - f(1)| = 1$; at step 2, $|f(0) - f(0.25)| = 0.06$; at step 3, $|f(0) - f(0.05)| = 0.0025$.

For part B of Table 6.1, the values for these different stopping criteria are:

1. $|\Delta x^*| < \varepsilon$. From step 1 to step 2, $|(0.25) - (-0.05)| = 0.30$; from step 2 to 3, $|(-0.05) - (0.01)| = 0.04$.
2. $|\Delta f(x^*)| < \varepsilon$. From step 1 to 2, $|(-2.94) - (-2.9975)| = 0.0575$; from step 2 to 3, $|(-2.9975) - (-2.9999)| = 0.0024$.
3. $|\Delta f(x^*)| \div (0.5) [| f(x^*_{k-1})| + |f(x^*_k)|] \leq \varepsilon$. From step 1 to step 2, $0.0575/2.96875 = 0.0194$; from step 2 to step 3, $0.0024/2.9987 = 0.0008$.
4. $|f(x^*) - f(x^m)| < \varepsilon$. At step 1, $|f(0.25) - f(-0.5)| = 0.19$; at step 2, $|f(-0.05) - f(0.10)| = 0.0075$; at step 3, $|f(0.01) - f(-0.02)| = 0.0003$.

Note that in this set of steps (part B of Table 6.1), the minimum, at $x = 0$, is not isolated exactly—in fact, the search intervals and the spacing between points in each interval were intentionally selected so as to miss $x = 0$, in order to introduce some variety into the values of the stopping criteria. Depending on which stopping rule and value of ε are chosen, further steps might be required.

Table 6.2 indicates one set of steps that would isolate the local minimum to $f(x) = (x - 1)(x + 1)^2 - 1$ at $x = 0.333$. It also suggests (in step 1) that $f(x)$ becomes increasingly negative as $x \to -\infty$, below some value of x, and that it becomes increasingly positive as $x \to +\infty$ above some point. The stopping criterion was the relative change measure for $f(x^*)$, with $\varepsilon = 10^{-5}$. This was achieved at step 7. At step 6, $|\Delta f(x^*)| \div (0.5) [|f(x^*_{k-1})| + |f(x^*_k)|] = 0.000032 (= 3.2 \times 10^{-5})$ and at step 7, the measure is $0.0000068 (= 0.68 \times 10^{-5})$.

Note that if the initial search interval had been, say, $(-6, 6)$ and if the spacing between evaluation points had been 2, the function values would have been $-176, -46, -4, -2, 8, 74, 244$, suggesting that $f(x)$ was a monotonically increasing function over that range of values for x; the twist in shape near the origin (giving the local maximum and minimum) would not be revealed. Effective use of search methods on "difficult" functions requires that you employ a number of different initial intervals in conjunction with a number of different spacings between evaluation points.

6.1.2 Sequential Methods

In the second major group of search techniques the choice of the evaluation points is made sequentially; each new choice of a value for x is based on information gathered from the preceding choices. These methods are effective for functions that are known to be unimodal, or for multimodal functions when the initial search interval is known to contain (or "bracket") one extreme point. The unimodality characteristic is exploited to eliminate large intervals of values of x over which it is clear that the maximum or minimum point cannot lie. For that reason, we first investigate the notion of bracketing a maximum or a minimum and then explore effective sequential search methods within such a bracketing interval.

TABLE 6.2   Equal-Interval Search for Minimum of $f(x) =$ $(x - 1)(x + 1)^2 - 1$

x_i	$f(x_i)$

Step 1. Search Interval: (–5, 4); Spacing between Evaluation Points: 1.0

x_i	$f(x_i)$
–5	–97
–4	–46
–3	–17
–2	–4
–1	–1
0	–2
1	–1
2	8
3	3
4	74

Step 2. Search Interval: (–2, 3); Spacing between Evaluation Points: 0.5

x_i	$f(x_i)$
–2	–4
–1.5	–1.625
–1	–1
–0.5	–1.375
0	–2
0.5	–2.125
1	–1
1.5	2.125
2	8
2.5	17.375
3	31

Step 3. Search Interval: (0, 1); Spacing between Evaluation Points: 0.25

x_i	$f(x_i)$
0	–2
0.25	–2.172
0.50	–2.125
0.75	–1.766
1	–1

Step 4. Search Interval: (0, 0.7); Spacing between Evaluation Points: 0.1

x_i	$f(x_i)$
0	–2
0.1	–2.089
0.2	–2.152
0.3	–2.183
0.4	–2.176
0.5	–2.125
0.6	–2.024
0.7	–1.867

TABLE 6.2 *continued*

x_i	$f(x_i)$
Step 5. Search Interval: (0.20, 0.40); Spacing between Evaluation Points: 0.02	
0.20	−2.1520
0.22	−2.1610
0.24	−2.1686
0.26	−2.1748
0.28	−2.1796
0.30	−2.1830
0.32	−2.1848
0.34	−2.1851
0.36	−2.1837
0.38	−2.1807
0.40	−2.1760
Step 6. Search Interval: (0.320, 0.360); Spacing between Evaluation Points: 0.005	
0.320	−2.18483
0.325	−2.18504
0.330	−2.18516
0.335	−2.18517
0.340	−2.18509
0.345	−2.18491
0.350	−2.18462
0.355	−2.18423
0.360	−2.18374
Step 7. Search Interval: (0.330, 0.340); Spacing between Evaluation Points: 0.001	
0.330	−2.185163
0.331	−2.185174
0.332	−2.185182
0.333	−2.185185
0.334	−2.185184
0.335	−2.185180
0.336	−2.185171
0.337	−2.185158
0.338	−2.185142
0.339	−2.185121
0.340	−2.185096

Bracketing a Maximum or a Minimum   In Chapter 5 we recognized that a *root* of $f(x) = 0$ was *bracketed* in an interval (a, b) when $f(a)f(b) < 0$. Bracketing a maximum or a minimum is a little more complicated; in particular, three values of x are required, along with the values of the function at each of those three points. Then a minimum, for example, is bracketed in the *interval* (x_0, x_2) when, given the *triplet* of points $x_0, x_1,$ and x_2 $(x_0 < x_1 < x_2), f(x_1) < f(x_0)$ and $f(x_1) < f(x_2)$. This is expressed compactly by writing $f(x_1) < \min[f(x_0), f(x_2)]$.

The initial bracketing interval may, as usual, reflect some specific knowledge or just a guess about the shape of $f(x)$, or it may be an entirely random choice. For example, for a minimum point:

1. Choose an x_0 and find $f(x_0)$.
2. Choose an $x_1 > x_0$ and find $f(x_1)$.
3. (a) If $f(x_1) < f(x_0)$, choose an $x_2 > x_1$.
    (i) If $f(x_2) > f(x_1)$, as in Figure 6.3a, the minimum is said to be bracketed by (x_0, x_1, x_2).
    (ii) If $f(x_2) \leq f(x_1)$, as in Figure 6.3b, forget x_2 and choose instead a new point, $x_{2'}$, that is larger than x_2. If $f(x_{2'}) > f(x_1)$ [again, as in Figure 6.3b], then $(x_0, x_{2'})$ is the initial bracketing interval and $(x_0, x_1, x_{2'})$ is the initial bracket-

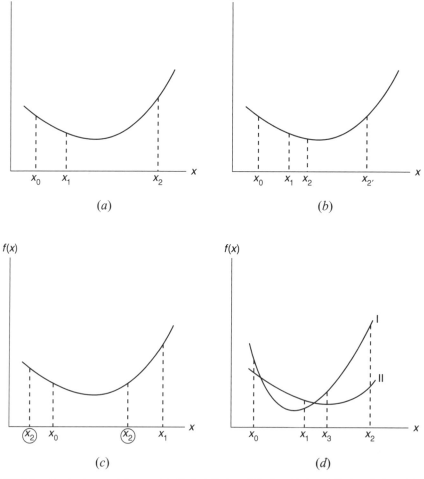

FIGURE 6.3   Bracketing a minimum. (a) $f(x_1) < f(x_0)$ and $f(x_2) > f(x_1)$. (b) $f(x_1) < f(x_0)$ and $f(x_2) \leq f(x_1)$. (c) $f(x_1) > f(x_0)$. (d) Information gained from x_3.

ing triplet. If $f(x_{2'}) \leq f(x_1)$, keep increasing x until the function finally turns up; that is, until an $x_{2''}$ is found for which $f(x_{2''}) > f(x_1)$. Then $(x_0, x_1, x_{2''})$ is the initial bracketing triplet and $(x_0, x_{2''})$ is the initial bracketing interval. Relabel $x_{2''}$ as x_2.

(b) If, instead, $f(x_1) > f(x_0)$ (Fig. 6.3c), either reverse the direction of search, so that the next x chosen is smaller than x_0, or else choose an x_2 between x_0 and x_1 and find $f(x_2)$, and so forth. (Two alternative possible locations for x_2 are circled in Fig. 6.3c.)

Once an initial interval (x_0, x_2) and an associated bracketing triplet (x_0, x_1, x_2) have been found, the problem is how to close in on the minimum point. The general idea is to choose a new point, x_3, so that it either (1) replaces one of the initial interval endpoints, leaving triplets (x_0, x_1, x_3) or (x_3, x_1, x_2) or (2) serves as the interior point of a new and smaller triplet, which would be either (x_0, x_3, x_1) or (x_1, x_3, x_2). We look first at the kind of information that is gained from a new point, x_3, and then investigate intelligent procedures for selecting the location of that point.

Suppose that an x_3 has been chosen and that the geometry is as shown in Figure 6.3d. Compare $f(x_3)$ and $f(x_1)$, the value of the function at the last interior point. For the function labeled I, $f(x_3) > f(x_1)$. In that case, the new bracketing interval is (x_0, x_3) and the associated bracketing triplet is (x_0, x_1, x_3). Alternatively, for a function with the general shape of II, $f(x_3) < f(x_1)$, and so the new bracketing interval is (x_1, x_2) with the associated triplet (x_1, x_3, x_2).[3] In either case, we next look for a new point, x_4, that will be chosen within the new interval, and then $f(x_4)$ is compared with the value of $f(x)$ for the current interior point of the triplet (x_1 for function I and x_3 for function II). And the procedure continues; that is, given the next new bracketing interval, an x_5 is chosen and is compared with the value of the function at the current interior point, and so forth.

Once a bracketing interval has been found, sequential search can proceed using information from either the *bracketing interval* alone (the two endpoints) or from the *bracketing triplet* (the two endpoints and also the interior point). We examine these options in turn.

Dichotomous Search Techniques that use the two endpoints only of the interval are illustrated by *dichotomous search*. The sequence proceeds as follows:

1. Define an initial interval (x_0, x_n) over which to search for an extreme point. In our illustrations, the point is a minimum, and the following description of the steps is written with that in mind. (You can easily modify the text to deal with maximum problems.) This interval choice may reflect some knowledge about the shape of the function or it may be chosen completely randomly. For simplicity of notation, suppose that $x_0 = 0$ and that $x_0 < x_n$, as in Figure 6.4. Thus the initial "interval of uncertainty" (about the location of the appropriate x) is $(0, x_n)$.

2. Find the *midpoint* of this interval of uncertainty, label it x_1, and evaluate $f(x_1 + \delta/2)$ and $f(x_1 - \delta/2)$, for some small $\delta > 0$. If, as in Figure 6.4, $f(x_1 - \delta/2) < f(x_1 + \delta/2)$, eliminate the interval from $(x_1 + \delta/2)$ to x_n from further consideration. This leaves $[0, (x_1 + \delta/2)]$

---

[3]It could just happen that $f(x_3) = f(x_1)$. If that is the case, note that the minimum will lie between x_1 and x_3 and hence both x_0 and x_2 could be replaced by the new interval endpoints.

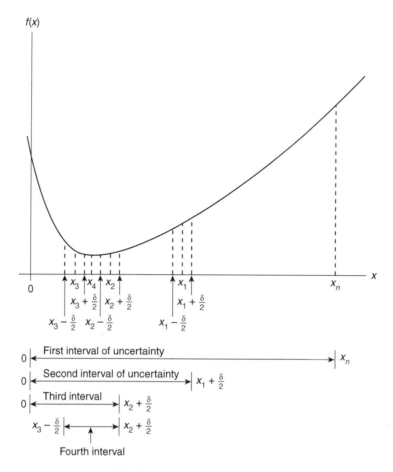

FIGURE 6.4 Dichotomous search.

as the new interval of uncertainty. Note that this is just slightly [in fact, $(\delta/2)$] longer than one-half of the original interval of uncertainty. This procedure is tantamount to estimating the derivative at x_1 using a central-difference approximation (Appendix 5.1).[4]

3. Next, find x_2, the midpoint of the new interval $[0, (x_1 + \delta/2)]$. Evaluate $f(x_2 + \delta/2)$ and $f(x_2 - \delta/2)$. If, again as in the figure, $f(x_2 - \delta/2) < f(x_2 + \delta/2)$, remove $[(x_2 + \delta/2), (x_1 + \delta/2)]$ from consideration as a possible location for x. The remaining interval of uncertainty is again just $\delta/2$ more than one-half the length of the interval found in the previous step. It is now $[0, (x_2 + \delta/2)]$.

4. The midpoint of the new interval is at x_3. Since, for this example, $f(x_3 - \delta/2) > f(x_3 + \delta/2)$, the interval $[0, (x_3 - \delta/2)]$ is next removed from consideration, leaving $[(x_3 - \delta/2), (x_2 + \delta/2)]$.

---

[4]Other finite-difference approximations to the derivative could also be used at x_1. If $f(x_1)$ replaces $f(x_1 - \delta/2)$ [so that $f(x_1)$ and $f(x_1 + \delta/2)$ are compared], we would be using a forward-difference approximation, and if $f(x_1)$ replaces $f(x_1 + \delta/2)$, it would be a backward-difference approximation. In any case, the function must be evaluated at *two* points at and/or near x_1.

5. Continue in this fashion. For this particular illustration, note that x_4, the midpoint of the remaining interval of uncertainty, is quite close to the value of x at which $f(x)$ has its minimum.

We continue to use the same two functions for illustrative purposes. In Table 6.3, dichotomous search has been used on $f(x) = x^2 - 3$, starting with an initial search interval $(-5, 4)$ and using a $\delta = 0.002$. In the case of a unimodal function such as this, if the initial interval had not bracketed the minimum point, dichotomous search would lead you single-mindedly toward the lower end of the initial search interval. When that happens, select a new beginning search interval that extends beyond the current one, in the direction that the search has been heading.

As with simultaneous search procedures, two of the most common stopping rules for sequential iterative search methods are based on differences in consecutive values of x^* or $f(x^*)$. These were denoted as $|\Delta x^*| < \varepsilon$ and $|\Delta f(x^*)| < \varepsilon$, respectively. In addition, in the case of dichotomous search, a stopping rule can also be based on the length of the interval of uncertainty, within which the extreme point is known to lie. For example, in Table 6.3, if the criterion had been to stop when the interval of uncertainty was less than 0.025, this would have allowed the search to stop at step 9, where x^* is known to lie within $(-0.00927, 0.010393)$. A reasonable estimate of x^* at that point would be the midpoint of that interval, namely, $x^* = 0.000562$ which, for many purposes, could be read as $x^* = 0$ (the correct minimum point).

The interval of uncertainty at step k is guaranteed to be exactly $(\delta/2)$ more than one-half of the interval in the $(k-1)$st step. (You can easily check that this is the case, except possibly for rounding, for the illustrations in both Tables 6.3 and 6.4.) Therefore, it is possible to know from the outset (given the choice of an initial search interval, I^0, and a value of δ) how many iterations will be needed to reduce the interval of uncertainty to less than some specified $\varepsilon > 0$. The relationship between the size of the interval of uncertainty at the kth step, and I^k, and I^0 and δ is

$$I^k = (0.5)^k (I^0) + \left\{ \left[\sum_{i=0}^{k-1} (0.5)^i \right] \left(\frac{\delta}{2} \right) \right\}$$

(Problem 6.1 asks you to derive this result.)

TABLE 6.3 Dichotomous Search for Minimum of $f(x) = x^2 - 3$ ($\delta = 0.002$)

Step	Search Interval	Length of Interval of Uncertainty	Midpoint (x_1)	$f[x_1 + (\delta/2)]$	$f[x_1 - (\delta/2)]$
0	(−5, 4)	9	−0.5	−2.75099	−2.74899
1	(−0.501, 4)	4.501	1.7495	0.06425	0.05725
2	(−0.501, 1.7505)	2.2515	0.62475	−2.60843	−2.61093
3	(−0.501, 0.62575)	1.1268	0.06238	−2.99598	−2.99623
4	(−0.501, 0.06338)	0.5644	−0.21881	−2.95255	−2.95168
5	(−0.21981, 0.06338)	0.2832	−0.07822	−2.99403	−2.99372
6	(−0.07992, 0.06338)	0.1433	−0.00827	−2.99994	−2.99991
7	(−0.00927, 0.06338)	0.07265	0.027055	−2.99921	−2.99932
8	(−0.00927, 0.028055)	0.037325	0.009393	−2.99989	−2.99992
9	(−0.00927, 0.010393)	0.019663	0.000562	−2.99999	−2.99999

Notice that another potentially logical stopping criterion for dichotomous search might appear to be one based on differences in the values of $f[x^* + (\delta/2)]$ and $f[x^* - (\delta/2)]$ at a given step; the smaller $|f[x^* + (\delta/2)] - f[x^* - (\delta/2)]|$, the more likely that a maximum or minimum has been isolated at x^*. For the illustration in Table 6.3, if the criterion had been that the absolute value of this difference be less than 0.00005, this would have allowed the search to stop at step 6. However, the example in Table 6.3 makes clear that this measure does not decrease monotonically with an increasing number of steps. For example, at step 3 it is 2.5×10^{-4}; at step 4, 8.7×10^{-4}, at step 6, 0.3×10^{-4} and at step 7, 1.1×10^{-4}.

In Table 6.4, dichotomous search is used on $f(x) = (x - 1)(x + 1)^2 - 1$. The initial search interval $(-5, 4)$ indeed brackets the minimum at $x = 0.333$; it also includes the maximum at $x = -1$. This turns out, in this example, not to be a problem because the initial dichotomization generates a new left-hand endpoint at -0.501, and the function is indeed unimodal within the $(-0.501, 4)$ interval. Here the search was stopped when the interval of uncertainty became less than 0.005, leading to an estimate of x^* of 0.334 (rounded to three decimals).

Golden Section Search   A different sequential procedure uses all three of the points in each of the current bracketing triplets. It addresses the issue of how "best" to choose each next value, x_{k+1}, given the current triplet that contains the three most recently chosen points, x_k, x_{k-1} and x_{k-2}. Notice that we cannot know in advance which two of these three values of x constitute the current bracketing interval endpoints and which one is the interior point. It depends on the shape of the function, as is illustrated by the two examples in Figure 6.3d. However, to simplify the exposition, we can imagine that the three points have been relabeled so that x_k represents the *interior point* of an interval (x_{k-2}, x_{k-1}). This means that x_k is the current best estimate of the value of x at which the function is minimized, and $f(x_k)$ is therefore the associated current best estimate of the minimum value of the function.

One straightforward rule is to locate the next point, x_{k+1}, in the middle of the *next bracketing interval*. This is known, reasonably, as the "bisection" rule. However, there are

TABLE 6.4   Dichotomous Search for Minimum of $f(x) = (x - 1)(x + 1)^2 - 1$ ($\delta = 0.002$)

Step	Search Interval	Length of Interval of Uncertainty	Midpoint (x_1)	$f[x_1 + (\delta/2)]$	$f[x_1 - (\delta/2)]$
0	$(-5, 4)$	9	-0.5	-1.37625	-1.37375
1	$(-0.501, 4)$	4.501	1.7495	4.67772	4.65435
2	$(-0.501, 1.7505)$	2.2515	0.62475	-1.98916	-1.99200
3	$(-0.501, 0.62575)$	1.1268	0.06238	-2.05910	-2.05738
4	$(0.06138, 0.62575)$	0.5644	0.34357	-2.18493	-2.18501
5	$(0.06138, 0.34457)$	0.2832	0.20298	-2.15388	-2.15294
6	$(0.20198, 0.34457)$	0.1426	0.27328	-2.17841	-2.17795
7	$(0.27228, 0.34457)$	0.0723	0.30843	-2.18405	-2.18386
8	$(0.30743, 0.34457)$	0.0371	0.32600	-2.18510	-2.18504
9	$(0.325, 0.34457)$	0.0196	0.34479	-2.18517	-2.18518
10	$(0.325, 0.33579)$	0.0108	0.33040	-2.18517	-2.18515
11	$(0.3294, 0.33579)$	0.0064	0.33260	-2.18518	-2.18517
12	$(0.3316, 0.33579)$	0.0042	0.33370	-2.18518	-2.18518

f(x)

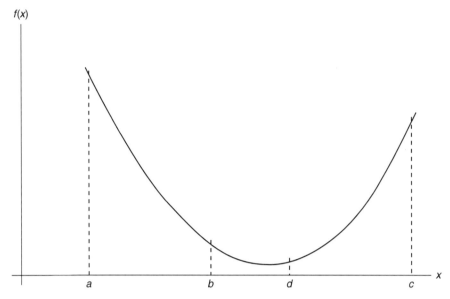

FIGURE 6.5   Initial points for golden section search.

two subintervals in the current bracketing interval, (x_{k-2}, x_k) and (x_k, x_{k-1}), so in which subinterval should x_{k+1} lie? One of the most popular and successful of the search procedures that do not rely on any derivative information at all is known as *golden section search*, for reasons that we will soon see. Consider the illustration in Figure 6.5. To simplify notation, let $x_{k-2} = a$, $x_k = b$ and $x_{k-1} = c$. For the moment, assume that the bracketing triplet (a, b, c), where $a < b < c$, is given.

Suppose that the next point, $x_{k+1} = d$, is to the right of (larger than) b. This means that the next bracketing triplet will be either (a, b, d) or (b, d, c), depending on whether $f(b) <$ min $[f(a), f(d)]$ or $f(d) <$ min $[f(b), f(c)]$. This, in turn, depends on where the minimum of $f(x)$ really lies, which is what we are trying to establish.

If the next triplet is (a, b, d), the length of the associated bracketing interval will be $(d - a)$; if it is (b, d, c), the length of the next bracketing interval will be $(c - b)$.[5] A reasonable goal would be for the next bracketing interval to be the shorter of these two; this would give us the maximum reduction in the interval of uncertainty. This suggests that one possible strategy (which might be called "ultra conservative") would be to choose $x_{k+1} = d$ so that the lengths of the two possible new bracketing intervals are the same.[6] The algebraic import of this is very simple: choose $x_{k+1} = d$ so that $(d - a) = (c - b)$, which means $d = a + c - b$. That tells us how to select d, once we have a, b and c.

Suppose, however, that the "previous" step had been the first. Our first bracketing interval was $x_0 = a$ and $x_1 = c$. Precisely how would the location of $x_2 = b$ have been determined? Since interval lengths change from step to step, we cannot use an absolute mea-

---

[5]In Figure 6.5, the next bracketing triplet turns out to be (b, d, c), since $f(d) <$ min $[f(b), f(c)]$. You can easily alter the following argument for the case in which $f(b) <$ min $[f(a), f(d)]$, which would be true if d were moved sufficiently further out toward c.

[6]What is the logic of this? It reduces (to zero) the risk of ending up with a selection for x_{k+1} that is in the *wrong* (longer) interval, because both those intervals will now be the same length.

sure of, say, the distance of the interior point from the left-hand end of its interval as the appropriate requirement from step to step. But we *can* use proportions. We want b to have been located in such a way that its *relative* location in *its* interval (a, c) is the same as the *relative* location of the next point, d, in *its* interval (b, c). These relative locations are $(b - a)/(c - a)$ and $(d - b)/(c - b)$, for b and d, respectively. Stated otherwise, these expressions indicate the *proportion* of the interval length that is to the left of b or d, respectively.

So our criterion for location of b is that

$$\frac{b - a}{c - a} = \frac{d - b}{c - b}$$

This, in conjunction with the "equal interval length" requirement, $d = a + c - b$, gives us two equations in two unknowns, b and d. We can assume that the initial interval (a, c) has been normalized to unit length; this is simply done by letting $a = 0$ and $c = 1$. Then substitution leads to $b^2 - 3b + 1 = 0$, so that $b = 2.618, 0.382$. Since $b < 1$ (it is the fraction of the unit distance from 0 to 1 at which b is located), it follows that $b = 0.382$.

Then, in the next step, $d = a + c - b = 0.618$. In d's bracketing interval (b, c), this means that d's relative location (to the right of the left-hand endpoint of its bracketing interval) is $(0.618 - 0.382)/(0.618) = 0.382$ (!). This will continue to be the case; each new point is 0.382 of the way into the larger subinterval, measured from the interior point in the current triplet. The idea is illustrated in Figure 6.6, which extends the evaluation points from Figure 6.5 and indicates their precise locations.

Tables 6.5 and 6.6 illustrate golden section search on the two illustrative functions. The initial bracketing triplet is such that both $f(b) < \min[f(a), f(c)]$ and $(b - a)/(c - a) = 0.382$. You should work through at least a few initial steps to get a feel for the process. Note that, as expected, the length of each new interval of uncertainty is $1 - 0.382 = 0.618$ times the length of the preceding interval (except for rounding). From Table 6.5 we see that at step 11 the minimum value of x has been trapped in the interval $(-0.029, 0.016)$, with a length of 0.045, and that the best estimate of the minimizing value of x is -0.001, at which $f(x) = -2.99999$. [Starting with step 9 the function values were expressed to five decimals, to indicate more precisely how the search closes in on the minimum of $f(x)$, which is -3.]

As with dichotomous search, it is possible to know in advance, given an initial interval of uncertainty, how many steps will be needed before the interval of uncertainty is less than a specified ε. Let I^k denote the length of the interval of uncertainty at step k, and I^0 be the length of the initial interval. Then it is easily shown that $I^k = (0.618)^k I^0$. To find the step for which $I^k = \varepsilon$ (and hence after which the interval of uncertainty will be less than ε), solve $(0.618)^k I^0 = \varepsilon$ for k, giving $k = \log_{0.618}(\varepsilon/I^0)$. For example, from Table 6.5, where $I^0 = 9$, let $\varepsilon = 0.05$, so that $k = \log_{0.618}(0.005556) = 10.79$. This suggests, and the results in Table 6.5 confirm, that the interval of uncertainty will be more than 0.05 at step 10 and less than 0.05 at step 11.

The (arbitrary) stopping criterion used for Table 6.5 was that the interval of uncertainty be less than 0.05. Alternative possibilities would include the following:

1. A criterion such as was used in Tables 6.1 and 6.2, based on the absolute value of the difference between $f(x)$ for the current best and next to best points. Here, at step 11, this difference is $|f(-0.001) - f(-0.012)| = 0.00014$.

2. A criterion based on the absolute value of the difference between two successive new values of x. Between steps 10 and 11 in Table 6.5, this value is $|(-0.029) - (-0.001)| = 0.028$.

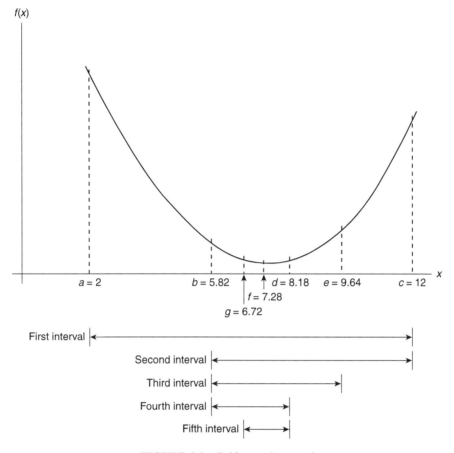

FIGURE 6.6   Golden section search.

3. A criterion based on the absolute value of the difference between two successive new values of $f(x)$. Again, between steps 10 and 11 in Table 6.5, this is $|(-2.99915) - (-2.99999)| = 0.00084$.

Suppose that the criterion had been that the interval of uncertainty be less than 0.10. This is satisfied at step 10 in Table 6.5. Note that here $f(b) = f(-0.012)$, the value of the function at the *interior* point of the current bracketing triplet, is better (less) than $f(d) = f(-0.029)$, the value of the function associated with the new point. That is because -0.029 becomes one of the *endpoints* in the next triplet, if you proceed to step 11. Thus the current $x = b$ would be the best estimate of x under that criterion. On the other hand, if the stopping criterion had been that the interval of uncertainty be less than 0.025, then step 8 would have been final. There $f(b) = f(-0.057) = -2.997$, whereas $f(d) = f(-0.012) = -2.99985$, because -0.012 would become the *interior* point in the next bracketing triplet. Hence, under this value of the stopping criterion on the length of the interval of uncertainty, the current $x = d$ is the best estimate of the minimizing value of x.

Table 6.6 contains similar results for $f(x) = (x - 1)(x + 1)^2 - 1$. In this case, for variety, iterations were stopped when the interval of uncertainty was less than 0.025, and initially

TABLE 6.5 Golden Section Search for Minimum of $f(x) = x^2 - 3$

Step	Bracketing Triplet $(a,b,c)'$	Function Values $[f(a),f(b),f(c)]'$	Length of Interval of Uncertainty	New x $(= d)$ {0.382 into max$[(a,b),(b,c)]$}	$f(d)$
0	−5	22	9	0.563	−2.683
	−1.562	−0.560			
	4	13			
1	−1.562	−0.560	5.562	1.880	0.053
	0.563	−2.683			
	4	13			
2	−1.562	−0.560	3.442	−0.249	−2.938
	0.563	−2.683			
	1.880	0.053			
3	−1.562	−0.560	2.125	−0.751	−2.436
	−0.249	−2.938			
	0.563	−2.683			
4	−0.751	−2.436	1.314	0.061	−2.996
	−0.249	−2.938			
	0.563	−2.683			
5	−0.249	−2.938	0.812	0.253	−2.936
	0.061	−2.996			
	0.563	−2.683			
6	−0.249	−2.938	0.502	−0.057	−2.997
	0.061	−2.996			
	0.253	−2.936			
7	−0.249	−2.938	0.310	−0.130	−2.983
	−0.057	−2.997			
	0.061	−2.996			
8	−0.130	−2.983	0.191	−0.012	−2.99985
	−0.057	−2.997			
	0.061	−2.996			
9	−0.057	−2.99675	0.118	0.016	−2.99974
	−0.012	−2.99985			
	0.061	−2.99627			
10	−0.057	−2.99675	0.073	−0.029	−2.99915
	−0.012	−2.99985			
	0.016	−2.99974			
11	−0.029	−2.99915	0.045	−0.001	−2.99999
	−0.012	−2.99985			
	0.016	−2.99974			

$f(x)$ was expressed to four decimal places. Here the best estimate of the minimizing x under that criterion is $x = 0.334$. As before, increased accuracy in the final estimate would result from a stricter stopping criterion and carrying more decimal places in the results.

A Comment on Terminology   The term *golden section* comes from the ancient Greeks who had a tremendous respect for aesthetically pleasing proportion, especially in architecture. For example, certain proportions that were found in the human body were also thought appropriate in the architecture of temples and public buildings. The golden sec-

TABLE 6.6   Golden Section Search for Minimum of $f(x) = (x - 1)(x + 1)^2 - 1$

Step	Bracketing Triplet $(a,b,c)'$	Function Values $[f(a),f(b),f(c)]'$	Length of Interval of Uncertainty	New $x (= d)$ {0.382 into max[$(a,b),(b,c)$]}	$f(d)$
0	−1	−1	3	0.854	−1.5018
	0.146	−2.1216			
	2	8			
1	−1	−2	1.854	−0.292	−1.6476
	0.146	−2.1216			
	0.854	−1.5018			
2	−0.292	−1.6476	1.146	0.416	−2.1710
	0.146	−2.1216			
	0.854	−1.5018			
3	0.146	−2.1216	0.708	0.583	−2.0450
	0.416	−2.1710			
	0.854	−1.5018			
4	0.146	−2.1216	0.437	0.313	−2.1844
	0.416	−2.1710			
	0.583	−2.0450			
5	0.146	−2.1216	0.270	0.249	−2.1716
	0.313	−2.1844			
	0.416	−2.1710			
6	0.249	−2.1716	0.167	0.352	−2.1845
	0.313	−2.1844			
	0.416	−2.1710			
7	0.313	−2.1844	0.103	0.376	−2.18147
	0.352	−2.1845			
	0.416	−2.1710			
8	0.313	−2.18437	0.063	0.337	−2.18516
	0.352	−2.18448			
	0.376	−2.18147			
9	0.313	−2.18437	0.039	0.328	−2.18513
	0.337	−2.18516			
	0.352	−2.18448			
10	0.328	−2.18513	0.024	0.334	−2.18518
	0.337	−2.18516			
	0.352	−2.18448			

tion (or *golden ratio*) was defined as follows: suppose that a line is divided into two un-equal segments with lengths a and b ($a < b$). If the division is made so that $a/b = b/(a + b)$—in words, so that the ratio of the smaller to the larger line segment is the same as the ratio of the larger segment to the entire line—then that ratio is known as the golden section. Specifically, suppose that the line has unit length; $a + b = 1$. Then the golden section division would be $a/b = b/1$, which means $b^2 = a$. But since $a = 1 - b$, $b^2 + b - 1 = 0$. Since (from the geometry of the problem) $b > 0$, the only solution to this quadratic equation is $b = 0.618$, and hence $a = 0.382$. (As we saw above, 0.382 was our measure of location of the next point into the new bracketing interval and 0.618 represented the decrease in the length of the interval of uncertainty at each step.) The Parthenon in Athens, for example, was built so that its height:width ratio—and many other pairs of its dimensions—is 0.618. Additional more elaborate and fascinating geometric constructions have also been found for the golden ratio.

There are interesting algebraic properties too. Consider the set of ratios $a/b = b/c = c/d = d/e = e/f$, where each fraction is a golden ratio; that is, where $x_i/x_j = x_j/(x_i + x_j)$ for each pair of adjacent ratios. A *golden section series* is one made up of the consecutive elements $a,b,c,d,e,f,\ldots$ from this set of golden ratios; that is, $a/b = 0.618$, and each term, beginning with c, is the sum of the previous two. For example, if $a = 1$, then $b = 1.618$, $c = 2.618$, $d = 4.236$, $e = 6.854$, $f = 11.090$, and so forth. Alternatively, if, say, $a = 0.708$, then $b = 1.146$, $c = 1.854$, $d = 3$, $e = 4.854$, and $f = 7.854$, and so on.

A related series, involving only *integers*, was investigated by Fibonacci (Leonardo of Pisa, Italian, 1180–1225). It begins

$$1, 1, 2, 3, 5, 8, 13, 21, 34, 55, 89, \ldots$$

where, after the initial pair of 1's, each successive term is the sum of the preceding two, as in the golden section series. In this case, the ratio between two successive terms approaches 0.618 as the series progresses:

$$\tfrac{1}{1} = 1, \tfrac{1}{2} = 0.5, \tfrac{2}{3} = 0.667, \tfrac{3}{5} = 0.6, \tfrac{5}{8} = 0.625,$$

$$\tfrac{8}{13} = 0.6154, \tfrac{13}{21} = 0.6190, \tfrac{21}{34} = 0.6176, \tfrac{34}{55} = 0.6182, \tfrac{55}{89} = 0.6180, \ldots$$

Because of this similarity to the elements that form the terms in the golden section series, iterative methods employing what is known as *Fibonacci search* have also been developed.

6.1.3   Parabolic Interpolation

Most iterative minimization methods for $f(x)$, along a line, use search methods up to a point and then start the process of approximating the function near to its minimum by a series of increasingly accurate curves—parabolas, or cubics, or perhaps fourth-order functions. The notion underpinning the simplest of these, parabolic approximation, is simple. It is always possible to fit a parabola through three (not collinear) points in a plane.

Assume that through golden section or similar search you have found three current values, x_{k-2}, x_{k-1}, x_k, that bracket a minimum and are themselves relatively close together.[7] Now fit a parabola through the points $f(x_k), f(x_{k-1})$, and $f(x_{k-2})$; this means, find the coefficients c_1, c_2, and c_3 that will identify the specific quadratic function

$$q(x) = c_1 x^2 + c_2 x + c_3$$

for which

$$c_1(x_k)^2 + c_2(x_k) + c_3 = f(x_k)$$

$$c_1(x_{k-1})^2 + c_2(x_{k-1}) + c_3 = f(x_{k-1})$$

$$c_1(x_{k-2})^2 + c_2(x_{k-2}) + c_3 = f(x_{k-2})$$

---

[7]Determination of when such points are "relatively close together" is itself an open question. One approach would be to switch to parabolic interpolation after the initial interval of uncertainty is reduced to some specified fraction of its original value.

Given the current known values of x_k, x_{k-1}, and x_{k-2} and the associated functional values $f(x_k), f(x_{k-1})$, and $f(x_{k-2})$, this is (generally) a simple problem of solving three *linear* equations in three unknowns, c_1, c_2, and c_3. Let the solution values be c_1^*, c_2^*, and c_3^*.[8]

Given the specific parabola $q(x) = c_1^* x^2 + c_2^* x + c_3^*$ for which $q'(x) = 2c_1^* x + c_2^*$, its extreme point is where $q'(x) = 0$, at $x^* = -c_2^*/2c_1^*$. This represents a minimum if $q''(x^*) > 0$, which means if $c_1^* > 0$. Then use x^* as the next estimate of the point where the minimum of $f(x)$ occurs, and proceed iteratively. The logic is applied to several triplets of points for the second of the two test functions; results are shown in Table 6.7. (Since the first function is itself a parabola, a parabolic estimate will always represent the function exactly. Try it.)

To illustrate the procedure, we make explicit the calculations required to fit the parabola at step 3 in Table 6.7. Here $x_k = 0.1912$, $x_{k-1} = 0.0909$ and $x_{k-2} = 2$; $f(x_k) = -2.1477$, $f(x_{k-1}) = -2.0819$, and $f(x_{k-2}) = 8$. The linear equation system is therefore (to four-decimal accuracy)

$$0.0366c_1 + 0.1912c_2 + c_3 = -2.1477$$

$$0.0083c_1 + 0.0909c_2 + c_3 = -2.0819$$

$$4c_1 + 2c_2 + c_3 = 8$$

The solution (straightforward matrix algebra) is $c_1^* = 3.2824$, $c_2^* = -1.5822$, and $c_3^* = -1.9652$, giving the parabola, $q(x)$, shown in Table 6.7 for step 3. Thus $q'(x) = 6.5648x - 1.5822$, and setting this equal to zero produces the result that $x^* = 0.2410$, as shown in the table.

The stopping rule that was used in Table 6.7 was that the change in $f(x^*)$ between an adjacent pair of steps be less than 10^{-4}. In this example, the right-hand end of the search interval remained at 2 and the interval of uncertainty remained larger than 1 throughout the iterations. Thus, a stopping criterion based on the size of the interval of uncertainty would not have been very satisfactory in this case. Again, alternative stopping decisions could have been based on the size of the change (improvement) in x^* from one step to the next.

6.1.4 Combined Techniques

Frequently the best iterative approach to a maximization or a minimization problem will begin with a fairly straightforward search, such as golden section, and then switch to a curve fitting method, such as parabolic interpolation, at some point when the search interval is "narrow," so that an assumption that the function has a particular shape (e.g., parabolic) in an interval near to the maximum or minimum is not unreasonable. The problem is then one of deciding when to switch, and there are several sophisticated approaches to this question. [You can find the details in more complete numerical methods texts, such as Press et al. (1986).]

---

[8]Recall that Brent's method (Chapter 5) uses a fitted parabola to approximate $f(x)$ near to a root. In that case, however, the function is an inverse parabola in the sense that x is expressed as a quadratic function of y, so that the parabola opens leftward or rightward, instead of upward or downward, as in the text here, when y is a quadratic function of x.

TABLE 6.7   Parabolic Interpolation for $(x-1)(x+1)^2 - 1$

Step	Bracketing Triplet $(a,b,c)'$	Function Values $[f(a),f(b),f(c)]'$	Fitted Parabola	Location of Minimum of Fitted Parabola $(x = d)$	$f(d)$
0	−1	−1	$2x^2+x-2$	−0.25	−1.7031
	0	−2			
	2	8			
1	−0.25	−1.7031	$2.7500x^2-0.5001x-2$	0.0909	−2.0819
	0	−2			
	2	8			
2	0	−2	$3.0909x^2-1.1819x-2$	0.1912	−2.1477
	0.0909	−2.0819			
	2	8			
3	0.0909	−2.0819	$3.2824x^2-1.5822x-1.9652$	0.2410	−2.1689
	0.1912	−2.1477			
	2	8			
4	0.1912	−2.1477	$3.4313x^2-1.9085x-1.9083$	0.2781	−2.1793
	0.2410	−2.1689			
	2	8			
5	0.2410	−2.1689	$3.5203x^2-2.1078x-1.8654$	0.2994	−2.1829
	0.2781	−2.1793			
	2	8			
6	0.2781	−2.1793	$3.5755x^2-2.2338x-1.8346$	0.3124	−2.18349
	0.2994	−2.1829			
	2	8			
7	0.2994	−2.1829	$3.6068x^2-2.3057x-1.8159$	0.3196	−2.18439
	0.3124	−2.1843			
	2	8			
8	0.3124	−2.18432	$3.6317x^2-2.3631x-1.8005$	0.3253	−2.18491
	0.3196	−2.18481			
	2	8			
9	0.3196	−2.18481	$3.6469x^2-2.3983x-1.7908$	0.3288	−2.18510
	0.3253	−2.18506			
	2	8			
10	0.32534	−2.185058	$3.6502x^2-2.4060x-1.7887$	0.3296	−2.18517
	0.32882	−2.185145			
	2	8			

6.1.5   Line Search with Derivatives

Suppose that $f(x)$ has first derivatives everywhere, at least in the neighborhood of a minimum, and that these are relatively easy and inexpensive to evaluate, so that they can be used in a search procedure. Naturally, if the function $f'(x)$ can be found, then the first-order condition $f'(x) = 0$ can be formulated; whether it is easily solved depends entirely on the complexity of the derivative function $f'(x)$ and hence on the complexity of $f(x)$ from which it is derived. The contribution of derivatives in line search methods is principally to indicate, by their *sign*, the *direction* in which to move toward the minimum point, from iteration k to iteration $k + 1$.

Consider the bracketing triplet a,b,c (where, as usual, $a < b < c$), and $f(b) < \min[f(a),$

$f(c)$]. If $f'(b) > 0$, then the next step toward a minimum should be leftward from b, into the interval (a, b). Conversely, when $f'(b) < 0$, the bracketed minimum is to the right of b, in the (b, c) interval.[9] Once this is established, something as simple as bisection of the relevant interval can be effective.

6.2 MINIMIZATION OF $f(\mathbf{X})$: NONDERIVATIVE METHODS

In this and the following section we consider several alternative kinds of approaches to multivariable minimization problems. For these iterative methods, it is useful to distinguish those that do not require derivatives at all from those that use them (actual or estimated), because the number of first and second partial (and cross-partial) derivatives—arranged in gradient vectors and Hessian matrices—grows rapidly with n, the number of variables. Nonderivative methods are examined in this section; derivative-based approaches are covered in Section 6.3.

For functions of two variables, *simultaneous* search methods are often impractical; for functions of three or more variables they become virtually impossible because of the difficulty of deciding on a sensible set of initial points [in n-dimensional space, for $f(x_1, \ldots, x_n)$] and, more importantly, of figuring out what you have learned after the functional evaluations at those points. On the other hand, *sequential* search methods may work well, depending on how cleverly you decide on the sequence of searches and interpret the results of those searches.

6.2.1 Test Functions

Throughout this and the next section, we illustrate each iterative technique on several test functions. As each method is presented, its performance on three functions will be explored. A standard approach (which will be followed here) is to report the number of iterations required by each of the techniques to satisfy some predetermined stopping criterion. This measure may fail to represent accurately the number of calculations, or their complexity, or the amount of information that is needed in each iteration. Often, the number of functional evaluations required is also used as a measure of relative effectiveness of alternative numerical methods. In addition, derivative-based methods require that values of derivatives (at least first, sometimes second, whether exact or approximated) be found at each step. Also, each iteration that requires finding the minimum value of a function along a particular line ("line searches") can employ approximation methods either for finding the roots of first-derivative equations (Chapter 5), which means that the first-derivative functions must be specified, or for minimizing the function directly (Section 6.1). As noted already in Chapter 5, computing time also can be used as a measure of relative efficiency, although it is (naturally) very dependent on the characteristics and configuration of the particular computer and the efficiency of the actual program(s) or software employed to implement the iterative methods.

The first three test functions are as follows:

(a) $f(\mathbf{X}) = x_1^2 + x_2^2 - 4x_1 - 5x_2 - 5$. This function is a bowl in three-dimensional space with circular contours in the x_1, x_2 plane. It is easily established from ordinary first- and

---

[9]This is precisely the same reasoning as is used in dichotomous search, where the derivative at each new point is approximated by one of the finite-difference procedures (Appendix 5.1).

second-order conditions ($\partial f / \partial x_1 = 0$, $\partial f / \partial x_2 = 0$, $\mathbf{H}$ positive definite) that a minimum of $f(\mathbf{X}^*) = -15.25$ occurs at $\mathbf{X}^* = \begin{bmatrix} 2 \\ 2.5 \end{bmatrix}$. Our objective (for all test functions) will be to approach $\mathbf{X}^*$ iteratively. This function is extremely simple; one reason for using it is so that comparisons can be made for each of the methods with a very slight (algebraic) variant, in (b), below.

(b) $f(\mathbf{X}) = x_1^2 + x_2^2 - 4x_1 - 5x_2 - x_1 x_2 - 5$. Including the $x_1 x_2$ interaction term changes the circular contours in the x_1, x_2 plane into (tilted) ellipses, and it affects some of the iterative procedures enormously, as we will see below. Again, application of calculus rules identifies a minimizing point at $\mathbf{X}^* = \begin{bmatrix} 4.33333 \\ 4.66667 \end{bmatrix}$, where $f(\mathbf{X}^*) = -25.33333$.

(c) $f(\mathbf{X}) = (x_1^2/4) + (x_2^2/9) - 0.8x_1 - x_2 - 0.3x_1 x_2 - 3$. Now the tilted elliptical contours are more pronounced. The function has a minimum value of -29.57896 at $\mathbf{X}^* = \begin{bmatrix} 22.63158 \\ 35.05263 \end{bmatrix}$. These functions are shown in Figures 6.7–6.9. After we have examined three nonderivative methods in this section we will summarize their respective performances on three additional test functions.

6.2.2 Simplex Methods

A very successful nonderivative approximation method for maximizing or minimizing a multivariable function is the *downhill simplex method* for minimization problems *(uphill simplex method* for maximization problems), with which the names Nelder and Mead are usually associated (Nelder and Mead, 1965). The "simplex" in the name refers to the geometric figure which forms the set of trial points at which the function is evaluated at each iteration; it also gives its name to the popular iterative computational method for linear programming problems (as we will see in Chapter 7). The geometric figure underpinning the approaches is the same, but the methods themselves are quite different. In the linear programming problem, the objective function is linear and the maximization or minimization is carried out subject to linear constraints (usually inequalities). The approach under consideration in this chapter is one that was designed for *unconstrained* problems with a *nonlinear* function to be maximized or minimized.

In *two*-dimensional space, a simplex is a triangle; officially it is the figure that results from forming the convex combination of (connecting with straight lines) *three* points, not all on the same *line*. In *three*-dimensional space, the convex combination of *four* points, not all on the same *plane* (three-dimensional linear subspace), forms a pyramid (tetrahedron). In general, a simplex is formed by the convex combination of $n + 1$ points (not all on the same n-dimensional hyperplane) in n-dimensional space. If the simplex is "regular," all the sides are the same length (forming, for $n = 2$, an *equilateral* triangle, and so on, in higher dimensions).

The downhill simplex method for minimization of a function of two variables uses a sequence of sets of three points each, where the points at each step are the corners (vertices) of a simplex. One variant always forms regular simplices (the plural of simplex) with sides of equal length; another allows for other than equilateral shapes. We concentrate only on the latter variant, which is the more general approach.

For $y = f(x_1, x_2)$ a sequence of values of the independent variables will be points on the x_1, x_2 plane. In general, an iterative method would evaluate $f(\mathbf{X})$ for a series of (judicious-

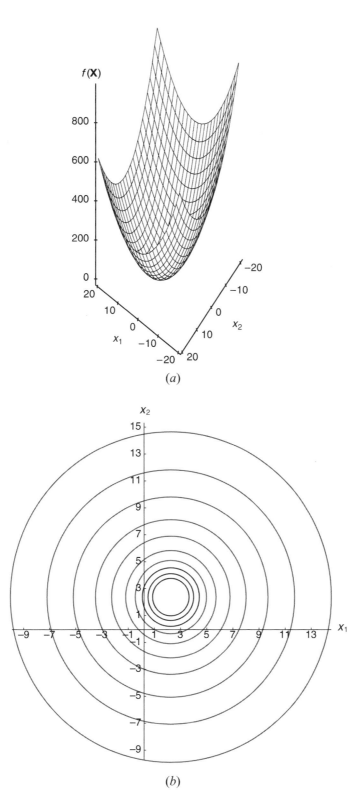

(a)

(b)

FIGURE 6.7 (a) Test function (a); $f(\mathbf{X}) = x_1^2 + x_2^2 - 4x_1 - 5x_2 - 5$; (b) contours for test function (a).

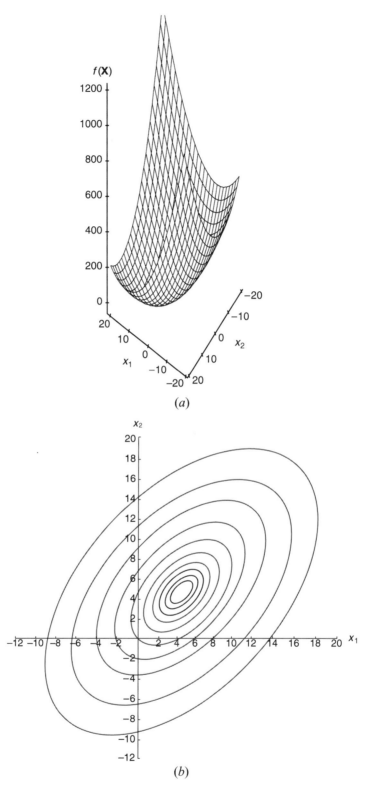

FIGURE 6.8 (*a*) Test function (b); $f(\mathbf{X}) = x_1^2 + x_2^2 - 4x_1 - 5x_2 - x_1x_2 - 5$; (*b*) contours for test function (b).

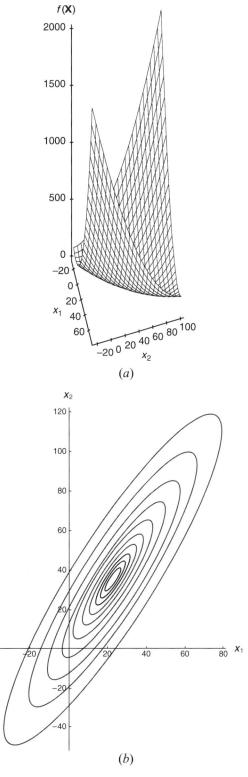

FIGURE 6.9 (*a*) Test function (c); $f(\mathbf{X}) = (x_1^2/4) + (x_2^2/9) - 0.8x_1 - x_2 - 0.3x_1x_2 - 3$; (*b*) contours for test function (c).

ly chosen) points, starting from an initial choice $\mathbf{X}^0 = [x_1^0, x_2^0]'$. By contrast, the downhill simplex method begins with not one but *three* initial points, $\mathbf{X}^0 = [x_1^0, x_2^0]'$, $\mathbf{X}^1 = [x_1^1, x_2^1]'$ and $\mathbf{X}^2 = [x_1^2, x_2^2]'$ on the x_1, x_2 plane, where these points are chosen so that they form a simplex (they cannot all be on the same line).

Steps in the Downhill Simplex Method The function is evaluated at each of the three initial points, giving $f(\mathbf{X}^0)$, $f(\mathbf{X}^1)$, and $f(\mathbf{X}^2)$. Choose the $\mathbf{X}^i (i = 0, 1, 2)$ for which $f(\mathbf{X}^i)$ is largest. For simplicity, suppose that the points have been numbered after the evaluation so that $\mathbf{X}^0$ gives the largest value to $f(\mathbf{X})$. This means that $\mathbf{X}^0$ is the *worst* initial choice of the three, for a minimization problem, since $f(\mathbf{X}^0) > \max[f(\mathbf{X}^1), f(\mathbf{X}^2)]$. Then replace $\mathbf{X}^0$ with a new point, $\mathbf{X}^3$, that is found by *projecting* $\mathbf{X}^0$ through the center of the side opposite it in the triangle.

This gives a new triangle (simplex) with corners $\mathbf{X}^1$, $\mathbf{X}^2$, and $\mathbf{X}^3$. Now evaluate $f(\mathbf{X})$ for each of *these* points [you already have $f(\mathbf{X}^1)$ and $f(\mathbf{X}^2)$] and repeat the procedure, replacing that one of the current $\mathbf{X}$'s that gives the largest value of $f(\mathbf{X})$ by a new point, $\mathbf{X}^4$, found by a similar kind of projection. Continue in this way, always replacing the currently worst corner by projecting it through the opposite side, until some stopping criterion is reached.

There are a number of subtle variations in these general rules that are designed to increase the speed of convergence to a solution or to deal with specific computational problems. A few of these are as follows:

(a) Given a current simplex with corners $\mathbf{X}^{k-2}$, $\mathbf{X}^{k-1}$, and $\mathbf{X}^k$, assume that the labelling has been done such that $\mathbf{X}^{k-2}$ is the worst, giving a larger value to $f(\mathbf{X})$ than either $\mathbf{X}^{k-1}$ or $\mathbf{X}^k$. Then $\mathbf{X}^{k-2}$ will be projected through the opposite side of the simplex, generating $\mathbf{X}^{k+1}$. However, suppose that $f(\mathbf{X}^{k+1}) > f(\mathbf{X}^{k-2})$—at the new point we are worse off than we were at $\mathbf{X}^{k-2}$, because we have started back up a hillside, rather than further down into the valley. One sensible rule at this point would be to reject the new $\mathbf{X}^{k+1}$ and instead look for a "better" point somewhere along the line connecting $\mathbf{X}^{k-2}$ and $\mathbf{X}^{k+1}$ (the unattractive new point); that is, look for an intermediate point that generates a better value of $f(\mathbf{X})$ than either the old ($\mathbf{X}^{k-2}$) or the new ($\mathbf{X}^{k+1}$) endpoints. This procedure would be an option only for cases in which *regular* simplices were not required at each iteration, which is one good reason for not using the equilateral simplex version of this procedure. This option is included in the simplex search program used on our test functions.

Another approach, given an unsatisfactory $\mathbf{X}^{k+1}$, is to forget about projecting $\mathbf{X}^{k-2}$ and instead select the *second worst* point in the $\mathbf{X}^{k-2}$, $\mathbf{X}^{k-1}$, $\mathbf{X}^k$ set and project it through its opposite side. Geometrically this is generating a *turn* in the search procedure away from the $\mathbf{X}^{k-2}$-to-$\mathbf{X}^{k+1}$ direction, which may have been unsatisfactory because it was essentially traversing a valley rather than leading down into it. This kind of turning maneuver may effectively redirect the iterations in more of a downhill direction.

(b) In contrast to the illustration in (a), suppose that the projection of $\mathbf{X}^{k-2}$ through the opposite side to $\mathbf{X}^{k+1}$, turns out to be a really good move. For example, perhaps $f(\mathbf{X}^{k+1})$ is far less than $f(\mathbf{X}^k)$, the currently lowest (best) value. Then it might be that a projection *even further out* along that same line would lead to an even better point; the new point could perhaps be moved out, say, twice as far from the edge of the current simplex. This approach, also, is possible only when nonregular simplices are allowed, and it is included in the simplex search that is used for our test functions.

(c) Suppose that a particular $\mathbf{X}^k$ has remained in the "current" simplex for a number of consecutive iterations. This means that it has *not been worst* for several steps; thus, it might, in fact, be "near" to a best point. A logical procedure in that case is to *contract* the simplex around this point. For example, with current vertices defined by $\mathbf{X}^{k-2}$, $\mathbf{X}^{k-1}$, and $\mathbf{X}^k$ (where $\mathbf{X}^k$ has persisted in the simplex for several iterations), contract the sides connecting $\mathbf{X}^k$ with $\mathbf{X}^{k-2}$ and with $\mathbf{X}^{k-1}$ to (say) one-half of their previous value and start a new iteration from this smaller simplex.

Numerical Illustrations   Table 6.8 contains results from application of the downhill simplex method to the three test functions. The relative change in $f(\mathbf{X}^*)$ stopping criterion has generally been used, with $\varepsilon = 10^{-5}$. Alternative stopping criteria are also illustrated in Table 6.8 for test function (a). For consistency, the (arbitrarily chosen) initial set of starting points [(0,0), (1,0), (0,1)] has been used for all three functions. Additional variants are also used for test function (c) in order to explore sensitivity to the choice of initial values.

Results for test function (a) are shown to six decimals in order to give a flavor of the way in which the solution is approached. (In general, fewer decimal places will be used in the remaining tables.) At each step, only the currently *best* point (of the three corners of the current simplex) is indicated in the table. For that reason, the same point may appear at more than one step; this happens several times in Table 6.8.

To be clear about the workings of this method, we examine in detail the first few steps reported in Table 6.8(a). The starting triplet has $f(0,0) = -5$, $f(1,0) = -8$ and $f(0,1) = -9$, so it is $\mathbf{X} = \begin{bmatrix} 0 \\ 1 \end{bmatrix}$ (the best point) that is recorded in the line for iteration 1. Since $\begin{bmatrix} 0 \\ 0 \end{bmatrix}$ is the worst of these three initial simplex corners, that is the corner that is projected through the middle of the opposite side (connecting $\begin{bmatrix} 1 \\ 0 \end{bmatrix}$ and $\begin{bmatrix} 0 \\ 1 \end{bmatrix}$), which means through $\begin{bmatrix} 0.5 \\ 0.5 \end{bmatrix}$. This gets us to $\begin{bmatrix} 1 \\ 1 \end{bmatrix}$, where $f(1,1) = -12$. Since this is indeed better than $f(0,1)$ (the best of the current corners), the projection is continued even further out along the direction from $\begin{bmatrix} 0 \\ 0 \end{bmatrix}$ to $\begin{bmatrix} 1 \\ 1 \end{bmatrix}$—as in the refinement in (b). In the algorithm used here, the distance from the midpoint to the reflected point is doubled, so we end up at $\begin{bmatrix} 1.5 \\ 1.5 \end{bmatrix}$, where $f(1.5, 1.5) = -14$. This is what is recorded as the outcome of Iteration 2. This initial simplex projection is shown in Figure 6.10a.

At this point, the simplex of interest has corners at $\begin{bmatrix} 0 \\ 1 \end{bmatrix}$, $\begin{bmatrix} 1 \\ 0 \end{bmatrix}$, and $\begin{bmatrix} 1.5 \\ 1.5 \end{bmatrix}$. The worst of these is $\begin{bmatrix} 1 \\ 0 \end{bmatrix}$, where $f(1, 0) = -8$, so projecting through $\begin{bmatrix} 0.75 \\ 1.25 \end{bmatrix}$—the midpoint of the line connecting the other two corners of the current simplex—we get to $\begin{bmatrix} 0.5 \\ 2.5 \end{bmatrix}$, where $f(0.5,$ 2.5) = -13$. This indeed improves on $\begin{bmatrix} 1 \\ 0 \end{bmatrix}$, but it is not better than the currently best, at $\begin{bmatrix} 1.5 \\ 1.5 \end{bmatrix}$ where $f(1.5, 1.5) = -14$, so we do not project out any further. This move is shown in

TABLE 6.8   Downhill Simplex Method Search for Minimum of Test Functions (a)–(c)

Iteration	x_1	x_2	$f(\mathbf{X})$
	Test Function (a): $f(\mathbf{X}) = x_1^2 + x_2^2 - 4x_1 - 5x_2 - 5$		
	[Starting Triplet: (0,0), (1,0), (0,1)]		
1	0.0	1.0	−9.0
2	1.5	1.5	−14.0
3	1.5	1.5	−14.0
4	2.0	3.0	−15.0
5	2.0	3.0	−15.0
6	2.0	3.0	−15.0
7	2.0	3.0	−15.0
8	2.250000	2.375000	−15.171875
9	2.062500	2.343750	−15.221680
10	2.062500	2.343750	−15.221680
11	2.062500	2.343750	−15.221680
12	2.027344	2.587891	−15.241528
13	1.967773	2.557129	−15.245698
14	2.030029	2.458130	−15.247345
15	1.984680	2.467499	−15.248709
16	1.987564	2.509971	−15.249746
17	1.987564	2.509972	−15.249746
18	1.987564	2.509972	−15.249746
19	2.003669	2.493186	−15.249940
20	2.003669	2.493186	−15.249940
21	1.995521	2.505468	−15.249950

	Test Function (a): Alternative Stopping Criteria			
ε	Iterations	x_1	x_2	$f(\mathbf{X})$
1×10^{-10}	35	2.00001	2.50005	−15.25000
1×10^{-8}	31	1.99998	2.50015	−15.25000
1×10^{-6}	24	1.99871	2.49997	−15.25000

Iteration	x_1	x_2	$f(\mathbf{X})$
	Test Function (b): $f(\mathbf{X}) = x_1^2 + x_2^2 - 4x_1 - 5x_2 - x_1x_2 - 5$		
1	0.0	1.0	−9.00
2	1.5	1.5	−16.25
3	1.5	1.5	−16.25
4	3.0	4.0	−24.00
5	3.0	4.0	−24.00
6	5.5	5.5	−24.25
7	4.12500	3.87500	−24.82813
8	3.90625	4.34375	−25.18457
9	4.75781	4.80469	−25.19269
10	4.43555	4.92383	−25.28304
11	4.25146	4.60400	−25.32783
12	4.25146	4.60400	−25.32783
13	4.25146	4.60400	−25.32783
14	4.25146	4.60400	−25.32783
15	4.30180	4.71267	−25.32877
16	4.35539	4.67316	−25.33295
17	4.35539	4.67316	−25.33295
18	4.35539	4.67316	−25.33295

TABLE 6.8   *continued*

Iteration	x_1	x_2	$f(\mathbf{X})$
	Test Function (b): $f(\mathbf{X}) = x_1^2 + x_2^2 - 4x_1 - 5x_2 - x_1x_2 - 5$ (*cont.*)		
19	4.31715	4.65124	−25.33308
20	4.33782	4.67586	−25.33327
21	4.34144	4.66835	−25.33328
	Test Function (c): $f(\mathbf{X}) = (x_1^2/4) + (x_2^2/9) - 0.8x_1 - x_2 - 0.3x_1x_2 - 3$ [Starting Triplet: (0,0), (1,0), (0,1)]		
1	0.00000	1.00000	−3.88889
2	1.50000	1.50000	−5.56250
3	1.50000	1.50000	−5.56250
4	3.00000	4.00000	−8.97222
5	3.00000	4.00000	−8.97222
6	7.50000	7.50000	−13.06250
7	7.75000	11.25000	−17.52813
8	16.87500	20.12500	−22.31467
9	21.93750	32.06250	−29.08770
10	21.93750	32.06250	−29.08770
11	21.93750	32.06250	−29.08770
12	21.93750	32.06250	−29.08770
13	21.93750	32.06250	−29.08770
14	22.18750	35.81250	−29.36426
15	24.37500	36.62500	−29.36675
16	22.60938	34.14063	−29.49248
17	22.83984	35.59766	−29.56915
18	22.83984	35.59766	−29.56915
19	22.83984	35.59766	−29.56915
20	22.28204	34.51190	−29.57262
21	22.52522	34.76115	−29.57598
22	22.62174	35.11709	−29.57827
23	22.62174	35.11709	−29.57827
24	22.63428	35.00144	−29.57861
25	22.63428	35.00144	−29.57861
26	22.65605	35.11211	−29.57884
27	22.65605	35.11211	−29.57884

Test Function (c): Alternative Starting Points

Starting Point	Iteration	x_1	x_2	$f(\mathbf{X})$
$\mathbf{S}_1 = [(1,2), (3,4), (5,1)]$	1	3.00	4.00	−8.97222
	25	22.58070	34.96699	−29.57879
$\mathbf{S}_2 = [(100,50), (30,20), (5,10)]$	1	5.00	10.00	−14.63889
	42	22.69047	35.11816	−29.57876
$\mathbf{S}_3 = [(0,0), (1000,0), (0,1000)]$	1	0.00	0.00	−3.0
	33	22.61347	35.00091	−29.57885
$\mathbf{S}_4 = [(100,200), (300,400), (500, 100)]$	1	100.00	200.00	661.44444
	31	22.61695	35.05610	−29.57888
$\mathbf{S}_5 = [(1000,2000), (3000,4000), (5000, 1000)]$	1	1000.00	2000.00	91641.44444
	36	22.68760	35.10467	−29.57874
$\mathbf{S}_6 = [(1000, 1000), (1001,1000), (1000, 1001)]$	1	1000.00	1001.00	59299.44444
	61	22.65884	35.11319	−29.57885

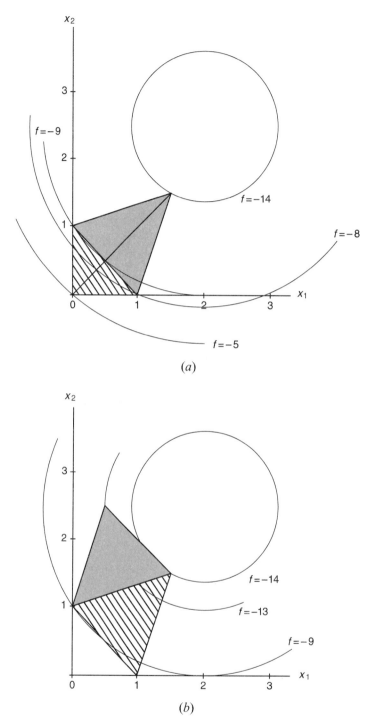

FIGURE 6.10 Initial steps in the downhill simplex method on test function (a) from [(0,0), (1,0), (0,1)] (for each iteration, the beginning simplex is crosshatched, and the projected simplex is shaded): (a) iteration 2, extended projection; (b) iteration 3, standard projection.

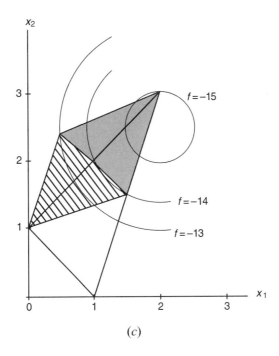

(c)

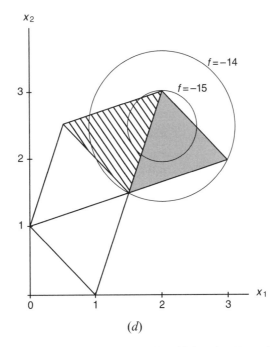

(d)

FIGURE 6.10   (c) iteration 4, standard projection; (d) iteration 5, standard projection.

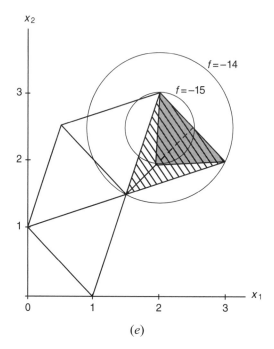

(e)

FIGURE 6.10 (e) iteration 6, shrunken projection.

Figure 6.10b. This means that the simplex after iteration 3 has corners at $\begin{bmatrix} 0 \\ 1 \end{bmatrix}$, $\begin{bmatrix} 1.5 \\ 1.5 \end{bmatrix}$, and $\begin{bmatrix} 0.5 \\ 2.5 \end{bmatrix}$ for which $\begin{bmatrix} 1.5 \\ 1.5 \end{bmatrix}$ is still the best of the three. That is why it is repeated in the line for iteration 3. The remaining panels in Figure 6.10 indicate several more steps in this downhill simplex search for the minimum of test function (a).

Note that, as is the case for many iterative methods and for many functions, the first few iterations make large inroads toward the minimizing solution, and then the final (sometimes very many) steps "polish" the solution values. For example, at iteration 13, the estimates, if rounded to three decimals, would be $x_1 = 1.968$ and $x_2 = 2.557$, at which point (also to three decimals) $f(\mathbf{X}) = -15.246$. Depending on the nature of the problem, this accuracy might be acceptable. Iterations 14–21 are needed to refine the estimate to satisfy the stopping criterion that the relative change in $f(\mathbf{X})$ from one iteration to the next be no more than 10^{-5}.

Test function (a) does not present much of a challenge to this iterative method, or many others, for that matter. It is included both because it is easy to visualize and because, as noted above, only a slight (algebraic) modification of the function will cause considerable trouble for some of the iterative methods. This is not the case, however, for the downhill simplex method, as results for test function (b) illustrate.

Results for test function (c) confirm that as functional complexity increases, there is an increase in the number of iterations, as would be expected. The final panel of Table 6.8 contains the first and the final iterations in the downhill simplex method for this function from several alternative starting points. What emerges from these results is that this technique seems to be quite insensitive to the position of the starting points relative to the actual minimum. Note that the minimizing solution is inside the simplices formed by new starting points $\mathbf{S}_1$ and $\mathbf{S}_3$; this appears to offer no great advantage either in terms of number of iterations or accuracy of the result. Other comparisons are possible. For example, $\mathbf{S}_6$ represents a rather large initial simplex, far from the origin, while $\mathbf{S}_7$ moves the starting triplet from part (a) of this table out to the position where (1000,1000) acts as the origin. One might have predicted, and in this particular example it is true, that it takes longer when starting with a "small" simplex far from the solution than when the simplex is "large." Simply put, projections across large triangles will tend to move around x_1, x_2 space in larger (at least initial) steps than projections across small triangles.

6.2.3 Sequential Univariate Search

The idea of a sequence of one-dimensional searches (line searches) underpins this and a number of other approaches. Sequential univariate search is the simplest of an entire group of techniques known as *direction set* methods, which we will examine in detail below. The concept is relatively simple and appealing.

Steps in the Sequential Univariate Search Procedure

(a) Consider $f(x_1, x_2)$ and begin by choosing an initial point, $\mathbf{X}^0 = [x_1^0, x_2^0]'$ with an associated $f(\mathbf{X}^0)$. Leaving x_2 fixed at x_2^0, minimize $f(x_1, x_2^0)$. This function can be denoted as $g^1(x_1)$ since it is a function of x_1 only. The minimum of this *univariate* function might be found using first-order conditions directly, if the derivative is easily found and the resulting equation dg^1/dx_1 [$= \partial f(x_1, x_2^0)/\partial x_1$] $= 0$ is easily solved for x_1. (Of course, some attention must also be paid to second-order conditions, namely, whether $d^2g^1/dx_1^2 > 0$, in order to be certain that a minimum and not a maximum point has been found.) Alternatively, any of the sequential search methods for a function of one variable from Section 6.1 could be used.

This process of finding the minimizing x_1, given x_2^0, can be described as searching along the x_1 direction from the initial point $\mathbf{X}^0$—that is, along the vector $\mathbf{I}_1 = [1,0]'$—for the minimum of the function along that line. Geometrically this point will be where the horizontal line is just *tangent* to a contour of the original function (the lowest-valued contour touched by the line). The procedure is called "line minimizing."

(b) By whatever means, assume that the minimizing x_1 has been found in step (a)—call it x_1^1. Then, introduce this value back into the original function $f(x_1, x_2)$, giving $f(x_1^1, x_2)$; denote this as $h^1(x_2)$, a function of x_2 only. Again, univariate search methods can be used to find the value of x_2 that minimizes $h^1(x_2)$. Let this value be x_2^1. This is the first value after x_2^0 in what will be a sequence of values of x_2. Geometrically, the line minimization in the vertical (x_2) direction—along the vector $\mathbf{I}_2 = [0,1]'$ from $[x_1^1, x_2^0]'$—will stop at the point where now the vertical line is tangent to a contour of the original function (again the lowest-valued contour touched by the line).

(c) Now return to (a), replace x_2^0 with x_2^1, and continue, finding the new value of x_1 that minimizes $f(x_1, x_2^1)$, which can be denoted $g^2(x_1)$. Let this value be x_1^2, and use it in (b) to find x_2^2, the value that minimizes $f(x_1^2, x_2)$ [$= h^2(x_2)$].

(d) Continue in this alternating sequence until some stopping criterion is reached.

The general idea is illustrated in Figures 6.11a and 6.11b for test functions (a) and (b). It is called a "direction set" method because search is first carried out along the direction

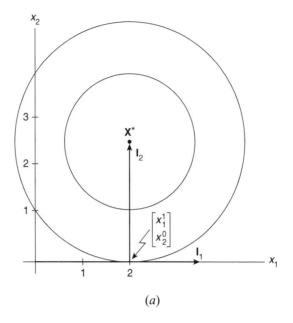

(*a*)

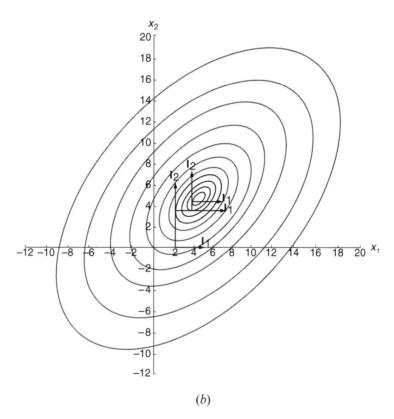

(*b*)

FIGURE 6.11   (*a*) Sequential univariate search on test function (a) from $[0, 0]'$; (*b*) initial steps in sequential univariate search on test function (b) from $[0, 0]'$.

indicated by $\mathbf{I}_1 = [1,0]'$ and then along the direction indicated by $\mathbf{I}_2 = [0,1]'$. These two unit vectors, the two columns of an order 2 identity matrix, define a complete *set* of directions that spans two-dimensional x_1,x_2 space. This approach can be quite adequate for a function whose contours are relatively circular. On the other hand, it can turn out to be very inefficient for a function with a long narrow valley, because it could require many very short steps. The following examples illustrate this.

Numerical Illustrations Table 6.9 contains results for sequential univariate search on the three test functions, from an arbitrary starting point of [0,0]. The stopping criterion, as before, is that the relative change in $f(\mathbf{X}^*)$ be no larger than $\varepsilon = 10^{-5}$. Again, for test function (c) alternative starting points are also used to explore sensitivity. Unlike the downhill simplex procedure, we now need only one initial pair, $\mathbf{X}^0 = [x_1^0, x_2^0]'$ to start the search.

The sequence for the bowl [test function (a)] is very brief. Note that for the bowl, sequential search will *always* move to the exact minimum point after one pair of steps, irrespective of the initial point. Since $\partial f / \partial x_1 = 2x_1 - 4$, the first step (line minimization) along the x_1 direction goes to $x_1 = 2$ (the initially selected value of x_2 does not contribute to $\partial f / \partial x_1$). Similarly, since $\partial f / \partial x_2 = 2x_2 - 5$, the first step (line minimization) along the x_2 direction goes to $x_2 = 2.5$—the value of x_1 that was found in the preceding step plays no role in the minimization with respect to x_2. The absence of an *interaction* term in $f(\mathbf{X})$, such as $x_1 x_2$, gives the circular shape to the contours shown in Figure 6.11 and explains why $\partial f / \partial x_1$ is independent of x_2 and why $\partial f / \partial x_2$ is independent of x_1.

Contours for test function (b) are shown in Figure 6.11b. Note that the efficiency of the procedure is decreased considerably. Instead of one pass through a direction set (two iterations, one each in the $\mathbf{I}_1$ and $\mathbf{I}_2$ directions), seven searches in the $\mathbf{I}_1$ direction and six in the $\mathbf{I}_2$ direction are now required (13 iterations), and sensitivity to alternative values of the stopping criterion is illustrated.

Finally, test function (c) requires 71 line minimizations before the same stopping criterion is satisfied. In this table, details are shown for only every 10 iterations, and search directions are omitted. Results for alternative starting points illustrate the general inferiority of this sequential approach to maximization and minimization problems. Recall that the downhill simplex method required only six more iterations when tackling test function (c) instead of (b) (Table 6.8).

For functions of more than two variables, if the minimizations are carried out sequentially from x_1 through x_n, then it is necessary initially to specify the $n-1$ values $x_2^0, \ldots, x_n^0$. This makes $f(\mathbf{X})$ a function of x_1 only and univariate minimization can proceed. The x_1^1 that solves this minimization problem, along with the original values of $x_3^0, \ldots, x_n^0$, are inserted into $f(\mathbf{X})$, creating a function of x_2 only, and the procedure continues exactly as described. The direction set, in this case, is made up of the n unit vectors $\mathbf{I}_1, \ldots, \mathbf{I}_n$ (the columns of an identity matrix of order n).

6.2.4 Conjugate Direction Methods

A particular kind of direction set method, with which Powell's name is usually attached (Powell, 1964), replaces the unit vectors used in sequential univariate search with a set of directions that generally lead to a minimum (or maximum) with fewer iterations. Test function (c), whose contours are elliptical and whose axes are not parallel to the x_1 and x_2 axes, serves to illustrate the advantages very nicely. In sequential univariate search we saw that the path is made up of a series of right-angle turns, creating a zigzag staircase kind of

TABLE 6.9 Sequential Univariate Search for Minimum of Test Functions (a)–(c)

k	$\mathbf{X}^k$	$f(\mathbf{X}^k)$	Search Direction from $\mathbf{X}^k$
	Test Function (a): $f(\mathbf{X}) = x_1^2 + x_2^2 - 4x_1 - 5x_2 - 5$		
0	$\begin{bmatrix} 0 \\ 0 \end{bmatrix}$	-5	$\mathbf{I}_1$
1	$\begin{bmatrix} 2 \\ 0 \end{bmatrix}$	-9	$\mathbf{I}_2$
2	$\begin{bmatrix} 2 \\ 2.5 \end{bmatrix}$	-15.25	$\mathbf{I}_1$
3	$\begin{bmatrix} 2 \\ 2.5 \end{bmatrix}$	-15.25	Stop
	Test Function (b): $f(\mathbf{X}) = x_1^2 + x_2^2 - 4x_1 - 5x_2 - x_1x_2 - 5$		
0	$\begin{bmatrix} 0 \\ 0 \end{bmatrix}$	-5	$\mathbf{I}_1$
1	$\begin{bmatrix} 2 \\ 0 \end{bmatrix}$	-9	$\mathbf{I}_2$
2	$\begin{bmatrix} 2 \\ 3.5 \end{bmatrix}$	-21.25	$\mathbf{I}_1$
3	$\begin{bmatrix} 3.75 \\ 3.5 \end{bmatrix}$	-24.3125	$\mathbf{I}_2$
4	$\begin{bmatrix} 3.75 \\ 4.375 \end{bmatrix}$	-25.07813	$\mathbf{I}_1$
5	$\begin{bmatrix} 4.1875 \\ 4.375 \end{bmatrix}$	-25.26953	$\mathbf{I}_2$
6	$\begin{bmatrix} 4.1875 \\ 4.59375 \end{bmatrix}$	-25.31738	$\mathbf{I}_1$
7	$\begin{bmatrix} 4.29688 \\ 4.59375 \end{bmatrix}$	-25.32935	$\mathbf{I}_2$
8	$\begin{bmatrix} 4.29688 \\ 4.64844 \end{bmatrix}$	-25.33234	$\mathbf{I}_1$
9	$\begin{bmatrix} 4.32422 \\ 4.64844 \end{bmatrix}$	-25.33308	$\mathbf{I}_2$
10	$\begin{bmatrix} 4.32422 \\ 4.66211 \end{bmatrix}$	-25.33327	$\mathbf{I}_1$
11	$\begin{bmatrix} 4.33105 \\ 4.66211 \end{bmatrix}$	-25.33332	$\mathbf{I}_2$
12	$\begin{bmatrix} 4.33105 \\ 4.66553 \end{bmatrix}$	-25.333329	$\mathbf{I}_1$
13	$\begin{bmatrix} 4.33276 \\ 4.66553 \end{bmatrix}$	-25.333332	Stop

TABLE 6.9 *continued*

k	$\mathbf{X}^k$	$f(\mathbf{X}^k)$	Search Direction from $\mathbf{X}^k$

Test Function (b): Alternative Stopping Criteria

ε	Iterations	x_1	x_2	$f(\mathbf{X})$
1×10^{-10}	21	4.33333	4.66666	−25.33333
1×10^{-8}	19	4.33332	4.66666	−25.33333
1×10^{-6}	15	4.33319	4.66660	−25.33333

Test Function (c): $f(\mathbf{X}) = (x_1^2/4) + (x_2^2/9) - 0.8x_1 - x_2 - 0.3x_1x_2 - 3$

k	$\mathbf{X}^k$	$f(\mathbf{X}^k)$
0	$\begin{bmatrix} 0 \\ 0 \end{bmatrix}$	−3.0
10	$\begin{bmatrix} 11.45454 \\ 19.96362 \end{bmatrix}$	−23.64494
20	$\begin{bmatrix} 18.73438 \\ 29.79142 \end{bmatrix}$	−28.85751
30	$\begin{bmatrix} 21.27271 \\ 33.21816 \end{bmatrix}$	−29.49124
40	$\begin{bmatrix} 22.15777 \\ 34.41299 \end{bmatrix}$	−29.56828
50	$\begin{bmatrix} 22.46637 \\ 34.82961 \end{bmatrix}$	−29.57765
60	$\begin{bmatrix} 22.57398 \\ 34.97487 \end{bmatrix}$	−29.57879
70	$\begin{bmatrix} 22.61149 \\ 35.02552 \end{bmatrix}$	−29.57893
71	$\begin{bmatrix} 22.61531 \\ 35.03006 \end{bmatrix}$	−29.57893

Test Function (c): Additional Starting Points

Starting Point	Total Number of Iterations
$[3, 4]'$	70
$[5, 10]'$	68
$[100, 200]'$	84
$[1000, 2000]'$	110
$[1000, 1001]'$	104

picture, in which each of the steps (1) is taken parallel to one of the two axes (and, therefore, perpendicular to the other) and (2) is generally very small. For a function with a long, narrow valley that does not run parallel to either axis, the method takes no account of the particular topography of the function, since it only allows 90° turns. The conjugate directions approach removes the right-angle restriction in a very effective way.

The Logic of Conjugate Direction Methods Simply put, the idea behind conjugate direction methods is that successive directions should be chosen so that they do not "undo" the optimality characteristics of the previously chosen directions. We concentrate on illustrations for functions of two variables with their associated contours in two-dimensional space. It will be useful at this point to examine some particular characteristics of quadratic functions. This is because Powell's method is based on properties of such functions; it works perfectly if the problem is to minimize a quadratic function, and it works less well the more $f(\mathbf{X})$ differs from a quadratic.

In Section 1.6 we used $\mathbf{X'AX}$ as the standard representation of a *quadratic form*, and we examined conditions on the matrix $\mathbf{A}$ that would determine the sign of the quadratic form. One could equally well use $\frac{1}{2}\mathbf{X'AX}$, where all the elements in $\mathbf{A}$ are simply twice as large as in the earlier expression. (There is no obvious motivation for doing this at the moment; there soon will be.) A general *quadratic function* may have, in addition, terms that are linear in the x's, as well as a constant. Therefore, a completely general representation of a quadratic function is

$$q(\mathbf{X}) = \tfrac{1}{2}\mathbf{X'AX} + \mathbf{B'X} + c$$

where $\mathbf{X} = [x_1, \ldots, x_n]'$, $\mathbf{A}$ is an $n \times n$ *symmetric* matrix, $\mathbf{B}$ is an n-element column vector, and c is a constant. With this expression, ordinary rules for differentiation will produce the result that the *gradient* of $q(\mathbf{X})$ is

$$\nabla q(\mathbf{X}) = \frac{\partial q(\mathbf{X})}{\partial x_i} = \mathbf{AX} + \mathbf{B}$$

which is a linear function of the x's. Further differentiation makes clear that the Hessian matrix of second partial and cross-partial derivatives is

$$\mathbf{H} = \frac{\partial^2 q(\mathbf{X})}{\partial x_i \partial x_j} = \mathbf{A}$$

This is the reason for doubling $\mathbf{A}$ and multiplying by $\frac{1}{2}$ in the general expression for $q(\mathbf{X})$; it produces the result that the Hessian is just exactly the symmetric coefficient matrix $\mathbf{A}$ in the original expression for $q(\mathbf{X})$.

Evaluating the gradient at $\mathbf{X}^k$ and at $\mathbf{X}^{k+1}$ gives

$$\nabla q(\mathbf{X}^k) = \mathbf{AX}^k + \mathbf{B} \qquad \text{and} \qquad \nabla q(\mathbf{X}^{k+1}) = \mathbf{AX}^{k+1} + \mathbf{B}$$

so that the *difference* in the value of the gradient at the two points is

$$[\nabla q(\mathbf{X}^{k+1}) - \nabla q(\mathbf{X}^k)] = \mathbf{A}[\mathbf{X}^{k+1} - \mathbf{X}^k] = \mathbf{H}[\mathbf{X}^{k+1} - \mathbf{X}^k]$$

In words, the Hessian matrix translates differences in $\mathbf{X}$ into differences in $\nabla q(\mathbf{X})$. This property of quadratic functions is at the heart of Powell's method.

Given a quadratic function $f(x_1, x_2)$, we initially examine the directions in the set $\{\mathbf{I}_1, \mathbf{I}_2\}$ from an initial point $\mathbf{X}^0$; that is, during the first two ($= n$) iterations in the conjugate direction method we search along the same directions as in the sequential univariate search procedure. Denote the line minimizer in the $\mathbf{I}_1$ direction by x_1^1 and the point $[x_1^1, x_2^0]'$ by $\mathbf{X}^{1*}$—suggesting an (initial) minimizing point with respect to the variable x_1. At that point the horizontal line is tangent to a contour of $f(\mathbf{X})$, as we know, and therefore ∇f will be orthogonal to the direction $\mathbf{I}_1$. Next, in sequential univariate search, a minimizer is found on the vertical line through $\mathbf{X}^{1*}$ ($\mathbf{I}_2$ direction, which is also the gradient direction at that point). At this minimizing point, $[x_1^1, x_2^1]'$, the vertical line will be tangent to a contour of $f(\mathbf{X})$; ∇f there will be orthogonal to the direction $\mathbf{I}_2$, and consequently at this point ∇f will no longer be orthogonal to the initial search direction $\mathbf{I}_1$. The search along $\mathbf{I}_2$ has generated a point at which the value in the x_1 direction is no longer optimal; that is why a new look along $\mathbf{I}_1$ is undertaken as the next step in sequential univariate search.

If the next point chosen, after $\mathbf{X}^{1*}$, were to continue also to be a minimum in the $\mathbf{I}_1$ direction, then ∇f there would have to be orthogonal to $\mathbf{I}_1$. Suppose that this next point is labeled $\mathbf{X}^{2*}$. Then it would be true that both $[\nabla f(\mathbf{X}^{1*})] \circ \begin{bmatrix} 1 \\ 0 \end{bmatrix} = 0$ and $[\nabla f(\mathbf{X}^{2*})] \circ \begin{bmatrix} 1 \\ 0 \end{bmatrix} = 0$, which means that

$$[\nabla f(\mathbf{X}^{2*}) - \nabla f(\mathbf{X}^{1*})] \circ \begin{bmatrix} 1 \\ 0 \end{bmatrix} = 0$$

But we have just seen that for quadratic functions the Hessian matrix translates changes in $\mathbf{X}$ into changes in the gradient of the function. This means that

$$\left[\mathbf{H}[\mathbf{X}^{2*} - \mathbf{X}^{1*}]\right] \circ \begin{bmatrix} 1 \\ 0 \end{bmatrix} = 0$$

or (since $\mathbf{H}$ is symmetric)

$$[\mathbf{X}^{2*} - \mathbf{X}^{1*}]' \mathbf{H} \begin{bmatrix} 1 \\ 0 \end{bmatrix} = 0$$

We now need an additional definition from matrix algebra. Given two n-element vectors, $\mathbf{V}_1$ and $\mathbf{V}_2$, and a positive definite $n \times n$ matrix, $\mathbf{M}$, when $\mathbf{V}_1' \mathbf{M} \mathbf{V}_2 = 0$, the two vectors are said to be "conjugate." (When $\mathbf{M}$ is an identity matrix, $\mathbf{I}$, or $\alpha \mathbf{I}$, for any $\alpha \neq 0$, the conjugate vectors will also be orthogonal.) Therefore, if $\mathbf{H}$ is positive definite, which means that if $f(\mathbf{X})$ is strictly convex and thus has a unique minimum, the vector $[\mathbf{X}^{2*} - \mathbf{X}^{1*}]$, connecting $\mathbf{X}^{2*}$ and $\mathbf{X}^{1*}$, is conjugate to $\begin{bmatrix} 1 \\ 0 \end{bmatrix}$. Any vector that is proportional to $[\mathbf{X}^{2*} - \mathbf{X}^{1*}]$ will also be conjugate to $\begin{bmatrix} 1 \\ 0 \end{bmatrix}$.

Steps in the Conjugate Direction Method: An Illustration for $f(x_1, x_2)$ Powell's discovery was that, for *quadratic* functions:

(a) If you build a set of directions made up of the *average directions* from a series of direction sets, this set of average directions has the *noninterfering* property; specifically, any point at which you choose to stop along the second average direction (call it $\mathbf{A}^1$) will still also be a minimizer along the first average direction (call it $\mathbf{A}^0$). In two-dimensional contour space, which is what you need for functions of two variables, there will only be *two* such average directions. [Remember that you need a third dimension to represent $f(x_1, x_2)$ itself. In three-dimensional space, for functions of three variables, you would find *three* average directions, etc.].

(b) These two average direction vectors turn out to have the mathematical property of conjugacy; namely, $(\mathbf{A}^1)'\mathbf{H}(\mathbf{A}^0) = 0$, where $\mathbf{H}$ is the Hessian matrix of $f(\mathbf{X})$. (Remember that for a quadratic function, the Hessian is made up of constant terms.)

We sketch the general idea for a quadratic function $f(x_1, x_2)$:

1. Start (arbitrarily) at $\mathbf{X}^0 = \begin{bmatrix} 1 \\ 1 \end{bmatrix}$ and use $\mathbf{D}^0 = \{\mathbf{I}_1, \mathbf{I}_2\}$ as the *initial* direction set, only to get things started. What is important to the conjugate direction method is the *average* direction, $\mathbf{A}^0$, that results from minimizing in turn along the two directions in this direction set, first along $\mathbf{I}_1$ and then along $\mathbf{I}_2$. Then, when $\mathbf{A}^0$ is determined, find the minimum along it and call that point $\mathbf{X}^1$ (Fig. 6.12a).

2. The next direction set, $\mathbf{D}^1$, replaces either $\mathbf{I}_1$ or $\mathbf{I}_2$ with $\mathbf{A}^0$. (We will explore later how this choice might be made.) For the present, assume that $\mathbf{I}_1$ is discarded and replaced

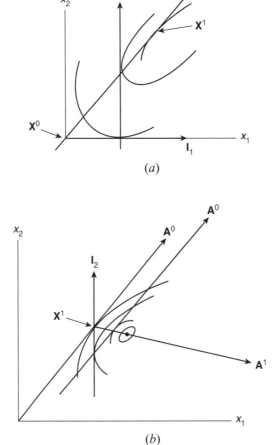

(a)

(b)

FIGURE 6.12 Conjugate directions for a quadratic function: (a) minimization along $\mathbf{A}^0$; (b) minimization along $\mathbf{A}^1$ (enlarged).

by $\mathbf{I}_2$ as the *first* direction in $\mathbf{D}^1$ and that $\mathbf{A}^0$ moves in to become the *second* direction; so $\mathbf{D}^1 = \{\mathbf{I}_2, \mathbf{A}^0\}$. Again, what is of interest is the *average* direction of movement after working through this new direction set, starting at $\mathbf{X}^1$, which means after minimizing first along $\mathbf{I}_2$ and then along $\mathbf{A}^0$. Call this new average direction $\mathbf{A}^1$.

3. Minimization along $\mathbf{A}^1$ will take you right to the minimum of $f(\mathbf{X})$ (Fig. 6.12*b*).

The fact that movement along the *n*th (in this example, the second) average direction leads directly to the minimum is a property of *quadratic* functions only. For quadratic functions of *n* variables, you will need *n* direction sets, from which you sequentially build a set of *n* conjugate directions $\{\mathbf{A}^0, \ldots, \mathbf{A}^{n-1}\}$, and minimization along the last of these will lead directly to the minimum. Powell's conjugate directions method is built on this quadratic approximation to $f(\mathbf{X})$; if $f(\mathbf{X})$ is not quadratic, more direction sets will generally be needed.

Steps in the Conjugate Direction Method: Generalization to $f(\mathbf{X})$

1. Start at an initial point, $\mathbf{P}_0^0 = [x_1^0, \ldots, x_n^0]'$, with an initial set of directions given by the unit vectors; $\mathbf{D}^0 = \{\mathbf{D}_1^0, \ldots, \mathbf{D}_n^0\} = \{\mathbf{I}_1, \ldots, \mathbf{I}_n\}$. Find the minimum along the $\mathbf{I}_1$ direction, given $x_2^0, \ldots, x_n^0$, move $\mathbf{P}_0^0$ to that minimum point, and label it $\mathbf{P}_1^0$. Then, with the new value of x_1 and the initial values of $x_3, \ldots, x_n$, find the minimum in the $\mathbf{I}_2$ direction, move $\mathbf{P}_1^0$ there and call it $\mathbf{P}_2^0$. Continue in this way, along each $\mathbf{I}_i$ direction in turn, using $\mathbf{P}_{i-1}^0$ as the starting point for that search. The final search, along $\mathbf{I}_n$, serves to locate $\mathbf{P}_n^0$.

2. Calculate the direction defined by $[\mathbf{P}_n^0 - \mathbf{P}_0^0]$, the vector that connects the starting point with the final point that was found in the last of the *n* line searches. This *single* vector summarizes the results of all *n* of the line minimizations just completed; it is one measure of the *average* direction of movement over the first set of *n* directions. It is, in that sense, the "best" direction of movement away from $\mathbf{P}_0^0$. Therefore, do a line search for the minimum along *this* average direction, $\mathbf{A}^0 = [\mathbf{P}_n^0 - \mathbf{P}_0^0]$, and make that minimizing point the starting location, $\mathbf{P}_0^1$, for the *next* set of *n* line minimizations. (The *super*script on $\mathbf{P}$ serves to identify which *set* of minimizations is being explored; the *sub*script indicates which line search within the set has just been completed.)

3. Construct a new set of search directions $\mathbf{D}^1$ by throwing out $\mathbf{D}_1^1$, continuing to use the *last* (*n*–1) directions of the previous set as the *first* (*n*–1) directions in the next set, and using $\mathbf{A}^0 = [\mathbf{P}_n^0 - \mathbf{P}_0^0]$ as the *n*th direction in this next set. That is, the new set of directions is

$$\mathbf{D}^1 = \{\mathbf{D}_1^1, \ldots, \mathbf{D}_n^1\} = \{\mathbf{D}_2^0, \mathbf{D}_3^0, \ldots, \mathbf{D}_n^0, \mathbf{A}^0\}$$

4. Beginning at $\mathbf{P}_0^1$, search along each of the directions in $\mathbf{D}^1$ in turn. Connect the final minimizing point, $\mathbf{P}_n^1$ (found along the final direction, $\mathbf{A}^0$) with $\mathbf{P}_0^1$, find the minimum along this direction, and denote it by $\mathbf{P}_0^2$. This will be the starting point for the next set of line searches.

5. Rebuild the direction set, discarding the first direction from the previous set, setting each $\mathbf{D}_i^2$ ($i = 1, \ldots, n-1$) equal to $\mathbf{D}_{i+1}^1$, and letting $\mathbf{D}_n^2 = \mathbf{A}^1 = [\mathbf{P}_n^1 - \mathbf{P}_0^1]$.

6. Continue in this way, searching along successive sets of *n* directions and then finding the minimum along the direction representing the average movement described by those *n* minimizations. In building the *k*th direction set, $\mathbf{D}_1^{k-1}$ is discarded, $\mathbf{D}_2^{k-1}, \ldots, \mathbf{D}_{n-1}^k$

become $\mathbf{D}_1^k, \ldots, \mathbf{D}_{n-1}^k$, and $\mathbf{D}_n^k$, the final direction in the new set, is given by $\mathbf{A}^{k-1} = [\mathbf{P}_n^{k-1} - \mathbf{P}_0^{k-1}]$. From the final point, $\mathbf{P}_n^k$, reached after n line searches, calculate the direction $\mathbf{A}^k = [\mathbf{P}_n^k - \mathbf{P}_0^k]$, and find the minimum along this direction; label that location $\mathbf{P}_0^{k+1}$ and continue until some stopping criterion is reached.

Variations in the Steps Most implementations of direction set methods use several variations on these basic ideas.[10] For example, it can happen that the directions in a particular direction set become (at least close to) linearly dependent. If, at step k, the direction set contained n vectors that were not linearly independent, those vectors would fail to span the space over which the searches at step k were carried out, and the desired minimum would not (in general) be found. For a function of two variables, the picture would be one in which the two vectors in the direction set are multiples of each other, and hence only minima along that one dimension (line) in two-dimensional space would be identified.

Also, if a function has a twisty valley, then a direction that is good for one part of the valley will turn out to be poor for the next part (after a twist). For this reason, one might *not* want to add the average direction from the last direction set—$\mathbf{A}^{k-1} = [\mathbf{P}_n^{k-1} - \mathbf{P}_0^{k-1}]$—as $\mathbf{D}_n^k$ in the next direction set. It could be that the average direction has already been adequately exploited by the moves in the $(k-1)$st direction set. In that case, one could (and implementations of Powell's method often do) simply decide to keep the direction set from the $(k-1)$st step and use it again for the kth step.

One fairly simple way to test whether an average direction has been "used up" during the $(k-1)$st step is to evaluate $f(\mathbf{X})$ at some further point, $\widetilde{\mathbf{P}}^{k-1}$, beyond $\mathbf{P}_n^{k-1}$, along that average direction. If $f(\widetilde{\mathbf{P}}^{k-1}) \geq f(\mathbf{P}_0^{k-1})$, then forget about any further use of the average direction from the $(k-1)$st step; start step k from $\mathbf{P}_0^k$ and instead simply reuse *all* the directions from the previous direction set. We will use this modification in our numerical illustrations, defining $\widetilde{\mathbf{P}}^{k-1} = [2\mathbf{P}_n^{k-1} - \mathbf{P}_0^{k-1}]$. (It is not difficult to show that $\widetilde{\mathbf{P}}^{k-1}$ is, indeed, on the line connecting $\mathbf{P}_n^{k-1}$ and $\mathbf{P}_0^{k-1}$. Try it.) Furthermore, if $\mathbf{A}^{k-1}$ is discarded, forget about minimizing along that line at the end of the $(k-1)$st set of directions. It can be assumed that the sequence of line searches through the $(k-1)$st direction set has already taken sufficient advantage of that direction. This will save an occasional line search, and we also use this modification in our numerical illustrations.

On the other hand, if $\mathbf{A}^{k-1}$ is to be introduced into the direction set for step k, there might be some more intelligent rule for deciding which direction to discard rather than simply always throwing out $\mathbf{D}_1^{k-1}$. One rule that has been found to be useful is to use $\mathbf{A}^{k-1}$ in place of the *best* step in the previous set, that is, along the $\mathbf{D}_i^{k-1}$ on which the function experienced its largest decrease. This is the variant that we will use in the examples in this text, although it may seem odd. If a direction was particularly good, why not keep it? The reason is that best directions can tend to become dominating influences in average directions, and keeping them in subsequent direction sets increases the possibility of linear dependence among vectors in those sets.

Numerical Illustrations In all illustrations of the conjugate direction method we will start at $\mathbf{X}^0 = \mathbf{P}_0^0 = \begin{bmatrix} 1 \\ 1 \end{bmatrix}$ and use the unit vectors (as in sequential univariate search) for the first direction set; $\mathbf{D}^0 = \{\mathbf{D}_1^0, \mathbf{D}_2^0\} = \{\mathbf{I}_1, \mathbf{I}_2\}$. Results are shown in Table 6.10, where k is

---

[10]Details can be found in, among others, Press et al. (1986).

TABLE 6.10 Conjugate Direction Search for Minimum of Test Functions (a)–(c)

k	$\mathbf{X}^k\,(=\mathbf{P}_0^k)$	$f(\mathbf{X}^k)$	$\mathbf{D}^k$
	Test Function (a): $f(\mathbf{X}) = x_1^2 + x_2^2 - 4x_1 - 5x_2 - 5$		
0	$\begin{bmatrix} 0 \\ 0 \end{bmatrix}$	-5	$[\mathbf{I}_1, \mathbf{I}_2]$

Minimization along $d_1^0 \Rightarrow \mathbf{P}_1^0 = \begin{bmatrix} 2.0 \\ 0 \end{bmatrix}, f(\mathbf{P}_1^0) = -9$

Minimization along $d_2^0 \Rightarrow \mathbf{P}_2^0 = \begin{bmatrix} 2.0 \\ 2.5 \end{bmatrix}, f(\mathbf{P}_2^0) = -15.25$

Average direction $\mathbf{A}^0 = (\mathbf{P}_2^0 - \mathbf{P}_0^0) = \begin{bmatrix} 2.0 \\ 2.5 \end{bmatrix}$

Minimize along average direction; add to next direction set? No.

1	$\begin{bmatrix} 2.0 \\ 2.5 \end{bmatrix}$	-15.25	$[\mathbf{I}_1, \mathbf{I}_2]$

Minimization along $d_1^1 \Rightarrow \mathbf{P}_1^1 = \begin{bmatrix} 2.0 \\ 2.5 \end{bmatrix}, f(\mathbf{P}_1^1) = -15.25$

Minimization along $d_2^1 \Rightarrow \mathbf{P}_2^1 = \begin{bmatrix} 2.0 \\ 2.5 \end{bmatrix}, f(\mathbf{P}_2^1) = -15.25$

Stop

	Test Function (b): $f(\mathbf{X}) = x_1^2 + x_2^2 - 4x_1 - 5x_2 - x_1x_2 - 5$		
0	$\begin{bmatrix} 0 \\ 0 \end{bmatrix}$	-5	$[\mathbf{I}_1, \mathbf{I}_2]$

Minimization along $d_1^0 \Rightarrow \mathbf{P}_1^0 = \begin{bmatrix} 2 \\ 0 \end{bmatrix}, f(\mathbf{P}_1^0) = -9.00$

Minimization along $d_2^0 \Rightarrow \mathbf{P}_2^0 = \begin{bmatrix} 2 \\ 3.5 \end{bmatrix}, f(\mathbf{P}_2^0) = -21.25$

Average direction $\mathbf{A}^0 = (\mathbf{P}_2^0 - \mathbf{P}_0^0) = \begin{bmatrix} 2 \\ 3.5 \end{bmatrix}$

Minimize along average direction; add to next direction set?

Yes $\Rightarrow \mathbf{P}_3^0 = \begin{bmatrix} 2.75676 \\ 4.82432 \end{bmatrix}, f(\mathbf{P}_3^0) = -22.57432$

1	$\begin{bmatrix} 2.75676 \\ 4.82432 \end{bmatrix}$	-22.57432	$\left\{\mathbf{I}_1, \begin{bmatrix} 2 \\ 3.5 \end{bmatrix}\right\}$

Minimization along $d_1^1 \Rightarrow \mathbf{P}_1^1 = \begin{bmatrix} 4.41216 \\ 4.82432 \end{bmatrix}, f(\mathbf{P}_1^1) = -25.31469$

Minimization along $d_2^1 \Rightarrow \mathbf{P}_2^1 = \begin{bmatrix} 4.32268 \\ 4.66773 \end{bmatrix}, f(\mathbf{P}_2^1) = -25.33321$

Average direction $\mathbf{A}^1 = (\mathbf{P}_2^1 - \mathbf{P}_0^1) = \begin{bmatrix} 1.56592 \\ -0.15659 \end{bmatrix}$

(continued)

TABLE 6.10   Conjugate Direction Search for Minimum of Test Functions (a)–(c) *(continued)*

k	$\mathbf{X}^k\ (=\mathbf{P}_0^k)$	$f(\mathbf{X}^k)$	$\mathbf{D}^k$

Test Function (b): $f(\mathbf{X}) = x_1^2 + x_2^2 - 4x_1 - 5x_2 - x_1x_2 - 5$ *(cont.)*

Minimize along average direction; add to next direction set?

$$\text{Yes} \Rightarrow \mathbf{P}_3^1 = \begin{bmatrix} 4.33333 \\ 4.66667 \end{bmatrix}, f(\mathbf{P}_3^1) = -25.33333$$

| 2 | $\begin{bmatrix} 4.33333 \\ 4.66667 \end{bmatrix}$ | -25.33333 | $\left\{ \begin{bmatrix} 1.56592 \\ -.15659 \end{bmatrix}, \begin{bmatrix} 2.0 \\ 3.5 \end{bmatrix} \right\}$ |

Minimization along $d_1^2 \Rightarrow \mathbf{P}_1^2 = \begin{bmatrix} 4.33333 \\ 4.66667 \end{bmatrix}, f(\mathbf{P}_1^2) = -25.33333$

Minimization along $d_2^2 \Rightarrow \mathbf{P}_2^2 = \begin{bmatrix} 4.33333 \\ 4.66667 \end{bmatrix}, f(\mathbf{P}_2^2) = -25.33333$

Stop

Test Function (c): $f(\mathbf{X}) = (x_1^2/4) + (x_2^2/9) - 0.8x_1 - x_2 - 0.3x_1x_2 - 3$

| 0 | $\begin{bmatrix} 0 \\ 0 \end{bmatrix}$ | -3 | $[\mathbf{I}_1, \mathbf{I}_2]$ |

Minimization along $d_1^0 \Rightarrow \mathbf{P}_1^0 = \begin{bmatrix} 1.6 \\ 0 \end{bmatrix}, f(\mathbf{P}_1^0) = -3.64$

Minimization along $d_2^0 \Rightarrow \mathbf{P}_2^0 = \begin{bmatrix} 1.6 \\ 6.66 \end{bmatrix}, f(\mathbf{P}_2^0) = -8.56840$

Average direction $\mathbf{A}^0 = (\mathbf{P}_2^0 - \mathbf{P}_0^0) = \begin{bmatrix} 1.6 \\ 6.66 \end{bmatrix}$

Minimize along average direction; add to next direction set?

$$\text{Yes} \Rightarrow \mathbf{P}_3^0 = \begin{bmatrix} 2.67836 \\ 11.14868 \end{bmatrix}, f(\mathbf{P}_3^0) = -9.64568$$

| 1 | $\begin{bmatrix} 2.67836 \\ 11.14868 \end{bmatrix}$ | -9.64568 | $\left\{ \mathbf{I}_1, \begin{bmatrix} 1.6 \\ 6.66 \end{bmatrix} \right\}$ |

Minimization along $d_1^1 \Rightarrow \mathbf{P}_1^1 = \begin{bmatrix} 8.28916 \\ 11.14868 \end{bmatrix}, f(\mathbf{P}_1^1) = -17.51608$

Minimization along $d_2^1 \Rightarrow \mathbf{P}_2^1 = \begin{bmatrix} 10.55663 \\ 20,58685 \end{bmatrix}, f(\mathbf{P}_2^1) = -22.27895$

Average direction $\mathbf{A}^1 = (\mathbf{P}_2^1 - \mathbf{P}_0^1) = \begin{bmatrix} 7.87827 \\ 9.43817 \end{bmatrix}$

Minimize along average direction; add to next direction set?

$$\text{Yes} \Rightarrow \mathbf{P}_3^1 = \begin{bmatrix} 22.63158 \\ 35.05263 \end{bmatrix}, f(\mathbf{P}_3^1) = -29.57895$$

| 2 | $\begin{bmatrix} 22.63158 \\ 35.05263 \end{bmatrix}$ | -29.57895 | $\left\{ \begin{bmatrix} 7.87827 \\ 9.43817 \end{bmatrix}, \begin{bmatrix} 1.6 \\ 6.66 \end{bmatrix} \right\}$ |

Minimization along $d_1^2 \Rightarrow \mathbf{P}_1^2 = \begin{bmatrix} 22.63158 \\ 35.05263 \end{bmatrix}, f(\mathbf{P}_1^2) = -29.57895$

Minimization along $d_2^2 \Rightarrow \mathbf{P}_2^2 = \begin{bmatrix} 22.63158 \\ 35.05263 \end{bmatrix}, f(\mathbf{P}_2^2) = -29.57895$

Stop

TABLE 6.10   *continued*

Test Function (c): Additional Starting Points		
Starting Point	Number of Direction Sets	Total Number of Iterations
$[1, 2]'$	3	8
$[100\ 50]'$	3	8
$[100, 200]'$	3	7
$[1000, 2000]'$	3	7
$[1000, 1000]'$	3	8

Summary for test functions (b) and (c): starting point $[0, 0]'$; number of direction sets 3; total number of iterations 8. (The last direction set and the two iterations in it were needed only as a stopping check; the minimum was found along the second average direction vector, $\mathbf{A}^1$.)

now used to identify the direction set. Each pass through a direction set requires at least two iterations (line minimizations) and sometimes three, depending on whether a minimization is carried out along the average direction created by the first two moves in the direction set. In all cases, the stopping criterion was again that the relative change in $f(\mathbf{X})$ be no larger than 10^{-5}.

For test function (a) the initial pair of minimizations—along the horizontal and then the vertical direction—leads immediately to $\mathbf{X}^* = \begin{bmatrix} 2 \\ 2.5 \end{bmatrix}$. This reproduces exactly the sequence that appeared when sequential univariate search was used on the same function. The final two iterations, in the next direction set, are needed in order to *confirm* that indeed a minimum point has been reached.

The sequence for test function (b) is more complicated, precisely because the contours have become tilted ellipses because of the presence of the multiplicative $x_1 x_2$ term in $f(\mathbf{X})$. Results are summarized in Table 6.10, but we also work through each step of the approach here, for anyone who would like more detail.

(a) Starting from $\mathbf{X}^0 = \mathbf{P}_0^0 = \begin{bmatrix} 0 \\ 0 \end{bmatrix}$, where $f(\mathbf{X}^0) = -5$, use the unit vectors, as in sequential univariate search, as the first direction set. The initial task therefore is to find the minimum along the horizontal line $x_2 = 0$. As with sequential univariate search, each line minimization in the conjugate direction procedure can be accomplished by either derivative-based methods or by some kind of sequential search procedure of the sort described in Section 6.1. Here, for simplicity, it is easy to substitute $x_2 = 0$ into $f(\mathbf{X})$ and solve $df/dx_1 = 0$ for x_1, giving $x_1 = 2$.

In the notation of the conjugate direction method, we have found $\mathbf{P}_1^0 = \begin{bmatrix} 2 \\ 0 \end{bmatrix}$, where $f(\mathbf{P}_1^0) = -9$, and hence the decrease in $f(\mathbf{X})$ generated by the search along $\mathbf{I}_1$ is 4. Next, minimizing along the $\mathbf{D}_2^0 (= \mathbf{I}_2)$ direction, at $x_1^1 = 2$, we find $x_2^1 = 3.5$, so that $\mathbf{P}_2^0 = \begin{bmatrix} 2 \\ 3.5 \end{bmatrix}$.

Now $f(\mathbf{P}_2^0) = -21.25$, and the (additional) decrease in the function along this direction from $\mathbf{P}_1^0$ is 12.25. Since $n = 2$, we have completed one pass through this first direction set, $\mathbf{D}^0$.

(b) Find $\widetilde{\mathbf{P}}^0$ and check whether $f(\widetilde{\mathbf{P}}^0) \geq f(\mathbf{P}_0^0)$. Here $\widetilde{\mathbf{P}}^0 = 2\mathbf{P}_2^0 - \mathbf{P}_0^0 = \begin{bmatrix} 4 \\ 7 \end{bmatrix}$, $f(\widetilde{\mathbf{P}}^0) = -19$, $f(\mathbf{P}^0) = -5$, and since $f(\widetilde{\mathbf{P}}^0) < f(\mathbf{P}^0)$, we want to find the minimum along the line defined by $\mathbf{A}^0 = [\mathbf{P}_2^0 - \mathbf{P}_0^0]$ and then add this new direction to the next direction set. Note that the new direction is not parallel to an axis, so both x_1 and x_2 are changed at the new minimum. The line that connects the origin and $\begin{bmatrix} 2 \\ 3.5 \end{bmatrix}$ is $x_2 = 1.75x_1$. Identify the minimizing point along that line as $\mathbf{P}_0^1$, the *next* starting point. In this example both x_1 and x_2 are changed; $\mathbf{P}_0^1 = [2.75676, 4.82432]'$ and there $f(\mathbf{P}_0^1) = -22.57432$.[11]

(c) Now rebuild the direction set, since we have decided to introduce $\mathbf{A}^0$ into the new set of directions. Throw away either $\mathbf{I}_1$ or $\mathbf{I}_2$, depending on which of those directions produced the largest decrease in $f(\mathbf{X})$ in the previous step. In this example the largest decrease came along $\mathbf{I}_2$; it is therefore replaced by $\mathbf{A}^0$, and the next direction set, $\mathbf{D}^1 = \{\mathbf{D}_1^1, \mathbf{D}_2^1\}$, is $\{\mathbf{I}_1, [\mathbf{P}_2^0 - \mathbf{P}_0^0]\} = \{[1, 0]', [2, 3.5]'\}$. Searches for minima along these two directions are now carried out, in sequence, starting from $\mathbf{P}_0^1$. Minimizing along $\mathbf{D}_1^1$ (the x_1 direction) means that we are now looking first for the best x_1 at this stage, given that $x_2^2 = 4.82432$; this leads to $x_1^2 = 4.41216$, and the first point found in this next set of searches is $\mathbf{P}_1^1 = [4.41216, 4.82432]'$, where $f(\mathbf{P}_1^1) = -25.31469$. The next move in this iteration is along the $\mathbf{D}_2^1$ direction, starting from $\mathbf{P}_1^1$. The line through $\mathbf{P}_1^1$ in that direction must have a slope of 1.75; it turns out to be[12] $x_2 = 1.75x_1 - 2.89696$, and minimization leads to $x_1^3 = 4.32268$ and $x_2^3 = 4.66773$, so $\mathbf{P}_2^1 = [4.32268, 4.66773]'$ and $f(\mathbf{P}_2^1) = -25.33321$. Again both x_1 and x_2 were changed by the last line search, in the $\mathbf{D}_2^1$ direction, since that direction was not parallel to an axis.

(d) We now once again identify the average direction of movement, this time from $\mathbf{P}_0^1$ to $\mathbf{P}_2^1$. This is, using the same logic as in step (b), $\mathbf{A}^1 = [\mathbf{P}_2^1 - \mathbf{P}_0^1] = [1.56592, -0.15659]'$. Do we want to line minimize along this direction? That depends on $f(\widetilde{\mathbf{P}}^1)$ and $f(\mathbf{P}_0^1)$; here $\widetilde{\mathbf{P}}^1 = [5.88860, 4.51114]'$. Evaluation of the function at those two points shows that $f(\widetilde{\mathbf{P}}^1) < f(\mathbf{P}_0^1)$, so we again choose to minimize along the average direction from step 1. This leads to $\mathbf{P}_3^1 = [4.33333, 4.66667]'$, where $f(\mathbf{P}_3^1) = -25.33333$.

(e) The next iteration begins from here, $\mathbf{P}_0^2 = \mathbf{P}_3^1$, and $\mathbf{D}^2 = \{\mathbf{D}_1^2, \mathbf{D}_2^2\} = \{[1.56592, -0.15659]', [2.0, 3.5]'\}$. Starting from $\mathbf{P}_0^2$ and minimizing first along $\mathbf{D}_1^2$ leads to $\mathbf{P}_1^2 = [4.33333, 4.66667]'$. From that point, a line search in the direction $\mathbf{D}_2^2$ generates $\mathbf{P}_2^2 = [4.33333, 4.66667]'$, and since $\mathbf{P}_0^2$, $\mathbf{P}_1^2$, and $\mathbf{P}_2^2$ are equal (to five decimal places), we can conclude that we have reached a decent approximation to the solution, and that at its minimum $f(\mathbf{X}) = -25.33333$.

Note, finally, that the Hessian for this function is $\mathbf{H} = \begin{bmatrix} 2 & -1 \\ -1 & 2 \end{bmatrix}$, and it is easily established that the two directions in the final direction set are indeed conjugate directions; namely, that $(\mathbf{D}_1^2)'\mathbf{H}(\mathbf{D}_2^2) = 0$. (Check it.) You might also check the conjugacy of the directions in the final direction set for test function (c), for which $\mathbf{H} = \begin{bmatrix} 0.5 & -0.3 \\ -0.3 & 0.222 \end{bmatrix}$. The steps to solution for test function (c) are indicated in Figure 6.13.

For the three test functions examined thus far, it is clear that the conjugate direction method reaches the minimizing values of $\mathbf{X}$ most directly. Sequential univariate search

---

[11]Some versions of the conjugate direction method would simply put $\mathbf{A}^0$, the average direction from the last set, into the next direction set and start the next n searches from $\mathbf{P}_2^0$, instead of finding the minimum along $\mathbf{A}^0$ and starting from there. Since $\mathbf{A}^0$ would be included as one of the new directions, minimization along it will be carried out at some point during the next n line minimizations, and the $(n+1)$st line minimization in direction set $\mathbf{D}^0$ would be avoided.

[12]The equation is found easily from the general formula for a line through a given point $[a,b]'$ with given slope s: $(y - b)/(x - a) = s$.

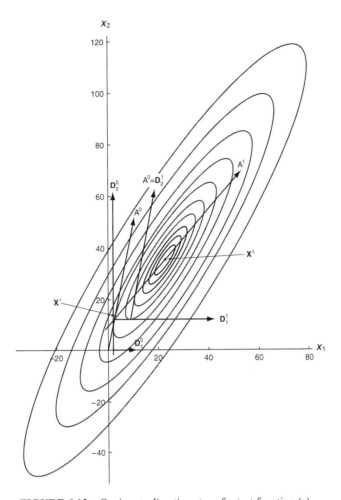

FIGURE 6.13   Conjugate direction steps for test function (c).

goes to the trouble of finding line minimizations along a series of often very inefficient lines. The downhill simplex method has the virtue of simplicity (no minimizations required). Conjugate direction search requires more calculations to select appropriate directions along which to search, but the payoff for such careful selection is apparent in the results in the three tables. Also, at least for test function (c), it is far less sensitive to choice of a starting point than either of the other two methods explored thus far.

6.2.5   Results for Additional Test Functions

We include some comparative results for three additional test functions in Table 6.11. The functions are

Test function (d): $f(\mathbf{X}) = (x_1^2/4) + (x_2^2/10) - 0.8x_1 - x_2 - 0.3x_1x_2 - 3$. At $\mathbf{X}^* = \begin{bmatrix} 46 \\ 74 \end{bmatrix}$ this function has a minimum value of -58.4.

TABLE 6.11   Results for Three Additional Test Functions

	Downhill Simplex	Sequential Univariate Search	Conjugate Directions[a]
Test Function (d): $f(\mathbf{X}) = (x_1^2/4) + (x_2^2/10) - 0.8x_1 - x_2 - 0.3x_1x_2 - 3$			
Number of iterations	28	48	8 (3)
X*	$\begin{bmatrix} 45.9388 \\ 73.8966 \end{bmatrix}$	$\begin{bmatrix} 45.6861 \\ 73.5292 \end{bmatrix}$	$\begin{bmatrix} 46.0000 \\ 74.0000 \end{bmatrix}$
$f(\mathbf{X}^*)$	−58.3999	−58.3975	−58.4000
Test Function (e): $f(\mathbf{X}) = (x_1^2/4) + (x_2^2/11) - 0.8x_1 - x_2 - 0.3x_1x_2 - 3$			
Number of iterations	33	373'	8 (3)
X*	$\begin{bmatrix} 488.9272 \\ 812.2130 \end{bmatrix}$	$\begin{bmatrix} 479.1665 \\ 796.1247 \end{bmatrix}$	$\begin{bmatrix} 490.0000 \\ 814.0001 \end{bmatrix}$
$f(\mathbf{X}^*)$	−605.9971	−605.7066	−606.0000
Test Function (f): $f(\mathbf{X}) = 100(x_2 - x_1^2)^2 + (1 - x_1)^2 + 0.01$			
Number of iterations	59	1580	32 (11)
X*	$\begin{bmatrix} 1.0001 \\ 1.0002 \end{bmatrix}$	$\begin{bmatrix} 0.9956 \\ 0.9911 \end{bmatrix}$	$\begin{bmatrix} 1.0000 \\ 1.0000 \end{bmatrix}$
$f(\mathbf{X}^*)$	0.0100	0.0100	0.0100
Iteration Counts for Alternative Stopping Criterion: $\varepsilon = 10^{-10}$			
Test Function			
(d)	46	210	8 (3)
(e)	51	1960	8 (3)
(f)	73	6740	35 (12)

[a]For conjugate directions, the first number is the total number of line minimizations; this is comparable to the total number of iterations in sequential univariate search. The number in parentheses represents the total number of direction sets used. The final direction set, with only two line searches, is needed only to ensure that the stopping criterion has been met.

Test function (e): $f(\mathbf{X}) = (x_1^2/4) + (x_2^2/11) - 0.8x_1 - x_2 - 0.3x_1x_2 - 3$. At $\mathbf{X}^* = \begin{bmatrix} 490 \\ 814 \end{bmatrix}$ this function has a minimum value of −606.

Test function (f): $f(\mathbf{X}) = 100(x_2 - x_1^2)^2 + (1 - x_1)^2 + 0.01$. At $\mathbf{X}^* = \begin{bmatrix} 1 \\ 1 \end{bmatrix}$ the function has a minimum value of 0.01.

Geometry is indicated in Figures 6.14–6.16. Functions (d) and (e) are included because they illustrate how a relatively slight algebraic change in the function can lead to rather dramatic changes in the contour geometry and hence in the location of the minimizing point and the number of iterations required by some of these numerical methods. In function (c) the x_2^2 term is divided by 9, in (d) by 10, and in (e) by 11. Function (f) is a variant of a class of widely used test functions due to Rosenbrock (hence they are known generically as "Rosenbrock's functions"). Their potentially irritating characteristic (for iterative methods) is the bananalike shape of their contours, which means that the best direction (along the valley floor) changes dramatically from one location to another.

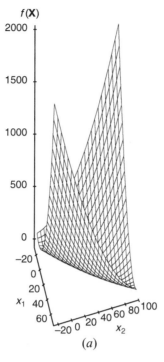

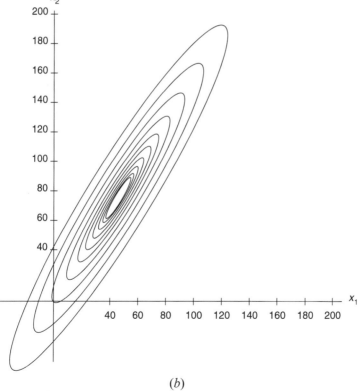

FIGURE 6.14   (a) Test function (d); $f(\mathbf{X}) = (x_1^2/4) + (x_2^2/10) - 0.8x_1 - x_2 - 0.3x_1x_2 - 3$; (b) contours for test function (d).

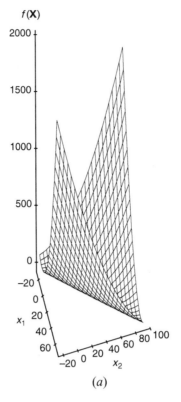

(a)

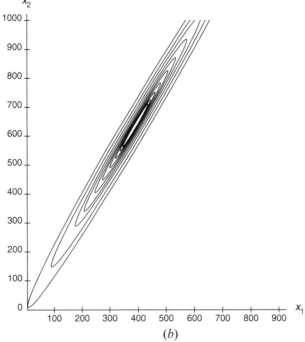

(b)

FIGURE 6.15   (a) Test function (e); $f(\mathbf{X}) = (x_1^2/4) + (x_2^2/11) - 0.8x_1 - x_2 - 0.3x_1x_2 - 3$; (b) contours for test function (e).

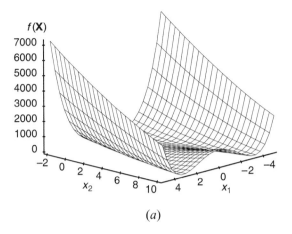

(a)

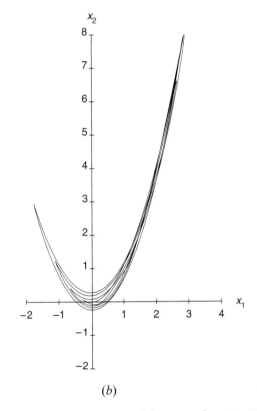

(b)

FIGURE 6.16   (a) Test function (f); $f(\mathbf{X}) = 100(x_2 - x_1^2)^2 + (1 - x_1)^2 + 0.01$; (b) contours for test function (f).

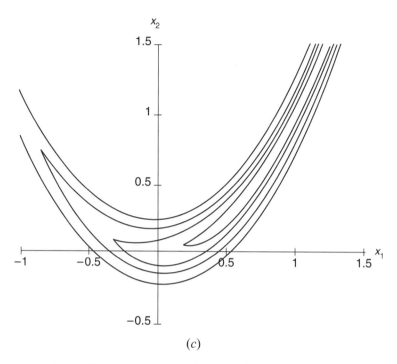

(c)

FIGURE 6.16 (c) magnified contours for test function (f).

It is clear that sequential univariate search is not a useful approach for anything but the simplest of problems. It begins to perform badly with test function (d), and by test function (e) its relative performance is very much worse. Compared with test function (d), the downhill simplex method requires a few (5) more iterations and conjugate directions needs no more at all. Finally, with test function (f), the fundamental mindlessness of sequential univariate search is demonstrated forcefully. It requires an agonizing number (1580) of generally very tiny steps.

The conjugate directions method is fine if line minimizations are not a problem, although it uses 4 times as many iterations for test function (f), which is *not* quadratic, as for the other test functions, which *are* quadratic. The downhill simplex method, where line minimizations play no role at all, is relatively unfazed even by test function (f), for which it requires less than twice as many iterations as for test function (e).

The same relative effectiveness of these three methods is reflected in the results for the much more stringent stopping criterion when $\varepsilon = 10^{-10}$. Here sequential univariate search is really miserable, the downhill simplex method now needs less than twice as many iterations and conjugate directions is hardly affected at all.

6.3 MINIMIZATION OF $f(\mathbf{X})$: DERIVATIVE METHODS

Two "classical" iterative methods for maximization and minimization go as far back as Cauchy (Augustin-Louis Cauchy, French, 1789–1857) and Newton. These techniques are built on the observation (Section 4.2) that the gradient of a function at a point indicates the direction of maximum increase of the function from that point and that the negative of the gradient points in the direction of maximum decrease. A number of more modern derivative-based approaches also make use of this important gradient property, but in combination with other clever ideas. For this reason, we describe the older approaches as "classical gradient methods," and we look at them first.

6.3.1 Classical Gradient Methods

In this section, we continue to concentrate on minimization problems; all that is needed for maximization problems is to change "negative gradient" to "gradient" in what follows.

The Method of Steepest Descent (Cauchy)   After the kth step in an iterative procedure seeking the minimum of $f(\mathbf{X})$, we know that the best *direction* to move is indicated by $-\nabla f(\mathbf{X}^k)$, which we will henceforth denote simply by $-\nabla^k$. Since for nonlinear functions the elements in the gradient vector change as we move about on the function $f(\mathbf{X})$, we need to decide how far to go in the currently best direction before reevaluating the gradient and (probably) changing course. That is, a *step length* needs to be determined. Put simply, the problem after the kth step is to find the next point, $\mathbf{X}^{k+1}$, as

$$\mathbf{X}^{k+1} = \mathbf{X}^k + \Delta\mathbf{X}^k = \mathbf{X}^k + \alpha^k[-\nabla^k] \tag{6.1}$$

where α^k is a scalar indicating, for the move from $\mathbf{X}^k$ to $\mathbf{X}^{k+1}$, how far to move along the direction of the negative gradient. Here is one simple approach to determination of an "optimal" step size. Using the second-order Taylor series approximation for the value of $f(\mathbf{X})$ at a point, $\mathbf{X}^{k+1}$, near to $\mathbf{X}^k$, we have

$$f(\mathbf{X}^{k+1}) = f(\mathbf{X}^k) + [\nabla^k]\circ[\mathbf{X}^{k+1} - \mathbf{X}^k] + \tfrac{1}{2}[\mathbf{X}^{k+1} - \mathbf{X}^k]'\mathbf{H}^k[\mathbf{X}^{k+1} - \mathbf{X}^k] \tag{6.2}$$

where $\mathbf{H}^k$ is the Hessian matrix evaluated at $\mathbf{X}^k$. Let $\Delta f^{k+1} = f(\mathbf{X}^{k+1}) - f(\mathbf{X}^k)$. Then, using (6.1), the difference between the value of the function at the next point and the current one can be expressed (after some rearrangement) as

$$\Delta f^{k+1} = \alpha^k[-\nabla^k]'[\nabla^k] + \tfrac{1}{2}(\alpha^k)^2 \{[-\nabla^k]'\mathbf{H}^k[-\nabla^k]\} \tag{6.3}$$

This difference in values of the function depends only on α^k, since ∇^k and $\mathbf{H}^k$ are made up of constants. For an iterative method that seeks to minimize $f(\mathbf{X})$, $f(\mathbf{X}^{k+1}) < f(\mathbf{X}^k)$, so Δf^{k+1} will be negative. The best choice of $\mathbf{X}^{k+1}$ will be the one for which Δf^{k+1} has as large a negative value as possible, so α^k should be chosen so as to *minimize* Δf^{k+1}. To do this, differentiate (6.3) with respect to α^k and set the resulting expression equal to zero:

$$\frac{d(\Delta f^{k+1})}{d\alpha^k} = [-\nabla^k]'[\nabla^k] + \alpha^k\{[-\nabla^k]'\mathbf{H}^k[-\nabla^k]\} = 0$$

from which (after eliminating redundant minus signs)

$$\alpha^k = \{[\nabla^k]'\mathbf{H}^k[\nabla^k]\}^{-1}[\nabla^k]'[\nabla^k] \tag{6.4}$$

Putting this into (6.1) generates the Cauchy rule for minimization problems:

$$\mathbf{X}^{k+1} = \mathbf{X}^k + \{[\nabla^k]'\mathbf{H}^k[\nabla^k]\}^{-1}[\nabla^k]'[\nabla^k]\,[-\nabla^k] \qquad (6.1')$$

For a minimum, $d^2(\Delta f^{k+1})/(d\alpha^k)^2 > 0$, which will always be true if $\mathbf{H}^k$ is positive definite; this will be the case if $f(\mathbf{X})$ is a convex function. Note in (6.4) that since the expression in curly brackets is a scalar, the -1 exponent simply denotes a reciprocal. In fact, the entire expression possibly looks more complicated than it is. The optimal step size is just the product of the gradient at $\mathbf{X}^k$ with itself (transposed), divided by a scalar involving the gradient and the Hessian at $\mathbf{X}^k$. For a maximization problem, $[-\nabla^k]$ in (6.1) is replaced by simply $[\nabla^k]$, the direction of maximum *increase* of $f(\mathbf{X})$.[13]

When the shape of $f(\mathbf{X})$ does not assure positive definiteness of the Hessian, this particular "optimal" gradient method is not particularly successful, and, in fact, there are alternatives that turn out to be much better in practice. The point of the illustration was simply to sketch one way in which the "best direction" property of the gradient could be exploited in an iterative procedure that needs to be given both a direction and a distance at each step.

Newton's Method   If a function is quite irregular, it is likely that the gradient direction at $\mathbf{X}^k$ will be best for only a very short part of any step, whereas some other direction from $\mathbf{X}^k$ will lead to a larger decrease in $f(\mathbf{X})$ for the same step length. For example, one might modify (6.1) to

$$\mathbf{X}^{k+1} = \mathbf{X}^k + \alpha^k[\mathbf{A}^k][-\nabla^k] \qquad (6.5)$$

where at each step a "weighting matrix," $\mathbf{A}^k$, is used to modify the negative gradient direction. [The procedure in (6.1), above, can be thought of as the special case in which $\mathbf{A}^k = \mathbf{I}$ for all k.]

Newton devised an approach along these lines by the simple expedient of choosing the next value, $\mathbf{X}^{k+1}$, so as to minimize $f(\mathbf{X}^{k+1})$ directly, by straightforward first-order derivative conditions. Differentiating the expression in (6.2) and setting it equal to zero gives

$$\frac{\partial f(\mathbf{X}^{k+1})}{\partial \mathbf{X}^{k+1}} = \nabla^k + \mathbf{H}^k(\mathbf{X}^{k+1} - \mathbf{X}^k) = \mathbf{0} \qquad (6.6)$$

Provided $\mathbf{H}^k$ is nonsingular, it follows that

$$\mathbf{X}^{k+1} = \mathbf{X}^k + (\mathbf{H}^k)^{-1}\,(-\nabla^k) \qquad (6.7)$$

From second-order conditions, $\mathbf{H}^k$ must be positive definite for a minimum. Thus, Newton's method has the form of (6.5) with $\alpha^k = 1$ and $\mathbf{A}^k = (\mathbf{H}^k)^{-1}$ at each step.

If we define the second term on the right-hand side of (6.7) as $\mathbf{D}^k$ (for both "distance" and "direction")

$$\mathbf{D}^k = -(\mathbf{H}^k)^{-1}\,\nabla^k \qquad (6.8)$$

---

[13]You might work out the consequences of that in the final Cauchy rule; it comes down to what is algebraically the *same* expression as in (6.1'), only the negative sign in the second term on the right has "migrated" leftward.

then the linear equation system that defines the unknown $\mathbf{D}^k$'s in terms of the known $\mathbf{H}^k$ and ∇^k is just $\mathbf{H}^k\mathbf{D}^k = -\nabla^k$, and one could also consider *iterative methods* for this system of *linear* equations (Chapter 2), at each step, instead of having to find an inverse each time, as in (6.7) and (6.8).

Numerical Illustrations Table 6.12 contains results for Cauchy's method on test functions (a)–(c). The stopping criterion is the usual one; relative change in $f(\mathbf{X}^*)$ from one step to the next must be no larger than 10^{-5}. Another stopping criterion that is often used in derivative-based methods is that the iterations should end when the elements of the gradient are "sufficiently close" to zero, indicating that there is no next direction in which to move because the current solution is optimal; that is, stop when $\max_i |f_i(\mathbf{X}^k)| \le \gamma$, for some specified, small value of $\gamma > 0$.

For test function (a), with its circular contours, the (constant) gradient, $\begin{bmatrix} -4 \\ -5 \end{bmatrix}$, *always*

points in the direction of the minimum point. Cauchy's method exploits this fact by finding the optimum step length along this (optimum) direction and hence reaches the true minimum in one step. For test functions (b) and (c), note that the first step makes enormous inroads toward the minimizing solution. As with test function (a), this is related to the geometry of the contours of these functions; you can see why by sketching the initial

gradients at $\begin{bmatrix} 0 \\ 0 \end{bmatrix}$ in Figures 6.7–6.9. Sensitivity to alternative values of ε is also illustrated

in Table 6.12 for test function (b). From the contours in Figure 6.9, it should also be clear why Cauchy's method displays the sensitivity that it does to starting points for test function (c), as reported in Table 6.12.

Results for Newton's method are shown in Table 6.13. They illustrate the fact that this iterative technique will always reach the minimum (or, with appropriate changes, the maximum) of a quadratic function in just one step. For such functions, it is completely insensitive to the choice of a starting point. The price you pay is that you must have not only the gradient vector but also the Hessian matrix of second partial and cross-partial derivatives must be known and its inverse must be calculated at each point. For functions that are not strictly quadratic, more iterations will generally be required. For functions with only a few variables, the Hessian matrix will be relatively small, and for functions that are near to quadratic, the number of iterations will be small also. However, for an irregular function of many variables, Newton's method requires a lot of information at each step.

6.3.2 Restricted Step Methods

Newton's method can be very fast and efficient, if the Hessian matrix and its inverse can be found at each $\mathbf{X}^k$ and if the Hessian is positive definite at each of those points. Later in this chapter we will explore some of the ways in which *estimates* of the Hessian (or its inverse) at each point can be used in what is often known as a quasi-Newton iterative approach. At present we want to investigate some different variants of Newton's method. These are useful when (1) the Hessian matrix is not positive definite at all of the $\mathbf{X}^k$ and/or (2) when the quadratic approximation to the function at step k [given by the second-order Taylor series expression for $f(\mathbf{X}^k)$, as in (6.2)] is not a good approximation for more than a very small area around $\mathbf{X}^k$. For simplicity, we refer to $\mathbf{D}^k = -(\mathbf{H}^k)^{-1}\,\nabla^k$ in (6.8) as a *full Newton step* (FNS).

TABLE 6.12   Cauchy's Method on Test Functions (a)–(c)

k	$\mathbf{X}^k$	$f(\mathbf{X}^k)$	∇^k	α
		Test Function (a): $f(\mathbf{X}) = x_1^2 + x_2^2 - 4x_1 - 5x_2 - 5$		
0	$\begin{bmatrix} 0 \\ 0 \end{bmatrix}$	-5	$\begin{bmatrix} -4 \\ -5 \end{bmatrix}$	0.5
1	$\begin{bmatrix} 2 \\ 2.5 \end{bmatrix}$	-15.25	$\begin{bmatrix} 0 \\ 0 \end{bmatrix}$	
		Test Function (b): $f(\mathbf{X}) = x_1^2 + x_2^2 - 4x_1 - 5x_2 - x_1 x_2 - 5$		
0	$\begin{bmatrix} 0 \\ 0 \end{bmatrix}$	-5	$\begin{bmatrix} -4 \\ -5 \end{bmatrix}$	0.97619
1	$\begin{bmatrix} 3.90476 \\ 4.88095 \end{bmatrix}$	-25.01190	$\begin{bmatrix} -1.07143 \\ 0.85714 \end{bmatrix}$	0.33607
2	$\begin{bmatrix} 4.26483 \\ 4.59290 \end{bmatrix}$	-25.32825	$\begin{bmatrix} -0.06323 \\ -0.07904 \end{bmatrix}$	0.97619
3	$\begin{bmatrix} 4.32656 \\ 4.67005 \end{bmatrix}$	-25.33325	$\begin{bmatrix} -0.01694 \\ 0.01355 \end{bmatrix}$	0.33607
4	$\begin{bmatrix} 4.33225 \\ 4.66550 \end{bmatrix}$	-25.33333	$\begin{bmatrix} -0.00100 \\ -0.00125 \end{bmatrix}$	0.97619

Test Function (b): Alternative Stopping Criteria

ε	Iterations	x_1	x_2	$f(\mathbf{X})$
1×10^{-10}	7	4.33333	4.66667	-25.33333
1×10^{-8}	6	4.33332	4.66665	-25.33333
1×10^{-6}	5	4.33323	4.66672	-25.33333

k	$\mathbf{X}^k$	$f(\mathbf{X}^k)$	∇^k	α
		Test Function (c): $f(\mathbf{X}) = (x_1^2/4) + (x_2^2/9) - 0.8x_1 - x_2 - 0.3x_1 x_2 - 3$		
0	$\begin{bmatrix} 0 \\ 0 \end{bmatrix}$	-3	$\begin{bmatrix} -0.8 \\ -1 \end{bmatrix}$	26.35714
1	$\begin{bmatrix} 21.08571 \\ 26.35714 \end{bmatrix}$	-24.61286	$\begin{bmatrix} 1.83571 \\ -1.46857 \end{bmatrix}$	1.46139
2	$\begin{bmatrix} 18.40303 \\ 28.50329 \end{bmatrix}$	-28.65107	$\begin{bmatrix} -0.14947 \\ -0.18684 \end{bmatrix}$	26.35714
3	$\begin{bmatrix} 22.34274 \\ 33.42794 \end{bmatrix}$	-29.40558	$\begin{bmatrix} 0.34299 \\ -0.27439 \end{bmatrix}$	1.46139
4	$\begin{bmatrix} 21.84150 \\ 33.82893 \end{bmatrix}$	-29.54655	$\begin{bmatrix} -0.02793 \\ -0.03491 \end{bmatrix}$	26.35714
5	$\begin{bmatrix} 22.57761 \\ 34.74907 \end{bmatrix}$	-29.57290	$\begin{bmatrix} 0.06409 \\ -0.05127 \end{bmatrix}$	1.46139
6	$\begin{bmatrix} 22.48396 \\ 34.82399 \end{bmatrix}$	-29.57782	$\begin{bmatrix} -0.00522 \\ -0.00652 \end{bmatrix}$	26.35714
7	$\begin{bmatrix} 22.62150 \\ 34.99591 \end{bmatrix}$	-29.57874	$\begin{bmatrix} 0.01197 \\ -0.00958 \end{bmatrix}$	1.46139
8	$\begin{bmatrix} 22.60400 \\ 35.00991 \end{bmatrix}$	-29.57891	$\begin{bmatrix} -0.00098 \\ -0.00122 \end{bmatrix}$	26.35714

TABLE 6.12 *continued*

Test Function (c): Additional Starting Points

Starting Point	Total Number of Iterations
$[1, 2]'$	18
$[100, 50]'$	6
$[100, 200]'$	34
$[1000, 2000]'$	57
$[1000, 1000]'$	23

Suppose that after a FNS from $\mathbf{X}^k$ to a new point $\mathbf{X}^{k+1}$ we find that $f(\mathbf{X}^{k+1}) > f(\mathbf{X}^k)$, either because the FNS carried us outside the realm of accuracy of the quadratic model or because $\mathbf{H}^k$ was not positive definite. It is not absolutely clear that one should always require $f(\mathbf{X}^{k+1}) < f(\mathbf{X}^k)$ at each step; in fact, we will soon explore an example in which Newton's method very effectively corrects for such a momentary lapse and continues quickly to the minimum point.

TABLE 6.13 **Newton's Method on Test Functions (a)–(c)**

k	$\mathbf{X}^k$	$f(\mathbf{X}^k)$	∇^k	$[\mathbf{H}^k]^{-1}$
		Test Function (a): $f(\mathbf{X}) = x_1^2 + x_2^2 - 4x_1 - 5x_2 - 5$		
0	$\begin{bmatrix} 0 \\ 0 \end{bmatrix}$	-5	$\begin{bmatrix} -4 \\ -5 \end{bmatrix}$	$\begin{bmatrix} 0.5 & 0 \\ 0 & 0.5 \end{bmatrix}$
1	$\begin{bmatrix} 2 \\ 2.5 \end{bmatrix}$	-15.25	$\begin{bmatrix} 0 \\ 0 \end{bmatrix}$	
		Test Function (b): $f(\mathbf{X}) = x_1^2 + x_2^2 - 4x_1 - 5x_2 - x_1x_2 - 5$		
0	$\begin{bmatrix} 0 \\ 0 \end{bmatrix}$	-5	$\begin{bmatrix} -4 \\ -5 \end{bmatrix}$	$\begin{bmatrix} 0.66667 & 0.33333 \\ 0.33333 & 0.66667 \end{bmatrix}$
1	$\begin{bmatrix} 4.33333 \\ 4.66667 \end{bmatrix}$	-25.33333	$\begin{bmatrix} 0 \\ 0 \end{bmatrix}$	
		Test Function (c): $f(\mathbf{X}) = (x_1^2/4) + (x_2^2/9) - 0.8x_1 - x_2 - 0.3x_1x_2 - 3$		
0	$\begin{bmatrix} 0 \\ 0 \end{bmatrix}$	-3	$\begin{bmatrix} -0.8 \\ -1 \end{bmatrix}$	$\begin{bmatrix} 10.52632 & 14.21053 \\ 14.21053 & 23.68422 \end{bmatrix}$
1	$\begin{bmatrix} 22.63159 \\ 35.05265 \end{bmatrix}$	-29.57896	$\begin{bmatrix} 0 \\ 0 \end{bmatrix}$	

Test Function (c): Additional Starting Points

Starting Point	Total Number of Iterations
$[1, 2]'$	1
$[100, 50]'$	1
$[100, 200]'$	1
$[1000, 2000]'$	1
$[1000, 1000]'$	1

Nonetheless, intuition suggests that if a FNS generates an $\mathbf{X}^{k+1}$ for which $f(\mathbf{X}^{k+1}) > f(\mathbf{X}^k)$, some kind of "correction" ought to be introduced. There are, broadly speaking, two approaches to making such a correction. One is to keep the FNS direction and change the step length; the other is to fix a step length (at something less than $\mathbf{D}^k$) and change the direction. The first is known as *backtracking* on a Newton step, the second is the *model trust region* approach. We examine these in turn.

Backtracking on the Newton Step   If, after a full Newton step, $\mathbf{D}^k = -(\mathbf{H}^k)^{-1} \nabla^k$, we find that $f(\mathbf{X}^{k+1}) > f(\mathbf{X}^k)$, we could choose to reject that $\mathbf{X}^{k+1}$ and instead do a line minimization along the direction $\mathbf{D}^k$ between $\mathbf{X}^k$ and $\mathbf{X}^k + \mathbf{D}^k$. This would introduce the need for some kind of iterative procedure (Section 6.1) to carry out the line minimization at this step. An alternative that has been found successful is to explore a few scalar reductions of $\mathbf{D}^k$—for example, evaluating $f(\mathbf{X}^{k+1})$ for two or three new values of $\mathbf{X}$, such as $\mathbf{X}^{k+1} = \mathbf{X}^k + (0.5)\mathbf{D}^k$, $\mathbf{X}^{k+1} = \mathbf{X}^k + (0.25)\mathbf{D}^k$, and $\mathbf{X}^{k+1} = \mathbf{X}^k + (0.1)\mathbf{D}^k$, and selecting the $\mathbf{X}^{k+1}$ for which $f(\mathbf{X}^{k+1})$ is smallest.

In order to illustrate how backtracking works, we need a function for which Newton's method does not produce the minimum point immediately after one iteration; that is, we need something other than a quadratic function. We use test function (f); results from application of Newton's method to this function are shown in Table 6.14. At $k = 1$, $\mathbf{X}^1 = \begin{bmatrix} -0.9950 \\ 0.9900 \end{bmatrix}$ and $f(\mathbf{X}^1) = 3.9901$; the full Newton step from there leads to $\mathbf{X}^2 = \begin{bmatrix} 0.9901 \\ -2.9604 \end{bmatrix}$ for which $f(\mathbf{X}^2) = 1552.97$. Since $f(\mathbf{X}^2)$ is very much larger than $f(\mathbf{X}^1)$, we decide to backtrack along the Newton step. Note that this is not necessarily a smart move, since by letting Newton proceed unhampered, the optimum is reached in just three more steps. Newton's method recognizes that $\mathbf{X}^2$ is not a good choice by virtue of the large size of the elements in ∇^2, which contribute to a relatively large correction at that point.

In any case, assuming that we use a rule that says reject $\mathbf{X}^{k+1}$ when $f(\mathbf{X}^{k+1}) > f(\mathbf{X}^k)$, we want to backtrack along the Newton step from $\mathbf{X}^1$ to $\mathbf{X}^2$. The results of this approach

TABLE 6.14   Full Newton Steps on Test Function (f) [Starting Point: (–1, –1)]

k	$\mathbf{X}^k$	$f(\mathbf{X}^k)$	∇^k	$\Delta\mathbf{X}$
0	$\begin{bmatrix} -1 \\ -1 \end{bmatrix}$	404.01	$\begin{bmatrix} -804 \\ -400 \end{bmatrix}$	$\begin{bmatrix} 0.0050 \\ 1.9900 \end{bmatrix}$
1	$\begin{bmatrix} -0.9950 \\ 0.9900 \end{bmatrix}$	3.9901	$\begin{bmatrix} -3.9999 \\ -0.0050 \end{bmatrix}$	$\begin{bmatrix} 1.9851 \\ -3.9504 \end{bmatrix}$
2	$\begin{bmatrix} 0.9901 \\ -2.9604 \end{bmatrix}$	1552.97	$\begin{bmatrix} 1560.72 \\ -788.15 \end{bmatrix}$	$\begin{bmatrix} 0.0000 \\ 3.9408 \end{bmatrix}$
3	$\begin{bmatrix} 0.9901 \\ 0.9804 \end{bmatrix}$	0.0101	$\begin{bmatrix} -0.0197 \\ 0.0000 \end{bmatrix}$	$\begin{bmatrix} 0.0099 \\ 0.0195 \end{bmatrix}$
4	$\begin{bmatrix} 1.0000 \\ 0.9999 \end{bmatrix}$	0.0100	$\begin{bmatrix} 0.0389 \\ -0.0195 \end{bmatrix}$	$\begin{bmatrix} 0.0000 \\ 0.0001 \end{bmatrix}$
5	$\begin{bmatrix} 1.0000 \\ 1.0000 \end{bmatrix}$	0.0100	$\begin{bmatrix} 0 \\ 0 \end{bmatrix}$	—

are shown in Table 6.15. At $k = 1$, by using $\mathbf{D}^1 = (0.1)[-(\mathbf{H}^k)^{-1}\nabla^k]$, we generate $\mathbf{X}^2 = \begin{bmatrix} -0.7965 \\ 0.5950 \end{bmatrix}$, for which $f(\mathbf{X}^2) = 3.3928 < f(\mathbf{X}^1) = 3.9901$. However, using this $\mathbf{X}^2$ as the starting point for further iterations, even an uninterrupted series of full Newton steps requires 11 more iterations (many more than were needed in Table 6.14, where full Newton steps were never rejected). Nonetheless, continuing with the backtracking approach from the new $\mathbf{X}^2$ requires a total of 17 more steps, and at three of them the FNS is not allowed. So while backtracking does get us to the optimal solution, it turns out not to be very efficient in this illustration.

Model Trust Regions   In this approach, based on work by Levenberg (1944) and Marquardt (1963), and with which those names are often associated, a region is defined around the current point, $\mathbf{X}^k$, within which it is felt that the quadratic approximation in (6.2) is "reasonably" accurate. (We will soon see what that means.) Since the quadratic function is a model of $f(\mathbf{X}^k)$, the effect is to establish a region within which the model can be *trusted*; hence the name. [Apparently the first application of this idea to optimization problems is to be found in Goldfeld et al. (1966).]

   One common approach is to establish a circular (two dimensions) or spherical (three or more dimensions) region by defining its radius. Then the problem at step k is defined as minimizing the quadratic approximation to $f(\mathbf{X})$ at $\mathbf{X}^k$ subject to the constraint that the new point, and hence the step length from $\mathbf{X}^k$ to it, lie within this region. For example, find the step from $\mathbf{X}^k$, $\mathbf{S}^k = \mathbf{X}^{k+1} - \mathbf{X}^k$, so as to

$$\text{Minimize} \quad q^k(\mathbf{S}^k) = f(\mathbf{X}^k) + [\nabla^k]'\mathbf{S}^k$$
$$+ (0.5)[\mathbf{S}^k]'\mathbf{H}^k\mathbf{S}^k \qquad (6.9)$$
$$\text{subject to} \quad \|\mathbf{S}^k\| \leq r^k$$

Here q^k represents the quadratic approximation to $f(\mathbf{X}^{k+1})$ around the point $\mathbf{X}^k$ (note that it is a function of $\mathbf{S}^k$ only), and $\|\mathbf{S}^k\|$ denotes the *length* (norm) of the vector, $[(\mathbf{S}^k)'\mathbf{S}^k]^{1/2}$. (This was the form of $\|\mathbf{S}^k\|$ that was proposed in the work of Levenberg and Marquardt; other writers have proposed different mathematical definitions of vector norms which we need not explore here.) Finally, r^k is the radius that has been selected for the trust region around $\mathbf{X}^k$. (How that radius is decided upon is another matter.)

   In structure, (6.9) is a constrained minimization problem of the sort that will be explored in Chapters 10 and 11. However, if the constraint were a strict equality, we could use the Lagrange multiplier approach of Chapter 4. Assume for the moment that this constraint is in fact

$$\|\mathbf{S}^k\| = [(\mathbf{S}^k)'\mathbf{S}^k]^{1/2} = r^k$$

or, squaring both sides

$$(\mathbf{S}^k)'\mathbf{S}^k = (r^k)^2$$

The associated Lagrangian function is

$$L = f(\mathbf{X}^k) + [\nabla^k]'\mathbf{S}^k + \tfrac{1}{2}[\mathbf{S}^k]'\mathbf{H}^k\mathbf{S}^k + \tfrac{1}{2}\lambda^k[(\mathbf{S}^k)'\mathbf{S}^k - (r^k)^2] \qquad (6.10)$$

TABLE 6.15   Backtracking on a Full Newton Step. [Test Function (f); Starting Point (−1, −1)]

k	$\mathbf{X}^k$	$f(\mathbf{X}^k)$	Should a FNS be taken? (If no, see footnotes)	If no, use $\mathbf{D}^k = -\alpha(\mathbf{H}^k)^{-1}\nabla^k$. Value of α chosen at this step	$\mathbf{D}^k$
0	$\begin{bmatrix} -1 \\ -1 \end{bmatrix}$	404.01	Y	—	$\begin{bmatrix} 0.0050 \\ 1.9900 \end{bmatrix}$
1	$\begin{bmatrix} -0.9950 \\ 0.9900 \end{bmatrix}$	3.9901	N	0.1	$\begin{bmatrix} 0.1985 \\ -0.3950 \end{bmatrix}$
2	$\begin{bmatrix} -0.7965 \\ 0.5950 \end{bmatrix}$	3.3928	Y	—	$\begin{bmatrix} 0.2022 \\ -0.2826 \end{bmatrix}$
3	$\begin{bmatrix} -0.5943 \\ 0.3123 \end{bmatrix}$	2.7189	Y	—	$\begin{bmatrix} 0.1738 \\ -0.1657 \end{bmatrix}$
4	$\begin{bmatrix} -0.4206 \\ 0.1467 \end{bmatrix}$	2.1191	Y	—	$\begin{bmatrix} 0.2018 \\ -0.1396 \end{bmatrix}$
5	$\begin{bmatrix} -0.2187 \\ 0.0071 \end{bmatrix}$	1.6612	Y	—	$\begin{bmatrix} 0.1332 \\ -0.0175 \end{bmatrix}$
6	$\begin{bmatrix} -0.0855 \\ -0.0104 \end{bmatrix}$	1.2198	Y	—	$\begin{bmatrix} 0.2385 \\ -0.0230 \end{bmatrix}$
7	$\begin{bmatrix} 0.1531 \\ -0.0335 \end{bmatrix}$	1.0511	Y	—	$\begin{bmatrix} 0.0684 \\ 0.0778 \end{bmatrix}$
8	$\begin{bmatrix} 0.2215 \\ 0.0444 \end{bmatrix}$	0.6183	N	0.5	$\begin{bmatrix} 0.2011 \\ 0.0914 \end{bmatrix}$
9	$\begin{bmatrix} 0.4225 \\ 0.1358 \end{bmatrix}$	0.5264	Y	—	$\begin{bmatrix} 0.0604 \\ 0.0938 \end{bmatrix}$
10	$\begin{bmatrix} 0.4830 \\ 0.2296 \end{bmatrix}$	0.2786	N	0.5	$\begin{bmatrix} 0.1494 \\ 0.1461 \end{bmatrix}$
11	$\begin{bmatrix} 0.6324 \\ 0.3757 \end{bmatrix}$	0.2034	Y	—	$\begin{bmatrix} 0.0631 \\ 0.1039 \end{bmatrix}$
12	$\begin{bmatrix} 0.6954 \\ 0.4797 \end{bmatrix}$	0.1043	N	0.5	$\begin{bmatrix} 0.0848 \\ 0.1199 \end{bmatrix}$
13	$\begin{bmatrix} 0.7802 \\ 0.5996 \end{bmatrix}$	0.0667	Y	—	$\begin{bmatrix} 0.0775 \\ 0.1301 \end{bmatrix}$
14	$\begin{bmatrix} 0.8577 \\ 0.7297 \end{bmatrix}$	0.0338	Y	—	$\begin{bmatrix} 0.0646 \\ 0.1169 \end{bmatrix}$
15	$\begin{bmatrix} 0.9224 \\ 0.8466 \end{bmatrix}$	0.0178	Y	—	$\begin{bmatrix} 0.0423 \\ 0.0829 \end{bmatrix}$
16	$\begin{bmatrix} 0.9647 \\ 0.9288 \end{bmatrix}$	0.0116	Y	—	$\begin{bmatrix} 0.0260 \\ 0.0520 \end{bmatrix}$
17	$\begin{bmatrix} 0.9907 \\ 0.9808 \end{bmatrix}$	0.0101	Y	—	$\begin{bmatrix} 0.0082 \\ 0.0169 \end{bmatrix}$

TABLE 6.15 *continued*

k	$\mathbf{X}^k$	$f(\mathbf{X}^k)$	Should a FNS be taken? (If no, see footnotes)	If no, use $\mathbf{D}^k = -\alpha(\mathbf{H}^k)^{-1}\nabla^k.$ Value of α chosen at this step	$\mathbf{D}^k$
18	$\begin{bmatrix} 0.9989 \\ 0.9977 \end{bmatrix}$	0.0100	Y	—	$\begin{bmatrix} 0.0011 \\ 0.0023 \end{bmatrix}$
19	$\begin{bmatrix} 1.0000 \\ 1.0000 \end{bmatrix}$	0.0100	Y	—	$\begin{bmatrix} 0.00001 \\ 0.00003 \end{bmatrix}$

Notes:

At $k = 1$, a full Newton step generates $\mathbf{X}^2 = \begin{bmatrix} 0.9901 \\ -2.9604 \end{bmatrix}$ for which $f(\mathbf{X}^2) = 1552.97 > f(\mathbf{X}^1)$, so that value for $\mathbf{X}^2$ is rejected.

At $k = 8$, a full Newton step generates $\mathbf{X}^9 = \begin{bmatrix} 0.6237 \\ 0.2272 \end{bmatrix}$ for which $f(\mathbf{X}^9) = 2.7696 > f(\mathbf{X}^8)$, so that value for $\mathbf{X}^9$ is rejected.

At $k = 10$, a full Newton step generates $\mathbf{X}^{11} = \begin{bmatrix} 0.7818 \\ 0.5218 \end{bmatrix}$ for which $f(\mathbf{X}^{11}) = 0.8570 > f(\mathbf{X}^{10})$, so that value for $\mathbf{X}^{11}$ is rejected.

At $k = 12$, a full Newton step generates $\mathbf{X}^{13} = \begin{bmatrix} 0.8650 \\ 0.7195 \end{bmatrix}$ for which $f(\mathbf{X}^{13}) = 0.1107 > f(\mathbf{X}^{12})$, so that value for $\mathbf{X}^{13}$ is rejected.

where λ^k is the Lagrange multiplier at the kth step and the final term is multiplied by $-\frac{1}{2}$ for numerical convenience in the first-order conditions. These conditions are

$$\frac{\partial L}{\partial \mathbf{S}^k} = \nabla^k + \mathbf{H}^k \mathbf{S}^k + \lambda^k \mathbf{S}^k = \mathbf{0} \tag{6.11}$$

and

$$\frac{\partial L}{\partial \lambda^k} = (\mathbf{S}^k)' \mathbf{S}^k - (r^k)^2 = 0 \tag{6.12}$$

The condition in (6.12) simply reproduces the constraint, as always when using a Lagrange multiplier approach for equality constrained problems. From (6.11), we see that

$$(\mathbf{H}^k + \lambda^k \mathbf{I})\mathbf{S}^k = -\nabla^k \qquad \text{or} \qquad \mathbf{S}^k = -(\mathbf{H}^k + \lambda^k \mathbf{I})^{-1}\nabla^k \tag{6.13}$$

which says that the Hessian matrix at $\mathbf{X}^k$ should be modified by adding a multiple of the identity matrix. Note that if $\lambda^k = 0$, we have precisely a full Newton step, as in (6.8). If λ^k (and hence $\lambda^k \mathbf{I}$) is very large relative to (the on-diagonal elements of) $\mathbf{H}^k$, $\mathbf{S}^k$ approaches $-(1/\lambda^k)\nabla^k$, the negative gradient (Cauchy) direction.

We explored results for *inequality* constrained optimization problems in Chapter 4, and we will do so again later, in Chapter 10, but they are needed now to complete our investigation of the model trust region approach. It turns out that optimal solution to the problem in (6.9)—that is, with the *inequality* constraint—can be characterized as

$$(\mathbf{H}^k + \lambda^k \mathbf{I})\mathbf{S}^k = -\nabla^k \qquad \text{or} \qquad \mathbf{S}^k = -(\mathbf{H}^k + \lambda^k \mathbf{I})^{-1}\nabla^k \qquad (6.14a)$$

$$\lambda^k[(\mathbf{S}^{k\prime}\mathbf{S}^k - (r^k)^2] = 0 \qquad (6.14b)$$

$$[(\mathbf{S}^k)^\prime \mathbf{S}^k - (r^k)^2] \le 0 \qquad (6.14c)$$

$$\lambda^k \ge 0 \qquad (6.14d)$$

From (6.14b) and (6.14d), if $\lambda^k \ne 0$ (which means $\lambda^k > 0$) then $(\mathbf{S}^k)^\prime \mathbf{S}^k = (r^k)^2$, the step length is as long as allowed by the radius of the trust region, and so the next point, $\mathbf{X}^{k+1}$, lies on the circumference of that region. On the other hand, if $(\mathbf{S}^k)^\prime \mathbf{S}^k < (r^k)^2$, then $\lambda^k = 0$, which means that $\mathbf{S}^k$ is a full Newton step.

Therefore a solution approach to the constrained minimization problem in (6.9) is as follows:

(a) Find the FNS, $\mathbf{S}^k = \mathbf{D}^k$, as in (6.8). If $(\mathbf{D}^k)^\prime \mathbf{D}^k \le (r^k)^2$, keep $\mathbf{D}^k$ as $\mathbf{S}^k$ and find $\mathbf{X}^{k+1} = \mathbf{X}^k + \mathbf{D}^k = \mathbf{X}^k - (\mathbf{H}^k)^{-1}\nabla^k$.

(b) If $(\mathbf{D}^k)^\prime \mathbf{D}^k > (r^k)^2$, search for the particular $\lambda^k \ge 0$ that generates an associated $\mathbf{S}^k$ for which $(\mathbf{S}^k)^\prime \mathbf{S}^k = (r^k)^2$. This is in itself an iterative search problem, and there are several approaches to finding a solution value for λ^k. Two of the most common of these produce step results that are known as the *hook step* [an attempt to describe the geometry (Moré, 1977)] and the *dogleg step* [also an attempt, less successful, to describe the geometry (Powell, 1970)], or *double dogleg step* (Dennis and Mei, 1979). [Details of these procedures can be found, for example, in Dennis and Schnabel (1983, Chapter 6).]

The following ratio is often used as an indicator of whether the quadratic model, as in (6.9), is adequate over the length of the entire step length, $\mathbf{S}^k$, from $\mathbf{X}^k$ to $\mathbf{X}^{k+1}$ [see, e.g., Fletcher (1987, Chapter 5)]:

$$R^k = \frac{f(\mathbf{X}^k) - f(\mathbf{X}^k + \mathbf{S}^k)}{f(\mathbf{X}^k) - q^k(\mathbf{S}^k)} \qquad (6.15)$$

In words, the numerator is the *actual* change in the value of $f(\mathbf{X})$ (usually a decrease, for a minimization problem), between $\mathbf{X}^k$ and $\mathbf{X}^{k+1} = \mathbf{X}^k + \mathbf{S}^k$, and the denominator is the difference between $f(\mathbf{X})$ and the *predicted* value of $f(\mathbf{X}^{k+1})$ that is given by the quadratic model in (6.9) (again, usually a decrease in minimization problems). The following rules of thumb are arbitrary but seem to have worked well [again, e.g., see Fletcher (1987); a slightly different version can be found in Dennis and Schnabel (1983, Chapter 6)]:

(a) If $R^k < 0.25$, the model is not predicting very well, so shrink the trust region radius for the next step to $\frac{1}{4}$ of its current value—that is, $r^k + 1 = (0.25)r^k$.

(b) if $R^k \ge 0.75$, correspondence between the actual function and the quadratic model of the function is judged to be "reasonable"; if $\|\mathbf{S}^k\| = r^k$, set $r^{k+1} = 2r^k$, otherwise (meaning when $\|\mathbf{S}^k\| < r^k$), set $r^{k+1} = r^k$.

(c) Finally, if $R^k \le 0$, set $\mathbf{X}^{k+1} = \mathbf{X}^k$ and, as in (a), set $r^{k+1} = (0.25)r^k$. This just says to ignore the newly found $\mathbf{X}^{k+1}$; go back to where you were $(\mathbf{X}^k)$, shrink the radius of the trust region, and try again.

Results from using these kinds of model trust region approaches to test function (f) are shown in Table 6.16. Again, as with the backtracking procedure on this same test function

TABLE 6.16   The Model Trust Region Approach on Test Function (f) [Starting Point (–1, –1)]

k	$\mathbf{X}^k$	$f(\mathbf{X}^k)$	∇^k	$\mathbf{H}^k$	$\Delta\mathbf{X}^k$	r^k	$\lVert S^k\rVert$	R^k	λ^k
0	$\begin{bmatrix}-1\\2\end{bmatrix}$	404.01	$\begin{bmatrix}-804\\-400\end{bmatrix}$	$\begin{bmatrix}1602 & 400\\400 & 200\end{bmatrix}$	$\begin{bmatrix}0.0050\\1.9900\end{bmatrix}$	2	1.99	1	0
1	$\begin{bmatrix}-0.9950\\0.9900\end{bmatrix}$	3.9901	$\begin{bmatrix}-3.9999\\-0.0050\end{bmatrix}$	$\begin{bmatrix}794.05 & 398.00\\398.00 & 200.00\end{bmatrix}$	$\begin{bmatrix}0.2179\\-0.4265\end{bmatrix}$	1	0.48	0.81	3.33[a]
2	$\begin{bmatrix}-0.7771\\0.5635\end{bmatrix}$	3.3312	$\begin{bmatrix}-16.1066\\-8.0764\end{bmatrix}$	$\begin{bmatrix}501.27 & 310.84\\310.84 & 200.00\end{bmatrix}$	$\begin{bmatrix}0.1958\\-0.2639\end{bmatrix}$	1	0.32	1.32	0
3	$\begin{bmatrix}-0.5813\\0.2996\end{bmatrix}$	2.6575	$\begin{bmatrix}-12.0766\\-7.6671\end{bmatrix}$	$\begin{bmatrix}287.67 & 232.52\\232.52 & 200.00\end{bmatrix}$	$\begin{bmatrix}0.1825\\-0.1737\end{bmatrix}$	1	0.25	1.33	0
4	$\begin{bmatrix}-0.3989\\0.1258\end{bmatrix}$	2.0776	$\begin{bmatrix}-8.1086\\-6.6576\end{bmatrix}$	$\begin{bmatrix}142.59 & 159.54\\159.54 & 200.00\end{bmatrix}$	$\begin{bmatrix}0.1827\\-0.1124\end{bmatrix}$	1	0.21	1.30	0
5	$\begin{bmatrix}-0.2162\\0.0134\end{bmatrix}$	1.6005	$\begin{bmatrix}-5.3181\\-6.6741\end{bmatrix}$	$\begin{bmatrix}52.74 & 86.47\\86.47 & 200.00\end{bmatrix}$	$\begin{bmatrix}0.1585\\-0.0351\end{bmatrix}$	1	0.16	1.34	0
6	$\begin{bmatrix}-0.0577\\-0.0218\end{bmatrix}$	1.1918	$\begin{bmatrix}-2.6951\\-5.0231\end{bmatrix}$	$\begin{bmatrix}14.71 & 23.08\\23.08 & 200.00\end{bmatrix}$	$\begin{bmatrix}0.1756\\0.0048\end{bmatrix}$	1	0.18	1.24	0
7	$\begin{bmatrix}0.1179\\-0.0169\end{bmatrix}$	0.8832	$\begin{bmatrix}-0.3098\\-6.1676\end{bmatrix}$	$\begin{bmatrix}25.56 & -47.16\\-47.16 & 200.00\end{bmatrix}$	$\begin{bmatrix}0.1231\\0.0599\end{bmatrix}$	1	0.14	1.35	0
8	$\begin{bmatrix}0.2410\\0.0429\end{bmatrix}$	0.6091	$\begin{bmatrix}-0.0582\\-3.0291\end{bmatrix}$	$\begin{bmatrix}54.51 & -96.39\\-96.39 & 200.00\end{bmatrix}$	$\begin{bmatrix}0.1884\\0.1059\end{bmatrix}$	1	0.22	0.88	0
9	$\begin{bmatrix}0.4294\\0.1489\end{bmatrix}$	0.4616	$\begin{bmatrix}4.9538\\-7.0980\end{bmatrix}$	$\begin{bmatrix}163.67 & -171.74\\-171.74 & 200.00\end{bmatrix}$	$\begin{bmatrix}0.0705\\0.0960\end{bmatrix}$	1	0.12	1.19	0
10	$\begin{bmatrix}0.4998\\0.2449\end{bmatrix}$	0.2626	$\begin{bmatrix}-0.0208\\-0.9800\end{bmatrix}$	$\begin{bmatrix}203.85 & -199.93\\-199.93 & 200.00\end{bmatrix}$	$\begin{bmatrix}0.1220\\0.1257\end{bmatrix}$	0.25	0.18	0.88	2.075[b]

From this point onward, full Newton steps are always successful in producing an $\mathbf{X}^{k+1}$ for which $f(\mathbf{X}^{k+1}) < f(\mathbf{X}^k)$. We indicate only the last two iterations.

k	$\mathbf{X}^k$	$f(\mathbf{X}^k)$	∇^k	$\mathbf{H}^k$	$\Delta\mathbf{X}^k$	r^k	$\lVert S^k\rVert$	R^k	λ^k
18	$\begin{bmatrix}0.9998\\0.9997\end{bmatrix}$	0.0100	$\begin{bmatrix}0.0006\\-0.0005\end{bmatrix}$	$\begin{bmatrix}801.72 & -399.93\\-399.93 & 200.00\end{bmatrix}$	$\begin{bmatrix}0.0002\\0.0003\end{bmatrix}$	0.25	0.0004	1.00	0
19	$\begin{bmatrix}1.0000\\1.0000\end{bmatrix}$	0.0100	$\begin{bmatrix}0\\0\end{bmatrix}$	$\begin{bmatrix}802 & -400\\-400 & 200\end{bmatrix}$	Stop	—	—	—	—

[a]At $k = 1$, a full Newton step ($\lambda = 0$) generates $\mathbf{X}^2 = \begin{bmatrix}0.9901\\-2.9604\end{bmatrix}$ for which $R^1 = -1548.98/3.96 = -391$ [$f(\mathbf{X}^2) =$ 1552.97], so that value for $\mathbf{X}^2$ is rejected, the trust region radius is reduced to one-quarter of its previous size, and the optimization problem in (6.15) is solved for that new radius.

[b]At $k = 10$, a full Newton step ($\lambda = 0$) generates $\mathbf{X}^{11} = \begin{bmatrix}0.7508\\0.5007\end{bmatrix}$ for which $R^{10} = -1.61$ [$f(\mathbf{X}^{11}) = 0.4687$], so that value for $\mathbf{X}^{11}$ is rejected, the trust region radius is reduced to one-quarter of its previous size, and the optimization problem in (6.15) is solved for that new radius.

(in Table 6.15), we find that for this particular example it would have been more efficient to use an unhampered series of full Newton steps.

6.3.3   Conjugate Gradient Methods

Attempts to combine the steepest descent properties of classical gradient methods with the positive ("noninterfering") characteristics of conjugate directions have led to what are known, logically, as *conjugate gradient methods*. The idea is simply to choose a direction at each iteration that will adequately modify the gradient direction at that point (the negative gradient direction for minimization problems) so that the new direction is conjugate to (and hence "noninterfering" with) the last direction. As with conjugate direction methods (Section 6.2.4), the conjugate gradient approach relies on quadratic approximations to $f(\mathbf{X})$. While the idea is simple, the derivation is somewhat complicated. It occupies the next few paragraphs.

The Logic of Conjugate Gradient Methods   Start at $\mathbf{X}^0$. Find $\boldsymbol{\nabla}^0$ and search along the direction given by $-\boldsymbol{\nabla}^0$ for the minimum point, $\mathbf{X}^1$. Evaluate $\nabla f(\mathbf{X}^1)$ ($= \boldsymbol{\nabla}^1$) but, instead of now finding the minimum point along $-\boldsymbol{\nabla}^1$, construct a direction, $\mathbf{D}^1$, that incorporates this negative gradient direction but with a modification so that $\mathbf{D}^1$ will be conjugate to the initial direction, $\mathbf{D}^0 = -\boldsymbol{\nabla}^0$. For example, we could represent $\mathbf{D}^1$ as $\mathbf{D}^1 = -\boldsymbol{\nabla}^1 + \mathbf{M}^1$, where the vector $\mathbf{M}^1$ contains the appropriate modifications. Since $\mathbf{D}^0$ is the only other n-element vector that we have at this point, we could imagine that the adjustment terms in $\mathbf{M}^1$ might be representable as some scalar modification (multiple) of it, as $\mathbf{M}^1 = w^0\mathbf{D}^0$. If this were the case, then

$$\mathbf{D}^1 = -\boldsymbol{\nabla}^1 + w^0\mathbf{D}^0 \tag{6.16}$$

Conjugacy of $\mathbf{D}^1$ and $\mathbf{D}^0$ requires $(\mathbf{D}^1)'\mathbf{A}\mathbf{D}^0 = 0$, for some matrix $\mathbf{A}$. Postmultiplying both sides of (6.16) by $\mathbf{A}\mathbf{D}^0$, we obtain

$$(\mathbf{D}^1)'\mathbf{A}\mathbf{D}^0 = 0 = (-\boldsymbol{\nabla}^1)'\mathbf{A}\mathbf{D}^0 + w^0[(\mathbf{D}^0)'\mathbf{A}\mathbf{D}^0]$$

from which

$$w^0 = \frac{(\boldsymbol{\nabla}^1)'\mathbf{A}\mathbf{D}^0}{(\mathbf{D}^0)'\mathbf{A}\mathbf{D}^0} \tag{6.17}$$

[as long as $(\mathbf{D}^0)'\mathbf{A}\mathbf{D}^0 \neq 0$] and so, from (6.16):

$$\mathbf{D}^1 = -\boldsymbol{\nabla}^1 + \left[\frac{(\boldsymbol{\nabla}^1)'\mathbf{A}\mathbf{D}^0}{(\mathbf{D}^0)'\mathbf{A}\mathbf{D}^0}\right]\mathbf{D}^0$$

This vector defines a direction of movement from $\mathbf{X}^1$ that modifies the steepest descent direction, $-\boldsymbol{\nabla}^1$, in order to remain conjugate to the direction in the immediately previous step, $\mathbf{D}^0$.

Recall from Section 6.2.4 that for $q(\mathbf{X}) = \frac{1}{2}\mathbf{X}'\mathbf{A}\mathbf{X} + \mathbf{B}'\mathbf{X} + c$, the Hessian matrix translates differences in $\mathbf{X}$ into differences in $\nabla q(\mathbf{X})$:

$$\boldsymbol{\nabla}^{k+1} - \boldsymbol{\nabla}^k = \mathbf{A}(\mathbf{X}^{k+1} - \mathbf{X}^k) \quad \text{or} \quad \mathbf{X}^{k+1} - \mathbf{X}^k = \mathbf{A}^{-1}(\boldsymbol{\nabla}^{k+1} - \boldsymbol{\nabla}^k) \tag{6.18}$$

where $\mathbf{A} = \mathbf{H} = \mathbf{H}^k$. If $f(\mathbf{X})$ is not quadratic, the same result holds from a Taylor series approximation of the function $\nabla f(\mathbf{X})$ at $\mathbf{X}^{k+1}$ around $\nabla f(\mathbf{X}^k)$, ignoring third (and higher) terms. The ith element in $(\mathbf{X}^{k+1} - \mathbf{X}^k)$ represents the distance between x_i^{k+1} and x_i^k, measured along the x_i axis. The vector itself represents both the *direction* from $\mathbf{X}^k$ to $\mathbf{X}^{k+1}$, $\mathbf{D}^k$, and a *distance* or *step size, s^k,* along that direction, so

$$\nabla^{k+1} = \nabla^k + s^k \mathbf{A} \mathbf{D}^k$$

In particular, for $k = 0$

$$\nabla^1 = \nabla^0 + s^0 \mathbf{A} \mathbf{D}^0 \qquad (6.19)$$

If, as is the case in gradient methods, successive gradients are to be orthogonal, $(\nabla^0)'\nabla^1 = 0$, then, multiplying (6.19) through by $(\nabla^0)'$, we have

$$(\nabla^0)'\nabla^0 + s^0(\nabla^0)'\mathbf{A}\mathbf{D}^0 = 0$$

and so

$$s^0 = \frac{(-\nabla^0)'\nabla^0}{(\nabla^0)'\mathbf{A}\mathbf{D}^0}$$

Consider the expression for w^0 in (6.17), using the fact from (6.19) that $\mathbf{A}\mathbf{D}^0 = (\nabla^1 - \nabla^0)/s^0$.

(a) For the *numerator* of (6.17), $(\nabla^1)'\mathbf{A}\mathbf{D}^0 = [(\nabla^1)'\nabla^1 - (\nabla^1)'\nabla^0]/s^0 = (\nabla^1)'(\nabla^1)/s^0$ (provided successive gradients are orthogonal, so that $(\nabla^1)'\nabla^0 = 0$).

(b) For the *denominator* in (6.17), $(\mathbf{D}^0)'\mathbf{A}\mathbf{D}^0 = [(\mathbf{D}^0)'\nabla^1 - (\mathbf{D}^0)'\nabla^0]/s^0$. Since a new point $\mathbf{X}^{k+1}$ represents the minimum along $\mathbf{D}^k$, $(\mathbf{D}^k)'\nabla^{k+1} = 0$; each new gradient must be orthogonal to the line along which the previous minimization was carried out. Therefore, $(\mathbf{D}^0)'\nabla^1 = 0$ so $(\mathbf{D}^0)'\mathbf{A}\mathbf{D}^0 = (-\mathbf{D}^0)'\nabla^0/s^0$, and, since $\mathbf{D}^0 = -\nabla^0$, $(\mathbf{D}^0)'\mathbf{A}\mathbf{D}^0 = (\nabla^0)'\nabla^0/s^0$. Therefore (and finally), w^0 in (6.17) becomes

$$w^0 = \frac{(\nabla^1)'\nabla^1}{(\nabla^0)'\nabla^0}$$

and (6.16) can be written as

$$\mathbf{D}^1 = -\nabla^1 + \left[\frac{(\nabla^1)'\nabla^1}{(\nabla^0)'\nabla^0} \right] \mathbf{D}^0 \qquad (6.16')$$

The value of this observation is that the first new direction, $\mathbf{D}^1$, can now be found from $\mathbf{D}^0$ *without knowing the matrix* $\mathbf{A}$. To summarize:

(a) Given $\mathbf{X}^0$, find $-\nabla^0 = \mathbf{D}^0$.

(b) Find the minimum along $\mathbf{D}^0$, denote it $\mathbf{X}^1$, and find ∇^1.

(c) Use the old direction along with the old and new gradients to find $\mathbf{D}^1$ as in (6.16'). (*Note*: Only two *gradients* and the old direction are needed.)

(d) Find the minimum along $\mathbf{D}^1$; denote it $\mathbf{X}^2$, and find ∇^2.

(e) Construct $\mathbf{D}^2 = -\nabla^2 + [(\nabla^2)'\nabla^2/(\nabla^1)'\nabla^1]\mathbf{D}^1$. (Again, only the last two gradients and the previous direction are used to determine the next direction.)

(f) Find the minimum along $\mathbf{D}^2$ and continue as above.

It turns out that this procedure creates a sequence of directions in which (i) not only is each direction conjugate to the immediately preceding one $[(\mathbf{D}^{k+1})'\mathbf{A}\mathbf{D}^k = 0]$ but in fact *all directions are mutually conjugate* $[(\mathbf{D}^k)'\mathbf{A}\mathbf{D}^j = 0$ for all $j \neq k]$ and (ii) not only is each gradient orthogonal to its immediate predecessor $[(\nabla^{k+1})'\nabla^k = 0]$ but in fact *all gradients are mutually orthogonal* $[(\nabla^k)'\nabla^j = 0$ for all $j \neq k]$.[14]

The rules in (a)–(f) use

$$w^k = \frac{(\nabla^{k+1})'\nabla^{k+1}}{(\nabla^k)'\nabla^k} \tag{6.20}$$

This representation of the iterative procedure is credited to Fletcher and Reeves (1964). An alternative, due to Polak and Ribiere (Polak, 1971) uses, instead

$$w^k = \frac{(\nabla^{k+1} - \nabla^k)'\nabla^{k+1}}{(\nabla^k)'\nabla^k} \tag{6.21}$$

For quadratic functions, $(\nabla^k)'\nabla^{k+1} = 0$, and so the distinction is unnecessary. However, for functions that are being *approximated* as quadratics, there is some evidence that the Polak–Ribiere version in (6.21) works better.

Numerical Illustrations Conjugate gradient search results for the three test functions are given in Table 6.17. The first step, in each case, is identical to that in Cauchy's method. (Do you see why?) The method is not quite as spectacular as Newton's approach, but neither does it require Hessians and their inverses. And it seems to handle a variety of starting points with relative ease. The steps to the solution for test function (c) are shown in Figure 6.17.

6.3.4   Quasi-Newton (Variable Metric) Methods

Newton's method, in which

$$\mathbf{X}^{k+1} = \mathbf{X}^k - (\mathbf{H}^k)^{-1}\nabla^k$$

[as in (6.7)] is very fast; it reaches the minimum of a quadratic function in just one iteration. There is, naturally, a price for this kind of performance. All first-partial derivative functions (needed for the gradient vector), all second-partial and all cross-partial derivative functions (needed for the Hessian matrix) must be known and evaluated, and then the inverse of the Hessian must be found—all this at each step. These extensive requirements have inspired research into ways for *approximating* the Hessian (or its inverse) at each step.

---

[14]Try showing that $(\nabla^2)'\nabla^1 = 0$ and that $(\mathbf{D}^2)'\mathbf{A}\mathbf{D}^1 = 0$. You can also easily show that $(\mathbf{D}^1)'\nabla^1 = (-\nabla^1)'\nabla^1$. Proving the more general case, say, that $(\nabla^2)'\nabla^0 = 0$ or that $(\mathbf{D}^2)'\mathbf{A}\mathbf{D}^0 = 0$, is not so simple.

TABLE 6.17 Conjugate Gradient Search for Minima of Test Functions (a)–(c)

k	$\mathbf{X}^k$	$f(\mathbf{X}^k)$	$\mathbf{D}^k$
	Test Function (a): $f(\mathbf{X}) = x_1^2 + x_2^2 - 4x_1 - 5x_2 - 5$		
0	$\begin{bmatrix} 0 \\ 0 \end{bmatrix}$	-5	$\begin{bmatrix} -4 \\ -5 \end{bmatrix}$
1	$\begin{bmatrix} 2 \\ 2.5 \end{bmatrix}$	-15.25	$\begin{bmatrix} 0 \\ 0 \end{bmatrix}$
	Test Function (b): $f(\mathbf{X}) = x_1^2 + x_2^2 - 4x_1 - 5x_2 - x_1x_2 - 5$		
0	$\begin{bmatrix} 0 \\ 0 \end{bmatrix}$	-5	$\begin{bmatrix} -4 \\ -5 \end{bmatrix}$
1	$\begin{bmatrix} 3.90476 \\ 4.88095 \end{bmatrix}$	-25.01190	$\begin{bmatrix} -1.25510 \\ 0.62755 \end{bmatrix}$
2	$\begin{bmatrix} 4.33333 \\ 4.66667 \end{bmatrix}$	-25.33333	$\begin{bmatrix} 0 \\ 0 \end{bmatrix}$
	Test Function (c): $f(\mathbf{X}) = (x_1^2/4) + (x_2^2/9) - 0.8x_1 - x_2 - 0.3x_1x_2 - 3$		
0	$\begin{bmatrix} 0 \\ 0 \end{bmatrix}$	-3	$\begin{bmatrix} -0.8 \\ -1 \end{bmatrix}$
1	$\begin{bmatrix} 21.08571 \\ 26.35714 \end{bmatrix}$	-24.61286	$\begin{bmatrix} -0.86016 \\ -4.83842 \end{bmatrix}$
2	$\begin{bmatrix} 22.63158 \\ 35.05263 \end{bmatrix}$	-29.57895	$\begin{bmatrix} 0 \\ 0 \end{bmatrix}$

Test Function (c): Additional Starting Points

Starting Point	Total Number of Iterations
$[1, 2]'$	2
$[100, 50]'$	2
$[100, 200]'$	2
$[1000, 2000]'$	2
$[1000, 1000]'$	2

One approach would be to use a finite-difference approximation to the Hessian (Appendix 5.1). This requires a number of functional evaluations (n for the gradient, $2n^2$ for the Hessian if the central difference method is used), the inverse would still be needed at each step, and there is no guarantee that such approximations maintain positive definite Hessians at each step. Positive definiteness is essential to ensure that Newton's method generates a minimizing, not a maximizing, step.

Updating the Estimate of the Inverse of the Hessian Matrix   An alternative approach has been developed in which the Hessian inverse is approximated at each step by *updating* the estimate of the Hessian inverse from the preceding step. To simplify notation, let $\mathbf{B}^k$ represent the approximation to the Hessian inverse that is generated at step k, $\mathbf{B}^k \approx (\mathbf{H}^k)^{-1}$. Because some of the expressions to follow are rather complicated, simplifying notation

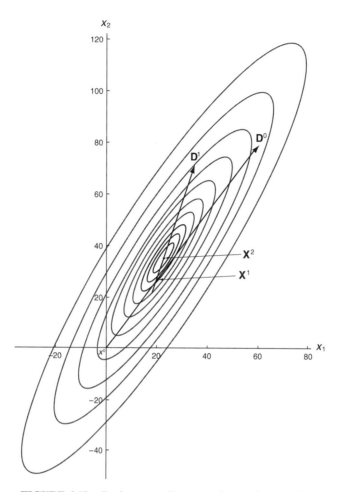

FIGURE 6.17 Conjugate gradient steps for test function (c).

will be very helpful. Let the **X** and ∇ differences be $\mathbf{x}^k \equiv \mathbf{X}^{k+1} - \mathbf{X}^k$ and $\mathbf{g}^k \equiv \nabla^{k+1} - \nabla^k$. Then (6.18) can be written more compactly as

$$\mathbf{g}^k = \mathbf{H}^k \mathbf{x}^k \qquad \text{or} \qquad \mathbf{x}^k = (\mathbf{H}^k)^{-1} \mathbf{g}^k \tag{6.22}$$

At step k, we have $\mathbf{X}^k$, ∇^k, and $\mathbf{B}^k$, the *estimate* of the Hessian inverse. Given these facts, $\mathbf{X}^{k+1}$ is found as

$$\mathbf{X}^{k+1} = \mathbf{X}^k - \mathbf{B}^k \nabla^k \tag{6.23}$$

and, with $\mathbf{X}^{k+1}$ established, ∇^{k+1} can be evaluated. Will it be true, as in (6.22), that $\mathbf{x}^k = \mathbf{B}^k \mathbf{g}^k$? Not likely. $\mathbf{B}^k$ is just an *estimate* of $(\mathbf{H}^k)^{-1}$, and from the way that $\mathbf{X}^{k+1}$ was found in (6.23), we know that $\mathbf{x}^k \equiv \mathbf{X}^{k+1} - \mathbf{X}^k = \mathbf{B}^k(-\nabla^k)$. Only if $\nabla^{k+1} = 0$ (which means that $\mathbf{X}^{k+1}$ is a minimizing point), so that $\mathbf{g}^k = \nabla^{k+1} - \nabla^k = -\nabla^k$, will

$$\mathbf{x}^k = \mathbf{B}^k \mathbf{g}^k \tag{6.24}$$

hold. As a consequence, quasi-Newton methods find the *next* **B**, $\mathbf{B}^{k+1}$, so that

$$\mathbf{x}^k = \mathbf{B}^{k+1}\mathbf{g}^k \tag{6.25}$$

This is sometimes referred to as the "quasi-Newton condition." That is, quasi-Newton methods "update" the estimate of the Hessian inverse one step later. $\mathbf{B}^{k+1}$ is required to have the property that $\mathbf{B}^k$ would have had if it had been the true $(\mathbf{H}^k)^{-1}$.

An acceptable updating process should not require an inordinate amount of computation, and it should preserve both the symmetry and the positive definiteness of the previous estimate, $\mathbf{B}^k$. Symmetry is not so difficult, even for finite-difference approximations to the Hessian. Positive definiteness is quite another matter. The now-popular rank 2 updates that we will examine below can be shown to generate positive definite matrices [e.g., see Fletcher (1987, pp. 54–55) for a proof]. And they are computationally effective and preserve symmetry as well.

Rank 1 Updates   In order to establish the basic ideas, we first look at a symmetric rank 1 update (Appendix 6.1) in which $\mathbf{B}^{k+1}$ would be constructed from $\mathbf{B}^k$ as

$$\mathbf{B}^{k+1} = \mathbf{B}^k + \alpha\mathbf{U}\mathbf{U}' \tag{6.26}$$

where **U** is an $n \times 1$ vector and α is a scalar. The issue is how to define α and **U**. Putting (6.26) into (6.25), we obtain

$$\mathbf{B}^k\mathbf{g}^k + \alpha\mathbf{U}\mathbf{U}'\mathbf{g}^k = \mathbf{x}^k \tag{6.27}$$

Whatever the elements of **U** are, $\alpha\mathbf{U}'\mathbf{g}^k$ is just a scalar, so from (6.27) **U** is proportional to $(\mathbf{X}^k - \mathbf{B}^k\mathbf{g}^k)$. In particular, if we let $\alpha\mathbf{U}'\mathbf{g}^k = 1$, then $\mathbf{U} = (\mathbf{x}^k - \mathbf{B}^k\mathbf{g}^k)$ and $\alpha = 1/\mathbf{U}'\mathbf{g}^k$ (which requires that $\mathbf{U}'\mathbf{g}^k \neq 0$). Putting all this into (6.26) gives[15]

$$\mathbf{B}^{k+1} = \mathbf{B}^k + \frac{(\mathbf{x}^k - \mathbf{B}^k\mathbf{g}^k)(\mathbf{x}^k - \mathbf{B}^k\mathbf{g}^k)'}{(\mathbf{x}^k - \mathbf{B}^k\mathbf{g}^k)'\mathbf{g}^k} \tag{6.28}$$

This is a specific symmetric rank 1 update that creates $\mathbf{B}^{k+1}$ from $\mathbf{B}^k$. It turns out to be reasonably successful, but it does not guarantee that updates remain positive definite. Also, problems arise if the denominator is zero ($\mathbf{B}^{k+1}$ is undefined) or near to zero (the updating term explodes).

Rank 2 Updates   A rank 2 updating scheme overcomes both of these difficulties. This would have the form (Appendix 6.1)

$$\mathbf{B}^{k+1} = \mathbf{B}^k + \alpha\mathbf{U}\mathbf{U}' + \beta\mathbf{V}\mathbf{V}' \tag{6.29}$$

In order to satisfy the quasi-Newton condition in (6.25), it must be true that

$$\mathbf{B}^k\mathbf{g}^k + \alpha\mathbf{U}\mathbf{U}'\mathbf{g}^k + \beta\mathbf{V}\mathbf{V}'\mathbf{g}^k = \mathbf{x}^k \tag{6.30}$$

---

[15]The result in (6.28) was reported (independently) in Broyden (1967), Davidon (1968), Fiacco and McCormick (1968), Murtagh and Sargent (1969), and Wolfe (1968).

Now there are two vectors, $\mathbf{U}$ and $\mathbf{V}$, and two scalars, α and β, to determine. After the kth iteration, we have $\mathbf{x}^k$, $\mathbf{g}^k$ and the $\mathbf{B}^k$ that was used at step k. Suppose that we let

$$\mathbf{U} = \mathbf{x}^k \qquad \text{and} \qquad \mathbf{V} = \mathbf{B}^k\mathbf{g}^k \tag{6.31}$$

As in the rank 1 updating scheme, let $\alpha\mathbf{U}'\mathbf{g}^k = 1$. Then, if we let $\beta\mathbf{V}'\mathbf{g}^k = -1$, we have $\alpha = 1/\mathbf{U}'\mathbf{g}^k$ and $\beta = -1/\mathbf{V}'\mathbf{g}^k$, and (6.29) becomes

$$\mathbf{B}^{k+1} = \mathbf{B}^k + \frac{\mathbf{x}^k(\mathbf{x}^k)'}{(\mathbf{x}^k)'\mathbf{g}^k} - \frac{\mathbf{B}^k\mathbf{g}^k(\mathbf{g}^k)'\mathbf{B}^k}{(\mathbf{g}^k)'\mathbf{B}^k\mathbf{g}^k} \tag{6.32}$$

This result was (independently) developed by Davidon (1959) and by Fletcher and Powell (1963) and for that reason is widely known as the "DFP" rank 2 symmetric update method. To express (6.32) more compactly, let

$$\mathbf{R} = [\mathbf{U} \mathbin{\vdots} \mathbf{V}] = [\mathbf{x}^k \mathbin{\vdots} \mathbf{B}^k\mathbf{g}^k]$$

$$\mathbf{S} = \begin{bmatrix} \alpha & 0 \\ 0 & \beta \end{bmatrix} = \begin{bmatrix} \dfrac{1}{\mathbf{U}'\mathbf{g}^k} & 0 \\ 0 & \dfrac{-1}{\mathbf{V}'\mathbf{g}^k} \end{bmatrix} \tag{6.33}$$

Then the DFP result in (6.32) can be written as

$$\mathbf{B}^{k+1} = \mathbf{B}^k + \mathbf{R}\mathbf{S}\mathbf{R}' \tag{6.34}$$

Updating the Estimate of the Hessian Matrix   Instead of estimating $(\mathbf{H}^k)^{-1}$ by $\mathbf{B}^k$, one could consider a sequence of updates on the Hessian itself, $\mathbf{H}^k$, not its inverse, and then find the inverse of this estimate using Sherman–Morrison–Woodbury (SMW) techniques. Let $\mathbf{C}^k \equiv (\mathbf{B}^k)^{-1} \approx \mathbf{H}^k$. Then, multiplying both sides of the quasi-Newton condition in (6.25) by $(\mathbf{B}^{k+1})^{-1}$ gives

$$(\mathbf{B}^{k+1})^{-1}\mathbf{x}^k = \mathbf{g}^k \qquad \text{or} \qquad \mathbf{C}^{k+1}\mathbf{x}^k = \mathbf{g}^k \tag{6.35}$$

Consider estimating $\mathbf{C}^{k+1}$ through a symmetric rank 2 update to $\mathbf{C}^k$:

$$\mathbf{C}^{k+1} = \mathbf{C}^k + \alpha\mathbf{U}\mathbf{U}' + \beta\mathbf{V}\mathbf{V}' \tag{6.36}$$

Putting (6.36) into (6.35) leads to

$$\mathbf{C}^k\mathbf{x}^k + \alpha\mathbf{U}\mathbf{U}'\mathbf{x}^k + \beta\mathbf{V}\mathbf{V}'\mathbf{x}^k = \mathbf{g}^k \tag{6.37}$$

Note that, compared with (6.30), $\mathbf{x}^k$ and $\mathbf{g}^k$ have been interchanged, as have $\mathbf{B}^k$ and $\mathbf{C}^k$. Therefore, parallel to (6.31), but with these interchanges, let $\mathbf{U} = \mathbf{g}^k$ and $\mathbf{V} = \mathbf{C}^k\mathbf{x}^k$. As before, also let $\alpha\mathbf{U}'\mathbf{x}^k = 1$ and $\beta\mathbf{V}'\mathbf{x}^k = -1$, so that $\alpha = 1/\mathbf{U}'\mathbf{x}^k$ and $\beta = -1/\mathbf{V}'\mathbf{x}^k$. Then (6.36) becomes

$$\mathbf{C}^{k+1} = \mathbf{C}^k + \frac{\mathbf{g}^k(\mathbf{g}^k)'}{(\mathbf{g}^k)'\mathbf{x}^k} - \frac{\mathbf{C}^k\mathbf{x}^k(\mathbf{x}^k)'\mathbf{C}^k}{(\mathbf{x}^k)'\mathbf{C}^k\mathbf{x}^k} \tag{6.38}$$

If we again define $\mathbf{R} = [\mathbf{U} \vdots \mathbf{V}]$ and $\mathbf{S} = \begin{bmatrix} \alpha & 0 \\ 0 & \beta \end{bmatrix}$, we now have

$$\mathbf{R} = [\mathbf{U} \vdots \mathbf{V}] = [\mathbf{g}^k \vdots \mathbf{C}^k \mathbf{x}^k]$$

$$\mathbf{S} = \begin{bmatrix} \alpha & 0 \\ 0 & \beta \end{bmatrix} = \begin{bmatrix} \dfrac{1}{\mathbf{U}'\mathbf{x}^k} & 0 \\ 0 & \dfrac{-1}{\mathbf{V}'\mathbf{x}^k} \end{bmatrix} \tag{6.39}$$

[Compare $\mathbf{R}$ and $\mathbf{S}$ in (6.33). Again, $\mathbf{x}^k$ and $\mathbf{g}^k$ have been interchanged and $\mathbf{C}^k$ replaces $\mathbf{B}^k$.] Using these definitions of $\mathbf{R}$ and $\mathbf{S}$, $\mathbf{C}^{k+1}$ in (6.38) can be represented compactly as

$$\mathbf{C}^{k+1} = \mathbf{C}^k + \mathbf{R}\mathbf{S}\mathbf{R}' \tag{6.40}$$

[Cf. (6.34).]

Finally, to get back to updates of the Hessian *inverse*, we can use SMW results on $\mathbf{C}^{k+1}$ ($\approx \mathbf{H}^{k+1}$) in (6.40) to find $\mathbf{B}^{k+1}$ $[\equiv (\mathbf{C}^{k+1})^{-1}] \approx (\mathbf{H}^{k+1})^{-1}$ as

$$\mathbf{B}^{k+1} = \mathbf{B}^k + \left[1 + \frac{(\mathbf{g}^k)'\mathbf{B}^k\mathbf{g}^k}{(\mathbf{x}^k)'\mathbf{g}^k} \right] \frac{\mathbf{x}^k(\mathbf{x}^k)'}{(\mathbf{x}^k)'\mathbf{g}^k} - \left[\frac{\mathbf{x}^k(\mathbf{g}^k)'\mathbf{B}^k + \mathbf{B}^k\mathbf{g}^k(\mathbf{x}^k)'}{(\mathbf{x}^k)'\mathbf{g}^k} \right] \tag{6.41}$$

This expression for $\mathbf{B}^{k+1}$ is credited simultaneously to Broyden (1970), Fletcher (1970), Goldfarb (1970), and Shanno (1970) and is consequently known as the "BFGS" rank 2 symmetric update for the inverse of the Hessian. It is the quasi-Newton variant that we will use in the numerical illustrations in this section.

There are many alternative expressions in the literature (all algebraically equivalent) for the DFP and BFGS procedures. For example, the variant that is used in the calculations for the numerical illustrations to follow in this section is a translation into APL of the FORTRAN program for an alternative version of BFGS that is given in Press et al. (1986). It is, in particular

$$\mathbf{B}^{k+1} = \left\{ \mathbf{B}^k + \frac{\mathbf{x}^k(\mathbf{x}^k)'}{(\mathbf{x}^k)'\mathbf{g}^k} - \frac{\mathbf{B}^k\mathbf{g}^k(\mathbf{g}^k)'\mathbf{B}^k}{(\mathbf{g}^k)'\mathbf{B}^k\mathbf{g}^k} \right\} +$$

$$(\mathbf{g}^k)'\mathbf{B}^k\mathbf{g}^k \left[\left(\frac{\mathbf{x}^k}{(\mathbf{x}^k)'\mathbf{g}^k} - \frac{\mathbf{B}^k\mathbf{g}^k}{(\mathbf{g}^k)'\mathbf{B}^k\mathbf{g}^k} \right) \left(\frac{\mathbf{x}^k}{(\mathbf{x}^k)'\mathbf{g}^k} - \frac{\mathbf{B}^k\mathbf{g}^k}{(\mathbf{g}^k)'\mathbf{B}^k\mathbf{g}^k} \right)' \right] \tag{6.42}$$

Note that the expression in curly brackets on the right is just the DFP result for $\mathbf{B}^{k+1}$, in (6.32). Problem 6.10 asks you to demonstrate the equivalence of (6.41) and (6.42).

Iterative estimates of the Hessian inverse, as in (6.42), can be used directly in Newton's method or they can be employed in restricted step approaches, such as backtracking, that employ line minimizations at each iteration along the estimated full Newton step direction. In the case of Newton's method, the rank 2 symmetric update, $\mathbf{B}^k$, simply replaces $(\mathbf{H}^k)^{-1}$ in (6.7) and (6.8), so that the iterative process is $\mathbf{X}^{k+1} = \mathbf{X}^k + \mathbf{B}^k(-\nabla^k)$ [cf. (6.7)] and, in particular, $\mathbf{D}^k = -(\mathbf{B}^k)\nabla^k$ [cf. (6.8)]. In the numerical illustrations, we refer to this method as "BFGS Newton."

Alternatively, an iterative updating procedure for estimating the Hessian inverse, as in (6.42), is frequently combined with a line minimization calculation at each iteration, along the *direction* given by the full Newton step, as in backtracking when the full New-

TABLE 6.18   BFGS Newton Method Search for Minimum of Test Functions (a)–(c)

k	$\mathbf{X}^k$	$f(\mathbf{X}^k)$	∇^k	$\mathbf{B}^k$
		Test Function (a): $f(\mathbf{X}) = x_1^2 + x_2^2 - 4x_1 - 5x_2 - 5$		
0	$\begin{bmatrix} 0 \\ 0 \end{bmatrix}$	-5	$\begin{bmatrix} -4 \\ -5 \end{bmatrix}$	$\begin{bmatrix} 0.95 & 0 \\ 0 & 0.95 \end{bmatrix}$
1	$\begin{bmatrix} 3.6 \\ 4.75 \end{bmatrix}$	-6.95	$\begin{bmatrix} 3.6 \\ 4.5 \end{bmatrix}$	$\begin{bmatrix} 0.7744 & -0.2195 \\ -0.2195 & 0.6756 \end{bmatrix}$
2	$\begin{bmatrix} 2 \\ 2.5 \end{bmatrix}$	-15.25	$\begin{bmatrix} 0 \\ 0 \end{bmatrix}$	
		Test Function (b): $f(\mathbf{X}) = x_1^2 + x_2^2 - 4x_1 - 5x_2 - x_1x_2 - 5$		
0	$\begin{bmatrix} 0 \\ 0 \end{bmatrix}$	-5	$\begin{bmatrix} -4 \\ -5 \end{bmatrix}$	$\begin{bmatrix} 1 & 0 \\ 0 & 1 \end{bmatrix}$
1	$\begin{bmatrix} 4 \\ 5 \end{bmatrix}$	-25	$\begin{bmatrix} -1 \\ 1 \end{bmatrix}$	$\begin{bmatrix} 1.21769 & 0.05782 \\ 0.05782 & 0.80442 \end{bmatrix}$
2	$\begin{bmatrix} 4.33279 \\ 4.66603 \end{bmatrix}$	-24.13782	$\begin{bmatrix} 2.06632 \\ -1.65306 \end{bmatrix}$	$\begin{bmatrix} 0.67018 & 0.33739 \\ 0.33739 & 0.67136 \end{bmatrix}$
3	$\begin{bmatrix} 4.33333 \\ 4.66667 \end{bmatrix}$	-25.33333	$\begin{bmatrix} -0.00046 \\ -0.00072 \end{bmatrix}$	$\begin{bmatrix} 0.66981 & 0.33727 \\ 0.33727 & 0.67158 \end{bmatrix}$
4	$\begin{bmatrix} 4.33333 \\ 4.66667 \end{bmatrix}$	-25.33333	$\begin{bmatrix} 0 \\ 0 \end{bmatrix}$	

Test Function (b):Alternative Stopping Criteria

ε	Iterations	x_1	x_2	$f(\mathbf{X})$
1×10^{-10}	5	4.33333	4.66667	-25.33333
1×10^{-8}	5	4.33333	4.66667	-25.33333
1×10^{-6}	4	4.33334	4.66667	-25.33333

k	$\mathbf{X}^k$	$f(\mathbf{X}^k)$	∇^k	$\mathbf{B}^k$
		Test Function (c): $f(\mathbf{X}) = (x_1^2/4) + (x_2^2/9) - 0.8x_1 - x_2 - 0.3x_1x_2 - 3$		
0	$\begin{bmatrix} 0 \\ 0 \end{bmatrix}$	-3	$\begin{bmatrix} -0.8 \\ -1.0 \end{bmatrix}$	$\begin{bmatrix} 1 & 0 \\ 0 & 1 \end{bmatrix}$
1	$\begin{bmatrix} 0.8 \\ 1 \end{bmatrix}$	-4.60889	$\begin{bmatrix} -0.7 \\ -1.01778 \end{bmatrix}$	$\begin{bmatrix} 10.41919 & 13.61020 \\ 13.61020 & 20.30740 \end{bmatrix}$
2	$\begin{bmatrix} 21.94588 \\ 31.19556 \end{bmatrix}$	-28.60184	$\begin{bmatrix} 0.81427 \\ -0.65142 \end{bmatrix}$	$\begin{bmatrix} 10.46553 & 14.46177 \\ 14.46177 & 22.64574 \end{bmatrix}$
3	$\begin{bmatrix} 22.84474 \\ 34.17158 \end{bmatrix}$	-29.42499	$\begin{bmatrix} 0.37090 \\ -0.25974 \end{bmatrix}$	$\begin{bmatrix} 9.81679 & 13.40736 \\ 13.40736 & 22.77504 \end{bmatrix}$
4	$\begin{bmatrix} 22.68613 \\ 35.11438 \end{bmatrix}$	-29.57880	$\begin{bmatrix} 0.00875 \\ -0.00264 \end{bmatrix}$	$\begin{bmatrix} 9.86918 & 13.28488 \\ 13.28488 & 22.38033 \end{bmatrix}$
5	$\begin{bmatrix} 22.63488 \\ 35.05728 \end{bmatrix}$	-29.57895	$\begin{bmatrix} 0.00026 \\ 0.00004 \end{bmatrix}$	$\begin{bmatrix} 10.51274 & 14.16758 \\ 14.16758 & 23.54840 \end{bmatrix}$

TABLE 6.18 *continued*

Test Function (c): Additional Starting Points	
Starting Point	Total Number of Iterations
$[1, 2]'$	7
$[100, 50]'$	7
$[100, 200]'$	10
$[1000, 2000]'$	11
$[1000, 1000]'$	11

ton step is unacceptable. Many of the most successful procedures in this group, which we will call "quasi-Newton Methods," have the following general structure: At iteration k (a) set $\mathbf{D}^k = -\mathbf{B}^k\nabla^k$; (b) line-search for the minimum along the direction $\mathbf{D}^k$, generating the new point $\mathbf{X}^{k+1} = \mathbf{X}^k + \alpha^k\mathbf{D}^k$; and (c) update $\mathbf{B}^k$ to $\mathbf{B}^{k+1}$ using a routine such as (6.42), return to (a), with $k = k + 1$, and continue until some stopping criterion is satisfied. (Initially, let $\mathbf{B}^0 = \mathbf{I}$.)

Numerical Illustrations Results for the BFGS Newton approach to minimization of test functions (a)–(c) are given in Table 6.18. Using the default procedure of setting $\mathbf{B}^0 = \mathbf{I}$ is not a successful strategy for test function (a) from an initial point $\mathbf{X}^0 = \begin{bmatrix} 0 \\ 0 \end{bmatrix}$, where $f(\mathbf{X}^0) = -5$. It leads to $\mathbf{D}^0 = \begin{bmatrix} 4 \\ 5 \end{bmatrix}$ and so $\mathbf{X}^1 = \begin{bmatrix} 4 \\ 5 \end{bmatrix}$, for which $f(\mathbf{X}^1) = -5$, so the difference between the values of $f(\mathbf{X})$ at these two points is identically zero and the iterations will terminate. For that reason, $\mathbf{B}^0$ was set at $(0.95)\mathbf{I}$ for test function (a), and the method required an extra step (as compared with Newton's method, in Table 6.13), to overcome this bad beginning and get to the optimum point. For test function (b) BFGS Newton is not as efficient as Newton itself, requiring four iterations as opposed to only one; but then, it also does not require finding the Hessian and its inverse at each step. Note, in comparing the steps in Tables 6.13 and 6.18, how the BFGS estimates of the Hessian inverse, $\mathbf{B}^k$, approach the exact results given in the $[\mathbf{H}^k]^{-1}$ column of Table 6.13. As might be expected, more iterations are needed for test function (c), and the BFGS Newton approach exhibits some sensitivity to the starting point used.

Results for the quasi-Newton method are given in Table 6.19. The sequence for test function (a) is the same as it was for Newton's method. (Is it clear why?). For test functions (b) and (c) quasi-Newton requires the same number of iterations as the conjugate gradient method, although the amount of calculation in each iteration is larger, since the Hessian inverse must be updated at each iteration. The quasi-Newton and conjugate gradient methods seem to be about equally (in)sensitive to alternative starting points.

6.3.5 Results for Additional Test Functions

Table 6.20 contains information on the relative performance of these five derivative-based search methods on test functions (d)–(f). The stopping criterion, again, is that the relative change in $f(\mathbf{X}^*)$ from one iteration to the next be no more than 10^{-5}. Cauchy's method fal-

TABLE 6.19   Quasi-Newton Method Search for Minimum of Test Functions (a)–(c)

k	$\mathbf{X}^k$	$f(\mathbf{X}^k)$	∇^k	$\mathbf{H}^k$	$\mathbf{D}^k = -\mathbf{H}^k\nabla^k$
		Test Function (a): $f(\mathbf{X}) = x_1^2 + x_2^2 - 4x_1 - 5x_2 - 5$			
0	$\begin{bmatrix} 0 \\ 0 \end{bmatrix}$	-5	$\begin{bmatrix} -4 \\ -5 \end{bmatrix}$	$\begin{bmatrix} 1 & 0 \\ 0 & 1 \end{bmatrix}$	$\begin{bmatrix} 4 \\ 5 \end{bmatrix}$
1	$\begin{bmatrix} 2 \\ 2.5 \end{bmatrix}$	-15.25	$\begin{bmatrix} 0 \\ 0 \end{bmatrix}$	Stop	—
		Test Function (b): $f(\mathbf{X}) = x_1^2 + x_2^2 - 4x_1 - 5x_2 - x_1x_2 - 5$			
0	$\begin{bmatrix} 0 \\ 0 \end{bmatrix}$	-5	$\begin{bmatrix} -4 \\ -5 \end{bmatrix}$	$\begin{bmatrix} 1 & 0 \\ 0 & 1 \end{bmatrix}$	$\begin{bmatrix} 4 \\ 5 \end{bmatrix}$
1	$\begin{bmatrix} 3.90476 \\ 4.88095 \end{bmatrix}$	-25.01190	$\begin{bmatrix} -1.07143 \\ 0.85714 \end{bmatrix}$	$\begin{bmatrix} 1.22698 & 0.06944 \\ 0.06944 & 0.81894 \end{bmatrix}$	$\begin{bmatrix} 1.25510 \\ -0.62755 \end{bmatrix}$
2	$\begin{bmatrix} 4.33333 \\ 4.66667 \end{bmatrix}$	-25.33333	$\begin{bmatrix} 0 \\ 0 \end{bmatrix}$	Stop	—
		Test Function (c): $f(\mathbf{X}) = (x_1^2/4) + (x_2^2/9) - 0.8x_1 - x_2 - 0.3x_1x_2 - 3$			
0	$\begin{bmatrix} 0 \\ 0 \end{bmatrix}$	-3	$\begin{bmatrix} -0.8 \\ -1.0 \end{bmatrix}$	$\begin{bmatrix} 1 & 0 \\ 0 & 1 \end{bmatrix}$	$\begin{bmatrix} 0.8 \\ 1.0 \end{bmatrix}$
1	$\begin{bmatrix} 21.08571 \\ 26.35714 \end{bmatrix}$	-24.61286	$\begin{bmatrix} 1.83571 \\ -1.46863 \end{bmatrix}$	$\begin{bmatrix} 0.52412 & 1.24087 \\ 1.24087 & 4.84573 \end{bmatrix}$	$\begin{bmatrix} 0.86016 \\ 4.83842 \end{bmatrix}$
2	$\begin{bmatrix} 22.63158 \\ 35.05263 \end{bmatrix}$	-29.57895	$\begin{bmatrix} 0 \\ 0 \end{bmatrix}$	Stop	—

Test Function (c): Additional Starting Points

Starting Point	Total Number of Iterations
$[1, 2]'$	2
$[100, 50]'$	2
$[100, 200]'$	2
$[1000, 2000]'$	2
$[1000, 1000]'$	2

TABLE 6.20   Number of Iterations for Three Additional Test Functions ($\mathbf{X}^0 = [0, 0]'$ throughout)

Cauchy	Newton	Conjugate Gradient	BFGS Newton	Quasi-Newton
		Test Function (d): $f(\mathbf{X}) = (x_1^2/4) + (x_2^2/10) - 0.8x_1 - x_2 - 0.3x_1x_2 - 3$		
13	1	2	7	2
		Test Function (e): $f(\mathbf{X}) = (x_1^2/4) + (x_2^2/11) - 0.8x_1 - x_2 - 0.3x_1x_2 - 3$		
87	1	2	9	2
		Test Function (f): $f(\mathbf{X}) = 100(x_2 - x_1^2)^2 + (1 - x_1)^2 + 0.01$		
1758	2	13	46	12

ters, Newton's method shines, conjugate gradient and quasi-Newton are about equally impressive [and in particular do very well overcoming the difficulties of test function (f)], and BFGS Newton is better than Cauchy only, which is not a strong recommendation.

6.4 SUMMARY

In this chapter we have explored a number of iterative routines that are designed to move toward a minimizing point (or, with simple modifications, a maximizing point) for $f(x)$ and for $f(\mathbf{X})$. These numerical methods provide an alternative to the classical calculus approach to maximization and minimization, which requires solving the one or more first-order condition equations for candidate points and then checking second-order conditions to separate points that are maxima from those that are minima (and both from points that are neither). These are the kinds of approaches that are employed in most computer software packages for solving unconstrained maximization and minimization problems.

For $f(x)$, successful iterative techniques generally combine features of sequential search (usually for the beginning steps) and more refined curve fitting, such as parabolic interpolation (for the final steps). For $f(\mathbf{X})$, the approaches need to be quite a bit more complex. The same is true for the classical calculus results, where first-order conditions change from a single equation to a set of equations and second-order conditions (testing for concavity or convexity of the function) change from establishing the sign of a second derivative to testing for positive or negative definiteness of a matrix. The downhill simplex method requires no derivatives at all and works well for $f(x_1, \ldots, x_n)$, at least when n is not too large.[16] Otherwise, conjugate direction methods are quite successful. If you can supply first-derivative functions and evaluate them at each successive point, then conjugate direction and quasi-Newton methods are both excellent choices and underpin a great many computer minimization routines. Variants abound, particularly with regard to the details of how the line minimizations are carried out at each iteration.

Iterative methods for optimization problems is an area in which an enormous amount of research has been and is being done, both in improving existing techniques and in introducing and testing new ones. Many important aspects of this work have had to be omitted from this already long chapter. Among the many omissions are discussions of

1. *Convergence Proofs.* Before a new method can be fully accepted as useful, it is necessary to demonstrate that the technique will, indeed, always converge, or if not always, then under what circumstances it does not. The proofs may be *global*, in the sense that convergence is shown even when the initial point, $\mathbf{X}^0$, is not "near to" the true solution, $\mathbf{X}^*$, or *local*, requiring that one start the method "near to" $\mathbf{X}^*$. In addition, it is general to specify conditions [for example, on $f(\mathbf{X})$] under which the convergence will be to a global maximum or minimum as opposed to a local solution only.

2. *Convergence Speed.* A very practical consideration is the speed of convergence for successful numerical methods. Convergence speed is generally discussed in terms of one or more alternative vector norms. The general definition of the norm of a vector is

$$\|\mathbf{V}\|_p = \left[\sum_{i=1}^n |v_i|^p \right]^{1/p}$$

---

[16]According to Dennis and Schnabel (1983, p. xii), n should be less than 5.

When $p = 1$, this is known as the L_1 norm, and it measures simply the sum of the absolute values of the elements in the vector. When $p = 2$, this is the L_2 norm (also called the Euclidian norm) which measures the length of the vector in n-dimensional Euclidian space.

Given $\mathbf{X}^*$ and two successive iterates, $\mathbf{X}^k$ and $\mathbf{X}^{k+1}$, two frequently used measures of speed of convergence are:

(i) If we define $\Delta^k = \mathbf{X}^k - \mathbf{X}^*$, then *linear* (or first-order) *convergence* is said to occur when

$$\frac{\|\Delta^{k+1}\|_1}{\|\Delta^k\|_1} \leq \alpha$$

(Fletcher (1987, p. 20) suggests that this represents a satisfactory speed of convergence only if $\alpha \leq 0.25$.) If it is also true that, for all k

$$\frac{\|\Delta^{k+1}\|_1}{\|\Delta^k\|_1} \to 0$$

convergence is termed *superlinear*.

(ii) If

$$\frac{\|\Delta^{k+1}\|_2}{\|\Delta^k\|_2} \leq \alpha$$

convergence is described as *quadratic*, or second-order. Since the L_2 norm measures length in $x_1, \ldots, x_n$ space, $\|\Delta^k\|_2$ measures the distance between $\mathbf{X}^k$ and $\mathbf{X}^*$.

3. *Alternative Ways of Assessing Relative Performance of One or More Test Functions.* The additional considerations at each iteration include the following:

(i)   Storage requirements. Methods that calculate (or estimate) successive gradients need to store those n values; if the Hessian, or an estimate, is used at each step, an additional n^2 values must be stored. This hardly seems important for what many might consider reasonably sized problems (say, completely arbitrarily, $n \leq 100$). However, note that with $n = 500$, the Jacobian contains 250,000 elements, and Fletcher (1987, p. 85) mentions problems in atomic structures with 3000 variables (a Hessian of 9 million elements!).

(ii) Estimates of what are called "housekeeping operations," such as numbers of multiplications, updates of variables that serve as counters, ordering values from largest to smallest, replacing last iteration values with current ones, and similar operations.

(iii) Comparisons of the numbers of functional evaluations required in each iteration, including $f(\mathbf{X})$ itself, the n elements (functions) of the gradient vector, and/or the n^2 elements (functions) in the Hessian matrix, if they are used.

4. *Roundoff Error.* Different numerical methods give rise to different possibilities for the buildup of roundoff error during the course of the calculations.

5. *Accuracy and Tolerance*. There may be several different places in a particular numerical method where a criterion is needed to terminate that part of the calculations. For example, if, as is frequently the case, line searches form a part of each step, they themselves usually employ an iterative scheme, for which a stopping criterion must be specified. The more exacting such criteria are, the longer each iteration will take. This need not be reflected in the number of iterations, but it could influence the computation time for that method.

These are the kinds of more advanced subjects that would be included in a book on numerical methods. The object of this chapter has been simply to introduce you to the logic and the workings of a number of representative kinds of iterative methods for unconstrained minimization problems. These are the building blocks of modern computer programs for such problems. After exposure to this material, you may be better able to understand a reference to fact that a particular software package for minimization of $f(\mathbf{X})$ uses Levenberg–Marquardt or Polak–Ribiere or quasi-Newton methods, for example, or why you need to search for an interval that brackets a minimum if you are running a program for minimizing $f(x)$ that uses golden section search.

APPENDIX 6.1   THE SHERMAN–MORRISON–WOODBURY FORMULA REVISITED

6A.1.1   Introduction

The basic Sherman–Morrison–Woodbury (SMW) results were examined in Appendix 5.2. For a nonsingular $n \times n$ matrix $\mathbf{A}$, these results relate a change in one or more elements of $\mathbf{A}$ to changes in the elements of $\mathbf{A}^{-1}$. When one element, a_{ij}, is changed to $a_{ij} + \Delta a_{ij}$, the new matrix, $\mathbf{A}^*$, can be represented as $\mathbf{A}^* = \mathbf{A} + \Delta\mathbf{A}$, where $\Delta\mathbf{A}$ is an $n \times n$ matrix with Δa_{ij} in the i,jth position and zeros elsewhere.

An efficient way to summarize the algebra is as follows. Let $\mathbf{U}$ be the ith column from an $n \times n$ identity matrix ($\mathbf{U} = \mathbf{I}_i$; this serves to identify the *row* in which the change occurs) and let $\mathbf{V}$ be the jth row from an $n \times n$ identity matrix, with Δa_{ij} replacing 1 in the jth position ($\mathbf{V} = \Delta a_{ij}\mathbf{I}'_j$; this indicates the *column* in which the change occurs as well as its magnitude). Then

$$\Delta\mathbf{A} = \mathbf{UV} \tag{6.1.1}$$

will be a null matrix except for the i,jth position, which will contain Δa_{ij}.

For example, if $\Delta\mathbf{A} = \begin{bmatrix} 0 & 0 & 0 \\ 0 & 0 & 3 \\ 0 & 0 & 0 \end{bmatrix}$, let $\mathbf{U} = \begin{bmatrix} 0 \\ 1 \\ 0 \end{bmatrix}$ and $\mathbf{V} = [0 \quad 0 \quad 3]$; then $\Delta\mathbf{A} = \mathbf{UV}$,

as in (6.1.1). The Sherman–Morrison result for $(\mathbf{A}^*)^{-1}$ in this situation [this was (5.2.1) in Appendix 5.2] is

$$(\mathbf{A}^*)^{-1} = \mathbf{A}^{-1} - \Delta(\mathbf{A}^{-1}) = \mathbf{A}^{-1} - \frac{(\mathbf{A}^{-1}\mathbf{U})(\mathbf{VA}^{-1})}{(1 + \mathbf{VA}^{-1}\mathbf{U})} \tag{6.1.2}$$

Note that the denominator, $(1 + \mathbf{VA}^{-1}\mathbf{U})$, is a scalar; the Sherman–Morrison expression is only valid as long as this denominator is not zero. In fact, when the denominator is zero, $\mathbf{A}$ is singular. (This was illustrated in Appendix 5.2.)

The Woodbury generalization extends this kind of result to a case in which more than one element in **A** is changed. If there are changes in one or more elements in each of m $(1 < m < n)$ different rows of **A**, **U** is expanded to an $n \times m$ matrix composed of the appropriate columns (m in number) from an $n \times n$ identity matrix, indicating, in turn, each row of **A** in which one or more changes occur. Similarly, **V** becomes an $m \times n$ matrix with the magnitudes of the changes, row by row, recorded in appropriate column positions and with zeros elsewhere. Using these definitions of **U** and **V**, it remains true that $\Delta \mathbf{A} = \mathbf{UV}$. Then the extended result [this was (5.2.2) in Appendix 5.2] is

$$(\mathbf{A}^*)^{-1} = \mathbf{A}^{-1} - \Delta(\mathbf{A}^{-1}) = \mathbf{A}^{-1} - [(\mathbf{A}^{-1}\mathbf{U})(\mathbf{I} + \mathbf{V}\mathbf{A}^{-1}\mathbf{U})^{-1}(\mathbf{V}\mathbf{A}^{-1})] \qquad (6.1.3)$$

where **I** is an $m \times m$ identity matrix. The extension to this case of the nonzero requirement for the denominator in (6.1.2) is that $(\mathbf{I} + \mathbf{V}\mathbf{A}^{-1}\mathbf{U})$ (along with **A**) in (6.1.3) must be nonsingular.

These results were useful and adequate for purposes of understanding the Broyden iterative updating approach for systems of nonlinear equations. However, the updates that arise in quasi-Newton iterative methods for maximization and minimization problems are more complicated. Results for some of these updating schemes, especially those for symmetric Hessian matrices, frequently employ alternative representations of these specific kinds of matrix updates. These, in turn, lead to slightly different expressions of the SMW results for the associated inverses. This is the subject matter of this appendix.

6A.1.2 Symmetric Rank 1 Changes

Given *any* pair of nonzero n-element vectors where **U** is $n \times 1$ and **V** is $1 \times n$, we have seen that the product **UV** is an $n \times n$ matrix whose rank is 1. We now concentrate on those particular rank 1 changes that are *symmetric*; that is, for which $\Delta a_{ij} = \Delta a_{ji}$. In order to do that, we need to define **U** and **V** differently and more generally than previously, in discussions of the SMW results in Appendix 5.2 and in Section 6A.1.1, above.

For example, suppose that $\mathbf{V} = \mathbf{U}'$, and consider, for simplicity, the case of $n = 3$. Then

$$\mathbf{UV} = \mathbf{UU}' = \begin{bmatrix} u_1^2 & u_1u_2 & u_1u_3 \\ u_2u_1 & u_2^2 & u_2u_3 \\ u_3u_1 & u_3u_2 & u_3^2 \end{bmatrix}$$

which, indeed, is a symmetric 3×3 rank 1 matrix.[17] In fact, whenever **V'** is proportional to (linearly dependent on) **U**, the product **UV** will be a symmetric rank 1 matrix. For example, with $\alpha \neq 0$, let

$$\mathbf{V}' = \begin{bmatrix} \alpha u_1 \\ \alpha u_2 \\ \alpha u_3 \end{bmatrix}$$

so that $\mathbf{V} = [\alpha u_1, \alpha u_2, \alpha u_3] = \alpha[u_1, u_2, u_3] = \alpha\mathbf{U}'$, and $\mathbf{UV} = \mathbf{U}\alpha\mathbf{U}' = \alpha\mathbf{UU}'$.

---

[17]These results—symmetry and a rank of 1—hold for the product **UU'** of any nonzero n-element column vector **U**.

Thus, if **A** happens to be an $n \times n$ symmetric matrix, and if $\Delta\mathbf{A} = \alpha\mathbf{UU}'$, which is a symmetric, rank 1 change, then the new matrix, $\mathbf{A}^* = \mathbf{A} + \Delta\mathbf{A}$, will also be symmetric. If **A** is nonsingular, its inverse, as well as that for $\mathbf{A}^*$, will also be symmetric.

For example, let **A** be the symmetric matrix $\mathbf{A} = \begin{bmatrix} 1 & 2 & 1 \\ 2 & 0 & 3 \\ 1 & 3 & 4 \end{bmatrix}$, $\mathbf{U} = \begin{bmatrix} 0 \\ 2 \\ 5 \end{bmatrix}$ and $\alpha = 1$.

Then $\mathbf{UU}' = \begin{bmatrix} 0 & 0 & 0 \\ 0 & 4 & 10 \\ 0 & 10 & 25 \end{bmatrix}$ (which is symmetric, as we know must be the case), so if $\Delta\mathbf{A}$

$= \alpha\mathbf{UU}'$, $\mathbf{A}^* = \mathbf{A} + \Delta\mathbf{A} = \begin{bmatrix} 1 & 2 & 1 \\ 2 & 4 & 13 \\ 1 & 13 & 29 \end{bmatrix}$, which is also symmetric, again as expected.

For any $\alpha \neq 0$, $\alpha\mathbf{UU}'$ is symmetric. For example, when $\alpha = 2$, $\alpha\mathbf{UU}' = \begin{bmatrix} 0 & 0 & 0 \\ 0 & 8 & 20 \\ 0 & 20 & 50 \end{bmatrix}$.

Thus, a generalized symmetric rank 1 matrix change is given by

$$\mathbf{U}\alpha\mathbf{U}' \qquad \text{or} \qquad \alpha\mathbf{UU}' \qquad\qquad (6.1.4)$$

6A.1.3   An Alternative SMW Expression

The General Case   We now examine a slight variation in the expression for a rank 1 change to **A**. (The variation will be less slight when the change is rank 2 or larger.) Again, given an $n \times 1$ **U** and a $1 \times n$ **V**, let $\mathbf{V}^* = \alpha\mathbf{V}$ (for $\alpha \neq 0$). Then $\Delta\mathbf{A} = \mathbf{UV}^* = \mathbf{U}(\alpha\mathbf{V})$ is an $n \times n$ rank 1 (but not, in general, symmetric) matrix. With $\mathbf{A}^* = \mathbf{A} + \Delta\mathbf{A} = \mathbf{A} + \mathbf{U}(\alpha\mathbf{V})$, the SMW approach from (6.1.2) gives

$$(\mathbf{A}^*)^{-1} = \mathbf{A}^{-1} - \left[\frac{(\mathbf{A}^{-1}\mathbf{U})(\alpha\mathbf{VA}^{-1})}{1 + (\alpha\mathbf{V})\mathbf{A}^{-1}\mathbf{U}} \right]$$

As before, the denominator of the expression in square brackets is a scalar and must be different from zero.

Since α is a nonzero scalar, both numerator and denominator of the term in square brackets on the right can be divided by it, giving

$$(\mathbf{A}^*)^{-1} = \mathbf{A}^{-1} - \frac{(\mathbf{A}^{-1}\mathbf{U})(\mathbf{VA}^{-1})}{(1/\alpha) + \mathbf{VA}^{-1}\mathbf{U}} \qquad\qquad (6.1.5)$$

This is simply an alternative way of expressing the relationship between $\mathbf{A}^{-1}$ and $(\mathbf{A}^*)^{-1}$ when the second term in a rank 1 change is expressed as $\alpha\mathbf{V}$. In order to suggest the structure that will be taken by the more general form of this kind of expression, when dealing with rank 2 (or higher) changes, we can rewrite (6.1.5) as

$$(\mathbf{A}^*)^{-1} = \mathbf{A}^{-1} - [(\mathbf{A}^{-1}\mathbf{U})(\alpha^{-1} + \mathbf{VA}^{-1}\mathbf{U})^{-1}(\mathbf{VA}^{-1})] \qquad\qquad (6.1.5')$$

The Symmetric Case   If $\mathbf{V}^* = \alpha\mathbf{V}$ and $\mathbf{V} = \mathbf{U}'$, then $\Delta\mathbf{A} = \mathbf{U}\mathbf{V}^* = \mathbf{U}\alpha\mathbf{U}' = \alpha\mathbf{U}\mathbf{U}'$, which is symmetric, and

$$(\mathbf{A}^*)^{-1} = \mathbf{A}^{-1} - \left[\frac{(\mathbf{A}^{-1}\mathbf{U})(\mathbf{U}'\mathbf{A}^{-1})}{(1/\alpha) + \mathbf{U}'\mathbf{A}^{-1}\mathbf{U}}\right] \tag{6.1.6}$$

or, alternatively

$$(\mathbf{A}^*)^{-1} = \mathbf{A}^{-1} - [(\mathbf{A}^{-1}\mathbf{U})(\alpha^{-1} + \mathbf{U}'\mathbf{A}^{-1}\mathbf{U})^{-1}(\mathbf{U}'\mathbf{A})] \tag{6.1.6'}$$

When $\mathbf{A}$ and $\Delta\mathbf{A}$ are symmetric (and $\mathbf{A}$ is nonsingular), then $\mathbf{A}^{-1}$ and also $(\mathbf{A}^*)^{-1}$ will be symmetric, too.

6A.1.4   Symmetric Rank 2 Changes

Consider now a *pair* of n-element column vectors, $\mathbf{U}_1$ and $\mathbf{U}_2$. Both $\alpha\mathbf{U}_1\mathbf{U}_1'$ $(\alpha \neq 0)$ and $\beta\mathbf{U}_2\mathbf{U}_2'$ $(\beta \neq 0)$ generate symmetric rank 1 matrices. The important point is that, when $\mathbf{U}_2$ is not proportional to (linearly dependent on) $\mathbf{U}_1$, and for nonzero scalars α and β

$$\Delta\mathbf{A} = \alpha\mathbf{U}_1\mathbf{U}_1' + \beta\mathbf{U}_2\mathbf{U}_2' = [\mathbf{U}_1 \vdots \mathbf{U}_2]\begin{bmatrix} \alpha\mathbf{U}_1' \\ \beta\mathbf{U}_2' \end{bmatrix} \tag{6.1.7}$$

generates a symmetric $n \times n$ matrix of rank 2.

For example, let $\mathbf{U}_1 = \begin{bmatrix} 1 \\ 2 \\ 3 \end{bmatrix}$, $\mathbf{U}_2 = \begin{bmatrix} 1 \\ 4 \\ 6 \end{bmatrix}$ (so that $\mathbf{U}_1$ and $\mathbf{U}_2$ are linearly independent), and let $\alpha = \beta = 1$; then

$$\alpha\mathbf{U}_1\mathbf{U}_1' = \begin{bmatrix} 1 & 2 & 3 \\ 2 & 4 & 6 \\ 3 & 6 & 9 \end{bmatrix} \quad \text{and} \quad \beta\mathbf{U}_2\mathbf{U}_2' = \begin{bmatrix} 1 & 4 & 6 \\ 4 & 16 & 24 \\ 6 & 24 & 36 \end{bmatrix}$$

(both symmetric rank 1 matrices), so that

$$\alpha\mathbf{U}_1\mathbf{U}_1' + \beta\mathbf{U}_2\mathbf{U}_2' = \begin{bmatrix} 2 & 6 & 9 \\ 6 & 20 & 30 \\ 9 & 30 & 45 \end{bmatrix}$$

which is symmetric and of rank 2. Note that for this case in which $\alpha = \beta = 1$, letting $\mathbf{U} = [\mathbf{U}_1 \vdots \mathbf{U}_2]$, $\mathbf{U}_1\mathbf{U}_1' + \mathbf{U}_2\mathbf{U}_2' = \mathbf{U}\mathbf{U}'$. (Try it.)

Other nonzero values of α and β serve only to alter the magnitudes in the $n \times n$ matrices $\mathbf{U}_1\mathbf{U}_1'$ and $\mathbf{U}_2\mathbf{U}_2'$, not the ranks of those matrices, so that a generalized symmetric rank 2 matrix change is given by the expression in (6.1.7). Notice that this expression can be rewritten as

$$\alpha\mathbf{U}_1\mathbf{U}_1' + \beta\mathbf{U}_2\mathbf{U}_2' = \mathbf{U}\mathbf{D}\mathbf{U}' \tag{6.1.8}$$

where $\mathbf{D}$ is the diagonal matrix $\begin{bmatrix} \alpha & 0 \\ 0 & \beta \end{bmatrix}$. This constitutes a compact expression for the generalized symmetric rank 2 matrix change.

6A.1.5 Another Alternative SMW Expression

In (6.1.6′) we presented a variant of the SMW formula because it suggests the general structure of the expression for more complicated symmetric changes, as in (6.1.8). The SMW result for a symmetric rank 2 change—that is, where $\Delta A = UDU'$, as in (6.1.8)—is

$$(A^*)^{-1} = A^{-1} - (A^{-1}U)[D^{-1} + U'A^{-1}U]^{-1}(U'A^{-1}) \qquad (6.1.9)$$

The nonsingularity requirement now is imposed on **A** (as always) and also on $(D^{-1} + U'A^{-1}U)$.

6A.1.6 Symmetric Rank n Changes ($n > 2$) and the Accompanying SMW Expression

The iterative methods for maximization and minimization problems that we investigate in this book make use of symmetric rank 1 and rank 2 changes only. To understand these methods, it is not necessary to investigate the algebra of higher-rank symmetric changes. However, the extra effort to include them at this point is small, and so the discussion is included, for completeness.

A symmetric rank m matrix change is easily shown to be an extremely straightforward generalization of the result in (6.1.8). All that is needed is that **U** now be defined as an $n \times m$ matrix made up of the columns $U_1, U_2, \ldots, U_m$:

$$U = [U_1 \vdots U_2 \vdots \cdots \vdots U_m]$$

and the $n \times n$ diagonal matrix **D** is redefined to contain elements $s_1, s_2, \ldots, s_m$ on its diagonal:

$$D = \begin{bmatrix} s_1 & 0 & \cdots & 0 \\ \vdots & s_2 & \cdots & 0 \\ \vdots & & \ddots & \\ \vdots & & & \\ 0 & 0 & \cdots & s_m \end{bmatrix}$$

(When describing a rank 2 change, we used α for s_1 and β for s_2.) Then, just as in (6.1.8), but with **U** and **D** defined as above, the generalized symmetric rank m matrix change is given by UDU'. Furthermore, when $\Delta A = UDU'$, the SMW expression for $(A^*)^{-1} = (A + \Delta A)^{-1}$ is exactly that given in (6.1.9), again with **U** and **D** defined as immediately above.

The SMW result is intended as a computational shortcut. Since it requires the inverse of the $m \times m$ matrix $(D^{-1} + U'A^{-1}U)$, it is clear that the result in (6.1.9) is only effective if m is considerably smaller than n; otherwise it is just as easy to find the new $n \times n$ inverse directly.

REFERENCES

Broyden, C. G., "Quasi-Newton Methods and Their Application to Function Minimization," *Math. Comput.* **21**, 368–381 (1967).

———, "The Convergence of a Class of Double Rank Minimization Algorithms," Parts I and II, *J. Inst. Math. Appl.* **6**, 76–90, 222–231 (1970).

Davidon, W. C., *Variable Metric Method for Minimization,* AEC Research and Development Report ANL-5990 (revised), 1959.

———, "Variance Algorithms for Minimization," *Comput. J.* **10,** 406–410 (1968).

Dennis, J. E., Jr., and Robert B. Schnabel, *Numerical Methods for Unconstrained Optimization and Nonlinear Equations,* Prentice-Hall, Englewood Cliffs, NJ, 1983.

Dennis, J. E., Jr., and H. H. W. Mei, "Two New Unconstrained Optimization Algorithms which Use Function and Gradient Values," *J. Optim. Theory Appl.* **28,** 453–482 (1979).

Fiacco, Anthony V., and Garth P. McCormick, *Nonlinear Programming,* Wiley, New York, 1968.

Fletcher, R., "A New Approach to Variable Metric Algorithms," *Comput. J.* **13,** 317–322 (1970).

———, *Practical Methods of Optimization,* Wiley, New York, 1987.

Fletcher, R., and M. J. D. Powell, "On the Modification of LDL^T Factorizations," *Math. Comput.* **29,** 1067–1087 (1963).

Fletcher, R., and C. M. Reeves, "Function Minimization by Conjugate Gradients," *Comput. J.* **7,** 149–154 (1964).

Goldfarb, D., "A Family of Variable Metric Methods Derived by Variational Means," *Math. Comput.* **24,** 23–26 (1970).

Goldfeld, S. M., R. E. Quandt, and H. F. Trotter, "Maximization by Quadratic Hill Climbing," *Econometrica* **34,** 541–551 (1966).

Levenberg, K., "A Method for the Solution of Certain Problems in Least Squares," *Quart. Appl. Math.* **2,** 164–168 (1944).

Marquardt, D., "An Algorithm for Least-Squares Estimation of Nonlinear Parameters," *SIAM J. Appl. Math.* **11,** 431–441 (1963).

Moré, J. J., "The Levenberg-Marquardt Algorithm: Implementation and Theory," in G. A. Watson, ed., *Numerical Analysis,* Lecture Notes in Mathematics, No. 630, Springer Verlag, Berlin, 1977.

Murtagh, B. A., and R. W. H. Sargent, "Second Derivative Methods," in R. Fletcher, ed., *Numerical Methods for Unconstrained Optimization,* Academic Press, London, 1969.

Nelder, J. A., and R. Mead, "A Simplex Method for Function Minimization," *Comput. J.* **8,** 351–367 (1965).

Polak, E., *Computational Methods in Optimization: A Unified Approach,* Academic Press, New York, 1971.

Powell, M. J. D., "An Efficient Method for Finding the Minimum of a Function of Several Variables without Calculating Derivatives," *Comput. J.* **7,** 155–162 (1964).

———, "A Hybrid Method for Nonlinear Equations," in P. Rabinowitz, ed., *Numerical Methods for Nonlinear Equations,* Gordon & Breach, London, 1970.

Press, William H., Brian P. Flannery, Saul A. Teukolsky, and William T. Vetterling, *Numerical Recipes. The Art of Scientific Computing,* Cambridge University Press, Cambridge, UK, 1986.

Shanno, D. F., "Conditioning of Quasi-Newton Methods for Function Minimization," *Math. Comput.* **24,** 647–656 (1970).

Wolfe, Philip, 1968. "Another Variable Metric Method," working paper [cited in Fletcher (1987, p. 51)].

PROBLEMS

6.1 For the dichotomous search technique, show that the interval of uncertainty at the kth step, I^k, is related to the initial search interval, I^0, and δ, the perturbation of the interval midpoint, by

$$I^k = (0.5)^k(I^0) + \left\{ \left[\sum_{i=0}^{k-1} (0.5)^i \right]\left(\frac{\delta}{2} \right) \right\}.$$

6.2   Complete the geometry for the downhill simplex search method in Figure 6.10 for the remaining iterations (after iteration 6), to see exactly how this procedure closes in on the minimum point at $\mathbf{X^*} = \begin{bmatrix} 2 \\ 2.5 \end{bmatrix}$.

6.3   Find the x (or x's) that minimize $f(x) = x^4 - 18x^2 + 15$.

(a) Solve the problem directly using standard first- and second-order conditions from Chapter 3.

(b) Solve the minimization problem directly using the gradient approach in (6.1) for $f(x)$. This is

$$x^{k+1} = x^k - \alpha^k f'(x) \text{ (for a minimization problem)}.$$

$$x^{k+1} = x^k + \alpha^k f'(x) \text{ (for a maximization problem)}.$$

Do your computational routine at least twice, starting at least once from an initial value of x^0 that is positive and at least once from an initial value of x_0 that is negative.

(c) Now approach the problem by using any iterative root-finding method that you wish (Chapter 5) on the equation that constitutes the first-order condition for either a maximum or a minimum for this function. Again, run through the iterative method at least twice, once with a bracket whose lower and upper bounds are both positive, a second time with a bracket whose lower and upper bounds are both negative.

6.4   Using the simple gradient method approach with $\alpha^k = 0.25$ for all k, find an approximation to the maximum of $f(\mathbf{X}) = 16 - 2(x_1 - 3)^2 - (x_2 - 7)^2$. Let $\mathbf{X^0} = \begin{bmatrix} 0 \\ 0 \end{bmatrix}$. Check your results by finding the maximum exactly, using the methods of Chapter 3. [This was Problem 3.6(d) of Chapter 3.]

6.5   Using the same gradient approach as in Problem 6.4, find the minimum of

(a) $f(\mathbf{X}) = x_1^2 + x_1 x_2 + x_2^2 - x_2$ using $\alpha^k = 0.5$ for all k.

(b) $f(\mathbf{X}) = x_1^2 + x_2^2 + (3x_1 + 4x_2 - 26)^2$ using $\alpha^k = 0.01$ for all k.

In both cases, compare your result with the exact answer. [These were Problems 3.6(a) and (b) in Chapter 3.]

6.6   A general description of an iterative method for multivariable functions is that each new point depends on the previous point in a simple additive way, as in (6.1):

$$\mathbf{X}^{k+1} = \mathbf{X}^k + \Delta\mathbf{X}^k$$

It is usual (although there are other options as well) to consider the second term on the right as made up of a *direction* component ($\mathbf{D}^k$) and a *step size* component (the scalar α^k), so

$$\Delta\mathbf{X}^k = \alpha^k \mathbf{D}^k \qquad \text{and} \qquad \mathbf{X}^{k+1} = \mathbf{X}^k + \alpha^k \mathbf{D}^k \qquad\qquad (*)$$

In *gradient-based* methods, $\mathbf{D}^k = \nabla f(\mathbf{X}^k)$ is used as the direction in maximization problems, since it points in the direction of maximum increase of the function from $\mathbf{X}^k$, and the negative of the gradient, $[-\nabla f(\mathbf{X}^k)]$ is used for $\mathbf{D}^k$ in minimization problems.

(a) Consider Cauchy's approach. Replace $\mathbf{D}^k$ in (∗) by the gradient (for a max problem) and by the negative of the gradient (for a min problem), express $f(\mathbf{X}^{k+1})$ with a second-order Taylor series, and derive the first-order conditions for maximizing the change (increase) in $\mathbf{X}$, $\mathbf{X}^{k+1} - \mathbf{X}^k$, for a max problem and for maximizing the change (decrease) in $\mathbf{X}$ for a min problem. Your result should show that the optimal α^k for a maximization problem will be the negative of the optimal α^k for a minimization problem. Check and explain second-order conditions in each of the problems.

(b) What would be the particularization of Cauchy's method, from part (a), to the problem of maximization or of minimization of a *function of a single variable*, $f(x)$?

(c) Show how Cauchy's method would work on the problem of maximizing $f(x) = -3x^2 + 6x + 100$. Use three different starting points: $x_0 = 0$, then $x_0 = -10$, then $x_0 = 6000$.

(d) Show how Cauchy's method would work on the problem of minimizing $f(x) = 3x^2 - 6x + 100$. Use three different starting points: $x_0 = 1$, then $x_0 = -1000$, then $x_0 = 10,000$.

6.7 Newton's method for a minimization problem was shown in (6.7). How would it work in practice on the problems in (c) and (d), in the previous question, for a function of one variable?

6.8 In the usual gradient approach for a maximization problem, we have

$$\mathbf{X}^{k+1} = \mathbf{X}^k + \Delta\mathbf{X}^k = \mathbf{X}^k + \alpha^k \nabla f(\mathbf{X}^k)$$

where α^k is the step length component of the change in $\mathbf{X}^k$ and $\nabla f(\mathbf{X}^k)$ is the direction component.

(a) Suppose, instead, that we define $\mathbf{X}^{k+1} = \mathbf{X}^k + \mathbf{C}^k$, where $\mathbf{C}^k$ is a vector of "changes from" k that incorporates *both* direction and length. Show how Newton's method emerges from maximization of the *change* in $f(\mathbf{X})$, from $\mathbf{X}^k$ to $\mathbf{X}^{k+1}$, through choice of $\mathbf{C}^k$. Consider second-order conditions as well as first-order conditions.

(b) Suppose that the step length component, α, is fixed at a constant value in the usual gradient approach; $\alpha^k = \bar{\alpha}$ for all k. Then what is an optimal choice of the elements in $\Delta\mathbf{X}^k$, the direction components? [Compare with the usual gradient method where the direction component is given by the gradient, $\Delta\mathbf{X}^k = \nabla f(\mathbf{X}^k)$ and then the step length at move k, α^k, is chosen optimally.] Illustrate your "optimal direction" rule with

$$f(\mathbf{X}) = -2x_1^2 + 12x_1 - x_2^2 + 14x_2 - 51$$

6.9 For the DFP rank 2 symmetric update in (6.32) it can be shown that if $(\mathbf{x}^k)'\mathbf{g}^k > 0$, then $\mathbf{B}^{k+1}$ will be positive definite if $\mathbf{B}^k$ is positive definite. Illustrate this property for the case in which $\mathbf{B}^k = \mathbf{I}$ and where

(a) $\mathbf{x}^k = \begin{bmatrix} 1 \\ 2 \\ 3 \end{bmatrix}$ and $\mathbf{g}^k = \begin{bmatrix} 1 \\ 0 \\ 3 \end{bmatrix}$, so that $(\mathbf{x}^k)'\mathbf{g}^k > 0$. Show that, following (6.32), $\mathbf{B}^{k+1} =$

$\begin{bmatrix} 1 & 0.2 & 0 \\ 0.2 & 1.4 & 0.6 \\ 0 & 0.6 & 1 \end{bmatrix}$ and then show that $\mathbf{B}^{k+1}$ is, indeed, positive definite.

(b) $\mathbf{x}^k = \begin{bmatrix} -1 \\ -2 \\ -3 \end{bmatrix}$ and $\mathbf{g}^k = \begin{bmatrix} 1 \\ 0 \\ 3 \end{bmatrix}$, so that $(\mathbf{x}^k)'\mathbf{g}^k < 0$. Find, for this case, that $\mathbf{B}^{k+1} =$

$\begin{bmatrix} 0.8 & -0.2 & -0.6 \\ -0.2 & 0.6 & -0.6 \\ -0.6 & -0.6 & -0.8 \end{bmatrix}$ and show that this is not a positive definite matrix.

6.10 Show that the alternative expressions for $\mathbf{B}^{k+1}$, in (6.41) and (6.42), are, in fact, equivalent.

If you have computer software that includes one or more of the numerical methods for unconstrained maximization or minimization that are explored in this chapter, then you can examine sensitivity of results (accuracy and efficiency, for example) to alternative starting points and stopping criteria on one or more test functions. Accuracy can be assessed, for functions whose exact maxima and/or minima are known, by some measure of the difference between the exact result and the result from the iterative procedure. Measures of the efficiency of a computational procedure might be number of iterations or computation time.

PART IV

APPLICATIONS: CONSTRAINED OPTIMIZATION IN LINEAR MODELS

Finding the x that provides an unconstrained maximum or minimum of a *linear* function $f(x)$ is impossible using classical calculus techniques because the function will have no finite maximum or minimum value, as long as it has a slope that is not zero. (If the slope is zero, then the function has the same value, both a maximum and a minimum, for all x's, and that is a completely uninteresting problem.)

If there are constraints on x, however, then the problem may become more interesting. If it is required that $x = a$, then there is no choice and, again, the problem is uninteresting. On the other hand, if it is required that $a \leq x \leq b$, then in general there *will* be a single value of x for which $f(x)$ will be largest and another value for which $f(x)$ will be smallest (again, if the slope of the line is not zero). Finding those critical values of x for a linear $f(x)$ is a (very simple) problem of constrained optimization in linear models.

When there is a *vector* of choice variables, **X**, an associated *linear* function $f(\mathbf{X})$, and one or more *linear* inequality constraints on the variables (or some linear inequalities and some linear equations), we are faced with a linear programming problem. Problems with this structure are examined in this part.

7

LINEAR PROGRAMMING: FUNDAMENTALS

This chapter is the first of three in which we examine techniques for constrained optimization in an all-linear model. These models require an extremely specialized and, it turns out, relatively simple approach. The essential mathematical characteristic is that both the objective function and all the constraints (usually inequalities) are *linear* functions of the decision variables (the x's). The *linear* in the name linear programming is therefore an exact description of the mathematical characteristics of the functions in the problem. *Programming* was used at the time the method was developed (in the United States during the 1940s in response to problems that arose during World War II) to denote "scheduling" or "planning" of activities (selection of values for the x's). This terminology has remained; it has nothing really to do with computer programming, although we will see that methods for solving linear programming problems (also called "linear programs") are done very efficiently on computers. This is fortunate, since real-world problems may be very large, having thousands of variables and/or constraints.

We saw in Chapter 4 how the conditions for *unconstrained* maximization or minimization problems with many variables could be formulated through extensions of the logic for a function of one variable, using the concept of the total differential. Second-order conditions depended on an understanding of the rules for determining the sign of a quadratic form, unless the concavity or convexity of the function could be established from the context of the problem. These sign rules become very complicated and difficult to apply for a function of many variables.

Then we looked at constraints. *Equality* constraints, removing the independence of the variables, required a further examination of the structure of quadratic forms for dependent variables. The basic approach, however, was essentially the same—first differentials had to be set equal to zero, the resulting equations solved, and then maximum points, minimum points, or those that were neither were distinguished by the sign of the second total differential. This required evaluation of the signs of leading principal minors of a bordered Hessian matrix. For functions of many variables, subject to several equality constraints, these rules become complicated, difficult, and tedious to apply. *Inequality* constraints posed a somewhat different problem, since the nature of the dependence between the variables is specified less exactly in an inequality than it is in an equation. The more complex results of Section 4.5 reflect this; it seems clear that for many-variable, many-constraint problems, the calculus-based approach of that section would generally be unworkable.

In fact, when both the objective function and all the constraints are *linear* functions of the unknowns, such calculus-based criteria for maximization and minimization cannot be used at all. This is because the first and second differential rules depend on the existence of first and second partial derivatives, where the first derivatives are themselves functions of the unknown x's—so that the first-order conditions, $\nabla f(\mathbf{X}) = \mathbf{0}$, constitute a set of equations in the x's. If all functions are linear, then all first partial derivatives are constants, all second partial and cross-partial derivatives are zero, and the criteria of Chapter 4 simply do not work.

In this chapter we explore the structure of a linear programming problem and see how *basic* solutions to systems of linear equations with more unknowns than equations (Chapter 2) play a crucial role in solutions to linear programs. We will see that linear programs always come in pairs, and in solving one problem we automatically solve the related problem as well. Finally, we will examine some ways of assessing the "robustness" of a solution—exploring the sensitivity of the solution to variations in some of the parameters of the problem.

There are many very effective computer software packages that solve linear programming problems. The best of these will generate solutions to both problems in the pair, and they will also compute and display sensitivity results. Therefore, it generally will not be necessary to actually solve linear programming problems by hand (just as you would not solve a large system of linear equations by hand). But as an intelligent user of linear programming computer software you ought to know what is being done for you, what the various kinds of results mean, and what is wrong when you are told that there is no solution or that there are many solutions to the linear program.

As an illustration of how a linear programming problem might arise, consider again the job retraining example of Chapter 1. The mayor of a city has introduced a pair of job retraining programs to help currently unemployed city residents upgrade their skill levels. Educational costs per student are $2500 for program 1 and $3500 for program 2; the mayor has a budget (allocated from the state) of $1 million to cover the costs of running these two programs next year. One of the measures of success of the programs is the income that will be earned by graduates since with their new skills they become employable. Annual income earned is estimated at $10,000 and $20,000 for graduates of programs 1 and 2, respectively. The city has a wage tax (2.5%), and $4,600,000 in new earned income (as a result of employment of graduates of the programs) would increase the city's tax revenue by $115,000; this figure is the mayor's target for new city revenue from that tax source. Letting x_1 and x_2 represent the numbers of students admitted next year's classes, the pair of linear equations describing this problem was [as in (1.5) in Chapter 1]

$$2500x_1 + 3500x_2 = 1,000,000$$
$$10,000x_1 + 20,000x_2 = 4,600,000$$

(7.1)

The solution to this pair of equations is $x_1 = 260$ and $x_2 = 100$, the numbers of students to admit to each of the two programs next year.

The left-hand side of the first equation is the total cost of admitting x_1 students to program 1 and x_2 students to program 2. The right-hand side is the budget allocation from the state. It might be more reasonable to state the relationship between costs and budget in terms of an *inequality*; letting the $1 million act as an *upper limit* on the amount to be spent. (The state would be happy if some of the money were returned and the city realizes that it cannot spend more than this allocation.) Similarly, it might be more realistic to pose the second constraint as an inequality also, suggesting that the mayor's target serves as a *lower limit* on the new tax base (the mayor would not complain if *more* income were earned). In this restatement of the problem, the two linear *equations* from the scenario in Chapter 1 become linear *inequalities*:

$$2500x_1 + 3500x_2 \leq 1{,}000{,}000$$

$$10{,}000x_1 + 20{,}000x_2 \geq 4{,}600{,}000 \tag{7.2}$$

A weak linear inequality defines a closed half-space (Section 3.5), and hence the two inequalities in (7.2) can be shown geometrically in x_1, x_2 space as the boundary defined by the equation plus everything on one side of it. In two dimensions the boundary is a line, and in Figure 7.1 we indicate the original solution-space geometry for the equation system (Fig. 7.1a) and the new geometry associated with the linear inequalities (Fig. 7.1b). Clearly, the solution to the original equations still satisfies the inequalities (where both are equations), but many other points (anywhere in the shaded region) also satisfy the inequalities. So, the mayor's problem, as currently expressed, has a great many possible solutions. Which one (or ones) are best?

One way to narrow down the solution possibilities is to specify some additional *objective*, in addition to those imposed by the constraints—that the budget not be exceeded and the tax target not be underfulfilled. For example, suppose the city also has a "temporary" 1.25% sales tax to raise funds for a new sports arena. It is estimated that on average employees with $10,000 in earnings spend about 35% of that amount in the city while those with $20,000 annual earnings spend about 30% (a higher proportion of out-of-city spending because of a higher proportion of suburban residency). Therefore, each graduate of program 1 might be expected to generate annually $(0.0125)(0.35)(\$10{,}000) = \$43.75$ in stadium-dedicated tax revenue while each graduate of program 2 would generate $75. So expected new annual sales tax collections earmarked for the stadium would be $43.75x_1 + 75x_2$. The mayor might indicate that the best program of all would be one that, while satisfying the constraints, also *maximized* this total sales tax revenue, $43.75x_1 + 75x_2$.

Then the problem would become much more subtle—find from among *all* the possible combinations of x_1 and x_2 that satisfy the inequalities in (7.2) the particular pair of values for which the *objective function*, $f(\mathbf{X}) = 43.75x_1 + 75x_2$, is maximized. Note, finally, that in this example it does not make sense for either x_1 or x_2 to be negative; that is, we would logically require that $x_1 \geq 0$ and $x_2 \geq 0$. What we now have is a full-blown linear programming problem:

$$\begin{aligned}
\text{Maximize} \quad & 43.75x_1 + 75x_2 \\
\text{subject to} \quad & 2500x_1 + 3500x_2 \leq 1{,}000{,}000 \\
& 10{,}000x_1 + 20{,}000x_2 \geq 4{,}600{,}000 \\
\text{and} \quad & x_1 \geq 0, x_2 \geq 0
\end{aligned} \tag{7.3}$$

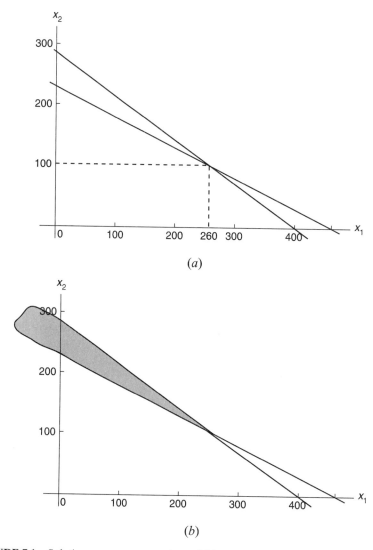

FIGURE 7.1   Solution-space representations of (a) equations (7.1) and (b) inequalities (7.2).

The *linear* structure of the objective function and each constraint allows us to express the entire problem with the convenience of matrix notation. For reasons soon to be explored, suppose that we multiply both sides of the second constraining inequality by -1. This reverses the sign of the inequality (if $7 \geq 4$, then $-7 \leq -4$) and gives us $-10{,}000x_1 - 20{,}000x_2 \leq -4{,}600{,}000$. To simplify the appearance but not alter the content of these constraints, divide both sides of the first by 500 and of the second by 10,000, so that the problem becomes

$$\text{Maximize} \quad 43.75x_1 + 75x_2$$
$$\text{subject to} \quad 5x_1 + 7x_2 \leq 2000 \tag{7.4}$$
$$-x_1 - 2x_2 \leq -460$$
$$\text{and} \quad x_1, x_2 \geq 0$$

Finally, let

$$\mathbf{P} = \begin{bmatrix} 43.75 \\ 75 \end{bmatrix}, \qquad \mathbf{X} = \begin{bmatrix} x_1 \\ x_2 \end{bmatrix}, \qquad \mathbf{A} = \begin{bmatrix} 5 & 7 \\ -1 & -2 \end{bmatrix}, \qquad \mathbf{B} = \begin{bmatrix} 2000 \\ -460 \end{bmatrix}$$

Then the entire linear programming problem can be stated as

$$\text{Maximize} \quad \mathbf{P'X}$$
$$\text{subject to} \quad \mathbf{AX} \le \mathbf{B} \tag{7.5}$$
$$\text{and} \quad \mathbf{X} \ge \mathbf{0}$$

In this chapter we will see that, precisely because of the linearities, the problem is enormously simplified; complications suggested by the rules for the inequality-constrained nonlinear case (Section 4.5) are completely avoided, and the actual calculations needed to solve a linear programming problem are the simplest kind of arithmetic, founded on our examination of linear equations systems in Chapter 2. (Although the calculations are simple, they may also be very numerous; this makes them obvious candidates for computer solution.)

Initially, we will review and extend some of the fundamental convex set theory from Section 3.5. We will see how this leads directly to theorems that enable us to relate the linear programming problem to one of selection from among *basic* solutions to a nonsquare linear *equation* system (with fewer equations than unknowns) formed from the constraints (Section 2.5). Because of the added presence of an objective function, efficient arithmetic procedures for this solution process can be developed. Then we will see that all linear programs come in pairs and that the two problems have certain mathematical connections that may provide additional insight into the optimal solution. Finally, we will explore the robustness of an optimal solution, indicating how sensitive the results are to changes in certain coefficients in the problem.

7.1 FUNDAMENTAL STRUCTURE, ALGEBRA, AND GEOMETRY

7.1.1 Illustrative Example

We will explore the algebra and geometry of the following example problem throughout this chapter. We use it instead of the variant of the job retraining example, above, because it is illuminating to have more constraints that unknowns. Specifically, consider the linear programming problem:

$$\text{Maximize} \quad \Pi = 2x_1 + 5x_2$$
$$\text{subject to} \quad x_1 + 2x_2 \le 10$$
$$3x_1 + 2x_2 \le 24$$
$$x_1 + 10x_2 \le 40 \tag{7.6}$$
$$\text{and} \quad x_1, x_2 \ge 0$$

A plausible "story" behind this illustration might be something like the following. A producer makes two goods, on which per unit profits of 2 and 5 are earned. This is why "Π" [Greek "P" (pi)] is often used to denote the objective function in a linear programming maximization problem. The manufacturing process requires the use of three inputs ("resources"), and the producer has fixed stocks of 10, 24, and 40, respectively, on hand and available for use. Each unit of the first good requires, for its production, one unit of resource 1, three units of resource 2, and one unit of resource 3; each unit of the second product takes two units of resource 1, two of resource 2, and 10 of resource 3. Finally, negative production makes no sense in this illustration. The producer's problem is to choose amounts of each of the two goods to produce, x_1 and x_2, so as to maximize profits, subject to upper limits set by the resource constraints and lower limits (namely, zero) set by the nonnegativity requirements on the x's.

You should be clear that, with straightforward definitions of **P** as the column vector of coefficients in the objective function, **X** as the column vector of unknowns, **A** as the coefficient matrix, exactly as with linear equations, and **B** as the column vector of right-hand sides, this problem has exactly the compact representation given in (7.5). The dimensions of **A** (3×2 in the illustration) completely define the size of the problem; rows of **A** represent the linear inequality constraints (exclusive of the nonnegativities) and columns of **A** represent the unknowns. The matrix representation in (7.5) is possible only because all functions in the problem (objective function and all constraints) are linear.

Thus, if we assign to **A** the dimensions $m \times n$—and, as a consequence, **P**, **X**, and **0** become n-element column vectors while **B** has m elements—we have in (7.5) the completely *general* linear programming maximization problem with n nonnegative variables subject to m linear inequality constraints. (Whenever reference is made to the number of constraints in a problem, it will exclude the nonnegativity requirements on the x's, unless otherwise specified.) Note that the objective is to *maximize* a linear function of the x's and the constraints all supply *upper bounds* on linear functions of those x's. If all coefficients in the **A** matrix and **P** vector are positive, then "less than or equal to" constraints are logical for the problem; they keep the variables, and hence the value of the objective function, from becoming infinitely large. More generally, as was true in the example in (7.3), some of the coefficients in a problem may be negative and/or some inequalities may initially run in the other direction. These inequalities [as in (7.3)] can be converted to "≤" form by multiplying them through by –1; hence $\mathbf{AX} \leq \mathbf{B}$ is perfectly general, and this form for the inequalities will be called the "standard form" for a *maximization* problem. We will consider minimization problems in a moment. (Inclusion of linear *equality* constraints along with linear inequalities will be discussed in the next chapter.)

Here are several less compact statements of the fundamental linear programming maximization problem that you might also find in the literature:

$$\text{Maximize} \quad \Pi = \sum_{j=1}^{n} p_j x_j$$

$$\text{subject to} \quad \sum_{j=1}^{n} a_{ij} x_j \leq b_i \quad (i = 1, \ldots, m) \tag{7.7}$$

$$\text{and} \quad x_j \geq 0 \quad (j = 1, \ldots, n)$$

or, in linear combination form (where $\mathbf{A}_j$ is the jth column of $\mathbf{A}$),

$$\text{Maximize} \quad \Pi = \sum_{j=1}^{n} p_j x_j$$

$$\text{subject to} \quad \sum_{j=1}^{n} \mathbf{A}_j x_j \leq \mathbf{B} \tag{7.8}$$

$$\text{and} \quad x_j \geq 0 \qquad (j = 1, \ldots, n)$$

or, in its most explicit form:

$$\text{Maximize} \quad \Pi = p_1 x_1 + p_2 x_2 + \cdots + p_n x_n$$

$$\text{subject to} \quad a_{11} x_1 + \cdots + a_{1n} x_n \leq b_1$$
$$\vdots \qquad \vdots \qquad \vdots \tag{7.9}$$
$$a_{m1} x_1 + \cdots + a_{mn} x_n \leq b_m$$

$$\text{and} \quad x_1 \geq 0, \ldots, x_n \geq 0$$

7.1.2   The Minimization Problem: Algebra

A problem that has a linear objective function to be *minimized*, subject to a set of constraints that set *lower* limits, as well as nonnegativities on the unknowns, is also a linear program. A classic example is known as the diet problem. Assume that you know the calorie contents, c_j, of unit amounts (however defined) of various food items. Suppose you are planning a menu (for a week)—deciding on the amounts of various foods to consume, y_j, during that week. Subject to certain *minimum* levels of other food essentials, such as vitamins and minerals, the objective may be to select a combination of foods that minimizes (weekly) calorie intake. Let $\mathbf{C}$ and $\mathbf{Y}$ be the (column) vectors of (known) unit calorie content and (unknown) amounts of food items and $\mathbf{E}$ be a vector of minimum levels of food essentials. Finally, let a matrix $\mathbf{K}$ contain elements k_{ij} that represent content of food essential i (vitamin B_1, for example) per unit of food item j (e.g., a cupful of rice). Then the diet problem is

$$\text{Minimize} \quad \mathbf{C}'\mathbf{Y}$$

$$\text{subject to} \quad \mathbf{K}\mathbf{Y} \geq \mathbf{E} \tag{7.10}$$

$$\text{and} \quad \mathbf{Y} \geq \mathbf{0}$$

[There are more disaggregate statements, parallel to those in (7.7)–(7.9) for the maximization problem.]

Any problem with the general form of (7.5) or (7.10) is a linear program. In both problems the unknowns are required to be nonnegative,[1] and a linear function of those vari-

---

[1] More correctly, this is *customarily* required. Allowing some or indeed all of the variables to take on negative values as well as positive would not destroy the important mathematical characteristics of linear programs, as subsequent sections will show; it would, however, alter the computational procedure. In any case, in a great majority of linear programming applications, negative values for the variables are meaningless in the context of the actual problem (you cannot consume less than *no* cupfuls of rice next week or admit fewer than *no* students to a particular job retraining program).

ables is to be maximized or minimized subject to a set of linear inequality constraints that set upper or lower limits, respectively.

7.1.3   The Maximization Problem: Geometry

The areas defined by each of the constraints in the linear program in (7.6) are shown in Figures 7.2a–7.2c. The area common to all constraints (their *intersection*) is also shown (Fig. 7.2d). Because of the $\mathbf{X} \geq \mathbf{0}$ constraints, the pictures are restricted to the nonnegative quadrant. (In higher-dimensional problems, the area within which no variables are negative is called the nonnegative *orthant*.)

Points in the shaded area inside the dark boundaries, as well as all points on the boundaries themselves, satisfy all three constraints in (7.6) and the nonnegativity requirements. The shaded area (including its boundaries) is the *feasible region* for the problem, and any $\mathbf{X}$ in the feasible region is called a *feasible solution* to the problem. Since the constraints are linear, the boundaries of the feasible region are straight lines.

If an additional linear constraint were added to the problem, it would either "cut away" at the current feasible region, adding a new segment to its boundary, or would have no effect whatever, if it were completely outside the present feasible region. Such a constraint is *redundant*; its addition does not change the feasible region at all. For example, try adding $x_1 + 5x_2 \leq 80$ to Figure 7.2d. Since the boundary set by this constraint lies completely above and to the right of the shaded feasible region, all feasible solutions to the original problem will automatically satisfy this new constraint also. Adding constraints to a problem, then, generally adds a new side to the feasible region (or else leaves it un-

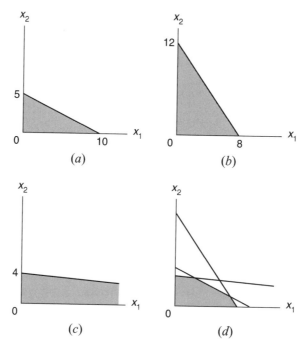

FIGURE 7.2   (*a*) Constraint 1; (*b*) constraint 2; (*c*) constraint 3; (*d*) all three constraints.

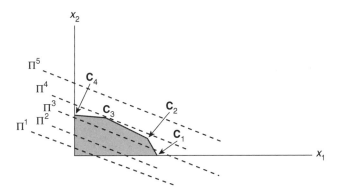

FIGURE 7.3 Objective function contour lines.

changed); adding variables to a problem increases the dimensionality of the solution space in which the feasible region is represented.

The only aspect of the problem in (7.6) that is missing from the geometry in Figure 7.2*d* is the objective function—$\Pi(\mathbf{X}) = 2x_1 + 5x_2$. Clearly, for any specific values x_1^0 and x_2^0, this function assumes a specific value, $\Pi(\mathbf{X}^0)$, or just $\Pi^0 = 2x_1^0 + 5x_2^0$. In slope-intercept form, $x_2^0 = -\frac{2}{5}x_1^0 + (\Pi^0/5)$. While the (vertical axis) intercept of the line, $(\Pi^0/5)$, obviously depends on Π^0 and hence on the particular values x_1^0 and x_2^0, the *slope* is given unambiguously by the coefficients of x_1 and x_2; here it is $-\frac{2}{5}$. Therefore, *contour lines* for various values of Π^0 can be represented in Figure 7.2*d* by a series of parallel lines with that slope. A set of five such lines, labeled $\Pi^1, \ldots, \Pi^5$ is shown in Figure 7.3.

Since the intercept on the x_2 axis is a direct function of Π, it is clear that the higher the intercept—the farther it is from the origin—the larger is Π. From among the five lines in Figure 7.3, Π^5 is the best from this point of view. But none of the combinations of x_1 and x_2 along that line lie in the feasible region. So the linear programming problem can be seen as one of finding an objective function contour line that is as far as possible from the origin (measured upward along the x_2 axis) while still having *at least one* point in the feasible region. Visually, in Figure 7.3, Π^4 is such a line. Objective function values associated with Π^1, Π^2, and Π^3 are clearly inferior; they intersect the x_2 axis below Π^4; Π^5 has a higher objective function value but is inadmissible because it violates the feasibility requirement.

The coordinates of the *corner point*, $\mathbf{C}_3$, at which the objective function is *tangent to* the feasible region, represent the values of x_1 and x_2 that maximize $2x_1 + 5x_2$ while also satisfying the constraints defining the feasible region. This then is the *optimal solution*. If you consider carefully the implications of Figure 7.3, you will see that the linearity of the objective function and all the constraints ensures that the optimal solution to a linear program with two unknowns will *always* be at a corner of the feasible region in two-dimensional solution space. The only slight variant of this result occurs when the objective function happens to be parallel to one of the constraints; in that case (in the two-variable example) two (adjacent) corners and all points along the edge joining them will be "equally optimal." (Imagine that the objective function contour lines are parallel to the edge connecting corners $\mathbf{C}_2$ and $\mathbf{C}_3$; then both $\mathbf{C}_2$ and $\mathbf{C}_3$ and all points on the boundary between them will be optimal.)[2] The very linearity that invalidated any approach utilizing calculus

---

[2]We consider this special case in more detail in Chapter 8.

and the Lagrange multiplier or Kuhn–Tucker results of Chapter 4 guarantees that a corner solution will always be the best for a linear program. Therefore, any solution procedure that examines *only* corners of the feasible region is assured of finding the optimum solution. This motivates us to examine the characteristics of these corners in more detail.[3]

It is useful to express a linear program in *augmented* form, in which each inequality is converted into an equation by the addition of a *slack* variable. Consider (7.6) again. Add to the left-hand side of the first constraint a *nonnegative* variable, s_1, which serves to take up the "slack" or balance the difference between the two sides. Since 10 is never less than $x_1 + 2x_2$, s_1 must necessarily be nonnegative. The first constraint is now $x_1 + 2x_2 + s_1 = 10$. Similarly, different slack variables, s_2 and s_3, can be added to the left-hand sides of constraints 2 and 3. Then the equality (augmented) form of the constraints is

$$x_1 + 2x_2 + s_1 = 10$$

$$3x_1 + 2x_2 + s_2 = 24$$

$$x_1 + 10x_2 + s_3 = 40$$

This is a set of three linear equations in five unknowns—the two original variables (x_1 and x_2) and the three additional slack variables (s_1, s_2, and s_3). The slack variables have been added for (what will turn out to be) mathematical convenience; they have nothing to contribute to the objective function of the problem. However, for purposes of symmetry, they could be thought of as appearing in the objective function with coefficients of zero. So the fully augmented form of the linear program in (7.5) can be written

$$\text{Maximize} \quad 2x_1 + 5x_2 + 0s_1 + 0s_2 + 0s_3$$
$$\text{subject to} \quad x_1 + 2x_2 + 1s_1 + 0s_2 + 0s_3 = 10$$
$$3x_1 + 2x_2 + 0s_1 + 1s_2 + 0s_3 = 24 \qquad (7.11)$$
$$x_1 + 10x_2 + 0s_1 + 0s_2 + 1s_3 = 40$$
$$\text{and} \quad x_1 \geq 0, x_2 \geq 0; \quad s_1 \geq 0, s_2 \geq 0, s_3 \geq 0$$

The matrix form of the augmented general linear programming maximization problem is therefore

$$\text{Maximize} \quad \mathbf{P'X + 0'S}$$
$$\text{subject to} \quad \mathbf{AX + IS = B} \qquad (7.12)$$
$$\text{and} \quad \mathbf{X \geq 0, S \geq 0}$$

where $\mathbf{S}$ is an m-element column vector of slack variables, $\mathbf{I}$ is an $m \times m$ identity matrix, the null vector associated with $\mathbf{X}$ contains n elements and that associated with $\mathbf{S}$ contains m elements. The important point about the augmented form of a linear programming

---

[3]For a linear program with three unknowns, the feasible region will be a convex set in three-dimensional solution space whose sides are defined by planes. Edges are defined by intersections of two planes and corners of this region will be defined by intersections of three planes. The objective function contours will also be planes, and so corners will also be where optimal solutions are located. In more than three dimensions, the principles are the same but the geometry is not available.

TABLE 7.1 Positive and Zero-Valued Variables at Corners of the Feasible Region

	Origin	Corner			
		C_1	C_2	C_3	C_4
Positive-valued variables	s_1, s_2, s_3	x_1, s_1, s_3	x_1, x_2, s_3	x_1, x_2, s_2	x_2, s_1, s_2
Zero-valued variables	x_1, x_2	x_2, s_2	s_1, s_2	s_1, s_3	x_1, s_3

problem is that the constraints will always be a set of m linear equations in $m + n$ variables, since one slack variable for each constraint (of which there are m) is added to the original x's (of which there are n). We know (from Section 2.5) that if the equation system is not inconsistent we can systematically eliminate "extra" variables by setting them equal to zero, so that $m \times m$ equation systems remain, to which a series of *basic solutions* can be found. Recall that these are solutions in which at least n of the $m + n$ variables are equal to zero.

Looking again at Figure 7.3, we consider the values of x_1 and x_2 and of the three slack variables, s_1, s_2 and s_3, at several of the corners of the feasible region. At the origin, $x_1 = x_2 = 0$ and each slack variable has to be strictly positive—$s_1 = 10$, $s_2 = 24$ and $s_3 = 40$, making up the difference between the left-hand sides (zero in each case when $x_1 = x_2 = 0$) and the right-hand sides. Consider C_1, a corner adjacent to the origin. Now x_1 has become positive and s_2 has turned to zero, since C_1 is on the line that is the equality form of constraint 2. (All along the edge defined by C_1C_2 in the figure, which is the boundary set by constraint 2, $s_2 = 0$.) Since this corner is defined as the intersection of $3x_1 + 2x_2 = 24$ and $x_2 = 0$ (the x_1 axis), it is clear that $x_1 = 8$; also at C_1, given the known values for x_1 and x_2, it is easily established that $s_1 = 2$ and $s_3 = 32$. Geometrically, C_1 is inside the boundaries that are set by the first and third constraints. (You can confirm this by extending the sides contributed by the first and third constraints until they intersect the x_1 axis, at 10 and 40, respectively.) Table 7.1 summarizes these kinds of results for all corners in Figure 7.3.

The very important point is that each of the corners has three of the five variables at positive value and two at a value of zero; hence each corner corresponds to a *basic solution* to the constraint equations in (7.11). Since there are three equations and five unknowns in this example, we know that altogether there are $C_3^5 = 10$ possible basic solutions. Yet the feasible region has only five corners, so in this case up to $10 - 5 = 5$ basic solutions are lost when we look only at the corners of the feasible region. The missing basic solutions are those in which at least one of the basic variables would have to be strictly *negative*. These five basic solutions would be associated with the following (infeasible) intersections in Figure 7.3:[4]

(a) Constraint 1 and the x_1 axis

(b) Constraint 3 and the x_1 axis

(c) Constraint 1 and the x_2 axis

(d) Constraint 2 and the x_2 axis

(e) Constraints 2 and 3

---

[4]You might examine at least some of these cases in detail, identifying which variables are zero, positive, and/or negative in each case. Remember that for points along the x_1 axis, $x_2 = 0$ and vice versa, while for points along a constraint line the slack variable *for that constraint* is zero.

Therefore, a possible method of solution for a linear programming problem would appear to be one in which all nonnegative basic solutions to the augmented constraint set—basic solutions in which no variable is negative—were found and the objective function evaluated for each. That solution for which the objective function was largest would then be chosen as the optimal solution.[5] We present a variant of this idea below, after making somewhat more precise the algebraic–geometric results that appear obvious in the case of two-dimensional solution space.

7.2 CONVEX SET THEORY AGAIN

Earlier in this book, in Section 3.5, we saw that

(a) A linear inequality defines a half-space; if the inequality is weak ($\leq$ or $\geq$), the half-space is closed (its boundary line—or plane or hyperplane—is included in the half-space).

(b) A closed half-space is a convex set (the straight line joining any two points in the set lies entirely within the set).

(c) Points that satisfy several linear inequalities simultaneously (points in the intersection of the half-spaces defined by the inequalities) form a convex set. Therefore, the x's satisfying the constraints in a linear programming maximization problem in standard form, $\mathbf{AX} \leq \mathbf{B}$ and $\mathbf{X} \geq \mathbf{0}$, form a closed convex set.

We have seen in Section 7.1 that corners of the feasible region are particularly important in linear programming problems, since the optimal solution is to be found among the corners. It will be useful to have an unambiguous definition of a "corner" (or vertex, or extreme point) of a convex set in solution space of any number of dimensions, where we cannot visualize the geometry.

EXTREME POINT   A point $\mathbf{P}_0$ is an extreme point of a convex set if and only if it does not lie between two other points in the set. That is, if and only if there do not exist two distinct points, $\mathbf{P}_1$ and $\mathbf{P}_2$, in the set such that

$$\mathbf{P}_0 = \alpha\mathbf{P}_1 + (1 - \alpha)\mathbf{P}_2$$

for $0 < \alpha < 1$.

An extreme point is one that cannot be expressed as a convex combination of some two other points in the set.

The following theorems serve to establish the connection between optimal solutions and extreme points (corners) of a feasible region for linear programming problems of any

---

[5]Problems with many constraints and many variables—with a many-sided and multidimensional feasible region—will have many corners and hence many nonnegative basic solutions, but computers can sort through this kind of linear algebra problem very efficiently. If you visualize well in three dimensions, imagine a set of three axes (the corner of a room often does well for the origin), introduce one constraint (a plane), then a second, then a third, and so on, counting the number of corners in each case. We will also see (Chapter 9) that in "large" problems the number of candidate corners can become enormous, and alternative computational procedures have been devised because of this.

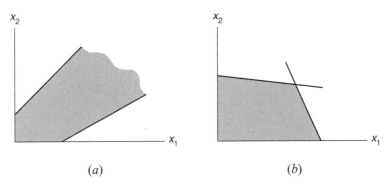

FIGURE 7.4   (*a*) Unbounded, closed convex region; (*b*) bounded, closed convex region.

size—meaning for feasible regions requiring any number of dimensions. They extend the logic of the two- and three-variable problems to cases where the geometry eludes us. Some of the proofs are omitted; they were presented in many texts in the 1950s and 1960s (the early days of linear programming). Others are included simply to indicate how easy it can be to use straightforward linear algebra to extend results that are apparent and obvious from the geometry of two-dimensional illustrations.

The first theorem deals with the number of extreme points that any linear programming problem could have. It is essential for the underlying computational method for linear programming problems, since that method proceeds via examination of basic solutions to the augmented constraints.

Theorem 7.1   The intersection of a finite number of closed half-spaces is a closed convex set *with a finite number of extreme points*.

This means that the feasible region to any linear programming problem has a finite number of corners. Consequently, if an optimal solution is always to be found at a corner, then you will never be required to search among an infinite number of possibilities (which would be an impossible task).[6]

If the intersection described in Theorem 7.1 is *bounded*, which means that any point in the set is at a finite distance from the origin, then the figure is a *convex polyhedron*. Figure 7.4 illustrates bounded and unbounded convex regions in two-dimensional space; note that both regions are *closed*, since the boundaries are included (the constraints are defined by weak inequalities; closedness and boundedness are seen to be very different concepts).

Theorem 7.2   Any point inside a closed and bounded convex set can be expressed as a convex combination of the extreme points of the set.

We need this result for the next theorem. Theorem 7.2 is illustrated in two dimensions in Figure 7.5, for a closed and bounded set made up of four straight line sides with four extreme points, $\mathbf{C}_1 \left(= \begin{bmatrix} x_{11} \\ x_{21} \end{bmatrix} \right)$ through $\mathbf{C}_4 \left(= \begin{bmatrix} x_{14} \\ x_{24} \end{bmatrix} \right)$. The point $\mathbf{P}$, inside the set, is on a line that starts at any corner and ends at a point, $\mathbf{S}$, on a side. (In the figure, $\mathbf{C}_1$ is the corner

---

[6]Although finite, the number of corners can be very large. See footnote 5, above.

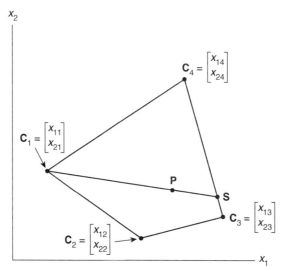

FIGURE 7.5   Illustration of Theorem 7.2.

used.) So $P = \alpha C_1 + (1 - \alpha)S$ for $\alpha > 0$. But $S = \beta C_3 + (1 - \beta)C_4$, for $\beta > 0$.[7] Substitution immediately identifies that $P = \alpha C_1 + 0C_2 + (1 - \alpha)\beta C_3 + (1 - \alpha)(1 - \beta)C_4$. It is easily seen that all coefficients on the C's are nonnegative, and you can quickly identify that the sum of these coefficients is exactly one, so P is indeed a convex combination of the extreme points.

Theorem 7.3   If the objective function takes on a maximum value at some point in the feasible region, then it takes on this value at an extreme point of the region.

There are two kinds of situations in which the objective function may not have a maximum value in the feasible region: if the feasible region is empty or if it is unbounded. Figure 7.6 illustrates these two possibilities in two-dimensional space:

(a) *Empty Feasible Region*   Suppose that the constraints in a linear programming problem are (i) $2x_1 + x_2 \leq 50$ and (ii) $-x_1 \leq -30$, along with $X \geq 0$. Constraint (i) sets the hypotenuse boundary for the triangle on the left in Figure 7.6a and constraint (ii) sets the vertical boundary for the shaded area on the right. No $X \geq 0$ can satisfy both constraints.

(b) *Unbounded Feasible Region*   If the constraints are (i) $-x_1 + x_2 \leq 10$ and (ii) $-x_1 + 2x_2 \leq 40$, along with $X \geq 0$, we have the feasible region indicated in Figure 7.6b. It extends infinitely far in the east–northeast direction. A downward-sloping objective function—such as $x_1 + x_2$—will have no finite maximum.

If a linear programming problem exhibits neither (a) nor (b), then the objection function will take on a maximum value somewhere in the feasible region and, according to Theorem 7.3, it will take on this value at an extreme point.

Suppose that $P'X$ is maximized at X^* and that the feasible region has corners $C_1 =$

---

[7] If P were on a side but not an extreme point, then it could be expressed as a convex combination of the two extreme points at the ends of the side.

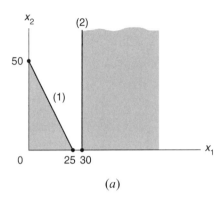

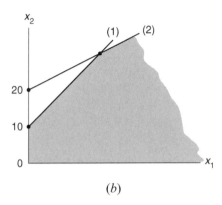

FIGURE 7.6   (*a*) Empty feasible region; (*b*) unbounded feasible region.

$\begin{bmatrix} x_{11} \\ x_{21} \end{bmatrix}, \dots, \mathbf{C}_k = \begin{bmatrix} x_{1k} \\ x_{2k} \end{bmatrix}$. If $\mathbf{X}^*$ is one of the corners, the theorem is valid. If not, then (from Theorem 7.2) $\mathbf{X}^*$ can be expressed as

$$\mathbf{X}^* = \sum_{i=1}^{k} \alpha_i \mathbf{C}_i \qquad \left(\alpha_i \geq 0 \text{ and } \sum_{i=1}^{k} \alpha_i = 1 \right)$$

Find that corner for which $\mathbf{P}'\mathbf{C}_i$ is maximized; denote it by $\mathbf{C}^*$, so that $\mathbf{P}'\mathbf{C}^* = \max [\mathbf{P}'\mathbf{C}_1, \dots, \mathbf{P}'\mathbf{C}_k]$. Note that, since $\mathbf{P}'\mathbf{C}^*$ is a constant and $\Sigma_{i=1}^{k} \alpha_i = 1$,

$$\sum_{i=1}^{k} \alpha_i \mathbf{P}'\mathbf{C}^* = \mathbf{P}'\mathbf{C}^* \sum_{i=1}^{k} \alpha_i = \mathbf{P}'\mathbf{C}^*$$

and so, because of the way $\mathbf{C}^*$ is defined, we have

$$\mathbf{P}'\mathbf{C}^* = \sum_{i=1}^{k} \alpha_i \mathbf{P}'\mathbf{C}^* \geq \sum_{i=1}^{k} \alpha_i \mathbf{P}'\mathbf{C}_i$$

Now, from the definition of $\mathbf{X}^*$, we find that

$$\mathbf{P}'\mathbf{X}^* = \mathbf{P}' \sum_{i=1}^{k} \alpha_i \mathbf{C}_i = \sum_{i=1}^{k} \alpha_i \mathbf{P}'\mathbf{C}_i$$

and examination of the last two equations shows that $\mathbf{P}'\mathbf{C}^* \geq \mathbf{P}'\mathbf{X}^*$. But $\mathbf{X}^*$ is a maximizer of the objective function, so $\mathbf{P}'\mathbf{X}^* \geq \mathbf{P}'\mathbf{X}$ for *all* feasible $\mathbf{X}$, including all the corners, $\mathbf{C}_1, \ldots, \mathbf{C}_k$. And in particular, $\mathbf{P}'\mathbf{X}^* \geq \mathbf{P}'\mathbf{C}^*$. Since both $\mathbf{P}'\mathbf{C}^* \geq \mathbf{P}'\mathbf{X}^*$ and $\mathbf{P}'\mathbf{X}^* \geq \mathbf{P}'\mathbf{C}^*$, it follows that $\mathbf{P}'\mathbf{X}^* = \mathbf{P}'\mathbf{C}^*$, which is what the theorem says; the objective function attains its maximum value at at least one extreme point.

Theorem 7.4   If the objective function takes on a maximum value at *more than one* extreme point of the feasible region, then it has that same maximum value at every convex combination of those extreme points.

In Figure 7.3, if the objective function contours happened to be parallel to one of the sides of the feasible region (e.g., the side defined by constraint 1), then two corners, $\mathbf{C}_2$ and $\mathbf{C}_3$ would be equally optimal, as would all points on their convex combination (the line joining them, which is the edge of the feasible region defined by constraint 1). In higher-dimension solution space, the convex combination may connect two extreme points, in which case the objective function is parallel to an edge of the feasible region, or it may connect three or more extreme points, in which case the objective function is parallel to a face of the feasible region. In either case, the linear programming problem is said to have *multiple optima* or *alternative optima*.

It is fairly easy to demonstrate that this theorem is valid. Suppose that the optima occur at extreme points $\mathbf{C}_1, \ldots, \mathbf{C}_j$ ($j \leq k$), so $\mathbf{P}'\mathbf{C}_1 = \cdots = \mathbf{P}'\mathbf{C}_j = \Pi^*$. Any convex combination of these equally optimal corners is defined as $\widetilde{\mathbf{C}} = \Sigma_{i=1}^{j} \alpha_i \mathbf{C}_i$ ($\alpha_i \geq 0$ and $\Sigma_{i=1}^{j} \alpha_i = 1$). Then $\mathbf{P}'\widetilde{\mathbf{C}} = \Sigma_{i=1}^{j} \mathbf{P}'\alpha_i \mathbf{C}_i = \Sigma_{i=1}^{j} \alpha_i \mathbf{P}'\mathbf{C}_i = \Pi^* \Sigma_{i=1}^{j} \alpha_i = \Pi^*$, which is what the theorem says.

The final two theorems deal with the augmented form of the linear programming problem in which m slack variables have been introduced to create equality constraints: m linear equations in $m + n$ unknowns. We need to be clear about the relationship between the feasible regions in the original and augmented problems. We saw in the previous section that the constraints in the augmented problem are [as in (7.12)] $\mathbf{AX} + \mathbf{IS} = \mathbf{B}$ and $\mathbf{X} \geq \mathbf{0}$, $\mathbf{S} \geq \mathbf{0}$, where the two null column vectors, associated with $\mathbf{X}$ and $\mathbf{S}$, contain n and m elements, respectively; $\mathbf{S}$ is an m-element column vector of slack variables, and $\mathbf{I}$ is an $m \times m$ identity matrix. To represent the constraints of the augmented problem efficiently, suppose that we create an $m \times (n + m)$ matrix in which $\mathbf{I}$ has been placed next to the original $\mathbf{A}$, so we have $[\mathbf{A} \,\vdots\, \mathbf{I}]$. This matrix contains all of the coefficients for the linear constraints $\mathbf{AX} + \mathbf{IS} = \mathbf{B}$. To keep the notation simple, denote this coefficient matrix for the equations in the augmented problem by $\mathbf{A}$. Similarly, denote the added slack variables as $s_1 = x_{n+1}, \ldots, s_m = x_{n+m}$ and then define an $(n + m)$-element column vector of all the unknowns $[\mathbf{X} \,\vdots\, \mathbf{S}]'$. Again, for notational simplicity, we continue to call this $\mathbf{X}$. Then the constraints of the augmented problem can be expressed as[8] $\mathbf{AX} = \mathbf{B}$ and $\mathbf{X} \geq \mathbf{0}$.

The feasible region defined by $\mathbf{AX} \leq \mathbf{B}$ and $\mathbf{X} \geq \mathbf{0}$ in the original form of the problem is the part of the intersection of the m half-spaces in $\mathbf{AX} \leq \mathbf{B}$ that is in the nonnegative orthant ($\mathbf{X} \geq \mathbf{0}$). In the augmented problem, each individual equation in $\mathbf{AX} = \mathbf{B}$ defines a

---

[8]The potential for confusion is outweighed by notational convenience in what follows in this chapter. Just remember that in augmented form, where the constraints are equations, $\mathbf{A}$ contains both the original coefficients and also the identity matrix associated with the slack variables, and $\mathbf{X}$ contains the original unknowns plus all of the slack variables.

(hyper)plane in the nonnegative orthant of $(n + m)$-dimensional space, and the set $\mathbf{AX} = \mathbf{B}$, along with $\mathbf{X} \geq \mathbf{0}$, identifies the intersection of the m (hyper)planes in the nonnegative orthant.

The *dimensionality* of a surface is, essentially, the number of dimensions needed to represent it. A plane in three-dimensional space really "takes up" only two dimensions; by moving the axes around, you could shift it so that it was contained in the x_1, x_2 plane. Similarly, a line in x_1, x_2 space has dimensionality 1. This is expressed formally by saying that a hyperplane in r-dimensional space is a linear *manifold* (vector space) of dimensionality $(r - 1)$.

It can be shown that the intersection of k linearly independent hyperplanes in r-dimensional space is a linear manifold of dimension $(r - k)$—for example, two linearly independent planes in three dimensions intersect in a line (dimensionality $3 - 2 = 1$), three linearly independent planes intersect in a point (dimensionality $3 - 3 = 0$). Therefore, m linearly independent hyperplanes in $(n + m)$-dimensional space intersect in a linear manifold of dimensionality n. This describes the feasible region of the augmented linear programming problem (when all constraints are independent). We already know that the dimensionality of the feasible region in the original form of the problem is just that of the solution space, namely, n. So the dimensionality of the feasible region in both forms of the problem is the same.

It is also possible to show the correspondence of an extreme point in both forms of the problem. The following small example is the largest that can be done without exceeding three-dimensional geometry. Suppose that the original problem has one constraint and two unknowns: $3x_1 + 2x_2 \leq 12$ and $\mathbf{X} = \begin{bmatrix} x_1 \\ x_2 \end{bmatrix} \geq \mathbf{0}$. Using x_3 for the (single) slack variable in this case, the augmented form of the problem is $3x_1 + 2x_2 + x_3 = 12$ and $\mathbf{X} = \begin{bmatrix} x_1 \\ x_2 \\ x_3 \end{bmatrix} \geq \mathbf{0}$.

Figure 7.7 illustrates this point. In the original problem, we are able to find that $x_3 = 12$ at C_0. In the augmented version of the problem, C_0 is transformed into C_3, where we see that $x_3 = 12$. This illustrates a general fact—examining extreme points of the feasible region in

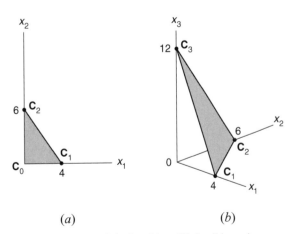

(a) (b)

FIGURE 7.7 (a) Feasible region, original problem; (b) feasible region, augmented problem.

the augmented problem (equality constraints) is the same as examining extreme points of the feasible region in the original problem (inequality constraints).

Now we are ready for the final two theorems. (As promised, proofs are omitted.)

Theorem 7.5   If a nonnegative basic solution can be found to the $\mathbf{AX} = \mathbf{B}$ constraints in the augmented problem, such as

$$\mathbf{A}_1 x_1^* + \cdots + \mathbf{A}_k x_k^* = \mathbf{B}$$

(where $k \leq m$ and $x_1^* > 0, \ldots, x_k^* > 0$), then the $(n + m)$-element column vector

$$\mathbf{X}^* = [x_1^*, \ldots, x_k^*, 0, \ldots, 0]'$$

is an extreme point of the feasible region for the augmented problem.

This says that to every nonnegative basic solution there corresponds an extreme point. The exact relevance of the theorem is that if we find a series of nonnegative basic solutions to $\mathbf{AX} = \mathbf{B}$ and also find the value of the objective function for each of those solutions, we are in fact evaluating the objective function at extreme points.

The only remaining question is whether *every* corner of the feasible region has a corresponding nonnegative basic solution. Theorem 7.6 provides an affirmative answer.

Theorem 7.6   If $\mathbf{X}^* = [x_1^*, \ldots, x_k^*, 0, \ldots, 0]'$ is an extreme point of the feasible region of the augmented problem, then the column vectors $\mathbf{A}_1, \ldots, \mathbf{A}_k$ from $\mathbf{A}$ that correspond to the positive x_i^* in $\mathbf{X}^*$ are linearly independent.

When $k = m$, these columns form a basis; when $k < m$, other columns from $\mathbf{A}$ must be added. (This would represent a degenerate basic solution.) Nonetheless, in either case, since (1) an optimum, if it exists, will be at an extreme point, of which there are a finite number; (2) each extreme point is represented by a nonnegative basic solution to $\mathbf{AX} = \mathbf{B}$; and (3) each nonnegative basic solution corresponds to an extreme point, then a solution procedure that evaluates the objective function for each nonnegative basic solution to $\mathbf{AX} = \mathbf{B}$ will always find the optimum and will do so in a finite number of steps. The *simplex method* is precisely such a procedure; it is the subject matter of the next section.[9]

7.3   THE SIMPLEX METHOD

Although the number of extreme points for any feasible region is finite, it still may be very large. Instead of evaluating the objective function for *all* corners, a more efficient procedure would, at any particular corner, next examine only a "better" corner—one for which it was known in advance that the objective function would have a larger value. This would have the effect, generally, of eliminating some extreme points from possible con-

---

[9]The set of all convex combinations of $n + 1$ points in n-dimensional space, not all on the same hyperplane (or plane, if $n = 3$; or line, if $n = 2$) is called a *simplex*. For $n = 2$, the two-dimensional simplex is a triangle and its interior. The term is used for the solution procedure in linear programming problems because of the geometry of the operations as they appear in vector space, not in the solution space that we have used in illustrating feasible regions.

sideration with each move from one corner to a better one, and once eliminated they would never again be worth considering. The method therefore requires the following:

(a) A procedure for finding a starting (initial) basic feasible solution, $\mathbf{X}^0$.
(b) A procedure identifying a better basic solution.
(c) A mechanism for "moving" from one basic solution, $\mathbf{X}^k$, to a better one[10], $\mathbf{X}^{k+1}$, for which $\mathbf{P'X}^{k+1} > \mathbf{P'X}^k$.
(d) A rule for stopping.

We see that this is an *iterative* procedure; it has the same sequence of steps as the methods in Chapters 2 and 5 for solving equations and in Chapter 6 for unconstrained maximization and minimization.

In small problems with two or three variables, the optimum can easily be found geometrically; for the general *n*-variable, *m*-constraint case we need a completely algebraic solution procedure. This is what the simplex method provides. In this chapter we examine a method for linear programming maximization problems. Because of relationships to be explored in Section 7.4, however, we will see that this also equips us to solve minimization problems.

7.3.1 The Simplex Criterion

If the origin is included among the extreme points of the feasible region, it provides a convenient starting point, since it is completely described by letting the slack variable for each constraint take on the entire right-hand side value.[11] This takes care of (a), above. For (b), the *simplex criterion* provides a means of ranking other corners relative to the current one in terms of their marginal contribution to the current objective function value. This will become clear in what follows.

Assume that a basic feasible solution has been found for the augmented problem

$$\text{Maximize} \quad \mathbf{P'X}$$

$$\text{subject to} \quad \mathbf{AX} = \mathbf{B}$$

$$\text{and} \quad \mathbf{X} \geq \mathbf{0}$$

where $\mathbf{A}$ and $\mathbf{X}$ are expanded, as in Section 7.2, to an $m + (n + m)$ matrix and an $(n + m)$-element vector, respectively. Slack variables are denoted $x_{n+1} = s_1, \ldots, x_{n+m} = s_m$, and a vector of m zeros has been appended to $\mathbf{P}$ in order to include the slack variables in the objective function. The variables can always be renumbered so that the first m are present in this basic solution, so the equation system can be represented as

$$\mathbf{A}_1 x_1 + \cdots + \mathbf{A}_m x_m = \mathbf{B} \tag{7.13}$$

---

[10]In Chapters 2 and 5 we have noted that $\mathbf{X}^{(k)}$ is sometimes used instead of $\mathbf{X}^k$, to distinguish it from an $\mathbf{X}$ vector in which each element is raised to the kth power. We prefer the more streamlined notation without parentheses; the meaning should always be clear from the context.

[11]We will see in Chapter 8 how an initial basic solution for the simplex method can be found when the origin is not part of the feasible region.

Denote this first basic solution as $\mathbf{X}^1 = [x_1^1, \ldots, x_m^1]'$. To avoid notational complications, assume that $\overline{\mathbf{P}}'$ represents the m-element row vector of objective function coefficients that are associated with exactly *these* m variables, so the objective function value that is associated with this basic solution is $\overline{\mathbf{P}}'\mathbf{X}^1$. The issue is what the effect on the value of this objective function would be if we introduced one of the $(m + n) - m = n$ currently excluded variables into the basic solution. Consider the case for any currently nonbasic variable, x_q. For the new solution to remain basic, one of the variables currently included in the basic solution—$x_1, \ldots, x_m$— must be removed.

Since $\mathbf{A}_1, \ldots, \mathbf{A}_m$ in (7.13) forms a basis in the m-dimensional vector space of $\mathbf{AX} = \mathbf{B}$, any other column in $\mathbf{A}$ that is not currently in the basis can be expressed as a linear combination of $\mathbf{A}_1, \ldots, \mathbf{A}_m$ (Section 1.5). Let $\mathbf{A}_q$ be the column in $\mathbf{A}$ associated with x_q. Then we know that y_{iq}'s can be found for which

$$\mathbf{A}_1 y_{1q} + \cdots + \mathbf{A}_m y_{mq} = \mathbf{A}_q \tag{7.14}$$

Assume that at least one of the y_{iq}'s is positive.

To maintain feasibility of the new basic solution, the new values of $x_1, \ldots, x_n$—call them $x_1^2, \ldots, x_m^2$—in conjunction with x_q must still satisfy the constraints, so

$$\mathbf{A}_1 x_1^2 + \cdots + \mathbf{A}_m x_m^2 + \mathbf{A}_q x_q = \mathbf{B} \tag{7.15}$$

must hold. They must also be nonnegative. We will consider that issue later. Since $x_1^1, \ldots,$ x_m^1 previously satisfied (7.13)—which means used up all the resources in $\mathbf{B}$—it is clear that the introduction of x_q must be accompanied by a *reduction* in the sizes of one or more of the elements in $\mathbf{X}^1$. The issue is which ones and by how much.[12]

Note that (7.14) expresses the resource-use column for x_q in terms of those for $x_1^1, \ldots,$ x_m^1, the current basic variables. Multiplying both sides of (7.14) by x_q gives

$$\mathbf{A}_1 y_{1q} x_q + \cdots + \mathbf{A}_m y_{mq} x_q = \mathbf{A}_q x_q \tag{7.16}$$

The right-hand side indicates total use of resources for any value of x_q. So $x_1^1, \ldots, x_m^1$ must be adjusted to release just this amount from their current consumption of resources, if the next solution is to remain feasible. Subtracting (7.16) from (7.13) (and rearranging) shows how much these decreases must be:

$$\mathbf{A}_1(x_1^1 - x_q y_{1q}) + \cdots + \mathbf{A}_m(x_m^1 - x_q y_{mq}) + \mathbf{A}_q x_q = \mathbf{B} \tag{7.17}$$

Note that this has exactly the structure of (7.15); it simply identifies exactly how the x_i^2 are related to the original x_i^1; $x_i^2 = [x_i^1 - x_q y_{iq}]$.

Since the next solution, in addition to being feasible, must also be *basic* (at most m variables nonzero), at least one of the $(x_i^1 - x_q y_{iq})$ terms $(i = 1, \ldots, m)$ in (7.17) must become zero. Also, no variable remaining in the basis can be negative. If $y_{iq} > 0$, then a large x_q could make $x_i^2 = [x_i^1 - x_q y_{iq}]$ negative and thus infeasible. This means that the value of

[12]If there were a currently excluded variable whose column in the $\mathbf{A}$ matrix contained all negative or zero entries (meaning that the activity associated with that variable either *produced*, rather than consumed, resources, or else used none at all) then that variable could be increased without limit from the point of view of the constraints. In that case, the feasible region would be *unbounded*.

the new variable, x_q, must be small enough so that no x_i^2 is negative. (The problem arises only for those y_{iq} that are positive.) Since the minimum value for any x_i^2 is zero, at that minimum

$$x_i^2 = [x_i^1 - x_q y_{iq}] = 0 \qquad \text{or} \qquad x_q = \frac{x_i^1}{y_{iq}}$$

This means that the *largest* value for x_q is the *smallest* of the ratios x_i^1/y_{iq}, from among those y_{iq} that are positive. That solution value for x_q will send to zero one variable previously in the basic solution; all others will still be positive. If two or more of the ratios, x_i^1/y_{iq}, are tied for smallest, then more than one previously included variable will become zero, and the new basic feasible solution will be degenerate (Section 2.5). We consider the computational implications of degeneracy in Chapter 8.

We now know *how* to introduce a new variable, x_q, into the basic feasible solution; the question that remains is whether we want to. The answer is yes, provided the objective function increases in value. Introduction of x_q means an objective function *increase* of $p_q x_q$; it also means a *decrease* in objective function value because $x_1^2, \ldots, x_m^2$ are, in general, smaller than they were without x_q, and one of them is in fact zero. Specifically, the new value of the objective function is

$$p_1 x_1^2 + \cdots + p_m x_m^2 + p_q x_q$$
$$= p_1[x_1^1 - x_q y_{1q}] + \cdots + p_m[x_m^1 - x_q y_{mq}] + p_q x_q \qquad (7.18)$$
$$= \overline{\mathbf{P}}' \mathbf{X}^1 + x_q \left(p_q - \sum_{i=1}^m p_i y_{iq} \right)$$

In particular, the *change* in the objective function associated with the introduction of x_q into the solution and making compensating changes to ensure that the solution remains both basic and feasible is given by the second term in (7.18). The parenthetical part of this term

$$p_q - \sum_{i=1}^m p_i y_{iq} \qquad (7.19)$$

is known as the *simplex criterion*. If it is positive, x_q could logically be brought into the basic solution, since to introduce it would cause an *increase* in the value of the objective function. If it is negative, introduction of x_q (with the required compensating changes in $x_1^1, \ldots, x_m^1$) would cause a *decrease* in objective function value, so x_q should not be introduced into the basic solution; and if it is zero, the objective function value would be unchanged by introduction of x_q.

From $x_i^2 = [x_i^1 - x_q y_{iq}]$,

$$\frac{[x_i^1 - x_i^2]}{x_q} = y_{iq} \qquad (7.20)$$

This shows that y_{iq} indicates exactly how much each currently included variable, x_i^1, must be changed per unit of x_q brought into the basis. So $\sum_{i=1}^m p_i y_{iq}$ is the *total reduction* in the value of the objective function because of the introduction of x_q at unit value and p_q is the increase associated with an $x_q = 1$. The difference, in (7.19), is the *unit* gain (or loss) in objective function value from introducing setting $x_q = 1$ in a new basic feasible solution. When this is multiplied by x_q, as in the second term in (7.18), this is the *total* gain (or loss).

Precisely because the objective function is linear in the x's, it is adequate to assess the impact of introducing a *unit* amount of a nonbasic variable, since there are no increasing or decreasing returns. If the unit gain exceeds the unit loss, the variable could be introduced into the basis.

In general, there will be several currently excluded variables for which the simplex criterion in (7.19) will be positive. An effective mechanism is to find (7.19) for each and then introduce into the new basis that variable for which this criterion is largest. This means moving from each $\mathbf{X}^k$ to the next, $\mathbf{X}^{k+1}$, by introducing *one* new variable into the set $x_1^k, \ldots, x_m^k$, and at the same time removing one of those variables, so that $\mathbf{X}^{k+1}$ remains basic. The mechanics of moving to the next basic feasible solution, part (c) of our requirements list, is provided by simplex method arithmetic.

7.3.2 Simplex Arithmetic

The steps in the simplex method solution procedure are:

1. Find an initial basic feasible solution.
2. Examine the simplex criterion for all variables that are excluded from that basis. Transform the solution in (1) into a new basic feasible solution by introducing the variable with the largest simplex criterion and removing one of the currently included variables.
3. Continue in this manner until, at $\mathbf{X}^k$, there is no excluded variable with a positive simplex criterion.

Each transformation from $\mathbf{X}^{k-1}$ to $\mathbf{X}^k$—bringing in a new variable and removing an old one—is called an *iteration*.

We illustrate the steps with the problem shown in (7.6). Converted to augmented form, the problem is

$$\text{Maximize} \quad \Pi = 2x_1 + 5x_2$$

$$\text{subject to} \quad x_1 + 2x_2 + s_1 = 10$$
$$3x_1 + 2x_2 + s_2 = 24 \qquad (7.21)$$
$$x_1 + 10x_2 + s_3 = 40$$
$$\text{and} \quad x_1, x_2;\ s_1, s_2, s_3 \geq 0$$

Rearranging constraints gives

$$s_1 = 10 - x_1 - 2x_2$$
$$s_2 = 24 - 3x_1 - 2x_2 \qquad (7.22)$$
$$s_3 = 40 - x_1 - 10x_2$$

TABLE 7.2   Simplex Table for Constraints in (7.22)

		x_1	x_2
s_1	10	-1	-2
s_2	24	-3	-2
s_3	40	-1	-10

This simple rearrangement puts the constraints in a particularly useful form. By setting $x_1 = x_2 = 0$, we obtain instantly an initial basic feasible solution, with $s_1 = 10$, $s_2 = 24$ and $s_3 = 40$. This is, of course, the origin, where $\mathbf{X} = \mathbf{0}$ and each slack variable takes on the entire right-hand side value.[13] The value of the objective function associated with this initial solution is zero.

The linear equations in (7.22) express values of variables in the current basic solution as linear functions of those that are not in the current basis (excluded because they were assigned a value of zero). We know that all basic solutions to this problem will have three (in general, m) variables included in the solution and two (in general, n) excluded from the solution. This suggests a simple format in which to record this and any further basic solutions. Table 7.2 represents such a *simplex table*.

The bottom line in Table 7.2, for example, summarizes exactly all the information in the third equation in (7.22); coefficients have simply been detached from their corresponding variables. Letting the double vertical line represent "equals," this line is read as "s_3 equals 40 minus 1 times x_1 minus 2 times x_2." And when the two variables whose labels are at the tops of the columns are set equal to zero (excluding them from the current basis), then we can immediately read the resulting basic solution as $s_1 = 10$, $s_2 = 24$ and $s_3 = 40$.

A *new* basic feasible solution will need another table of exactly this same shape: three rows (one for each equation) and three columns (two for the nonbasic variables and one for the current values of the basic variables). The difference in labeling will be that one of the variables indicated along the top will now appear at the left-hand side (moved from nonbasic to basic) and one of the currently included variables, with a left-hand side label, will be shifted to the top (moved from basic to nonbasic).

If information about the objective function could also be indicated in this grid arrangement, then it would be complete. This is easily accomplished. For this problem, where $\Pi = 2x_1 + 5x_2$, we can think of Π in exactly the role of a slack variable, by rewriting as

$$-2x_1 - 5x_2 + \Pi = 0 \tag{7.23}$$

So Π acts as a measure of the difference between the (negative of) a current value of the objective function (for a particular set of x_1 and x_2 values) and zero. This can easily be added to the information in Table 7.2; we do it at the top, as shown in Table 7.3.

The top line in Table 7.3 can be read just like any of the others. At this initial basic feasible solution, with $x_1 = x_2 = 0$, the value of the objective function is 0 (left-most column,

---

[13]If the origin is not in the feasible region, one or more of the s_i's at this point will be negative. [This would happen if one or the right-hand side elements in (7.21) were negative.] For such problems, a modification of beginning steps is required; this is taken up in Chapter 8.

TABLE 7.3   Complete Simplex Table for the Problem in (7.21)

		x_1	x_2
Π	0	2	5
s_1	10	-1	-2
s_2	24	-3	-2
s_3	40	-1	$-10^\star$

top row), and $p_1 = 2$ and $p_2 = 5$ give the gains expected if a unit of either the first or the second activity is introduced into the basis. These are the simplex criteria for x_1 and x_2 when viewed from the initial basic feasible solution at the origin.

To move to a new basic feasible solution means putting either x_1 or x_2 into the basis and removing one of the s's. This is the algebraic equivalent of moving to an *adjacent* corner of the feasible region. There are three parts to the procedure:

1. Decide which nonbasic variable (top row) to bring into the basis.
2. Decide which basic variable (left column) to remove from the basis.
3. Express the new basic variables as linear functions of the new nonbasic variables. This means finding new numbers for the bottom three rows of Table 7.3. In addition, find the value of the simplex criterion for each of the variables excluded from the new basis. This means, find new numbers for the top row.

Repeat until all top-row elements, except the leftmost, are negative (or until some are negative and some zero). When that happens, there is no point in changing basic feasible solutions any further; no change would increase the value of the objective function, and the solution at that point is optimal.

We illustrate these simplex arithmetic steps for the problem in Table 7.3.

Variable to Enter the Basis   From Table 7.3, it is clear that either x_1 or x_2 would improve the value of the objective function. A reasonable rule is to choose as the variable to enter the basis that one for which the top row coefficient is largest. (If several are tied, any one of them will do.) This does not necessarily lead to the largest gain in objective function value, but it is a simple and logical rule.[14] So in our problem we want x_2 to be in the next basic solution. This means that the x_2 label moves to the left-hand side in the simplex table representing the next solution, replacing one of the three s_i's. Because of the nature of the procedure for interchanging variables from one solution to the next, the operation is known as *pivoting*, and the column of the variable coming into the basis is the *pivot column*.

---

[14]It is easy to imagine an example in which this rule does not lead to the largest objective function increase. Suppose that the problem is to maximize $100x_1 + 2x_2$ subject to $500x_1 + x_2 \le 500$ and $\mathbf{X} \ge \mathbf{0}$. Starting at the origin and choosing x_1 as the variable to enter the basis will lead to the corner where $\mathbf{X}^1 = \begin{bmatrix} 1 \\ 0 \end{bmatrix}$ and $\Pi(\mathbf{X}^1) = 100$. If, instead, x_2 were introduced, the next basic feasible solution would be $\mathbf{X}^2 = \begin{bmatrix} 0 \\ 500 \end{bmatrix}$ where $\Pi(\mathbf{X}^2) = 1000$, even though x_2 has the smaller simplex criterion value.

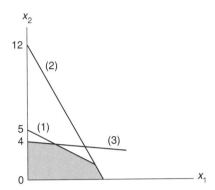

FIGURE 7.8   Intercepts of constraints 1–3 on the x_2 axis.

Variable to Leave the Basis   In this problem, with its two-dimensional solution-space geometry, it is easy to identify the variable that must leave the basic solution when x_2 enters. The geometry is repeated in Figure 7.8, where the intercepts of the three constraints on the x_2 axis are shown explicitly.

Since each unit of x_2 in the basis will increase the objective function value by five units, we would like to make x_2 as large as possible in the next solution. That means that we want to move upward along the x_2 axis until we hit a constraint—the constraint that has the lowest intercept on that axis. From Figure 7.8, we recognize that this will be at the point where $x_2 = 4$. This is the *corner* of the feasible region defined by the intersection of the x_2 axis and the third constraint. And since $s_3 = 0$ anywhere on the third constraint, it is precisely that slack variable that leaves the basis. In simplex method terminology, the s_3 row in Table 7.3 is the *pivot row*. And the element at the intersection of the pivot column and pivot row is the *pivot element* (here this is –10, which is starred). In the next simplex table, then, the labels for x_2 and s_3 will be interchanged, and the process of deriving a new simplex table from the current one is called a *pivot operation*.

Note that the intercepts for constraints 1 and 2 on the x_2 axis are at 5 and 12, respectively. Note, also, that the values of the three intercepts can be found directly from Table 7.3, as the values of the left-column elements divided by *the negatives of* the elements in the x_2 column: $\frac{10}{2}$, $\frac{24}{2}$, and $\frac{40}{10}$.[15]

The rule for selecting the pivot row is the same at any step, even though the geometric logic is sometimes more complicated than just moving out along an axis. The rule remains: for each negative element in the pivot column, find the absolute value of the ratio of left-column element to pivot column element. The pivot row will be the one for which this ratio is smallest. (If the ratios for two or more rows are tied for smallest, it makes no difference which is chosen.)

Expressing New Basic Variables as Linear Functions of Nonbasic Variables   For our illustration, we want to create a new basic feasible solution in which s_1, s_2, and x_2 are included in the basis and x_1 and s_3 are excluded (set equal to zero). Specifically, we want to

---

[15]You should see why this is the case by recalling exactly how the feasible region sides are drawn from the constraints in the original problem. The negative elements in both the x_1 and x_2 columns of Table 7.3 were originally positive in the statements of the constraints; their signs changed because we expressed each of the slack variables in terms of the x's.

create a new simplex table, like Table 7.3, only now with the x_2 and s_3 labels interchanged. Mathematically, this means that we want to express s_1, s_2, and x_2 as linear functions of x_1 and s_3; $s_1 = \ell_1(x_1,s_3)$, $s_2 = \ell_2(x_1,s_3)$, and $x_2 = \ell_3(x_1,s_3)$. Since we have seen that we can also consider Π as a kind of slack variable for the objective function, we also want to find a new linear expression for it: $\Pi = \ell_0(x_1,s_3)$.

Consider first the linear equation connecting s_3 with x_1 and x_2. This is exactly the information in the pivot row in Table 7.3; $s_3 = 40 - x_1 - 10x_2$. It is very straightforward algebra to rewrite this so that x_2 is alone on one side of the equation:

$$x_2 = \tfrac{1}{10}(40) - \tfrac{1}{10}x_1 - \tfrac{1}{10}s_3 \tag{7.24}$$

This is exactly the linear function ℓ_3 that we need. This means, in turn, that the new coefficients in the x_2 row of the next simplex table will be: 4, $-\tfrac{1}{10}$, and $-\tfrac{1}{10}$. This much of the new simplex table is shown in Table 7.4.

Now that we have identified this linear function, in (7.24), we are in a position to find $\ell_1(x_1,s_3)$, $\ell_2(x_1,s_3)$, or $\ell_0(x_1,s_3)$. We first do ℓ_1 for illustration. From the s_1 row of Table 7.3 [or from the original constraints, in (7.22)], we know that $s_1 = 10 - x_1 - 2x_2$. But in (7.24) we have an expression for x_2 in terms of x_1 and s_3, so we need only substitute, giving

$$s_1 = 10 - x_1 - 2[\tfrac{1}{10}(40) - \tfrac{1}{10}x_1 - \tfrac{1}{10}s_3]$$

and all that remains is to collect terms. This gives

$$s_1 = [10 - 2(\tfrac{1}{10})(40)] + [-1 + 2(\tfrac{1}{10})]x_1 + [2(\tfrac{1}{10})s_3] \tag{7.25}$$

or

$$s_1 = 2 - \tfrac{8}{10}x_1 + \tfrac{2}{10}s_3$$

which is exactly what we want for $\ell_1(x_1,s_3)$.

Exactly the same kind of substitution, this time of (7.24) into the expression for s_2 in Table 7.3, leads to

$$s_2 = 16 - \tfrac{28}{10}x_1 + \tfrac{2}{10}s_3 \tag{7.26}$$

This is the $\ell_2(x_1,s_3)$ that we need.

TABLE 7.4   New Simplex Table with Transformed Labels and New Pivot Row Elements

		x_1	s_3
Π			
s_1			
s_2			
x_2	4	$-\tfrac{1}{10}$	$-\tfrac{1}{10}$

Finally, consider $\ell_0(x_1, s_3)$. In Table 7.3 we have

$$\Pi = 0 + 2x_1 + 5x_2$$

and replacing x_2 with the expression in (7.24) gives

$$\Pi = 0 + 2x_1 + 5[\tfrac{1}{10}(40) - \tfrac{1}{10}x_1 - \tfrac{1}{10}s_3]$$

or, rearranging,

$$\Pi = [0 + 5(\tfrac{1}{10})(40)] + [2 - 5(\tfrac{1}{10})]x_1 - 5(\tfrac{1}{10})s_3 \qquad (7.27)$$

which is

$$\Pi = 20 + \tfrac{3}{2}x_1 - \tfrac{1}{2}s_3$$

This is $\ell_0(x_1, s_3)$, the new linear function relating Π to the two currently nonbasic variables.

All this information is shown in Table 7.5, the complete simplex table for the second basic feasible solution, $\mathbf{X}^2$. This corresponds to the first corner selected after the initial basic feasible solution, $\mathbf{X}^1 = \begin{bmatrix} s_1 \\ s_2 \\ s_3 \end{bmatrix} = \begin{bmatrix} 10 \\ 24 \\ 40 \end{bmatrix}$, at the origin. As with Table 7.4, we can immediately identify the basic feasible solution at this point by setting nonbasic variables equal to zero. Here this means $x_1 = 0$ and $s_3 = 0$, so $x_2 = 4$, $s_1 = 2$, and $s_2 = 16$—in our notation for successive basic feasible solutions, this is $\mathbf{X}^2 = \begin{bmatrix} x_2 \\ s_1 \\ s_2 \end{bmatrix} = \begin{bmatrix} 4 \\ 2 \\ 16 \end{bmatrix}$—which can easily be verified by substitution into the constraint equations in (7.22). Moreover, $\Pi = 20$, again as can be easily verified by substitution of $x_1 = 0$, $x_2 = 4$ into the objective function.

The very important point is that these top-row coefficients multiplying the new nonbasic variables (x_1 and s_3) are precisely the simplex criteria for those variables. That is, they represent the value to the objective function of introducing a unit of x_1 or of s_3, *from the vantage point of the new corner*, where $x_1 = 0$ and $x_2 = 4$.

Consider the new coefficient on x_1, as shown in (7.27). It is $[2 - 5(\tfrac{1}{10})]$, indicating that if a unit of x_1 were brought into a basic solution, it would no longer be worth the original 2.

TABLE 7.5   Complete Simplex Table for Second Basic Feasible Solution

		x_1	s_3
Π	20	$\frac{3}{2}$	$-\frac{1}{2}$
s_1	2	$-\frac{8}{10}$ ★	$\frac{2}{10}$
s_2	16	$-\frac{28}{10}$	$\frac{2}{10}$
x_2	4	$-\frac{1}{10}$	$-\frac{1}{10}$

From Figure 7.8 we see exactly why. If x_1 is to become positive, x_2 must be reduced, in order to free up some of resource 3, since at $x_2 = 4$, all 40 units are consumed. To move into positive x_1 territory, we move along the downward-sloping edge of the feasible region defined by constraint 3.

More precisely, each unit of x_1 uses one unit of resource 3; each unit of x_2 uses 10 units of resource 3. At the current corner, $x_2 = 4$ consumes all of that resource. In order to release enough of that resource to allow *one unit* of x_1, it is therefore necessary to *reduce* x_2 by $\frac{1}{10}$th of a unit. But since each unit of x_2 contributes 5 to the objective function, reducing x_2 from 4 to 3.9 means an objective function loss of $5(\frac{1}{10})$units. So the *net* effect on the objective function of introducing x_1 at unit value is $[2 - 5(\frac{1}{10})] = \frac{3}{2}$, exactly as shown in (7.27) and in the top row of Table 7.5.

This is exactly the simplex criterion, in (7.19), for this problem. Here x_2, s_1 $(= x_3)$, and s_2 $(= x_4)$ are currently in the basis, so (7.19) when considering introducing x_1 is

$$p_1 - (p_2 y_{21} + p_3 y_{31} + p_4 y_{41})$$

where $p_1 = 2, p_2 = 5$, and $p_3 = p_4 = 0$ (objective function coefficients for slack variables), and

$$y_{21} = \frac{[x_2^1 - x_2^2]}{x_1} \text{ [from (7.20)]} = \frac{1}{10}$$

(change needed in x_2 for a *unit* change in x_1). This works out to exactly $2 - 5(\frac{1}{10})$.

Generalized Simplex Arithmetic If you look carefully at the results in (7.24), you will see that the *pivot element* in Table 7.3 [solution $\mathbf{X}^1$] is replaced by its reciprocal in the corresponding position in Table 7.5 (solution $\mathbf{X}^2$). Other elements in the pivot row in Table 7.3 appear in corresponding positions in Table 7.5 after having been divided by *the negative of* the pivot element.

Rules for transforming *pivot column* elements from one table to the next are illustrated by the results in (7.25) and (7.26). In the pivot column, new elements are seen to be the old element divided by the pivot [in both cases, $(-2)/(-10)$]. *All other elements* (not in the pivot row or column) are more complicated. Consider the -1, in the s_1 row, x_1 column in Table 7.3. It has become $-\frac{8}{10}$ through the transformation $\{-1 - [(-2)(-1)/(-10)]\}$. In words: {old element $-$ [(pivot column, s_1 row element)(pivot row, x_1 column element)/(pivot element)]}. This general rule is also illustrated in the transformations from Table 7.3 to Table 7.5 of all other elements that are located neither in the pivot row nor the pivot column. Table 7.6 summarizes these rules.

In Problem 7.10 you are asked to derive these simplex arithmetic rules for the general case, starting with an arbitrary simplex table matrix with elements a_{ij}, assuming a particu-

TABLE 7.6 Summary of Simplex Arithmetic

Old Basis	New Basis
Pivot element	Reciprocal of pivot element
Pivot-column elements	Old element/pivot element
Pivot-row elements	$-$(Old element)/pivot element
All other elements	Old element $-$ [("corner product")/pivot element]

lar pivot column and row, and expressing the new basic variables in terms of the new non-basic variables. The results in (7.24)–(7.27) are just specific examples of those general rules.

Deciding to Stop   When a solution $\mathbf{X}^k$ is found for which all top-row elements (except the leftmost) are negative or zero, there is at that point no excluded (nonbasic) variable that would increase the objective function value if introduced into the basis, so $\mathbf{X}^k$ represents an optimal solution, $\mathbf{X}^*$, and further pivot operations are unnecessary. In our example, we see in Table 7.5 that x_1 has a positive simplex criterion (it is the only excluded variable that does, in this small example). So the x_1 column now becomes the new pivot column (for the next pivot operation). The appropriate ratios are now $\frac{20}{8}$, $\frac{160}{28}$, and $\frac{40}{1}$ (absolute values of the ratios of first column to pivot column elements, for those pivot elements that are negative—in this case, all of them), and so s_1 is identified as the new pivot row and $-\frac{8}{10}$ is the new pivot element (again, it is starred).

Table 7.7 contains the next simplex table. The basic feasible solution is now $x_1 = 2\frac{1}{2}$, $x_2 = 3\frac{3}{4}$, $s_2 = 9$; $s_1 = s_3 = 0$. (We leave all results as fractions, since that is how they are most likely to appear if you do this problem by hand, following the rules in Table 7.6.) Now $\Pi = 23\frac{3}{4}$, and both currently excluded variables would *reduce* this value if introduced into the current basis, so this solution is optimal. This is $\mathbf{C}_3$ in Figure 7.3, and we see that simplex arithmetic is consistent with geometric insight. Constraints 1 and 3 are met as strict equalities (their slack variables are zero), and there are 9 "extra" units of resource 2.

For this small problem, in (7.6), with two unknowns and three constraints, we found the optimal solution after examining only three basic solutions (Tables 7.3, 7.5, and 7.7). We saw earlier that there are up to $C_3^5 = 10$ possible basic solutions for this problem, and also, from the geometry (Fig. 7.3), that there are five basic and *feasible* solutions (corners of the feasible region, including the origin). The simplex method looks at the feasible subset of all basic solutions, and (usually) not at all of them (although we will meet a notable exception in Chapter 9).

In larger problems, this efficiency of the simplex method is much more impressive. For example, with 20 unknowns and four inequality constraints, there could be as many as $C_4^{24} = 10,626$ basic solutions. Some of these may not exist (linearly dependent columns in the 4×4 coefficient matrix), and many will be infeasible. The simplex method never even considers looking at an infeasible solution (the pivot rules force nonnegativity of the next solution). Problem 7.6 contains the data ($\mathbf{P}$, $\mathbf{A}$, and $\mathbf{B}$) for a 20-variable, four-constraint problem. Out of the potential 10,626 basic solutions, the simplex method finds the optimal solution after examining only *three* (after the origin).

TABLE 7.7   Complete Simplex Table for Optimal Basic Feasible Solution

		s_1	s_3
Π	$23\frac{3}{4}$	$-\frac{15}{8}$	$-\frac{1}{8}$
x_1	$2\frac{1}{2}$	$-\frac{10}{8}$	$\frac{1}{4}$
s_2	9	$\frac{28}{8}$	$-\frac{1}{2}$
x_2	$3\frac{3}{4}$	$\frac{1}{8}$	$-\frac{1}{8}$

7.4 DUALITY

7.4.1 Mathematical Relationships and Interpretation

Suppose that the producer whose problem we looked at in Section 7.3 were considering retirement and wants to advertise in the local paper in order to sell the entire stock of resources on hand. How much should be charged? That is, what might be a reasonable value (prices to advertise) for these resources?

Let y_1 be the to-be-determined value for one unit of resource 1, with y_2 and y_3 defined similarly. Then $10y_1 + 24y_2 + 40y_3$ is the total value of the resources. In order to make the advertisement an attractive one, the producer should try to keep this total cost to a potential buyer low; if it is ridiculously high, there will be no response to the advertisement. Of course, prices could be set to zero (give the resources away); that would attract customers but would not be very advantageous for the producer.

To avoid setting prices that are too low, the producer might reason as follows. The physical resource content of a unit of product 1 is given by the elements in the first column of the **A** matrix for the problem in (7.6), namely, one unit of resource 1, three of resource 2, and one of resource 3. Therefore, the *value* of a unit of product 1 in terms of the resources used to produce it is $y_1 + 3y_2 + y_3$. The producer receives a return of 2 for each unit of product 1 produced. So it might be reasonable to price these resources in such a way that the valuation of the resource content of a unit of product 1 is *no less than* the amount gained by producing that unit. This means $y_1 + 3y_2 + y_3 \geq 2$.

The same kind of reasoning pertains to product 2. The resource content of each unit of product 2 is 2 units of both resources 1 and 2 and 10 units of resource 3; the return from producing a unit of product 2 is 5, so it would be logical to also require of y_1, y_2, and y_3 that they satisfy $2y_1 + 2y_2 + 10y_3 \geq 5$. Finally, assuming that the producer is not interested in paying someone to take the resources away, $y_1 \geq 0$, $y_2 \geq 0$, and $y_3 \geq 0$.

The producer's resource valuation problem is then

$$\text{Minimize} \quad 10y_1 + 24y_2 + 40y_3$$
$$\text{subject to} \quad y_1 + 3y_2 + y_3 \geq 2$$
$$2y_1 + 2y_2 + 10y_3 \geq 5 \tag{7.28}$$
$$\text{and} \quad y_1 \geq 0, y_2 \geq 0, y_3 \geq 0$$

Here are the two problems, (7.6) and (7.28), side by side:

$$\text{Maximize} \quad 2x_1 + 5x_2 \qquad\qquad \text{Minimize} \quad 10y_1 + 24y_2 + 40y_3$$
$$\text{subject to} \quad x_1 + 2x_2 \leq 10 \qquad\qquad \text{subject to} \quad y_1 + 3y_2 + y_3 \geq 2$$
$$3x_1 + 2x_2 \leq 24 \qquad\qquad\qquad 2y_1 + 2y_2 + 10y_3 \geq 5$$
$$x_1 + 10x_2 \leq 40 \qquad\qquad\qquad\qquad\qquad\qquad\qquad \tag{7.29}$$
$$\text{and} \quad x_1, x_2 \geq 0 \qquad\qquad\qquad \text{and} \quad y_1, y_2, y_3 \geq 0$$

The symmetry is even more evident when the problems are written in general matrix notation:

$$\begin{array}{ll} \text{Maximize} \quad \mathbf{P'X} & \text{Minimize} \quad \mathbf{B'Y} \\ \text{subject to} \quad \mathbf{AX} \leq \mathbf{B} & \text{subject to} \quad \mathbf{A'Y} \geq \mathbf{P} \qquad\qquad (7.30) \\ \text{and} \quad \mathbf{X} \geq \mathbf{0} & \text{and} \quad \mathbf{Y} \geq \mathbf{0} \end{array}$$

Specifically

1. One is a maximization problem, the other a minimization problem.
2. The constraints in the maximum problem set *upper* bounds (aside from nonnegativity of the variables), in the minimum problem they set *lower* bounds.
3. The coefficient matrices are transposes of one another. If one problem has m constraints and n unknowns, the associated problem has n constraints and m unknowns.
4. The objective function coefficients of one problem become the right-hand sides of the constraints of the other problem.

When two problems are related in this mathematical way, they are said to be *dual* to one another. Somewhat confusingly, in such a pair, the problem that is of primary concern (here the maximization problem) is called the *primal* problem and the other (in this case, the minimization problem, which was somewhat artificially constructed) is known as the *dual*.[16]

The important point about dual linear programming problems is this. Because of certain mathematical relationships that exist between the *optimal values* for the variables in a pair of dual problems, the optimal values of the dual variables may be of interest even though the dual *problem* itself may appear to be somewhat artificial. Moreover, it turns out that the final (optimal) simplex table (as in Table 7.7) contains the optimal values for *both* problems. In solving one problem, you have automatically solved them both.[17]

In order to examine some important mathematical relationships between a pair of dual linear programs, as in (7.29) or (7.30), it is useful once again to express the problems in augmented form. Here, as usual, $\mathbf{S}$ represents the column vector of nonnegative slack variables for the maximization problem; correspondingly, let $\mathbf{T}$ be a column vector of nonnegative *surplus* variables for the minimization problem. (In minimization problems with constraints that set lower bounds, notice that these new slack variables must be *subtracted* from the left side of each constraint, to absorb the surplus on that side.) We use "Δ" [Greek "D" (delta)] for the *dual* objective function.[18] So (7.30) becomes

$$\begin{array}{ll} \text{Maximize} \quad \Pi = \mathbf{P'X} & \text{Minimize} \quad \Delta = \mathbf{B'Y} \\ \text{subject to} \quad \mathbf{AX} + \mathbf{S} = \mathbf{B} & \text{subject to} \quad \mathbf{A'Y} - \mathbf{T} = \mathbf{P} \qquad\qquad (7.31) \\ \text{and} \quad \mathbf{X} \geq \mathbf{0}, \mathbf{S} \geq \mathbf{0} & \text{and} \quad \mathbf{Y} \geq \mathbf{0}, \mathbf{T} \geq \mathbf{0} \end{array}$$

---

[16]In any pair of dual problems, the "more important" of the two is the primal and the other is the dual. This means that either a maximization or a minimization problem *could* be the primal in a set of dual problems; it depends entirely on the context in which they arise. Note that the dual of the dual problem is just the primal.

[17]For this reason, our discussion of a simplex method approach for maximization problems is not as restrictive as it might appear. If you have a minimization linear program to solve, you could write down its dual (maximization) problem and solve it using the rules from Section 7.3. The solution to the primal (minimization) problem will also have been generated. We are about to see how to find it.

[18]The "Π" which represented "profit" in the earlier sample maximization problem, is now seen also to stand for "primal."

In what follows, we will assume that the maximization problem is our primal (the one of primary interest) and the minimization problem is therefore the dual.

We now need to do some algebra, in matrix form. It will lead, in a few steps, to important conclusions, even if the goal is not immediately obvious.

(a) Multiply each constraint i ($i = 1, \ldots, m$) in the maximization problem by its associated y_i and each constraint j ($j = 1, \ldots, n$) in the dual by an associated x_j. In matrix terms, this gives

$$\mathbf{Y'AX} + \mathbf{Y'S} = \mathbf{Y'B} \tag{7.32a}$$

and

$$\mathbf{X'A'Y} - \mathbf{X'T} = \mathbf{X'P} \tag{7.32b}$$

from the constraints in the primal and dual problems in (7.31).

(b) The expression on the right-hand side of (7.32a) is just the dual objective function, Δ, and the expression on the right-hand side of (7.32b) is Π, the primal objective function.[19] Therefore

$$\Pi = \mathbf{X'A'Y} - \mathbf{X'T} \quad \text{and} \quad \Delta = \mathbf{Y'AX} + \mathbf{Y'S}$$

and, since $\mathbf{X'A'Y} = \mathbf{Y'AX}$ (for reasons mentioned in the last footnote), it follows that

$$\Delta - \Pi = \mathbf{Y'S} + \mathbf{X'T} \tag{7.33}$$

Since feasibility in both the primal and the dual problems requires nonnegativity of all variables (including slacks), Theorem PD1 (primal–dual theorem 1) follows immediately.

Theorem PD1   For any *feasible* solutions to a pair of dual linear programming problems, as in (7.30), $\Delta \geq \Pi$. This is clearly true, since all elements in each of the products making up each of the terms on the right-hand side of (7.33) are nonnegative.

This theorem is important because it shows that the value of the objective function in the minimization problem puts an *upper limit* on the objective function for the maximization problem. At the same time, the value of the maximization problem's objective function establishes a *lower limit* on the value of the objective function for the minimization problem. Theorem PD2 follows from this.

Theorem PD2   If you have a pair of *feasible* solutions, $\mathbf{X}^f$ and $\mathbf{Y}^f$, with objective function values $\Pi^f = \mathbf{P'X}^f$ and $\Delta^f = \mathbf{B'Y}^f$, for which $\Pi^f = \Delta^f$, then these solutions are optimal: $\mathbf{X}^f = \mathbf{X}^*$ and $\mathbf{Y}^f = \mathbf{Y}^*$.

---

[19]Suppose $n = 3$. Then $\mathbf{P'X} = p_1 x_1 + p_2 x_2 + p_3 x_3$; $\mathbf{X'P} = x_1 p_1 + x_2 p_2 + x_3 p_3$, and the two products are obviously the same. This is always true for any pair of vectors containing the same number of elements. Both $\mathbf{P'X}$ and $\mathbf{X'P}$ generate the same scalar, and both are the same as the *scalar product* (or *inner product* or *dot product*) $\mathbf{P} \circ \mathbf{X}$ (Section 1.5).

This follows immediately from Theorem PD1. If the maximization problem objective function has hit its ceiling and, at the same time, the minimization problem objective function has hit its floor, then no improvement in either objective function is possible and the solutions are optimal.

Here is another theorem that follows from the result in (7.33).

Theorem PD3 A pair of feasible solutions has $\Pi' = \Delta'$ if and only if $\mathbf{Y'S} = 0$ and $\mathbf{X'T} = 0$. Explicitly, $\mathbf{Y'S} = y_1 s_1 + \cdots + y_m s_m$ and $\mathbf{X'T} = x_1 t_1 + \cdots + x_n t_n$. So $\mathbf{Y'S} = 0$ if and only if $y_i s_i = 0$ for *all* $i = 1, \ldots, m$ and $\mathbf{X'T} = 0$ if and only if $x_j t_j = 0$ for *all* $j = 1, \ldots, n$. In this form, the theorem has an interesting interpretation. Each primal slack variable is

$$s_i = b_i - \sum_{j=1}^{n} a_{ij} x_j$$

which indicates the difference between the available amount of resource i and the amount used by a particular solution (set of x_j's). So, from Theorem PD3, for an optimal solution

$$y_i^* \left(b_i - \sum_{j=1}^{n} a_{ij} x_j^* \right) = 0 \qquad (i = 1, \ldots, m) \tag{7.34}$$

Similarly

$$t_j = \sum_{i=1}^{n} a_{ij} y_i - p_j$$

so t_j is a measure of the difference between the valuation of resources in a unit of j and its objective function return. Again, from Theorem PD3, at an optimal solution

$$x_j^* \left(\sum_{i=1}^{n} a_{ij} y_i^* - p_j \right) = 0 \qquad (j = 1, \ldots, n) \tag{7.35}$$

Specifically

1. If $y_i^* > 0$, then $s_i^* = 0$. If, in the optimal solution, the valuation placed on resource i is positive, this means that in that solution all of the ith resource was used up.
2. If $x_j^* > 0$, then $t_j^* = 0$. If a positive amount of j is produced in the optimal solution, then the imputed value of resources embodied in one unit of j is just equal to p_j, the per unit objective function return.
3. If $s_i^* > 0$, then $y_i^* = 0$. A resource that is not fully used up in an optimal solution is valued at zero; it is in excess supply.
4. If $t_j^* > 0$, then $x_j^* = 0$. If the imputed value of resources in a unit of j is greater than its per unit return, product j is not produced in the optimal solution.

Note that the inference cannot be made in the reverse direction—if a particular s_i^*, t_j^*, y_i^*, or x_j^* is zero, then the optimal value of the associated variable may or may not be zero. The relationships $y_i^* s_i^* = 0$ and $x_j^* t_j^* = 0$ require that *at least one* of the variables in each pair of terms must be zero; this does not rule out that both are zero.

Here is a theorem about the existence of optimal solutions to a pair of dual linear programs.

Theorem PD4 If both problems in a pair of dual linear programs have *feasible* solutions, then both have *optimal* solutions.

This sometimes makes possible an easy investigation of the existence question, which may be worth exploring before spending time and effort trying to find an optimal solution. For example, the origin is often easily found to be a feasible solution to a maximization problem. If a problem has no feasible solution its feasible region is empty, which means that the constraints are contradictory—as we saw in regard to Theorem 7.3, above. Once feasibility for both problems is established, optimality follows from Theorem PD1, since feasibility means that there is a ceiling (Δ) for Π and a floor (Π) for Δ.

Finally, under certain conditions[20]

Theorem PD5 $\partial \Pi^* / \partial b_i = y_i^*$. In words, the optimal value of the ith dual variable is a measure of the amount by which the optimal value of the primal objective function can be expected to change for a small (marginal) change in the amount of the ith resource.

This follows from Theorem PD2. Since

$$\Pi^* = \mathbf{P'X^*} = \mathbf{B'Y^*} = y_1^* b_1 + \cdots + y_m^* b_m$$

$\partial \Pi^* / \partial b_i = y_i^*$. Thus the dual variables in a linear programming problem have characteristics that parallel those of the Lagrange multipliers that were first introduced in Section 4.2. There is one associated with each constraint, and their optimal values measure the contribution of marginal changes in the right-hand sides of the constraints. (We will explore a similar connection for the case of duality in nonlinear programming problems in Chapter 11.)

This interpretation of the optimal values of dual variables as marginal values of primal constraint quantities is one of the major reasons for interest in the linear programming dual. There are often problems in which the marginal worth of each of the resources to the primal objective function is an important piece of information—for example, in making decisions about whether (and if so, for what price) to acquire additional units of one or more of the resources. These dual values, y_i^*, sometimes called "shadow prices," represent, in our story, the *maximum* price that the producer would be willing to pay for an additional unit of resource i. It is important to note that the marginal valuation property holds regardless of whether one chooses to construct a story around the dual problem.

At the same time, the marginal nature of the valuation also means that it can be misleading, since it is valid only over a particular range of change in resource amount. For this reason, we turn now first to locating the optimal dual values in the final simplex tables and then to establishing ranges of relevance for each of them.

---

[20]Provided the proper right- and left-sided limits exist and are equal so that the derivative is defined (Chapter 3). We will see in Chapter 8 that degeneracy of an optimal solution may invalidate this theorem for some of the constraints.

7.4.2 Dual Values in the Simplex Tables

For the general maximization problem in (7.30) the **A** matrix has dimensions $m \times n$; the unknowns are $x_1, \ldots, x_n$, and there are m constraints. It is clear that the dual, a minimization problem, has m unknowns, $y_1, \ldots, y_m$, one for each primal constraint; and it has n constraints, one for each unknown in the primal. And since the coefficient matrix in one problem is just the transpose of the matrix in the other, this suggests that Table 7.3 really contains all the information for the minimization problem as well.

Suppose that we convert the illustrative minimization problem in (7.29) to augmented form, as in (7.31). Here the constraints would be

$$y_1 + 3y_2 + y_3 - t_1 = 2$$

and

$$2y_1 + 2y_2 + 10y_3 - t_2 = 5$$

This is a pair of linear equations in five unknowns, and so we know that basic solutions to this minimization linear programming problem will have three of the five unknowns set equal to zero.

Expressing the negatives of the slack variables as linear functions of the y's, we obtain

$$-t_1 = 2 - y_1 - 3y_3 - y_3$$

and

$$-t_2 = 5 - 2y_1 - 2y_2 - 10y_3$$

Suppose that we attach the labels $-t_1$ and $-t_2$ to the x_1 and x_2 columns, respectively, and add the labels y_1, y_2 and y_3 to the s_1, s_2, and s_3 rows in Table 7.3. This is done in Table 7.8. The inner set of labels identifies variables in the primal problem, as usual. Variables in a basic solution for the primal are listed along the left side; they are expressed as linear functions of the nonbasic variables, listed along the top. The outer set of labels identifies variables in the dual problem. Variables in a basic solution to the dual are listed along the top; they are expressed as linear functions of nonbasic (dual) variables, listed along the left side.

So by "rotating" our view of the initial simplex table, and adding labels for the dual problem variables, we have a representation of dual constraints. Letting the double horizontal line represent "equals," the last column would be read as

$$-t_2 = 5 - 2y_1 - 2y_2 - 10y_3$$

(which is exactly the second dual constraint).

Finally, if we attach a label (Δ) for the objective function in the dual problem, we see that all appropriate coefficients appear in exactly the right places down the leftmost column. So, the information that $\Delta = 10y_1 + 24y_2 + 40y_3$ is contained down that column, using the dual labels from the outer set. [The reasoning is the same as that behind (7.23) for II.]

TABLE 7.8   Initial Simplex Table for Both Primal and Dual Problems

		Δ	$-t_1$	$-t_2$
			x_1	x_2
	Π	0	2	5
y_1	s_1	10	-1	-2
y_2	s_2	24	-3	-2
y_3	s_3	40	-1	-10

Tables 7.8–7.10 repeat the information from Tables 7.3, 7.5, and 7.7, only now the dual variable labels are interchanged, along with those for the primal problem, at each iteration. The basic solutions at each step are indicated.

(a) *Table 7.8.*   This is the initial basic feasible solution for the primal problem, where $x_1 = 0, x_2 = 0, s_1 = 10, s_2 = 24, s_3 = 40$ and $\Pi = 0$. The associated basic solution for the dual problem is $y_1 = 0, y_2 = 0, y_3 = 0, -t_1 = 2, -t_2 = 5$ and $\Delta = 0$. (You should check to see that in fact this solution satisfies the two constraints of the augmented dual.) While this is clearly a *basic* solution to the dual (three of the five variables equal zero), it is not *feasible* since t_1 and t_2 are both negative.

(b) *Table 7.9.*   Here the next basic *and* feasible solution for the primal is $x_1 = 0, x_2 = 4, s_1 = 2, s_2 = 16, s_3 = 0$ and $\Pi = 20$. The associated basic solution for the dual is $y_1 = 0, y_2 = 0, -y_3 = -\frac{1}{2}, -t_1 = \frac{3}{2}, t_2 = 0$ and $\Delta = 20$. (Again, you might check that this solution satisfies the dual constraints.) And again, this is not a dual feasible solution, since $t_1 < 0$.

(c) *Table 7.10.*   Now the primal solution is $x_1 = 2\frac{1}{2}, x_2 = 3\frac{3}{4}, s_1 = 0, s_2 = 9, s_3 = 0$, and $\Pi = 23\frac{3}{4}$. The top-row negatives tell us that this is indeed an optimal solution. Associated with this is a dual solution that is both basic and (now) also feasible, namely $y_1 = \frac{15}{8}, y_2 = 0, y_3 = \frac{1}{8}, t_1 = 0, t_2 = 0$, and $\Delta = 23\frac{3}{4}$. This illustrates the fact that, in this version of the simplex method, only when the primal *optimal* solution is reached will be corresponding dual basic solution be *feasible* as in Table 7.10.

TABLE 7.9   Second Simplex Table for Both Primal and Dual Problems

		Δ	$-t_1$	$-y_3$
			x_1	s_3
	Π	20	$\frac{3}{2}$	$-\frac{1}{2}$
y_1	s_1	2	$-\frac{8}{10}$	$\frac{2}{10}$
y_2	s_2	16	$-\frac{28}{10}$	$\frac{2}{10}$
t_2	x_2	4	$-\frac{1}{10}$	$-\frac{1}{10}$

TABLE 7.10   Third (Optimal) Simplex Table for Both Primal and Dual Problems

		Δ	$-y_1$	$-y_3$
			s_1	s_3
	Π	$23\frac{3}{4}$	$-\frac{15}{8}$	$-\frac{1}{8}$
t_1	x_1	$2\frac{1}{2}$	$-\frac{10}{8}$	$\frac{1}{4}$
y_2	s_2	9	$\frac{28}{8}$	$-\frac{1}{2}$
t_2	x_2	$3\frac{3}{4}$	$\frac{1}{8}$	$-\frac{1}{8}$

7.4.3   The Optimal Solution to the Dual

The question then is whether this solution will be an optimal one for the dual. It is clear that the correct variables will appear in the dual solution, because of the labeling procedure that we have adopted. A variable that is *in* the dual basis—whose label is in the top row—automatically has its corresponding primal variable *excluded* from the primal basis (the left-hand column). Thus it is ensured that Theorem PD3 will hold. We have $\Pi = \Delta$ and since the dual solution is also feasible (for the first time), it is optimal (Theorem PD2).

7.5   SENSITIVITY ANALYSIS

"Sensitivity" or "postoptimality" analysis considers how much an optimal solution might be changed if some of the original data of the problem were different. In the standard linear programming maximization problem, the fixed numbers are the elements in **P** (the objective function coefficients), **B** (the right-hand sides of the linear constraints), and **A** (the coefficients that connect the unknown *x*'s to those right-hand-side values). In this section we will explore just two specific kinds of sensitivity questions—those in which there are (1) changes in right-hand sides of constraints or (2) changes in objective function coefficients. Most linear programming computer software includes numerical results for these two kinds of changes, along with the optimal values of both primal and dual variables, in the optimal solution output display.

7.5.1   Changes in the Right-Hand Side of a Constraint

In Theorem PD5 we saw that the optimal value of a dual variable, y_i^*, measures the marginal effect on the optimal value of the primal objective function of a small (infinitesimal) change in the right-hand side of the *i*th primal constraint; $\partial \Pi^*/\partial b_i = y_i^*$. As it is a partial derivative, this is strictly a *ceteris paribus* concept; everything else in $\Pi^* = y_1^* b_1 + \cdots + y_m^* b_m$ [specifically, the values of all the rest of the b_j $(i \neq j)$] must remain unchanged.
  The geometry of the illustrative problem (essentially, Fig. 7.3) is repeated here, as Figure 7.9; it includes Π^4, the objective function contour line that identifies the optimal corner. We saw that $y_2^* = 0$, which is consistent with the figure; a small change in the amount of b_2, which means in the position of constraint 2, will not alter the optimum at all. However, a small change (consider, for now, an increase) in either b_1 or b_3 *will* have the effect

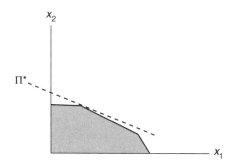

FIGURE 7.9   Feasible region and optimal objective function contour for problem 7.6.

of moving the optimal extreme point outward or upward, away from the origin. This will result in an *increased* primal objective function value.

If only (1) moves outward, the optimal corner will creep out along the extension of (3), until such time as it bumps into (2), also extended. Further outward movement of (1) will not affect the feasible region. That region will then be completely defined by (2) and (3). The dual valuation on resource 1, $y_1^* = \frac{15}{8}$, is thus valid along that range of increase of b_1 up to the point in Figure 7.9 where another corner is hit—until constraint 2, in this case, becomes binding. It turns out that the optimal simplex table, Table 7.10, contains all the information that we need to establish the *range of increase* over which $\frac{15}{8}$ is a valid measure of the per unit worth of b_1.

Algebraically, adding units of capacity to b_1 is equivalent to allowing the slack variable in that primal constraint, s_1, to become negative. This means that variables that are currently in the primal basis and that have *positive* coefficients in the s_1 ($= -y_1$) column will become *smaller* for increasingly negative values of s_1.[21] Therefore, among the *positive* coefficients in the s_1 column we must find the smallest ratio of leftmost column to s_1 column elements, since this will be precisely the variable in the current optimum basis that hits zero first. In this case, we compare $9 \div \frac{28}{8} = \frac{18}{7}$ (s_2 row) and $\frac{15}{4} \div \frac{1}{8} = 30$ (x_2 row) and select the smaller (in general, with more than two choices, the smallest). Thus, s_2 will become zero before any of the other currently included basic variables as s_1 becomes more and more negative; it will do so at $s_1 = -2\frac{4}{7}$, and so $2\frac{4}{7}$ is the maximum that b_1 can *increase* before another corner (constraint) is hit in the feasible region. Referring to Figure 7.9, this will happen when (1) moves out until constraints 2 and 3 intersect. Thus in this case the *marginal* valuation on resource 1 is valid for a *unit* increase in b_1 or even if b_1 goes up by 2 units, but not if it increases by 3 or more units.

You could solve several new linear programs, identical to (7.6) except for a change in the 10 in the first constraint.

(a) If the 10 becomes 11 (within the range of validity of y^*), the new optimal solution is $\mathbf{X}^* = \begin{bmatrix} 3\frac{3}{4} \\ 3\frac{5}{8} \end{bmatrix}$, with $\Pi^* = 25\frac{5}{8}$, which is exactly the old optimal value of Π ($23\frac{3}{4}$) plus $(y_1^*)(\Delta b_1) = (1\frac{7}{8})(1)$.

---

[21]Recall that in any simplex table the coefficients express basic primal variables as linear functions of nonbasic ones. From Table 7.10, for example, $s_2 = 9 + \frac{28}{8}s_1 - \frac{1}{2}s_3$. Beyond a particular *negative* value of s_1, s_2 would turn negative, if s_3 remains zero. This critical value is just where $9 + \frac{28}{8}s_1 = 0$; namely when $s_1 = -9 \div \frac{28}{8} = -\frac{18}{7}$.

(b) If the 10 becomes 12 (still within the valid range for y_1^*), the new optimal solution is $\mathbf{X}^* = \begin{bmatrix} 5 \\ 3\frac{1}{2} \end{bmatrix}$, with $\Pi^* = 27\frac{1}{2}$. Again, this is the original value of Π^* plus $(y_1^*)(\Delta b_1) = (1\frac{7}{8})(2)$.

(c) If the 10 changes to 13 (beyond the $2\frac{4}{7}$ range for y_1^*), the new optimal solution is $\mathbf{X}^* = \begin{bmatrix} 5\frac{5}{7} \\ 3\frac{3}{7} \end{bmatrix}$ at which point $\Pi^* = 28\frac{4}{7}$; this is *less* than $23\frac{3}{4} + (y_1^*)(\Delta b_1) = 23\frac{3}{4} + (1\frac{7}{8})(3) = 29\frac{3}{8}$.

The geometry of the three new positions for constraint 1 and hence for the optimal solution is shown in the various parts of Figure 7.10. The changes in (a) and (b), for which $y_1^*(\Delta b_1)$ provides a correct measure of change in the objective function, involve movement of the optimal extreme point out along the line representing the constraint 3 boundary. The last move, in (c), takes constraint 1 beyond the point of intersections of the lines for constraints 2 and 3; hence the first constraint is now *redundant* since the feasible region defined by constraints 2 and 3 lies entirely within the half-space of constraint 1. Thus outward movement along the line for constraint 3 is stopped short by the presence of constraint 2 and the increase in the objective function is less than in the move from the original problem to position (a) or from (a) to (b). Since constraint 2 was not binding in the original solution ($s_2^* > 0$), that resource was given no value ($y_2^* = 0$), and so we have no in-

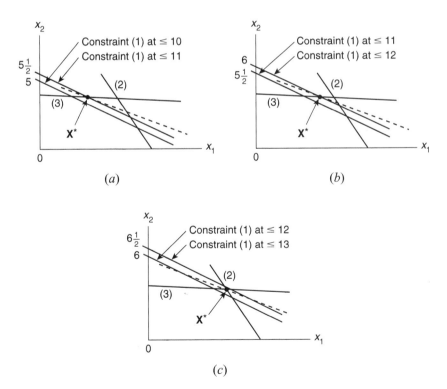

FIGURE 7.10   Three (unit) changes in location of constraint 1: (*a*) unit change from 10 to 1l; (*b*) unit change from 11 to 12; (*c*) unit change from 12 to 13.

formation, from the optimal simplex table, on the value of resource 2, once it does become critical.

The practical implication of a result like this is straightforward. If a supplier appears and offers an additional unit of resource 1 to the producer in the story behind the illustration in (7.6), that producer should be willing to pay up to $y_1^* = 1\frac{7}{8}$ for the unit of resource 1. As we have seen, $1\frac{7}{8}$ represents the amount the producer could gain, in terms of the objective function, by resolving the linear program in (7.6) with the right-hand side of the first constraint now set at 11. Similarly, for two additional units of resource 1, the producer would be willing to pay up to $2(y_1^*) = 3\frac{3}{4}$. However, an additional 3 units would not be worth $3(y_1^*) = 5\frac{5}{8}$ to the producer, but rather only $(2\frac{4}{7})(y_1^*) = (2\frac{4}{7})(1\frac{7}{8}) = 4.8214$, since beyond the upper limit of 12.5714 $(= 12\frac{4}{7} = 10 + 2\frac{4}{7})$, additional units of resource 1 are worthless. Indeed, the producer should be willing to pay only up to 4.8214 for *any* additional amount of n units, where $n > 3$.

Similar reasoning supports the observation that each unit of *decrease* in the right-hand side of the first constraint, from 10 down to 8, *diminishes* the optimal objective function value by $1\frac{7}{8}$. When the right-hand side is 9, the linear programming problem has an optimal objective function value of $21\frac{7}{8}$; if the right-hand side is 8, the optimal objective function value is 20. (You might solve these two new linear programs, first with $b_1 = 9$ and then with $b_1 = 8$, as a check.) If the right-hand side of constraint 1 is *less than* 8, the optimal value of the objective function decreases by *more than* $1\frac{7}{8}$ per unit. This is because the optimal corner now slides down the x_2 axis instead of (further) leftward along constraint 3, since the boundary set by $x_1 \geq 0$ (that is, the vertical axis) is hit. (You might also solve another linear programming problem, now with $b_1 = 7$, and compare the optimal value of the objective function with what it was for $b_1 = 8$.) The simplex method cannot see around corners; it can, however, identify for each y_i^* the range $[(l_i, u_i)$, for *lower* and *upper* limit] over which each unit of increase in b_i is worth y_i^* per unit and each unit of decrease is worth $-y_i^*$ per unit. In the case of y_1^*, this range is (8, 12.5714).

A completely analogous calculation shows that the range over which $y_3^* = \frac{1}{8}$ is a correct valuation on *increases* in resource 3 is 10 units; for each one-unit increase in the right-hand side of constraint 3—up to 10 units more (up to a total of 50)—the value of the objective function will increase by $\frac{1}{8}$. From Figure 7.10, we see that this is the range over which constraint 3 contributes something to the boundary of the feasible region. Above $x_1 + 10x_2 \leq 50$ the constraint is redundant; the feasible region is then defined *entirely* by constraints 1 and 2, and so additional increases in the right-hand side of constraint 3 contribute nothing. Similarly, each unit of *decrease* of b_3, down to 22, diminishes the optimal value of the objective function by $\frac{1}{8}$. The range for y_3^* is therefore (22, 50).

Finally, the geometry of Figure 7.10 shows that outward movement of the boundary defined by constraint 2 will never change the location of the present optimal solution; $y_2^* = 0$ for *any* increase in b_2 (the upper limit is $+\infty$). However, at some point, decreases in b_2 *will* change the optimal corner—from the present point defined by the intersection of constraints 1 and 3 to a point defined by the intersection of constraints 2 and 3. The algebra is a little different; the result is that *decreases* down to 15 do nothing to the optimal value of the objective function, so $l_2 = 15$. Below 15, further decreases in b_2 cut away at the feasible region in such a way that the optimal objective function value is decreased; y_2 becomes a nonzero number, but the simplex method cannot find that number by looking only at the *current* optimal simplex table. The range over which $y_2^* = 0$ is (15, $+\infty$).

We see, then, that the linearity of the sides of the feasible region has two influences on the marginality interpretation of the optimal dual variables, $y_i^* = \partial\Pi^*/\partial b_i$. First, it means that there may be a range much larger than the infinitesimal one implied by the partial derivative over which y_i^* is valid. Second, however, at some point the dual valuations become totally useless, since beyond that range the constraints either become redundant or define a new optimal corner in conjunction with new constraints that have become binding.[22]

7.5.2 Changes in an Objective Function Coefficient

The coefficients in the objective function (the elements of **P**) govern the slope of the representation of objective function contours in solution space. In two-variable problems, the slope of those contours is $-(p_1/p_2)$. (In three-variable problems, the orientation of the objective function *plane* in three-dimensional solution space is a more complicated function of p_1, p_2, and p_3, and the representation is even more complex for n-variable problems when $n > 3$.) An increase in p_1 or a decrease in p_2 would increase the absolute value of the slope; a decrease in p_1 or an increase in p_2 would decrease the absolute value of the slope.

Small changes in the slope of the objective function would simply change the "tilt" of the linear contour on the *current optimal corner*. Larger changes in the slope would cause the optimal solution to *jump* to an adjacent corner. Imagine the objective function contour in Figure 7.10 becoming flatter—that means that the (negative) slope is smaller in absolute value (for example, if p_2 were 10 instead of 5). But the optimal corner would still be defined by the intersection of constraints 1 and 3. That means that the optimal values, x_1^* and x_2^*, would not change, although the optimal value of the objective would, because p_2 has increased.

Suppose that p_2 increased instead to 20; then the slope of the objective function contours would be $-\frac{1}{10}$. This means that these contours would be exactly parallel to the boundary defined by constraint 3. In that case, there would be multiple optima, although the current corner would still be one of the optimal solutions. However, if p_2 increased beyond 20, say, to 21, then the objective function contours, with a slope of $-\frac{2}{21}$ would identify an optimal solution at the point where constraint 3 intersects the vertical axis. Similarly, after a certain increase in the absolute value of the slope, the optimal solution would jump to the corner defined by constraints 1 and 2.

The ranges given for objective function coefficients indicate the lower and upper limits on the values of each of those coefficients, beyond which the optimal solution would change to a different corner. In this problem, since a jump to the left occurs when the slope becomes $-\frac{1}{10}$, this means that if p_1 decreases to beyond 0.5 [where $-(p_1/p_2) = -\frac{1}{10}$] or (as we just saw) if p_2 increases to beyond 20 (again where the slope is $-\frac{1}{10}$), the optimal corner changes. The lower limit for p_1 and the upper limit for p_2 are thus defined by constraint 3.

---

[22]The total differential (Chapter 3) can be of some help in assessing the impacts of *simultaneous* changes in the right-hand sides of several constraints. Suppose that the right-hand sides of both constraints 1 and 3 increase by 1 unit. Using the total differential

$$d\Pi^* = \frac{\partial \Pi^*}{\partial b_1}\,(db_1) + \frac{\partial \Pi^*}{\partial b_3}\,(db_3) = y_1^*(1) + y_3^*(1) = \tfrac{15}{8} + \tfrac{1}{8} = 2$$

Geometrically, the optimal corner is moving out along a line with a slope that lies between those of constraints 1 and 3 in Figure 7.10. Eventually either constraint 2 or the x_2 axis will be hit. If the original problem is solved again, this time with ≤ 11 for constraint 1 and ≤ 41 for constraint 3, the new optimum is $x_1^* = 3\frac{1}{2}$, $x_2^* = 3\frac{3}{4}$, and $\Pi^* = 25\frac{3}{4}$, which is the old value, $23\frac{3}{4}$, plus the 2 from the evaluation of $d\Pi^*$.

Similarly, the slope of constraint 1 is $-\frac{1}{2}$. This means that p_1 could increase up to 2.5 or p_2 could decrease down to 4, without disturbing the optimality of the current corner. Therefore the range for p_1 is (0.5, 2.5) and the range for p_2 is (4, 20). Note that these ranges are valid when *only that coefficient* is changing. If both p_1 and p_2 change, then the effect on the slope is ambiguous; it depends on the relative sizes of the two changes. In terms of the producer's story, these ranges indicate that inaccuracy in the estimate of either of the per unit profits, p_1 or p_2, would not necessarily lead to a wrong decision regarding x_1 and x_2.

7.6   SUMMARY

In exploring the structure and solution methods for linear programming problems, we have seen how this particular variation of the general inequality-constrained optimization problem of Chapter 4 is saved by the very characteristics that invalidate the calculus approach, via either Lagrange multipliers or the Kuhn–Tucker method. Precisely because (geometrically) the objective function and the sides of the feasible region are linear or (algebraically) the augmented problem consists of a set of linear equations with more unknowns than equations, we can use an efficient solution procedure for examining objective function values associated with basic solutions to the equations. This is the simplex method.

We have seen that linear programs always come in pairs, and that after the one that is of interest in a particular situation (the primal) has been formulated and solved, there may or may not be relevant information in the optimal values of the associated (dual) problem.

In Chapter 8 we investigate several additional topics in linear programming. Some are particular issues that can arise in the simplex calculations owing to special characteristics of the problem that are easily shown geometrically. Others are cases where the algebraic structure of a variant of the form that we have explored in this chapter allows a special (simpler) solution procedure. Then, in Chapter 9, we explore an entirely different method for solving linear programming problems. This approach shoots through the feasible region rather than moving around the edges.

PROBLEMS

7.1   Maximize $3x_1 + 2x_2 + 15x_3$ subject to $x_1 + x_2 + x_3 \leq 12$ and $2x_1 + x_2 + 5x_3 \leq 18$, as well as x_1, x_2 and $x_3 \geq 0$.

7.2   Minimize $6x_1 + 8x_2$ subject to $2x_1 + 7x_2 \geq 12$ and $2x_1 + x_2 \geq 1$, as well as nonnegativity of the two variables.

7.3   Solve the following problem:

$$\text{Minimize} \quad 5x_1 + 9x_2$$
$$\text{subject to} \quad 3x_1 + 4x_2 \geq 2$$
$$x_1 + 4x_2 \geq 6$$
$$\text{and} \quad x_1, x_2 \geq 0$$

7.4   Maximize $[1 \quad 8 \quad 10 \quad 9]\mathbf{X}$, subject to $[2 \quad 0 \quad 1 \quad 0]\mathbf{X} \leq 10$, $[0 \quad 2 \quad 2 \quad 5]\mathbf{X} \leq$
30 and $\mathbf{X} \geq 0$, where $\mathbf{X} = [x_1 \quad x_2 \quad x_3 \quad x_4]'$.

7.5   Maximize $x_1 + 2x_2$ subject to $-x_1 + x_2 \leq 6$ and $-x_1 + 2x_2 \leq 16$, along with $x_1, x_2 \geq$
0.

7.6   (a)   You could do this problem by hand, but it will be a lot easier if you have simplex method computer software. Here is a set of data for a linear programming
problem with 20 unknowns and four constraints:

$$\mathbf{P} = [6 \quad 7 \quad 8 \quad 9 \quad 1 \quad 2 \quad 3 \quad 4 \quad 5 \quad 6 \quad 7 \quad 8 \quad 9 \quad 1 \quad 2 \quad 3 \quad 4 \quad 5 \quad 6 \quad 7]'$$

$$\mathbf{A} = \begin{bmatrix} 3 & 6 & 5 & 7 & 8 & 9 & 7 & 6 & 5 & 6 & 7 & 8 & 7 & 6 & 5 & 4 & 6 & 5 & 8 & 3 \\ 2 & 8 & 6 & 4 & 5 & 3 & 2 & 6 & 1 & 2 & 3 & 4 & 5 & 6 & 7 & 8 & 5 & 8 & 6 & 8 \\ 6 & 8 & 4 & 5 & 3 & 2 & 1 & 5 & 8 & 7 & 6 & 5 & 4 & 3 & 2 & 1 & 2 & 2 & 7 & 9 \\ 5 & 6 & 7 & 8 & 9 & 2 & 3 & 4 & 6 & 7 & 8 & 3 & 4 & 5 & 6 & 7 & 2 & 3 & 4 & 5 \end{bmatrix}$$

$$\mathbf{B}' = [100 \quad 200 \quad 300 \quad 400]$$

Find the optimal solution—it is $x_1^* = 11.11$, $x_{20}^* = 22.22$, $s_3^* = 33.36$ and $s_4^* = 233.35$,
with $\Pi^* = 222.2$. In particular, note how many pivots are needed, despite the fact
that there are as many as $C_4^{24} = 10{,}626$ potential basic solutions. (If you begin at the
origin, you should find the optimal solution after three pivots.)

   (b)   Expand the data in part (a) to, say, 20 unknowns and six constraints. Now there
are potentially $C_6^{26} = 230{,}230$ basic solutions. See how many pivots the simplex
method needs.

   (c)   If you can do it easily, try a problem with 20 unknowns and 10 constraints, for
which there are over 30 million potential basic solutions! See how efficient the
simplex method is in this case. *Note*: For one completely general maximization
problem of this size, the simplex method took only *three* pivots, starting from
the origin, to reach the optimal solution. (The elements of $\mathbf{A}$ were random numbers between 1 and 9; those in $\mathbf{B}$ were random numbers between 1 and 4, then
multiplied by 100.)

7.7   Write down the dual linear program for Problems 7.1–7.5. In each case
   (a)   What is the optimal solution to this dual problem?
   (b)   Show that all of the "complementary slackness" relations of Theorem PD3 hold.

7.8   Consider two linear programs that differ only in that one of the constraints in one of
them is a constant, k, times the corresponding constraint in the other problem. How
would the optimal values of the dual variables in these two problems differ?

7.9   Consider a maximization problem where per unit profits in the objective function
are expressed in hundreds of dollars (e.g., $2x_1$ in the objective function means that a
unit amount of product 1 generates \$200 in profits). Suppose that you reexpressed
objective function profits in thousands of dollars.
   (a)   How would this affect the optimal values of the variables in the primal?
   (b)   How would the optimal values of the dual variables be affected?

7.10 The general m-constraint, n-variable linear programming maximization problem
was shown in (7.9) in the text. Assume all $b_i > 0$. Derive the simplex method rules

(as in Table 7.6) for moving from the initial basic feasible solution to a better basic feasible solution, assuming that the initial pivot element is located in row k ($k \leq m$) and column ℓ($\ell \leq n$).

7.11 Show how the conversion of a primal maximization linear programming problem from inequality to equality form (i.e., conversion to the augmented problem) ensures the nonnegativity of the dual variables.

7.12 Show that the dual to the following linear programming problem has no feasible solution:

$$\text{Maximize} \quad 2x_1 + 3x_2 + x_3$$
$$\text{subject to} \quad -5x_1 - x_2 + x_3 \leq 20$$
$$-x_1 + 2x_2 \leq 50$$
$$\text{and} \quad x_1, x_2 \geq 0.$$

7.13 A pleasure boat manufacturer produces two basic lines: a standard model (the Cruiser) and a more luxurious version (the Clipper). These are sold to dealers at a profit of $2000 per Clipper and $1000 per Cruiser. A Clipper requires 150 person-hours of labor for assembly, 50 person-hours for painting and finishing, and 10 person-hours for checking out and testing. A Cruiser needs 60, 40, and 20 person-hours, respectively, for the three tasks. During each production run, there are 30,000 person-hours available in the assembly shops, 13,000 in the painting and finishing shops, and 5000 in the checking and testing division. How many of each model should be scheduled for each production run in order to realize the greatest possible profit?

7.14 In Problem 7.13, what would be the most profitable production program if both Clippers and Cruisers were sold to dealers at the same profit of $1500 per boat?

7.15 Regions A and B can both product Miniautos, but using different production technologies. It is estimated that each unit produced in A contributes $300 to that region's income, while each unit in B contributes $700 to B's income. Skilled labor is freely mobile between the regions. Production of a Miniauto in A requires 30 person-hours of skilled labor while in B it requires 100 person-hours. The total pool of skilled labor available is 3000 person-hours per day. Unskilled labor is also required, and it is immobile between the regions. Production of one Miniauto in A requires 20 person-hours of unskilled labor; in B it requires 15 person-hours. Total available supplies of unskilled labor are 800 person-hours per day in A and 360 in B. Finally, production of each Miniauto requires one engine; these are produced at a factory on the border between the two regions and are available the next day in either region. The factory produces 60 engines per day.

(a) An interregional council wants to set production in both regions so as to maximize the total regional income (income in A plus income in B). What should the council recommend?

(b) What is the value to the two-region system of
   (i) An additional person-hour of skilled labor?
   (ii) An additional 40 person-hours of skilled labor?

(c) If an additional 8 person-hours of unskilled labor per day could be made available in either A or B from outside the system, to which region should it be allocated?

7.16 A small developing country has abundant supplies of bauxite (from which aluminum is made) and coal. Aluminum and coal, as luck would have it, are the primary inputs into the production process of the country's three most important state-controlled export industries. The inputs of aluminum and coal per unit of product in the three industries are as shown below:

	Industry		
	A	B	C
Coal	6 tons	6 tons	8 tons
Aluminum	10 lb	4 lb	12 lb

The nation's extractive capacity (for bauxite and coal) and its aluminum production capacity are such that aluminum can be made available for industrial production at the rate of 360 lb per day, while coal is mined at the rate of 270 tons per day. Measured in dollars, finished products from the three industries yield the following revenues for the country through international trade: $600 per unit from A, $400 per unit from B, and $200 per unit from C.

(a) Is it optimal for all three industries to engage in daily production? Why or why not? (Answer this question without doing any simplex method calculations at all.)

(b) What level of daily production in each industry will maximize the country's total revenues from international sales of the products of A, B, and C?

(c) You later discover that production in each of the industries also uses recycled plastic. In particular, a unit of product A requires 20 lb of recycled plastic, in B a unit of product requires 14 lb and a unit of C's product requires 16 lb of recycled plastic. The federal government of the country sets a target (for environmental purposes) that the three industries should consume at least 800 lb of recycled plastic per day. Would a conflict arise between revenue maximization and environmental cleanup? How much plastic can be recycled if revenues are maximized?

Answer the remaining parts of this question without reworking the entire problem; that is, use information available from the original optimal solution. This is relatively easy to do if you have used computer software for this problem

(d) If, for employment reasons, the government insisted that all three industries produce at least one unit of output per day, by how much would maximal total revenues [from part (b)] be changed, or would they remain unaffected?

(e) Relative to the results in part (b), what is the economic cost to the country of a strike by bauxite miners that reduces daily availability of aluminum to 350 lb per day? Why would it be necessary to rework the problem if the strike reduced daily aluminum availability to 150 lb?

(f) How would your answer to part (b) change (if at all) if the revenue per unit of C sold in the international market increased to $500? What about $750?

7.17 Consider the transportation network (e.g., a set of oil pipelines) shown below, indicating connections between point A (in the oil field) and point B (the port, from which oil is shipped) through intermediate pumping stations J, K, L, and M. Numbers above the lines indicate *capacities* on those network segments, in terms of maximum number of barrels of oil (in thousands) per day that could move over the route (e.g., 20,000 barrels per day is the maximum over the route from A to J). The pipeline is currently empty, and an empty tanker is expected to arrive at the port tomorrow.

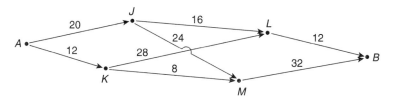

(a) Formulate and solve a linear programming problem that would identify the maximum amount of oil that could be moved from the fields to the port tomorrow, and also identify the extent to which each segment of the pipeline would be used.

(b) If you could add exactly five units of capacity on one and only one link, at no cost, where would you choose to put it and why? What would be the effect on the amount of oil that could be shipped?

(c) Suppose that the picture and the numbers were the same, but now the numbers indicate the *costs* of moving one barrel of oil along each link in the network. Formulate and solve a linear programming problem that would identify the least-cost path along which to move one barrel of oil, from the field (A) to the port (B).

8

LINEAR PROGRAMMING: EXTENSIONS

In Chapter 7 we examined the fundamental structure and computational approach to linear programming problems. In this chapter we investigate several additional topics, including special features that a problem may have which require sometimes simple alterations in the simplex method and sometimes very considerable changes or even entirely new solution approaches.

Problems whose objective function and/or constraints happen to allow more than one optimal solution, or which exclude the origin from the feasible region (so that the initial basic solution that we found in the simplex method of Chapter 7 is not possible), or in which some of the constraints are equations instead of inequalities, all are easily and automatically handled by most linear programming computer software. As we are about to see, these variations really affect only the shape of the feasible region or the relationship between the objective function contours and the sides of the feasible region. It is nonetheless useful to know exactly how the underlying geometry of the problem and algebra of the simplex method are modified (Sections 8.1–8.3).

Other variants (e.g., the transportation and assignment problems in Sections 8.4 and 8.5) have very specific and unique algebraic features that permit alternative and generally very fast solution methods. Again, linear programming software packages often automatically take advantage of these features, sometimes by supplying an entirely different computing approach for these problem types.

The computational implications of adding a whole-number requirement to the problem (*discrete* or *integer* linear programming problems) are dramatic and are explored in Section 8.6. One solution method builds on the simplex method and attacks the problem by adding a series of cleverly constructed linear constraints; this is the "cutting plane" method. A second and completely different approach to integer programming problems has also been developed; this is the branch-and-bound technique.

Finally, some ways of introducing an amount of uncertainty into the rigid linear programming structure are explored in Section 8.7. These include hierarchical optimization and goal programming approaches. The former is designed to accommodate the fact that real-world problems may have more than one objective. The latter does that as well, in a framework that also allows for flexibility in the specification of right-hand side values for constraints.

8.1 MULTIPLE OPTIMA AND DEGENERACY

8.1.1 Primal Multiple Optima

Consider the following linear programming problem:

$$
\begin{aligned}
\text{Maximize} \quad & 6x_1 + 4x_2 \\
\text{subject to} \quad & -2x_1 + x_2 \leq 2 \\
& x_1 - x_2 \leq 2 \\
& 3x_1 + 2x_2 \leq 9 \\
\text{and} \quad & \mathbf{X} \geq \mathbf{0}
\end{aligned}
\tag{8.1}
$$

Using the simplex arithmetic from Chapter 7, we find the simplex table shown in Table 8.1 after two iterations (try it, for practice).

Looking at the simplex criterion elements in the top row, we see that while neither of the excluded variables would *increase* the value of the objective function (no positive elements), introduction of s_2 into the basis would leave the objective function unchanged. If we decide to do that, iteration rules compel us to remove s_1 from the basis when bringing s_2 in. The new simplex table is shown in Table 8.2. (You should be convinced that pivoting in the s_1 column of Table 8.2 will lead back exactly to Table 8.1.)

The two alternative optima given in these tables are identified in Table 8.3. In both cases, the objective function value is 18. Although $s_3^* = 0$ in both solutions, s_1 is included in one optimal basis and s_2 is included in the other. This suggests that the Table 8.1 solution represents one endpoint of constraint 3 (its intersection with constraint 2) and Table 8.2 represents the other endpoint of constraint 3 (its intersection with constraint 1). These points are "equally optimal," which suggests that the objective function contours must be parallel to constraint 3.

Looking at the problem, in (8.1), we see that this is the case. It is also clear from the so-lution-space geometry in Figure 8.1, where the solutions are identified as I and II, respectively. When parallelism of this kind occurs, there will be more than one *basic* optimal so-

TABLE 8.1 Simplex Table for the Problem in (8.1) after Two Iterations

		s_2	s_3
Π	18	0	-2
s_1	$6\frac{3}{5}$	$-1\frac{2}{5}$	$-\frac{1}{5}$
x_1	$2\frac{3}{5}$	$-\frac{2}{5}$	$-\frac{1}{5}$
x_2	$\frac{3}{5}$	$\frac{3}{5}$	$-\frac{1}{5}$

TABLE 8.2   Simplex Table for the Problem in (8.1) after Three Iterations

		s_1	s_3
Π	18	0	-2
s_2	$4\frac{5}{7}$	$-\frac{5}{7}$	$-\frac{1}{7}$
x_1	$\frac{5}{7}$	$\frac{2}{7}$	$-\frac{1}{7}$
x_2	$3\frac{3}{7}$	$-\frac{3}{7}$	$-\frac{2}{7}$

TABLE 8.3   Alternative Basic Optima for the Problem in (8.1)

	Included in Optimal Basis	Excluded from Optimal Basis
Table 8.1	$x_1 = 2\frac{3}{5}, x_2 = \frac{3}{5}, s_1 = 6\frac{3}{5}$	$s_2 = 0, s_3 = 0$
Table 8.2	$x_1 = \frac{5}{7}, x_2 = 3\frac{3}{7}, s_2 = 4\frac{5}{7}$	$s_1 = 0, s_3 = 0$

lution and also an infinite number of *nonbasic* optima, namely all points on the convex combination of solutions I and II, along the edge of the feasible region defined by constraint 3.

Therefore, the appearance of one or more zeros in a top-row simplex table with otherwise negative elements means that the primal problem has multiple (or alternative) optima. The relevance of the simplex method—which looks only at basic solutions—is not affected, however, since among the multiple optima there will be at least two (for a problem with two unknowns) that are at corners of the feasible region and hence are basic.

Now consider the solutions to the dual problem as shown in Tables 8.1 and 8.2. These are collected together in Table 8.4. In either case, the optimal solution is $\mathbf{Y}^* = \begin{bmatrix} 0 \\ 0 \\ 2 \end{bmatrix}$ and

$\mathbf{T}^* = \begin{bmatrix} 0 \\ 0 \end{bmatrix}$, *even though* Tables 8.1 and 8.2 have a different pair of dual variables in the ba-

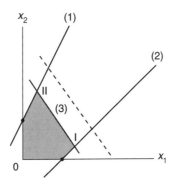

FIGURE 8.1   Multiple optima for problem in (8.1).

TABLE 8.4   Dual Optimal Solutions from Tables 8.1 and 8.2

	Included in Optimal Basis	Excluded from Optimal Basis
Table 8.1	$y_2 = 0, y_3 = 2$	$y_1 = 0, t_1 = 0, t_2 = 0$
Table 8.2	$y_1 = 0, y_3 = 2$	$y_2 = 0, t_1 = 0, t_2 = 0$

sis (top-row labels). This is because the dual optimal solutions are *degenerate* (one or more variables in the basis at value zero). Thus a primal problem with multiple optima (one or more zeros in an otherwise negative top row) has an associated dual problem with a degenerate optimal solution.

8.1.2   Primal Degenerate Optimum

Consider this linear programming problem:

$$\text{Maximize} \quad x_1 + 2x_2$$
$$\text{subject to} \quad x_2 \leq 3$$
$$x_1 \leq 2 \tag{8.2}$$
$$3x_1 + 2x_2 \leq 12$$
$$\text{and} \quad \mathbf{X} \geq \mathbf{0}$$

After one iteration, we have the simplex table shown in Table 8.5.

It is clear that x_1 should come in to the next basis, but the criterion for choosing the variable to leave (choice of the pivot row) indicates that s_2 and s_3 are equally good. In both cases, the relevant ratio is equal to 2, and the x_2 row in Table 8.5 is not a candidate, since its x_1 column element is not negative. The two possible pivot choices are labeled "a" and "b," and the results of these choices are shown in the next simplex tables, Tables 8.6(a) and 8.6(b). Either choice produces an optimal solution in one additional pivot—from Table 8.5 to Table 8.6(a) or to Table 8.6(b). The primal optima are recorded in Table 8.7. It is clear that they are (i) degenerate and (ii) the same in both cases.

Geometrically, in moving *from* Table 8.5, introducing x_1 into the solution that already includes x_2, constraints 2 and 3 are hit simultaneously. This is the geometric implication of

TABLE 8.5   Simplex Table for the Problem in (8.2) after One Iteration

		x_1	s_1
Π	6	1	-2
x_2	3	0	-1
s_2	2	-1^a	0
s_3	6	-3^b	2

TABLE 8.6 **Alternative Simplex Tables for the Problem in (8.2) after Two Iterations**

		s_2	s_1
Π	8	-1	-2
x_2	3	0	-1
x_1	2	-1	0
s_3	0	3	2

(a)   Pivot a

		s_3	s_1
Π	8	$-\frac{1}{3}$	$-\frac{4}{3}$
x_2	3	0	-1
s_2	0	$\frac{1}{3}$	$-\frac{2}{3}$
x_1	2	$-\frac{1}{3}$	$\frac{2}{3}$

(b)   Pivot b

the equality of the ratios in Table 8.5 from the s_2 and s_3 rows. Normally, in a pivoting operation, one moves along an edge of the feasible region until a corner is reached—along a constraint until it meets a new one. Here, however, *two* new constraints define the same stopping point. Either one would be sufficient to define the corner. This is clear from the solution-space geometry, in Figure 8.2.

Table 8.5 represents the corner at I in Figure 8.2. Introduction of x_1 at that point requires movement out along constraint 1, and *either* constraint 2 or 3 would serve to define the corner at II, which is the stopping point in the x_1 direction. From the point of view of the problem, however, only constraint 3 is redundant; its removal would not alter the feasible region but removal of constraint 2 would change the shape of the region.

Although the optimal values of the primal variables are unaffected by the pivot choice in Table 8.5, it is clear from Tables 8.6(a) and 8.6(b) that the optimal values of the dual variables *do* differ. These are collected together in Table 8.8. Therefore, although degeneracy at a primal optimum point does not affect the solution values in the primal, it *does* mean that the dual has multiple optima. It is important to keep this in mind in applications in which optimal dual variables (shadow prices) are used in a normative sense and compared with real-world prices or valuations; the dual variables that are used as the norm for

TABLE 8.7 **Optima for the Problem in (8.2)**

	Included in Optimal Basis	Excluded from Optimal Basis
Table 8.6(a)	$x_1 = 2, x_2 = 3, s_3 = 0$	$s_1 = 0, s_2 = 0$
Table 8.6(b)	$x_1 = 2, x_2 = 3, s_2 = 0$	$s_1 = 0, s_3 = 0$

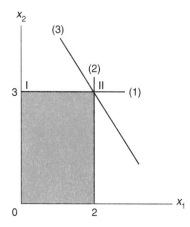

FIGURE 8.2 Degenerate optimum for problem in (8.2)

comparison will depend on the rule used to choose among the equally valid pivot elements, as in Table 8.5.

At a degenerate optimal corner, such as II in Figure 8.2, the effects of marginal increases and decreases are *different* for some of the constraints; that is, their right- and left-sided derivatives at that point will differ and so $\partial\Pi^*/\partial b_i$ will not exist. For example, a unit *increase* in the right-hand side of constraint 1 (upward movement) leads to an increase in the objective function that is smaller than the (absolute value of the) decrease associated with a unit *decrease* in the right-hand side of that constraint. Or, again, an *increase* in constraint 3 does nothing to the objective function value whereas a *decrease* reduces Π^*. Thus, Theorem PD5 of the last chapter is strictly true only for nondegenerate optima.

Degeneracy and multiple optima are seen to be opposite sides of a coin, but their implications are quite different. Because of the complete symmetry between a primal problem and its dual, a linear programming problem has multiple optima if and only if the optimal solution to the associated dual problem is degenerate.

8.2 ARTIFICIAL VARIABLES

The simplex method of Chapter 7 was valid for maximization problems with nonnegative numbers on the right-hand sides of the constraints. The duality properties of a linear programming problem and the structure on the simplex tables assure that a minimization

TABLE 8.8 Dual Alternative Optima for the Problem in (8.2)

	Included in Optimal Basis	Excluded from Optimal Basis
Table 8.6(a)	$y_1 = 2, y_2 = 1$	$y_3 = 0, t_1 = 0, t_2 = 0$
Table 8.6(b)	$y_1 = \frac{4}{3}, y_3 = \frac{1}{3}$	$y_2 = 0, t_1 = 0, t_2 = 0$

problem can be handled by solving its dual.[1] The requirement that right-hand sides be nonnegative simply ensures that the origin is a corner of the feasible region, so an initial basic feasible solution has all x's at zero value and all s's equal to the right-hand sides of the constraints. However, it is easy to imagine a problem in which the origin will not be part of the feasible region. An example is provided by the maximization problem in (7.3) in the last chapter (Fig. 7.1b).

When this is the case, the problem can be resolved by addition of one or more *artificial* variables. Computer software packages for linear programming problems automatically generate an initial feasible basis using artificial variables techniques, or using variants of what is known as a two-phase procedure. The following example serves simply to illustrate the logic of one of the methods most frequently used to overcome the infeasible-origin problem.

Consider the problem in (7.6) in Chapter 7, with the direction of the third inequality reversed:

$$\text{Maximize} \quad 2x_1 + 5x_2$$
$$\text{subject to} \quad x_1 + 2x_2 \leq 10$$
$$3x_1 + 2x_2 \leq 24$$
$$x_1 + 10x_2 \geq 40$$
$$\text{and} \quad \mathbf{X} \geq \mathbf{0}$$

or, multiplying the third constraint by -1, we obtain

$$\text{Maximize} \quad 2x_1 + 5x_2$$
$$\text{subject to} \quad x_1 + 2x_2 \leq 10$$
$$3x_1 + 2x_2 \leq 24 \tag{8.3}$$
$$-x_1 - 10x_2 \leq -40$$
$$\text{and} \quad \mathbf{X} \geq \mathbf{0}$$

Figure 8.3 indicates what has happened to the feasible region (compare with Fig. 7.2d).
Writing the constraints in augmented form, we have

$$x_1 + 2x_2 + s_1 = 10$$
$$3x_1 + 2x_2 + s_2 = 24 \tag{8.4}$$
$$-x_1 - 10x_2 + s_3 = -40$$

and so when $x_1 = x_2 = 0$, $s_1 = 10$, $s_2 = 24$, but $s_3 = -40$, violating the nonnegativity requirements on $\mathbf{S}$.

---

[1]Actually, it is not difficult to modify the simplex arithmetic rules from Chapter 7 so that a minimization problem can be handled directly. You might try this, noting that sign differences will be necessary because nonnegative surplus variables must now be subtracted from, not added to, the left-hand sides of "greater than or equal to" inequalities—the standard form for a linear programming minimization problem.

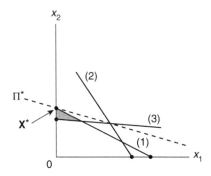

FIGURE 8.3 Solution-space geometry for problem in (8.3)

Suppose, for the moment, that we let $s_3 = 0$ also and alter the third constraint further by *subtracting* a new and artificial variable, s_a, so that the constraint becomes

$$-x_1 - 10x_2 + s_3 - s_a = -40$$

Then an initial basic *and feasible* solution will be $x_1 = x_2 = 0$; $s_1 = 10$, $s_2 = 24$, $s_3 = 0$, and $s_a = 40$. Having included a new and artificial variable in a constraint, we must also include it in the objective function. One way to ensure that this variable will not remain in further basic solutions is to insert it in the objective function (of this maximization problem) with a *large* and *negative* coefficient. This will ensure that it is immediately removed from the initial basic feasible solution. All we need to do is give it a coefficient of $-K$, where we understand K to be an immense number.[2]

The final version of the augmented problem is then

$$\text{Maximize} \quad 2x_1 + 5x_2 + 0s_1 + 0s_2 + 0s_3 - Ks_a$$
$$\text{subject to} \quad x_1 + 2x_2 + s_1 = 10$$
$$3x_1 + 2x_2 + s_2 = 24 \tag{8.5}$$
$$-x_1 - 10x_2 + s_3 - s_a = -40$$
$$\text{and} \quad x_1, x_2; s_1, s_2, s_3, s_a \geq 0$$

We know that x_1, x_2, and s_3 will be zero in the initial simplex table; they will be the column labels. We also know from constraint 3 that

$$s_a = 40 - x_1 - 10x_2 + s_3$$

and this provides us with exactly the coefficients for the s_a row in the initial simplex table. Moreover, since s_a is totally artificial from the point of view of the objective function, this rearrangement of constraint 3 allows us to eliminate s_a from the objective function. It becomes

$$2x_1 + 5x_2 + 0s_1 + 0s_2 + 0s_3 - K(40 - x_1 - 10x_2 + s_3)$$

---

[2]Often M is used, which has led to the name "the big M method" for this procedure.

TABLE 8.9   Initial Simplex Table for the Problem in (8.5)

		x_1	x_2	s_3
Π	$-40K$	$(2+K)$	$(5+10K)$	$(0-K)$
s_1	10	-1	-2	0
s_2	24	-3	-2	0
s_a	40	-1	-10	1

or

$$(2 + K)x_1 + (5 + 10K)x_2 + 0s_1 + 0s_2 + (0 - K)s_3 - 40K$$

In the initial simplex table, with $s_1 = 10$, $s_2 = 24$, and $s_a = 40$, the value of the objective function is $-40K$.

All this is recorded in Table 8.9. Since $-40K$ is not a very good maximum, and since both x_1 and x_2 have positive coefficients for the objective function at this point, we now use the usual rules of simplex arithmetic to select a pivot column and row. In the first iteration, x_2 will identify the pivot column and s_a will be removed, by the usual smallest-ratio rules. In the second pivot operation, x_1 comes into the next basis, replacing s_1. Finally, a third pivot brings s_3 into the basis and removes x_1 (even though it just came into the basis on the previous pivot step). At this point, we have the simplex table shown in Table 8.10. (You might carry out these pivot operations to see exactly how it is that the K's disappear completely from everywhere except the s_a column.) So we have an optimal solution in which $x_1^* = 0$, $x_2^* = 5$; $s_1^* = 0$, $s_2^* = 14$, $s_3^* = 10$, $s_a^* = 0$, and $\Pi^* = 25$. This solution is indicated in Figure 8.3.

This kind of artificial-variable technique can be used for a linear programming maximum problem with any number of constraints having negative right-hand side values. If the ordinary rules of pivoting produce a unique optimal solution in which not all artificial variables have been removed from the basis, then the original problem has no feasible solution.

An essentially identical approach is the *two-phase* procedure. After adding artificial variables, exactly as above, to each constraint with a negative right-hand side—denoting by $s_{a(i)}$ the artificial variable for constraint i—the phase I problem seeks to minimize the sum of the artificial variables, $\Sigma_i s_{a(i)}$, subject to the (enlarged) constraints of the original problem. Since these variables are also required to be nonnegative, the minimum of this sum is zero, except in the rare case where there is no feasible solution at all. When this optimal phase I solution is reached, the resulting values for the x's and s's constitute an initial basic feasible solution to the original linear program. Phase II then begins with this solution as a

TABLE 8.10   Optimal Simplex Table for the Problem in (8.5)

		s_1	s_a	x_1
Π	25	$-\frac{5}{2}$	$-K$	$-\frac{1}{2}$
s_3	10	-5	1	-4
s_2	14	1	0	-2
x_2	5	$-\frac{1}{2}$	0	$-\frac{1}{2}$

starting point for the simplex calculations on the original problem. In other words, phase I establishes an alternative to the (excluded) origin from which to begin the real work.

8.3 EQUATIONS AS CONSTRAINTS

We saw in Chapter 4 that problems of maximization or minimization of a function subject to constraints that are equations can be approached using Lagrange multipliers. We also saw that inequalities are a different matter, and the development of the simplex arithmetic in Chapter 7 depended on constraints in the form of weak linear inequalities. In this section we note that linear *equations* as constraints can also be handled by the simplex method. This means that the calculations procedures of Chapter 7 can be used for a problem with some inequality and some equality constraints, thus increasing the generality of the method. Again, we consider two related approaches. And again it should be noted that linear programming software for modern computers handles equality constraints automatically. In fact, constraints in the form of equalities as well as those that set lower bounds and others that set upper bounds can all be accommodated simultaneously. In most software packages, all that is needed is that the user specify, for each constraint, whether it reads "≤", "≥," or "=".

The first approach is extremely simple. It involves merely converting each linear equation to *two* linear inequalities. Consider $2x_1 + 3x_2 = 10$; this is equivalent to requiring $2x_1 + 3x_2 \leq 10$ and $2x_1 + 3x_2 \geq 10$ simultaneously. If this were a constraint in a maximization problem, we could write the second inequality as $-2x_1 - 3x_2 \leq -10$ and proceed as in Section 8.2.

One consequence for the dual problem may be of interest, depending on the use that is made of the optimal values of the dual variables. Since there is a dual variable associated with each primal inequality constraint, each primal equality constraint (when written in double-inequality form) will generate *two* dual variables. In fact, however, the two can be combined into one (which is logical, since they are associated with what is fundamentally one primal constraint), but this dual variable will be unrestricted in sign. This is easily illustrated by a small example. (You might carry out a similar kind of demonstration for the general case.)

Let the primal linear programming problem be

$$\text{Maximize} \quad x_1 + 6x_2$$
$$\text{subject to} \quad 3x_1 + 4x_2 \leq 5$$
$$2x_1 + 8x_2 = 7 \qquad (8.6)$$
$$\text{and} \quad \mathbf{X} \geq \mathbf{0}$$

In all-inequality form, this becomes

$$\text{Maximize} \quad x_1 + 6x_2$$
$$\text{subject to} \quad 3x_1 + 4x_2 \leq 5$$
$$2x_1 + 8x_2 \leq 7 \qquad (8.7)$$
$$-2x_1 - 8x_2 \leq -7$$
$$\text{and} \quad \mathbf{X} \geq \mathbf{0}$$

Following the general rules in Chapter 7, the dual to this problem is

$$\text{Minimize} \quad 5y_1 + 7y_2 - 7y_3$$
$$\text{subject to} \quad 3y_1 + 2y_2 - 2y_3 \geq 1$$
$$4y_1 + 8y_2 - 8y_3 \geq 6 \tag{8.8}$$
$$\text{and} \quad y_1, y_2, y_3 \geq 0$$

But this is the same as

$$\text{Minimize} \quad 5y_1 + 7(y_2 - y_3)$$
$$\text{subject to} \quad 3y_1 + 2(y_2 - y_3) \geq 1$$
$$4y_1 + 8(y_2 - y_3) \geq 6$$
$$\text{and} \quad y_1, y_2, y_3 \geq 0$$

Therefore, if we define $y_d = y_2 - y_3$ (the subscript reminds us that it is the difference between two variables), the problem becomes

$$\text{Minimize} \quad 5y_1 + 7y_d$$
$$\text{subject to} \quad 3y_1 + 2y_d \geq 1$$
$$4y_1 + 8y_d \geq 6 \tag{8.9}$$
$$\text{and} \quad y_1 \geq 0 \text{ but } y_d \gtreqless 0$$

since the difference between two nonnegative variables (y_2 and y_3) can be of either sign, or zero. So, in the final form of the dual problem, (8.9), we have one dual variable for each original primal constraint, in (8.6), but the variable associated with the primal equation is unrestricted in sign.

The second approach to equality constraints is exactly the two-phase procedure described in Section 8.2. Artificial variables can be added to each *equation*; that is, it can be treated as if it were an inequality. Then in phase I one seeks to minimize the sum of the added artificial variables, subject to the revised constraint equations that include these variables. Except in unusual circumstances, the minimum will be zero, and this optimum to the phase I problem becomes the initial basic feasible solution for the phase II simplex computations on the problem of original interest.

8.4   THE TRANSPORTATION PROBLEM

8.4.1   Fundamental Structure

The name *transportation problem* denotes a class of linear programs with a particular structure that allows computational streamlining. It is a framework of quite wide applicability; it can be introduced in the following setting, which explains its name. Consider the problem of distributing units of a homogeneous product from places where they are produced—the origins, m in number—to places where they are consumed—the destinations, n in number. Suppose that each origin i ($i = 1, \ldots, m$) has a known supply or number of units produced, a_i, and each destination j ($j = 1, \ldots, n$) has a known demand or number of units needed, b_j, and that production has already been planned so that demand is exactly met, so

$$\sum_{i=1}^{m} a_i = \sum_{j=1}^{n} b_j$$

The problem is to decide which destinations should be served from which origins and in what amounts. Assume that the object is to distribute items in such a way that the total cost of transporting them is minimized. Let x_{ij} denote the amount of product sent from origin i to destination j, and t_{ij} be the *unit* transport cost associated with the i–j route. The objective function is then to minimize

$$T = \sum_{i=1}^{m} \sum_{j=1}^{n} t_{ij} x_{ij}$$

It is important to note that the objective function in a transportation problem of this sort, while linear in the x_{ij}, nonetheless allows a realistic kind of "nonlinearity" in transportation costs. This is because the t_{ij} represent the total costs of sending a unit amount over the entire length of the route from i to j. For example, suppose that origin 1 is the same distance on a roadmap from destination 2 as it is from destination 5, but that there is a six-lane highway connecting origin 1 with destination 2 while the route from origin 1 to destination 5 requires crossing a river on a ferry boat. Then, although the number of miles between the origin and the two destinations may be the same, t_{12} would very likely be less than t_{15}, since the latter involves a longer shipping time (waiting for the ferry, slower time crossing the river) and extra expenses (cost of the ferry trip) for each unit transported. Similarly, a destination may be the same road distance from two origins, but one of the origins could be on an island, from which a combination of ship–truck transportation was required, as opposed to truck only from the other origins. (If goods are shipped by air, then the presence of water may not be significant, but you can think of other plausible stories; there may be a mountain range between one of the origins and some of the destinations, etc.)

In addition to nonnegativity requirements on the x_{ij}, there are two classes of constraints. One concerns each of the origins, the other each of the destinations. Specifically, shipments *out of* any origin must equal the supply at that origin:

$$\sum_{j=1}^{n} x_{ij} = a_i \qquad \text{(for each origin } i\text{)}$$

and shipments *into* any destination must equal the demand at that destination:

$$\sum_{i=1}^{m} x_{ij} = b_j \qquad \text{(for each destination } j\text{)}$$

So the complete problem is

$$\text{Minimize} \quad T = \sum_{i=1}^{m} \sum_{j=1}^{n} t_{ij} x_{ij}$$

$$\text{subject to} \quad (1) \sum_{j=1}^{n} x_{ij} = a_i \qquad (i = 1, \ldots, m)$$

$$(2) \sum_{i=1}^{m} x_{ij} = b_j \qquad (j = 1, \ldots, n) \tag{8.10}$$

$$\text{and} \quad x_{ij} \geq 0$$

The constraint sets (1) and (2) in (8.10) have a particularly simple structure. Not only are they equations, but also all the coefficients on the x_{ij}'s are either 0 (absence of that particular x_{ij} in that constraint) or 1 (presence of that x_{ij}). This is more easily seen in a small example; suppose there are two origins (say, 1 = New York and 2 = Philadelphia) and three destinations (1 = Chicago, 2 = Seattle, and 3 = San Francisco). The constraints are therefore

$$
\begin{aligned}
x_{11} + x_{12} + x_{13} & & & = a_1 \\
& x_{21} + x_{22} + x_{23} & & = a_2 \\
x_{11} & + x_{21} & & = b_1 \\
\phantom{x_{11}} x_{12} & + x_{22} & & = b_2 \\
\phantom{x_{11}} x_{13} & + x_{23} & & = b_3
\end{aligned}
\tag{8.11}
$$

The first equation, for example, says that shipments from New York to Chicago, Seattle, and San Francisco (the three possible destinations) must exhaust New York's production. The last constraint says that the amounts coming into San Francisco from New York and from Philadelphia (the only origins in the problem) must just satisfy San Francisco's demand.

Consider the linear equations in (8.11) in $\mathbf{AX} = \mathbf{B}$ form. The matrix of coefficients has the following particularly simple structure:

$$
\mathbf{A}_{(5\times6)} =
\begin{bmatrix}
1 & 1 & 1 & 0 & 0 & 0 \\
0 & 0 & 0 & 1 & 1 & 1 \\
1 & 0 & 0 & 1 & 0 & 0 \\
0 & 1 & 0 & 0 & 1 & 0 \\
0 & 0 & 1 & 0 & 0 & 1
\end{bmatrix}
$$

The rank of this matrix cannot exceed 5 (the smaller of its two dimensions). However, removing *any* column leaves a 5×5 submatrix whose determinant is zero; this is because the rows are linearly dependent. For example

$$(\text{row } 1) = (\text{row } 3) + (\text{row } 4) + (\text{row } 5) - (\text{row } 2)$$

This is true in **A** and in any 5×5 submatrix in **A**; in fact, $\rho(\mathbf{A}) = 4$. Since $a_1 + a_2 = b_1 + b_2 + b_3$, the same linear dependence exists in the augmented matrix $[\mathbf{A} \,\vdots\, \mathbf{B}]$ as in **A**, and therefore (8.11) represents a consistent set of *four* linearly independent equations. Thus this simplest form of the transportation problem consists of mn variables and $(m + n - 1)$ independent constraint equations. Except in the trivial cases in which $m = 1$ or $n = 1$ (or both), there will be more unknowns than equations; hence, as usual, we must search the *basic* solutions in order to find the optimum. An optimal solution will contain at most $(m + n - 1)$ variables in the basis at positive value. Variants of the simplex method, programmed to deal exclusively with equations, can be used. A more general statement of the linear programming transportation problem allows for the several additional and realistic possibilities:

(a) Total supply in the system might not equal total demand: $\sum_{i=1}^{m} a_i \neq \sum_{j=1}^{n} b_j$. If there is more production capacity than demand ($\sum_{i=1}^{m} a_i > \sum_{j=1}^{n} b_j$) there will be units of product left unshipped at one or more origins, and the origin constraints become

$$\sum_{j=1}^{n} x_{ij} \leq a_i \qquad (i = 1, \ldots, m)$$

If there is more demand than supply ($\sum_{i=1}^{m} a_i < \sum_{j=1}^{n} b_j$), there will be unfilled demand at one or more destinations, and the destination constraints in this case are

$$\sum_{i=1}^{m} x_{ij} \leq b_j \qquad (j = 1, \ldots, n)$$

In the completely general transportation problem, both sets of constraints can be written as inequalities, as long as $\sum_{i=1}^{m} a_i \geq \sum_{j=1}^{n} b_j$—$\sum_{j=1}^{n} x_{ij} \leq a_i$ for the origins and $\sum_{i=1}^{m} x_{ij} \geq b_j$ for the destinations. The objective function for this more general problem is then exactly as in (8.10). As long as $t_{ij} > 0$ for all i and j, the destination constraints will always be equalities at an optimal solution. (You would never ship more than the minimum needed at each destination, as long as transportation costs are positive.)

You might consider the dual to this generalized version of the transportation problem. The variables in the dual (maximization) problem are necessarily spatially oriented; they are associated with primal capacity or demand constraints that are specific to a particular point—an origin (i) or a destination (j). It is possible to interpret the dual objective function and the constraint relations in this dual problem in terms of the spatial economics concept of location rents.

(b) One or more of the arcs, connecting an origin, i, and a destination, j, may have an upper limit, or *capacity*, k_{ij}, on the amount of shipment that it can accommodate. This is known as the *capacitated transportation problem*. For example, there may be a maximum possible flow per hour in a pipeline (determined by the diameter of the pipe, the characteristics of the material being sent through it, and the force with which it is pumped), or a maximum number of flights per hour over an air route (because of takeoff and/or landing capabilities at airports, required spacing between aircraft for safety reasons, etc.). The alteration to the statement of the general problem is simply that a set of constraints $x_{ij} \leq k_{ij}$ is added, for those i–j pairs whose arcs have such upper limits.

(c) A third variant is known as the *transshipment* problem; sometimes this is also called the *transportation problem with warehousing*. That provides an easy illustration. Suppose that there are intervening nodes to which goods can be shipped from origins and from which they can then be further shipped to destinations. Warehouses serve exactly this kind of function. In a different context, airline passengers flying from Philadelphia to Seattle will often not be flown nonstop but rather via an intermediate airport, Chicago or Denver, for example, where there may also be a change of airplane.

In addition to the original t_{ij} [unit shipment costs from each origin (i) to each destination (j)], we now must also have (1) the unit costs, t_{ik}, of shipping from each origin to each transshipment point (k) and the unit costs, t_{kj}, of shipping from each transshipment point to each destination and (2) the capacities of the transshipment points. If these are warehouses, the capacities would likely be a reflection of physical volume constraints (e.g., cubic yardage); if they are airports, they might be in terms of aircraft landings and/or takeoffs per hour.

Many computer programs are designed specifically to solve linear programming transportation problems with any or all of these extensions. In general they solve both primal and dual problems and, in addition, usually give sensitivity measures such as the effects on the objective function optimal solution of a unit change in origin capacities or destination demands (dual variables). Generally, also, the ranges of these dual variables are calculated, as well as ranges for objective function coefficients (here unit transportation costs) over which the optimal solution remains unchanged.

8.4.2 Numerical Illustration

Here is a small sample problem that will let us examine the kinds of results that are obtained from current computer software for the general transportation problem.[3] There are two origins (O) and three destinations (D), with supplies $a_1 = 10$, $a_2 = 15$, demands $b_1 = 8$, $b_2 = 12$, $b_3 = 5$ and unit transportation costs as shown in the following matrix:

	D_1	D_2	D_3
O_1	30	40	20
O_2	40	10	60

(8.12)

In fact, a very efficient and compact way to illustrate all the relevant data in a transportation problem is to expand the cost matrix by adding (1) one more column, labeled a_i, in which to record the supplies of each origin and (2) one more row, labeled b_j, in which to record the demands at each destination. For this example

	D_1	D_2	D_3	a_i
O_1	30	40	20	10
O_2	40	10	60	15
b_j	8	12	5	

(8.13)

---

[3]These results were generated using the Eastern Software Products program "TSA88," for transportation (TS) and assignment (A) problems.

The full problem, then, is

$$\text{Minimize} \quad T = 30x_{11} + 40x_{12} + 20x_{13} + 40x_{21} + 10x_{22} + 60x_{23}$$

$$\text{subject to} \quad x_{11} + x_{12} + x_{13} = 10$$

$$x_{21} + x_{22} + x_{23} = 15$$

$$x_{11} + x_{21} = 8 \qquad\qquad (8.14)$$

$$x_{12} + x_{22} = 12$$

$$x_{13} + x_{23} = 5$$

$$\text{and} \quad x_{ij} \geq 0$$

The optimal solution to this problem has a minimized total shipping cost, T^*, of 490, associated with the following shipments: $x_{11}^* = 5$, $x_{13}^* = 5$, $x_{21}^* = 3$, and $x_{22}^* = 12$. [Note that, as we expect, there are $m + n - 1 = 4$ variables, out of a possible mn ($= 6$), in the optimal basic solution.] It is visually effective to present the optimal shipments in a grid that is the same size as we used above to record all data of the problem, as in (8.13). By leaving the supply (last) column and the demand (last) row, we can easily check to see how each origin's capacity is used and how each destination's demand is satisfied. This is done in Table 8.11.

We designate dual variables for the origin constraints by u_i's and for the destination constraints by v_j's. If, as in the present example, $\sum_{i=1}^{m} a_i = \sum_{j=1}^{n} b_j$, then the dual variables for the origin supplies are only valid measures of the impacts of unit capacity *increases* (decreases would generate a problem with no feasible solution, since there would not be enough supply to satisfy all demand) and of unit demand *decreases* (increases in demand would, again, create the same kind of problem in which there would be one or more destinations with unsatisfied demand). In minimization problems, the dual variable represents the amount by which the optimal value of the objective function would *decrease* for a unit change in the right-hand side of the constraint. The ranges for this problem are shown in Table 8.12.

It is easy to confirm these values by rerunning the problem with appropriate changes in one of the supplies or one of the demands. For example, if $a_1 = 11$, the minimized total cost associated with satisfying all demand and not exceeding total supply (in fact, leaving one unit unshipped from origin 2) is 480, which, indeed, is a reduction of 10, as indicated

TABLE 8.11   Optimal Shipments in the Sample Transportation Problem

	D_1	D_2	D_3	a_i
O_1	5	0	5	10
O_2	3	12	0	15
b_j	8	12	5	

TABLE 8.12   Optimal Values of Dual Variables and Their Ranges

Dual Variable	Range
$u_1^* = 10$	+3
$u_2^* = 0$	+∞
$v_1^* = 40$	−3
$v_2^* = 10$	−12
$v_3^* = 30$	−3

TABLE 8.13   Optimal Shipments in the Capacitated Transportation Problem

	D_1	D_2	D_3	a_i
O_1	3	2	5	10
O_2	5	10	0	15
b_j	8	12	5	

by u_1. An increase of a_1 to 12 would generate a cost decrease of 20, and an increase of a_1 to 13 would shrink the total shipping cost to 460.

Increases in the supply from origin 1 beyond 13 are useless, since the cost-minimizing shipping pattern has all of destination 1 and 3 demand met from origin 1, which is the cheaper supplier to both of those destinations (refer to the table of unit shipping costs). At the same time, all of destination 2's demand is satisfied by shipments from origin 2, and it is the cheaper alternative for that destination. This is reflected in the fact that the valid range for u_1^* is +3 units. Changes from the present 10 up to a maximum of 13 will each generate cost decreases of 10 units; changes beyond 13 are useless, from the point of view of permitting further decreases in the total transportation cost. You should also be clear why $u_2^* = 0$ and why the range for that variable has no upper limit. The dual values associated with destination demand decreases have a similar interpretation; you should solve this problem again for one or two of the unit changes in the b_j's and confirm the value and range of the v_j's given in the table.

If we add an upper limit of 10 on shipments from origin 2 to destination 2—$x_{22} \leq 10$—then the optimal transportation pattern is as shown in Table 8.13. In this case, $T^* = 570$. Since x_{22}^* was 12 in the uncapacitated version of the problem, it is logical that the capacity constraint would be binding in the optimal solution to the capacitated problem. The increased costs are attributable to the rearrangements of the original optimal solution that are necessary in order to deal with the decrease of x_{22}^* from 12 to 10.

Finally, suppose that there are two transshipment points in the problem, T_1 and T_2. We add the unit transportation costs from each origin to each transshipment point and from each such point to each of the destinations to the original cost matrix in (8.12). This is shown in (8.15), where the two 10's in the lower right-hand corner represent assumed unit shipping costs from one transshipment point to the other.

	D_1	D_2	D_3	T_1	T_2
O_1	30	40	20	12	23
O_2	40	10	60	15	8
T_1	15	20	6		10
T_2	25	5	30	10	

(8.15)

Figure 8.4 suggests one possible topography for this particular example. Assume that the first transshipment point (e.g., warehouse) has a capacity of 20 and the second of 10. Then the optimal shipments for this modification of the original problem are shown in Table 8.14; minimized transportation costs are now $T^* = 435$. Comparison with the optimal solution to the original problem, with neither capacity constraints nor transshipment points, illustrates the kinds of cost savings that transshipment may make possible. (Naturally, the numbers are completely specific to this numerical example, but the general principles are the same in any real-world problem.)

The transportation problem is but one (quite simple) example of a general class known as *network optimization* problems. These are problems in which geographically distinct points (nodes) are connected by some kind of network (arcs), as in the transportation problem example, and the problem is to optimize some kind of activity on the network. A few examples are as follows:

1. Variants of the *traveling salesperson* problem in which the object is to find a routing from an origin (say Chicago) to a destination (perhaps Chicago again, if it is the salesperson's home base) by which to visit each of a set of cities (nodes) located on a transportation route (arcs) in such a way that total distance traveled, or total time taken to make the whole tour, is minimized.

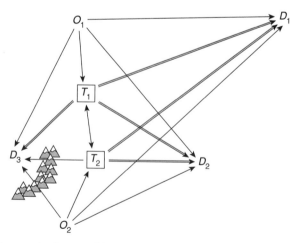

FIGURE 8.4 Topography for transshipment problem. (Double lines are limited access highways. There is a range of mountains between D_3 and O_2 or T_2.)

TABLE 8.14   Optimal Shipments in the Transshipment Problem

	D_1	D_2	D_3	T_1	T_2	a_i
O_1				10		10
O_2		12		3		15
T_1	8		5			20
T_2						10
b_j	8	12	5	20	10	

2. *Maximum flow* problems, in which, for example, a complicated set of pipelines connects many oil wells ("sources") to many oil storage facilities ("sinks") through a series of intermediate pumping stations (acting somewhat like transshipment points); the object is to find the routing of oil from the sources to the sinks that maximizes the flow (e.g., per day) through the network.

3. *Shortest route* (or *minimum distance*) problems are those in which an origin (say, New York) and a destination (say, San Francisco) on a network are connected by a great many alternative series of links (e.g., rail), and the object is to find the shortest route (total elapsed distance) through all the possible links from origin to destination.

8.5   THE ASSIGNMENT PROBLEM

8.5.1   Fundamental Structure

In this section we describe a particular form of the transportation problem that has broader applicability than its "assignment problem" name might initially suggest. This will become clear as we look at the structure of the problem.

Suppose that n jobs are to be completed and m individuals are available to do them. In the most straightforward case, $m = n$. Suppose that each person is capable of performing each task, but their efficiencies differ. This difference is reflected in the known costs—or times, or some other measure—for each person, i, at each job, j; let these known figures be denoted by c_{ij}. The object is to assign individuals to jobs so that total cost (or time), T, to complete all jobs is minimized. Let x_{ij} be a variable that can take on only two values: 0 if individual i is *not* assigned to job j and 1 if the person *is* assigned. This is termed a 0–1 *variable*, and the problem represents one kind of *integer linear programming* problem; we will consider these in detail in Section 8.6. For the moment, we concentrate on the structure of the problem.

The objective is to minimize $T = \sum_{i=1}^{m} \sum_{j=1}^{n} c_{ij} x_{ij}$. Without constraints (if the c_{ij} are strictly positive), the minimum is zero, achieved when no assignments are made and no jobs get done. To avoid this, we require that

1. Each job has one person assigned to it:

$$\sum_{i=1}^{m} x_{ij} = 1 \qquad \text{(for each job, } j\text{)}$$

This ensures that no job is left undone; it does not prevent the assignment of one person to more than one task, however.

2. To avoid this, we also must require that each person be assigned to only one task:

$$\sum_{j=1}^{n} x_{ij} = 1 \qquad \text{(for each individual, } i\text{)}$$

The complete assignment problem is therefore

$$\text{Minimize} \quad T = \sum_{i=1}^{m} \sum_{j=1}^{n} c_{ij} x_{ij}$$

$$\text{subject to} \quad (1) \; \sum_{i=1}^{m} x_{ij} = 1 \qquad \text{(for each } j\text{)}$$

$$(2) \; \sum_{j=1}^{n} x_{ij} = 1 \qquad \text{(for each } i\text{)} \qquad (8.16)$$

$$\text{and} \quad x_{ij} = 0 \text{ or } 1 \text{ for all } i \text{ and } j$$

This last constraint could be written as $x_{ij} \geq 0$, $x_{ij} \leq 1$ and x_{ij} integer. In the assignment problem, just as in the transportation problem, it will turn out that the optimal values of all x_{ij} will be whole numbers as long as the right-hand sides of the constraints are themselves whole numbers. Therefore we need not be concerned with this integer requirement yet; we will explore the more general case of integer linear programming problems in Section 8.6.

If a particular individual, i, is not able to do a specific job, j, then a very large value for that c_{ij} can simply be entered into the data of the problem. If there are more tasks than people ($m < n$), then some jobs will remain undone in the optimal solution; if there are more people than jobs ($m > n$), then some people will remain unassigned in an optimal solution. There are completely straightforward ways of modifying the basic problem in (8.16) to deal with each of these kinds of variations. A standard approach is to create either dummy tasks (when there are more people than jobs), to which the ($m - n$) unused people will be assigned in an optimal solution, or dummy persons (an unfortunate term, when there are more jobs than people), who will be assigned to the ($n - m$) jobs that will remain unfilled in an optimal solution and to enter equal person-job assignment costs for those dummies.

8.5.2 Numerical Illustration

Five workers ($W_1, \ldots, W_5$) are available to work on different machines in a factory. The costs associated with each worker–machine combination are given in (8.17). An "X" indicates that a worker is unable to operate that machine, and "M_N" represents a new machine that is being considered to replace one of the five current machines.

An optimal solution can be represented in a matrix of the same size as (8.17) in which a "0" in cell i, j means that $x_{ij} = 0$ (worker i is not assigned to machine j) and a "1" means $x_{ij} = 1$ (worker i is assigned to machine j). For the problem in (8.16), the optimal assignment pattern turns out to be as shown in Table 8.15; the minimized total costs of this assignment are $T^* = 121$, and the new machine replaces machine 3 (since no one is assigned to machine 3 in the optimal solution).

	M_1	M_2	M_3	M_4	M_5	M_N
W_1	12	3	6	X	5	9
W_2	4	11	X	5	X	3
W_3	8	2	10	9	7	5
W_4	X	7	8	6	12	10
W_5	5	8	9	4	6	X

$$(8.17)$$

The assignment problem format has been extended to a variety of real-world problems, including plant location: assignment of plants (i) to geographic locations (j). This problem, in turn, is a specific example of a larger family of problems known as (*public*) *facility location* problems. These include problems of deciding on the optimal location of clinics, libraries, fire stations, and so on, that serve a geographically dispersed population. In the case of clinics or libraries, to which individuals must travel to receive the service, the objective might be to minimize some measure of total travel time (or cost) for the potentially served population; in the case of fire stations, from which units (fire trucks or aid cars) are dispatched to the location of a fire or medical need, the objective might be to locate the facility so as to minimize the average travel time, or the maximum travel time, for example. There is a wide literature on programming approaches to facility location problems, but we cannot explore the topic further here.

8.6 INTEGER PROGRAMMING

In this section we consider the consequences of including in any linear programming problem the added requirement that the solution values of the variables be whole numbers; this makes the problem a *discrete* or *integer* program. We only examine *linear* programming problems that include this additional requirement.

In all the preceding examples of Chapters 7 and 8 we stopped the simplex calculations when the criterion on top-row elements indicated that an optimum had been reached. We accepted, for example, in Table 7.7, the result that one or more of the x_i^* were not whole numbers. Clearly there are situations in which a noninteger result is not disturbing—for

TABLE 8.15 Optimal Assignments for the Problem in (8.17)

	M_1	M_2	M_3	M_4	M_5	M_N
W_1	0	0	0	0	1	0
W_2	0	0	0	0	0	1
W_3	0	1	0	0	0	0
W_4	0	0	0	1	0	0
W_5	1	0	0	0	0	0

example, if x_1^* and x_2^* are dimensions of a rectangular field, then to find that $x_1^* = 2.5$ and $x_2^* = 3.75$ is completely useful information. If the optimal number of cases of product j that should be produced during the coming quarter, x_j^*, turns out to be 1234.2, one is not too uncomfortable (because the relative error is so small) in rounding to 1234, or perhaps even to 1230.[4] Suppose, however, that x_j^* represents the optimal number of dams to build above a certain point on a river for hydroelectric generation purposes, and that in a particular problem $x_j^* = 2.589$. Now clearly the *relative* error involved in rounding (to either 2 or 3) is not small, and one would be far more comfortable if the problem could be *required* from the beginning to generate a whole-number answer for x_j^*.

Sometimes, then, noninteger answers are acceptable because the units of measurement (feet and dollars) are easily interpreted in fractions, at least up to some point. In other cases, noninteger answers are tolerable because the relative error incurred by rounding is small.[5] Sometimes, however, only whole numbers make sense. This was true of the dam example and is generally true of problems in which optimal solution values are relatively small and represent highly indivisible units. It is also the case for other classes of problems. We saw that many forms of the assignment problem (Section 8.5) only made sense if the variables took on values of 0 or 1 only. *Capital budgeting* problems also require 0–1 variables. Here there are upper limits on the amount of financing available for several projects. Letting $x_j = 0$ when project j is not undertaken (financed) and $x_j = 1$ if that project does receive financing, the objective often is to maximize some aggregate benefit from the projects, $\Sigma_{j=1}^n b_j x_j$ (where b_j is a measure of the benefit from project j) subject to a budget constraint $\Sigma_{j=1}^n c_j x_j \le B$, where c_j is the cost of project j and B represents the total budget amount available for all of the projects. It also turns out that several kinds of nonlinear problems can be solved by the addition of integer-valued variables.

8.6.1   The Geometry of Integer Programs

A simple two-variable example will illustrate the particulars of the integer programming problem and indicate (at least geometrically) what the solution procedure must accomplish. When only whole-number nonnegative solution values of x_1 and x_2 are allowed, the character of the feasible region alters drastically, since, within the bounds set by the constraints, only the integer points (*lattice points*) are feasible. Thus the majority of the space enclosed by the constraints becomes infeasible. Figure 8.5 illustrates a feasible region in the noninteger as well as integer cases.

It is clear from Figure 8.5b that the fundamental and extremely useful linear programming property that the optimum is always at a corner of the feasible region defined by the linear inequalities now no longer holds. In general the integer optimum will be *interior* to this feasible region. (In some cases, but not in general, the simplex arithmetic of Chapter 7 will lead to an optimal solution that happens also to be all integer.) By adding objective function contours to Figure 8.5b, we can also see in general why rounding—downward, in this example, so as not to violate the constraints—is unsatisfactory. Now consider Figure

---

[4]The accuracy of a problem's input data—the a_{ij}'s, b_i's, and p_j's—is obviously relevant. Extremely precise solutions are not very meaningful in a problem constructed from relatively imprecise data.

[5]Some computer programs for solving integer problems impose an upper bound (in one case, $2^8 = 256$) on the value that any integer variable can have. This is done because of the enormous amount of storage space that some integer programming solution routines need. The reasoning is that with variables whose values can be that large, rounding noninteger answers will not cause much error.

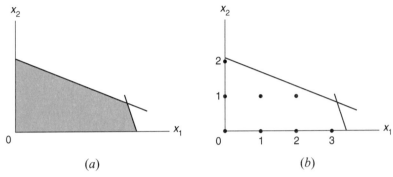

FIGURE 8.5 (*a*) Feasible region, noninteger problem (shaded area); (*b*) feasible region, integer problem (black dots).

8.6. The optimal noninteger solution is approximately $\mathbf{X}^* = \begin{bmatrix} 3.25 \\ 0.75 \end{bmatrix}$. Rounding to the next lowest integers would give $\mathbf{X}^R = \begin{bmatrix} 3 \\ 0 \end{bmatrix}$, which is clearly inferior—in terms of the objective function value—to $\mathbf{X}^i = \begin{bmatrix} 0 \\ 2 \end{bmatrix}$. What is needed, therefore, is an algebraic procedure for moving from a noninteger optimum such as $\mathbf{X}^* = \begin{bmatrix} 3.25 \\ 0.75 \end{bmatrix}$ to the associated integer-valued optimum, $\mathbf{X}^i$, when we do not have the luxury of an accurate geometric picture in front of us.

8.6.2 Complete Enumeration

It might seem that a process of *complete enumeration* would be a reasonable computational approach to integer linear programming problems. For example, in a problem with two unknowns, suppose that the constraints are $x_1 \le 6$, $x_2 \le 5$ and $x_1, x_2 \ge 0$. This means that x_1 must be one of the integers $0,1,2, \ldots, 6$ and x_2 must be one of the integers $0,1,2, \ldots,$ 5. If there are no further constraints, then it is easy to see that the optimal solution must be one of 42 [= (6+1)(5+1)] lattice points. In general, if $x_1 \le N_1$ and $x_2 \le N_2$ (and $x_1, x_2 \ge 0$),

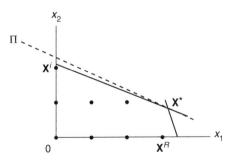

FIGURE 8.6 Noninteger versus integer optimum.

where N_1 and N_2 are integers, there are $(N_1 + 1)(N_2 + 1)$ candidate lattice points representing allowable integer values for the two variables.

For modern computers, identification of each of the feasible lattice points and evaluation of the linear objective function at each of them would not seem to be much of a task, provided the number of such points is "reasonable." And that is exactly where the trouble lies; in many real-world problems the number of integer points will not be "reasonable" at all. Here are some examples of the way in which the number of points grows alarmingly:

1. If each x_i in a problem ($i = 1, \ldots, n$) is confined to the set for which $x_i \geq 0$, $x_i \leq 1$ and integer (a 0–1 problem), then there are 2^n lattice points; denote the number of lattice points by L. Then, for 5 variables, $L = 32$; for 10 variables, $L = 1024$; for 20 variables, $L = 1,048,576$; for 30 variables, $L = 1,073,741,824$; and with 50 variables, $L = 1.13 \times 10^{15}$ (1.13 quadrillion). At some point, the number of evaluations becomes immense.

2. Most integer programs are not 0–1 problems. In a problem in which each x_i ($i = 1, \ldots, n$) is confined to the set for which $x_i \geq 0$ and $x_i \leq N_i$, the total number of feasible integer lattice points is $\Pi_{k=1}^{n} (N_k + 1)$. (The symbol Π is shorthand for "the product of" just as Σ represents "the sum of.") To simplify the illustrations, suppose that all unknowns have the same upper bound, which can be denoted simply by N; $N_1 = \cdots = N_n = N$. Table 8.16 shows the total number of feasible lattice points for a few representative values of n (the number of unknowns) and N (the integer upper bound for each of the unknowns).

These numbers indicate the way in which the number of eligible lattice points grows astonishingly with the number of unknowns and/or the number of integer values that each variable can take. Of course, there may be additional constraints that will cut away at the set of feasible lattice points. For example, in a two-variable problem, with $x_1 \leq 6$, $x_2 \leq 10$, and nonnegativity, there are 77 eligible lattice points; if there is, in addition, a constraint requiring $x_1 + 2x_2 \leq 12$, then the number of feasible lattice points is reduced by about 39%, to 47. And additional constraints (as long as they are not redundant) would reduce the number of relevant lattice points even further. However, as we will see when we convert some of the results from Table 8.16 into computation time, even drastic reduction because of additional constraints may not be very helpful at all.

Suppose, for the sake of illustration, that it were possible to do *one billion* lattice point identifications and objective function evaluations *per second*. In Table 8.17 the results from Table 8.16 are converted into estimated computation times; a "–" indicates computa-

TABLE 8.16   The Number of Feasible Lattice Points for Problems of Alternative Sizes[a]

Number of Variables (n)	Upper Limit on Each Variable (N)					
	2	4	6	8	10	20
2	9	25	49	81	121	441
3	27	125	343	729	1,331	9,261
4	81	625	2,401	6,561	14,641	194,481
5	243	3,125	16,807	59,049	161,051	*4.0m*
10	59,049	*9.7m*	*282.4m*	*3.4b*	*25.9b*	*1.6×10^{13}*
15	*14.3m*	*30.5b*	*4.7×10^{12}*	*2.0×10^{14}*	*4.1×10^{15}*	*6.8×10^{19}*
20	*3.4b*	*9.5×10^{13}*	*7.9×10^{16}*	*1.2×10^{19}*	*6.7×10^{20}*	*2.7×10^{26}*

[a]"*m*" denotes million (10^6), "*b*" denotes billion (10^9) and numbers in italics are rounded *downward*.

TABLE 8.17 Estimates of Computation Times for Lattice Points in Table 8.16^a

Number of Variables (n)	Upper Limit on Each Variable (N)					
	2	4	6	8	10	20
2	—	—	—	—	—	—
3	—	—	—	—	—	—
4	—	—	—	—	—	—
5	—	—	—	—	—	—
10	—	—	—	3.4 s	25.9 s	4.4 h
15	—	30.5 s	78.3 min	55.6 h	47.5 days	2.1×10^3 years
20	3.4 s	26.4 h	2.5 years	3.8×10^2 years	2.1×10^4 years	8.5×10^9 years

^{a}Estimates based on *1 billion* lattice point identifications, objective function evaluations and comparisons per second. All numbers have been rounded downward.

tion time less than one second. (In reading the results in the lower right part of the table, it is worth remembering that 10^2 years is one century!)

Even if there are additional constraints, cutting down the feasible region considerably, they are not likely to be of much help, in general. For example, with 15 unknowns, each with an upper integer limit of 10, if additional constraints on the x's cut away *99%* of the feasible lattice points in Table 8.16, this would still amount to 4.1×10^{13} points, with an estimated calculation time of 1.3 years—not an encouraging prospect.

Again, the point of the illustrations in Tables 8.16 and 8.17 is not the *exact* values of the lattice point counts or the estimated computation time (based on the assumption of 1 billion evaluations per second). Rather, the idea is to illustrate that complete enumeration approaches will not be reasonable for any but the smallest of problems.

8.6.3 Gomory's Cutting-Plane Method

Algebra and Geometry In Figure 8.7 we have reproduced the feasible region from Figure 8.5b with the added dotted lines that connect the "outermost" feasible lattice points. The convex region formed by the x_1 and x_2 axes (as lower bounds) and these dotted lines (as upper bounds) has two useful properties: (1) it includes, on its boundaries or in its interior, all the feasible lattice points from Figure 8.5b, and (2) each of its extreme points is also a lattice point. Hence (and this is generally true, but we do not prove it) the optimal

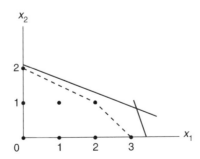

FIGURE 8.7 Reduced feasible region for integer problem.

integer solution to the problem whose inequality constraints define the boundaries in Figure 8.7 will occur at an extreme point of the feasible region whose boundaries are the dotted lines in the same figure. Thus, a solution method for integer programs would be one that removed, through the addition of *constraints*, the "excess" feasible region—the area between the solid and the dotted boundaries. This is, in effect, what we do; the procedure was developed by R. E. Gomory and is known as the Gomory method for integer programming problems.[6] The added constraints are often called "cutting lines" or, more generally, with more unknowns and hence more dimensions in solution space, "cutting planes."

We will consider here only the *all-integer* linear programming problem, in which all x_j^* are required to be whole numbers. A natural variation is the *mixed-integer* linear programming problem in which only certain of the variables must be integer at the optimum. You might visualize how the geometry in Figure 8.7 would be altered if, for example, x_1 were required to be integer-valued but x_2 were not. Most integer programming computer software deals equally well with an all-integer or a mixed-integer problem. A further complication would be to require all (or some) integer results in a *nonlinear programming problem*. Nonlinear programs are the subject matter of Section V (Chapters 10 and 11), and we do not pursue the matter further here (or later, either).

The first step in solving a linear program with the added requirement that all solution values be whole numbers is to solve the problem as if the integer requirement did not exist. It is always possible that the solution will happen to be all integers; if so, your work is finished. If not, one or more new constraints must be added. We now indicate how this is done.

Consider a maximization problem that has generated the optimal noninteger simplex table shown in (8.18). The two basic (and optimal) variables are labeled x_1^* and x_2^* for notational convenience; they represent any combination of original and slack variables—except, presumably, both slacks, which would represent the origin and a rather uninteresting maximum. Similarly, x_1^{nb} and x_2^{nb} represent the two (nonbasic) variables that are excluded from the optimal basis. Since the optimal values are $x_1^* = a_{10}$ and $x_2^* = a_{20}$, either a_{10} or a_{20} (or both) must be noninteger; otherwise the optimum would be integer-valued and the integer programming problem would be solved.

		x_1^{nb}	x_2^{nb}
Π	a_{00}	a_{01}	a_{02}
x_1^*	a_{10}	a_{11}	a_{12}
x_2^*	a_{20}	a_{21}	a_{22}

(8.18)

For illustration, assume that only a_{10} is fractional. Recall the way in which information is read from any row of any table in the simplex calculations. Each line expresses an included (basic) variable as a linear function of the excluded (nonbasic) variables and a constant. For example, from the x_1 line in (8.18)

---

[6]The earliest Gomory reference seems to be Gomory (1958), which was reprinted in Graves and Wolfe (1963).

$$x_1^* = a_{10} + a_{11}x_1^{nb} + a_{12}x_2^{nb} \tag{8.19}$$

or, rearranging, we obtain

$$a_{10} = x_1^* - a_{11}x_1^{nb} - a_{12}x_2^{nb} \tag{8.20}$$

Since it is the existence of a fractional part in a_{10} that spoils the solution, we consider the two aspects of a_{10}—its whole-number, or integer, part, I_{10}, and its nonnegative fractional part, f_{10}.[7] In fact, we break up each of the coefficients, a_{1j}, in the x_1 row, into an integer part (I_{1j}) and a nonnegative fractional part (f_{1j}).[8] So (8.20) can be expressed (where the negative signs are absorbed, where necessary, into the I_{ij})

$$(I_{10} + f_{10}) = x_1^* + (I_{11} + f_{11})x_1^{nb} + (I_{12} + f_{12})x_2^{nb} \tag{8.21}$$

Subtracting I_{10} from both sides of (8.21) and rearranging, we have

$$f_{10} = x_1^* + (f_{11}x_1^{nb} + f_{12}x_2^{nb}) + (I_{11}x_1^{nb} + I_{12}x_2^{nb}) - I_{10} \tag{8.22}$$

If we let

$$F = f_{11}x_1^{nb} + f_{12}x_2^{nb}$$

then (8.22) is

$$f_{10} - F = x_1^* + I_{11}x_1^{nb} + I_{12}x_2^{nb} - I_{10} \tag{8.23}$$

Since we are concerned with linear programming problems in which all x_j are required to be integer, all slack variables are also integers. The reason is that any of the original constraints in the problem that involved fractional a_{ij}'s or a fractional right-hand side, b_i, can be converted to all-integer a_{ij} and b_i by multiplication of both sides of the inequality by the lowest common denominator for the fractional parts. Thus, whether x_1^* represents an original (x_j) or a slack (s_i) variable, an all-integer solution will make the right-hand side of (8.23) an integer. Then, of course, the left-hand side of (8.23) must also be an integer. But since $0 \le f_{10} < 1$ and $F \ge 0$ (by definition), the left-hand side of (8.23) can be an integer *only if*

$$F \ge f_{10} \tag{8.24}$$

[This is a *necessary* condition for the left-hand side of (8.23) to be a whole number; it is not specific enough to also be *sufficient*.]

It is precisely from (8.24) that we form an additional constraint for the problem, to be added to the optimal simplex table in (8.18) to generate further pivoting. Replacing F by its definition in terms of the x_i^{nb}'s, (8.24) is

$$\sum_{i=1}^{2} f_{1i}x_i^{nb} \ge f_{10} \tag{8.25}$$

---

[7]Since a_{10} is always nonnegative, so is f_{10}.

[8]If an a_{ij} ($j \ne 0$) is negative, say, $-3\frac{1}{4}$, it can be thought of as $-4 + \frac{1}{4}$, so its integer part is -4 and its *nonnegative* fractional part is $\frac{1}{4}$.

If we convert this to an equality, by (as usual) subtracting a nonnegative surplus variable from the left-hand side, we have

$$\sum_{i=1}^{2} f_{1i}x_i^{nb} - s_1^G = f_{10} \tag{8.26}$$

(where "s_1^G" reminds us that this is the first Gomory constraint). Finally, as is usual in simplex arithmetic, we express the new surplus variable as a linear function of the nonbasic variables:

$$s_1^G = -f_{10} + \sum_{i=1}^{2} f_{1i}x_i^{nb} = -f_{10} + f_{11}x_1^{nb} + f_{12}x_2^{nb} \tag{8.27}$$

This becomes an additional constraint that must hold if an integer solution is found; it is in exactly the correct form to add to (8.18). This is done here:

		x_1^{nb}	x_2^{nb}
Π	a_{00}	a_{01}	a_{02}
x_1^*	a_{10}	a_{11}	a_{12}
x_2^*	a_{20}	a_{21}	a_{22}
s_1^G	$-f_{10}$	f_{11}	f_{12}

$$\tag{8.28}$$

Since $-f_{10}$ is negative, the primal solution in (8.28) is now infeasible. Pivoting on a *positive* element in the s_1^G row will remove this infeasibility, since the simplex rules require dividing *pivot-row* elements by the negative of the pivot element. Thus the pivot-row selection has been made first. The ordinary simplex method decisions of Chapter 7 have been reversed.

It appears that (8.28) has been viewed as if turned on its side; it is just that, and the procedure is in fact called the *dual simplex method*. The pivot column is therefore that one, among those with *positive* elements in the pivot row, for which $|a_{0j}/a_{ij}|$ is minimum.[9] Then pivoting follows exactly the rules of Chapter 7. The new simplex table (after pivoting) may still contain noninteger solution values—in the leftmost column. If so, another Gomory constraint is added, with coefficients derived exactly as described above. In general, if there are several noninteger a_{i0}, a good rule seems to be to select from among those associated with x_j variables (and not slacks) that one with the largest fractional part.

---

[9]This prevents us from destroying dual feasibility by turning a top-row coefficient positive. (Recall the negative signs that are attached to the top-row labels when reading the dual solution.) The (primal) simplex method of Chapter 7 maintains *primal* feasibility at all times and pivots until dual feasibility is also achieved (negative top-row elements). The dual simplex method maintains dual feasibility and pivots until primal feasibility is achieved.

Numerical Illustration of Gomory's Method   Suppose the problem is to

$$\text{Maximize} \quad 2x_1 + x_2$$

$$\text{subject to} \quad 3x_1 + x_2 \leq 7$$

$$3x_1 + 4x_2 \leq 12 \tag{8.29}$$

$$\text{and} \quad x_1, x_2 \geq 0 \text{ and integer}$$

You should work through the two pivot steps that are necessary to produce the optimal noninteger simplex result indicated in Table 8.18.

This optimal solution, with $\mathbf{X}^* = \begin{bmatrix} 1\frac{7}{9} \\ 1\frac{2}{3} \end{bmatrix}$, clearly does not satisfy the integer require-

ment on the variables. Therefore we introduce a new (Gomory) constraint; we could choose either the x_1 or the x_2 row from Table 8.18. Using the rule of thumb mentioned above (this example will illustrate that this is not always the best move) would create

$$x_1^* = 1\tfrac{7}{9} - \tfrac{4}{9}s_1 + \tfrac{1}{9}s_2 \tag{8.19'}$$

or, rearranging

$$1\tfrac{7}{9} = x_1^* + \tfrac{4}{9}s_1 - \tfrac{1}{9}s_2 \tag{8.20'}$$

Extracting nonnegative fractional parts, we obtain

$$(1 + \tfrac{7}{9}) = x_1^* + (0 + \tfrac{4}{9})s_1 + (-1 + \tfrac{8}{9})s_2 \tag{8.21'}$$

and, rearranging

$$\tfrac{7}{9} = x_1^* + (\tfrac{4}{9}s_1 + \tfrac{8}{9}s_2) + [0s_1 + (-1)s_2] \tag{8.22'}$$

Therefore the Gomory constraint to be added is

$$\tfrac{4}{9}s_1 + \tfrac{8}{9}s_2 \geq \tfrac{7}{9} \tag{8.25'}$$

TABLE 8.18   Optimal Noninteger Simplex Table for the Problem in (8.29)

		s_1	s_2
Π	$5\frac{2}{9}$	$-\frac{5}{9}$	$-\frac{1}{9}$
x_1	$1\frac{7}{9}$	$-\frac{4}{9}$	$\frac{1}{9}$
x_2	$1\frac{2}{3}$	$\frac{1}{3}$	$-\frac{1}{3}$

With a nonnegative surplus variable, s_1^G, subtracted from the left-hand side (and rearranged), we have

$$s_1^G = -\tfrac{7}{9} + \tfrac{4}{9}s_1 + \tfrac{8}{9}s_2 \tag{8.27'}$$

In (8.27') the new surplus variable has been expressed as a linear combination of the two excluded variables in the optimal noninteger solution (Table 8.18) and a constant. Thus the new constraint can be added directly to that table, as is done here:

		s_1	s_2
Π	$5\tfrac{2}{9}$	$-\tfrac{5}{9}$	$-\tfrac{1}{9}$
x_1	$1\tfrac{7}{9}$	$-\tfrac{4}{9}$	$\tfrac{1}{9}$
x_2	$1\tfrac{2}{3}$	$\tfrac{1}{3}$	$-\tfrac{1}{3}$
s_1^G	$-\tfrac{7}{9}$	$\tfrac{4}{9}$	$\tfrac{8}{9}*$

(8.30)

Since all that we need are the *coefficients* relating s_1^G to s_1 and s_2 and a constant term, the entire derivation can be done in simple tabular form, showing only the coefficients. From the x_1 row in Table 8.18, we obtain

$$1\tfrac{7}{9} \quad -\tfrac{4}{9} \quad \tfrac{1}{9} \tag{8.19''}$$

Now change the signs of all elements except the leftmost

$$1\tfrac{7}{9} \quad \tfrac{4}{9} \quad -\tfrac{1}{9} \tag{8.20''}$$

and find the nonnegative fractional parts:

$$\tfrac{7}{9} \quad \tfrac{4}{9} \quad \tfrac{8}{9} \tag{8.22''}$$

Finally, change the sign of the leftmost element:

$$-\tfrac{7}{9} \quad \tfrac{4}{9} \quad \tfrac{8}{9} \tag{8.27''}$$

These are exactly the coefficients for the s_1^G row, as in (8.27'), which is added to Table 8.18 to produce (8.30).

Pivoting *must* be done in the bottom row. Among positive elements in this row, examine the absolute values of a_{0j}/a_{ij}, and select as pivot column the one for which this ratio is smallest. In this case, it is in the s_2 column; hence the lower right element has been starred and must act as the pivot. Using the rules of Chapter 7, pivoting produces the following simplex table:

		s_1	s_1^G
Π	$5\frac{3}{24}$	$-\frac{1}{2}$	$-\frac{1}{8}$
x_1	$1\frac{21}{24}$	$-\frac{1}{2}$	$\frac{1}{8}$
x_2	$1\frac{9}{24}$	$\frac{1}{2}$	$-\frac{3}{8}$
s_2	$\frac{7}{8}$	$-\frac{1}{2}$	$\frac{9}{8}$

$$(8.31)$$

where $x_1 = 1\frac{21}{24}$ and $x_2 = 1\frac{9}{24}$, so at least one further Gomory constraint must be introduced into the problem. Working again with the row with the larger fractional part, we choose the x_1 row. Following the abbreviated derivation above, we start with

$$1\tfrac{21}{24} \quad -\tfrac{1}{2} \quad \tfrac{1}{8}$$

Then

$$1\tfrac{21}{24} \quad \tfrac{1}{2} \quad -\tfrac{1}{8}$$

and

$$\tfrac{21}{24} \quad \tfrac{1}{2} \quad \tfrac{7}{8}$$

lead to

$$-\tfrac{21}{24} \quad \tfrac{1}{2} \quad \tfrac{7}{8}$$

and so the starting simplex table for the next iteration will be

		s_1	s_1^G
Π	$5\frac{3}{24}$	$-\frac{1}{2}$	$-\frac{1}{8}$
x_1	$1\frac{21}{24}$	$-\frac{1}{2}$	$\frac{1}{8}$
x_2	$1\frac{9}{24}$	$\frac{1}{2}$	$-\frac{3}{8}$
s_2	$\frac{7}{8}$	$-\frac{1}{2}$	$\frac{9}{8}$
s_2^G	$-\frac{21}{24}$	$\frac{1}{2}$	$\frac{7}{8}$*

$$(8.32)$$

The pivot element has been starred, and you should work through one more pivot operation to find the final solution as shown in Table 8.19, where the optimal integer solution is identified as $\mathbf{X}^* = \begin{bmatrix} 2 \\ 1 \end{bmatrix}$, at which point $\Pi^* = 5$. Note in particular that in this example

TABLE 8.19   Optimal Integer Simplex Table for the Problem in (8.29)

		s_1	s_2^G
Π	5	$-\frac{3}{7}$	$-\frac{1}{7}$
x_1	2	$-\frac{4}{7}$	$\frac{1}{7}$
x_2	1	$\frac{5}{7}$	$-\frac{3}{7}$
s_2	2	$-\frac{8}{7}$	$\frac{9}{7}$
s_1^G	1	$-\frac{4}{7}$	$\frac{8}{7}$

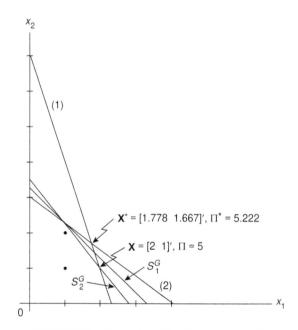

FIGURE 8.8   Geometry of the Gomory constraints.

the optimal integer solution would not have been found by simply rounding downward from the optimal noninteger solution in Table 8.18.[10]

Figure 8.8 indicates the geometry of this problem, including the two added Gomory constraints. Note that to include these in a solution space figure, with x_1 and x_2 on the axes, we must reexpress these added constraints as linear functions of x_1 and x_2. The first new constraint [in (8.25′)] was

$$\tfrac{4}{9}s_1 + \tfrac{8}{9}s_2 \geq \tfrac{7}{9}$$

But from the original problem, we know that $s_1 = 7 - 3x_1 - x_2$ and $s_2 = 12 - 3x_1 - 4x_2$.

---

[10]You might also want to investigate what happens if the initial Gomory cutting plane is built from the information in the x_2 row of Table 8.18 rather than the x_1 row, as in the example. It happens that in that case only one Gomory constraint is needed, so the rule that advises us to use the row for the variable with the largest fractional part in its solution is not necessarily always the most efficient. But it is not easy to devise rules that will *always* give the best advice about which row to choose.

Straightforward substitution and algebra show that this new constraint is really $4x_1 + 4x_2 \leq 13$. Similarly, the second Gomory constraint [as in the bottom row of (8.32)] was

$$\tfrac{1}{2}s_1 + \tfrac{7}{8}s_1^G \geq \tfrac{21}{24}$$

Again, substitution [this time for s_1 and s_1^G (from (8.27′)] will give $5x_1 + 4x_2 \leq 14$. These are the s_1^G and s_2^G constraints shown in Figure 8.8.

As a brief digression, we note that imposition of one or more integer requirements on a primal problem influences the character of the associated dual. Since each primal constraint has an associated dual variable, it is clear that the requirement of an integer solution to the primal will add variables to the dual problem. For example, the noninteger solution shown in Table 8.18 gives $y_1^* = \tfrac{5}{9}$ and $y_2^* = \tfrac{1}{9}$, because both constraints are binding in the optimal noninteger solution. However, in the integer solution, $y_1^* = \tfrac{3}{7}$ and $y_2^* = 0$ (it is absent from the dual optimal solution in Table 8.19, because in the integer solution the second constraint is no longer binding). Moreover, we now have $s_2^G = \tfrac{1}{7}$, a value of the dual variable associated with the second added Gomory constraint. These added dual variables, and the fact that many primal constraints may be met as inequalities in the optimal integer solution and hence have zero-valued dual variables, present fundamental and interesting problems for the interpretation of the dual to an integer programming problem.

8.6.4 Branch-and-Bound Methods

An alternative to the Gomory approach to solving integer linear programming problems is known as the *branch-and-bound* technique. [The term seems to have originated with Little et al. (1963).] The theory that underpins the method is probably most easily understood with a numerical example; we use the problem in (8.29) again. In essence, the branch-and-bound procedure generates a treelike structure to identify and solve a set of increasingly constrained subproblems, derived from the original integer linear program. This approach identifies only a subset of the integer lattice points and selects the best from among them.

Problem I: Initial Branching on x_1 Recall that the optimal (noninteger) solution was $\mathbf{X}^* = \begin{bmatrix} 1.778 \\ 1.667 \end{bmatrix}$, where $\Pi^* = 5.222$. (In this example we will express fractions to three decimals only; that is adequate for the purposes of the illustration.) Consider x_1; an optimal *and integer* solution for x_1 cannot lie between $x_1 = 1$ and $x_1 = 2$, as does $x_1^* = 1.778$. So, in fact, it must be true that either $x_1 \leq 1$ or $x_1 \geq 2$. We can therefore divide the feasible region for this problem into two subregions, as defined by these two (new) constraints on x_1. This explicitly excludes any values of x_1 in the (open) interval $1 < x_2 < 2$ and is known as *branching* on x_1. Define these two new (sub)problems as I(A) and I(B). We consider them in turn.

Problem I(A) The geometry of Problem I(A) is shown in Figure 8.9; this is the original problem with the added constraint $x_1 \leq 1$. It is a new linear program, for which simplex methods would be entirely appropriate. In this small example, however, we can find the optimal solution easily from the geometry of the problem. It is at $\mathbf{X}_{I(A)}^* = \begin{bmatrix} 1 \\ 2.25 \end{bmatrix}$, where Π = 4.25 and where the value for x_2 violates the integer requirement. Therefore, from the point of view of this new problem, an optimal and integer solution for x_2 cannot lie between 2 and 3. Reasoning as above when we initially branched on x_1, it must now be true

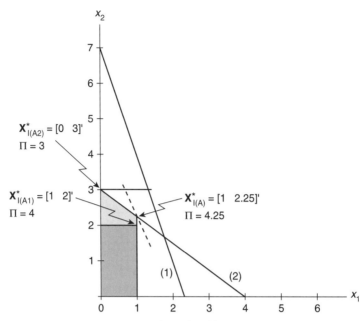

FIGURE 8.9   Branch-and-bound problems I(A), I(A1), and I(A2).

that either $x_2 \leq 2$ or $x_2 \geq 3$. At this point, within Problem I(A) defined by the initial branch on x_1, we create a second branch, now on x_2.

PROBLEM I(A) WITH A SECOND BRANCH, NOW ON x_2 We can denote the two new (sub)problems as I(A1) and I(A2), respectively. The geometry is indicated in Figure 8.9 also. Again, each of these problems could be solved in the usual way via the simplex method. Here the geometry makes the optimal solutions obvious:

Problem I(A1)   The optimal $\mathbf{X}$ is $\mathbf{X}^*_{I(A1)} = \begin{bmatrix} 1 \\ 2 \end{bmatrix}$, where $\Pi = 4$.

Problem I(A2)   Here the optimal solution is $\mathbf{X}^*_{I(A2)} = \begin{bmatrix} 0 \\ 3 \end{bmatrix}$, where $\Pi = 3$.

Problem I(B)   This is the original problem with the added constraint $x_1 \geq 2$. The geometry is shown in Figure 8.10, and the optimal solution is again obvious from the geometry. It is $\mathbf{X}^*_{I(B)} = \begin{bmatrix} 2 \\ 1 \end{bmatrix}$, where $\Pi = 5$. Since this optimal solution happens also to be an integer point, there is no need to impose additional bounds on either variable and to explore any further along this branch on x_1.

Figure 8.11 summarizes the treelike structure of the branch and bound approach to this problem, from the initial branch on x_1. There are nine lattice points in the feasible region:

$$(0,0) \; (0,1) \; (0,2) \; (0,3)$$

$$(1,0) \; (1,1) \; (1,2)$$

$$(2,0) \; (2,1)$$

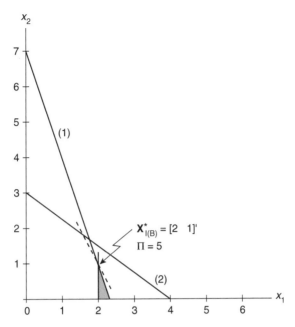

FIGURE 8.10 Branch-and-bound problem I(B).

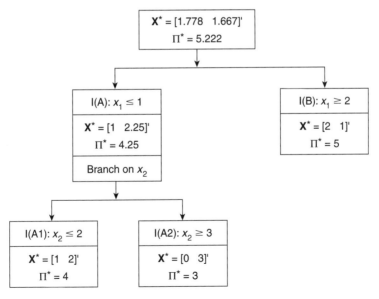

FIGURE 8.11 Branch-and-bound summary: initial branch on x_1. [*Note*: If Problem I(B) were solved first, then after finding the solution to Problem I(A), it would not be necessary to branch on x_2, since the optimal solution to Problem I(B) provides a better solution that is also integer.]

Note that only three of these nine are examined explicitly since, in this problem

(0,3) is superior to any other lattice point with $x_1 = 0$.
(1,2) is superior to any other lattice point with $x_1 = 1$.
(2,1) is superior to the only other lattice point with $x_1 = 2$.

This kind of efficient elimination of inferior lattice points is the strength of branch-and-bound techniques. There are many variations of this approach which attempt to incorporate clever ways for choosing the variables on which to branch in each part of the tree. However, the basic idea is adequately illustrated by this small example.

Problem II: Initial Branching on x_2 We could have initially branched on x_2 instead of x_1. Since the optimal noninteger solution for x_2 is 1.667, the two new subproblems are again defined through introduction of an upper bound of 1 and a lower bound of 2, but this time on x_2. Problems II(A) and II(B) contain the new constraints $x_2 \leq 1$ and $x_2 \geq 2$, respectively.

Problem II(A) The geometry for this problem is shown in Figure 8.12, and the solution is clearly $\mathbf{X}^*_{\text{II(A)}} = \begin{bmatrix} 2 \\ 1 \end{bmatrix}$, where, as we already know [from the solution to Problem I(B)], $\Pi = 5$.

Problem II(B) The geometry for this subproblem is shown in Figure 8.13; the solution is easily found, using the simplex method or the observation, from the geometry, that it is at the intersection of the equation $x_2 = 2$ and $3x_1 + 4x_2 = 12$, the boundary set by the the sec-

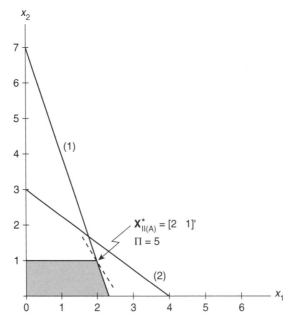

FIGURE 8.12 Branch-and-bound Problem II(A).

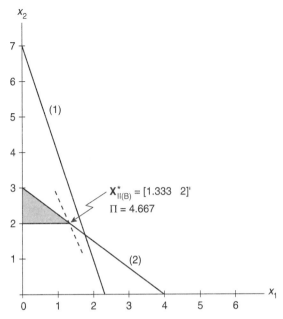

FIGURE 8.13   Branch-and-bound Problem II(B).

ond constraint. It is $\mathbf{X}^*_{\text{II(B)}} = \begin{bmatrix} 1.333 \\ 2 \end{bmatrix}$, where $\Pi = 4.667$. Since x_1 is not an integer at this

point, we would introduce two additional subproblems, branching now on x_1. However, exploration of these additional subproblems is unnecessary in this example, since we know that the optimal integer solutions to them will necessarily have an objective function value that is *no more than* the current noninteger solution, where $\Pi = 4.667$. (Adding binding constraints can never increase an objective function value.) And since we have already found, from Problem II(A), an integer solution with $\Pi = 5$, that value serves as a lower *bound* on our explorations along other branches. Whenever we reach an optimal noninteger solution with $\Pi \leq 5$, we can eliminate further exploration along that branch. That is the *bounding* part of the branch-and-bound technique; it has been suggested (Wagner, 1975, p. 488) that "branch and prune" might be more a more appropriate name.

For illustration only, we continue.

PROBLEM II(B) WITH A SECOND BRANCH, NOW ON x_1 Reasoning as before, we see that either $x_1 \leq 1$ or $x_1 \geq 2$. Call these problems II(B1) and II(B2), respectively. The geometry is in Figure 8.14.

Problem II(B1)   The solution to this new linear program is $\mathbf{X}^*_{\text{II(B1)}} = \begin{bmatrix} 1 \\ 2.25 \end{bmatrix}$, where Π
$= 4.25$. Since this solution now has a noninteger value for x_2, we next
should branch yet again on x_2, establishing two further subproblems within Problem II(B1), one with $x_2 \leq 2$ and the other with $x_2 \geq 3$.

Problem II(B1,1)   Add $x_2 \leq 2$ to Problem II(B1). Since we are currently on the main branch defined by $x_2 \geq 2$, the only possibility is $x_2 = 2$. In conjunction with $x_1 \leq 1$, the optimal solution is $\begin{bmatrix} 1 \\ 2 \end{bmatrix}$, where $\Pi = 4$.

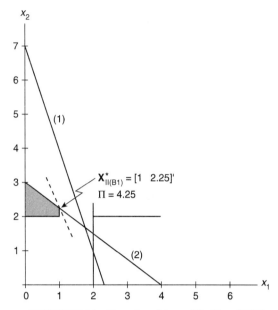

FIGURE 8.14   Branch-and-bound Problem II(B1).

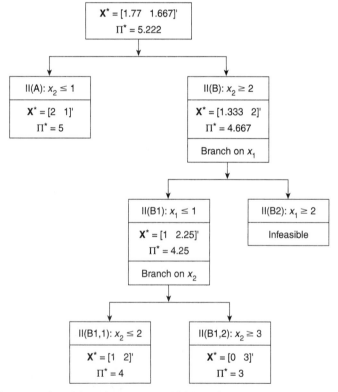

FIGURE 8.15   Branch-and-bound summary: initial branch on x_2. [*Note*: If Problem II(A) were solved first, then after finding the solution to Problem II(B), it would not be necessary to branch on x_1, since the optimal solution to Problem II(A) provides a better solution that is also integer.]

Problem II(B1,2)   Add $x_2 \geq 3$ to Problem II(B1). The only feasible (and hence also optimal) point that remains now is $\begin{bmatrix} 0 \\ 3 \end{bmatrix}$, where $\Pi = 3$.

Problem II(B2)   This is the problem formed when we add the requirement $x_1 \geq 2$ to Problem II(B). But Problem II, from the beginning, was identified with the constraint $x_2 \geq 2$. The geometry makes clear that these two requirements define an area that lies outside the feasible region established by the original constraints on the problem. Thus this particular subproblem has no feasible solution.

Figure 8.15 summarizes the structure of the branch-and-bound search from the initial branch on x_2. It is clear that the same three lattice points (out of a total of nine) have been isolated. In general, there will be different numbers of subproblems to be investigated depending on which variable is chosen for the initial branching. In a problem with many variables, it is not easy to know from the outset which branches are more efficient than others.

8.7   HIERARCHICAL OPTIMIZATION AND GOAL PROGRAMMING

In this section we explore some of the ways in which the linear programming framework has been extended to accommodate the fact that many real-world problems may not fit its fairly rigid structure. One complaint may be that at least some of the underlying relationships are inherently nonlinear. In that case, nonlinear programming models may be appropriate. These are the subject matter of Part V (Chapters 10 and 11).

Even in cases where linearity is not a bad assumption for the functions in a problem, there may be reasons to doubt the usefulness of the straightforward linear programming framework in Chapter 7. For example, there may be more than one potential objective function. As an illustration, in planning land use in a developing region it may be reasonable to want to try to maximize new employment generated and at the same time to minimize some measure(s) of environmental damage. One way out of this kind of dilemma is to formulate a pair of problems such as the following:

Problem 1 has an objective to maximize employment and a constraint that sets an upper limit on environmental impact (along with possibly many other constraints);

Problem 2 has an objective to minimize environmental impact and a constraint that sets a lower limit on employment generation (along with the same other constraints as in Problem 1).

One difficulty with this kind of approach is deciding on the right-hand sides of the environmental impact constraint in Problem 1 and the employment generation constraint in Problem 2. And if, as would generally be the case, the optimal solutions to the two problems are distinguishably different, then we are left with the issue of knowing which to prefer.[11]

Another approach to this *multiobjective* programming problem is available if the objectives can be *ranked* in order of preference. (That means, for Problems 1 and 2, above, the

---

[11]Some kinds of sensitivity analysis might be useful in cases like this. For example, changing the environmental and the employment targets in Problems 1 and 2, respectively, and finding new solutions might indicate some kinds of tradeoffs between the two objectives.

ability to identify whether employment generation or environmental protection is the more important objective. Note that it does not require that one specify *how much* more important one is than the other.) If such rankings are possible, then *hierarchical programming* (or *hierarchical optimization*) techniques can be used. We will examine how these work in this section.

Another complaint about the linear programming structure is the rigidity of the constraints. In real-world problems, the upper or lower bounds on the right-hand sides of constraints may be somewhat "loose"—more like ideal targets than absolutely definite figures. One approach is to introduce probabilistic reasoning into the programming structure, essentially specifying a probability for each constraint that the constraint be met. For example, for the ith constraint such a statement might be

$$\Pr\{a_{i1}x_1 + \cdots + a_{in}x_n \le b_i\} \ge 0.95$$

This says that the optimal $\mathbf{X}$ should be chosen so that the probability that constraint i is met is at least 95%. Each constraint can have a different probability associated with it, and in this way the "importance" of the constraints can be reflected—higher probabilities are attached to more important constraints. This kind of problem structure requires a different solution approach to deal with the fact that the constraints are expressed in this probabilistic way.

Alternatively, there is the approach of *goal programming*, in which a certain amount of "deviation" from the strict requirements of each constraint $a_{i1}x_1 + \cdots + a_{in}x_n \le b_i$ is possible. We will also look at this variation on the problem structure in this section.

8.7.1 Hierarchical Optimization: Structure

The assumption here is that there is more than one objective function and that they are conflicting in the sense that they have different optimal solutions. This means that the objective function contour lines are sloped in such a way that different corners of the feasible region are optimal for different objectives. It is also assumed that these objectives can be ranked in order of preference. If there are k objectives, they can always be numbered so that the ranking is

$$f^1(\mathbf{X}) \,\mathscr{P}\, f^2(\mathbf{X}) \,\mathscr{P}\, \cdots \,\mathscr{P}\, f^k(\mathbf{X})$$

where "$\mathscr{P}$" indicates "is preferred to."

The hierarchical approach can be applied equally well in a nonlinear programming situation. Our interest here is in illustrating its contribution in a linear programming framework. Assume that we represent the constraints by simply specifying that $\mathbf{X} \in F$ (read "$\mathbf{X}$ is an element of F," where F indicates the feasible region, including nonnegativity). The first in what will be a series of linear programming problems is then

Problem HO(1)

$$\text{Maximize} \quad f^1(\mathbf{X})$$

$$\text{subject to} \quad \mathbf{X} \in F$$

(where "HO" denotes hierarchical optimization). Let the solution to this problem be $\mathbf{X}^{*(1)}$.

The next problem has as its objective function the second most important objective, along with a new constraint that depends on the solution of Problem HO(1).

Problem HO(2)

$$\text{Maximize} \quad f^2(\mathbf{X})$$

$$\text{subject to} \quad \mathbf{X} \in F$$

$$\text{and} \quad f^1(\mathbf{X}) \geq \beta_1 f^1[\mathbf{X}^{*(1)}]$$

where $0 \leq \beta_1 \leq 1$.

The right-hand side of this new constraint represents the *minimum amount* of the original objective function optimal value, $f^1[\mathbf{X}^{*(1)}]$, that must be retained in the solution to Problem HO(2). For example, if $\beta = 0.85$, the constraint requires that the value of $f^1(\mathbf{X})$ in the new solution be no less than 85% of its optimal value for Problem HO(1). Put alternatively, $(1 - \beta)f^1[\mathbf{X}^{*(1)}]$ indicates the *maximum amount* of the original objective function optimal value that can be traded off in order to achieve something of the second objective. So β_1 is called a *tradeoff coefficient* between objective functions 1 and 2; it must be specified by the analyst, and that may or may not be easy to do.[12] The solution to this problem is $\mathbf{X}^{*(2)}$.

The next problem takes up the cause of the third most important objective.

Problem HO(3)

$$\text{Maximize} \quad f^3(\mathbf{X})$$

$$\text{subject to} \quad \mathbf{X} \in F$$

$$f^1(\mathbf{X}) \geq \beta_1 f^1[\mathbf{X}^{*(1)}]$$

$$\text{and} \quad f^2(\mathbf{X}) \geq \beta_2 f^2[\mathbf{X}^{*(2)}]$$

where $0 \leq \beta_i \leq 1$ ($i = 1,2$). Here β_2 represents the same kind of tradeoff coefficient, indicating the least amount of $f^2[\mathbf{X}^{*(2)}]$ that must be retained as one attempts to address less important objectives.

Finally, then, we have the last linear programming problem in the hierarchy.

Problem HO(k)

$$\text{Maximize} \quad f^k(\mathbf{X})$$

$$\text{subject to} \quad \mathbf{X} \in F$$

$$f^1(\mathbf{X}) \geq \beta_1 f^1[\mathbf{X}^{*(1)}]$$

$$f^2(\mathbf{X}) \geq \beta_2 f^2[\mathbf{X}^{*(2)}]$$

$$\vdots$$

$$\text{and} \quad f^{k-1}(\mathbf{X}) \geq \beta_{k-1} f^{k-1}[\mathbf{X}^{*(k-1)}]$$

where $0 \leq \beta_i \leq 1$ ($i = 1, \ldots, k - 1$).

---

[12]If the objective function in the first problem was minimized, then the added constraint would be $f^1(\mathbf{X}) \leq [1 + (1 - \beta_1)]f^1[\mathbf{X}^{*(1)}]$.

As an additional wrinkle, it is possible that the tradeoff coefficients for a particular objective might change over the sequence of programming problems. In Problem HO(2), β_1 was the proportion of the maximum possible value of objective function 1 that the analyst was unwilling to give up in the name of objective 2. Perhaps when it comes to objective 3, that coefficient would be higher (since objective 3 is judged less important than objective 2). In that case, we would have β_{12} in Problem HO(2), a different β_{13} in Problem HO(3), . . . , and β_{1k} in Problem HO(k).

8.7.2 Hierarchical Optimization: Numerical Illustration

We return to the job retraining example at the beginning of Chapter 7. Recall that x_1 and x_2 were numbers of students to admit to each of two job retraining programs in an urban area for the coming year; the first constraint was budgetary (funding from the state) and the second reflected the mayor's interest in expected increased revenue from city wage taxes. The problem was shown in (7.4) and is repeated here.

$$\text{Maximize} \quad f(\mathbf{X}) = 43.75x_1 + 75x_2$$
$$\text{subject to} \quad 5x_1 + 7x_2 \leq 2000$$
$$x_1 + 2x_2 \geq 460 \tag{8.33}$$
$$\text{and} \quad x_1, x_2 \geq 0$$

Figure 8.16 repeats the geometry for this problem (this was Figure 7.1b in the last chapter).

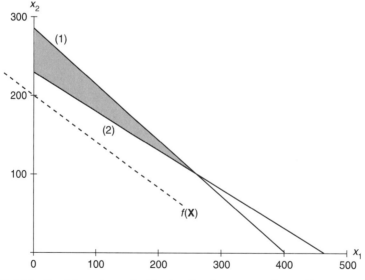

FIGURE 8.16   Feasible region and objective function contour for problem in (8.33) [Problem HO(1′)].

Although this problem was not solved in Chapter 7, you can easily find that $\mathbf{X}^* = \begin{bmatrix} 0 \\ 285.71 \end{bmatrix}$. This is a situation in which fractional solutions (students) do not make much sense. With variables whose values are numbers of this magnitude, the usual argument is that rounding (downward in this case, to maintain feasibility with the budget constraint) does not introduce "serious" error. If we do that, we have $\mathbf{X}^* = \begin{bmatrix} 0 \\ 285 \end{bmatrix}$ and $f(\mathbf{X}^*) = $ \$21,375. In fact, in this problem, the optimal integer programming solution is $\mathbf{X}^* = \begin{bmatrix} 2 \\ 284 \end{bmatrix}$, for which $f(\mathbf{X}^*) = \$21,387.50$, so the rounded solution produces an optimized objective function that is 0.06% less.[13]

Suppose, however, that the mayor sees the job retraining effort as having more than one objective. The first and most important objective function is that for sales tax revenue for the new sports arena which was reflected in the problem in (8.33): $f^1(\mathbf{X}) = 43.75x_1 + 75x_2$. Additionally, but less important than $f^1(\mathbf{X})$, the mayor would like to have as many enrollees in the two programs as possible; in the notation of hierarchical optimization problems, the second objective is to maximize $f^2(\mathbf{X}) = x_1 + x_2$. Finally, past experience has indicated that students in either program are likely to need individual tutoring during the course of the program. On average, Program 1 students need about 20 hours of tutoring and Program 2 students need about 10. This is undertaken by instructors and leaves them less time for their other obligations, so the mayor would like the expected number of tutoring hours to be as small as possible. The third objective, therefore, is to *minimize* $f^3(\mathbf{X}) = 20x_1 + 10x_2$.

Problem HO(1′)   This is just the problem in (8.33), for which, as we have seen, the optimal solution (rounded downward) is $\mathbf{X}^{*(1)} = \begin{bmatrix} 0 \\ 285 \end{bmatrix}$. Figure 8.16 indicates the boundaries of the feasible region and an objective function contour. For this solution, $f^1[\mathbf{X}^{*(1)}] = \$21,375$ (expected new sales tax revenue for the new stadium). We might also note that $f^2[\mathbf{X}^{*(1)}] = 285$ (total enrollment) and $f^3[\mathbf{X}^{*(1)}] = 2850$ tutoring hours.

Problem HO(2′)   In order to formulate the second of the series of hierarchical programs, we have to specify the maximum amount of $f^1[\mathbf{X}^{*(1)}] = \$21,375$ that might be given up in order to increase the number of students in the two programs. Suppose that $\beta_1 = 0.90$, meaning that the mayor is willing to give up no more than 10 percent of the \$21,375 in an effort to increase enrollment. So Problem HO(2′) is

$$\begin{aligned} \text{Maximize} \quad & f^2(\mathbf{X}) = x_1 + x_2 \\ \text{subject to} \quad & 5x_1 + 7x_2 \leq 2000 \\ & x_1 + 2x_2 \geq 460 \\ & 43.75x + 75x \geq 19237.50 \\ \text{and} \quad & x_1, x_2 \geq 0 \end{aligned} \qquad (8.34)$$

---

[13]Perhaps an entering class of two students is unrealistic. In that case, we might also need to have a minimum threshold size, say, 15, for each class. But that adds another complication, as long as either class could also have no students. Then x_1 or x_2 could be zero but if not, they have a lower bound of 15. Solving this kind of problem is a little tricky, and it goes beyond our interests here. But it can be tackled by proper transformation of the variables.

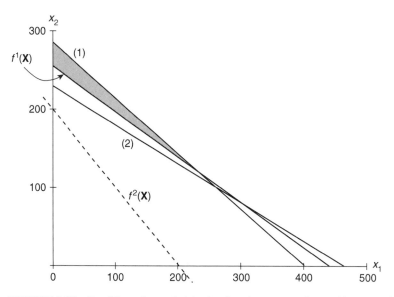

FIGURE 8.17 Feasible region and objective function contour for Problem HO(2').

The geometry is shown in Figure 8.17. The new constraint is simply one of the contour lines of the objective function from Problem HO(1'). Changing β_1 in (8.34) changes location of the new constraint, but not its slope. It is clear that at $\beta_1 = 1.00$, the contour will be exactly at the optimal solution from Problem HO(1'), and with no willingness to give up any of $f^1[(\mathbf{X}^{*(1)}]$ there is no possibility of considering the second objective at all. The new feasible region would simply be one point, $\mathbf{X}^{*(1)}$. As β_1 decreases, indicating increasing willingness to trade off objective 1 for objective 2, the constraint moves downward and the new feasible region increases.

The optimal solution to this problem is $\begin{bmatrix} 223.09 \\ 126.36 \end{bmatrix}$. Rounding downward, we will use

$\mathbf{X}^{*(2)} = \begin{bmatrix} 223 \\ 126 \end{bmatrix}$ with $f^2[\mathbf{X}^{*(2)}] = 349$. As might be expected because of the trade off with

the first objective function, total enrollment has increased [from 285 in the solution to Problem HO(1')]. The value of the third objective function, $f^3[\mathbf{X}^{*(2)}]$ (tutoring hours), is now 5720. This is much worse than $f^3[\mathbf{X}^{*(1)}]$ but that is irrelevant for now, since the tutoring hours objective is ranked lowest of the three. Tax revenue has dropped to $f^1[\mathbf{X}^{*(2)}] = 19,206.25$.

It is easily established that if the third constraint were absent from (8.34), so that we were solving the original problem with the budget and revenue constraints only, and with $f^2(\mathbf{X})$ as the objective function, the optimal solution, which happens to be integer, is at $\mathbf{X} = \begin{bmatrix} 260 \\ 100 \end{bmatrix}$. By trying alternative tradeoff coefficients in (8.34) you can find how much of $f^1[\mathbf{X}^{*(1)}]$ needs to be given up in order to achieve this best of all solutions for $f^2(\mathbf{X})$.[14]

---

[14]It turns out to be between 11% and 12%. You could also identify this figure by noting that at $\mathbf{X} = \begin{bmatrix} 260 \\ 100 \end{bmatrix}$, $f^1(\mathbf{X}) = 18,875$, which is 88.3% of $f^1[\mathbf{X}^{*(1)}] = 21,375$.

Problem HO(3′)   Suppose that the maximum allowable tradeoff on the second objective (enrolled students) is also 10 percent, so that $\beta_2 = 0.90$. Then the last program in the hierarchy is

$$\text{Minimize}\quad f^3(\mathbf{X}) = 20x + 10x_2$$

$$\text{subject to}\quad 5x_1 + 7x_2 \leq 2000$$

$$x_1 + 2x_2 \geq 460 \tag{8.35}$$

$$43.75x_1 + 75x_2 \geq 19237.50$$

$$x_1 + x_2 \geq 314.1$$

$$\text{and}\quad x_1, x_2 \geq 0$$

The geometry is indicated in Figure 8.18. As expected, the new constraint has added another side to the feasible region; this time it is a contour line of $f^2(\mathbf{X})$. Again its exact location depends on the right-hand side of the constraint; higher values of β_2 move the new side upward, shrinking the size of the feasible region. The optimal solution is $\begin{bmatrix} 99.35 \\ 214.75 \end{bmatrix}$, so we use $\mathbf{X}^{*(3)} = \begin{bmatrix} 99 \\ 214 \end{bmatrix}$. At this point, $f^1[\mathbf{X}^{*(3)}] = \$20{,}381.25$ [an increase over Problem HO(2′)], $f^2[\mathbf{X}^{*(3)}] = 313$ students enrolled and $f^3[\mathbf{X}^{*(3)}] = 4120$ tutoring hours.

8.7.3   An Alternative Hierarchical Optimization Approach

A different way of accommodating the hierarchical nature of a problem like this is to compress all the requirements into one larger linear programming problem.

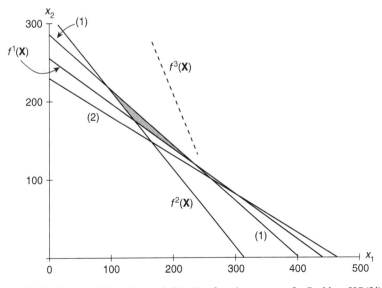

FIGURE 8.18   Feasible region and objective function contour for Problem HO(3′).

Objective Function   Consider each objective function alone, one at a time, in conjunction with only the original constraints of the problem. Denote the optimal values of the objective functions in each of these single-objective problems by $f^{1}*(\mathbf{X})$, $f^{2}*(\mathbf{X})$, $\ldots$, $f^{k*}(\mathbf{X})$.[15] For each $f^{i*}(\mathbf{X})$, calculate the absolute value of the *relative deviation* of any other value, $f^{i}(\mathbf{X})$, from it:

$$\frac{|[f^{i*}(\mathbf{X}) - f^{i}(\mathbf{X})]|}{f^{i*}(\mathbf{X})}$$

(Multiplied by 100, these are percentage deviations.) Then the objective in this new problem is to *minimize D*, the sum of these relative deviations of each objective function from its best possible optimum:

$$D = \sum_{i=1}^{k} \frac{|[f^{i*}(\mathbf{X}) - f^{i}(\mathbf{X})]|}{f^{i*}(\mathbf{X})}$$

The *absolute values* of the deviations are important, so that positive deviations for functions being maximized do not cancel out negative deviations for functions being minimized. There are straightforward ways to deal with *absolute value programs*, but it is also easy to build in the effect of taking absolute values by using the expressions

$$\frac{f^{i*}(\mathbf{X}) - f^{i}(\mathbf{X})}{f^{i*}(\mathbf{X})}$$

for *maximization* objectives, where we know that $f^{i*}(\mathbf{X}) \geq f^{i}(\mathbf{X})$ [no value of $f^{i}(\mathbf{X})$ can be larger than its "best" maximum] and

$$\frac{f^{j}(\mathbf{X}) - f^{j*}(\mathbf{X})}{f^{j*}(\mathbf{X})}$$

for *minimization objectives*, where we know equally well that $f^{j*}(\mathbf{X}) \leq f^{j}(\mathbf{X})$ [no value of $f^{j}(\mathbf{X})$ can be smaller than its "best" minimum]. Note that there will be trouble if any $f^{i*}(\mathbf{X}) = 0$; for that reason, this is a less-preferred hierarchical optimization approach.

Constraints   Thus far, no account has been taken of the ranking of the objective functions. We do that by introducing a new set of constraints specifying that the relative discrepancy for objective 1 must be less than that for objective 2, which in turn must be less than that for objective 3, and so on.

In addition, then, to the original constraints of the problem (which we continue to denote simply as $\mathbf{X} \in F$), we must build a representation of the hierarchical nature of the objectives. This is done in this version of a hierarchical optimization approach by specifying that

---

[15]Note that for f^{1} only it will always be true that $f^{1}*(\mathbf{X})$ is the same as $f^{1}[\mathbf{X}*^{(1)}]$, the optimal value of the objective function from Problem HO(1), above. In the solutions to Problems HO(2), $\ldots$, HO(k), the values $f^{2}[\mathbf{X}*^{(2)}]$, $\ldots, f^{k}[\mathbf{X}*^{(k)}]$ will generally be "less optimal" (smaller in maximization problems, larger in minimization problems) than their associated $f^{2}*(\mathbf{X}), \ldots, f^{k*}(\mathbf{X})$, because of the added constraints in Problems HO(2), $\ldots$, HO(k).

(a) The relative deviation of the first objective function must be strictly less than that of the second:

$$\frac{|[f^{1*}(\mathbf{X}) - f^1(\mathbf{X})]|}{f^{1*}(\mathbf{X})} < \frac{|[f^{2*}(\mathbf{X}) - f^2(\mathbf{X})]|}{f^{2*}(\mathbf{X})}$$

(b) The relative deviation of the second objective function must be strictly less than that of the third:

$$\frac{|[f^{2*}(\mathbf{X}) - f^2(\mathbf{X})]|}{f^{2*}(\mathbf{X})} < \frac{|[f^{3*}(\mathbf{X}) - f^3(\mathbf{X})]|}{f^{3*}(\mathbf{X})}$$

(c) Continuing in this way, the last of these hierarchical constraints requires that the relative deviation of the $(k-1)$st objective function be strictly less than that of the kth:

$$\frac{|[f^{(k-1)*}(\mathbf{X}) - f^{(k-1)}(\mathbf{X})]|}{f^{(k-1)*}(\mathbf{X})} < \frac{|[f^{k*}(\mathbf{X}) - f^k(\mathbf{X})]|}{f^{k*}(\mathbf{X})}$$

Compactly, then, all of these new constraints can be written as

$$\frac{|[f^{i*}(\mathbf{X}) - f^i(\mathbf{X})]|}{f^{i*}(\mathbf{X})} < \frac{|[f^{(i+1)*}(\mathbf{X}) - f^{(i+1)}(\mathbf{X})]|}{f^{(i+1)*}(\mathbf{X})}$$

for $i = 1, \ldots, (k-1)$. Again, the absolute-value problem can be taken care of by reversing the order of subtraction for *minimization* objectives. When written in that way, the absolute-value signs can be dropped.

Note that the constraints as shown contain strict inequalities. This is done to reflect the hierarchy of objectives. But the simplex method approach is built on the notion of a feasible region that includes its boundaries (the "=" part of a "≤" or "≥" constraint). An easy way out of this problem is to add a very small amount, say, 0.0001, to the left-hand sides of the constraints, and thereby convert them to weak inequalities.

8.7.4 Numerical Illustration of this Alternative Approach

For the three objectives in our numerical illustration, it is easily established that $f^{1*}(\mathbf{X}) = 21{,}375$, $f^{2*}(\mathbf{X}) = 360$, and $f^{3*}(\mathbf{X}) = 2300$.[16] In the objective function for this new version of the problem we take account of absolute-value requirements by changing the order of subtraction for f^3 (the minimization objective), and we multiply everything by 10,000, to get rid of the small numbers created by division by 21,375, 360, and 2300, respectively, in the three terms of the objective function. (Multiplication by 100 creates percentages from proportions; multiplication by 100 again just changes the scale so that not all data are to the right of the decimal point.)

---

[16]These are optimal values of the respective objective functions (rounded appropriately, when necessary) in three programming problems with only the two original constraints. The optima are, respectively: $\mathbf{X}^* = \begin{bmatrix} 0 \\ 285 \end{bmatrix}$ [from Problem HO(1′) and rounded, as we have seen], $\mathbf{X}^{**} = \begin{bmatrix} 260 \\ 100 \end{bmatrix}$ (without rounding, also as we have seen) and $\mathbf{X}^{***} = \begin{bmatrix} 0 \\ 230 \end{bmatrix}$ (without rounding).

In this case, then, we have

$$D = [10{,}000 - 20.468x_1 - 35.087x_2]$$
$$+ [10{,}000 - 27.778x_1 - 27.778x_2]$$
$$+ [86.956x_1 + 43.478x_2 - 10{,}000] \tag{8.36}$$

The three bracketed terms represent the percentage deviations of each of the three objective functions (multiplied by 100). Algebraically, this works out to

$$D = 10{,}000 + 38.720x_1 - 19.387x_2$$

and the constant plays no role in the optimization.

There will be two new constraints reflecting the hierarchy. Changing the order of subtraction for $f^3(\mathbf{X})$ and adding 0.0001 to the left-hand sides, to create weak inequalities, we have

$$\frac{[f^{1*}(\mathbf{X}) - f^1(\mathbf{X})]}{f^{1*}(\mathbf{X})} + 0.0001 \leq \frac{[f^{2*}(\mathbf{X}) - f^2(\mathbf{X})]}{f^{2*}(\mathbf{X})}$$

and

$$\frac{[f^{2*}(\mathbf{X}) - f^2(\mathbf{X})]}{f^{2*}(\mathbf{X})} + 0.0001 \leq \frac{[f^3(\mathbf{X}) - f^{3*}(\mathbf{X})]}{f^{3*}(\mathbf{X})}$$

In this problem, again multiplying by 10,000 to get rid of very small numbers, these work out to be

$$7.309x_1 - 7.309x_2 \leq -1$$

and

$$-114.734x_1 - 71.256x_2 \leq -20{,}001$$

Collecting all of this together, the alternative linear programming problem to deal with the three conflicting and ranked objective functions is given in (8.37). The geometry of this problem is sketched in Figure 8.19. Constraints 1 and 2 are the original constraints in the problem. Constraints 3 and 4 are entirely new and do not have a straightforward interpretation in terms of the original unknowns (numbers of students in each of the two programs).

$$
\begin{aligned}
\text{Minimize} \quad & D = 10{,}000 + 38.720x_1 - 19.387x_2 \\
\text{subject to} \quad & 5x_1 + 7x_2 \leq 2{,}000 \\
& x_1 + 2x_2 \geq 460 \\
& 7.309x_1 - 7.309x_2 \leq -1 \\
& -114.734x_1 - 71.256x_2 \leq -20{,}001 \\
& \text{and } \mathbf{X} \geq \mathbf{0}
\end{aligned}
\tag{8.37}
$$

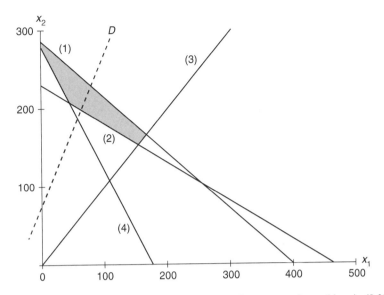

FIGURE 8.19 Feasible region and objective function contour for problem in (8.37).

The optimal solution is the same as for Problem HO(1). Rounded downward, it is $\mathbf{X}^* = \begin{bmatrix} 0 \\ 285 \end{bmatrix}$. The percentage deviations from each objective function's optimal values are zero for $f^1(\mathbf{X})$, 21% for $f^2(\mathbf{X})$ (285 vs. 360 enrollees), and 24% for $f^3(\mathbf{X})$ (2850 vs. 2300 tutoring hours). It is the sum of these three deviations that has been minimized, at 45%, in the problem in (8.37) (multiplied by 100).

Earlier solutions in the hierarchical optimization sequence of programs were $\mathbf{X}^{*(2)} = \begin{bmatrix} 223 \\ 126 \end{bmatrix}$, for which D in (8.37) is 162% (and the *order* of relative deviations is also wrong—the deviation for objective function 1 is worse than for objective function 2). At $\mathbf{X}^{*(3)} = \begin{bmatrix} 99 \\ 214 \end{bmatrix}$, the order is preserved but the total value (D) is 97%. Similarly, at $\begin{bmatrix} 0 \\ 230 \end{bmatrix}$ and $\begin{bmatrix} 260 \\ 100 \end{bmatrix}$ (the other corners of the original feasible region), the order is wrong (in both cases) and D is 55% and 181%, respectively.

8.7.5 Goal Programming: Structure

In the goal programming approach, conflicting objectives are turned into constraints—but constraints that are potentially "loose." For example, at the beginning of this section we suggested a land use problem in a developing region in which one objective is to maximize new employment generated by the development and another was to minimize some measure of environmental damage caused by the development. We noted that one problem in converting objectives to constraints was deciding on the right-hand sides. This is what goal programming addresses by suggesting that these numbers should be viewed as "goals" (or "targets") and that they may be treated in a way that introduces an element of flexibility.

Conflicting Objectives Become Constraints   Suppose that $f^1(\mathbf{X})$ is the linear function relating employment generation to projects chosen (the x's) and $f^2(\mathbf{X})$ is some other linear function relating environmental impact to those same projects. In addition, there may be other linear constraints that would have to hold in a problem with either of these objectives: maximize $f^1(\mathbf{X})$ subject to $\mathbf{X} \in C$ or minimize $f^2(\mathbf{X})$ over that same convex constraint set, $\mathbf{X} \in C$. Then the conversion from objectives to constraints proceeds as follows:

1. Decide for $f^1(\mathbf{X})$ on some "reasonably attractive" level of new employment; call it g_1 (for the goal for the first objective). Then, initially, we would want to require that $\mathbf{X}$ be chosen so that $f^1(\mathbf{X})$ was not lower than this goal; so $f^1(\mathbf{X}) \geq g_1$ becomes an additional constraint.

2. Similarly, decide for $f^2(\mathbf{X})$ on some also "reasonably admissible" level of environmental impact, g_2. Then $\mathbf{X}$ should be chosen subject to a constraint that this level not be exceeded, so $f^2(\mathbf{X}) \leq g_2$ is another new constraint.

3. To reflect the fact that we generally cannot hope to be extremely precise about the exact size of either g_1 or g_2, we allow for some possibility that $f^1(\mathbf{X})$ might actually be less than g_1 (which means that g_1 should have been lower than was specified) and that $f^2(\mathbf{X})$ might be more than g_2 (which means that the g_2 might have been higher than specified). This is accomplished by using "deviational variables." These are much like slack or surplus variables, except that there are two in each "loose" constraint.

In ordinary linear programming we would convert the employment objective turned constraint, $f^1(\mathbf{X}) \geq g_1$, to an equation by subtracting a nonnegative surplus variable from the left-hand side; $f^1(\mathbf{X}) - s_1 = g_1$, where $s_1 \geq 0$ represents the amount of employment *above* the floor established by g_1. Instead, in the goal programming approach, we define exactly this kind of surplus variable by $d_1^+ \geq 0$. This is called a *positive deviational* variable; it represents the amount by which $f^1(\mathbf{X})$ is *above* its target, g_1. Similarly, $d_1^- \geq 0$ is like a slack variable; it is a *negative deviational* variable for goal 1, representing the amount by which $f^1(\mathbf{X})$ is *below* its target, g_1.[17] So the employment inequality constraint becomes $f^1(\mathbf{X}) - d_1^+ + d_1^- = g_1$.

For any particular solution, it is not possible that both $d_1^+ \neq 0$ and $d_1^- \neq 0$ (although both may be equal to zero, if $f^1(\mathbf{X}) = g_1$). If there is a feasible solution, $\mathbf{X}$, in which $d_1^+ > 0$ and $d_1^- = 0$, this simply means that $f^1(\mathbf{X}) > g_1$ (inside the boundary set by g_1). On the other hand, if $d_1^- > 0$ and $d_1^+ = 0$, then $f^1(\mathbf{X}) < g_1$. The original "$\geq$" constraint is violated; this is the same as "relaxing" g_1 and allowing a lower value of employment. This is how the goal programming approach introduces flexibility into a constraint.

In the same way, d_2^+ and d_2^- are the pair of nonnegative deviational variables for the environmental impact goal. The converted constraint is

$$f^2(\mathbf{X}) - d_2^+ + d_2^- = g_2$$

Here, since g_2 established an *upper* bound in the inequality constraint, d_2^+ is the variable that allows flexibility. If for some solution, $d_2^- > 0$ and $d_2^+ = 0$, then $f^2(\mathbf{X}) < g_2$; the solution is inside the boundary set by g_2. On the other hand, if $d_2^+ > 0$ and $d_2^- = 0$, then $f^2(\mathbf{X}) >$

---

[17]The "negative" in the name of d_1^- refers to the amount by which the left-hand side of the constraint is less than the right. When $d_1^- \neq 0$, then $f^1(\mathbf{X}) - g_1$ is negative. In particular, it does *not* refer to the sign of d_1^-. Both d_1^- and d_1^+ must be *nonnegative*.

g_2 and the solution violates the upper boundary in the "$\leq$" constraint (which is the same as increasing g_2, which means relaxing it).

In general, then, the converted goal constraint for the ith objective function will be

$$f^i(\mathbf{X}) - d_i^+ + d_i^- = g_i$$

If $f^i(\mathbf{X})$ was to be maximized, then the added inequality constraint was $f^i(\mathbf{X}) \geq g_i$. In this case, d_i^+ serves exactly the role of an ordinary surplus variable in a lower bound inequality constraint, and d_i^- is a nontraditional kind of slack variable that allows $f^i(\mathbf{X})$ to be *less than* g_i. On the other hand, if $f^i(\mathbf{X})$ was a function to be minimized, then the added constraint was $f^i(\mathbf{X}) \leq g_i$. Now d_i^- is like an ordinary slack variable in an upper bound inequality constraint, and d_i^+ is the new and different variable that allows $f^i(\mathbf{X})$ to be *more than* g_i.

Since the terminology can be a bit confusing at first, here are simple illustrations. Suppose $g_1 = 10$, so $f^1(\mathbf{X}) \geq 10$ becomes $f^1(\mathbf{X}) - d_1^+ + d_1^- = 10$. If we find an $\mathbf{X}$ for which $f^1(\mathbf{X}) = 12$, then $d_1^+ = 2$ and $d_1^- = 0$ and the original constraint is satisfied; if for some other $\mathbf{X}$, $f^1(\mathbf{X}) = 7$, then $d_1^+ = 0$ and $d_1^- = 3$ and the original constraint is violated. Similarly, let $g_2 = 400$, so $f^2(\mathbf{X}) \leq 400$ becomes $f^2(\mathbf{X}) - d_2^+ + d_2^- = 400$. If $f^2(\mathbf{X}) = 500$, then $d_2^+ = 100$ and $d_2^- = 0$ (the original constraint is violated); if $f^2(\mathbf{X}) = 360$, then $d_2^+ = 0$ and $d_2^- = 40$ (the original constraint is satisfied).

Thus far, then, our original land development problem has been turned into the following. Find an $\mathbf{X}$ so that

$$\mathbf{X} \in C$$
$$f^1(\mathbf{X}) - d_1^+ + d_1^- = g_1$$
$$f^2(\mathbf{X}) - d_2^+ + d_2^- = g_2$$

It may also be that some or all of the original constraints, in $\mathbf{X} \in C$, should be converted to "double slack" form, to allow for flexibility in their right-hand sides. If so, it would be accomplished in exactly the same way. However, if the original constraints are not to be considered flexible, then the usual single slack or surplus variable approach is appropriate, adding a nonnegative d^- to the left-hand side of a "$\leq$" constraint and subtracting a nonnegative d^+ from the left-hand side of a "$\geq$" constraint. In any event, note that now there is no objective function left.

A New Kind of Objective Function Suppose that, as in the hierarchical optimization structure, it is possible to *rank* the objectives, $f^1(\mathbf{X})$ and $f^2(\mathbf{X})$, that have been turned into constraints. As in the hierarchical optimization discussion, assume that they are numbered in order of importance, so $f^1(\mathbf{X}) \mathcal{P} f^2(\mathbf{X})$. A straightforward way to try to avoid $f^1(\mathbf{X}) < g_1$ (new employment below its stipulated floor) would be to make d_1^- as small as possible. If $d_1^- = 0$ (at its lower limit), then $f^1(\mathbf{X})$ will be at or above g_1. Similarly, a way to try to keep $f^2(\mathbf{X})$ less than g_2 is to make d_2^+ as small as possible. When $d_2^+ = 0$ (its lower limit), $f^2(\mathbf{X})$ will be at or below g_2. So, because $f^1(\mathbf{X}) \mathcal{P} f^2(\mathbf{X})$, we want to first of all minimize d_1^- and then, when that is accomplished, minimize d_2^+, at least to the extent possible.

Exactly how this is accomplished in a solution algorithm need not concern us very much. Some approaches use an objective such as

$$\text{Minimize} \quad P_1 d_1^- + P_2 d_2^+$$

where P_1 and P_2 indicate first and second "priority levels;" that is, they simply reflect an ordering of importance. Then a simplexlike method would have to deal with this kind of *ordinal* objective function. Other approaches reject the notion of summation in the objective function and use other means to force a sequence of optimizations, dealing with the individual deviational variables in order—much in the spirit of hierarchical optimization, only without any tradeoffs during the sequence of problems. The easiest way to see how this works out is to look at the geometry of a two-variable problem.

8.7.6 Goal Programming: Numerical Illustration

We continue with the problem from the hierarchical optimization illustration earlier in this section. If the two constraints from the original problem, in (8.33), were truly flexible, that would mean that the original budget *could* be exceeded and that wage tax receipts *could* be less than the original minimum proposed by the mayor. In that case, these constraints would be

$$5x_1 + 7x_2 + d_1^- - d_1^+ = 2000$$

$$x_1 + 2x_2 + d_2^- - d_2^+ = 460$$

On the other hand, if we want to ensure that $5x_1 + 7x_2 \leq 2000$ and $x_1 + 2x_2 \geq 460$, we need only insist that $d_1^+ = 0$ and $d_2^- = 0$, by simply leaving them out of the formulation. Assume, for this illustration, that we have done that. This means that the feasible region defined by the first two constraints is exactly as in Figure 8.16.

Suppose that we have established $g_1 = \$20{,}000$ (wage tax receipts), $g_2 = 300$ (enrollees) and $g_3 = 4000$ (tutoring hours). The first two are lower limits and the third is an upper limit.[18] Then the objectives turned constraints are

$$43.75x_1 + 75x_2 + d_3^- - d_3^+ = 20{,}000$$

$$x_1 + x_2 + d_4^- - d_4^+ = 300$$

$$20x_1 + 10x_2 + d_5^- - d_5^+ = 4{,}000$$

Continuing with the assumption that the three objectives are numbered in order of preference, we want first to minimize d_3^-, the amount that $43.75x_1 + 75x_2$ is below 20,000. (If d_3^- can be forced all the way down to zero, then we are assured of being on or *above* the target of sales tax receipts for the new stadium.) Second, we want to minimize d_4^-, the amount by which the number of enrollees is below 300. (Again, if this can be set to zero, we are on or *above* the target of number of students enrolled.) Finally, after these two objectives are dealt with, we would also like to minimize d_5^+, the amount by which needed tutoring hours exceed 4000. If it were possible to also make $d_5^+ = 0$, we would be on or *below* the target for maximum tutoring hours.

Here are all five constraints. The first two have been altered to reflect our assumption that these are inflexible.

---

[18]Again, it is not always easy to come up with target numbers like these. One approach is to find the optimum for each objective subject to the two original constraints only. We did this already earlier in this section. Then, just as we did *sequentially* in the hierarchical optimization procedure, we could set g_1 and g_2 at some percentage below (for maximization objectives) and g_3 at some percentage above (for minimization objectives) these optima.

$$5x_1 + 7x_2 + d_1^- = 2{,}000$$

$$x_1 + 2x_2 - d_2^+ = 460$$

$$43.75x_1 + 75x_2 + d_3^- - d_3^+ = 20{,}000 \qquad (8.38)$$

$$x_1 + x_2 + d_4^- - d_4^+ = 300$$

$$20x_1 + 10x_2 + d_5^- - d_5^+ = 4{,}000$$

These five constraints are shown in Figure 8.20a. The region defined by the first two (in-flexible) constraints is shaded. Arrows indicate the half-spaces in which each of the deviational variables is positive.

The sequence of optimization problems is easily visualized geometrically. Considering first the objective of minimizing d_3^-, this will be accomplished anywhere in the closed half-space on or above constraint 3 (where $d_3^- = 0$; vertical shading). Next, d_4^- is minimized at zero anywhere on or above constraint 4 (another closed half-space, this time with horizontal shading). Finally, minimizing d_5^+ (at zero) is accomplished anywhere on or below constraint 5 (another closed half-space, here with 45° shading). The area in which all three of these unwanted deviational variables are equal to zero is magnified in Figure 8.20b. In this case, any $\mathbf{X}$ inside or on the boundaries of this elongated diamond shaped region is on the "desired" side of each of the last three constraints. So then we can optimize further by, say, selecting within *this* area that point for which $f^1(\mathbf{X})$ is maximized; that would be at $\mathbf{X} = \begin{bmatrix} 50 \\ 250 \end{bmatrix}$.

The more interesting cases for which goal programming is appropriate are those in which the targets are such that not all unwanted deviational variables can simultaneously be zero (where there is no area like the diamond shape in Fig. 8.20). This kind of situation would arise, for example, if we raised g_2 to 340 in the problem shown in (8.38). The geometry is shown in Figure 8.21a and the area of interest is magnified in Figure 8.21b. Subject to the first two constraints in (8.38), there is no area in which d_3^-, d_4^-, and d_5^+ are all zero.

Following the same geometric logic as in the preceding problem, we again start with the closed half-space where $d_3^- = 0$. Now, however, the area in which $d_4^- = 0$ is entirely outside the shaded feasible region defined by the first two constraints. Since larger values of d_4^- are associated with points farther away (here southwest) from constraint 4, it is clear from the figure that once we have minimized d_3^- at zero, the smallest possible value of d_4^- will be associated with the point at the lower right corner of the feasible region, where $\mathbf{X} = \begin{bmatrix} 144 \\ 181 \end{bmatrix}$ (rounded downward). Thus the "region" in which d_3^- and d_4^- are minimized is reduced to a single point, and nothing can be done about the third goal. At this "best" point, $d_3^- = 0$ but $d_4^- = 15$ and $d_5^+ = 690$.

You can easily try alternative values for one or more of the goals or alternative rankings for the three original objective functions and see how the optimal solution will be changed. At the extreme, if each goal is set at its optimal value from the problem with only the first two constraints in (8.38),[19] then all that is left of the original feasible region defined by those two constraints are its three corners. Depending on which goal has first

---

[19]These points were identified in footnote 16 above.

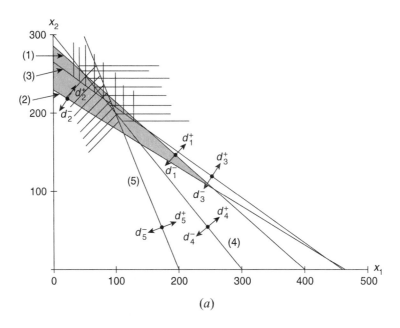

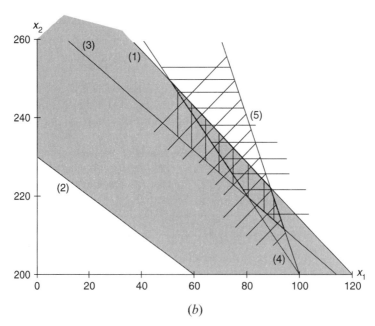

FIGURE 8.20 Geometry of problem in (8.38): (*a*) the five constraints; (*b*) magnification of Figure 8.20*a*.

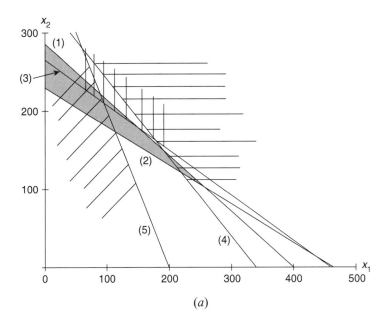

(a)

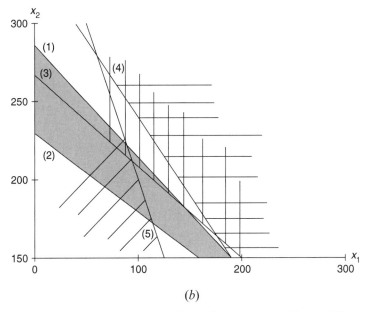

(b)

FIGURE 8.21   Alternative goal program: (a) the five constraints; (b) magnification of Figure 8.21a.

priority, the unwanted deviational variable for that constraint will be zero and no attention will be able to be given to the remaining two goals.

8.8   SUMMARY

In this chapter, we have considered the effect on a linear program of peculiarities that could be present in the data of a particular problem (multiple optima, degeneracy, the origin not a feasible basis, equations as constraints). Some of these introduce only modest changes in simplexlike computations. Others may influence our interpretation of an optimal solution (e.g., multiple optima).

We have also investigated the implications of peculiarities in the fundamental structure of a problem (transportation and assignment models, integer requirements on the solution values, multiple objectives, uncertainty about goals as expressed in right-hand sides of constraints). We have seen that sometimes (transportation and assignment problems) simplifications of solution methods are possible. For integer programs, somewhat more subtle solution procedures are required, but computer software is available to deal with them. With multiple objective problems, several alternative formulations are available. Usually only an ordering of objectives is needed; sometimes specific tradeoffs between objectives are also required. All of these variations will also have accompanying dual problems. We have not explored them or their interpretations.

There is still a great deal in the linear programming area that has been omitted—for example, the use of these models to solve problems of the network flow variety, the theory and results for parametric and stochastic programming models, and the more complicated dynamic programming models for multistage optimization. These variations are covered in more detailed and specialized books.

The potentially immense size of some real-world linear programming problems has led to the development of alternatives to the simplex method. Many of these newer approaches do not creep around the edge of the feasible region at all but rather try to take direct paths through it.

One example of such a class of potentially large problems is known as the "traveling salesperson" problem in which the task is to find a path through a set of points on a network (e.g., cities on a map) that minimizes total travel time, or total distance traveled, or some other measure of "cost." For a problem with as few as, say, 20 cities, there are $20! \cong 2.433 \times 10^{18}$ possible itineraries (i.e., ways to *order* the cities, from the first to the last on the itinerary). A computer that could evaluate the costs associated with *one billion* possible orderings per second would still require more than 77 years to evaluate the costs associated with all the possible itineraries.

This prototype traveling salesperson problem is closely related to the kinds of optimal vehicle scheduling problems faced by companies that distribute packages throughout a city (or, indeed, the entire country, or worldwide). It also resembles the kinds of scheduling problems that are faced by airlines serving multiple cities.

Currently a great deal of work is being done on developing and refining these newer approaches—known collectively as *interior point methods*. They are closer in spirit to some of the techniques for *nonlinear* programming problems that we will examine in Part V. (We will also see, in Chapter 10, that some computational approaches to nonlinear programming problems create approximating *linear* programming problems so that a variant of the simplex method can be used to find a solution. Many interior point approaches are just the reverse; they solve a linear programming problem by creating and solving an associated *nonlinear* program.) We explore some of these newer techniques in Chapter 9.

REFERENCES

Gomory, Ralph E., "An Algorithm for Integer Solutions to Linear Programs," paper first issued as Princeton-IBM Mathematics Research Project, Technical Report 1, Nov. 17, 1959. [Reprinted in Graves and Wolfe (1963), pp. 269–302.]

Graves, Robert L., and Philip Wolfe, eds., *Recent Advances in Mathematical Programming*, McGraw-Hill, New York, 1963.

Little, John D. C., Katta G. Murty, Dura W. Sweeney, and Karel Caroline, "An Algorithm for the Travelling Salesman Problem," *Oper. Res.* **11**, 972–989 (1963).

Wagner, Harvey M., *Principles of Operations Research*, 2nd ed., Prentice-Hall, Englewood Cliffs, N.J., 1975.

PROBLEMS

8.1 Find all solutions to the following linear program:

$$\text{Maximize} \quad 6x_1 + 9x_2$$
$$\text{subject to} \quad x_1 + 2x_2 \leq 14$$
$$3x_1 + 2x_2 \leq 28$$
$$2x_1 + 3x_2 \leq 22$$
$$\text{and} \quad x_1, x_2 \geq 0$$

8.2 Find the optimal solution to the following problem and its dual:

$$\text{Maximize} \quad 8x_1 + 15x_2 + 3x_3$$
$$\text{subject to} \quad 2x_1 + 6x_2 + 5x_3 \leq 30$$
$$4x_1 + 5x_2 + 10x_3 \leq 39$$
$$2x_1 + 4x_2 + 3x_3 \leq 24$$
$$\text{and} \quad x_1, x_2, x_3 \geq 0$$

If more than one solution exists to either the primal or the dual, find all solutions.

8.3

$$\text{Maximize} \quad 2x_1 + x_2$$
$$\text{subject to} \quad 15x_1 + 9x_2 \leq 170$$
$$-2x_1 + 4x_2 \geq 16$$
$$x_1 \geq 2$$
$$\text{and} \quad x_1, x_2 \geq 0$$

8.4

$$\text{Maximize} \quad 10x_1 + 7x_2$$

$$\text{subject to} \quad 4x_1 + 3x_2 \leq 26$$

$$x_1 \leq 6$$

$$2x_1 + 4x_2 \leq 20$$

$$\text{and} \quad x_1, x_2 \geq 0 \text{ and integer}$$

Explore both Gomory and branch-and-bound approaches to the solution.

8.5 Add the requirement that x_1, x_2, and x_3 be integers to Problem 7.1 in Chapter 7.

8.6 A state government has $40,000 to allocate among less-developed counties to help to finance public works projects for unemployment relief. Six counties (A, B, C, D, E, and F) qualify as eligible on the basis of their current levels of unemployment. The projects will require state funding of $15,000, $15,000, $12,000, $18,000, $19,000, and $17,000, respectively, in counties A–F. Relative locations of the counties are as follows (where counties G and H are not eligible for project location because their current levels of unemployment are not high enough):

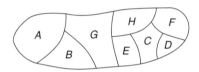

Each eligible county can receive at most one project. Estimates of new jobs that would be created by these projects are 100, 130, 120, 125, 110, and 135, in A through F, respectively. In addition, for environmental reasons, projects cannot be undertaken in contiguous (adjacent) counties. Formulate and solve an integer linear programming problem that would tell the state where to invest, if the aim is to maximize the total number of new jobs created.

8.7 A company wants to decide how to allocate its budget for the coming quarter. Two proposals have been presented (call them projects A and B), but there is a capital constraint. Project A requires $2000 and project B needs $3000. The company has $4000 available. Project A generates a net profit of $6000 and project B a net profit of $5000. Set this problem up an in integer program and solve it using either the Gomory or branch-and-bound procedure. [*Hint:* Include the constraints $x_A \leq 1$ and $x_B \leq 1$ and the requirement that x_A and x_B be integer, where x_A and x_B represent the number of "units" of projects A and B that are adopted.] Note that while this problem can very easily be solved intuitively (virtually by inspection) or graphically, there is need for some formal method such as this when there are many projects and many constraints.

The following transportation and assignment problems are best solved using computer software designed specifically for those problem types, or they can be done with any linear programming software that accommodates problems with equality constraints.

8.8 Find the optimal shipping transportation problem connecting two origins and three destinations:

$$\text{Minimize} \quad 10x_{11} + 20x_{12} + 12x_{13} + 15x_{21} + 10x_{22} + 18x_{23}$$

subject to origin capacities of 30 and 50, respectively, and destination demands of 20, 45, and 15, respectively.

8.9 Determine the optimal shipments for a problem with the following data (where cell entries indicate unit transport costs):

	D_1	D_2	D_3	D_4	D_5	D_6	Capacities
O_1	2	1	4	0	1	2	18
O_2	3	2	0	4	2	5	23
O_3	1	3	3	5	3	0	6
Demands	8	6	6	10	7	10	

8.10 An ice cream distributor has four trucks and one person to sell from each of them. One truck will be sent to each of four tourist areas in the city during the summer season. Expected weekly sales (in dollars) are shown below for trucks $1, \ldots, 4$ (they differ in amount of storage space for ice cream and in their decoration) in each of the areas A, B, C, and D. What assignment of trucks to areas will maximize profit? (Assume that operating costs are the same for all trucks and that they do not vary with location.)

	Area			
Truck	A	B	C	D
1	600	900	400	1000
2	1200	1000	800	1100
3	500	700	400	600
4	1800	1000	1100	1400

8.11 Five waiters $(W_1, \ldots, W_5)$ are available to work in five branches of a restaurant chain. The entries in the matrix below indicate the daily costs associated with each waiter–location assignment. An X indicates that a waiter cannot work at the indicated location (e.g., because of travel-time constraints). W_r represents costs of a waiter who is available as a replacement for any one of the current waiters. Find the best combination of waiters and the optimal assignment of them to locations. What is the total cost of this assignment?

	L_1	L_2	L_3	L_4	L_5
W_1	120	40	80	X	50
W_2	30	110	20	70	80
W_3	60	X	100	80	90
W_4	X	50	90	60	40
W_5	50	X	70	120	60
W_r	90	30	50	100	X

8.12 What is the effect on the minimum transportation costs in Problem 8.9 of

(a) A decrease of one unit in demand at destination 1 (with all other facts remaining unchanged)

(b) An increase of one unit in supply at origin 1 (again, with all other facts from the question unchanged; in particular, with demand at D_1 equal to 8).

8.13 How, if at all, is the optimal solution to Problem 8.9 changed if you later discover that, because of road repair, there is a capacity limit of 8 on the shipments that can be sent from origin 1 to destination 4?

8.14 Suppose that two warehouses are built to serve the network depicted in Problem 8.9. Unit transport costs from origin 1 to the two warehouses are 0.5 and 0.5; from origin 2 they are 1.0 and 1.0; from origin 3 they are 0.5 and 0.5, respectively. Unit transport costs from warehouses 1 and 2 to destinations D_1 through D_6 are as shown in the following table:

	D_1	D_2	D_3	D_4	D_5	D_6
W_1	1	2	1	2	1	2
W_2	0.5	1	1	2	0.75	1

Both warehouses have a capacity of transshipping up to 30 units, and shipments cannot be made from either warehouse to the other. How are the optimal shipments from Problem 8.9 changed by the construction of these warehouses? What happens to minimum total shipping costs?

8.15 In region 1 there are two sparkling water bottling plants, with production capacities of 100 and 200 cases per day, respectively. In the same region, three cities are markets for the sparkling water; they have daily demands of 100, 100, and 200 cases, respectively. Sparkling water is also bottled in region 2, at three bottling plants with daily production capacities of 150, 150, and 200 cases, respectively. Two urban areas in region 2 have daily demands of 250 and 50 cases of sparkling water, respectively. For political reasons, region 1 refuses to export sparkling water to region 2; however, bottling plants in region 2 willingly export to region 1. The *intra*regional transportation costs per case are as follows:

	Region 1				Region 2	
	City 1	City 2	City 3		City 1	City 2
Plant 1	50	60	20	Plant 1	40	30
Plant 2	30	40	30	Plant 2	40	20
				Plant 3	50	30

Costs of shipping a case of sparkling water from *any* bottling plant in region 2 to *any* market in region 1 are 100.

(a) What pattern of shipments from factories in both regions will minimize total shipping costs and satisfy demands in all markets?

(b) What happens to optimal shipping patterns and total transportation cost if it becomes twice as expensive to ship sparkling water into market 1 in region 2 because a bridge has washed out and all goods destined for that market must now be taken into the city on barges?

(c) Suppose that shipments from plant 1 in region 2 to city 1 in region 2 cannot exceed 50 cases per day because of shipping restrictions through a tunnel on the highway between that plant and that city. How would you now answer part (a)?

(d) What is the optimal shipping plan and what are the costs associated with it if the tunnel described in part (c) collapses, meaning that there can be *no* shipments from plant 1 in region 2 to city 1 in region 2?

(e) In which market in which region would a one-case increase in demand cause the greatest increase in total systemwide shipping costs? What would be the size of that increase?

(f) At which bottling plant in which region would a one-case increase in daily production capacity cause the greatest decrease in total systemwide shipping costs? What would be the size of that decrease?

8.16 A company wants to decide where to locate three new branch plants. There are three potential sites (in three different states). Each site can accommodate only one plant. The estimates of annual labor costs (in $ millions) differ for each plant–site combination because of different wage rates in each state and different labor-skill mixes needed by each plant. The company's estimates of these annual costs are

	Location		
	A	B	C
Plant 1	10	5	9
Plant 2	4	3	2
Plant 3	5	3	4

(a) The company asks for your opinion on which plant should be located in each location. Without using any computer software at all, how do you know that it is impossible for each plant to be located at its least-cost site? Also, how do you know that it is impossible to allocate to each site its least-cost plant?

(b) Now, solving the problem in (a), what do you tell the company to do? What are the expected annual labor costs that would result from your suggestion?

(c) What would you tell the company if it turned out to be possible to locate two plants at B?

(d) How would you change your recommendation to the company, if at all, from the one that you gave in part (b), if there were a fourth location, D, where one plant could be sited, with estimated annual labor costs of 4 for plants 1 and 3 (plant 2 could not locate at D because labor with the skills needed by plant 2 is not available at D)?

8.17 Problems 7.2, 7.3, 7.5, 7.11, and 7.13 in Chapter 7 have only two unknowns, so solution-space geometry is easily sketched. For each of these five problems:

(a) Formulate two additional possible objective functions, specify an ordering of these objectives, and solve the multiple objective linear programming problem with a sequence of three hierarchical optimization problems. (You will also need to specify tradeoffs between objectives for the second and third problems in each sequence.)

(b) Explore alternative solutions and the effects on solution-space geometry caused by different choices for

   (i) Objective function orderings

   (ii) Tradeoff coefficients

8.18 (a) Approach the problems in Problem 8.17(a) using the alternative of a single linear programming problem, as in (8.37). Since the feasible regions are all two-dimensional, you can find the solutions geometrically in each case.

(b) Introduce the alternatives that you chose in Problem 8.17(b) for the formulation in part (a) of this problem.

8.19 (a) For each of the cases in Problem 8.17(a), convert the problem to a goal programming format and solve these versions of the problems. (You will need to specify the goals in each case.)

(b) Explore in this goal programming framework the geometric and numerical implications of the alternatives chosen in Problem 8.17(b).

9

LINEAR PROGRAMMING: INTERIOR POINT METHODS

The simplex method is founded on two mathematical observations that connect the algebra and the geometry of the problem: (1) an optimal solution (if one exists) will be at one or more corners of the convex feasible region in n-dimensional solution space (where n is the number of unknowns) and (2) each of those corners corresponds to a nonnegative basic solution to the linear equations of the *augmented* constraint set. Exploring successively better corners can be visualized as moving along the boundaries of the feasible region. In many cases, the simplex method is amazingly efficient in sorting through basic solutions for the best among those that are nonnegative.[1] It is no wonder that this approach enjoyed enormous popularity during the entire last half of the twentieth century.

For some kinds of problems (essentially those that are very very large) the simplex method can be slow. This disadvantage generally does not show up in textbook examples because typically they are too small. However, some clever (if perhaps not very realistic) linear programming problems have been devised for which a simplex method approach, starting at the origin, would visit each and every corner before finding the optimal solution at the very last available vertex. Here is an illustration [adapted from Fletcher (1987, p. 184); also Chvátal (1983, Chapter 4)]. It is attributed to Klee and Minty (1972):

$$\text{Maximize} \quad x_1 + 10x_2 + 100x_3$$

$$\text{subject to} \quad x_1 + 20x_2 + 200x_3 \leq 10^4 = 10{,}000$$

$$x_2 + 20x_3 \leq 10^2 = 100 \tag{9.1}$$

$$x_3 \leq 10^0 = 1$$

$$\text{and} \quad \mathbf{X} \geq \mathbf{0}$$

[1]For example, a linear programming maximization problem with 20 unknowns and 10 constraints could have up to C_{20}^{30} (more than 30 million) basic solutions. For one particular sample problem of this size, the simplex method required only *three* pivots (examined three basic feasible solutions after starting at the origin) to reach the optimal solution. (This is mentioned in Problem 7.6 in Chapter 7.)

The feasible region for this problem in three-dimensional solution space has eight vertices. These are indicated in Figure 9.1, which represents the essential geometry of the problem but with constraint coefficients changed in order to avoid large differences in scale—such as an intercept of 10,000 on the x_1 axis for constraint 1 and an intercept of 1 on the x_3 axis for constraint 3.

In fact, if we look again at the problem in (9.1), we realize that the geometry in three-dimensional solution space would look like a kind of "distorted" cube. The third constraint, $x_3 \leq 1$, is a straightforward upper limit on x_3—so that, in conjunction with non-negativity, we have $0 \leq x_3 \leq 1$. The two planes that define these limits, $x_3 = 0$ and $x_3 = 1$, are parallel.

Given the upper limit on x_3, the role played by it in the second constraint will probably be *relatively* unimportant, because the right-hand side of that constraint is 100 times larger than the one in constraint 3. So constraint 2 (in conjunction with nonnegativity) effectively brackets x_2 in the range $0 \leq x_2 \leq 100$. The plane defining $x_2 = 100$ is parallel to that defined by the lower limit, $x_2 = 0$.

Finally, then, the roles played by both x_3 and x_2 will be (again, relatively) unimportant in constraint 1 whose right-hand side is 100 times that in constraint 2 and 10,000 times that in constraint 3; this means, approximately, that $0 \leq x_1 \leq 10,000$. And the planes defined by $x_1 = 10,000$ and $x_1 = 0$ are, again, parallel. So the feasible region is essentially a long, narrow box—a distorted cube whose faces have dimensions 1×100 in the $x_1 = 0$ plane, $1 \times 10,000$ in the $x_2 = 0$ plane, and $100 \times 10,000$ in the $x_3 = 0$ plane. So it is clear that the feasible region has eight (2^3) vertices.

The initial simplex table for this problem is in Table 9.1, and you can see how the simplex method might be tricked. The initial basic feasible solution is at the origin. Using the rule that the variable with the largest positive top-row element should be introduced into the next basic solution, the simplex method would find x_3 to be the most attractive currently excluded variable (by far) and so would bring it into the

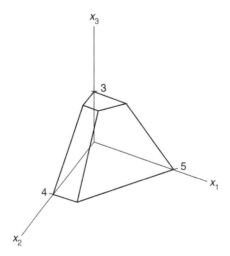

FIGURE 9.1   Characteristics of feasible region for a problem such as that presented in (9.1).

TABLE 9.1   Complete Simplex Table for the Problem in (9.1)

		x_1	x_2	x_3
Π	0	1	10	100
s_1	10,000	−1	−20	−200
s_2	100	0	−1	−20
s_3	1	0	0	−1

next basis (replacing s_3), even though its value in the new basic solution can only be x_3 = 1.[2]

But there are further pathologies to the problem. If you solve this linear program by hand (Problem 9.1 asks you to do just that), starting with the array in Table 9.1 and using the simplex method rules from Section 7.3, you will find that *all* remaining vertices are examined; only after the seventh pivot will you arrive at $\mathbf{X^*} = \begin{bmatrix} 10,000 \\ 0 \\ 0 \end{bmatrix}$ (a solution that is fairly obvious when you look at the peculiar structure of this problem).

The example in (9.1) is simply the $n = 3$ illustration of the following general problem:

$$\text{Maximize} \quad \sum_{j=1}^{n} 10^{(j-1)} x_j$$

$$\text{subject to} \quad x_i + 2\sum_{j>i} 10^{(j-i)} x_j \le 10^{(2n-2i)} \quad (i = 1, \ldots, n) \quad (9.2)$$

$$\text{and} \quad \mathbf{X} \ge \mathbf{0}$$

The feasible region will always have 2^n vertices, and the simplex method will visit them all.[3] Because of this observation on how the simplex method would work on this kind of "worst case" problem in (9.2), it is said to be an *exponential time* algorithm. The required

---

[2]The choice of x_3 to enter the next basic solution is based on the "largest positive top-row coefficient" rule that we used throughout Chapter 7. This rule is easy to apply; the choice is made by inspection (simple ranking). There are alternative but more complicated rules. One might be called the "largest objective function change" rule. In that case, you would examine *each* of the currently nonbasic variables with positive top-row coefficients. For each of these, you would need to find the current basic variable that would be replaced in the next solution (i.e., find the pivot element in each of the candidate columns) and then calculate the change in objective function value that would be associated with that new variable, using the last of the four simplex rules in Table 7.6. (It happens, in this case, that using that particular rule would lead to the optimal solution in one pivot!) There are numerous other variants of these entering-variable selection rules. All of them require more computation than the simple one that we used in Chapter 7, and for that reason, the largest top-row coefficient rule, even if not always efficient, is very frequently used in practice.

[3]Problem 9.2 asks you to write down the $n = 4$ case and to solve it by hand, again using the simplex method and the rules of Chapter 7. There are $2^4 = 16$ vertices, and you will visit them all, starting at the origin, before stopping at the optimum. While the pivots take a lot of paper, the arithmetic is easy, since the pivot elements are always −1, and the coefficient matrix is upper triangular (zeros below the main diagonal). In fact, all that changes from one simplex table to the next are the *signs* of some of the numbers and one pair of top-row and left-column labels.

TABLE 9.2 Iterations and Estimated Computing Time for Exponential versus Polynomial Algorithms[a]

	n			
	50	100	500	1000
Exponential Time Algorithms (2^n)				
Iterations	1.13×10^{15}	1.27×10^{30}	3.27×10^{150}	1.07×10^{301}
Computing time	35.8 years	400 trillion centuries	—[b]	—[b]
Polynomial Time Algorithms (n^3)				
Iterations	125,000	1,000,000	1.25×10^8	1×10^9
Computing time	0.125 s	1 s	2.08 min	16.67 min

[a]Assuming 1 million iterations per second.
[b]You could work out the entries for these two cells, at 1 million iterations per second (3.1536×10^{13} iterations per century), but the point may be adequately made by the figure under the $n = 100$ column.

number of iterations (pivots), and hence also the required number of arithmetic operations and the computational time, can be bounded from above by an exponential function of n, namely, $2^n - 1$. By contrast, as we will see in the sections that follow, many current variants of interior point methods can be shown to have an upper bound (again, for iterations and computing time) that is a polynomial function of n (or some other measure of the *size* of the problem)—for example, n^3. These algorithms are therefore known as *polynomial time* algorithms, and the motivation for the recent and current work in interior point methods for linear programs is because of the potentially enormous differences between these two kinds of upper bounds.

Table 9.2 contains a few illustrations showing how quickly you might get into trouble with an algorithm that requires 2^n iterations as opposed to n^3. For a problem like that in (9.2), with $n = 50$ (the first column in the table), $2^n = 1.13 \times 10^{15}$ (this is 1.13 quadrillion).[4] If we imagine that the simplex method on a fast computer could execute 1 *million* pivots per second, it would take more than 35 years to reach the (final) optimal solution. For the moment, suppose that interior point method iterations can also be done at 1 million per second. In that case, an interior point method would need no more than 0.125 s for the same problem. And a problem with $n = 50$ is "small" by modern standards (as we will see in Section 9.5). The point is made even more forcefully in the remaining columns of Table 9.2, for larger values of n. So there is some reason to be concerned about the performance of the simplex method in very large and peculiarly structured problems. On the other hand, it is not obvious how likely it is that you would meet such a problem in a real-world situation. Later in this chapter we will see that for very large but not so peculiarly structured problems, the simplex method often remains very competitive in terms of computation time.

9.1 INTRODUCTION TO INTERIOR POINT METHODS

In an attempt to overcome this potential weakness in the simplex method, a number of alternative procedures have been developed. They are known collectively as interior point

---

[4]The number of iterations is actually $2^n - 1$, but with numbers of this magnitude, 2^n is good enough.

methods. The first of these new and modern algorithms to receive widespread attention was described in Karmarkar (1984). This and other interior point methods adopt an approach that is completely different from the simplex algorithm.[5] They are iterative procedures that, as you would expect from the name, move through the *interior* of the feasible region rather than around the edges; in fact, they generally intentionally *avoid* the boundaries. In the years since 1984, when Karmarkar's first paper appeared, many researchers have worked on developing "better" interior point methods for linear programs. There are literally thousands of papers in which variations, major and minor, are presented.[6]

Criteria for assessing the merits of these alterative iterative algorithms are usually based on measures of the *complexity* of the method. Put simply, one attempts to find an estimate of the order of magnitude of the number of arithmetic operations that are required in order to arrive at a solution to a problem (frequently calculated as the product of the number of iterations and the average number of operations in each iteration, or some similar measure of computation time or effort).[7] These measures are usually expressed as an upper bound on that number, and they are a function of the *size* of the problem. This, in turn, requires that there be an agreed upon definition of problem size, which turns out, usually, to be rather more subtle than just a count of unknowns (n) and/or constraints (m). (We used n in the simple illustrations in Table 9.2, as an indication of size.)

One commonly accepted measure of the size, L, of a linear program is $L \cong mn + \lceil \log_2 |k| \rceil$. In this case, n is the number of decision *plus* slack variables (so that the constraints in the problem are m linear *equations* in n unknowns) and k is the product of all of the nonzero elements in the problem (in **P**, **A** and **B**, in the notation of Chapters 7 and 8).[8] For this measure, these matrices and vectors are assumed to contain only integers (which can always be accomplished by appropriate, if tedious, scaling) and $\lceil c \rceil$ denotes (when c is not integer) c rounded up to the nearest integer. The second term reflects the number of bits needed to represent the problem in binary notation for a computer.

Most of the currently successful work on interior point methods is with algorithms for which the number of iterations is of order nL or $(\sqrt{n})L$—written $O(nL)$ and $O((\sqrt{n})L)$ in "big-O" notation.[9] A large number of these interior point methods also require, on aver-

---

[5]The first work on interior point methods for linear programs is often credited to Dikin (1967). This was what is called an *affine scaling* method. This work was not widely known at that time; it was uncovered in the flurry of research that followed Karmarkar's paper. The first polynomial time method for linear programs is generally attributed to Khachian (1979)—a short paper that was originally in Russian and was based on earlier work by other Russians. This was known as the *ellipsoid algorithm*. Descriptions of this algorithm can be found in many books and articles, including Chvátal (1983, pp. 443–454) or Karloff (1991, Chapter 4). The ellipsoid algorithm has been eclipsed by other interior point methods, developed since 1984 and based on or at least stimulated by Karmarkar's work. Very active research continues into improvements on and variants of these newer algorithms. There are many descriptions both of the original approach and its variants in the literature. See, for example, Arbel (1993), Panik (1996, Chapter 13), Saigal (1995, Chapter 5), Sierksma (1996, Chapter 5), or Vanderbei (1996).

[6]den Hartog (1994, p. 7, note 3) claims that there were about 2000 papers dealing with interior point methods by 1992, 8 years after Karmarkar's first publication.

[7]Often these numbers are calculated for worst-case scenarios, for example, problems with particularly troublesome features, like the linear program in (9.2).

[8]In the linear programming literature you will find that some authors use n for the number of decision variables in a linear program with inequality constraints and others use n to denote the total number of variables in an augmented problem, where the constraints are equations (so that n is the total number of decision plus slack variables).

[9]Informally, if the number of iterations, NI, is $O(n^3)$, this means $NI \leq n^3$.

TABLE 9.3   Arithmetic Operations and Estimated Computing Time for Polynomial Algorithms with Differing Complexities[a]

n:	50	100	500	1000
L:	1450	5700	145,000	570,000
Polynomial (n^3L)				
Arithmetic operations	1.8×10^8	5.7×10^9	1.8×10^{13}	5.7×10^{14}
Computing time	0.18 s	5.7 s	5.03 h	6.6 days
Polynomial ($n^{3.5}L$)				
Arithmetic operations	1.3×10^9	5.7×10^{10}	4.1×10^{14}	1.8×10^{16}
Computing time	1.28 s	57 s	4.7 days	208 days
Polynomial (n^4L)				
Arithmetic operations	9.1×10^9	5.7×10^{11}	9.1×10^{15}	5.7×10^{17}
Computing time	9.06 s	9.5 min	104 days	18 years

[a]Assuming 1 billion arithmetic operations per second and $n = 2m$.

age, $O(n^3)$ arithmetic operations per iteration (essentially to solve a system of linear equations in n unknowns using something like Gaussian elimination), so the overall measures of complexity (number of iterations times number of operations per iteration) are $O(n^4L)$ or $O(n^{3.5}L)$ for many interior point methods.

Frequently a "new" interior point algorithm will be a variation of an existing algorithm but for which the upper bound complexity measure has been reduced—so that, for example, an $O(n^4L)$ algorithm is refined to become $O(n^{3.5}L)$ or one that is $O(n^{3.5}L)$ is reduced to $O(n^3L)$. These differences are by no means trivial. Table 9.3 gives you an idea of their magnitude. In the calculations for L in this table it is assumed that a problem in inequality form has the same number of decision variables as constraints, so that in augmented form the number of decision plus slack variables, denoted by n, is equal to $2m$. This means, for example, that in the $n = 500$ case, $mn = 125,000$.[10] Also, it is assumed that 1 *billion* arithmetic calculations (not iterations) can be performed in one second.[11]

As an example of the impact that variations in complexity can have for interior point methods, in the $n = 1000$ column in Table 9.3, we see that for an $O(n^4L)$ algorithm the number of arithmetic calculations would be of the order of 570 quadrillion, while for $O(n^{3.5}L)$ and $O(n^3L)$ algorithms the numbers are 18 quadrillion and 570 trillion, respectively, with estimates calculation times of 18 years, 208 days and 6.6 days each. And a problem with $n = 1000$ is not particularly "large" these days; for $n > 1000$, the differences in total numbers of arithmetic operations and, consequently, in computation times, will be even more dramatic.

[10]For an estimate of the $\lceil \log_2 |k| \rceil$ term, it is assumed that of the total nm elements in $\mathbf{A}$, 90% are zero and half of those nonzero elements are equal to 1, so that they contribute nothing to the overall product. The remaining numbers were drawn randomly from the integers in the range from 1 to 20 and their product was calculated; this can be a very very large number. For example, in the $n = 500$ case, it was estimated to be $2^{20,000}$, which means that the value used for L was $125,000 + 20,000 = 145.000$.

[11]For the simplex method, using an estimate of $O(n^3)$ arithmetic operations in each pivot, the comparable overall complexity measure would be $O(2^{0.5n}n^3)$. As we will see later in this chapter, many large real-world problems are not as ill-conditioned as the example in (9.1), and the simplex method often performs much better than this worst-case upper bound complexity measure would suggest.

9.2   AFFINE SCALING INTERIOR POINT METHODS

Interior point methods for linear programs have been classified in several ways; sometimes the distinctions are not very clearcut. In this and the next section we concentrate on two methods that are representative of the most successful approaches at the present time; they are generally known as *affine scaling* and *path-following* procedures. We explore the features of affine scaling algorithms in this section and path-following methods in Section 9.3. Later in this section we will see exactly where the term "affine scaling" comes from and, later still, that virtually *all* approaches to solving a linear programming problem could be characterized as "path-following."

For the remainder of this chapter, it will be useful to concentrate on linear programs that are expressed in augmented form, so that the constraints are linear equations. Following the notation adopted in Section 7.2, we represent these augmented constraints as $\mathbf{AX} = \mathbf{B}$, where it is understood that the $\mathbf{A}$ here is really $[\mathbf{A} \vdots \mathbf{I}]$ from the original problem, $\mathbf{X}$ now represents $\begin{bmatrix} \mathbf{X} \\ \hline \mathbf{S} \end{bmatrix}$ from the original problem, where $\mathbf{S}$ is the m-element vector of slack variables and $\mathbf{P}$ has been expanded to include a zero coefficient for each slack variable (recall footnote 7 in Chapter 7). The prototype problem is

$$\text{Maximize} \quad \mathbf{P'X}$$
$$\text{subject to} \quad \mathbf{AX} = \mathbf{B} \tag{9.3}$$
$$\text{and} \quad \mathbf{X} \geq \mathbf{0}$$

The largest augmented problem that can be conveniently represented geometrically is one with two decision variables and one constraint; this requires three dimensions in augmented solution space, with axes for the original variables, x_1 and x_2, and a third axis for the slack variable, x_3. We explore the essential features of the affine scaling method with a specific numerical illustration of a problem of this size:

$$\text{Maximize} \quad x_1 + x_2$$
$$\text{subject to} \quad x_1 + 2x_2 \leq 10 \tag{9.4}$$
$$\text{and} \quad x_1 \geq 0, x_2 \geq 0$$

The solution-space geometry for the feasible region for this problem is very straightforward; it is shown in Figure 9.2a. The objective function contours in that geometry would be lines with a slope of -1. It is obvious that the optimal solution is at $x_1 = 10$ and $x_2 = 0$.

In augmented form, we have

$$\text{Maximize} \quad x_1 + x_2 + 0x_3$$
$$\text{subject to} \quad x_1 + 2x_2 + x_3 = 10 \tag{9.5}$$
$$\text{and} \quad x_1 \geq 0, x_2 \geq 0, x_3 \geq 0$$

Let $\mathbf{P} = \begin{bmatrix} 1 \\ 1 \\ 0 \end{bmatrix}$, $\mathbf{A} = [1 \quad 2 \quad 1]$, $\mathbf{B} = 10$, and $\mathbf{X} = \begin{bmatrix} x_1 \\ x_2 \\ x_3 \end{bmatrix}$; then the augmented problem is exactly as described in (9.3). The geometry of this version of the problem is indicated in Fig-

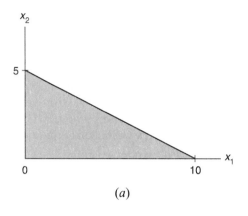

(a)

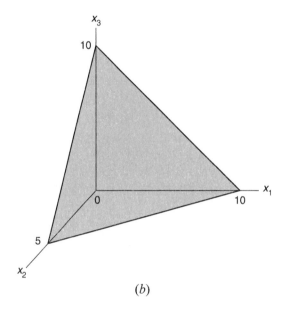

(b)

FIGURE 9.2   (a) Feasible region for problem presented in (9.4); (b) feasible region for problem presented in (9.5).

ure 9.2b. The objective function contours on the $x_1 + 2x_2 + x_3 = 10$ plane would be parallel lines created by a series of parallel planes that are orthogonal to the x_1x_2 plane (x_3 plays no role in defining these contours).

We explored iterative methods for unconstrained maximization problems in Chapter 6, where we saw that the essential elements are: (a) how (where) to start (selection of an initial point, $\mathbf{X}^0$); (b) how to move from the current point, $\mathbf{X}^k$, to the next point, $\mathbf{X}^{k+1}$, where $\mathbf{X}^{k+1} = \mathbf{X}^k + \Delta\mathbf{X}^k$, and (c) when to stop. Interior point methods adopt this same iterative approach, but it is made more complicated by the presence of both equality and nonnegativity (inequality) constraints. Since finding $\Delta\mathbf{X}^k$ [step (b)] is generally the most complicated of the three, we consider it first and in most detail.

9.2.1 Direction of Movement

Suppose that we start with a feasible solution that is *interior* to the feasible region, meaning in the augmented problem in (9.5) that all three unknowns are strictly positive—away from their nonnegativity boundaries. (We will soon see why this is a good idea.) Let this initial point be $\mathbf{X}^0 = \begin{bmatrix} 1 \\ 1 \\ 7 \end{bmatrix}$, where $\mathbf{AX}^0 = 10$. As we know from the discussion of iterative methods in Part III, movement to a next point, $\mathbf{X}^1$, involves, a choice of a *direction*, $\mathbf{D}^0$, in which to move and a decision on the *step length*, α^0, along that direction, so $\mathbf{X}^1 = \mathbf{X}^0 + \Delta\mathbf{X}^0 = \mathbf{X}^0 + \alpha^0\mathbf{D}^0$.

Remember (Section 4.2) that the *gradient* of a function points in the direction of maximum increase for the function, and this is therefore a reasonable direction in which to move in a maximization problem. As we saw in Chapter 6, this logic underpins an entire family of *gradient methods* for unconstrained maximization and minimization. We are about to see that it is also at the heart of many interior point methods for *linear programming* problems. (If the problem has a minimization objective, the direction of movement is indicated by the negative gradient.) The gradient of a linear function, $\mathbf{P}'\mathbf{X}$, is just the vector of coefficients, $\mathbf{P}$ (we continue to assume that the gradient is always expressed as a column vector). For our augmented problem in (9.5), $\nabla = \begin{bmatrix} 1 \\ 1 \\ 0 \end{bmatrix} = \mathbf{P}$, so it would be maximally beneficial to find $\mathbf{X}^1$ as $\mathbf{X}^0 + \alpha^0\nabla$. Without worrying about the step size for the moment, it is clear that we cannot move *at all* along the gradient vector from $\mathbf{X}^0$ because at that point the gradient points outward, away from the surface of the feasible region (Fig. 9.3a). With $\mathbf{X}^0 = \begin{bmatrix} 1 \\ 1 \\ 7 \end{bmatrix}$ it is clear that *nothing* can be added to x_1 and/or x_2 unless a compensating amount is taken away from x_3; but that is impossible if we are trying to move in the gradient direction, since we would have $\mathbf{X}^1 = \begin{bmatrix} 1 \\ 1 \\ 7 \end{bmatrix} + \alpha^0 \begin{bmatrix} 1 \\ 1 \\ 0 \end{bmatrix}$, and *any* positive value for α^0 will render $\mathbf{X}^1$ infeasible.

The solution proposed by the affine scaling approach (and other interior point methods also) is to *project* the gradient vector onto the feasible region in augmented solution space. Geometrically, this is tantamount to dropping a line from the end of the gradient so that it meets the feasible region surface orthogonally. The question is: how to accomplish this projection algebraically.

Both $\mathbf{X}^0$ and $\mathbf{X}^1$ must satisfy the constraint equations, $\mathbf{AX}^0 = \mathbf{B}$ and $\mathbf{AX}^1 = \mathbf{B}$, so subtracting, and since $\Delta\mathbf{X}^0 = \mathbf{X}^1 - \mathbf{X}^0$, it follows that $\mathbf{A}(\Delta\mathbf{X}^0) = \mathbf{0}$. This makes clear that the elements in $\Delta\mathbf{X}^0$ are solutions to a set of homogeneous linear equations. In our current example, this is just one homogeneous linear equation in three unknowns:

$$[1 \quad 2 \quad 1] \begin{bmatrix} \Delta x_1^0 \\ \Delta x_2^0 \\ \Delta x_3^0 \end{bmatrix} = 0$$

We know from Chapter 2 that there are infinitely many solutions to such an equation; in three-dimensional space (with Δx_1^0, Δx_2^0, and Δx_3^0 on the three axes), the equation de-

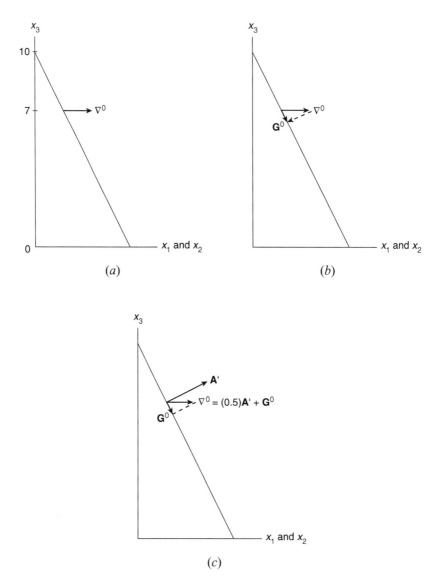

FIGURE 9.3 (*a*) Feasible region for problem presented in (9.5), viewed from the side, and ∇^0; (*b*) the projected gradient, $\mathbf{G}^0$; (*c*) ∇^0 as a linear combination of $\mathbf{G}^0$ and $\mathbf{A}'$.

scribes a single plane that goes through the origin. This collection of all possible solutions, $\Delta\mathbf{X}^0$, to $\mathbf{A}(\Delta\mathbf{X}^0) = \mathbf{0}$ is known as the *null space* of the matrix $\mathbf{A}$. Using that terminology, it is usual to say that this initial direction vector, $\Delta\mathbf{X}^0$, must lie in the null space of $\mathbf{A}$. We are interested in the computational implications of that requirement.

For the moment, our concern is with identifying the projection of the gradient, ∇, onto the augmented constraint $\mathbf{AX} = \mathbf{B}$. Denote the *projected gradient* at this initial point by $\mathbf{G}^0$

$= \begin{bmatrix} g_1^0 \\ g_2^0 \\ g_3^0 \end{bmatrix}$; we will use this as the *direction* component, $\mathbf{D}^0$, of $\Delta\mathbf{X}^0$ (Fig. 9.3*b*). We can worry about the step size, α^0, later. So we have $\mathbf{A}(\Delta\mathbf{X}^0) = \mathbf{A}(\alpha^0\mathbf{G}^0) = \alpha^0\mathbf{A}(\mathbf{G}^0) = 0$, and since we can divide by the scalar α^0, we see that $\mathbf{A}(\mathbf{G}^0) = 0$. We know (Chapter 4, Section 4.2) that $\mathbf{A}'$ is *orthogonal* to the constraint plane. So at our beginning point, $\mathbf{X}^0 = \begin{bmatrix} 1 \\ 1 \\ 7 \end{bmatrix}$, we have a pair of linearly independent[12] vectors, $\mathbf{V} = \begin{bmatrix} 1 \\ 1 \\ 0 \end{bmatrix}$ and $\mathbf{A}' = \begin{bmatrix} 1 \\ 2 \\ 1 \end{bmatrix}$ (Fig. 9.3c).

We are looking for the algebraic description of $\mathbf{G}^0$. This vector will be orthogonal to $\mathbf{A}'$, and the three vectors $\mathbf{A}'$, $\mathbf{V}$, and $\mathbf{G}^0$ will lie in a plane that is orthogonal to the constraint. Any one of these vectors can be described as a linear combination of the other two. For example, using w_1 and w_2 for the two scalars (weights) in the linear combination.

$$\mathbf{V} = \mathbf{A}'(w_1) + \mathbf{G}^0(w_2)$$

We can always "normalize" by letting $w_2 = 1$ (Fig. 9.3c again). Using w for the remaining scalar, this leaves

$$\mathbf{V} = \mathbf{A}'(w) + \mathbf{G}^0$$

For our illustrative example, this represents three linear equations in four unknowns— w and the three elements of $\mathbf{G}^0$. Explicitly, this is

$$\begin{bmatrix} 1 \\ 1 \\ 0 \end{bmatrix} = \begin{bmatrix} 1 \\ 2 \\ 1 \end{bmatrix}(w) + \begin{bmatrix} g_1^0 \\ g_2^0 \\ g_3^0 \end{bmatrix} = \begin{bmatrix} 1 & 1 & 0 & 0 \\ 2 & 0 & 1 & 0 \\ 1 & 0 & 0 & 1 \end{bmatrix}\begin{bmatrix} w \\ g_1^0 \\ g_2^0 \\ g_3^0 \end{bmatrix}$$

In addition, however, we know that $\mathbf{AG}^0 = 0$ or

$$\begin{bmatrix} 1 & 2 & 1 \end{bmatrix}\begin{bmatrix} g_1^0 \\ g_2^0 \\ g_3^0 \end{bmatrix} = 0$$

This adds one more linear equation in three of the four unknowns. Putting these four linear equations all together, we have

$$\begin{bmatrix} 0 & 1 & 2 & 1 \\ 1 & 1 & 0 & 0 \\ 2 & 0 & 1 & 0 \\ 1 & 0 & 0 & 1 \end{bmatrix}\begin{bmatrix} w \\ g_1^0 \\ g_2^0 \\ g_3^0 \end{bmatrix} = \begin{bmatrix} 0 \\ 1 \\ 1 \\ 0 \end{bmatrix} \tag{9.6}$$

[12]The vectors $\mathbf{V}$ and $\mathbf{A}'$ will always be linearly independent in a two-variable, one-constraint problem. Think of the way in which $\mathbf{V}$ ($= \mathbf{P}$) and $\mathbf{A}'$ are formed in the augmented linear programming problem; p_3 must always be zero (it is the weight on the slack variable in the objective function) and a_3 must always be one (the coefficient on the slack variable in the constraint). It is not difficult to generalize this observation to linear programming problems of any size.

Using partitioned matrix notation, (9.6) can be seen to be

$$
\left[\begin{array}{c|c} 0 & \mathbf{A} \\ \hline \mathbf{A}' & \mathbf{I} \end{array}\right]\left[\begin{array}{c} w \\ \hline \mathbf{G}^0 \end{array}\right] = \left[\begin{array}{c} 0 \\ \hline \mathbf{\nabla} \end{array}\right]
$$

Following the notation in Section 1.3 on inverses of partitioned matrices, we use **R** for the partitioned inverse matrix

$$
\left[\begin{array}{c} w \\ \hline \mathbf{G}^0 \end{array}\right] = \left[\begin{array}{c|c} \mathbf{R}_{11} & \mathbf{R}_{12} \\ \hline \mathbf{R}_{21} & \mathbf{R}_{22} \end{array}\right]\left[\begin{array}{c} 0 \\ \hline \mathbf{\nabla} \end{array}\right]
$$

so that

$$
w = \mathbf{R}_{11}(0) + \mathbf{R}_{12}(\mathbf{\nabla})
$$

and

$$
\mathbf{G}^0 = \mathbf{R}_{21}(0) + \mathbf{R}_{22}(\mathbf{\nabla})
$$

From the results in Equations (1.23) and (1.24) in Chapter 1, we can find that for the equations in (9.6) $\mathbf{R}_{12} = (\mathbf{AA}')^{-1}\mathbf{A}$ and $\mathbf{R}_{22} = [\mathbf{I} - \mathbf{A}'(\mathbf{AA}')^{-1}\mathbf{A}]$, which means

$$
w = [(\mathbf{AA}')^{-1}\mathbf{A}](\mathbf{\nabla})
$$

and

$$
\mathbf{G}^0 = [\mathbf{I} - \mathbf{A}'(\mathbf{AA}')^{-1}\mathbf{A}](\mathbf{\nabla}) \tag{9.7}
$$

Finally, from (9.7)

$$
\mathbf{M} = \mathbf{R}_{22} = [\mathbf{I} - \mathbf{A}'(\mathbf{AA}')^{-1}\mathbf{A}] \tag{9.8}
$$

is the matrix that projects $\mathbf{\nabla}$ onto the null space of A; for that reason, it is known as a *projection matrix.*

For the illustration in (9.5), you can find that

$$
\mathbf{M} = \begin{bmatrix} 0.8333 & -0.3333 & -0.1667 \\ -0.3333 & 0.3333 & -0.3333 \\ -0.1667 & -0.3333 & 0.8333 \end{bmatrix}
$$

and so $\mathbf{G}^0 = \mathbf{M}\mathbf{\nabla} = \begin{bmatrix} 0.5 \\ 0 \\ -0.5 \end{bmatrix}$. You can also easily find that $w = 0.5$, and indeed $\mathbf{\nabla} = \mathbf{A}'(w) + \mathbf{G}^0$; here this is

$$
\begin{bmatrix} 1 \\ 1 \\ 0 \end{bmatrix} = \begin{bmatrix} 1 \\ 2 \\ 1 \end{bmatrix}(0.5) + \begin{bmatrix} 0.5 \\ 0 \\ -0.5 \end{bmatrix}
$$

as in Figure 9.3*c.*

9.2.2 Step Size

In gradient-based iterative methods for unconstrained maximization problems (Chapter 6, Section 6.3), the step-size decision at each iteration is based on some optimizing criterion. For example, in the method of steepest ascent (Cauchy's method), α^k is chosen so as to maximize Δf^k, the change in function value from $\mathbf{X}^k$ to $\mathbf{X}^{k+1}$. By contrast, in Newton's method, the value of $f(\mathbf{X}^{k+1})$—expressed as a second-order Taylor series approximation at $\mathbf{X}^k$—is maximized directly; this leads to an iterative procedure in which the scalar α^k is replaced by a matrix—the inverse of the Hessian at each step. With nonnegativity constraints, the step-size choice may be more restricted; attention must be paid to the boundaries.

The maximum possible step size, α^0, from $\mathbf{X}^0$, along the projected gradient, $\mathbf{G}^0$, depends on both the location of $\mathbf{X}^0$ and the sizes and signs of the elements in $\mathbf{G}^0$. If $\mathbf{X}^0$ is close to a boundary—for example, near to $x_i = 0$ (the nonnegativity boundary for x_i)—and if g_i^0 is negative (meaning that x_i^0 should decrease), then the maximum movement that would be allowed along $\mathbf{G}^0$ is given by $|x_i^0/g_i^0|$. If several elements in $\mathbf{G}^0$ are negative, then the maximum α^0 is min $\{|x_i^0/g_i^0|\}$ for all i for which $g_i^0 < 0$.

We can illustrate in ordinary solution space (because it requires only two dimensions), even though the application will eventually be carried out in augmented solution space. This means that the gradient, $\nabla^0 = \begin{bmatrix} p_1 \\ p_2 \end{bmatrix}$, *not* its projection, $\mathbf{G}^0$, will be used. Figure 9.4 repeats the geometry of Figure 9.2a with the addition of two sets of initial values, first at $\mathbf{X}^0 = \begin{bmatrix} 1 \\ 1 \end{bmatrix}$ and then at $\mathbf{X}^0 = \begin{bmatrix} 2 \\ 3 \end{bmatrix}$. If either $p_1 < 0$ or $p_2 < 0$ (which is not the case in our numer-

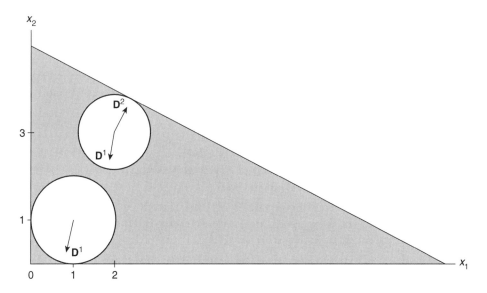

FIGURE 9.4   Gradient directions from $\begin{bmatrix} 1 \\ 1 \end{bmatrix}$ and $\begin{bmatrix} 2 \\ 3 \end{bmatrix}$ when $\nabla^0 = \begin{bmatrix} -2 \\ -10 \end{bmatrix}$ (indicated by $\mathbf{D}^1$) and $\nabla^0 = \begin{bmatrix} 6 \\ 12 \end{bmatrix}$ (indicated by $\mathbf{D}^2$).

ical illustration, but in general either, or both, could be negative), then the largest decrease that either x_1 or x_2 could undergo is

$$\min \left\{ \left| \frac{x_1^0}{p_1} \right|, \left| \frac{x_2^0}{p_2} \right| \right\}$$

So, for example, if $p_1 = -2$ and $p_2 = -10$, the largest step size possible along $\begin{bmatrix} -2 \\ -10 \end{bmatrix}$ from $\begin{bmatrix} 1 \\ 1 \end{bmatrix}$ would be $\alpha^0 = \min \{\frac{1}{2}, \frac{1}{10}\} = \frac{1}{10}$, since then $\mathbf{X}^1 = \begin{bmatrix} 1 \\ 1 \end{bmatrix} + \frac{1}{10} \begin{bmatrix} -2 \\ -10 \end{bmatrix} = \begin{bmatrix} 0.8 \\ 0 \end{bmatrix}$, and x_2 has reached its lower limit.

Geometrically, the maximum possible change in $\mathbf{X}$, from $\mathbf{X}^0$, can be seen to be equal to the radius of the largest circle that can be centered at $\mathbf{X}^0$ and inscribed in the feasible region. This will be a circle that is tangent to one boundary (or more) and interior to the others. In this illustration, the circle will have a radius of 1 and will be tangent to both the x_1 and x_2 axes.

On the other hand, if our initial point had been $\begin{bmatrix} 2 \\ 3 \end{bmatrix}$, then since $\mathbf{X}^1 = \begin{bmatrix} 2 \\ 3 \end{bmatrix} + \alpha^0 \begin{bmatrix} -2 \\ -10 \end{bmatrix}$, we could have taken a (longer) step of $\alpha^0 = \min \{1, \frac{3}{10}\} = \frac{3}{10}$ along the gradient before one of the variables becomes zero (here again it would be x_2 that reaches zero first). Suppose, however, that $\nabla^0 = \begin{bmatrix} 6 \\ 12 \end{bmatrix}$; then the boundary formed by $x_1 + 2x_2 = 10$ will be hit first, not one of the axes. These ideas are sketched in Figure 9.4.

In order to deal effectively with *all* constraints—in this case, $x_1 + 2x_2 \leq 10$ as well as $x_1 \geq 0$ and $x_2 \geq 0$—we move to augmented solution space. This is because, in the augmented problem, maintaining feasibility with respect to the $x_1 + 2x_2 \leq 10$ constraint (transformed into $x_1 + 2x_2 + x_3 = 10$) means maintaining nonnegativity of its slack variable, x_3. Increases in x_1 and/or x_2 must be accompanied by a decrease in x_3, which also has a lower limit of zero. So feasibility is maintained by selecting an $\mathbf{X}^1 = \begin{bmatrix} x_1^1 \\ x_2^1 \\ x_3^1 \end{bmatrix}$ that

(a) satisfies the *equation* $x_1 + 2x_2 + x_3 = 10$ and (b) for which $x_1^1 \geq 0$, $x_2^1 \geq 0$ and $x_3^1 \geq 0$.

In augmented solution space, the direction of movement is along the *projected* gradient, $\mathbf{G}^0$, and the largest step from a given $\mathbf{X}^0$ would be governed by which of the *three* variables, x_1, x_2, or x_3 was first driven to zero in moving along that direction. To continue with the illustration, only now in augmented solution space, suppose that we are at $\mathbf{X}^0 = \begin{bmatrix} 2 \\ 3 \\ 2 \end{bmatrix}$ and that $\mathbf{G}^0 = \begin{bmatrix} 1 \\ 2 \\ -5 \end{bmatrix}$. Then the step size is $\alpha^0 = 0.4$, since $\mathbf{X}^1 = \mathbf{X}^0 + \alpha^0 \mathbf{G}^0 = \begin{bmatrix} 2 \\ 3 \\ 2 \end{bmatrix} +$

$(0.4) \begin{bmatrix} 1 \\ 2 \\ -5 \end{bmatrix} = \begin{bmatrix} 2.4 \\ 3.8 \\ 0 \end{bmatrix}$, and x_3 has reached its lower limit (meaning in turn that $x_1 = 2.4$ and $x_2 = 3.8$ satisfy $x_1 + 2x_2 = 10$, in original solution space, exactly).

This kind of observation lies behind the notion that a better kind of starting point would be one that was more centered in the feasible region—for example, equidistant from the three nonnegativity boundaries. This logic is carried out in augmented solution space through a *scaling* of the problem.

9.2.3   Scaling

In our illustrative problem in augmented solution space, we had $\mathbf{X}^0 = \begin{bmatrix} 1 \\ 1 \\ 7 \end{bmatrix}$. One way to

have an initial point "centered" would be if it were at $\begin{bmatrix} 1 \\ 1 \\ 1 \end{bmatrix}$, for the simple reason that each

x_i (including x_3, the slack variable for the original problem) is then one unit away from its

nonnegativity boundary. One easy way to create an $\tilde{\mathbf{X}}^0 = \begin{bmatrix} 1 \\ 1 \\ 1 \end{bmatrix}$ from $\mathbf{X}^0 = \begin{bmatrix} 1 \\ 1 \\ 7 \end{bmatrix}$ would be

to define a *new* set of variables, $\tilde{\mathbf{X}}$, in which $\tilde{x}_3 = \frac{1}{7} x_3$ while x_1 and x_2 are unchanged (so
that $\tilde{x}_1 = x_1$ and $\tilde{x}_2 = x_2$).

This *scaling* of $\mathbf{X}^0$ to $\tilde{\mathbf{X}}^0$ is easily seen to be accomplished as

$$\tilde{\mathbf{X}}^0 = (\hat{\mathbf{X}}^0)^{-1} \mathbf{X}^0 \tag{9.9}$$

(Remember that putting a hat over a vector creates a diagonal matrix from that vector and
that the inverse of a diagonal matrix is another diagonal matrix containing the reciprocals
of the original elements.) In order to simplify notation, so that we do not have to keep
track of a series of both $\tilde{\mathbf{X}}$'s and $\hat{\mathbf{X}}$'s, let

$$\mathscr{D}^0 = \hat{\mathbf{X}}^0$$

so that

$$\tilde{\mathbf{X}}^0 = (\mathscr{D}^0)^{-1} \mathbf{X}^0 \tag{9.10}$$

In our example, only x_3 is transformed when it becomes $\tilde{x}_3$, but in general all elements of
any $\tilde{\mathbf{X}}^k$ may differ from the corresponding $\mathbf{X}^k$. This conversion of $\mathbf{X}^0$ into $\tilde{\mathbf{X}}^0$ in (9.9) or
(9.10) is known as an *affine transformation* or an *affine scaling* of $\mathbf{X}^0$.[13]

Having scaled the variables, we need to reexpress both the objective function and the
constraints in terms of these new variables. From (9.9) and (9.10), $\mathbf{AX}^0 = \mathbf{A}\mathscr{D}^0\tilde{\mathbf{X}}^0$ so, let-
ting $\tilde{\mathbf{A}} = \mathbf{A}\mathscr{D}^0$, the constraints in terms of the scaled variables are $\tilde{\mathbf{A}}\hat{\mathbf{X}}^0 = \mathbf{B}$. Also $\mathbf{P}'\mathbf{X}^0 =$
$\mathbf{P}'\mathscr{D}^0\tilde{\mathbf{X}}^0$ so, letting $\mathbf{P}'\mathscr{D}^0 = \tilde{\mathbf{P}}'$, $\tilde{\mathbf{P}} = \mathscr{D}^0\mathbf{P}$ and the objective function for the initial point in
the scaled problem is $\tilde{\mathbf{P}}'\tilde{\mathbf{X}}^0$. These scalings can be used to transform *any* $\mathbf{X}^k$ into an asso-

ciated $\tilde{\mathbf{X}}^k$, provided all $x_j^k > 0$. In this example, $\tilde{\mathbf{A}} = \begin{bmatrix} 1 & 2 & 7 \end{bmatrix}$ and $\tilde{\mathbf{P}} = \begin{bmatrix} 1 \\ 1 \\ 0 \end{bmatrix}$, so that the
affine scaled version of the original problem in (9.5) is[14]

$$\text{Maximize} \quad \tilde{\mathbf{P}}'\tilde{\mathbf{X}} = \tilde{x}_1 + \tilde{x}_2 + 0\tilde{x}_3$$

$$\text{Subject to} \quad \tilde{\mathbf{A}}\tilde{\mathbf{X}} = \tilde{x}_1 + 2\tilde{x}_2 + 7\tilde{x}_3 = 10 \tag{9.11}$$

$$\text{and} \quad \tilde{\mathbf{X}} \geq \mathbf{0}$$

---

[13]Formally, an affine transformation of one n-element vector, $\mathbf{X}$, into another n-element vector, $\mathbf{Y}$, is represented as $\mathbf{Y} = \mathbf{AX} + \mathbf{K}$, where $\mathbf{A}$ is an $n \times n$ matrix and $\mathbf{K}$ is an n-element vector of constants. In the particular case
when $\mathbf{K} = \mathbf{0}$, we have $\mathbf{Y} = \mathbf{AX}$, which is still an affine transformation but is also called a *linear* transformation.
[14]Since both $x_1^0 = 1$ and $x_2^0 = 1$, it turns out that there is *no* change in p_1 and p_2 in the objective function or in a_1
or a_2 in the constraint in this first scaling of the problem. That feature will change, as we are about to see, when
all unknowns have current values other than 1.

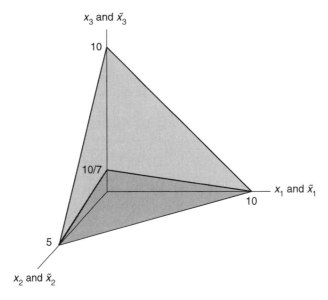

FIGURE 9.5   Feasible regions for problems presented in (9.5) and (9.11).

and for this problem we have $\tilde{\mathbf{X}}^0 = \begin{bmatrix} 1 \\ 1 \\ 1 \end{bmatrix}$. Figure 9.5 indicates the geometry in augmented solution space for both the original and the scaled versions of our illustrative problem.

The object is now to proceed with gradient projection and step-size selection, but to do it all in this scaled version of the problem in which the smallest allowable step size has been increased by the centering procedure.[15] So the same story now is told for the scaled problem. At $\tilde{\mathbf{X}}^0$, the gradient is $\tilde{\nabla}^0 = \tilde{\mathbf{P}}^0 = \begin{bmatrix} 1 \\ 1 \\ 0 \end{bmatrix}$, which points outward from the feasible triangle in Figure 9.5.[16] So we need to project $\tilde{\nabla}^0$ onto $\tilde{\mathbf{A}}\tilde{\mathbf{X}} = 10$. In this case, the projection matrix, (9.8), will be $\tilde{\mathbf{M}} = [\mathbf{I} - \tilde{\mathbf{A}}'(\tilde{\mathbf{A}}\tilde{\mathbf{A}}')^{-1}\tilde{\mathbf{A}}]$ and $\tilde{\mathbf{G}} = \tilde{\mathbf{M}}(\tilde{\nabla})$. The logic is identical to that in the original story before scaling; the only difference is all the "~"s (tildes). You can easily find that

$$\tilde{\mathbf{M}}^0 = \begin{bmatrix} 0.9815 & -0.0370 & -0.1296 \\ -0.0370 & 0.9259 & -0.2593 \\ -0.1296 & -0.2593 & 0.0926 \end{bmatrix}$$

---

[15]Imagine the point $\tilde{\mathbf{X}}^0 = \begin{bmatrix} 1 \\ 1 \\ 1 \end{bmatrix}$ on the surface of the triangle that represents the feasible region for the augmented *and scaled* problem in Figure 9.5. Since we are now in three dimensions, the circle centered on $\tilde{\mathbf{X}}^0$ has become a sphere, and with a radius of 1 it is *simultaneously* tangent to the three *planes*, $\tilde{x}_1 = 0$, $\tilde{x}_2 = 0$, and $\tilde{x}_3 = 0$, that serve as lower boundaries for $\tilde{x}_1$, $\tilde{x}_2$, and $\tilde{x}_3$.

[16]In this first iteration, $\tilde{\nabla}^0 = \nabla^0$ and $\tilde{\mathbf{P}}^0 = \mathbf{P}^0$, only because $x_1^0 = 1$ and $x_2^0 = 1$ in our illustration. As we will soon see, this is seldom the case.

so

$$\tilde{\mathbf{G}}^0 = \begin{bmatrix} 0.9444 \\ 0.8889 \\ -0.3889 \end{bmatrix}$$

and

$$\tilde{\mathbf{X}}^1 = \tilde{\mathbf{X}}^0 + (\alpha^0)\,\tilde{\mathbf{G}}^0 = \begin{bmatrix} 1 \\ 1 \\ 1 \end{bmatrix} + (\alpha^0) \begin{bmatrix} 0.9444 \\ 0.8889 \\ -0.3889 \end{bmatrix}$$

In this scaled problem, since only $\tilde{g}_3^0 < 0$, there is only one ratio of interest and $\alpha^0 = |x_3^0/g_3^0| = |1/-0.3889| = 2.5714$. It turns out to be a good idea in general not to take a full α-length step at any iteration, since that could put the next point right on a boundary which would interfere with creation of the subsequent step. One rule of thumb is to use only a proportion, ρ, of the allowable α step (where, usually, $0.9 \le \rho < 1.0$). If we employ that rule with $\rho = 0.95$, then we find

$$\tilde{\mathbf{X}}^1 = \tilde{\mathbf{X}}^0 + \rho(\alpha^0)\,\tilde{\mathbf{G}}^0 = \begin{bmatrix} 1 \\ 1 \\ 1 \end{bmatrix} + (0.95)(2.5714) \begin{bmatrix} 0.9444 \\ 0.8889 \\ -0.3889 \end{bmatrix} = \begin{bmatrix} 3.3071 \\ 3.1714 \\ 0.05 \end{bmatrix}$$

Since $\tilde{\mathbf{X}}^0 = (\mathcal{D}^0)^{-1}\mathbf{X}^0$, we can undo the scaling and return to the *unscaled* problem to find $\mathbf{X}^1 = (\mathcal{D}^0)\,\tilde{\mathbf{X}}^1$, so

$$\mathbf{X}^1 = \begin{bmatrix} 1 & 0 & 0 \\ 0 & 1 & 0 \\ 0 & 0 & 7 \end{bmatrix} \begin{bmatrix} 3.3071 \\ 3.1714 \\ 0.05 \end{bmatrix} = \begin{bmatrix} 3.3071 \\ 3.1714 \\ 0.35 \end{bmatrix}$$

In particular, $x_1^1 = 3.3071$ and $x_2^1 = 3.1714$, as compared to $x_1^0 = 1$ and $x_2^0 = 1$. The objective function value is now $\mathbf{P}'\mathbf{X}^1 = 6.4785$; we started with $\mathbf{P}'\mathbf{X}^0 = 2$.

Subsequent steps follow exactly these same rules:

(a) Given $\mathbf{X}^k$, form the diagonal matrix $\mathcal{D}^k = \hat{\mathbf{X}}^k$. Then $\tilde{\mathbf{X}}^k = (\mathcal{D}^k)^{-1}\mathbf{X}^k = \mathbf{1}$ (where, as usual, $\mathbf{1}$ represents a column vector of ones), $\tilde{\mathbf{A}}^k = \mathbf{A}(\mathcal{D}^k)$ and $\tilde{\mathbf{V}}^k = (\mathcal{D}^k)\mathbf{P}$.[17]

(b) Find the matrix that projects the current gradient onto the augmented feasible region of the scaled problem:

$$\tilde{\mathbf{M}}^k = \{\mathbf{I} - (\tilde{\mathbf{A}}^k)'[(\tilde{\mathbf{A}}^k)(\tilde{\mathbf{A}}^k)']^{-1}(\tilde{\mathbf{A}}^k)\}$$

so that the projected gradient is $\tilde{\mathbf{G}}^k = \tilde{\mathbf{M}}^k(\tilde{\mathbf{V}}^k)$.

(c) To establish a step size, identify $\alpha^k = \min \{|\tilde{x}_i^k/\tilde{g}_i^k|\}$ for all $\tilde{g}_i^k$ that are negative.

(d) The new point in the scaled problem is then

$$\tilde{\mathbf{X}}^{k+1} = \tilde{\mathbf{X}}^k + \Delta\tilde{\mathbf{X}}^k = \tilde{\mathbf{X}}^k + (0.95)\alpha^k\tilde{\mathbf{G}}^k$$

---

[17]Note that at each step, it is the *original* $\mathbf{A}$ and $\mathbf{P}$ that are scaled for the current step.

(e) Finally, the new point in the unscaled problem is

$$\mathbf{X}^{k+1} = (\mathcal{D}^k)\,\tilde{\mathbf{X}}^{k+1}$$

and this becomes $\mathbf{X}^k$ in step (a) for the next iteration, unless some stopping criterion is met by $\mathbf{X}^k$. We will explore stopping (and starting) rules below.

Table 9.4 shows how this approach works out for the sample problem in (9.4). (Using only four decimal places masks activity that serves to distinguish the last few steps more clearly than can be seen in the table.)

It is clear that, as predicted, the simplex method would have worked better on this small problem. Starting at $\mathbf{X}^0 = \begin{bmatrix} 0 \\ 0 \end{bmatrix}$ and ignoring the augmented form of the problem, at most

TABLE 9.4   An Affine Scaling Approach to the Linear Programming Problem in (9.4)

k	$\mathbf{X}^k$	$\tilde{\mathbf{X}}^k$	$\tilde{\mathbf{G}}^k$	$(0.95)\alpha^k$	$\Delta\tilde{\mathbf{X}}^k$	$\Delta\mathbf{X}^k$	$\mathbf{P}'\mathbf{X}^k$
0	$\begin{bmatrix} 1 \\ 1 \\ 7 \end{bmatrix}$	$\begin{bmatrix} 1 \\ 1 \\ 1 \end{bmatrix}$	$\begin{bmatrix} 0.9444 \\ 0.8889 \\ -0.3889 \end{bmatrix}$	2.4429	$\begin{bmatrix} 2.3071 \\ 2.1714 \\ -6.95 \end{bmatrix}$	$\begin{bmatrix} 2.3071 \\ 2.1714 \\ -6.65 \end{bmatrix}$	2
1	$\begin{bmatrix} 3.3071 \\ 3.1714 \\ 0.35 \end{bmatrix}$	$\begin{bmatrix} 1 \\ 1 \\ 1 \end{bmatrix}$	$\begin{bmatrix} 1.3049 \\ -0.6687 \\ -0.2119 \end{bmatrix}$	1.4027	$\begin{bmatrix} 1.8539 \\ -0.95 \\ -0.3010 \end{bmatrix}$	$\begin{bmatrix} 6.1311 \\ -3.0129 \\ -0.1054 \end{bmatrix}$	6.4785
2	$\begin{bmatrix} 9.4382 \\ 0.1586 \\ 0.2446 \end{bmatrix}$	$\begin{bmatrix} 1 \\ 1 \\ 1 \end{bmatrix}$	$\begin{bmatrix} 0.0116 \\ -0.1582 \\ -0.2443 \end{bmatrix}$	3.8881	$\begin{bmatrix} 0.0453 \\ -0.6150 \\ -0.9500 \end{bmatrix}$	$\begin{bmatrix} 0.4274 \\ -0.0975 \\ -0.2324 \end{bmatrix}$	9.5968
3	$\begin{bmatrix} 9.8657 \\ 0.0610 \\ 0.0122 \end{bmatrix}$	$\begin{bmatrix} 1 \\ 1 \\ 1 \end{bmatrix}$	$\begin{bmatrix} 0.0008 \\ -0.0610 \\ -0.0122 \end{bmatrix}$	15.5645	$\begin{bmatrix} 0.0120 \\ -0.9500 \\ -0.1904 \end{bmatrix}$	$\begin{bmatrix} 0.1183 \\ -0.0580 \\ -0.0023 \end{bmatrix}$	9.9267
4	$\begin{bmatrix} 9.9840 \\ 0.0031 \\ 0.0099 \end{bmatrix}$	$\begin{bmatrix} 1 \\ 1 \\ 1 \end{bmatrix}$	$\begin{bmatrix} 0.0000 \\ -0.0031 \\ -0.0099 \end{bmatrix}$	95.9282	$\begin{bmatrix} 0.0011 \\ -0.2928 \\ -0.9500 \end{bmatrix}$	$\begin{bmatrix} 0.0112 \\ -0.0009 \\ -0.0094 \end{bmatrix}$	9.9871
5	$\begin{bmatrix} 9.9952 \\ 0.0022 \\ 0.0005 \end{bmatrix}$	$\begin{bmatrix} 1 \\ 1 \\ 1 \end{bmatrix}$	$\begin{bmatrix} 0.0000 \\ -0.0022 \\ -0.0005 \end{bmatrix}$	440	$\begin{bmatrix} 0.0004 \\ -0.9500 \\ -0.2179 \end{bmatrix}$	$\begin{bmatrix} 0.0042 \\ -0.0021 \\ -0.0001 \end{bmatrix}$	9.9974
6	$\begin{bmatrix} 9.9994 \\ 0.0001 \\ 0.0004 \end{bmatrix}$	$\begin{bmatrix} 1 \\ 1 \\ 1 \end{bmatrix}$	$\begin{bmatrix} 0.0000 \\ -0.0001 \\ -0.0004 \end{bmatrix}$	2,453	$\begin{bmatrix} 0.0000 \\ -0.2648 \\ -0.9500 \end{bmatrix}$	$\begin{bmatrix} 0.0004 \\ 0.0000 \\ -0.0004 \end{bmatrix}$	9.9995
7	$\begin{bmatrix} 9.9998 \\ 0.0001 \\ 0.0000 \end{bmatrix}$	$\begin{bmatrix} 1 \\ 1 \\ 1 \end{bmatrix}$	$\begin{bmatrix} 0.0000 \\ -0.0001 \\ -0.0002 \end{bmatrix}$	11,971	$\begin{bmatrix} 0.0000 \\ -0.9500 \\ -0.2318 \end{bmatrix}$	$\begin{bmatrix} 0.0002 \\ -0.0001 \\ 0.0000 \end{bmatrix}$	9.9999
8	$\begin{bmatrix} 10.0000 \\ 0.0000 \\ 0.0000 \end{bmatrix}$	$\begin{bmatrix} 1 \\ 1 \\ 1 \end{bmatrix}$	—	—	Stop	—	—

TABLE 9.5   The Affine Scaling Method on Problem (7.21) in Chapter 7

k	$(\mathbf{X}^k)'$	$\mathbf{P}'\mathbf{X}^k$
0	[1, 1]	7
1	[2.0791, 3.6471]	22.3937
2	[2.7367, 3.6160]	23.5533
3	[2.4908, 3.7454]	23.7086
4	[2.5100, 3.7445]	23.7427
5	[2.4997, 3.7498]	23.7485
6	[2.5004, 3.7498]	23.7497
7	[2.5000, 3.7500]	23.7500

two pivot operations would be needed to reach the optimal corner, $\mathbf{X}^* = \begin{bmatrix} 10 \\ 0 \end{bmatrix}$, in original solution space. Table 9.5 shows how affine scaling works on the sample problem that was given in (7.21) in Chapter 7.

9.2.4   Starting and Stopping

Starting   The initial scaling operation requires that we form $\mathcal{D}^0 = \hat{\mathbf{X}}^0$ and then use it to generate $\tilde{\mathbf{X}}^0 = (\mathcal{D}^0)^{-1}\mathbf{X}^0$ [as in (9.10)]. This means tha $\hat{\mathbf{X}}^0$ must be nonsingular. For a diagonal matrix, this means that all elements must be nonzero. If $\mathbf{X}^0$ is *strictly inside* the feasible region in original solution space, this is guaranteed. In many problems, it is possible to use $x_1^0 = 1, \ldots, x_n^0 = 1$, in conjunction with $x_{n+1}^0, \ldots, x_{n+m}^0$ at appropriate values. This is just for convenience; the $x_i^0 = 1$ $(i = 1, \ldots, n)$ are easily tested for feasibility in original solution space. (Specifically, is $\mathbf{A1} < \mathbf{B}$, so that all slack variables are also positive?) If the point *is* feasible, then values of the slack variables $(x_{n+1}^0, \ldots, x_{n+m}^0)$ are easily found as $\mathbf{B} - \mathbf{A1}$.

If the constraints of the problem are violated by this $\mathbf{X}^0$, then there are many "start-up" procedures designed to create an initial feasible solution in which $\mathbf{X}^0 > \mathbf{0}$. These details are not important for an intuitive understanding of interior point methods, and they can be found in advanced linear programming texts, including those cited in the references to this chapter. However, it is worth mentioning that finding an initial feasible solution *can* turn out to be a very time-consuming part of the calculations. As we saw in Part III for iterative methods in general, it is also true that the performance of an affine scaling (or other interior point) algorithm may be sensitive to the starting point that is used.

Stopping   Any number of stopping rules can be used. In Chapter 6 we looked at several.

(a) One of these is based on the relative change in the objective function from one iteration to the next. Using the earlier notation $\Pi^k = \mathbf{P}'\mathbf{X}^k$, this rule for the linear programming case might be

$$\frac{|\Pi^{k+1} - \Pi^k|}{|\Pi^k|} < \varepsilon$$

or

$$\frac{|\Pi^{k+1} - \Pi^k|}{(0.5)(|\Pi^{k+1}| + |\Pi^k|)} < \varepsilon$$

for some small value of ε, such as $\varepsilon = 10^{-6}$. (Notice that to invoke this rule in our illustration, we need to have information beyond the four decimal places that are shown in Tables 9.4 and 9.5.)

(b) Another stopping rule was based on the size of relative changes in the unknown vector itself, such as

$$\frac{\|\mathbf{X}^{k+1} - \mathbf{X}^k\|}{\|\mathbf{X}^k\|} < \varepsilon$$

or

$$\frac{\|\mathbf{X}^{k+1} - \mathbf{X}^k\|}{(0.5)(\|\mathbf{X}^{k+1} + \mathbf{X}^k\|)} < \varepsilon$$

where $\|\mathbf{X}^k\|$ denotes some vector *norm*, such as $\sqrt{(x_1^k)^2 + \cdots + (x_n^k)^2}$.

(c) For gradient-based methods, a stopping rule can be made to depend on the closeness of $|\nabla^k|$ to a null vector (or the same test can be imposed on the projected gradient, $|\mathbf{G}^k|$). Again, then, for some small, preselected value of ε, we could choose to stop when

$$|\nabla^k| < \varepsilon \qquad \text{or} \qquad |\mathbf{G}^k| < \varepsilon$$

(d) Other kinds of stopping rules are based on the result in Theorem PD1 in Chapter 7. For a primal maximization problem and its dual minimization problem, the theorem says that, for any pair of *feasible* solutions to both primal and dual problems, the value of the dual objective function provides an upper limit to the value of the primal objective function. Some interior point methods calculate an estimate of a solution to the dual problem at each iteration (in addition to a solution to the primal; more about this in the sections that follow). Provided the estimated dual solution is also feasible, then the dual objective function at the kth iteration, $\mathbf{B}'\mathbf{Y}^k$, bounds $\mathbf{P}'\mathbf{X}^k$ from above, and a stopping criterion can be based on what is known as the "duality gap":

$$[|\mathbf{B}'\mathbf{Y}^k| - |\mathbf{P}'\mathbf{X}^k|] < \varepsilon$$

or, using earlier notation, where $\Delta^k = \mathbf{B}'\mathbf{Y}^k$, we have

$$[|\Delta^k| - |\Pi^k|] < \varepsilon$$

9.3   PATH-FOLLOWING INTERIOR POINT METHODS

It is clear that the affine scaling method in Section 9.2 generated a "path" from the initial point, $\mathbf{X}^0 = \begin{bmatrix} 1 \\ 1 \\ 7 \end{bmatrix}$, to the final optimal point, $\mathbf{X}^8 = \begin{bmatrix} 10 \\ 0 \\ 0 \end{bmatrix}$—namely, the series of points indicated in the $\mathbf{X}^k$ column of Table 9.4. In that sense, it could legitimately be called a *path-following* algorithm. Indeed, the same is true for *any* iterative method that travels from $\mathbf{X}^0$ to $\mathbf{X}^*$ via a series of intermediate points.[18] However, in the literature on interior point methods, the path-following label is usually reserved for the approach that we will explore in this section.

We begin, again, with the augmented form of the sample problem, in (9.5). It is repeated here. What we are going to do is create a problem in which the objective is to maximize a *nonlinear* function subject to only linear *equality* constraints. The reason? So that then the Lagrange multiplier approach (Chapter 4) can be used. It will turn out to be useful to replace x_3 by the more usual notation for a slack variable; in this case, since there is only one constraint, we simply use s.

$$\text{Maximize} \quad x_1 + x_2 + 0s$$

$$\text{subject to} \quad x_1 + 2x_2 + s = 10$$

$$\text{and} \quad x_1 \geq 0, x_2 \geq 0, s \geq 0$$

The first step is to remove the nonnegativities (the only remaining *inequality* constraints) by creating what is known as a *barrier function*. This function serves to encourage avoidance of the boundaries that are established by the $\mathbf{X} \geq \mathbf{0}$ and $s \geq 0$ constraints. One generally successful form is a *logarithmic barrier function* expressed as

$$f(x_1, x_2, s) = x_1 + x_2 + \mu(\ln x_1 + \ln x_2 + \ln s)$$

where μ is some prespecified "small" positive number (we return to this below). Since $\lim_{x \to 0} \ln x = -\infty$, it is clear that for any $\mu > 0$, the logarithmic term becomes extremely small as x_1, x_2, and/or s approach zero, their lower limits. And the smaller μ, the closer $f(\mathbf{X})$ is to the objective function in (9.5). With an objective of *maximizing* $f(\mathbf{X})$, this means that the nonnegativity boundaries will be avoided.

So now the revised version of the problem in (9.5) is to

$$\text{Maximize} \quad f(x_1, x_2, s) = x_1 + x_2 + \mu(\ln x_1 + \ln x_2 + \ln s)$$

$$\text{subject to} \quad x_1 + 2x_2 + s = 10$$

and for this we can use Lagrange. The strategy is to solve a *series* of problems, with decreasing values of μ (but always choosing $\mu > 0$). The set of optimal solutions to this series of problems is called the *central path* for the problem, and this is the "path" that is referred to in the title to this section.

The Lagrangian function is

$$L = x_1 + x_2 + \mu(\ln x_1 + \ln x_2 + \ln s) - \lambda(x_1 + 2x_2 + s - 10)$$

---

[18]It is also true for the simplex method, although in that case the points are all on the boundaries (at the corners) of the feasible region, not interior to it.

Standard first-order conditions are

$$\frac{\partial L}{\partial x_1} = x_1 + \mu - \lambda x_1 = 0$$

$$\frac{\partial L}{\partial x_2} = x_2 + \mu - 2\lambda x_2 = 0$$

$$\frac{\partial L}{\partial s} = \frac{\mu}{s} - \lambda = 0$$ (9.12)

$$\frac{\partial L}{\partial \lambda} = x_1 + 2x_2 + s = 10$$

For a given value of μ, this is a set of (not very) nonlinear equations in x_1, x_2, s, and λ, and any of the iterative methods from Chapter 5 could be applied. Indeed, one of the features that distinguishes one from another of the many available variations on path-following interior point methods is in how these kinds of first-order conditions are handled. It is not much of an issue in this small problem, but it can be extremely important in large problems which, again, are the sort for which interior point approaches may have an advantage over the simplex method.

For this reason, before we examine the solutions to these first-order conditions for our sample problem for a series of values of μ, it is instructive to generalize the characteristics of a path-following approach for a problem of any size. We explore simultaneously the first-order conditions for a pair of primal and dual augmented linear programs:

$$\begin{aligned} \text{Maximize} \quad & \mathbf{P'X} + \mathbf{0S} \\ \text{subject to} \quad & \mathbf{AX} + \mathbf{IS} = \mathbf{B} \\ \text{and} \quad & \mathbf{X} \geq \mathbf{0}, \mathbf{S} \geq \mathbf{0} \end{aligned}$$ (9.13)

and

$$\begin{aligned} \text{Minimize} \quad & \mathbf{B'Y} + \mathbf{0T} \\ \text{subject to} \quad & \mathbf{A'Y} - \mathbf{IT} = \mathbf{P} \\ \text{and} \quad & \mathbf{Y} \geq \mathbf{0}, \mathbf{T} \geq \mathbf{0} \end{aligned}$$ (9.14)

You can determine that the barrier functions will be

$$f(\mathbf{X,S}) = \mathbf{P'X} + \mu\left(\sum_{j=1}^{n} \ln x_j + \sum_{i=1}^{m} \ln s_i\right)$$

and

$$f(\mathbf{Y,T}) = \mathbf{B'Y} - \mu\left(\sum_{i=1}^{m} \ln y_i + \sum_{j=1}^{n} \ln t_j\right)$$

Using $\mathbf{Y}$ for the Lagrange multipliers in the primal problem and $\mathbf{X}$ for the Lagrange multipliers in the dual problem, the corresponding Lagrangian optimization problems will be

$$\text{Maximize } L^{\mathrm{P}}(\mathbf{X},\mathbf{S},\mathbf{Y};\ \mu) = \mathbf{P}'\mathbf{X} + \mu\left(\sum_{j=1}^{n} \ln x_j + \sum_{i=1}^{m} \ln s_i\right) - \mathbf{Y}'(\mathbf{AX} + \mathbf{S} - \mathbf{B}) \quad (9.15)$$

and

$$\text{Minimize } L^{\mathrm{D}}(\mathbf{Y},\mathbf{T},\mathbf{X};\ \mu) = \mathbf{B}'\mathbf{Y} - \mu\left(\sum_{i=1}^{m} \ln y_i + \sum_{j=1}^{n} \ln t_j\right) - \mathbf{X}'(\mathbf{A}'\mathbf{Y} - \mathbf{T} - \mathbf{P}) \quad (9.16)$$

Using "hats," as usual, to create diagonal matrices out of vectors, you can work out that the first-order conditions for maximization of L^{P} and minimization of L^{D} are compactly expressed as, respectively:

$$\mathbf{P} + \mu(\hat{\mathbf{X}})^{-1}\mathbf{1} - \mathbf{A}'\mathbf{Y} = \mathbf{0} \quad (9.17a)$$

$$\mu(\hat{\mathbf{S}})^{-1}\mathbf{1} - \mathbf{Y} = \mathbf{0} \quad (9.17b)$$

$$\mathbf{AX} + \mathbf{S} = \mathbf{B} \quad (9.17c)$$

and

$$\mathbf{B} - \mu(\hat{\mathbf{Y}})^{-1}\mathbf{1} - \mathbf{AX} = \mathbf{0} \quad (9.18a)$$

$$-\mu(\hat{\mathbf{T}})^{-1}\mathbf{1} + \mathbf{X} = \mathbf{0} \quad (9.18b)$$

$$\mathbf{A}'\mathbf{Y} - \mathbf{T} = \mathbf{P} \quad (9.18c)$$

From (9.17b), $\hat{\mathbf{S}}\hat{\mathbf{Y}}\mathbf{1} = \mu\mathbf{1}$ or $\mathbf{S} = \mu(\hat{\mathbf{Y}})^{-1}\mathbf{1}$, so (9.18a) is just (9.17c); from (9.18b), $\hat{\mathbf{T}}\hat{\mathbf{X}}\mathbf{1} = \mu\mathbf{1}$ or $\mathbf{T} = \mu(\hat{\mathbf{X}})^{-1}\mathbf{1}$, so (9.17a) is just (9.18c). Putting all of this together, we have

$$\mathbf{AX} + \mathbf{S} = \mathbf{B} \quad (9.19a)$$

$$\mathbf{A}'\mathbf{Y} - \mathbf{T} = \mathbf{P} \quad (9.19b)$$

$$\hat{\mathbf{T}}\hat{\mathbf{X}}\mathbf{1} = \mu\mathbf{1} \quad (9.19c)$$

$$\hat{\mathbf{S}}\hat{\mathbf{Y}}\mathbf{1} = \mu\mathbf{1} \quad (9.19d)$$

Note that (9.19a) and (9.19b) are simply the constraints in the primal and dual problems in (9.13) and (9.14). Additionally, when $\mu = 0$, (9.19c) and (9.19d) are $s_i y_i = 0$ ($i = 1, \ldots, m$) and $t_j x_j = 0$ ($j = 1, \ldots, n$), precisely the complementary slackness conditions characterizing an optimal solution, as in Theorem PD3 in Chapter 7. Finally, if $\mathbf{X} \geq \mathbf{0}$ and $\mathbf{Y} \geq \mathbf{0}$, then, from (9.19c) and (9.19d), again, both $\mathbf{S}$ and $\mathbf{T}$ will also be nonnegative.

Rewriting (9.19), we have

$$f^1(\mathbf{X,Y,S,T}) = \mathbf{AX} + \mathbf{S} - \mathbf{B} = \mathbf{0}$$

$$f^2(\mathbf{X,Y,S,T}) = \mathbf{A'Y} - \mathbf{T} - \mathbf{P} = \mathbf{0}$$

$$f^3(\mathbf{X,Y,S,T}) = \mathbf{\hat{T}\hat{X}1} - \mu\mathbf{1} = \mathbf{0}$$

$$f^4(\mathbf{X,Y,S,T}) = \mathbf{\hat{S}\hat{Y}1} - \mu\mathbf{1} = \mathbf{0}$$

(9.20)

or, compactly

$$\mathbf{F(X,Y,S,T)} = \begin{bmatrix} f^1(\mathbf{X,Y,S,T}) \\ f^2(\mathbf{X,Y,S,T}) \\ f^3(\mathbf{X,Y,S,T}) \\ f^4(\mathbf{X,Y,S,T}) \end{bmatrix} = \mathbf{0}$$

(9.21)

This is the set of nonlinear equations that constitute the first-order conditions for the Lagrangian optimization problems in (9.15) and (9.16).

If we were to use the Newton–Raphson approach (Chapter 5, Section 5.2) to finding a solution for this set of equations, we would invoke the second-order Taylor series approximation, with $\mathbf{V} = \begin{bmatrix} \mathbf{X} \\ \mathbf{Y} \\ \mathbf{S} \\ \mathbf{T} \end{bmatrix}$ and $\Delta\mathbf{V}^k = \mathbf{V}^{k+1} - \mathbf{V}^k = \begin{bmatrix} \Delta\mathbf{X} \\ \Delta\mathbf{Y} \\ \Delta\mathbf{S} \\ \Delta\mathbf{T} \end{bmatrix}$. This approximation is

$$\mathbf{F(V}^{k+1}) = \mathbf{F(V}^k) + \mathbf{J(V}^k)(\Delta\mathbf{V}^k)$$

[as in (5.21) in Chapter 5], from which, for $\mathbf{F(V}^{k+1}) = \mathbf{0}$, we have

$$\Delta\mathbf{V}^k = -[\mathbf{J(V}^k)]^{-1}\mathbf{F(V}^k)$$

(9.22)

[(5.22) in Chapter 5], or

$$\mathbf{J(V}^k)\Delta\mathbf{V}^k = -\mathbf{F(V}^k)$$

(9.23)

The Jacobian for this problem is

$$\begin{bmatrix} \dfrac{\partial f^1}{\partial \mathbf{X}} & \dfrac{\partial f^1}{\partial \mathbf{Y}} & \dfrac{\partial f^1}{\partial \mathbf{S}} & \dfrac{\partial f^1}{\partial \mathbf{T}} \\ \vdots & \vdots & \vdots & \vdots \\ \vdots & \vdots & \vdots & \vdots \\ \dfrac{\partial f^4}{\partial \mathbf{X}} & \dfrac{\partial f^4}{\partial \mathbf{Y}} & \dfrac{\partial f^4}{\partial \mathbf{S}} & \dfrac{\partial f^4}{\partial \mathbf{T}} \end{bmatrix} = \begin{bmatrix} \mathbf{A} & \mathbf{0} & \mathbf{I} & \mathbf{0} \\ \mathbf{0} & \mathbf{A'} & \mathbf{0} & -\mathbf{I} \\ \mathbf{\hat{T}} & \mathbf{0} & \mathbf{0} & \mathbf{\hat{X}} \\ \mathbf{0} & \mathbf{\hat{S}} & \mathbf{\hat{Y}} & \mathbf{0} \end{bmatrix}$$

(9.24)

so

$$
\begin{bmatrix}
\mathbf{A} & \mathbf{0} & \mathbf{I} & \mathbf{0} \\
\mathbf{0} & \mathbf{A'} & \mathbf{0} & -\mathbf{I} \\
\hat{\mathbf{T}} & \mathbf{0} & \mathbf{0} & \hat{\mathbf{X}} \\
\mathbf{0} & \hat{\mathbf{S}} & \hat{\mathbf{Y}} & \mathbf{0}
\end{bmatrix}
\begin{bmatrix}
\Delta\mathbf{X} \\
\Delta\mathbf{Y} \\
\Delta\mathbf{S} \\
\Delta\mathbf{T}
\end{bmatrix}
=
\begin{bmatrix}
\mathbf{B} - \mathbf{AX} - \mathbf{S} \\
\mathbf{P} - \mathbf{A'Y} + \mathbf{T} \\
\mu\mathbf{1} - \hat{\mathbf{T}}\hat{\mathbf{X}}\mathbf{1} \\
\mu\mathbf{1} - \hat{\mathbf{S}}\hat{\mathbf{Y}}\mathbf{1}
\end{bmatrix}
\tag{9.25}
$$

and $\mathbf{V}^{k+1} = \mathbf{V}^k + \alpha^k \Delta\mathbf{V}^k$.

The step parameter, α, is needed to ensure that nonnegativity is preserved for all of the elements in $\mathbf{X}^{k+1}$ and in $\mathbf{Y}^{k+1}$. There is more than one way to decide on the "best" step length at each iteration, and some algorithms take only *one* step in each Newton–Raphson iteration. Details can be found in many references, including Vanderbei (1996) or den Hertog (1994).

The results in (9.25) are really a set of Kuhn–Tucker (or Karush–Kuhn–Tucker) first-order conditions for the problems in (9.13) and (9.14) (Chapter 4, Section 4.5), only here they are applied to a pair of problems in which the inequality constraints have been converted to equations. For this reason, (9.25) is often referred to in the literature as the *KKT system* for the path-following algorithms.

Rather than finding the inverse of the Jacobian at each step in (9.22) or (9.25), there are numerous alternative iterative methods for solving this system of nonlinear equations: modified Newton and quasi-Newton approaches, including a rank 1 updating scheme due to Broyden, with or without the Sherman–Morrison–Woodbury (SMW) approximation. These techniques and others are covered in Chapter 5, Section 5.2. The precise way in which a technique is implemented can have a major impact on the resulting complexity of the algorithm, and this is the source of differentiation among many of the path-following interior point methods for linear programs.

Rather than deal with all of the equations in (9.25) directly, they may be simplified in the following way. From the third and fourth equations, respectively, $\Delta\mathbf{T}$ and $\Delta\mathbf{S}$ are found as

$$
\Delta\mathbf{T} = (\hat{\mathbf{X}})^{-1}(\mu\mathbf{1} - \hat{\mathbf{T}}\hat{\mathbf{X}}\mathbf{1} - \hat{\mathbf{T}}\,\Delta\mathbf{X})
\tag{9.26}
$$

$$
\Delta\mathbf{S} = (\hat{\mathbf{Y}})^{-1}(\mu\mathbf{1} - \hat{\mathbf{S}}\hat{\mathbf{Y}}\mathbf{1} - \hat{\mathbf{S}}\,\Delta\mathbf{Y})
\tag{9.27}
$$

Substituting $\Delta\mathbf{S}$ into the first equation in (9.25) and $\Delta\mathbf{T}$ into the second equation gives

$$
\mathbf{A}\,\Delta\mathbf{X} - (\hat{\mathbf{Y}})^{-1}\,\hat{\mathbf{S}}\,\Delta\mathbf{Y} = \mathbf{B} - \mathbf{AX} - \mu(\hat{\mathbf{Y}})^{-1}\mathbf{1}
$$

$$
\mathbf{A'}\Delta\mathbf{Y} + (\hat{\mathbf{X}})^{-1}\,\hat{\mathbf{T}}\,\Delta\mathbf{X} = \mathbf{P} - \mathbf{A'Y} + \mu(\hat{\mathbf{X}})^{-1}\mathbf{1}
$$

In matrix form, putting the second equation first, this is

$$
\begin{bmatrix}
(\hat{\mathbf{X}})^{-1}\hat{\mathbf{T}} & \mathbf{A'} \\
\mathbf{A} & -(\hat{\mathbf{Y}})^{-1}\hat{\mathbf{S}}
\end{bmatrix}
\begin{bmatrix}
\Delta\mathbf{X} \\
\Delta\mathbf{Y}
\end{bmatrix}
=
\begin{bmatrix}
\mathbf{P} - \mathbf{A'Y} + \mu(\hat{\mathbf{X}})^{-1}\mathbf{1} \\
\mathbf{B} - \mathbf{AX} - \mu(\hat{\mathbf{Y}})^{-1}\mathbf{1}
\end{bmatrix}
$$

This is known as the *reduced KKT system* for the problem. This, too, could be solved completely, using some iterative procedure.

However, to indicate the nature of the calculations that are involved, suppose that we solve the first set of equations for $\Delta\mathbf{X}$:

$$\Delta\mathbf{X} = [\hat{\mathbf{X}}(\hat{\mathbf{T}})^{-1}][\mathbf{P} - \mathbf{A}'\mathbf{Y} + \mu(\hat{\mathbf{X}})^{-1}\mathbf{1} - \mathbf{A}'\Delta\mathbf{Y}] \tag{9.28}$$

and substitute this result into the second set of equations, which will then contain only $\Delta\mathbf{Y}$ as unknowns:

$$-[(\hat{\mathbf{Y}})^{-1}\,\hat{\mathbf{S}} + \mathbf{A}\hat{\mathbf{X}}(\hat{\mathbf{T}})^{-1}\mathbf{A}']\Delta\mathbf{Y} = \mathbf{B} - \mathbf{AX} - \mu(\hat{\mathbf{Y}})^{-1}\mathbf{1} - [\mathbf{A}\hat{\mathbf{X}}(\hat{\mathbf{T}})^{-1}][\mathbf{P} - \mathbf{A}'\mathbf{Y} + \mu(\hat{\mathbf{X}})^{-1}\mathbf{1}] \tag{9.29}$$

This is in the form $(\mathbf{M} + \mathbf{ANA}')\Delta\mathbf{Y} = \mathbf{K}$. Note that $\mathbf{M} = (\hat{\mathbf{Y}})^{-1}\,\hat{\mathbf{S}}$ is a diagonal matrix that contains strictly positive elements (as long as all variables are strictly positive, which will be true of procedures that remain interior to the feasible region). It is therefore positive definite (and symmetric, as are all diagonal matrices). The same is true of $\mathbf{N} = \hat{\mathbf{X}}(\hat{\mathbf{T}})^{-1}$, and this assures that $\mathbf{ANA}'$ is positive definite; it is also symmetric, as is any matrix in the form $\mathbf{ADA}'$, where $\mathbf{D}$ is diagonal. (Problem 9.3 asks you to show that these statements are true in general.) Finally, then, $\mathbf{M} + \mathbf{ANA}'$ is a symmetric positive definite matrix of the form $\mathbf{A}\Theta\mathbf{A}'$. This means that a specific kind of factorization can be used to solve for $\Delta\mathbf{Y}$ in (9.29), rather than finding the inverse of $\mathbf{A}\Theta\mathbf{A}'$ directly. This is known as a *Cholesky factorization*; it has come to play an important role in many interior point methods, so we explore it briefly in Section 9.4.

With $\Delta\mathbf{Y}$ found from (9.29), substitution into (9.28) will give $\Delta\mathbf{X}$, and further substitution into (9.26) and (9.27) will give $\Delta\mathbf{S}$ and $\Delta\mathbf{T}$.[19] All of this is for one chosen value of μ. After step k, a $\mu^{k+1} < \mu^k$ is selected and the problem is resolved. An obvious question is how long does this continue? One stopping criterion makes use of the result in (7.33) in Chapter 7, namely, that for a pair of *feasible* primal and dual solutions:

$$\Delta - \Pi = \mathbf{B}'\mathbf{Y} - \mathbf{P}'\mathbf{X} = \mathbf{Y}'\mathbf{S} + \mathbf{X}'\mathbf{T}$$

This is sometimes called the *duality gap*. It can be expressed as $\mathbf{1}'\hat{\mathbf{S}}\hat{\mathbf{Y}}\mathbf{1} + \mathbf{1}'\hat{\mathbf{T}}\hat{\mathbf{X}}\mathbf{1}$, which hardly seems like an improvement, except that when these variables obey the results in (9.19c) and (9.19d), this means that the duality gap is simply $\Delta - \Pi = (m + n)\mu$.

Again, alternative expressions of the basic problems will generate different expressions for this gap. For example, if the maximization problem in (9.13) is written as

$$\text{Maximize} \quad \mathbf{P}'\mathbf{X}$$
$$\text{subject to} \quad \mathbf{AX} = \mathbf{B}$$
$$\text{and} \quad \mathbf{X} \geq \mathbf{0}$$

then n, the number of decision variables in this version of the problem, is what was $n + m$ in the expressions in (9.15). Here, then, the duality gap would be represented as $n\mu$. In any case, a stopping rule can be tied to the size of the duality gap.

---

[19]There are many alternative derivations of $\Delta\mathbf{X}$ and $\Delta\mathbf{Y}$ (and, sometimes, $\Delta\mathbf{S}$ and $\Delta\mathbf{T}$) results in the literature, and at first reading there seem to be many and differing sets of results. Much of the potential for confusion comes from the fact that some authors use a minimization problem as the primal (with x's as decision variables); for others, a maximization problem constitutes the primal (but again with x's for decision variables). Additionally, many authors express their primal in augmented form and hence do not consider explicitly a set of slack (or surplus) variables for that problem. That, in turn, means that their dual variables are unrestricted in sign.

There are many further subtleties, such as rules for selection of a sequence of values of μ to ensure that the objective function value changes by at least some small amount in each successive version of the problem. These and other details (of which there are many) are nicely covered in Vanderbei (1996).

Returning to our sample problem in (9.5), if we define $t_1 = \mu/x_1$ and $t_2 = \mu/x_2$, so that $\mathbf{T} = \mu(\hat{\mathbf{X}})^{-1}\mathbf{1}$, the first-order conditions in (9.12) can be reexpressed as

$$x_1 + 2x_2 + s - 10 = 0$$
$$y - t_1 - 1 = 0$$
$$2y - t_2 - 1 = 0$$
$$t_1 x_1 - \mu = 0 \tag{9.30}$$
$$t_2 x_2 - \mu = 0$$
$$sy - \mu = 0$$

The Jacobian matrix, in (9.24), is

$$\mathbf{J}(\mathbf{X}, y, s, \mathbf{T}) = \begin{bmatrix} 1 & 2 & 0 & 1 & 0 & 0 \\ 0 & 0 & 1 & 0 & -1 & 0 \\ 0 & 0 & 2 & 0 & 0 & -1 \\ t_1 & 0 & 0 & 0 & x_1 & 0 \\ 0 & t_2 & 0 & 0 & 0 & x_2 \\ 0 & 0 & s & y & 0 & 0 \end{bmatrix}$$

Here $\Delta\mathbf{V} = \begin{bmatrix} \Delta\mathbf{X} \\ \Delta\mathbf{Y} \\ \Delta\mathbf{S} \\ \Delta\mathbf{T} \end{bmatrix} = [\Delta x_1, \Delta x_2, \Delta y, \Delta s, \Delta t_1, \Delta t_2]'$. For a 6×6 Jacobian, the problem can be

solved directly in one step by finding the inverse and using (9.22), where

$$\mathbf{F}(\mathbf{V}^k) = [(10 - x_1^k - 2x_2^k - s^k), (1 - y + t_1), (1 - 2y + t_2), (\mu - x_1 t_1), (\mu - x_2 t_2), (\mu - sy)]'$$

Table 9.6 indicates the points along the central path that are traced out by a series of values for μ. Each of these points results from an *exact* solution to $\Delta\mathbf{V}^k = -[\mathbf{J}(\mathbf{V}^k)]^{-1}\mathbf{F}(\mathbf{V}^k)$, using $\mathbf{J}(\mathbf{V}^k)^{-1}$, rather than an *approximation* via Newton–Raphson or some other iterative approach. In order to illustrate the elements along the central path for this problem, results in the table start with $\mu = 1$ million and end with $\mu = 0.00001$. Figure 9.6 sketches that central path. Note that the duality gap at each iteration is, indeed, $(m + n)\mu$.

9.4   ADDITIONAL VARIATIONS AND IMPLEMENTATION ISSUES

9.4.1   Regular Simplices; Circles and Ellipses

The feasible region for our illustrative problem in augmented solution space is that part of the $x_1 + 2x_2 + x_3 = 10$ plane that is in the nonnegative orthant. It is a (two-dimensional)

TABLE 9.6 A Path-Following Approach to the Linear Programming Problem in (9.4)

μ	x_1	x_2	s	Primal Objective (II)	Duality Gap (Δ − II)	y	t_1	t_2	Dual Objective (Δ)
1×10^6	3.3333	1.6667	3.3333	5	3,000,000	300,000.5	299,999.5	600,000	3,000,005
1×10^5	3.3334	1.6667	3.3333	5.0001	300,000	30,000.5	29,999.5	60,000	300,005.0001
1×10^4	3.3339	1.6667	3.3328	5.0006	30,000	3000.5001	2999.5001	6000.0001	30,005.0006
1×10^3	3.3389	1.6667	3.3278	5.0056	3000	300.5006	299.5006	600.0011	3005.0056
100	3.3892	1.6664	3.2781	5.0556	300	30.5056	29.5056	60.0111	305.0556
10	3.9138	1.6366	2.8129	5.5505	30	3.5550	2.5550	6.1101	35.5505
5	4.5161	1.5555	2.3729	6.0716	15	2.1072	1.1072	3.2143	21.0716
1	7.5366	0.7903	0.8829	8.3269	3	1.1327	0.1327	1.2654	11.3269
0.5	8.6312	0.4481	0.4726	9.0793	1.5	1.0579	0.0579	1.1159	10.5793
0.25	9.2821	0.2372	0.2434	9.5193	0.75	1.0269	0.0269	1.0539	10.2693
0.10	9.7051	0.0980	0.0990	9.8030	0.3	1.0103	0.0103	1.0206	10.1030
0.01	9.9701	0.0100	0.0100	9.9800	0.03	1.0010	0.0010	1.0020	10.0100
0.001	9.9970	0.0010	0.0010	9.9980	0.003	1.0001	0.0001	1.0002	10.0010
1×10^{-4}	9.9997	0.0001	0.0001	9.9998	0.0003	1.0000	0.0001	1.0000	10.0001
1×10^{-5}	10.0000	0.0000	0.0000	10.0000	0	1.0000	0.0000	1.0000	10.0000

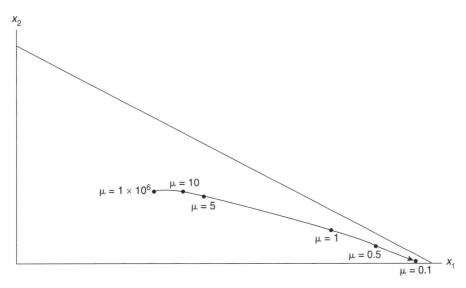

FIGURE 9.6 X^* satisfying (9.12) for various values of μ (points on the central path).

simplex (triangle)—the convex combination of three points not on the same line in two-dimensional space [the three points that are the axis intercepts $(10,0,0)$, $(0,5,0)$ and $(0,0,10)$].[20] The plane defined by $x_1 + x_2 + x_3 = k$ in the nonnegative orthant is a *regular* two-dimensional simplex in three-dimensional space; its axis intercepts are $(k,0,0)$, $(0,k,0)$, $(0,0,k)$. It is an *equilateral* triangle in the plane in which it lies; the length of each side is $\sqrt{2k^2} = 1.414k$.[21] The *center* of this simplex is at $(k/3,k/3,k/3)$. This point is equidistant from the three nonnegativity constraints for x_1, x_2, and x_3; in this case, from the planes $x_1 = 0$, $x_2 = 0$ and $x_2 = 0$ that are defined by pairs of axes.

The original simplex feasible region, $x_1 + 2x_2 + x_3 = 10$, could be converted into a regular simplex by a suitable transformation of variables. Define three new variables as $x_1' = 0.3x_1$, $x_2' = 0.6x_2$, and $x_3' = 0.3x_3$, then the constraint in x_1', x_2', x_3' space is $x_1' + x_2' + x_3' = 3$. With this kind of *scaling*, it is clear that *any* simplex can be converted into a regular simplex. If we choose $k = n$ (which means 3, as we did in this case), then the center will be at $(1,1,1)$.

By scaling at each step in the affine procedure so that the starting point is at $\widetilde{\mathbf{X}}^k = \begin{bmatrix} 1 \\ 1 \\ 1 \end{bmatrix}$,

we are in effect identifying a *sphere*, centered at $\widetilde{\mathbf{X}}^k$ and with a radius of 1, so that it is tangent to the three axis planes. By making $\Delta\mathbf{X}^k$ not quite as long as the radius of that sphere, we are certain that all three variables will continue to be nonnegative at the next point, $\widetilde{\mathbf{X}}^{k+1}$. The center of this sphere (by definition) lies on the augmented constraint; so the in-

[20]Simplex geometry is at the heart of the downhill (uphill) simplex method, a nonderivative iterative approach for minimization (maximization) of $f(\mathbf{X})$, as we saw in Chapter 6. For linear programming problems, the simplex method is a successful iterative solution technique, although the term refers to the geometry of the procedure in the vector-space representation of the problem, not in the more usual solution-space picture.

[21]A regular two-dimensional simplex for which $k = 1$ is known as a *standard* regular two-dimensional simplex. The generalizations are $\Sigma_{j=1}^n x_j = k$ and $\Sigma_{j=1}^n x_j = 1$ are regular and standard $(n-1)$-dimensional simplices in n-dimensional space.

tersection of the sphere and the constraint (the simplex plane in augmented solution space) is a *circle* with the same radius. Then projecting the gradient onto the simplex and requiring that $\Delta \mathbf{X}^k < 1$ means that in effect the triangular feasible region is being approximated by the circle.

Karmarkar's method (and other early variants of alternative interior point methods) were concerned with problems in which the feasible region was transformed into a *regular* simplex. The main reason is that regular simplices (equilateral triangles) are "better" approximated by spheres (circles) than are arbitrary simplices.[22]

Returning to the problem in (9.11), the equation for the sphere centered on $\tilde{\mathbf{X}} = \begin{bmatrix} 1 \\ 1 \\ 1 \end{bmatrix}$ and with a radius of 1 is

$$(\tilde{x}_1 - 1)^2 + (\tilde{x}_2 - 1)^2 + (\tilde{x}_3 - 1)^2 = 1 \tag{9.31}$$

The trace of this sphere on the plane $\tilde{x}_1 + 2\tilde{x}_2 + 7\tilde{x}_3 = 10$ is a circle centered at $\tilde{\mathbf{X}} = \begin{bmatrix} 1 \\ 1 \\ 1 \end{bmatrix}$ with a radius of 1.

Suppose now that we investigate what happens to a sphere when we transform $\tilde{\mathbf{X}}$ into the original variables, $\mathbf{X}$, via $\tilde{\mathbf{X}} = (\mathcal{D}^0)^{-1}\mathbf{X}$. In order to make the geometric point more clearly, suppose that the initial point was $\mathbf{X}^0 = \begin{bmatrix} 2 \\ 3 \\ 7 \end{bmatrix}$; this is done only so that *all three* variables differ between scaled and unscaled versions of the problem. This means that

$$\mathcal{D}^0 = \begin{bmatrix} 2 & 0 & 0 \\ 0 & 3 & 0 \\ 0 & 0 & 7 \end{bmatrix}$$

and so

$$\begin{bmatrix} \tilde{x}_1 \\ \tilde{x}_2 \\ \tilde{x}_3 \end{bmatrix} = \begin{bmatrix} \frac{1}{2} & 0 & 0 \\ 0 & \frac{1}{3} & 0 \\ 0 & 0 & \frac{1}{7} \end{bmatrix} \begin{bmatrix} x_1 \\ x_2 \\ x_3 \end{bmatrix}$$

If we replace $\tilde{x}_1$, $\tilde{x}_2$, and $\tilde{x}_3$ in (9.31), with, respectively, $x_1/2$, $x_2/3$, and $x_3/7$, we have

$$\left(\frac{x_1}{2} - 1\right)^2 + \left(\frac{x_2}{3} - 1\right)^2 + \left(\frac{x_3}{7} - 1\right)^2 = 1 \tag{9.32}$$

---

[22]A measure of "roundedness" of a two-dimensional figure is given by the ratio of the radius (R) of the smallest circumscribed circle (containing the figure) to the radius (r) of the largest inscribed circle (contained in the figure). (In more than two dimensions, the inscribed and circumscribed figures are spheres or hyperspheres.) For a circle, $R/r = 1$, and that is the best ratio you can have. Other two-dimensional figures will have $R/r > 1$. For a two-dimensional simplex with sides of length a, b, and c, it can be shown that $r = \sqrt{(s-a)(s-b)(s-c)/s}$ and $R = abc/4rs$, where $s = (a + b + c)/2$ (the *semiperimeter*). In Problem 9.4 you are asked to show that for any given value of s, r is maximized when $a = b = c = \frac{2}{3}s$. This means that for a triangle with a given perimeter the inscribed circle with the largest radius (and hence the largest area) can be created when the triangle is equilateral. In that case, $r^* = 0.1925s$, $R^* = 0.3848s$, so $R^*/r^* = 2$, and the regular simplex is "better" approximated by a circle than is any other simplex with the same perimeter. Problem 9.5 asks you to illustrate these facts for three specific triangles with a perimeter of 12.)

This is an *ellipsoid* in the three-dimensional solution space of the unscaled problem, and its trace on the unscaled constraint $x_1 + 2x_2 + x_3 = 10$ is an ellipse.

9.4.2 Alternative Scalings

In Karmarkar's original work, the transformations from original to scaled problems were a little more complicated than the simple affine scaling that we looked at in Section 9.2. He worked with problems in which one of the constraints had the form of a *standard* regular simplex, and for that reason his centering (relative to that constraint) at step k was generated by the scaling

$$\widetilde{x}_j = \frac{x_j}{x_j^k} \div \sum_{\ell=1}^{n} \frac{x_\ell}{x_\ell^k}$$

Letting $\mathcal{D}^k = \hat{\mathbf{X}}^k$, this means

$$\widetilde{\mathbf{X}}^k = [(\mathcal{D}^k)^{-1}\mathbf{X}] \div [\mathbf{1}'(\mathcal{D}^k)^{-1}\mathbf{X}]$$

and the objective function in the scaled problem at step k is

$$\text{Maximize} \quad [\mathbf{P}'(\mathcal{D}^k)^{-1}\mathbf{X}] \div [\mathbf{1}'(\mathcal{D}^k)^{-1}\mathbf{X}]$$

This is, officially, a linear *fractional* program, but it turns out to be adequate to concentrate on maximization of the numerator only. This is because among the constraints is $\Sigma_{j=1}^{n}$ $\widetilde{x}_j = 1$ (the standard regular n-dimensional simplex) and $\widetilde{\mathbf{X}} \geq \mathbf{0}$, so the denominator in the objective function will always be positive.

9.4.3 Computational Issues

A variant of the rules in (a)–(e) near the end of Sectin 9.2.3 is often employed in affine scaling approaches. Using $\Delta\widetilde{\mathbf{X}}^k = \rho(\alpha^k)\,\widetilde{\mathbf{G}}^k$, from (d), we concentrate on the $\widetilde{\mathbf{G}}^k$ component. From (b) in that list, and since $\widetilde{\mathbf{A}}^k = \mathbf{A}(\mathcal{D}^k)$ and $(\mathcal{D}^k)' = \mathcal{D}^k$, we obtain

$$\widetilde{\mathbf{G}}^k = \widetilde{\mathbf{M}}^k(\widetilde{\mathbf{V}}^k) = \{\mathbf{I} - (\widetilde{\mathbf{A}}^k)'[(\widetilde{\mathbf{A}}^k)(\widetilde{\mathbf{A}}^k)']^{-1}(\widetilde{\mathbf{A}}^k)\}(\mathcal{D}^k\mathbf{P})$$

$$= \mathcal{D}^k\mathbf{P} - (\mathbf{A}\mathcal{D}^k)'[(\mathbf{A}\mathcal{D}^k)(\mathbf{A}\mathcal{D}^k)']^{-1}\mathbf{A}(\mathcal{D}^k)^2\mathbf{P}$$

$$= \mathcal{D}^k\{\mathbf{P} - \mathbf{A}'[\mathbf{A}(\mathcal{D}^k)^2\mathbf{A}']^{-1}\mathbf{A}(\mathcal{D}^k)^2\mathbf{P}\}$$

Suppose that we define

$$\mathbf{v}^k = [\mathbf{A}(\mathcal{D}^k)^2\mathbf{A}']^{-1}\mathbf{A}(\mathcal{D}^k)^2\mathbf{P} \qquad (9.33)$$

so that

$$\widetilde{\mathbf{G}}^k = \mathcal{D}^k[\mathbf{P} - \mathbf{A}'\mathbf{v}^k]$$

and, in the original solution space,

$$\mathbf{D}^k = \mathcal{D}^k\widetilde{\mathbf{G}}^k = (\mathcal{D}^k)^2[\mathbf{P} - \mathbf{A}'\mathbf{v}^k]$$

To simplify further, let

$$\mathbf{z}^k = \mathbf{P} - \mathbf{A}'\mathbf{v}^k \tag{9.34}$$

then

$$\mathbf{D}^k = (\mathscr{D}^k)^2 \mathbf{z}^k$$

which is the direction of movement at the kth iteration in unscaled solution space.

Rearranging (9.33), we have

$$[\mathbf{A}(\mathscr{D}^k)^2\mathbf{A}']\mathbf{v}^k = \mathbf{A}(\mathscr{D}^k)^2\mathbf{P}$$

which is a set of linear equations in the unknowns, $\mathbf{v}^k$. Rather than finding the inverse of $[\mathbf{A}(\mathscr{D}^k)^2\mathbf{A}']$ at each step, we can find $\mathbf{v}^k$ by iterative methods, such as Gaussian elimination, Gauss–Jordan or Gauss–Seidel methods (Chapters 2 and 5), or factorization of $[\mathbf{A}(\mathscr{D}^k)^2\mathbf{A}']$, as in an **LU** decomposition (Chapter 2). Since $\mathscr{D}^k$ is a diagonal matrix, we know that $[\mathbf{A}(\mathscr{D}^k)^2\mathbf{A}']$ is *symmetric*, and there are additional factorization possibilities. The particular iterative method used may have a large impact on the running time of the interior point algorithm. Remember that interior point methods have been developed because of potential simplex method trouble with *large* problems; that means, for problems in which m and n are large. That means, in turn, that the coefficient matrix $\mathbf{A}$ is also large, which is why alternatives to finding the inverse of $[\mathbf{A}(\mathscr{D}^k)^2\mathbf{A}']$ are attractive.

In fact, as we saw in the previous section, $[\mathbf{A}(\mathscr{D}^k)^2\mathbf{A}']$ is also *positive definite* as long as $\mathbf{A}$ is of full rank and each element of $\mathbf{X}^k$ is *interior* to the feasible region—so that each element in $\mathscr{D}^k = (\hat{\mathbf{X}}^k)$ is strictly positive. That, in turn, means that at least one additional alternative representation is possible; this is a Cholesky factorization, to which we turn at the end of this section.

Finally, the first-order conditions for a minimum of a quadratic form

$$f(\mathbf{X}) = \tfrac{1}{2}\mathbf{X}'\mathbf{A}\mathbf{X} - \mathbf{B}'\mathbf{X} + c$$

are

$$\nabla f = \mathbf{A}\mathbf{X} - \mathbf{B} = \mathbf{0} \qquad \text{or} \qquad \mathbf{A}\mathbf{X} = \mathbf{B}$$

So, using iterative methods directly on the problem of minimizing $f(\mathbf{X})$ (Chapter 6) will lead to the $\mathbf{X}$ that is the solution to $\mathbf{A}\mathbf{X} = \mathbf{B}$. In the present case, the quadratic form to be minimized is

$$f(\mathbf{v}^k) = \tfrac{1}{2}(\mathbf{v}^k)'[\mathbf{A}(\mathscr{D}^k)^2\mathbf{A}']\mathbf{v}^k - [\mathbf{A}(\mathscr{D}^k)^2\mathbf{P}]'\mathbf{v}^k$$

and the logic is the same as in Chapter 6.

It also turns out that, at each iteration, $\mathbf{v}^k$ in (9.33) is an *estimate* of the solution vector for the dual linear programming problem. This, in turn, provides a link to the last of the possible stopping criteria, in rule (d) in Section 9.2.4. Moreover, the vector $\mathbf{z}^k$

in (9.34) bears a close relationship to the top row at the current iteration in the standard simplex method. Details concerning these connections can be found in many references.

9.4.4 Cholesky Factorizations

Because of the importance of Cholesky (André-Louis Cholesky, French, 1875–1918) factorizations in both affine scaling and path-following algorithms, we explore them here. If a square matrix, $\mathbf{M}$, is both symmetric and positive definite, then it is possible to perform a decomposition like $\mathbf{M} = \mathbf{LU}$ (Chapter 2), only now with $\mathbf{U} = \mathbf{L}'$, so that $\mathbf{M} = \mathbf{LL}'$, where $\mathbf{L}$ is lower triangular,

$$\mathbf{L} = \begin{bmatrix} \ell_{11} & 0 & \ldots & 0 \\ \ell_{21} & \ell_{22} & \ldots & 0 \\ \vdots & \vdots & & \vdots \\ \ell_{n1} & \ell_{n2} & \ldots & \ell_{nn} \end{bmatrix}$$

and so $\mathbf{L}'$ is upper triangular. If $\mathbf{M}$ is the coefficient matrix for a system of linear equations, $\mathbf{MX} = \mathbf{B}$, then these can be written as $\mathbf{LL}'\mathbf{X} = \mathbf{B}$; by letting $\mathbf{Z} = \mathbf{L}'\mathbf{X}$, they can also be expressed as $\mathbf{LZ} = \mathbf{B}$. But since $\mathbf{L}$ is lower triangular, these equations (in the z's) are easily solved without an inverse by *forward substitution* (Chapter 2, Section 2.7). For example, the first equation is just

$$\ell_{11}z_1 = b_1$$

from which $z_1 = b_1/\ell_{11}$. The second equation is

$$\ell_{21}z_1 + \ell_{22}z_2 = b_2$$

and with z_1 already found, z_2 is now determined; and so forth. Given these values ($\overline{\mathbf{Z}}$, say), then $\mathbf{X}$ is found (again without an inverse) by *backward substitution* in $\mathbf{L}'\mathbf{X} = \overline{\mathbf{Z}}$. Starting with $\ell_{nn}x_n = \bar{z}_n$, then, knowing x_n, we have

$$\ell_{n-1,n-1}x_{n-1} + \ell_{n-1,n}x_n = \bar{z}_{n-1}$$

and so on.

This decomposition of $\mathbf{M}$ into $\mathbf{LL}'$ is known as a *Cholesky factorization* or *Cholesky decomposition* of $\mathbf{M}$; it is also known as a *triangular factorization* of the matrix, and $\mathbf{L}$ is sometimes called the *Cholesky triangle*. It is worth noticing that a Cholesky factorization requires only half the storage of an $\mathbf{LU}$ factorization.

Here is how the Cholesky elements are found for a 3×3 positive definite and symmetric matrix $\mathbf{M}$. That means that we are looking for the six nonzero elements in

$$\mathbf{L} = \begin{bmatrix} \ell_{11} & 0 & 0 \\ \ell_{21} & \ell_{22} & 0 \\ \ell_{31} & \ell_{32} & \ell_{33} \end{bmatrix}$$

Written out explicitly, this is

$$\mathbf{M} = \mathbf{LL'} = \begin{bmatrix} \ell_{11} & 0 & 0 \\ \ell_{21} & \ell_{22} & 0 \\ \ell_{31} & \ell_{32} & \ell_{33} \end{bmatrix} \begin{bmatrix} \ell_{11} & \ell_{21} & \ell_{31} \\ 0 & \ell_{22} & \ell_{32} \\ 0 & 0 & \ell_{33} \end{bmatrix}$$

$$= \begin{bmatrix} (\ell_{11})^2 & \ell_{11}\ell_{21} & \ell_{11}\ell_{31} \\ \ell_{21}\ell_{11} & (\ell_{21})^2 + (\ell_{22})^2 & \ell_{21}\ell_{31} + \ell_{22}\ell_{32} \\ \ell_{31}\ell_{11} & \ell_{31}\ell_{21} + \ell_{32}\ell_{22} & (\ell_{31})^2 + (\ell_{32})^2 + (\ell_{33})^2 \end{bmatrix}$$

It is easy to establish a sequential solution procedure for the elements of $\mathbf{L}$ by proceeding down columns (it could equally well proceed along rows). Starting at the top of column 1 in $\mathbf{M}$ and in $\mathbf{LL'}$, we have

$m_{11} = (\ell_{11})^2$, so $\ell_{11} = \sqrt{m_{11}}$

$m_{21} = \ell_{21}\ell_{11}$, so $\ell_{21} = m_{21}/\sqrt{m_{11}}$

$m_{31} = \ell_{31}\ell_{11}$, so $\ell_{31} = m_{31}/\sqrt{m_{11}}$

$m_{22} = (\ell_{21})^2 + (\ell_{22})^2$, and knowing ℓ_{21}, then ℓ_{22} can be found

$m_{32} = \ell_{31}\ell_{21} + \ell_{32}\ell_{22}$, and knowing ℓ_{31}, ℓ_{21}, and ℓ_{22}, then ℓ_{32} can be found

$m_{33} = (\ell_{31})^2 + (\ell_{32})^2 + (\ell_{33})^2$, and knowing ℓ_{31} and ℓ_{32}, then ℓ_{33} can be found

Solving for the elements in column 1 of $\mathbf{L}$ requires finding $\sqrt{m_{11}}$; for this to be possible, m_{11} must be positive. If you work out the role of the elements in $\mathbf{M}$ in the solution for ℓ_{22}, you find

$$\ell_{22} = \sqrt{(m_{11}m_{22} - m_{12}^2)/m_{11}}$$

The expression under the square root is exactly $|\mathbf{M}_2|/|\mathbf{M}_1|$, the ratio of the second leading principal minor of $\mathbf{M}$ to the first leading principal minor (Chapter 1, Section 1.6). Problem 9.7 asks you to verify that

$$(\ell_{33})^2 = \frac{|\mathbf{M}_3|}{|\mathbf{M}_2|}$$

The same applies for the diagonal elements in an $\mathbf{L}$ matrix of any size:

$$(\ell_{jj})^2 = \frac{|\mathbf{M}_j|}{|\mathbf{M}_{j-1}|} \quad (j \geq 2)$$

So the requirement that $\mathbf{M}$ be positive definite (for Cholesky factorization) serves to ensure that these leading principal minors will be positive [this was given in the rule in (1.40) in Chapter 1].

Here is a numerical illustration. Let

$$\mathbf{M} = \begin{bmatrix} 2 & 1 & 2 \\ 1 & 2 & 3 \\ 2 & 3 & 5 \end{bmatrix}$$

which is certainly symmetric and for which you can easily verify that $|\mathbf{M}_1| = 2 > 0$, $|\mathbf{M}_2| = 3 > 0$ and $|\mathbf{M}_3| = 1 > 0$. Then, to four decimal places

$$\mathbf{L} = \begin{bmatrix} 1.4142 & 0 & 0 \\ 0.7071 & 1.2247 & 0 \\ 1.4142 & 1.6330 & 0.5774 \end{bmatrix}$$

and you can check that $\mathbf{LL'} = \mathbf{M}$.

The following example illustrates what is known as the problem of *fill-in* with Cholesky factorizations. Consider the matrix $\mathbf{A}$, which could be the coefficient matrix from an augmented linear program with four constraints and four decision variables:

$$\mathbf{A} = \begin{bmatrix} 1 & 2 & 3 & 2 & 1 & 0 & 0 & 0 \\ 0 & 1 & 0 & 0 & 0 & 1 & 0 & 0 \\ 0 & 0 & 2 & 0 & 0 & 0 & 1 & 0 \\ 0 & 0 & 0 & 1 & 0 & 0 & 0 & 1 \end{bmatrix}$$

Suppose that

$$\mathbf{M} = <[1 \quad 2 \quad 3 \quad 4]>$$

and

$$\mathbf{N} = <[2 \quad 3 \quad 4 \quad 1 \quad 2 \quad 3 \quad 4 \quad 1]>$$

Then

$$\mathbf{M} + \mathbf{ANA'} = \mathbf{A\Theta A'} = \begin{bmatrix} 57 & 6 & 24 & 2 \\ 6 & 8 & 0 & 0 \\ 24 & 0 & 23 & 0 \\ 2 & 0 & 0 & 6 \end{bmatrix}$$

where

$$\Theta = <[3 \quad 3 \quad 4 \quad 1 \quad 2 \quad 5 \quad 7 \quad 5]>$$

and the Cholesky factorization of $\mathbf{A\Theta A'}$ is, again to four decimals:

$$\mathbf{L} = \begin{bmatrix} 7.5498 & 0 & 0 & 0 \\ 0.7947 & 2.7145 & 0 & 0 \\ 3.1789 & -0.9307 & 3.4682 & 0 \\ 0.2649 & -0.0776 & -0.2636 & 2.4196 \end{bmatrix}$$

The original $\mathbf{A}$ matrix from the linear program is somewhat sparse; it contains 11 nonzero elements out of 32. Using the percentage of nonzero elements in a matrix to the total possible as a simple measure of "density," $\mathbf{A}$ has a density of 34%. For the $\mathbf{A\Theta A'}$

matrix, the density has increased—10 nonzero elements out of 16; density 62.5%. Finally, **L** has a density of 100% (counting density of the lower triangular part only, since the upper right triangle *must* contain only zeros).

Suppose now that we interchange rows 1 and 4 of the **A** matrix, which simply means that we are altering the *order* in which the constraints are written down. While this is easily done in this case, it can be accomplished formally using a permutation matrix **P** (Chapter 1, Section 1.6), where in this case[23]

$$\mathbf{P} = \begin{bmatrix} 0 & 0 & 0 & 1 \\ 0 & 1 & 0 & 0 \\ 0 & 0 & 1 & 0 \\ 1 & 0 & 0 & 0 \end{bmatrix}$$

so that

$$\mathscr{A} = \mathbf{PA} = \begin{bmatrix} 0 & 0 & 0 & 1 & 0 & 0 & 0 & 1 \\ 0 & 1 & 0 & 0 & 0 & 1 & 0 & 0 \\ 0 & 0 & 2 & 0 & 0 & 0 & 1 & 0 \\ 1 & 2 & 3 & 2 & 1 & 0 & 0 & 0 \end{bmatrix}$$

and

$$\mathscr{A}\Theta\mathscr{A}' = \begin{bmatrix} 6 & 0 & 0 & 2 \\ 0 & 8 & 0 & 6 \\ 0 & 0 & 23 & 24 \\ 2 & 6 & 24 & 57 \end{bmatrix}$$

This is just $\mathbf{P}(\mathbf{A}\Theta\mathbf{A}')\mathbf{P}'$ since $\mathscr{A}\Theta\mathscr{A}' = (\mathbf{PA})\Theta(\mathbf{PA})' = \mathbf{P}(\mathbf{A}\Theta\mathbf{A}')\mathbf{P}'$ [and $(\mathbf{PA})' = \mathbf{A}'\mathbf{P}'$]. The Cholesky factorization, $\mathscr{L}$, of $\mathscr{A}\Theta\mathscr{A}'$ is

$$\mathscr{L} = \begin{bmatrix} 2.4495 & 0 & 0 & 0 \\ 0 & 2.8284 & 0 & 0 \\ 0 & 0 & 4.7958 & 0 \\ 0.8165 & 2.1213 & 5.0043 & 5.1759 \end{bmatrix}$$

Notice what has happened to the density of the Cholesky triangle; with $\mathscr{A}\Theta\mathscr{A}'$, the corresponding $\mathscr{L}$ has a density of 70%. So the order in which the constraints of a linear programming problem are written down (the order of the rows in the **A** matrix) can have an influence the density of the Cholesky matrices.[24] Since sparseness is a quality to be desired in matrices that are part of an algorithm (less computer storage, fewer arithmetic op-

---

[23]It is customary to use **P** for a permutation matrix (Chapter 1); the context should not cause any confusion with the objective function coefficient vector in a linear programming maximization problem.

[24]This effect of ordering on fill-in is not very dramatic in this small example, but it can make a huge difference in algorithms for large linear programs in which Cholesky factorizations are used.

erations per iteration, etc.), it is often useful to try to find an "appropriate" rearrangement of constraints before invoking an interior point algorithm that employs Cholesky factorizations to address a linear program. This is part of what is known as "preprocessing" a problem. The trouble is that finding an appropriate permutation of constraints for such preprocessing can itself be a time- and space-consuming task, potentially defeating at least some of the advantage of the faster algorithm.

9.5   RELATIVE PERFORMANCE: SIMPLEX VERSUS INTERIOR POINT METHODS ON REAL-WORLD PROBLEMS

The linear programming problems in (9.1) and (9.2) were designed to test the simplex method under a worst-case scenario and, indeed, the simplex method would be unreasonably slow for such problems. On the other hand, the real issue is how the simplex method performs, compared with interior point methods, on "representative" or "average" real-world problems. That is the environment in which most practitioners will find themselves working. And it is presently not completely clear that interior point methods hold the advantage over the simplex method in those cases.

In part, this is because the frenzy of work on interior point methods that began with Karmarkar's 1984 paper also spurred renewed interest in the simplex method. In particular, by taking advantage of advances in computer speed, size, and technology, in conjunction with advanced linear algebra methods, it became possible to write ever-faster and more efficient simplex method algorithms that were competitive (sometimes superior) to alternative interior point methods.

Bixby (1994) presents a set of results on a group of fairly large "test problems" that have been used by the linear programming community to try out new algorithms. A version of the simplex method is used that contains many refinements to take advantage of capacities and other characteristics of up-to-date computers. Of eight test problems, half were best done by the simplex algorithm and half by a version of the barrier (path-following) interior point approach. So it is clear that interior point methods are by no means always superior to a refined simplex algorithm.

What is the size of problems that are currently considered to be representative of "large" real-world linear programs? Lustig et al. (1994, p. 10) define "small" linear programs as those in which $m + n < 2000$ and "moderate" problems as those for which $2000 \leq m + n <$ 10,000. (This means that the illustrations in Tables 9.2 and 9.3 were very much on the small side.) A set of empirical results for "large" problems collected from academia and industry are discussed by Bixby (1994, p. 20), who characterizes as "large models" a set of 28 real-world linear programming problems in which $m + n$ ranges from 3480 to 416,000.[25]

One very interesting fact about these (and, presumably, other real-world linear programming problems) is that they are quite "sparse." The average density of the **A** matrices for these 28 problems is only 0.04%. So, although the average problem in this set had 22,306 rows and 79,830 columns, only about 712,400 of the 1,781,000,000 cells in that average problem's coefficient matrix would be nonzero.

Table 9.7 presents some characteristics and some computational results for interior point methods and for the simplex method on the seven largest problems—those for

---

[25]The ranges for numbers of rows (m) and columns (n) across these 28 problems were: $42 \leq m \leq 56{,}790$ and $2{,}390 \leq n \leq 415{,}950$. The criterion for inclusion was that the **A** matrix contain at least 100,000 nonzero entries.

TABLE 9.7   Empirical Results for Large Real-World Problems Using Interior Point Methods (IPM) and the Simplex Method (SM)

Problem	m	n	mn (billions)	Density (%)	Computing Time[a] IPM	Computing Time[a] SM
1	56,794	139,121	7.90	0.008	**76.4 min**	76.6 min
2	43,687	164,831	7.20	0.01	**8.1 h**	16.4 h
3	43,387	107,164	4.65	0.004	**19.1 min**	72.5 min
4	33,874	105,728	3.58	0.006	—[b]	**93.4 min**
5	41,366	78,750	3.26	0.06	—[b]	**5.2 min**
6	27,349	97,710	2.67	0.01	**16.2 min**	32.8 min
7	22,513	99,785	2.25	0.02	5.1 h	**83.7 min**

[a]The better method is indicated in boldface.
[b]In these cases, the IPM method was discontinued after it became clear that it was inferior.
Source: Derived from data in Bixby (1994, Tables XII and XIV, pp. 20, 21).

which $mn > 2$ billion. What emerges from the figures in this table is that neither method is superior in a majority of the cases examined. It seems that both simplex and interior point methods may have a place in the future of linear programming computations.

9.6   SUMMARY

Research into alternative interior point methods for linear programming problems, especially very large problems, is ongoing and vigorous. And there are already many alternative expositions and derivations of interior point methods. These often begin from different statements of the basic linear programming problem (maximization vs. minimization, augmented vs. inequalities) and employ alternative notation, and one is sometimes misled into thinking that there are substantial differences among the methods when in fact they are actually quite similar.

It has been shown that many interior point methods can be expressed as a combination of a *centering* direction and an *optimizing* direction. In that kind of view, the differences become ones of emphasis only—for example, which part of the combination gets the greatest weight.

Differences among the largest current computers mean that one algorithm may solve a particular problem in the least time on one machine and another algorithm may be faster for the same problem on a different machine. Therefore, fast algorithms are often constructed to exploit specific features of the computer on which they will be tested.

A key aspect of modern linear programming algorithms for large problems is their *preprocessing* capabilities—for example, elimination of redundant constraints. As a very simple example, suppose that among the constraints in a problem are

$$x_1 + x_2 \leq 10$$
$$2x_1 + 3x_2 \leq 20$$
$$0 \leq x_1 \leq 5$$
$$0 \leq x_2 \leq 4$$

Then clearly the first constraint is redundant. The trick is to identify such redundancies in large problems and eliminate them, thus reducing the size of the problem. Additionally, rearrangement of the data in a problem may be very useful, such as reordering constraints to take advantage of sparsity for linear algebra operations (as in Cholesky factorizations).

Ultimately, it seems likely that an ideal linear programming method will use a combination of both interior and edge (simplex) steps—perhaps initially moving quickly through the interior of the feasible region and then jumping to a nearby edge and using simplexlike reasoning for the final run into the optimal corner.

REFERENCES

Arbel, Ami, *Exploring Interior-Point Linear Programming: Algorithms and Software*, The MIT Press, Cambridge, MA, 1993.

Bixby, Robert E., "Progress in Linear Programming," *ORSA J. Comput.* **6**, 15–22 (1994).

Chvátal, V., *Linear Programming*, Freeman, New York, 1983.

den Hertog, D., *Interior Point Approach to Linear, Quadratic and Convex Programming. Algorithms and Complexity,* Kluwer Academic Publishers, Boston, 1994.

Dikin, I., "Iterative Solution of Problems of Linear and Quadratic Programming," *Sov. Math. Dokl.* **8**, 674–675 (1967).

Fletcher, R., *Practical Methods of Optimization*, Wiley, New York, 1987.

Karloff, Howard, *Linear Programming*, Birkhäuser, Boston, 1991.

Karmarkar, Narenda K., "A New Polynomial Time Algorithm for Linear Programming," *Combinatorica* **4**, 373–395 (1984).

Khachian, L. G., "A Polynomial Algorithm for Linear Programming" (in Russian), *Dok. Akad. Nauk SSSR* **224**, 1093–1096 (1979). (Translated in *Sov. Math. Dokl.* **20**, 191–194.)

Klee, V., and G. J. Minty, "How Good is the Simplex Algorithm?" in O. Shisha, ed., *Inequalities III*, Academic Press, New York, 1972, pp. 159–175.

Lustig, Irvin J., Roy E. Marsten, and David F. Shanno, "Interior Point Methods for Linear Programming: Computational State of the Art," *ORSA J. Comput.* **6**, 1–14 (1994).

Panik, Michael J., *Linear Programming: Mathematics, Theory and Algorithms*, Kluwer Academic Publishers, Boston, 1996.

Saigal, Romesh, *Linear Programming. A Modern Integrated Analysis*, Kluwer Academic Publishers, Boston, 1995.

Sierksma, Gerard, *Linear and Integer Programming. Theory and Practice*, Marcel Dekker, New York, 1996.

Vanderbei, Robert J., *Linear Programming: Foundations and Extensions*, Kluwer Academic Publishers, Boston, 1996.

PROBLEMS

9.1 Use the simplex method from Chapter 7 to solve the problem in (9.1), whose initial simplex table is shown in Table 9.1, by hand. Note that, indeed, you will find the basic feasible solutions associated with all seven feasible region vertices that remain after the origin.

9.2 Write out the complete linear programming problem in (9.2) for $n = 4$. Starting at the origin, using the simplex rules from Chapter 7 and the largest-top-row-element criterion for the variable to enter the next basis, find the optimal solution. This will take 15 pivots, but the arithmetic for each pivot is relatively easy. Note how the largest-top-row-element criterion forces you to introduce (into the next basis), and later remove, and later reintroduce, then later remove, and so on, x_4 a total of 4 times (eight different pivots), x_3 twice (four different pivots), and x_2 once (two pivots). Only x_1 is introduced only once and never subsequently removed.

9.3 (a) Show that any $n \times n$ diagonal matrix, **D**, with $d_{jj} > 0$ ($j = 1, \ldots, n$) is positive definite, using the leading principal minors rule in (1.40) of Chapter 1.

(b) For the case of $m < n$, where **A** is an $m \times n$ matrix, show that **AA'** is (i) symmetric and (ii) positive definite if $\rho(\mathbf{A}) = m$. [In the context of an augmented linear program, $\rho(\mathbf{A}) = m$ means that there are no redundant constraint equations.]

(c) Show that **ADA'** is also (i) symmetric and (ii) positive definite under the same conditions as in (a) and (b).

(d) Explore what happens to **AA'** when $\rho(\mathbf{A}) < m$, for a specific numerical example that you construct, in which $m = 3$ and $n = 5$.

9.4 Find the values of a, b and c that

$$\text{Maximize} \quad \sqrt{\frac{(s-a)(s-b)(s-c)}{s}}$$

for a fixed value of s, say $\bar{s}$.

9.5 Find r and R, and hence the "roundedness" ratio R/r, using the results explained in footnote 22 (in Section 9.4.1), for the following three triangles, each with a perimeter of 12 units:

Case 1: $a = b = c = 4$

Case 2: $a = b = 5$, $c = 2$

Case 3: $a = b = 3.25$, $c = 5.5$

Sketch each of the triangles. The areas of each of these triangles can be shown to be $A_t = rs$. A good measure of the quality of the approximation of the triangular feasible region by the inscribed circle is the ratio of the area of the circle, A_c, to the area of the triangle in each case. Show that $A_c/A_t = 0.605$, 0.428 and 0.416, respectively, for cases 1–3, and verify that these give the same rankings of roundness (best to worst) as do the three R/r ratios.

9.6 (a) One solution to the problem of finding an initial point in either an affine scaling or interior point method is to use

$$\tilde{\mathbf{X}}^0 = (\mathbf{A}_R^{-1})\mathbf{B}$$

where $\mathbf{A}_R^{-1} = (\mathbf{A}')(\mathbf{AA}')^{-1}$ is the *right inverse* of **A**, as in (2.7) in Chapter 2. Find this starting point for the constraint in our augmented sample problem: $x_1 + 2x_2 + x_3 = 10$. Use several alternative points satisfying the constraint to demonstrate that the $\tilde{\mathbf{X}}^0$ found in this way is closer to the origin than any other point on the constraint.

(b) Show that an $\tilde{\mathbf{X}}^0$ found as in (a), above, will be the *center* of a constraint if it has the form of a regular simplex.

9.7 Verify that in the Cholesky factorization of $\mathbf{M}$ into $\mathbf{L}$ and $\mathbf{L}'$, $(\ell_{33})^2 = |\mathbf{M}_3|/|\mathbf{M}_2|$, where $|\mathbf{M}_j|$ is the jth leading principal minor of $\mathbf{M}$. (Leading principal minors are discussed in Chapter 1.)

PART V

APPLICATIONS: CONSTRAINED OPTIMIZATION IN NONLINEAR MODELS

In this part we consider approaches to the optimization problem in which there is at least one *nonlinear* function in the set $\{f(\mathbf{X}), h^i(\mathbf{X})\}$. This constitutes a programming problem for which the techniques of Part IV, where the underlying linearity was exploited, cannot be used. Of all the types of optimization problems considered in this book, these are the most complicated.

If $f(\mathbf{X})$ and all constraints are linear, the methods of Part IV do nicely. If $f(\mathbf{X})$ is nonlinear and there are no constraints, then the classical calculus methods (Part II, Chapter 3) are available, as are the more modern iterative approaches (Part III, Chapter 6). When $f(\mathbf{X})$ is nonlinear and there are constraints, the classical methods (Part II, Chapter 4) generally break down for all except the smallest problems, especially if the constraints are inequalities and not just equations. That is why the methods covered in this part have been developed and are important for solving real-world constrained optimization problems.

10

NONLINEAR PROGRAMMING: FUNDAMENTALS

The advantage of linear programs lies precisely in their linearity. It is because both the objective function and each of the constraints are linear functions of the unknowns that all the fundamental results relating linear programs, convex sets, and basic solutions to systems of linear equations hold. These results, in turn, make possible the computational approach of the simplex method (Chapters 7 and 8), a powerful package of arithmetic rules that efficiently finds the way to an optimum solution. Linearity is also exploited in various ways in interior point methods for linear programs (Chapter 9)—so that, for affine scaling methods, the projected gradient lies in the feasible region [a (hyper)plane], and for path-following approaches the first-order conditions on the Lagrangian function are a set of (almost) linear equations.

Clearly, the linearity requirements are stringent, but the rewards are great. As soon as one or more of the constraints and/or the objective function is a nonlinear function, the logical foundations of the simplex calculations are undermined and interior point methods become more complicated. In this and the following chapter, we explore the implications of nonlinearities on both the nature of the optimal solution and on methods for finding it. We will see that there are no *universally* valid solution procedures for general nonlinear programming problems. Considering the complexity of the results in Chapter 4 (especially Section 4.5), this is hardly surprising.

Initially we investigate the geometry of nonlinearities in the programming problem and see how the simplex arithmetic is thereby invalidated. Then we examine the additional complexities that are introduced into the results for the general constrained optimization problem of Chapter 4 when the requirement of nonnegative variables—which is common in programming problems—is added to the constraint set. The role of concavity and convexity assumptions on the objective function and the constraints is also explored in this section, where *quasi*concavity and *quasi*convexity (Appendix 3.1) are employed. The Kuhn–Tucker constraint qualification is discussed and illustrated in Section 10.3. Finally, in Section 10.4, we see how the Kuhn–Tucker results can be arrived at in a different way. This alternative route to the same outcome employs the notion of a *saddle point* of a function. This view of the nonlinear programming problem provides an additional connection with linear programming problems. In addition, there are techniques for finding saddle points in some cases; thus, the saddle point connection may be of computational help.

10.1 STRUCTURE OF THE NONLINEAR PROGRAMMING PROBLEM

10.1.1 Algebra

Parallel to the general linear program structure [as in (7.7) in Chapter 7], the nonlinear programming maximization problem can be represented as

$$
\begin{aligned}
\text{Maximize} \quad & f(\mathbf{X}) \\
\text{subject to} \quad & g^1(\mathbf{X}) \le b_1 \\
& \quad \vdots \\
& g^m(\mathbf{X}) \le b_m \\
\text{and} \quad & \mathbf{X} \ge \mathbf{0}
\end{aligned}
\tag{10.1}
$$

where, as usual, $\mathbf{X} = [x_1, \ldots, x_n]'$. As with linear programs, the nonnegativity requirement on $\mathbf{X}$ is a new feature, compared with the problems that we looked at in Chapter 4. This is a description of the general *mathematical programming* maximization problem; when $f(\mathbf{X})$ and all $g^i(\mathbf{X})$ are linear, then the problem is a linear program; otherwise it is a nonlinear program.

More generally, one or more of the constraints may set lower bounds, $g^j(\mathbf{X}) \ge b_j$, and one or more others may be *equalities*, $g^k(\mathbf{X}) = b_k$. The former can be converted to the style in (10.1) by multiplying both sides by (–1). Equalities, as we saw in Chapter 8 (Section 8.3) can be treated as a *pair* of inequalities. So the format in (10.1) is, in fact, completely general.

As we have also seen in earlier chapters, it is convenient to rewrite constraints as

$$
h^i(\mathbf{X}) \equiv g^i(\mathbf{X}) - b_i
$$

so that (10.1) becomes

$$
\begin{aligned}
\text{Maximize} \quad & f(\mathbf{X}) \\
\text{subject to} \quad & h^i(\mathbf{X}) \le 0 \qquad (i = 1, \ldots, m) \\
\text{and} \quad & \mathbf{X} \ge \mathbf{0}
\end{aligned}
\tag{10.2}
$$

Finally, letting $\mathbf{h}(\mathbf{X}) \equiv [h^1(\mathbf{X}), \ldots, h^m(\mathbf{X})]'$, we have

$$
\begin{aligned}
\text{Maximize} \quad & f(\mathbf{X}) \\
\text{subject to} \quad & \mathbf{h}(\mathbf{X}) \le \mathbf{0} \\
\text{and} \quad & \mathbf{X} \ge \mathbf{0}
\end{aligned}
\tag{10.2$'$}
$$

Similarly, the general nonlinear programming minimization problem is

$$\text{Minimize} \quad f(\mathbf{X})$$
$$\text{subject to} \quad \mathbf{h}(\mathbf{X}) \geq \mathbf{0} \tag{10.3}$$
$$\text{and} \quad \mathbf{X} \geq \mathbf{0}$$

10.1.2 Geometry

As we saw in Section 4.5, in an inequality constrained problem the optimal solution will either be completely inside the feasible region defined by the constraints or be somewhere on the boundary, where one or more of the constraints is binding.

Nonlinear Objective Function, Linear Constraints   The feasible region from the linear programming problem in (7.6) is repeated here in Figure 10.1, in which contours for several possible nonlinear objective functions have been added (dashed lines). When all the constraints in a problem are linear functions, the simplex method for finding nonnegative basic solutions is always available. Whether these solutions are of any value is another matter.

In Figures 10.1*a* and 10.1*b*, assume that higher values of the objective function are associated with contours that are farther from the origin. If the objective function, $f(\mathbf{X})$, is a

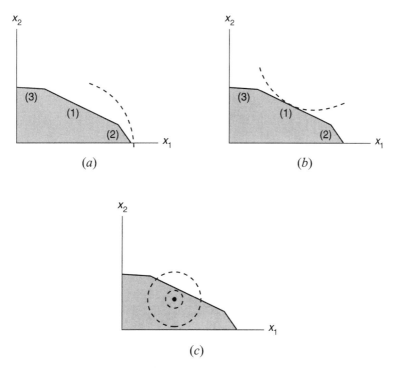

FIGURE 10.1   (*a*) Linear constraints, concave objective function contours; (*b*) linear constraints, convex objective function contours; (*c*) interior optimal solution.

bowl with a *minimum* that is located southwest of the origin, the contours in the positive quadrant of x_1, x_2 space will be concave, as in Figure 10.1*a*. In that case, the *maximum* of $f(\mathbf{X})$ subject to the constraints will occur at a corner of the feasible region, just as in a linear programming problem; hence the maximum will be identical to one of the nonnegative basic solutions to the linear constraining equations of the augmented problem, and the simplex method would find it.

If the objective function is an *inverted bowl* whose *maximum* is located in the positive quadrant and outside the feasible region, the contours will be convex, as in Figure 10.1*b*. The objective function contours could also be changed so that the maximum is, in fact, at a corner, but this would not *generally* be the case. Since the simplex method examines corners only, it would be totally invalid for a problem whose geometry was like that in Figure 10.1*b*. The optimum solution for that problem will have *four* variables positive (x_1, x_2, as well as s_2, s_3, the slack variables for the second and third constraints), instead of *three*, as required for a basic solution.

Finally, if the geometry were as shown in Figure 10.1*c*, where the optimal solution is in the *interior* of the feasible region, that solution is even less "basic," since now all *five* variables (both x's and all three slacks) would be positive.

Linear Objective Function, Nonlinear Constraints   With nonlinear constraints, the complications are rather different but the implication is the same—invalidation of a simplex method approach to a solution. Figure 10.2 illustrates for a problem with one linear and one nonlinear constraint, in conjunction with a linear objective function (dashed line). The feasible region is convex, but in addition to extreme points at the three corners ($\mathbf{C}_1$ through $\mathbf{C}_3$) where constraints intersect (as in the case of linear constraints), *all* points along the curved boundary from $\mathbf{C}_1$ to $\mathbf{C}_2$ are also extreme points (they cannot be expressed as a convex combination of any pair of points in the feasible region). Even if you invented an analog to the simplex method to find intersections of these constraints (when not all of them are linear), you would overlook the true optimal solution since it is not at such an intersection.

Nonlinear constraints, with or without a linear objective function, can present a different sort of difficulty also; they may define a nonconvex feasible region. This is illustrated in Figure 10.3. Nonconvexity of the feasible region can mislead solution procedures into thinking, wrongly, that a *local* optimum is in fact the *global* optimum. Assume that the optimum solution to the problem illustrated in Figure 10.3 is at $\mathbf{C}_2$, on the objective function contour that is farthest from the origin. A simplexlike procedure that started from the ori-

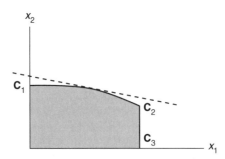

FIGURE 10.2   Nonlinear constraint, linear objective function (convex feasible region).

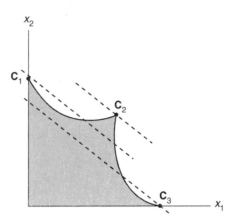

FIGURE 10.3   Nonlinear constraints, linear objective function (nonconvex feasible region).

gin and moved from corner to corner would be at either C_1 or C_3 after the first iteration. Suppose it was C_1. Viewed from that corner, the introduction of a small amount of x_2 into the current solution would *decrease* the value of the objective function. The myopia of the simplex method (looking at the effect of introducing a *very small* amount of x_2) cannot see past the immediate decrease to the ultimate increase in the objective function. With convex feasible regions (always present in linear programs), this problem cannot arise.

Nonlinear Objective Function, Nonlinear Constraints   When both nonlinear constraints and a nonlinear objective function appear in the same problem, the complexities illustrated in the figures are reinforced, and the simplex method is totally inadequate as a *general* solution procedure. As we saw in Section 4.5, when inequality constraints were first considered (without the added complication of $\mathbf{X} \geq \mathbf{0}$), in an inequality-constrained maximization or minimization problem the optimum solution may be one or more of the following:

1. At an extreme point of the feasible region, of which there may be an infinite number that are not defined by intersections of constraints (Fig. 10.2); or at one of a finite number of extreme points, which, however, cannot be reached from adjacent corners if one is using marginal (shortsighted, simplexlike) calculations (Fig. 10.3).
2. On the boundary of the feasible region but not at an extreme point (Fig. 10.1*b*).
3. In the interior of the feasible region—not on a boundary at all (Fig. 10.1*c*).

For these reasons, we explore in the next section a systematic method for solution of nonlinear programming problems. This is, as in Section 4.5, the Kuhn–Tucker approach.

10.2   THE KUHN–TUCKER APPROACH TO NONLINEAR PROGRAMMING PROBLEMS

We examined the Kuhn–Tucker technique for inequality-constrained maximization and minimization problems in Section 4.5, where there was no requirement that the variables be nonnegative. Results for those problems are repeated here, for comparison with the new results that we will derive for the problems in (10.2) and (10.3) (recall that $h_j^i \equiv \partial h^i/\partial x_j$). For the present, we continue to assume that problems are "well behaved", meaning that they meet the criteria of the Kuhn–Tucker constraint qualification. This means that the constraint gradients are not linearly dependent at an optimal point (Section 4.5).

The necessary conditions for $\mathbf{X}^* = [x_1^*, \dots, x_n^*]'$ to be a relative maximum of $f(\mathbf{X})$ subject to $h^i(\mathbf{X}) \le 0$ are

$$f_j - \sum_{i=1}^{m} \lambda_i [h_j^i] = 0 \qquad (j = 1, \dots, n) \tag{10.4a}$$

$$\lambda_i [h^i(\mathbf{X})] = 0 \qquad (i = 1, \dots, m) \tag{10.4b}$$

$$h^i(\mathbf{X}) \le 0 \qquad (i = 1, \dots, m) \tag{10.4c}$$

$$\lambda_i \ge 0 \qquad (i = 1, \dots, m) \tag{10.4d}$$

[This was (4.34) in Chapter 4.] Furthermore, we saw that if $f(\mathbf{X})$ is a *concave* function and all $h^i(\mathbf{X})$ are *convex* or *quasiconvex*, these conditions are also sufficient for a maximum. And if $f(\mathbf{X})$ is *strictly* concave, the maximum is an *absolute*, or *global*, one.

Expressed compactly, using the Jacobian matrix $\mathbf{J} = [\partial h^i/\partial x_j]$, along with $\mathbf{h}(\mathbf{X}) \equiv [h^1(\mathbf{X}), \dots, h^m(\mathbf{X})]'$ and $\mathbf{\Lambda} \equiv [\lambda_i, \dots, \lambda_m]'$, the conditions in (10.4) can be expressed as

$$\nabla f = \mathbf{J}' \mathbf{\Lambda} \tag{10.5a}$$

$$\mathbf{\Lambda} \circ [\mathbf{h}(\mathbf{X})] = 0 \tag{10.5b}$$

$$\mathbf{h}(\mathbf{X}) \le \mathbf{0} \tag{10.5c}$$

$$\mathbf{\Lambda} \ge \mathbf{0} \tag{10.5d}$$

[This was (4.34′) in Chapter 4.] Parallel conditions for the general minimization problem in (10.3) were given in (4.40) and (4.40′).

10.2.1   Kuhn–Tucker Conditions for Programming Problems that Include $\mathbf{X} \ge \mathbf{0}$

For the maximization problem in (10.2), the new nonnegativity constraints, $x_1 \ge 0, \dots, x_n \ge 0$, can be converted to equations in the same way as the original $h^i(\mathbf{X}) \le 0$ requirements were rewritten in Chapter 4. For a maximization problem in *standard form* (all constraints set upper limits), we can think of the nonnegativities as

$$h^{m+1}(\mathbf{X}) = -x_1 \le 0, \dots, h^{m+n}(\mathbf{X}) = -x_n \le 0$$

Following the logic of Section 4.5, add m squared slack variables, s_i^2, to the $h^i(\mathbf{X}) \le 0$ constraints and n (new) squared slack variables, s_{m+j}^2, to the $h^{m+j}(\mathbf{X}) \le 0$ (nonnegativity) constraints. Now the associated Lagrangian function contains two sets of additional terms—to account for the original constraints, $h^1(\mathbf{X}), \dots, h^m(\mathbf{X})$, and for the nonnegativity requirements in $h^{m+1}(\mathbf{X}), \dots, h^{m+n}(\mathbf{X})$, so

$$L = f(\mathbf{X}) - \sum_{i=1}^{m} \lambda_i[h^i(\mathbf{X}) + s_i^2] - \sum_{j=1}^{n} \lambda_{m+j}[h^{m+j}(\mathbf{X}) + s_{m+j}^2]$$

The procedure is the usual one with Lagrangian functions; all first partial derivatives are set equal to zero, as if the variables were independent.

$$\frac{\partial L}{\partial x_j} = f_j - \sum_{i=1}^{m} \lambda_i h_j^i - \sum_{j=1}^{n} \lambda_{m+j} h_j^{m+j} = 0 \qquad (j = 1, \ldots, n) \tag{10.6a}$$

$$\frac{\partial L}{\partial \lambda_i} = -[h^i(\mathbf{X}) + s_i^2] = 0 \qquad (i = 1, \ldots, m)$$

$$\frac{\partial L}{\partial \lambda_{m+j}} = -[h^{m+j}(\mathbf{X}) + s_{m+j}^2] = 0 \qquad (j = 1, \ldots, n) \tag{10.6b}$$

$$\frac{\partial L}{\partial s_i} = -2\lambda_i s_i = 0 \qquad (i = 1, \ldots, m)$$

$$\frac{\partial L}{\partial s_{m+j}} = -2\lambda_{m+j} s_{m+j} = 0 \qquad (j = 1, \ldots, n) \tag{10.6c}$$

There are $2m + 3n$ equations in (10.6), n for the original variables (x's), $m + n$ for the slack variables (s_i's and s_{m+j}'s), and $m + n$ for the Lagrange multipliers (λ_i's and λ_{m+j}'s). Reasoning exactly as in Section 4.5, the requirements in (10.6c) establish that, for *each i*, either $\lambda_i = 0$ or $s_i = 0$, or both, and for *each j*, either $\lambda_{m+j} = 0$ or $s_{m+j} = 0$, or both. If $s_i = 0$, then from (10.6b), $h^i(\mathbf{X}) = 0$; if $s_{m+j} = 0$, then, also from (10.6b), $h^{m+j}(\mathbf{X}) = 0$. Then the slack variables can be eliminated from the first-order conditions in (10.6) through replacement of both (10.6b) and (10.6c) with

$$\lambda_i[h^i(\mathbf{X})] = 0 \qquad (i = 1, \ldots, m)$$

$$\lambda_{m+j}[h^{m+j}(\mathbf{X})] = 0 \qquad (j = 1, \ldots, n) \tag{10.6d}$$

Since $h^{m+j}(\mathbf{X}) = -x_j$, the second set of requirements in (10.6d) is that $-\lambda_{m+j} x_j = 0$, or, eliminating the minus sign, just $\lambda_{m+j} x_j = 0$.

As before, once the (squared) slack variables have been eliminated from the first-order conditions, the *direction* of the inequality constraints has been lost. Note also that $h_j^{m+j} = -1$. Conditions (10.6) can now be restated in terms of the original x_j variables (without any slack variables) and the added $m + n$ Lagrange multipliers:

$$f_j - \sum_{i=1}^{m} \lambda_i h_j^i + \lambda_{m+j} = 0 \qquad (j = 1, \ldots, n) \tag{10.6'a}$$

$$\lambda_i[h^i(\mathbf{X})] = 0 \qquad (i = 1, \ldots, m)$$

$$-\lambda_{m+j} x_j = 0 \qquad (j = 1, \ldots, n) \tag{10.6'b}$$

$$h^i(\mathbf{X}) \le 0 \qquad (i = 1, \ldots, m)$$

$$\mathbf{X} \ge \mathbf{0} \tag{10.6'c}$$

The complete set of Kuhn–Tucker necessary conditions for the problem in (10.2) is as follows:

The necessary conditions for $\mathbf{X}^* = [x_1^*, \ldots, x_n^*]'$ to be a relative maximum of $f(\mathbf{X})$ subject to $h^i(\mathbf{X}) \leq 0$ and $x_j \geq 0$ are that nonnegative λ_i and λ_{m+j} (10.7) exist such that they and $\mathbf{X}^*$ satisfy (10.6′).

These conditions are also sufficient if $f(\mathbf{X})$ is concave and all of the constraints are convex or quasiconvex. Since the new $x_j \geq 0$ constraints are linear in the x's (and hence both concave and convex), the added requirement on constraints for sufficiency is still on the functions $h^i(\mathbf{X})$ only. And if $f(\mathbf{X})$ is strictly concave, the maximum is an absolute one. We explore sufficiency issues in more detail at the end of this section.

These conditions are often expressed differently in the literature, in a way that eliminates the Lagrange multipliers associated with the nonnegativities (the λ_{m+j}) from explicit consideration. This is reasonable, since it is likely to be the marginal valuations on the original (substantive) constraints $[h^i(\mathbf{X}) \leq 0]$ and not on the nonnegativity requirements that are of real interest in a problem.

Rewrite (10.6′a) as

$$f_j - \sum_{i=1}^{m} \lambda_i h_j^i = -\lambda_{m+j}$$

and substitute this into the second set of equations in (10.6′b):

$$x_j \left(f_j - \sum_{i=1}^{m} \lambda_i h_j^i \right) = 0$$

The nonnegativity requirement on the λ_{m+j} in (10.7) is equivalent to specifying, in (10.6′a), that

$$f_j - \sum_{i=1}^{m} \lambda_i h_j^i \leq 0$$

Therefore, another alternative expression of the Kuhn–Tucker necessary conditions (without any λ_{m+j}) for the problem in (10.2) is

The necessary conditions for $\mathbf{X}^* = [x_1^*, \ldots, x_n^*]'$ to be a relative maximum of $f(\mathbf{X})$ subject to $h^i(\mathbf{X}) \leq 0$ and $\mathbf{X} \geq \mathbf{0}$ are that $\mathbf{X}^*$ satisfy

$$f_j - \sum_{i=1}^{m} \lambda_i h_j^i \leq 0 \tag{10.8a}$$

$$x_j \left(f_j - \sum_{i=1}^{m} \lambda_i h_j^i \right) = 0 \tag{10.8b}$$

$$h^i(\mathbf{X}) \leq 0 \tag{10.8c}$$

$$\lambda_i [h^i(\mathbf{X})] = 0 \tag{10.8d}$$

$$\mathbf{X} \geq \mathbf{0}; \lambda_1 \geq 0, \ldots, \lambda_m \geq 0 \tag{10.8e}$$

Again, using compact vector and gradient notation, these results can be shown as

$$\nabla f \leq \mathbf{J}'\mathbf{\Lambda} \tag{10.9a}$$

$$\mathbf{X} \circ [\nabla f - \mathbf{J}'\mathbf{\Lambda}] = 0 \tag{10.9b}$$

$$\mathbf{h}(\mathbf{X}) \leq \mathbf{0} \tag{10.9c}$$

$$\mathbf{\Lambda} \circ [\mathbf{h}(\mathbf{X})] = 0 \tag{10.9d}$$

$$\mathbf{X} \geq \mathbf{0}, \mathbf{\Lambda} \geq \mathbf{0} \tag{10.9e}$$

The general results for the nonlinear programming minimization problem, in (10.3), require only a reversal of the inequalities in (10.9a) and (10.9c). (You should be clear why this is true.) For completeness, they are included here:

The necessary conditions for $\mathbf{X}^* = [x_1^*, \dots, x_n^*]'$ to be a relative minimum of $f(\mathbf{X})$ subject to $h^i(\mathbf{X}) \geq 0$ and $\mathbf{X} \geq \mathbf{0}$ are that $\mathbf{X}^*$ satisfy

$$f_j - \sum_{i=1}^{m} \lambda_i h_j^i \geq 0 \tag{10.10a}$$

$$x_j \left(f_j - \sum_{i=1}^{m} \lambda_i h_j^i \right) = 0 \tag{10.10b}$$

$$h^i(\mathbf{X}) \geq 0 \tag{10.10c}$$

$$\lambda_i [h^i(\mathbf{X})] = 0 \tag{10.10d}$$

$$\mathbf{X} \geq \mathbf{0}; \lambda_1 \geq 0, \dots, \lambda_m \geq 0 \tag{10.10e}$$

If $f(\mathbf{X})$ is convex and the constraints $h^i(\mathbf{X})$ are concave or quasiconcave, then these are also sufficient conditions for a minimum, which will be an absolute one if $f(\mathbf{X})$ is strictly convex.

In compact terms, the Kuhn–Tucker necessary conditions for the problem in (10.3) are

$$\nabla f \geq \mathbf{J}'\mathbf{\Lambda} \tag{10.11a}$$

$$\mathbf{X} \circ [\nabla f - \mathbf{J}'\mathbf{\Lambda}] = 0 \tag{10.11b}$$

$$\mathbf{h}(\mathbf{X}) \geq \mathbf{0} \tag{10.11c}$$

$$\mathbf{\Lambda} \circ [\mathbf{h}(\mathbf{X})] = 0 \tag{10.11d}$$

$$\mathbf{X} \geq \mathbf{0}, \mathbf{\Lambda} \geq \mathbf{0} \tag{10.11e}$$

10.2.2   Computational Implications

The computational demands generated by direct application of these conditions can be enormous. Consider the results in (10.8). Specifically

> There are $m + n$ variables (λ_i's and x's) that must satisfy $m + n$ nonlinear equations [in (10.8b) and (10.8d)] in conjunction with $2(m + n)$ inequalities [in (10.8a), (10.8c), and (10.8e)]. Included in the inequalities are the m original constraints [in (10.8c)] and $m + n$ nonnegativities on the x's and λ's [in (10.8e)].

The nonlinear equations in (10.8b) and (10.8d) require that the product of two terms be zero; if one of the two multiplicative terms in each equation is not zero, then the other term must be zero. The logic parallels but is more complicated than that in Section 4.5 for the example with two inequality constraints:

1. Consider (10.8d).

   a. For each i, if $\lambda_i \neq 0$, then $h^i(\mathbf{X})$ must be zero and the λ_i must be positive [again, from the nonnegativities in (10.8e)].

   b. Or, for each i, if $h^i(\mathbf{X}) \neq 0$, then the associated λ_i must be zero and the resulting $h^i(\mathbf{X})$ must be negative [from (10.8c)].

   We saw in Section 4.5 that these two options for each $\lambda_i[h^i(\mathbf{X})] = 0$ require that an exploration be carried out along each individual constraint [when $h^i(\mathbf{X}) = 0$] and also that each constraint be ignored (when $\lambda_i = 0$).

2. At the same time, consider (10.8b).

   a. For each j, if $x_j \neq 0$, then the corresponding $f_j - \Sigma_{i=1}^n \lambda_i h_j^i = 0$ [for $j = 1, \ldots, n$, this creates n equations but also assures that the inequalities in (10.8a) are satisfied] and, from (10.8e), the x_j that result must be positive (n nonnegativity requirements on the x's).

   b. Or, for each j, if $(f_j - \Sigma_{i=1}^n \lambda_i h_j^i) \neq 0$, then the corresponding $x_j = 0$ [ensuring that the nonnegativities on the x's in (10.8e) are met] and the bracketed term must be negative [from (10.8a)].

It should be clear that each of the $x_j (f_j - \Sigma_{i=1}^m \lambda_i h_j^i) = 0$ conditions now forces an exploration along each of the new boundaries (which are simply the x_j axes) set by the nonnegativity requirement on each x_j [as in 2(b), directly above] and also when each $x_j \neq 0$ [as in 2(a), directly above]. Note that in this latter case, the requirement is $f_j - \Sigma_{i=1}^n \lambda_i h_j^i = 0$, exactly as in (10.4a), for the original problem without nonnegativities.

There are therefore 2^m combinations for zero or nonzero λ_i's in (1) and each of these possibilities can theoretically be paired with each of the 2^n possible combinations of zero or nonzero x_j's in (2). This gives a total of $2^n 2^m (= 2^{n+m})$ possible combinations for the x's and λ's together. As we saw at the beginning of Chapter 9, 2^k can get large very quickly as k increases. For a (relatively small) problem with 10 variables and 10 constraints, $2^{20} = 1{,}048{,}576$; with 30 variables and 20 constraints, $2^{50} = 1.13 \times 10^{15}$ (1.13 quadrillion). Some of the combinations will very likely turn out to be impossible, as we will see in the numerical example that follows, but that is only established in the process of investigating each one.

10.2.3   Numerical Illustrations

Example 1

$$\text{Maximize}\quad f(\mathbf{X}) = -x_1^2 - x_2^2 - 4x_1 + 6x_2$$
$$\text{subject to}\quad h^1(\mathbf{X}) = x_1 + x_2 - 10 \leq 0$$
$$h^2(\mathbf{X}) = 3x_1 + x_2 - 18 \leq 0 \tag{10.12}$$
$$\text{and}\quad x_1 \geqq 0, x_2 \geqq 0$$

An unconstrained maximum occurs at $\mathbf{X}^u = \begin{bmatrix} -2 \\ 3 \end{bmatrix}$, where $f(\mathbf{X}^k) = 13$. This is shown in Figure 10.4, where $\mathbf{X}^u$ is flagged; the feasible region is shaded and several (circular) contours of the objective function are indicated. The constrained maximum is at $\mathbf{X}^* = \begin{bmatrix} 0 \\ 3 \end{bmatrix}$, where $f(\mathbf{X}^*) = 9$.

The full set of Kuhn–Tucker requirements in (10.8) is

$$-2x_1 - 4 - \lambda_1 - 3\lambda_2 \leq 0$$
$$-2x_2 + 6 - \lambda_1 - \lambda_2 \leq 0 \tag{10.13a}$$

$$x_1(-2x_1 - 4 - \lambda_1 - 3\lambda_2) = 0$$
$$x_2(-2x_2 + 6 - \lambda_1 - \lambda_2) = 0 \tag{10.13b}$$

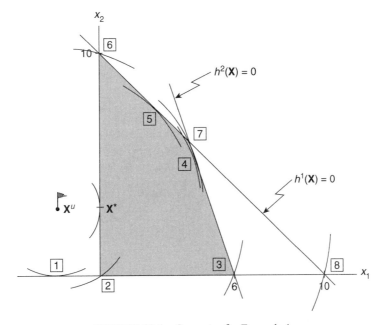

FIGURE 10.4   Geometry for Example 1.

$$x_1 + x_2 - 10 \leq 0$$
$$3x_1 + x_2 - 18 \leq 0 \tag{10.13c}$$

$$\lambda_1(x_1 + x_2 - 10) = 0$$
$$\lambda_2(3x_1 + x_2 - 18) = 0 \tag{10.13d}$$

$$x_1 \geq 0, x_2 \geq 0, \lambda_1 \geq 0, \lambda_2 \geq 0 \tag{10.13e}$$

Table 10.1 contains details of how these Kuhn–Tucker conditions identify the $2^{n+m} = 16$ potential candidate points for the proper constrained maximum point. In terms of the logic discussed in Chapter 4, Section 4.5.4 (identifying the number of ways of imposing none, one, two, or more constraints as equations), the possibilities are:

$C_0^4 = 1$ way of choosing none of the four constraints $[h^1(\mathbf{X}), h^2(\mathbf{X})$ and the two nonnegativities] as equations (finding the unconstrained maximum point, at $\mathbf{X}^u$);

$C_1^4 = 4$ ways of imposing one constraint at a time as an equation (examining the *sides* of the feasible region)

$C_2^4 = 6$ ways of imposing two constraints at a time as equations (examining the *corners* of the feasible region)

$C_3^4 = 4$ ways of (trying to) impose three constraints at a time as equations (which is impossible when $n = 2$)[1]

$C_4^4 = 1$ way of (again, trying to) impose all four constraints at a time as equations (which is also impossible when $n = 2$)

Example 2 Here is a problem with three unknowns and one constraint, so there are again $2^4 = 16$ possibilities:

$$\text{Maximize} \quad -x_1^2 + 20x_1 - x_2^2 + 20x_2 - x_3^2 + 20x_3$$
$$\text{subject to} \quad h(\mathbf{X}) = x_1 + x_2 + x_3 - 12 \leq 0 \tag{10.14}$$
$$\text{and} \quad \mathbf{X} \geq \mathbf{0}$$

The geometry of three-dimensional solution space is indicated in Figure 10.5. The feasible region is a four-sided solid (a tetrahedron); the sides are made up of parts of the three axis planes (e.g., the x_2, x_3 plane that is defined by $x_1 = 0$) and the constraint itself, $x_1 + x_2 + x_3 = 12$, which is the tilted plane [a regular simplex (Chapter 9)] in the figure. The objective function is a hypersphere in four-dimensional $x_1, x_2, x_3, h(\mathbf{X})$ space, with an unconstrained maximum at $\mathbf{X}^u = \begin{bmatrix} 10 \\ 10 \\ 10 \end{bmatrix}$, which violates the constraint; its "contours" are spheres in solution space, centered on $\mathbf{X}^u$. The constrained maximum is at $\mathbf{X}^* = \begin{bmatrix} 4 \\ 4 \\ 4 \end{bmatrix}$.

---

[1]The exception, as noted in Section 4.5, is if more than two (in two-dimensional solution space), or three (in three-dimensional solution space), or more constaints intersect in the same point. For example, if three equations with two unknowns should happen to intersect in a unique point, one of the inequality constraints would be redundant in the sense that the feasible region would remain unchanged if that constraint were removed from the problem.

TABLE 10.1   Alternatives Generated by the Conditions in (10.13)

Assumptions for (10.13d)	Assumptions for (10.13b)	Implications of Assumptions	C_0^4	C_1^4	C_2^4	C_3^4	C_4^4	Geometry
			Which of the 16 Kinds of Possibilities this Represents					
(1) $\lambda_1 = 0$, $\lambda_2 = 0$	(i) $x_1 = 0$, $x_2 = 0$	$6 \le 0$, from (10.13a), so impossible	X					Intersection of x_1 and x_2 axes (the origin; [2] in Fig. 10.4)
	(ii) $x_1 = 0$, $-2x_2 + 6 = 0$	$x_2 = 3$, and $\mathbf{X} = \begin{bmatrix} 0 \\ 3 \end{bmatrix}$ satisfies all of (10.13)		X				Tangency of $f(\mathbf{X})$ contour with x_2 axis ($\mathbf{X}^*$)
	(iii) $x_2 = 0$, $-2x_1 - 4 = 0$	$x_1 = -2$, which violates (10.13e)		X				Tangency of $f(\mathbf{X})$ contour with x_1 axis (point [1])
	(iv) $-2x_2 + 6 = 0$, $-2x_1 - 4 = 0$	$\mathbf{X} = \begin{bmatrix} -2 \\ 3 \end{bmatrix}$ and again, $x_1 < 0$ violates (10.13e)	X					Unconstrained maximum point ($\mathbf{X}^u$)
(2) $\lambda_1 = 0$, $3x_1 + x_2 - 18 = 0$	(i) $x_1 = 0$, $x_2 = 0$	$\mathbf{X} = \begin{bmatrix} 0 \\ 0 \end{bmatrix}$ does not satisfy $3x_1 + x_1 = 18$				X		Impossible; intersection of $h_2(\mathbf{X}) = 0$ with *both* axes
	(ii) $x_1 = 0$, $-2x_2 + 6 - \lambda_2 = 0$	$x_2 = 18$; $\lambda_2 = -30$ violates (10.13e)			X			Intersection of $h_2(\mathbf{X}) = 0$ and x_2 axis (not shown)
	(iii) $x_2 = 0$, $-2x_1 - 4 - 3\lambda_2 = 0$	$x_1 = 6$; $\lambda_2 = -16/3$ violates (10.13e)			X			Intersection of $h_2(\mathbf{X}) = 0$ and x_1 axis (point [3])
	(iv) $-2x_2 + 6 - \lambda_2 = 0$, $-2x_1 - 4 - 3\lambda_2 = 0$	$-x_1 + 3x_2 = 11$ and $3x_1 + x_2 = 18$ lead to $\mathbf{X} = \begin{bmatrix} 4.3 \\ 5.1 \end{bmatrix}$; $\lambda_2 = -4.2$ violates (10.13e)		X				Tangency of $f(\mathbf{X})$ contour with $h_2(\mathbf{X}) = 0$ (point [4])

(continued)

547

TABLE 10.1   Alternatives Generated by the Conditions in (10.13) (continued)

Assumptions for (10.13d)	Assumptions for (10.13b)	Implications of Assumptions	Which of the 16 Kinds of Possibilities this Represents					Geometry
			C_0^4	C_1^4	C_2^4	C_3^4	C_4^4	
(3) $\lambda_2 = 0$, $x_1 + x_2 - 10 = 0$	(i) $x_1 = 0, x_2 = 0$	$\mathbf{X} = \begin{bmatrix} 0 \\ 0 \end{bmatrix}$ does not satisfy $x_1 + x_2 = 10$			X			Impossible; intersection of $h_1(\mathbf{X}) = 0$ with *both* axes
	(ii) $x_1 = 0, -2x_2 + 6 - \lambda_1 = 0$	$x_2 = 10; \lambda_1 = -14$ violates (10.13e)		X				Intersection of $h_1(\mathbf{X}) = 0$ and x_2 axis (point [6])
	(iii) $x_2 = 0, -2x_1 - 4 - \lambda_1 = 0$	$x_1 = 10; \lambda_1 = -24$ violates (10.13e)		X				Intersection of $h_1(\mathbf{X}) = 0$ and x_1 axis (point [8])
	(iv) $-2x_2 + 6 - \lambda_1 = 0,$ $-2x_1 - 4 - \lambda_1 = 0$	$-x_1 + x_2 = 5$ and $x_1 + x_2 = 10$ lead to $\mathbf{X} = \begin{bmatrix} 2.5 \\ 7.5 \end{bmatrix}$; $\lambda_1 = -9$ violates (10.13e)	X					Tangency of $f(\mathbf{X})$ contour with $h_1(\mathbf{X}) = 0$ (point [5])
(4) $3x_1 + x_2 - 18 = 0$, $x_1 + x_2 - 10 = 0$	(i) $x_1 = 0, x_2 = 0$	$\mathbf{X} = \begin{bmatrix} 0 \\ 0 \end{bmatrix}$ does not satisfy either equation					X	Impossible; intersection of $h_1(\mathbf{X}) = 0, h_2(\mathbf{X}) = 0$ and *both* axes
	(ii) $x_1 = 0,$ $-2x_2 + 6 - \lambda_1 - \lambda_2 = 0$	$\lambda_1 + \lambda_2 = -6$ implies at least one $\lambda_i < 0$ violating (10.13e)				X		Impossible; intersection of $h_1(\mathbf{X}) = 0, h_2(\mathbf{X}) = 0$ and x_2 axis
	(iii) $x_2 = 0, -2x_1 - 4 - \lambda_1 - 3\lambda_2 = 0$	$\lambda_1 + 3\lambda_2 = -12$ implies at least one $\lambda_i < 0$ violating (10.13e)				X		Impossible; intersection of $h_1(\mathbf{X}) = 0, h_2(\mathbf{X}) = 0$ and x_1 axis
	(iv) $-2x_2 + 6 - \lambda_1 - \lambda_2 = 0,$ $-2x_1 - 4 - \lambda_1 - 3\lambda_2 = 0$	$\lambda_1 = -3, \lambda_2 = -3$ violates (10.13e)			X			Intersection of $h_1(\mathbf{X}) = 0$ and $h_2(\mathbf{X}) = 0$ (point [7])

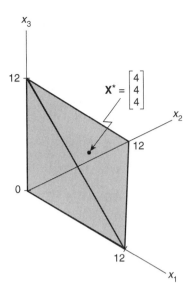

FIGURE 10.5 Geometry for Example 2.

Note that with $m = 1$ and $n = 3$

C_0^4 still identifies the case of choosing none of the four constraints [$\mathbf{h}(\mathbf{X})$ plus three nonnegativities].

C_1^4 represents the cases in which only one constraint is met as an equality, which now identifies the four *faces* of the feasible region.

C_2^4, with two constraints as equations (intersecting pairs of planes), defines the six *edges* of the feasible region.

C_3^4 are the four *corners* of the feasible region, where three constraints at a time intersect.

C_4^4 identifies the case in which all four constraints are simultaneously met as equations, which is impossible.

The Kuhn–Tucker conditions are

$$-2x_1 + 20 - \lambda \le 0$$
$$-2x_2 + 20 - \lambda \le 0 \qquad \text{(10.15a)}$$
$$-2x_3 + 20 - \lambda \le 0$$

$$x_1(-2x_1 + 20 - \lambda) = 0$$
$$x_2(-2x_2 + 20 - \lambda) = 0 \qquad \text{(10.15b)}$$
$$x_3(-2x_3 + 20 - \lambda) = 0$$

$$x_1 + x_2 + x_3 \le 12 \qquad \text{(10.15c)}$$

$$\lambda(x_1 + x_2 + x_3 - 12) = 0 \tag{10.15d}$$

$$\mathbf{X} \geq \mathbf{0}, \lambda \geq 0 \tag{10.15e}$$

Now there are only two cases under (10.15d)—$\lambda = 0$ and $x_1 + x_2 + x_3 = 12$. Each of these possibilities must be examined in conjunction with *eight* (2^3) possible cases under (10.15b). You can work through all 16 cases to see that in fact all faces, edges, and corners of the feasible region are accounted for in the process, and that only $\mathbf{X}^* = \begin{bmatrix} 4 \\ 4 \\ 4 \end{bmatrix}$ passes all the tests.

10.2.4 Constraint Gradient Geometry Once Again

In order to explore the vector geometry of Kuhn–Tucker necessary conditions for the complete nonlinear programming problems in (10.2) and (10.3), it is useful to return to (10.6′) and (10.7) and to generalize those results. Recall that the nonnegativities for a maximum problem were expressed as $-x_j \leq 0$, so as to have all the constraints expressed as upper bounds. As a (very simple, very linear) function of all x's, this can be expressed as

$$0x_1 + \cdots + 0x_{j-1} + (-1)x_j + 0x_{j+1} + \cdots + 0x_n \leq 0 \tag{10.16}$$

Using standard notation in which $\mathbf{I}_j$ represents the jth column from an identity matrix of appropriate size (here $n \times n$), (10.16) can be expressed in inner product form, in terms of the entire vector of x's, as

$$h^{m+j}(\mathbf{X}) = [-\mathbf{I}_j] \circ \mathbf{X} \leq 0 \tag{10.16′}$$

For $n = 2$, the two nonnegativity constraints in this form are $h^{m+1}(\mathbf{X}) = \begin{bmatrix} -1 \\ 0 \end{bmatrix} \circ \begin{bmatrix} x_1 \\ x_2 \end{bmatrix}$ and $h^{m+2}(\mathbf{X}) = \begin{bmatrix} 0 \\ -1 \end{bmatrix} \circ \begin{bmatrix} x_1 \\ x_2 \end{bmatrix}$. As we expect for *lower bound* constraints, the gradient for the non-negativity constraint on x_1, $\nabla h^{m+1} = \begin{bmatrix} -1 \\ 0 \end{bmatrix}$ is orthogonal to the x_2 axis and points leftward, out of the feasible region; a similar interpretation holds for the nonnegativity constraint on x_2: $\nabla h^{m+2} = \begin{bmatrix} 0 \\ -1 \end{bmatrix}$ is orthogonal to the x_1 axis and points downward, away from the feasible region. The geometric interpretations are similar for any number of unknowns.

If we rearrange (10.6′a), so that the λ terms are on the right-hand side, we have

$$f_j = \lambda_1 h_j^1 + \cdots + \lambda_m h_j^m - \lambda_{m+j} \qquad (j = 1, \ldots, n) \tag{10.6″a}$$

and using gradient notation

$$\nabla f = \lambda_1 \nabla h^1 + \lambda_2 \nabla h^2 + \cdots + \lambda_m \nabla h^m + \lambda_{m+1}(-\mathbf{I}_1) + \lambda_{m+2}(-\mathbf{I}_2) + \cdots + \lambda_{m+n}(-\mathbf{I}_n)$$

or

$$\nabla f = \sum_{i=1}^{m} \lambda_i \, \nabla h^i + \sum_{j=1}^{n} \lambda_{m+j} \nabla h^{m+j}$$

Letting $\Lambda = [\lambda_1, \ldots, \lambda_{m+n}]'$ [note that this includes *all* of the λ's, both on the original constraints $h^i(\mathbf{X}) \leq 0$ and also on the nonnegativity constraints $h^{m+j}(\mathbf{X}) \leq 0$] and, as before, $\mathbf{h}(\mathbf{X}) = [h^1(\mathbf{X}), \ldots, h^m(\mathbf{X})]'$, we can express $(10.6')$ as

$$\nabla f = \sum_{i=1}^{m} \lambda_i \, \nabla h^i + \sum_{j=1}^{n} \lambda_{m+j} \nabla h^{m+j} \tag{10.17a}$$

$$\Lambda \circ \mathbf{h}(\mathbf{X}) = 0 \tag{10.17b}$$

$$\mathbf{h}(\mathbf{X}) \leq \mathbf{0} \tag{10.17c}$$

$$\mathbf{X} \geq \mathbf{0}, \Lambda \geq \mathbf{0} \tag{10.17d}$$

In terms of *linear combinations* (Chapter 1), the requirement of an optimal solution as stated in (10.17a) is that, at $\mathbf{X}^*$, the gradient of the objective function must be expressbile as a linear combination of the gradients, at $\mathbf{X}^*$, of the substantive constraints of the problem [the $h^i(\mathbf{X})$'s] as well as the nonnegativity requirements [the $h^{m+j}(\mathbf{X})$'s, and herein lies the difference from the results in Section 4.5].

Again, computational insights come from the details in (10.17b) and (10.17c):

1. From (10.17b), if a particular ∇h^i is included in the linear combination in (10.17a)—which means if the associated λ_i is not zero—it must be true that the corresponding constraint is binding: $h^i(\mathbf{X}^*) = 0$. Similarly, if $\lambda_{m+j} \neq 0$, then the nonnegativity constraint on x_j is binding, namely, $x_j^* = 0$. Thus, the conditions in (10.17a) together with (10.17b) require that ∇f at $\mathbf{X}^*$ be expressible as a linear combination of the gradients of the *binding* (or *active*) constraints at $\mathbf{X}^*$.

2. From (10.17d), all the λ's must be nonnegative, so the requirement in (10.17a) is that, at $\mathbf{X}^*$, the gradient of the objective function must be expressible as a *nonnegative* linear combination of the gradients of the active constraints at $\mathbf{X}^*$. This algebra, in turn, has an immediate geometric interpretation.

We return to the geometry for Example 1 in solution space (Fig. 10.6). Gradient directions associated with the objective function and each of the active constraints are shown at each corner (A, C, E, and G) of the feasible region as well as at several non-corner locations (B, D, and F). The optimal solution is at B in the figure. (Only the direction, not the actual length, of ∇f is shown at each point. You should consider other nonoptimal points around the boundary of the feasible region and be convinced that ∇f cannot be expressed as a nonnegative linear combination of the relevant ∇h^i's at those points.)

The beginning portions of each of the relevant cones at each of the corners are indicated by shading. The complete cone extends infinitely far in the two directions indicated by the constraint gradients, since in each case it is formed from *all* nonnegative linear combinations of the two gradient vectors. At noncorner boundary locations, where only one

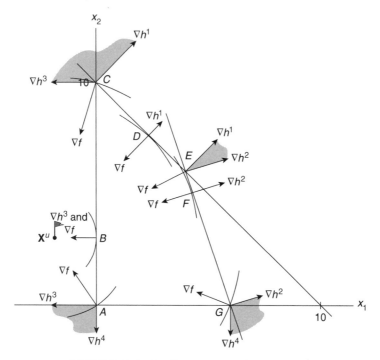

FIGURE 10.6 Gradient geometry for Example 1.

constraint is binding, there will only be one active-constraint gradient, and the cone will degenerate into a line.

It is easily seen that, at each corner of the feasible region, ∇f does not lie in the cone formed by the gradients of the two active constraints at that corner. Similarly, at nonoptimal noncorner points (D and F), ∇f is not a nonnegative multiple of ∇h for that constraint. Algebraic details regarding these gradients at A through G are given in Table 10.2.

For this problem, conditions (10.17a), which include the Lagrange multipliers for the nonnegativity constraints, are

$$\nabla f = \lambda_1 \begin{bmatrix} 1 \\ 1 \end{bmatrix} + \lambda_2 \begin{bmatrix} 3 \\ 1 \end{bmatrix} + \lambda_3 \begin{bmatrix} -1 \\ 0 \end{bmatrix} + \lambda_4 \begin{bmatrix} 0 \\ -1 \end{bmatrix}$$

At the optimal point, B, $\nabla f^* = \begin{bmatrix} -4 \\ 0 \end{bmatrix}$, and $\lambda_1^* = \lambda_2^* = \lambda_3^* = 0$ because $h^1(\mathbf{X}) \neq 0$, $h^2(\mathbf{X}) \neq 0$, and $h^3(\mathbf{X}) \neq 0$, and so

$$\begin{bmatrix} -4 \\ 0 \end{bmatrix} = (0) \begin{bmatrix} 1 \\ 1 \end{bmatrix} + (0) \begin{bmatrix} 3 \\ 1 \end{bmatrix} + \lambda_3 \begin{bmatrix} -1 \\ 0 \end{bmatrix} + (0) \begin{bmatrix} 0 \\ -1 \end{bmatrix}$$

meaning that $\lambda_3^* = 4$.

TABLE 10.2   Gradient Characteristics for Example 1

Boundary Location	A	B	C	D	E	F	G
X	$\begin{bmatrix} 0 \\ 0 \end{bmatrix}$	$\begin{bmatrix} 0 \\ 3 \end{bmatrix}$	$\begin{bmatrix} 0 \\ 10 \end{bmatrix}$	$\begin{bmatrix} 2.5 \\ 7.5 \end{bmatrix}$	$\begin{bmatrix} 4 \\ 6 \end{bmatrix}$	$\begin{bmatrix} 5 \\ 3 \end{bmatrix}$	$\begin{bmatrix} 6 \\ 0 \end{bmatrix}$
∇f	$\begin{bmatrix} -4 \\ 6 \end{bmatrix}$	$\begin{bmatrix} -4 \\ 0 \end{bmatrix}$	$\begin{bmatrix} -4 \\ 14 \end{bmatrix}$	$\begin{bmatrix} -9 \\ -9 \end{bmatrix}$	$\begin{bmatrix} -12 \\ -6 \end{bmatrix}$	$\begin{bmatrix} -14 \\ 0 \end{bmatrix}$	$\begin{bmatrix} -16 \\ 6 \end{bmatrix}$
Active constraints	h^3, h^4	h^3	h^1, h^3	h^1	h^1, h^2	h^2	h^2, h^4
Gradients of constraints (∇h^i)	$\begin{bmatrix} -1 \\ 0 \end{bmatrix}\begin{bmatrix} 0 \\ -1 \end{bmatrix}$	$\begin{bmatrix} -1 \\ 0 \end{bmatrix}$	$\begin{bmatrix} 1 \\ 1 \end{bmatrix}\begin{bmatrix} -1 \\ 0 \end{bmatrix}$	$\begin{bmatrix} 1 \\ 1 \end{bmatrix}$	$\begin{bmatrix} 1 \\ 1 \end{bmatrix}\begin{bmatrix} 3 \\ 0 \end{bmatrix}$	$\begin{bmatrix} 3 \\ 0 \end{bmatrix}$	$\begin{bmatrix} 3 \\ 0 \end{bmatrix}\begin{bmatrix} 0 \\ -1 \end{bmatrix}$

10.2.5   The Linear Programming Case and Dual Variables

The optimal solution to a *linear* programming problem can also be characterized using the Kuhn–Tucker conditions in (10.17). We return to the problem given in (7.6) in Chapter 7. It is reproduced here as Example 3.

Example 3

$$\text{Maximize} \quad f(\mathbf{X}) = 2x_1 + 5x_2$$

$$\text{subject to} \quad x_1 + 2x_2 \le 10$$

$$3x_1 + 2x_2 \le 24$$

$$x_1 + 10x_2 \le 40$$

$$\text{and} \quad x_1 \ge 0, x_2 \ge 0$$

Kuhn–Tucker Geometry for Example 3   The optimal solution to this problem was found via the simplex method in Chapter 7; it is $x_1^* = 2.5$ and $x_1^* = 3.75$. The optimal values of the dual variables (denoted as y_1, y_2, and y_3) were $y_1^* = 1.875$, $y_2^* = 0$, and $y_3^* = 0.125$. Aspects of the problem are illustrated in Figure 10.7. Since the constraints are all linear, each ∇h_i will be a vector of constants. An objective function contour (dashed line) and its gradient, $\nabla f(\mathbf{X}) = \begin{bmatrix} 2 \\ 5 \end{bmatrix}$, are also included in the figure, along with the (initial parts of the) cones formed by the gradients of the active constraints at each corner.

The Kuhn–Tucker conditions in (10.17) for this linear programming problem are written out explicitly below. (We intentionally use the Kuhn–Tucker conditions in the form that includes all the Lagrange multipliers, including those on the nonnegativity constraints.)

$$\begin{bmatrix} 2 \\ 5 \end{bmatrix} = \lambda_1 \begin{bmatrix} 1 \\ 2 \end{bmatrix} + \lambda_2 \begin{bmatrix} 3 \\ 2 \end{bmatrix} + \lambda_3 \begin{bmatrix} 1 \\ 10 \end{bmatrix} + \lambda_4 \begin{bmatrix} -1 \\ 0 \end{bmatrix} + \lambda_5 \begin{bmatrix} 0 \\ -1 \end{bmatrix} \tag{10.18a}$$

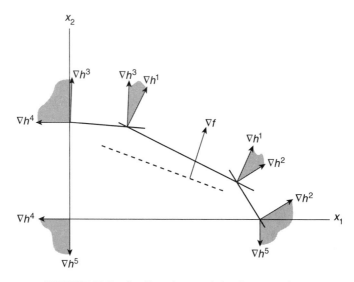

FIGURE 10.7   Gradient characteristics for Example 3.

$$\lambda_1(x_1 + 2x_2 - 10) = 0$$

$$\lambda_2(3x_1 + 2x_2 - 24) = 0$$

$$\lambda_3(x_1 + 10x_2 - 40) = 0 \qquad (10.18b)$$

$$-\lambda_4 x_1 = 0$$

$$-\lambda_5 x_2 = 0$$

$$x_1 + 2x_2 \leq 10$$

$$3x_1 + 2x_2 \leq 24 \qquad (10.18c)$$

$$x_1 + 10x_2 \leq 40$$

$$x_1, x_2 \geq 0;\ \lambda_1, \lambda_2, \lambda_3, \lambda_4, \lambda_5 \geq 0 \qquad (10.18d)$$

The conditions in (10.18a) have, once again, exactly the geometric interpretation that the gradient of the objective function at an optimal point be expressible as a nonnegative linear combination of the normals to the constraints that are active at that point. Using that criterion, it is apparent from Figure 10.7 that the optimal solution to this linear programming problem will be at the corner defined by the intersection of constraints 1 and 3. (You can move ∇f around to the other corners to be convinced that it does not lie in the cones defined by the constraints at those corners.)

This serves to illustrate the consistency between the geometry of the Kuhn–Tucker conditions and the simplex method linear programming result. Next, we explore where the algebra of the Kuhn–Tucker conditions leads us in this problem.

Kuhn–Tucker Algebra for Example 3 The two (m) linear equations in the five (n) unknowns (λ's) in (10.18a) can easily be represented in standard matrix form. Let $\mathbf{M}$ be the 2×5 matrix made up of the columns of constraint gradients:[2]

$$\mathbf{M} = \begin{bmatrix} 1 & 3 & 1 & -1 & 0 \\ 2 & 2 & 10 & 0 & -1 \end{bmatrix}$$

Then, with $\mathbf{\Lambda}' = [\lambda_1, \lambda_2, \lambda_3, \lambda_4, \lambda_5]$, the system is

$$\mathbf{M\Lambda} = \nabla f \qquad\qquad (10.18a')$$

where, furthermore, $\mathbf{\Lambda} \geq \mathbf{0}$. We know, from the two-dimensional solution space in Figure 10.7, that exactly two constraint gradients define a cone at each corner of the feasible region. That means that we need to find nonnegative *basic solutions* (Chapter 2) to the equations in (10.18a'), in which three ($n - m$) of the λ's are set equal to zero.

We also know, in a two-variable linear programming problem, that basic solutions to the augmented constraint set are associated with points of intersection of pairs of constraints. Moreover, the nonnegative basic solutions to those same equations identify each corner of the feasible region in solution space. For this linear program, the augmented constraint set contains the following three equations in five unknowns:

$$x_1 + 2x_2 + s_1 = 10$$

$$3x_1 + 2x_2 + s_2 = 24 \qquad\qquad (10.19)$$

$$x_1 + 10x_2 + s_3 = 40$$

If we examine all 10 basic solutions[3] to the Kuhn–Tucker requirements in (10.18a'), we identify exactly the same 10 solution-space locations as when all basic solutions are found to the linear equation system in (10.19). Table 10.3 summarizes the particulars of these 10 solutions; each corresponds to one of the 10 points indicated in Figure 10.8, which replicates the essentials of Figure 10.7 but extends each of the constraints to intersections that lie outside of the feasible region.

In each case, solution values for the two λ's found as a basic solution lead, via the other conditions in (10.18), to associated values for x_1 and x_2. (The three *nonbasic* λ's in any basic solution are, by definition, equal to zero.) The λ's in the basic solution must be nonnegative in order to satisfy (10.18d). When a basic solution with nonnegative λ's is found, the x's associated with that solution must be checked in each constraint in (10.18c). In this case, the expected optimal solution is identified (point 5 in the figure).

Kuhn–Tucker Conditions and Duality in Linear Programming Rewrite (10.18a), replacing λ_4 and λ_5 by the notation that was used in Chapter 7 for *surplus* variables in the *dual* constraints, namely, t_1 and t_2:

---

[2]This matrix is just the transpose of the *Jacobian* matrix to the system of five linear functions $h^1(\mathbf{X}) = x_1 + x_2 - 10, \ldots, h^5(\mathbf{X}) = 0x_1 - x_2$ that constitute the constraints in this problem.
[3]Recall that the maximum number of possible basic solutions to m linear equations in n unknowns ($m < n$) is C_n^m. In the example in (10.18a'), all $C_2^5 = 10$ basic solutions can be found, because all possible *pairs* of columns in $\mathbf{M}$ are linearly independent.

TABLE 10.3   Characteristics of Basic Solutions to (10.18a′) and (10.19)

X′	Λ′	Label in Figure 10.8 and Geometric Description	Constraint(s) Violated
[0, 0]	[0, 0, 0, −2, −5]	1 The origin	$\lambda_4 \geq 0, \lambda_5 \geq 0$
[0, 4]	[0, 0, 0.5, −1.5, 0]	2 Intersection of constraint 3 and x_2 axis	$\lambda_4 \geq 0$
[0, 5]	[2.5, 0, 0, 0.5, 0]	3 Intersection of constraint 1 and x_2 axis	Constraint 3
[0, 12]	[0, 2.5, 0, 5.5, 0]	4 Intersection of constraint 2 and x_2 axis	Constraints 1 and 3
[2.5, 3.75]	[1.875, 0, 0.125, 0, 0]	5 Intersection of constraints 1 and 3	NONE
[5.714, 3.429]	[0, 0.536, 0.393, 0, 0]	6 Intersection of constraints 2 and 3	Constraint 1
[7, 1.5]	[2.75, −0.25, 0, 0, 0]	7 Intersection of constraints 1 and 2	$\lambda_2 \geq 0$
[8, 0]	[0, 0.667, 0, 0, −3.667]	8 Intersection of constraint 2 and x_1 axis	$\lambda_5 \geq 0$
[10, 0]	[2, 0, 0, 0, −1]	9 Intersection of constraint 1 and x_1 axis	Constraint 2
[40, 0]	[0, 0, 2, 0, 15]	10 Intersection of constraint 3 and x_1 axis	Constraints 1 and 2

$$\lambda_1 + 3\lambda_2 + \lambda_3 - t_1 = 2$$
$$2\lambda_1 + 2\lambda_2 + 10\lambda_3 - t_2 = 5$$
(10.18a′)

Also, using s_1 through s_3 for the slack variables in the three constraints of the primal linear programming problem, conditions (10.18b) can be rewritten as

$$\lambda_1(-s_1) = 0, \ \lambda_2(-s_2) = 0, \ \lambda_3(-s_3) = 0; \ (-t_1)x_1 = 0, \ (-t_2)x_2 = 0 \qquad (10.18b')$$

and multiplication of each equation by −1 would remove all the negative signs.

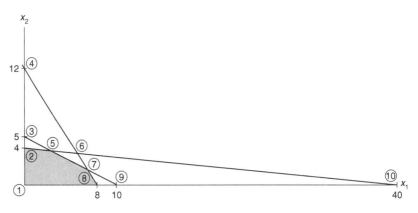

FIGURE 10.8   Locations of basic solutions to (10.19).

What we have in (10.18a′) is precisely the equality form for the constraints from the linear programming problem that is the *dual* to the original maximization problem. In Chapter 7, when we introduced this dual problem, we used y's for the dual variables; here we see that the λ's play exactly the role that the y's did in the discussion in Chapter 7. Moreover, the conditions in (10.18b′)—again with λ's here where y's were used before—express precisely the *complementary slackness* properties (Theorem PD3 in Chapter 7) that characterized optimal solutions to a pair of primal and dual linear programming problems. Finally, the λ's that we seek in forming the linear combinations must be nonnegative just as dual variables were required to be nonnegative in Chapter 7. The only aspect of the original dual problem that is missing from this Kuhn–Tucker derivation is an explicit statement of the dual objective function.

10.2.6 Quasiconcavity, Quasiconvexity and Sufficiency

For the constrained maximum problem in (10.2), if all $h^i(\mathbf{X})$ are convex and if $f(\mathbf{X})$ is concave (strictly or not), then we have seen that the Kuhn–Tucker necessary conditions in (10.9) are also sufficient. Arrow and Enthoven (1961) first proposed relaxations of these convexity/concavity requirements along the following lines:

1. *Relaxation of convexity conditions on $h^i(\mathbf{X})$.* If all of the $h^i(\mathbf{X})$ are *quasi*convex, the feasible region will still be a convex set, and so, with a concave objective function, the conditions in (10.8) will continue to be sufficient as well as necessary for a maximum.[4]

2. *Relaxation of concavity conditions on $f(\mathbf{X})$.* If $f(\mathbf{X})$ is *quasi*concave, along with quasiconvex constraints, then the Kuhn–Tucker conditions will be sufficient as well as necessary for a constrained maximum at $\mathbf{X^*}$ if either of the following holds:[5]

$$\text{(a) } \nabla f(\mathbf{X^*}) \neq 0 \qquad \text{or} \qquad \text{(b) } f(\mathbf{X}) \text{ is concave} \qquad (10.20)$$

Condition (b), of course, is just a "tightening" of quasiconcavity,[6] and puts us back to the conditions in (1). Condition (a) is designed to rule out certain kinds of inflection points.[7]

With quasiconcave objective functions, a problem can arise. (This was mentioned in Appendix 3.1.) Figure 3.1.2b illustrated a strictly quasiconcave function over $[c,d]$ with a zero-valued derivative at both x^* and x^{**}; the former is a point of inflection and the latter is the true maximum for the function.[8] The Kuhn–Tucker necessary conditions identify (among others) points at which the derivative or the gradient of the function is zero, so those conditions could not also be *sufficient* for a constrained maximum of a quasicon-

---

[4]We have made use of this fact—that quasiconvex constraints $h^i(\mathbf{X}) \leq 0$ define a convex feasible region—throughout our investigation of inequality-constrained problems, beginning in Section 4.5.

[5]The Arrow–Enthoven conditions are more detailed than the two listed here, but a for great many real-world problems the characterization given here is adequate.

[6]Recall that a concave function is always quasiconcave; a quasiconcave function may or may not be concave.

[7]Since Arrow and Enthoven chose to express the inequality constraints in a maximization problem as $g^i(\mathbf{X}) \geq 0$, the convexity of the feasible region is assured in their form of the problem if all $g^i(\mathbf{X})$ are quasiconcave functions (Appendix 3.1). For this reason, their work was termed "quasiconcave programming." Recall that we use the standard form of upper limits in maximization problems—$g^i(\mathbf{X}) \leq b_i$ or $h^i(\mathbf{X}) \leq 0$—and so the feasible region is a convex set when all $h^i(\mathbf{X})$ are quasiconvex.

[8]If a function has a point of inflection, where the shape of the function changes from concave (convex) to convex (concave), it need not have—but can have, and that is what causes trouble—a zero-valued derivative there.

cave objective function unless there were additional conditions that would weed out such inflection points.

Arrow and Enthoven's conditions were designed to do just that. In doing so, as in condition (a) in Eq. (10.20), above, they also eliminate a true maximum inside the constraints. (That is why they are sufficient conditions.) Subsequently, a number of specialized and refined definitions of quasiconcave and quasiconvex functions were proposed to eliminate inflection points with zero-valued gradients. For example, Mangasarian (1969) provides mathematical definitions of *pseudoconcave* and *pseudoconvex* functions that eliminate the possibility of such inflection points.[9] The definition is as follows:

The function $f(x)$, defined over a convex set $\mathbf{X}$, is *pseudoconcave* if, for $x' \neq x''$:

$$f(x'') > f(x') \quad \text{implies} \quad f'(x')(x'' - x') > 0 \tag{10.21}$$

[Compare the definition of quasiconcavity in (3.1.5) in Appendix 3.1, where both inequalities are weak.] The extension to $f(\mathbf{X})$ is obvious [cf. (3.1.5') in Appendix 3.1; change both inequalities in that definition from weak to strong]. *Pseudoconvexity* would be of interest in inequality-constrained minimization problems. It is defined similarly; just reverse the two inequalities to read "less than." (Again, you could compare the definitions in Appendix 3.1, this time for quasiconvexity.)

You might work through the geometry of why this definition disallows an inflection point with a zero-valued derivative but does not eliminate a function with a true maximum (where, also, the derivative is zero). Schematically, this kind of function falls between convex and strictly quasiconvex:

$$\text{Strictly convex} \Rightarrow \text{convex} \Rightarrow \text{pseudoconvex} \Rightarrow$$

$$\text{strictly quasiconvex} \Rightarrow \text{quasiconvex}$$

Other definitions provide additional specialized variations on the notions of concavity and convexity. Many of these are designed to identify functions for which (1) a local optimum is assured to be the global optimum and/or (2) a global optimum is unique.

To pursue these refinements would carry us too far afield, but you should be aware that they exist (and why they exist). Using the notion of pseudoconcavity, the Kuhn–Tucker conditions can be extended as follows:

If the objective function, $f(\mathbf{X})$, is pseudoconcave and all constraints, $h^i(\mathbf{X})$
are quasiconvex, then the Kuhn–Tucker necessary conditions in (10.8) are
also *sufficient* for a maximum to the problem in (10.2). 	(10.22)

Numerical Illustrations   Here are a few examples of how the Kuhn–Tucker conditions sort out candidates for a constrained maximum point in the face of quasiconcavity of $f(x)$:

1. Maximize $f(x) = e^{-(x-2)^2}$ subject to $x \leq 3$ and $x \geq 0$. [This function was shown in Figure 3.18c in the text of Chapter 3 and in Figure 3.1.7 in Appendix 3.1; it is repeated here as Figure 10.9.] The Kuhn–Tucker requirements in (10.8) are

---

[9]See also Vajda (1974) and Zoutendijk (1976). Other references are indicated in footnote 27 in Appendix 3.1.

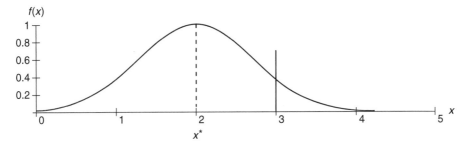

FIGURE 10.9 $f(x) = e^{-(x-2)^2}$.

$$-2(x-2)e^{-(x-2)^2} - \lambda = 0 \qquad\qquad (10.23a)$$

$$x[-2(x-2)e^{-(x-2)^2} - \lambda] = 0 \qquad\qquad (10.23b)$$

$$x - 3 \leq 0 \qquad\qquad (10.23c)$$

$$\lambda(x-3) = 0 \qquad\qquad (10.23d)$$

$$x \geq 0, \lambda \geq 0 \qquad\qquad (10.23e)$$

Starting with (10.23d):

(i) if $\lambda = 0$, we are led, via (10.23a) and (10.23b), to two options—$x = 0$, which fails (10.23a), or $x = 2$, which satisfies all remaining conditions; or

(ii) on the other hand, when $x = 3$ in (10.23d), (10.23b) generates $\lambda = -0.736$, which fails (10.23e).

So in this case, where the objective function is strictly quasiconcave (but also pseudo-concave), the Kuhn–Tucker conditions, (10.23), turn out to be also sufficient to identify the constrained maximum point at $x^* = 2$.

It is also easily established that, for example, if the constraint had been $x \leq 1$ (so that the constrained maximum is at $x = 1$), the Kuhn–Tucker conditions in (10.8) would force examination of the points $x = 0,1,2$ (the left boundary, right boundary, and unconstrained maximum points, respectively). Both $x = 0$ and $x = 2$ are (appropriately) rejected by some part of the Kuhn–Tucker conditions and $x = 1$ is accepted. (You can easily work through these possibilities.) Again, then, the conditions in (10.8) turn out to be sufficient as well for this strictly quasiconcave objective function.

2. Maximize $f(x) = (x - 1)^3$ subject to $x \leq 2$ as well as $x \geq 0$. Again, $f(x)$ is strictly quasiconcave, but now not pseudoconcave. You can easily write down the conditions in (10.8) and find that (10.8d) leads to $x = 0,1,2$. Now, however, while $x = 0$ is (appropriately-ly) rejected, both $x = 1$ (a point of inflection) and $x = 2$ (the true constrained maximum point) satisfy all the other necessary conditions in (10.8).

This illustrates that for an objective function that is strictly quasiconcave, but that is not pseudoconcave because it has the kind of inflection point exhibited in Figure 10.10, the Kuhn–Tucker conditions cannot be sufficient also.

3. Maximize $f(x) = -0.1(x + 1)^3(x - 2)^2 + 1$, with $x \leq 3$ and $x \geq 0$. (This is essentially

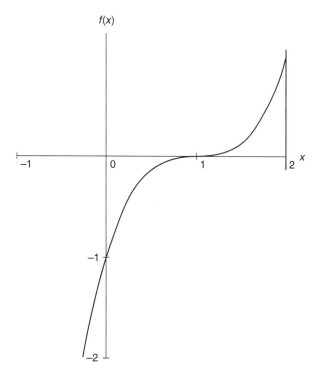

FIGURE 10.10 $f(x) = (x - 1)^3$.

the negative of the function shown in Figure 3.1.8 in Appendix 3.1. It is repeated here as Fig. 10.11.) Once more, $f(x)$ is strictly quasiconcave but not pseudoconcave.

The necessary conditions in (10.8) are, again, easily written down, but not so easily solved without some kind of computer algebra help. In this case, if you work through the implications of (10.8d), five values for x will be identified:

(i) $x = -1$, which fails nonnegativity in (10.8e); it is an inflection point, with $f' = 0$, but it is below the $x \geq 0$ boundary

(ii) $x = 0$, the left-hand boundary, which fails (10.8a)

(iii) $x = 0.5$, where $f' = 0$ again, and which fails nothing (it is the maximum point, with or without the constraints);

(iv) $x = 2$, where $f' = 0$ once more, and which again fails nothing, but that is wrong [it is, again, a point of inflection where $f'(x) = 0$]

(v) $x = 3$, the right-hand boundary, for which $\lambda = -24$, violating nonnegativity in (10.8e).

It is doubtful that one will generally know in advance whether the objective function in a maximization problem is pseudoconcave. One approach is simply to apply the Kuhn–Tucker conditions in (10.8) and, if more than one candidate appears (as in the last two illustrations), evaluate $f(\mathbf{X})$ at each and choose the best—the one for which $f(\mathbf{X})$ is largest.

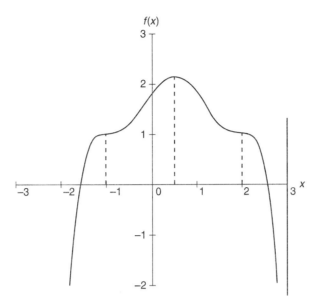

FIGURE 10.11 $f(x) = (-0.1)(x + 1)^3(x - 2)^3 + 1$.

Illustrations from Economics   Constrained optimization problems in economics provided the inspiration for Arrow and Enthoven's investigation of quasiconvexity and quasiconcavity. For example, economists often find it reasonable (and adequate) to assume that utility functions are only quasiconcave, not necessarily concave. Similarly, production functions may often be assumed to be quasiconcave but not concave. Relaxed assumptions about functional shape, such as these, often mean that the resulting theory can encompass a wider variety and more realistic set of real-world situations. We explore a production function example.

A Cobb–Douglas Production Function   The usual Cobb–Douglas production function relates output to use of labor and capital inputs as $f(\mathbf{X}) = Ax_1^\alpha x_2^\beta$, where x_1 and x_2 represent the amounts of labor and capital, respectively, employed in production, and where $A > 0$ and $0 < \alpha, \beta < 1$. When $\alpha + \beta > 1$, the production function reflects the property of increasing returns to scale, doubling the amount of each input more than doubles total output.

We examine $f(\mathbf{X})$ for concavity or convexity, using the rules on leading principal minors of the Hessian matrix, for the case in which positive amounts of both labor and capital are employed in production. For notational simplicity, let $\gamma = f(\mathbf{X})$ and note that $\gamma > 0$ for the case in which $x_1 > 0$ and $x_2 > 0$. The Hessian matrix for this function is

$$\mathbf{H} = \begin{bmatrix} f_{11} & f_{12} \\ f_{21} & f_{22} \end{bmatrix} = \begin{bmatrix} \alpha(\alpha - 1)\gamma x_1^{-2} & \alpha\beta\gamma x_1^{-1}x_2^{-1} \\ \alpha\beta\gamma x_1^{-1}x_2^{-1} & \beta(\beta - 1)\gamma x_2^{-2} \end{bmatrix}$$

For *concavity*, $\mathbf{H}$ must be negative semidefinite, which means that $|\mathbf{H}_1| < 0$ and $|\mathbf{H}_2| = |\mathbf{H}| > 0$; for *convexity*, the requirement is that $\mathbf{H}$ be positive semidefinite, meaning $|\mathbf{H}_1| > 0$ and $|\mathbf{H}_2| = |\mathbf{H}| > 0$. In this example, $|\mathbf{H}_1|$ is unambiguously negative. (Why?) This means that convexity is ruled out. What about concavity? Here $|\mathbf{H}_2| = \alpha\beta[1 - (\alpha + \beta)]\gamma^2 x_1^{-2} x_2^{-2}$, which is negative when $x_1 > 0$, $x_2 > 0$ and $\alpha + \beta > 1$ (the increasing-returns case). Therefore $f(\mathbf{X})$ is not a concave function when $\alpha + \beta > 1$.

Using results from Appendix 3.1, particularly Definition 4, for quasiconcavity, we form the bordered matrix for this function, as in (3.1.9):

$$\mathbf{D} = \begin{bmatrix} 0 & f_1 & f_2 \\ \hline f_1 & f_{11} & f_{12} \\ f_2 & f_{21} & f_{22} \end{bmatrix}$$

which is this case is

$$\mathbf{D} = \begin{bmatrix} 0 & \alpha\gamma x_1^{-1} & \beta\gamma x_2^{-1} \\ \hline \alpha\gamma x_1^{-1} & & \\ \beta\gamma x_2^{-1} & & \mathbf{H} \end{bmatrix}$$

Sufficient conditions for quasiconcavity [from (3.1.12) in Appendix 3.1] place sign conditions on the leading principal minors of $\mathbf{D}$ [remember that in (3.1.12) $\mathbf{D}_1$ is defined as the determinant of the 2×2 matrix in the upper left]. For this example, the requirements are that $|\mathbf{D}_1| < 0$ and $|\mathbf{D}_2| > 0$. Here is it not very difficult to confirm that these conditions are in fact satisfied; $|\mathbf{D}_1| = -(\alpha\gamma x_1^{-1})^2 < 0$ and $|\mathbf{D}_2| = |\mathbf{D}| = \alpha\beta[\alpha + \beta]\gamma^3 x_1^{-2} x_2^{-2} > 0$. Thus $f(\mathbf{X})$ satisfies *sufficient* conditions for a *quasiconcave* function.

A Cost-Minimization Illustration   Suppose that this Cobb–Douglas production function appeared as a constraint in the following example of a least-cost production problem (Arrow and Enthoven, 1961). Given positive unit prices for both labor and capital—$p_1 > 0$ and $p_2 > 0$—let $\mathbf{P} = [p_1, p_2]'$ and $\mathbf{X} = [x_1, x_2]'$, and consider the problem of choosing amounts of labor (x_1) and capital (x_2) to employ with the aim of minimizing total (labor and capital) costs of production, subject to a minimum level of total output (and nonnegativity of labor and capital use). This problem could be posed as follows:

Minimize $C(\mathbf{X}) = \mathbf{P}'\mathbf{X}$

subject to $f(\mathbf{X}) = A x_1^\alpha x_2^\beta \geq \overline{Q}$ or $h(\mathbf{X}) = A x_1^\alpha x_2^\beta - \overline{Q} \geq 0$ (10.24)

and $\mathbf{X} \geq \mathbf{0}$

where $\overline{Q} > 0$ (meaning that there must be a *positive* amount of output produced). The objective function in this problem is linear in x_1 and x_2 and hence both concave and convex (although neither strictly so). Under increasing returns, as we have just seen, the constraint function, $f(\mathbf{X})$, will not be concave, but rather quasiconcave; and it is differentiable.

For this constrained minimization problem, the Arrow–Enthoven results in (10.20) become either

(a) $\nabla C(\mathbf{X}^*) \neq \mathbf{0}$ or   (b) $C(\mathbf{X})$ is convex   (10.25)

The Kuhn–Tucker necessary conditions in (10.8) will be both necessary and sufficient for a minimum if one of the conditions in (10.25) is satisfied. But here (a) in (10.25) is clearly met, since $\partial C/\partial x_1 = p_1 > 0$ everywhere (and also $\partial C/\partial x_2 = p_2 > 0$ everywhere, although only one such inequality is required).

Numerical Example   Suppose in the problem in (10.24) that $\mathbf{P} = [3 \quad 10]'$, $f(\mathbf{X}) = (1.5)x_1^{(0.6)}x_2^{(0.8)}$, and $\overline{Q} = 100$. Application of the Kuhn–Tucker conditions generates the following requirements:

$$3 - \lambda[(0.9)x_1^{(-0.4)}x_2^{(0.8)}] \geq 0$$
$$10 - \lambda[(1.2)x_1^{(0.6)}x_2^{(-0.2)}] \geq 0 \tag{10.26a}$$

$$x_1\{3 - \lambda[(0.9)x_1^{(-0.4)}x_2^{(0.8)}]\} = 0$$
$$x_2\{10 - \lambda[(1.2)x_1^{(0.6)}x_2^{(-0.2)}]\} = 0 \tag{10.26b}$$

$$(1.5)x_1^{(0.6)}x_2^{(0.8)} \geq 100 \tag{10.26c}$$

$$\lambda[(1.5)x_1^{(0.6)}x_2^{(0.8)} - 100] = 0 \tag{10.26d}$$

$$\mathbf{X} \geq 0, \lambda \geq 0$$

From (10.26d), (i) $\lambda = 0$ or (ii) $(1.5)x_1^{(0.6)}x_2^{(0.8)} = 100$. Looking first at (i), the implications of $\lambda = 0$ are that $x_1 = 0$ and $x_2 = 0$ [from (10.26b)], which violates (10.26c). Under (ii) there are four cases to examine, generated by the requirements in (10.26b):

1. $x_1 = 0$ and $x_2 = 0$. We have already seen that this violates (10.26c). In fact, even if only $x_1 = 0$ or $x_2 = 0$ (not both), the requirements in (10.26c) will be violated. These are the outcomes of the next two possible cases, under (10.26b).

2. $x_1 = 0$ and $\{10 - \lambda[(1.2)x_1^{(0.6)}x_2^{(-0.2)}]\} = 0$. Again, since $x_1 = 0$, (10.26c) is not met.

3. $\{3 - \lambda[(0.9)x_1^{(-0.4)}x_2^{(0.8)}]\} = 0$ and $x_2 = 0$. Again, $x_2 = 0$ violates (10.26c).

4. Finally, then, we are left in (10.26b) with $\{3 - \lambda[(0.9)x_1^{(-0.4)}x_2^{(0.8)}]\} = 0$ and $\{10 - \lambda[(1.2)x_1^{(0.6)}x_2^{(-0.2)}]\} = 0$.

In conjunction with $(1.5)x_1^{(0.6)}x_2^{(0.8)} = 100$, we have three (nonlinear) equations in the three unknowns: x_1, x_2, and λ. The solution (using software for solving systems of nonlinear equations) is $x_1 = 33.8987$, $x_2 = 13.5595$, and $\lambda = 1.6950$, which can be verified to satisfy all of the *necessary* conditions in (10.26). Because $C(\mathbf{X})$ in (10.24) is linear and $h(\mathbf{X})$ is quasiconcave, the conditions in (10.26) are also *sufficient* for this constrained minimization problem.

Reflections   For problems in which $f(\mathbf{X})$ and all $h^i(\mathbf{X})$ are specific numerical functions, and for which a numerical solution is sought, it may be straightforward to examine all functions for concavity or concavity (strict or otherwise)—for example, using the leading principal minors tests. For large problems, however, these determinental tests can be complicated and tedious. For quasiconcavity or quasiconvexity, somewhat similar rules are available (Definitions 4a and 4b in Appendix 3.1), but they do not identify strict versions of either. For further distinctions—pseudoconcavity or pseudoconvexity, for example—there do not seem to be any rules that are (relatively) easy to apply.

However, if application of Kuhn–Tucker conditions in a numerical problem identifies more than one candidate for a constrained maximum (e.g., as in Figs. 10.10 and 10.11), you can evaluate $f(\mathbf{X})$ for each such point and choose the one for which $f(\mathbf{X})$ is largest. [If more than one point gives the same "largest" value for $f(\mathbf{X})$, then you have to decide whether there is any basis for choosing one over another.] If you are using some kind of numerical method in such a problem, you could have trouble; you could climb to a local hilltop that was inferior to others, and you might not realize that there were other hilltops. That is why it is useful to be able to establish that necessary conditions are also sufficient.

In theoretical discussions, the functions may be specified only in a *general* way (as in the illustrations from economics) or *assumed* to have particular shape characteristics. In those cases, the Kuhn–Tucker necessary requirements may provide useful characterizations of optimal behavior, but it is also important to have some assurance that those necessary requirements have in fact identified the right kind of extreme point (e.g., a maximum and not a minimum or something else). So it is helpful, again, to know if sufficiency can be established, even in the face of generalized functional forms.

10.3   THE KUHN-TUCKER CONSTRAINT QUALIFICATION

In this section we illustrate some of the ways in which it is possible for a constrained optimum to exist at a point that fails to satisfy all the Kuhn–Tucker conditions. In particular, a constrained maximum or minimum may exist at a point $\mathbf{X}^*$ where $\nabla f(\mathbf{X}^*)$ cannot be expressed as a nonnegative linear combination of the gradients of the active constraints at $\mathbf{X}^*$, violating the conditions in (10.17a). Such problems fail to meet Kuhn and Tucker's constraint qualification (or regularity condition), and the necessary conditions for inequality constrained problems, like those in Section 10.2, are not valid.

If we concentrate on the $n = 2$ case with (for simplicity) linear constraints, the geometry of (10.17a) is easy to grasp; it was illustrated in Figure 10.6. We consider two cases: when the unconstrained optimum is feasible and when it is not.

10.3.1   Unconstrained Optimum Inside or on the Boundaries of the Feasible Region

1. When the unconstrained optimal point, $\mathbf{X}^*$, is interior to the feasible region (no active constraints at $\mathbf{X}^*$), $\nabla f(\mathbf{X}^*) = \mathbf{0}$ and the requirement in (10.17a) is easily met by setting all $m+n$ of the λ's equal to zero. [You might think through why all the other conditions in (10.17) are also met at such an interior optimum.]

2. If $\mathbf{X}^*$ is exactly on the edge of the feasible region—on a *single* constraint, $h^j(\mathbf{X}) = 0$, then, again, $\nabla f(\mathbf{X}^*) = \mathbf{0}$ and setting all λ's equal to zero (including λ_j) will satisfy (10.17a), whatever the value of $\nabla h^j(\mathbf{X}^*)$.

3. If $\mathbf{X}^*$ is precisely at a point defined by the intersection of *two* constraints, $h^j(\mathbf{X})$ and $h^k(\mathbf{X})$ (at a *corner* of the feasible region), $\nabla f(\mathbf{X}^*) = \mathbf{0}$ and again setting all λ's equal to zero (including λ_j and λ_k) will satisfy (10.17a), irrespective of the values of $\nabla h^j(\mathbf{X}^*)$ and $\nabla h^k(\mathbf{X}^*)$.

This logic extends to more complicated problems with more variables and/or nonlinear constraints. In all cases in which the constraints turn out to be unnecessary, there is no need for concern with the Kuhn–Tucker constraint qualification. Using the necessary conditions in (10.17) will always lead you to the optimal solution—the trouble is that you do not know in advance whether the constraints are, in fact, unnecessary.

10.3.2 Unconstrained Optimum Outside the Feasible Region

Continuing the $n = 2$ case, the constrained optimal point, $\mathbf{X}^*$, will be located either on a single constraint (as at point B in Fig. 10.6) or at the intersection of two constraints (as at $\mathbf{X}^* = \begin{bmatrix} 2.5 \\ 3.75 \end{bmatrix}$ in Fig. 10.7). We explore these possibilities in turn.

1. Let $h^k(\mathbf{X})$ be the active constraint at $\mathbf{X}^*$, where $\nabla f(\mathbf{X}^*) \neq \mathbf{0}$, since the unconstrained optimum is outside the feasible region. From (10.17a), we should have

$$\nabla f(\mathbf{X}^*) = \lambda_k^* [\nabla h^k(\mathbf{X}^*)]$$

for $\lambda_k^* \geq 0$, and a problem would arise in trying to find a λ_k^* to satisfy this equation if $\nabla h^k(\mathbf{X}^*) = 0.$[10] This situation is illustrated in Example 4, in section 10.3.3.

2. If the optimal point is defined by the intersection of two constraints—$h^j(\mathbf{X}) = 0$ and $h^k(\mathbf{X}) = 0$—then (10.17a) requires that

$$\nabla f(\mathbf{X}^*) = \lambda_j^* [\nabla h^j(\mathbf{X}^*)] + \lambda_k^* [\nabla h^k(\mathbf{X}^*)]$$

for $\lambda_j^*, \lambda_k^* \geq 0$. With $\nabla f(\mathbf{X}^*) \neq \mathbf{0}$, a problem arises if $\nabla h^j(\mathbf{X}^*)$ and $\nabla h^k(\mathbf{X}^*)$ are *linearly dependent* (so that they define only a line, not a cone, in two-dimensional solution space; they do not *span* that space) and $\nabla f(\mathbf{X}^*)$ is linearly independent of them. Note that this also includes the possibility that one of the two gradients is a null vector.[11] There is no way that $\nabla f(\mathbf{X}^*)$ can be described as a linear combination of two such constraint gradients at $\mathbf{X}^*$. These kinds of possibilities are illustrated in Examples 5 and 6, in Sections 10.3.3 and 10.3.4.

10.3.3 Numerical Illustrations: Convex Feasible Regions

The following two examples illustrate the fact that the Kuhn–Tucker conditions can be confounded even in problems with convex feasible regions.[12]

Example 4 The kind of difficulty suggested in parapraph 1 in Section 10.3.2 occurs in the following problem, whose fundamental geometry is given in Figure 10.12:

---

[10]Recall that for a maximization problem in standard form, the gradients of the constraints, evaluated along the boundaries, point *outward* from the feasible region. When the unconstrained maximum is outside the feasible region, $\nabla f(\mathbf{X})$ evaluated along the boundaries will also point outward, so it is *nonnegative* λ's that are required, as in (10.17).

[11]It is also conceivable that *both* $\nabla h^j(\mathbf{X}^*) = 0$ and $\nabla h^k(\mathbf{X}^*) = \mathbf{0}$, so that in fact the two gradients define only a point, not even a line.

[12]Convexity of a feasible region allows one to conclude that necessary conditions for a constrained maximum or minimum point are also sufficient conditions. In this section we are examining situations in which even necessary conditions fail to identify an optimal point.

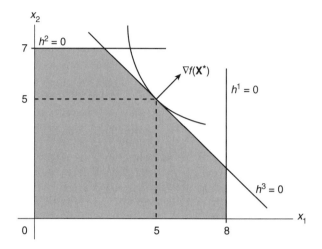

FIGURE 10.12 Geometry for Example 4.

$$\text{Maximize} \quad f(\mathbf{X}) = -(x_1 - 10)^2 - (x_2 - 10)^2 + 25$$

$$\text{subject to} \quad h^1(\mathbf{X}) = x_1 - 8 \leq 0$$

$$h^2(\mathbf{X}) = x_2 - 7 \leq 0$$

$$h^3(\mathbf{X}) = -(10 - x_1 - x_2)^3 \leq 0$$

$$\text{and} \quad x_1, x_2 \geq 0$$

Here $f(\mathbf{X})$ has an unconstrained hilltop at $\mathbf{X}^u = \begin{bmatrix} 9 \\ 9 \end{bmatrix}$, outside the feasible region, and its contours are circles in the x_1, x_2 plane. The boundaries created by the first two constraints in x_1, x_2 space are simply the lines $x_1 = 8$ and $x_2 = 7$. The boundary given by the third constraint is the *line* $x_1 + x_2 = 10$ [the locus of points for which $h^3(\mathbf{X}) = 0$].

The feasible region is seen to be a convex set all of whose sides plot as straight lines, but this is *not* because all constraints are themselves linear functions of x_1 and x_2; $h^3(\mathbf{X})$ is a nonlinear function of x_1 and x_2, although it happens to have a linear contour in the x_1, x_2 plane. It is clear, from the geometry, that only $h^3(\mathbf{X})$ will be binding at the optimal solution. Knowing that, we can easily solve $10 - x_1 - x_2 = 0$ for either unknown, substitute into $f(\mathbf{X})$, and find that an unconstrained optimum occurs at $\mathbf{X}^* = \begin{bmatrix} 5 \\ 5 \end{bmatrix}$.

For this problem

$$\nabla h^3(\mathbf{X}) = \begin{bmatrix} 3(10 - x_1 - x_2)^2 \\ 3(10 - x_1 - x_2)^2 \end{bmatrix}$$

and so, at $\mathbf{X}^*$, $\nabla h^3(\mathbf{X}^*) = \begin{bmatrix} 0 \\ 0 \end{bmatrix}$. In fact, it is clear that $\nabla h^3 = \mathbf{0}$ everywhere along the constraint formed by $10 - x_1 - x_2 = 0$. Even though an identifiable outward pointing normal (gradient) to the line could be *visualized* at any point along the $h^3(\mathbf{X}) = 0$ boundary, *alge-*

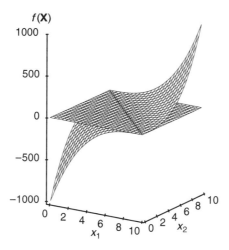

FIGURE 10.13 $h^3(\mathbf{X}) = -(10 - x_1 - x_2)^3$.

braic evaluation of that gradient anywhere along that boundary gives the null vector.[13] Since $\nabla f(\mathbf{X}^*) = \begin{bmatrix} 8 \\ 8 \end{bmatrix}$, it is impossible to find a $\lambda_3 \geq 0$ that will satisfy $\nabla f(\mathbf{X}^*) = \lambda_3$ $\nabla h^3(\mathbf{X}^*)$, which here means $\begin{bmatrix} 8 \\ 8 \end{bmatrix} = \lambda_3 \begin{bmatrix} 0 \\ 0 \end{bmatrix}$. [Only the direction, not complete length, of $\nabla f(\mathbf{X}^*)$ is shown in Figure 10.12.]

The problem is that $h^3(\mathbf{X})$ has a peculiar three-dimensional shape that gives it a null gradient in the x_1, x_2 plane, where $h^3(\mathbf{X}) = 0$. We can tell that something is amiss from the Hessian matrix, since

$$\nabla^2 h^3(\mathbf{X}) = \begin{bmatrix} -6(10 - x_1 - x_2) & -6(10 - x_1 - x_2) \\ -6(10 - x_1 - x_2) & -6(10 - x_1 - x_2) \end{bmatrix}$$

and at all points along the line $x_1 + x_2 = 10$, this is a null matrix. In particular, that means that $h^3(\mathbf{X})$ is neither a concave nor a convex function.

Figure 10.13 illustrates the three-dimensional geometry for this constraint. We see that the line $x_1 + x_2 = 10$ represents a locus of *inflection points* for $h^3(\mathbf{X})$. One object of the Kuhn–Tucker constraint qualification is to weed out optimization problems like this where the gradient of the constraint that is binding at the optimal solution is a null vector.[14]

---

[13]The same kind of problem was illustrated with a different numerical example in Chapter 4, Section 4.2.6 in the context of equality constrained problems.

[14]Note that if $h^3(\mathbf{X})$ were altered to, say, $h^3(\mathbf{X}) = -(10 - x_1 - x_2)^2 + c$, where c is any constant, then the function will be shifted downward or upward (depending on whether c is positive or negative) but the intersection with the x_1, x_2 plane will still be a straight line. For example, if $c = 27$, then $h^3(\mathbf{X})$ slices the x_1, x_2 plane in the line $x_1 + x_2 = 7$. At *any* point on this line, $\nabla h^3(\mathbf{X}) = \begin{bmatrix} 27 \\ 27 \end{bmatrix}$, so $\lambda_3^* = \frac{8}{27}$ satisfies (10.17a)—$\begin{bmatrix} 8 \\ 8 \end{bmatrix} = \lambda_3^* \begin{bmatrix} 27 \\ 27 \end{bmatrix}$. It is only when a constraint $h(\mathbf{X})$ intersects the x_1, x_2 plane in a line (or curve) along which $\nabla h(\mathbf{X}) = \mathbf{0}$ that the Kuhn–Tucker conditions are in trouble.

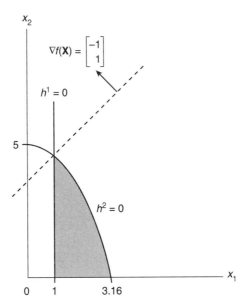

FIGURE 10.14   Geometry for Example 5.

Example 5   Here is a problem in which the constrained optimal solution is at a corner of a convex feasible region, where two ($= n$) constraints are binding. However, one of the constraint gradients is $\begin{bmatrix} 0 \\ 0 \end{bmatrix}$, and so again the Kuhn–Tucker requirements in (10.17) cannot be met:

$$\text{Maximize} \quad f(\mathbf{X}) = -x_1 + x_2$$
$$\text{subject to} \quad h^1(\mathbf{X}) = -x_1 \leq -1$$
$$h^2(\mathbf{X}) = -(5 - 0.5x_1{}^2 - x_2)^3 \leq 0$$
$$\text{and}^{15} \quad h^3(\mathbf{X}) = -x_2 \leq 0$$

The geometry is shown in Figure 10.14, and it is apparent that the optimal solution is at the corner of the feasible region where $h^1(\mathbf{X})$ and $h^2(\mathbf{X})$ intersect, at $\mathbf{X}^* = \begin{bmatrix} 1 \\ 4.5 \end{bmatrix}$

From Figure 10.14 we know that $h^1(\mathbf{X})$ and $h^2(\mathbf{X})$ are the binding constraints at the optimum, so the relevant gradients for this problem are

$$\nabla f(\mathbf{X}) = \begin{bmatrix} -1 \\ 1 \end{bmatrix}, \nabla h^1(\mathbf{X}) = \begin{bmatrix} -1 \\ 0 \end{bmatrix} \text{ and } \nabla h^2(\mathbf{X}) = \begin{bmatrix} 3x_1 D^2 \\ 3D^2 \end{bmatrix}$$

where $D = (5 - 0.5x_1^2 - x_2)$. The boundary provided by $h^2(\mathbf{X})$ in the positive quadrant of x_1, x_2 space is given by $-(5 - 0.5x_1^2 - x_2)^3 = 0$, which means by the curve $x_2 = 5 - 0.5x_1^2$. At

---

[15]The function $h^2(\mathbf{X})$ was also used in Section 4.2 to illustrate the constraint gradients difficulty in equality-constrained problems. Note that nonnegativity of x_1 is assured from the first constraint, $x_1 \geq 1$.

any point along this curve, including $\mathbf{X}^*$, $\nabla h^2(\mathbf{X}) = \begin{bmatrix} 0 \\ 0 \end{bmatrix}$. The optimality requirement on gradient vectors for this problem requires finding $\lambda_1 \geq 0$ and $\lambda_2 \geq 0$ that satisfy

$$\begin{bmatrix} -1 \\ 1 \end{bmatrix} = \lambda_1 \begin{bmatrix} -1 \\ 0 \end{bmatrix} + \lambda_2 \begin{bmatrix} 0 \\ 0 \end{bmatrix}$$

This is, of course, impossible; $\nabla f(\mathbf{X}^*)$ is linearly independent of the only nonnull gradient, $\nabla h^1(\mathbf{X}^*)$.[16]

10.3.4 Numerical Illustration: Nonconvex Feasible Regions with Cusps at Optimal Corners

The kind of irregularity noted in paragraph 2 in Section 10.3.2 can also arise when the feasible region has one or more very specific "cusps" along its boundaries. If the optimal solution happens to lie at one of these cusps, the Kuhn–Tucker necessary conditions will fail to hold. This is illustrated in Example 6.

Example 6   Figure 10.15 represents the (nonconvex) feasible region for a two-variable problem with the constraints:

$$h^1(\mathbf{X}) = x_2 - 2(1 - x_1)^3 - 1 \leq 0$$

$$h^2(\mathbf{X}) = -x_2 - 2(1 - x_1)^3 + 1 \leq 0$$

$$x_1, x_2 \geq 0$$

The contours of the two constraint functions in x_1, x_2 space are given by

$$x_2 = 2(1 - x_1)^3 + 1 \qquad \text{and} \qquad x_2 = -2(1 - x_1)^3 + 1$$

respectively. The functions intersect at $\mathbf{X} = \begin{bmatrix} 1 \\ 1 \end{bmatrix}$, where both have a slope of 0, so they are tangent to each other there. This creates the particular kind of cusp that is shown in the figure. Suppose that the objective for this problem is to maximize $f(\mathbf{X}) = 3x_1 + x_2$, whose contours in x_1, x_2 space are straight lines with a slope of -3. Then the optimal solution will be at this tangency point, $\mathbf{X}^* = \begin{bmatrix} 1 \\ 1 \end{bmatrix}$ in Figure 10.15. One dashed-line contour for this objective function has been included in the figure.

The tangency of the active constraints at $\mathbf{X}^*$ is also indicated by the constraint gradients there:

$$\nabla h^1(\mathbf{X}^*) = \begin{bmatrix} 0 \\ 1 \end{bmatrix} \qquad \text{and} \qquad \nabla h^2(\mathbf{X}^*) = \begin{bmatrix} 0 \\ -1 \end{bmatrix}$$

---

[16]As in the previous example, the Hessian matrix for the troublesome constraint [here $h^2(\mathbf{X})$] contains only zeros at all points along the $h^2(\mathbf{X}) = 0$ boundary. If you have access to computer graphics software, you might take a look at the function $h^2(\mathbf{X})$ in three-dimensional space. It is rather more complicated than its contour in the x_1, x_2 plane might suggest.

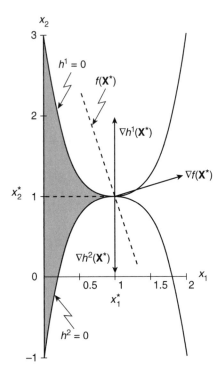

FIGURE 10.15   Geometry for Example 6.

The first points due north, the second due south. In contrast, $\nabla f = \begin{bmatrix} 3 \\ 1 \end{bmatrix}$ everywhere and does not lie on the line defined by the two constraint gradients (Fig. 10.15). Clearly, there is no linear combination of $\nabla h^1(\mathbf{X^*})$ and $\nabla h^2(\mathbf{X^*})$ that is equal to ∇f^*, so at this constrained optimum point the Kuhn–Tucker necessary conditions fail—the (nonnull) gradients of the active constraints at the optimal corner are *linearly dependent*, and ∇f^* there is *linearly independent* of them.

An attempt to apply the conditions in (10.17) to this problem uncovers the algebraic nature of the trouble. These conditions are as follows:

$$3 - 6(1 - x_1)^2\,\lambda_1 - 6(1 - x_1)^2\lambda_2 + \lambda_3 = 0$$

$$1 - \lambda_1 + \lambda_2 + \lambda_4 = 0$$
(10.27a)

$$\lambda_1[x_2 - 2(1 - x_1)^3 - 1] = 0$$

$$\lambda_2[-x_2 - 2(1 - x_1)^3 + 1] = 0$$
(10.27b)

$$\lambda_3 x_1 = 0$$

$$\lambda_4 x_2 = 0$$

$$x_2 - 2(1 - x_1)^3 - 1 \leq 0$$

$$-x_2 - 2(1 - x_1)^3 + 1 \leq 0 \tag{10.27c}$$

Nonnegativity of the x's and λ's   (10.27c)

At $\mathbf{X}^* = \begin{bmatrix} 1 \\ 1 \end{bmatrix}$, conditions (10.27b) require that $\lambda_3 = \lambda_4 = 0$. Putting these results into the pair of equations in (10.27a):

$$\begin{array}{c} 0\lambda_1 - 0\lambda_2 = 3 \\ \lambda_1 - \lambda_2 = 1 \end{array} \quad \text{or} \quad \begin{bmatrix} 0 & 0 \\ 1 & -1 \end{bmatrix} \begin{bmatrix} \lambda_1 \\ \lambda_2 \end{bmatrix} = \begin{bmatrix} 3 \\ 1 \end{bmatrix}$$

This is easily seen to be an inconsistent pair of linear equations in λ_1 and λ_2; the coefficient matrix $[\nabla h^1(\mathbf{X}^*) \vdots \nabla h^2(\mathbf{X}^*)] = \begin{bmatrix} 0 & 0 \\ 1 & -1 \end{bmatrix}$ is singular (its rank is 1) while the rank of the augmented matrix $[\nabla h^1(\mathbf{X}^*) \vdots \nabla h^2(\mathbf{X}^*) \vdots \nabla f(\mathbf{X}^*)] = \begin{bmatrix} 0 & 0 & \vdots & 3 \\ 1 & -1 & \vdots & 1 \end{bmatrix}$ is 2, so there are no solutions. These results on ranks indicate precisely the linear *dependence* of the constraint gradients and the linear *independence* of ∇f from them. Although the problem has an optimal solution at $\mathbf{X}^*$ that is well defined geometrically, the Kuhn–Tucker necessary conditions do not hold there.

10.3.5 Practical Considerations

Kuhn and Tucker (1950) excluded these kinds of troublesome situations by *restricting* their necessary conditions to only those problems whose constraints did not exhibit such linear dependence of gradients at the optimal point.[17]

A good deal of effort has been undertaken to modify, simplify and otherwise express in alternative ways the Kuhn–Tucker constraint qualification. [A very thorough summary of some of this work is given in Takayama (1985, especially pp. 93–104).] Some of the most powerful and useful results were brought together by Arrow, Hurwicz, and Uzawa (as noted in Takayama (1985, pp. 97–98)]. Among these are the following conditions—if any one of them holds, the constraint qualification conditions are assured, and a constrained optimum will satisfy the Kuhn–Tucker necessary conditions;[18] in all cases, the problem under consideration is the general inequality-constrained maximization problem in standard form

Maximize $f(\mathbf{X})$

subject to $h^i(\mathbf{X}) \leq 0$

(with nonnegativities included in the constraints, so $i = 1, \ldots, m, m+1, \ldots, m+n$):

---

[17]Their discussion was framed in terms of "test vectors" and "qualifying arcs," which do not provide very accessible rules for judging the constraints in any particular problem. Chiang (1984, pp. 734–738) contains some illustrations of these concepts.

[18]Kuhn and Tucker stated their general nonlinear programming problem as: Maximize $g(x)$, subject to $f^j(x) \geq 0$ and $x \geq 0$. Many other writers interchange f and g, using: Maximize $f(x)$, subject to $g^j(x) \geq 0$ (including nonnegativity of the x's in these constraints). Since we generally express maximization problem constraints as $h^i(\mathbf{X}) \leq 0$, with nonnegativities separately, as $h^{m+j}(\mathbf{X}) = -x_j \leq 0$, requirements that the functions $g^j(x)$ in $g^j(x) \geq 0$ be *convex* translate to the requirements that $h^i(\mathbf{X})$ in $h^i(\mathbf{X}) \leq 0$ be *concave*.

(a) All $h^i(\mathbf{X})$ are linear functions.

(b) The feasible region is convex, with an interior point, and $\nabla h^i(\mathbf{X}^*) \neq \mathbf{0}$ for all i associated with binding constraints at the maximum point $\mathbf{X}^*$.

(c) Let $\mathbf{M}(\mathbf{X}^*)$ be the matrix whose columns are the gradients $\nabla h^i(\mathbf{X}^*)$ for those i that are associated with binding constraints at $\mathbf{X}^*$. Suppose that there are k such binding constraints at $\mathbf{X}^*$; then $\rho[\mathbf{M}(\mathbf{X}^*)] = k$.

(d) All $h^i(\mathbf{X})$ are concave functions.

We consider each of these in turn.

(a) We have seen several feasible regions defined by linear constraints in the linear programs of Part IV. A three-constraint (plus nonnegativity) illustration is given in Figure 10.16, for the following constraints:

$$h^1(\mathbf{X}) = x_2 - 5 \leq 0$$

$$h^2(\mathbf{X}) = x_1 + x_2 - 8 \leq 0$$

$$h^3(\mathbf{X}) = x_1 - 6 \leq 0$$

$$h^4(\mathbf{X}) = -x_1 \leq 0$$

$$h^5(\mathbf{X}) = -x_2 \leq 0$$

Each constraint has a constant (nonnull) gradient everywhere, as indicated in Figure 10.16, and irrespective of the shape of the objective function or of the location of the unconstrained maximum point, the linear dependence of gradients problem simply cannot arise.

(b) This condition rules out the kinds of problem cases that arose Examples 4 and 5, even though the feasible regions in those examples were convex. The condition that the feasible region have an interior point, which means that the feasible region is itself not simply a single point, is often expressed by saying that there exists an $\overline{\mathbf{X}}$ such that $h^i(\overline{\mathbf{X}}) <$

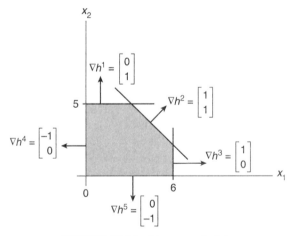

FIGURE 10.16 Linear constraints.

0, for all i.[19] This condition, itself, is needed to exclude a similar kind of null gradients problem in applying conditions (10.17a).

For example, suppose that the problem is

$$\text{Maximize}\quad x_1 + x_2$$

$$\text{subject to}\quad h^1(\mathbf{X}) = x_1^2 \leq 0$$

$$h^2(\mathbf{X}) = x_2^2 \leq 0$$

$$\text{and}\quad x_1 \geq 0, x_2 \geq 0$$

Both $h^1(\mathbf{X})$ and $h^2(\mathbf{X})$ are convex, and the feasible region is a convex set, but in conjunction with the nonnegativity requirements on x_1 and x_2, it reduces to the single point $\mathbf{X} = \begin{bmatrix} 0 \\ 0 \end{bmatrix}$; thus $\mathbf{X}^* = \begin{bmatrix} 0 \\ 0 \end{bmatrix}$. For this problem, the gradients are

$$\nabla f(\mathbf{X}) = \begin{bmatrix} 1 \\ 1 \end{bmatrix},\ \nabla h^1(\mathbf{X}) = \begin{bmatrix} 2x_1 \\ 0 \end{bmatrix},\ \nabla h^2(\mathbf{X}) = \begin{bmatrix} 0 \\ 2x_2 \end{bmatrix}$$

$$\nabla h^3(\mathbf{X}) = \begin{bmatrix} -1 \\ 0 \end{bmatrix},\ \nabla h^4(\mathbf{X}) = \begin{bmatrix} 0 \\ -1 \end{bmatrix}$$

In attempting to invoke (10.17a) at $\mathbf{X} = \begin{bmatrix} 0 \\ 0 \end{bmatrix}$, we have

$$\begin{bmatrix} 1 \\ 1 \end{bmatrix} = \lambda_1 \begin{bmatrix} 0 \\ 0 \end{bmatrix} + \lambda_2 \begin{bmatrix} 0 \\ 0 \end{bmatrix} + \lambda_3 \begin{bmatrix} -1 \\ 0 \end{bmatrix} + \lambda_4 \begin{bmatrix} 0 \\ -1 \end{bmatrix}$$

and it is impossible to find nonnegative λ's to satisfy this equation, because $\nabla h^1(\mathbf{X}^*)$ and $\nabla h^2(\mathbf{X}^*)$ are both null vectors.

(c) This condition rules out the kind of problem case represented in Example 6, where $\rho[\mathbf{M}(\mathbf{X}^*)] = 1$, not 2 ($= m$). In fact, the constraints in Examples 4 and 5 are disallowed by this condition also. This is because, for example, a two-element null vector is linearly dependent on *any* two-element (nonnull) vector, from the basic definition of linear dependence (Chapter 1).

(d) For a maximization problem, we have seen (Chapter 4 and earlier in this chapter) that if each $h^i(\mathbf{X})$ constraint is *convex* (or *quasiconvex*), the feasible region is assured to be a convex set. If, in addition, $f(\mathbf{X})$ is concave, then the Kuhn–Tucker necessary conditions are also sufficient for a constrained maximum. Of interest here is the reverse situation in which the constraint functions are *concave* in a maximization problem—so the feasible region will not, in general, be convex.

Here is an example with three *concave* constraints (two of them are linear) in addition to nonnegativity of x_1 and x_2. We use it only to illustrate (not prove) the effectiveness of this condition in ensuring that the problems of linear dependence of gradients, which con-

---

[19]In various forms this is known as Slater's condition or Karlin's condition, after their authors, who expressed them in slightly alternative ways. See Takayama (1985, Chapter 1, Section B) for much more detail.

found the Kuhn–Tucker necessary conditions, will not cause problems. The constraint set, in $h^i(\mathbf{X}) \leq 0$ form, is given by

$$h^1(\mathbf{X}) = x_2 - 5$$

$$h^2(\mathbf{X}) = -x_1^2 - x_2^2 + 10x_1 + 10x_2 - 46$$

[alternatively, $h^2(\mathbf{X}) = -(x_1 - 5)^2 - (x_2 - 5)^2 + 4$]

$$h^3(\mathbf{X}) = 1.913x_1 - 3.513x_2$$

$$h^4(\mathbf{X}) = -x_1 \quad \text{and} \quad h^5(\mathbf{X}) = -x_2$$

This feasible region is shown in Figure 10.17; it is decidedly nonconvex. This is a consequence of $h^2(\mathbf{X})$, which is strictly concave (the leading principal minors of the Hessian matrix are alternatively negative and positive). In the geometry of $x_1, x_2, h^2(\mathbf{X})$ space, $h^2(\mathbf{X})$ looks like an overturned bowl; the circle centered at $\begin{bmatrix} 5 \\ 5 \end{bmatrix}$ with a radius of 2 identifies the locus of points for which $h^2(\mathbf{X}) = 0$. Part of this circle forms the $h^2(\mathbf{X})$ boundary in Figure 10.17, and values of $\mathbf{X}$ that are on the circle or outside it represent points for which $h^2(\mathbf{X}) \leq 0$.

Suppose that the objective for this problem is to maximize $f(\mathbf{X}) = x_1 + x_2$. Geometrically, it is easy to see that the optimal solution will be at point F in Figure 10.17, where $\mathbf{X}^* = \begin{bmatrix} 9.182 \\ 5 \end{bmatrix}$, and ∇f there is indeed inside the cone defined by ∇h^1 and ∇h^3. Gradient directions are shown at various points on the boundary of the feasible region. Constraints 2 and 3 are tangent at point D, with collinear gradients, but that is irrelevant to the Kuhn–Tucker conditions, since D is not an optimal point when $f(\mathbf{X})$ is linear.

Suppose, instead, that the objective function is strictly concave (e.g., an overturned bowl). There will still be no problem with the collinear constraint gradients at D. The only

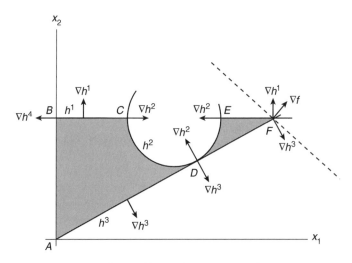

FIGURE 10.17   Concave constraints.

way that D could be the constrained maximum point is if an objective function contour were exactly tangent to $h^3(\mathbf{X}) = 0$ at D [which means also tangent to the $h^2(\mathbf{X}) = 0$ circle at that point].[20] But tangency of the $f(\mathbf{X})$ contour at D means collinearity of $\nabla f(\mathbf{X})$, $\nabla h^2(\mathbf{X})$, and $\nabla h^3(\mathbf{X})$ there, and so the requirements in (10.17a) will not be confounded by the linear dependence of the constraint gradients at D. If the constrained maximum is at D and the hilltop is southeast of D, then $\nabla f(\mathbf{X}^*) = \lambda_3^* \nabla h^3(\mathbf{X}^*)$, for $\lambda_3^* > 0$; if the hilltop is northwest of D, then $\nabla f(\mathbf{X}^*) = \lambda_2^* \nabla h^2(\mathbf{X}^*)$, for $\lambda_2^* > 0$.

As a practical matter, only conditions (a) and (d) immediately above are *global* characterizations of the constraints in a problem. They are easily checked—linearity by inspection, concavity via, say, leading principal minors tests on the Hessian matrices for each constraint. Conditions (b) and (c) immediately above are *local* in the sense that they specify characteristics on gradients of constraints that are active *at the optimal point*.

10.4 SADDLE POINT PROBLEMS

In this section we examine the characteristics of saddle points of a function. (A saddle point is illustrated in Fig. 3.33 in Chapter 3.) We do this because there are potentially interesting connections between saddle points and optimal solutions to mathematical programming problems—both linear programs and also certain nonlinear programs. Several results are presented in the form of standard theorems, without proofs. These relationships may provide clues for computational procedures for nonlinear programming problems; they also suggest a possible interpretation of Kuhn–Tucker results and of the Lagrange multipliers that emerged in solutions to inequality-constrained problems, in Section 4.5 and earlier in this chapter.

10.4.1 Saddle Points Defined

Consider a function, ϕ, of $n + m$ variables, in which we distinguish between the first n variables, $\mathbf{X} = [x_1, \ldots, x_n]'$, and the remaining variables, $\mathbf{Y} = [y_1, \ldots, y_m]'$. So

$$\phi(\mathbf{X},\mathbf{Y}) \equiv \phi(x_1, \ldots, x_n; y_1, \ldots, y_m)$$

This function is said to have a *saddle point* at $(\mathbf{X}^0,\mathbf{Y}^0)$ if and only if

$$\phi(\mathbf{X}^0,\mathbf{Y}) \geq \phi(\mathbf{X}^0,\mathbf{Y}^0) \geq \phi(\mathbf{X},\mathbf{Y}^0) \tag{10.28}$$

The right-hand inequality says that $\mathbf{X}^0$ maximizes $\phi(\mathbf{X},\mathbf{Y}^0)$; for a given $\mathbf{Y} = \mathbf{Y}^0$, $\mathbf{X}^0$ is a maximizer. The left-hand inequality says that $\mathbf{Y}^0$ minimizes $\phi(\mathbf{X}^0,\mathbf{Y})$; for a given $\mathbf{X} = \mathbf{X}^0$, $\mathbf{Y}^0$ is a minimizer. This is sometimes stated as follows: $\mathbf{X}^0$ is chosen so as to $\max_{\mathbf{X}} [\min_{\mathbf{Y}} \phi(\mathbf{X},\mathbf{Y})]$ and $\mathbf{Y}^0$ is chosen so as to $\min_{\mathbf{Y}} [\max_{\mathbf{X}} \phi(\mathbf{X},\mathbf{Y})]$. From this point of view, the $\mathbf{X}$'s are said to be *maximinimizing* points, and the $\mathbf{Y}$'s are *minimaximizing* points.

Consider the problem of finding a saddle point (if one exists) in the nonnegative orthant only—that is, of finding an $\mathbf{X}^0 \geq \mathbf{0}$ and $\mathbf{Y}^0 \geq \mathbf{0}$ such that (10.28) holds over all $\mathbf{X},\mathbf{Y}$

---

[20] Try moving an imaginary hilltop around to the southeast of D or also above D, inside the circle. Only if the $f(\mathbf{X})$ contour is tangent will D be the optimal point.

$\geq \mathbf{0}$. This can be called the "nonnegative saddle point problem." Using the usual gradient notation, let

$$\nabla \phi_{\mathbf{X}^0} = \left[\frac{\partial \phi}{\partial x_1^0}, \dots, \frac{\partial \phi}{\partial x_n^0} \right]', \qquad \nabla \phi_{\mathbf{Y}^0} = \left[\frac{\partial \phi}{\partial y_1^0}, \dots, \frac{\partial \phi}{\partial y_m^0} \right]'$$

with all partial derivatives evaluated at $(\mathbf{X}^0, \mathbf{Y}^0)$. If $\phi(\mathbf{X}, \mathbf{Y})$ is differentiable, so that the elements of these gradients can be found, then a set of necessary conditions for a nonnegative saddle point are given by the following.

Theorem SP1   For $(\mathbf{X}^0, \mathbf{Y}^0)$ to be a saddle point of $\phi(\mathbf{X}, \mathbf{Y})$, it is necessary that

   (a) $\nabla \phi_{\mathbf{X}^0} \leq \mathbf{0}$, $[\nabla \phi_{\mathbf{X}^0}] \circ \mathbf{X}^0 = 0$
   (b) $\nabla \phi_{\mathbf{Y}^0} \geq \mathbf{0}$, $[\nabla \phi_{\mathbf{Y}^0}] \circ \mathbf{Y}^0 = 0$

for $\mathbf{X}^0 \geq \mathbf{0}$, $\mathbf{Y}^0 \geq \mathbf{0}$.

   The maximizing properties of $\mathbf{X}^0$ when $\mathbf{Y} = \mathbf{Y}^0$ [from (10.28)] are reflected in condition (a), which says that $\partial \phi / \partial x_j^0 \leq 0$ for $x_1, \dots, x_n$. If for some x_j, on the contrary, $\partial \phi / \partial x_j^0 > 0$, it would be possible to increase x_j^0 (without violating the nonnegativity requirement on x_j) and also the value of ϕ. And if for some x_k, $\partial \phi / \partial x_k^0 < 0$, one would prefer an even lower value of x_k (decreasing x_k will increase ϕ). The only reason not to make x_k smaller is if it has reached its lower limit of zero. This means that if $\partial \phi / \partial x_k^0 < 0$, then $x_k^0 = 0$ and if $\partial \phi / \partial x_k^0 = 0$, then $x_k^0 \geq 0$. These are exactly the possibilities allowed by part (a) of Theorem SP1. The interpretation of (b) is similar, except that the y's are minimizers of $\phi(\mathbf{X}, \mathbf{Y})$ for $\mathbf{X} = \mathbf{X}^0$, and hence $\partial \phi / \partial y_i^0$ can never be negative.
   Sufficient conditions for a nonnegative saddle point are given in the following theorem. As we expect from earlier considerations of sufficient conditions, the issue is one of the shape of $\phi(\mathbf{X}, \mathbf{Y})$ near $(\mathbf{X}^0, \mathbf{Y}^0)$.

Theorem SP2   If

   (a) $\phi(\mathbf{X}, \mathbf{Y}^0)$ is a *concave* function of $\mathbf{X}$
   (b) $\phi(\mathbf{X}^0, \mathbf{Y})$ is a *convex* function of $\mathbf{Y}$

then the conditions of Theorem SP1 are both necessary *and* sufficient for $(\mathbf{X}^0, \mathbf{Y}^0)$ to be a saddle point of $\phi(\mathbf{X}, \mathbf{Y})$.

   These conditions have the usual kinds of concavity/convexity interpretations. In condition (a), the geometric requirement is that $\phi(\mathbf{X}, \mathbf{Y}^0)$ lie below its tangent plane for $\mathbf{X}$ in the neighborhood of $\mathbf{X}^0$ and in (b) the requirement is that $\phi(\mathbf{X}^0, \mathbf{Y})$ lie above its tangent plane for $\mathbf{Y}$ in the neighborhood of $\mathbf{Y}^0$.
   The function $f(\mathbf{X}) = 1 + x_1^2 - x_2^2$ was shown in Figure 3.33 of Chapter 3. There is a saddle point at $\begin{bmatrix} 0 \\ 0 \end{bmatrix}$, and you can easily verify that the necessary conditions of Theorem SP1 are met there. Furthermore, $f(x_1, 0) = 1 + x_1^2$ and $f'(x_1, 0) = 2 > 0$, which means that $f(x_1, 0)$ is (everywhere) a convex function of x_1; also $f(0, x_2) = 1 - x_2^2$ and $f'(0, x_2) = -2 < 0$,

identifying $f(0, x_2)$ as an everywhere-concave function of x_2. In the notation of Theorem SP2, x_1 plays the role of y and x_2 plays the role of x, and thus $\begin{bmatrix} 0 \\ 0 \end{bmatrix}$ satisfies the necessary and sufficient conditions for a saddle point to $f(\mathbf{X}) = 1 + x_1^2 - x_2^2$.

10.4.2 Saddle Points and Linear Programming Problems

We saw in Chapter 7 that a pair of dual linear programming problems [as in (7.30)] can be represented as

	Primal		**Dual**	
Maximize	$\Pi = \mathbf{P}'\mathbf{X}$	Minimize	$\Delta = \mathbf{B}'\mathbf{Y}$	
subject to	$\mathbf{AX} \leq \mathbf{B}$	subject to	$\mathbf{A}'\mathbf{Y} \geq \mathbf{P}$	(10.29)
and	$\mathbf{X} \geq \mathbf{0}$	and	$\mathbf{Y} \geq \mathbf{0}$	

Theorem SP3 indicates the connection between an optimal solution to the primal problem in (10.29) and a saddle point to an associated Lagrangian function.

Theorem SP3 $\mathbf{X}^0$ is an optimal solution to the maximizing (primal) linear programming problem in (10.29) if and only if a $\mathbf{Y}^0$ ($\geq \mathbf{0}$) exists such that $(\mathbf{X}^0, \mathbf{Y}^0)$ is a saddle point of the Lagrangian function

$$L(\mathbf{X},\mathbf{Y}) = \mathbf{P}'\mathbf{X} - \mathbf{Y}'(\mathbf{AX} - \mathbf{B})$$

for all $\mathbf{X} \geq \mathbf{0}$, $\mathbf{Y} \geq \mathbf{0}$.

This says that an optimal solution to the linear programming maximization problem is identical to the $\mathbf{X}^0$ vector in the *nonnegative saddle point* for the function $L(\mathbf{X},\mathbf{Y})$. The requirements for this saddle point can be found by using Theorem SP1 directly, for $L(\mathbf{X},\mathbf{Y}) = \phi(\mathbf{X},\mathbf{Y})$. The conditions in (b) of that theorem (you might want to check all of this) generate precisely (i) the requirements that $\mathbf{AX}^0 \leq \mathbf{B}$ (which means that the optimal $\mathbf{X}$'s must be feasible for the primal) and (ii) the "complementary slackness" conditions from Theorem PD3 in Chapter 7 that characterize *optimal* values of slack variables in the primal, s_i^0, and dual variables, y_i^0, by the condition that $s_i^0 y_i^0 = 0$ (for all i).

Similarly, the conditions in part (a) of Theorem SP1, applied to $L(\mathbf{X},\mathbf{Y})$, will generate (i) the requirements that the $\mathbf{Y}^0$'s satisfy the dual constraints and (ii) the other "complementary slackness" optimality conditions that $t_j^0 x_j^0 = 0$, for all j, where t_j^0 are the optimal values of the nonnegative slack variables in the dual constraints. Finally, since $L(\mathbf{X},\mathbf{Y}^0)$ is *linear* in the $\mathbf{X}$'s (and hence concave, although not strictly so) and since $L(\mathbf{X}^0,\mathbf{Y})$ is *linear* in the $\mathbf{Y}$'s (and so convex, but also not strictly so), these conditions are both necessary and sufficient for a nonnegative saddle point.

The impact of this result is that a maximization problem subject to linear inequality and nonnegativity constraints on the variables has been converted to a much less constrained (nonnegativity only) saddle point problem. To the extent that there are numerical methods for finding saddle points and since the constraints in a maximization or minimization problem virtually always complicate the calculations, one *could* be interested in this correspondence between linear programming optimal solutions and saddle points. This is not likely, however, since the simplex method and interior point methods have proved to be powerful approaches for linear programs. The important point is that there is a similar kind of connection between saddle points and nonlinear programming problems, where the computational implications may be of more use.

10.4.3 Saddle Points and Nonlinear Programming Problems

The general nonlinear programming problem in (10.2) was

$$\text{Maximize } f(\mathbf{X})$$
$$\text{subject to } h^i(\mathbf{X}) \le 0 \qquad (i = 1, \ldots, m) \tag{10.30}$$
$$\text{and } \mathbf{X} \ge \mathbf{0}$$

Create an associated Lagrangian function

$$L(\mathbf{X},\mathbf{Y}) = f(\mathbf{X}) - \sum_{i=1}^{m} y_i[h^i(\mathbf{X})] \tag{10.31}$$

The correspondence between nonnegative saddle points and nonlinear programming problems is given in the following two theorems (for which differentiability of the objective function and each constraint is required).

Theorem SP4   For $\mathbf{X}^0$ to be a solution to the nonlinear maximization problem in (10.30), it is necessary that $\mathbf{X}^0$ and some $\mathbf{Y}^0$ satisfy the conditions of Theorem SP1, for $\phi(\mathbf{X},\mathbf{Y}) \equiv L(\mathbf{X},\mathbf{Y})$ in (10.31).

Applying the conditions of Theorem SP1 for this case, we find (in all cases, the requirements are to hold at $\mathbf{X}^0$ and $\mathbf{Y}^0$)

(a) $\nabla L_{\mathbf{X}} \le \mathbf{0}$, which means $\partial L/\partial x_j \le 0$, or

$$f_j - \sum_{i=1}^{m} y_i h_j^i \le 0 \qquad (j = 1, \ldots, n)$$

and, from $[\nabla L_{\mathbf{X}}] \circ \mathbf{X} = 0$

$$\left(f_j - \sum_{i=1}^{m} y_i h_j^i \right) x_j = 0 \qquad (j = 1, \ldots, n)$$

(b) $\nabla L_{\mathbf{Y}} \ge \mathbf{0}$, which means $\partial L/\partial y_i \ge 0$, or

$$h^i(\mathbf{X}) \le 0 \qquad (i = 1, \ldots, m)$$

and, from $[\nabla L_{\mathbf{Y}}] \circ \mathbf{Y} = 0$

$$[h^i(\mathbf{X})] y_i = 0 \quad (i = 1, \ldots, m)$$

for all $\mathbf{X}, \mathbf{Y} \ge \mathbf{0}$ (the *nonnegative* saddle point problem).

Written out in this way, we see that the saddle point requirements (in Theorem SP4) are identical to the Kuhn–Tucker conditions from Section 10.2, with the λ's from there represented here as y's. In the earlier development, via Lagrange multipliers, the relationship between the optimal solution to the maximization problem and a nonnegative saddle point to the associated Lagrangian function was not explicit.

Moreover, in view of Theorem SP3 for linear programs, one might suspect that the y's of Theorem SP4—the Lagrange multipliers of Section 10.2—may be precisely the variables in a program that is *dual* to the maximization problem in (10.30). We will explore this further in the next chapter, where duality for nonlinear programs is treated.

Furthermore, if we interpret the nonlinear problem in (10.30) as one of profit (or revenue) maximization subject to resource constraints (as we did for the linear programming maximization problem in Chapter 7), then the optimality conditions of Theorem SP4 have logical interpretations. In this case, $\partial f/\partial x_j$ represents the marginal profit (or marginal revenue) of product j; $\partial h^i/\partial x_j$ gives the *input* of resource i needed to produce a marginal amount of product j, and y_i may be thought of (as before) as a shadow price or marginal valuation on resource i. Then conditions (a) require that, at an optimal solution, the net marginal profit (or the net marginal revenue) for each j be zero or negative, and if negative, then no j is produced ($x_j^0 = 0$). Conditions (b) require that resource constraints be met and if the ith resource is not fully used [if $h^i(\mathbf{X}^0) < 0$, which means $g^i(\mathbf{X}^0) < b_i$], its corresponding valuation is zero ($y_i^0 = 0$).

These are very similar to interpretations given to some of the primal-dual optimal relationships in linear programs—for example, Theorem PD3 of Chapter 7. This again suggests that the y_i here, from the nonnegative saddle point problem for $\phi(\mathbf{X},\mathbf{Y})$, might be thought of as the variables in a programming problem dual to (10.30).

Parallel to the sufficient conditions for a nonnegative saddle point in Theorem SP2, we have

Theorem SP5 For $\mathbf{X}^0$ to be a solution to the nonlinear maximization problem in (10.30), it is *sufficient* that $\mathbf{X}^0$ and some $\mathbf{Y}^0$ satisfy condition (a) of Theorem SP2, for $\phi(\mathbf{X},\mathbf{Y}) \equiv L(\mathbf{X},\mathbf{Y})$ in (10.31).

Consider (a) of Theorem SP2. The issue here is *concavity* of

$$L(\mathbf{X},\mathbf{Y}^0) = f(\mathbf{X}) - \sum_{i=1}^m y_i^0[h^i(\mathbf{X})] = f(\mathbf{X}) + \sum_{i=1}^m y_i^0[-h^i(\mathbf{X}^0)]$$

If each $-h^i(\mathbf{X})$ is concave and if $f(\mathbf{X})$ is concave, then since all $y_i \geq 0$, $L(\mathbf{X},\mathbf{Y}^0)$ will be a nonnegative linear combination of concave functions and hence will itself be concave. If each $-h^i(\mathbf{X})$ is *concave*, then each $h^i(\mathbf{X})$ will be *convex*, so $L(\mathbf{X},\mathbf{Y}^0)$ will be concave when $f(\mathbf{X})$ is a concave function and all $h^i(\mathbf{X})$ are convex functions of the x's.

Part (b) of Theorem SP2 requires that $L(\mathbf{X}^0,\mathbf{Y})$ be convex. Here

$$L(\mathbf{X}^0,\mathbf{Y}) = f(\mathbf{X}^0) - \sum_{i=1}^m y_i[h^i(\mathbf{X}^0)]$$

and this is *linear* in the y_i and hence automatically *convex* (as well as concave, although neither strictly so). Thus condition (b) of Theorem SP2 is certain to be satisfied by $L(\mathbf{X}^0,\mathbf{Y})$ and does not need to be made explicit in Theorem SP5.

To summarize, when the objective function, $f(\mathbf{X})$, is concave and when each constraint, $h^i(\mathbf{X})$, is convex, the conditions of Theorem SP4 are both necessary and sufficient for a solution to the general nonlinear programming maximization problem in (10.30). The object here was to give a new interpretation to primal and dual variables in a pair of linear programming problems, as maximinimizing and minimaximizing variables in a saddle point problem. This, in turn, suggests possible parallels for duals in nonlinear programming problems. It also illustrates how a constrained optimization problem, such as (10.30), involving (in general) n variables, m nonlinear constraints, and the nonnegativity of all the variables, can be converted into an equivalent saddle point problem in $(n + m)$ variables constrained only by nonnegativity requirements. Since some computational methods can locate saddle points, this provides a solution procedure for those nonlinear programming problems that satisfy the concavity–convexity requirements.

10.5   SUMMARY

Extension of the Kuhn–Tucker reasoning to nonlinear problems that include nonnegativity requirements on the variables was seen to involve no new principles but considerable additional complexity in the statement of necessary conditions. In the form of the results for a maximization problem, in (10.6′) or (10.8), we have a number of nonlinear equations to solve and inequalities to satisfy—in addition to nonnegativity requirements on the variables. The specific multiplicative nature of many of the equations, in the general form of $a_i b_i = 0$, suggests an exploration scheme in which the implications of first $a_i = 0$ and then $b_i = 0$ are investigated for each i. This quickly leads to an enormous number of possibilities, as suggested in Table 10.1.

The geometric implications of nonnegativity constraints were also examined, in order to understand why the Kuhn–Tucker conditions change as they do when $\mathbf{X} \geq \mathbf{0}$ is added to a problem. In addition, the role of concavity/convexity conditions on both $f(\mathbf{X})$ and each $h^i(\mathbf{X})$ constraint was explored in order to establish conditions under which the Kuhn–Tucker necessary conditions were also sufficient, and under which the extreme point found via the necessary conditions was a global, not just local, optimum. Quasiconcavity and quasiconvexity were presented as relaxations of the requirements of (ordinary) concavity and convexity, and some of the situations under which these less stringent characteristics were adequate to establish the sufficiency of the Kuhn–Tucker conditions were explored.

Next, the logic (both algebraic and geometric) of Kuhn and Tucker's constraint qualification was examined. In their important work, Kuhn and Tucker simply assumed that the constraints in their problems satisfied that qualification. Finally, saddle point problems were explored and connected to both linear and nonlinear programs. We saw that this provides insights into the Lagrange multipliers of Section 4.5 and suggests properties for nonlinear programming duality theory. It also indicates one direction for a computational approach to nonlinear problems. We explore several computational techniques for nonlinear programming problems in the next chapter.

REFERENCES

Arrow, Kenneth, and Alain C. Enthoven, "Quasi-Concave Programming," *Econometrica* **24**, 779–800 (1961).

Chiang, Alpha, *Fundamental Methods of Mathematical Economics*, 3rd ed., McGraw-Hill, New York, 1984.

Kuhn, Harold W., and Albert W. Tucker, "Nonlinear Programming," in Jerzy Neyman, ed., *Proceedings of the Second Berkeley Symposium on Mathematical Statistics and Probability*, University of California Press, Berkeley, 1950, pp. 481–492. (Reprinted in Peter Newman, ed., *Readings in Mathematical Economics, Vol. I, Value Theory*, Johns Hopkins Press, Baltimore, 1968.)

Mangasarian, Olvi L., *Nonlinear Programming*, McGraw-Hill, New York, 1969.

Takayama, Akira, *Mathematical Economics,* 2nd ed., Cambridge University Press, New York, 1985.

Vajda, S., *Theory of Linear and Nonlinear Programming*, Longman, London, 1974.

Zoutendijk, Guus, *Mathematical Programming Methods*, North-Holland, Amsterdam, 1976.

PROBLEMS

10.1 Find the maximum of $f(\mathbf{X}) = -x_1^2 - x_2^2 - x_3^2 + 4x_1 + 6x_2$ subject to $x_1 + x_1 \leq 2$ and $2x_1 + 3x_2 \leq 12$ and $\mathbf{X} \geq \mathbf{0}$, using the Kuhn–Tucker results in (10.8). Establish the sufficiency of conditions (10.8) by testing $f(\mathbf{X})$ for concavity or convexity.

10.2 Minimize $6x_1^2 + 5x_2^2$ subject to $x_1 + 5x_2 \geq 3$ and $\mathbf{X} \geq \mathbf{0}$. Use the Kuhn–Tucker results, as in Problem 10.1.

10.3 Minimize the same function as in Problem 10.2, subject to $\mathbf{X} \geq \mathbf{0}$ and $x_1 + 5x_2 \leq 3$, again using the Kuhn–Tucker approach, as in Problem 10.1.

10.4 For each of the following problems, formulate the Lagrangian function as in (10.31) and find its saddle point. (Note that if you have solved Problems 10.1 through 10.3, the work is essentially completed.) Maximize

(a) $f(\mathbf{X}) = -x_1^2 - x_2^2 - x_3^2 + 4x_1 + 6x_2$ subject to $x_1 + x_2 \leq 2$ and $2x_1 + 3x_2 \leq 12$ and $\mathbf{X} \geq \mathbf{0}$

(b) $f(\mathbf{X}) = -6x_1^2 - 5x_2^2$ subject to $-x_1 - 5x_1 \leq -3$ and $\mathbf{X} \geq \mathbf{0}$

(c) $f(\mathbf{X}) = -6x_1^2 - 5x_2^2$ subject to $x_1 + 5x_2 \leq 3$ and $\mathbf{X} \geq \mathbf{0}$

10.5 Consider Problems 7.1, 7.3, and 7.4 in Chapter 7. Find both primal and dual solutions, and show that the gradient of the objective function at the optimal point can be expressed as a nonnegative linear combination of the gradients to the constraints that are binding at that point, where the optimal values of the dual variables are the weights in the linear combination.

10.6 For each numbered numerical example in this Chapter (Examples 1–6), form the matrix $\mathbf{M}(\mathbf{X}^*)$ whose columns are the gradients, $\nabla h^i(\mathbf{X}^*)$, for those i that are associated with binding constraints at $\mathbf{X}^*$. (Let k = the number of constraints, including nonnegativities, that are binding at $\mathbf{X}^*$.) Verify that $\rho[\mathbf{M}(\mathbf{X}^*)] = k$ for the first three examples.

10.7 The daily emission of airborne pollutants from two petrochemical facilities is given (in thousands of cubic feet) by

$$E = 2x_1^2 - 80x_1 + 8x_2^2 + 1000$$

where x_1 and x_2 represent hours of operation per day of the two facilities.

(a) What amount of daily operation (in hours) of each of the two facilities would be optimal in the sense of minimizing daily pollution emission?

(b) How would you answer part (a), above, if, in addition, there were an upper limit of 12 h per day of operation at facility 1?

(c) You later discover (as is usually the case) that the E function was incorrectly estimated; actually it should be

$$E = 2x_1^2 - 80x_1 + 8x_2^2 - 4x_1x_2 + 1100$$

How would you now answer the question in part (a)?

Use the full set of Kuhn–Tucker conditions in answering at least one part of this question.

11

NONLINEAR PROGRAMMING: DUALITY AND COMPUTATIONAL METHODS

It is clear from Chapter 10 that nonlinearities in a programming problem largely eliminate any *generally* applicable solution procedures. For that reason, there are a number of special approaches, many of which depend on particular characteristics of the functions in the problem—for example, concavity and convexity. Usually, the objective function must be concave (sometimes only quasiconcave) for a maximization problem, or convex (or quasiconvex) for a minimization problem, and the constraints must be such that their intersection is a convex set.

In this chapter we investigate some of the most widely used computational techniques for nonlinear programs. Variants of these kinds of approaches are incorporated in many computer software packages for solving nonlinear programming problems. Some of these are akin to gradient methods; the added feature is that they must alter direction as constraints become binding, as was also true of interior point methods for linear programs (Chapter 9). Other methods rely on various kinds of linear approximations to a nonlinear problem for which variations on the simplex method can be applied. Nonlinear but unconstrained approximating models have also been successful.

Development of increasingly fast and accurate computational approaches for nonlinear programming problems, driven by the increasing speed and capacity of modern computers, is at the forefront of the mathematical programming field. (In Chapter 9 we saw that this was true for interior point approaches to linear programming problems as well.) Even though the computational details will change and the solution algorithms will be refined as computers continue to develop, the fundamental logic of and distinctions between the various approaches will remain valid.

We also explore, briefly, the notion of duality for the nonlinear programming case. The parallels to duality in linear programming are of some interest in their own right, and some of the duality results turn out to play a role in computational techniques.

Finally, we present the details of quadratic programming problems. In a mathematical sense, these are the "most linear" of the nonlinear programming models, and computational procedures have been developed to exploit this characteristic.

11.1   GRADIENTLIKE COMPUTATIONAL METHODS FOR NONLINEAR PROGRAMS

In Section 10.2 we saw that application of the first-order conditions for maxima and minima to the general nonlinear programming problem generated a complex set of Kuhn–Tucker conditions [(10.8) or (10.9)] containing inequalities and nonlinear equations. Using these conditions directly for optimal solutions can often be an extremely difficult and involved procedure. We have also seen, in Chapter 6, that the properties of the gradient to a function suggested a rather broad class of techniques for *unconstrained* maximization or minimization problems. The underlying logic of these gradient methods was quite simple: move toward the optimum along the "best" route possible. Gradient methods for unconstrained problems differ from one another principally in (1) whether they modify the gradient in defining the "best" direction and (2) the length of step that they take along the currently chosen direction before reevaluating the gradient and establishing a new direction.

For *constrained* maximization or minimization problems, the logic of a gradientlike approach is easy enough to express: move in the "best" direction until you are forced to alter it because you have hit a constraint. It is the mathematical implementation of an alteration to the gradient that turns out to be rather complex. Gradients also underpin many of the interior point methods for linear programs in Chapter 9—some using a *projected* gradient (to maintain feasibility) and some that do not step all the way to a boundary.

As usual, the general outline of a gradient-based procedure is as follows:

1. Begin with a feasible $\mathbf{X}^0$. This may be—but seldom is—a difficult initial choice, depending on the complexity of the constraints.

2. Move along the direction of the gradient, $\nabla f(\mathbf{X}^0)$, for a maximization problem $[-\nabla f(\mathbf{X}^0)$ for a minimization problem] through selection of a step size, α^0, to a new *and feasible* point, $\mathbf{X}^1$, for which $f(\mathbf{X}^1) > f(\mathbf{X}^0)$ $[f(\mathbf{X}^1) < f(\mathbf{X}^0)$ for a minimization problem].

3. Decide when to stop.

Regarding step 2 (above), the feasibility requirement could mean that it might not be possible to move to the *unconstrained maximum* point along the gradient direction because that point may be outside the feasible region. It could also mean that it is necessary to completely modify the gradient direction because movement along that direction leads from $\mathbf{X}^0$ *immediately* out of the feasible region. [This would be the case if $\mathbf{X}^0$ were on the boundary and $\nabla f(\mathbf{X}^0)$ pointed outward from that boundary.]

In this section we explore two methods, steeped in the gradient tradition, that deal with these issues. These techniques become interesting and particularly useful when the nonlinear optimization problem contains many variables and many constraints. The constraints are linear in the illustrations that follow; nonlinear constraints add mathematical complexity (e.g., linear approximations via Taylor's series are sometimes employed), but the underlying principles are adequately illuminated in problems with linear constraints.

When there is only one constraint that is binding at the current point, $\mathbf{X}^k$, the problem is relatively simple. The more usual case, which is at the same time more complex, occurs in a problem involving many variables and many constraints. Then it is likely that $\mathbf{X}^k$ is on a boundary of the feasible region that is defined by the intersection of several binding constraints. The essential problem is to find a way to represent the set of points defined by this intersection.[1] Because the effect of constraints is (generally) to cause a modification of the gradient direction, these gradientlike computational approaches for nonlinear programs are often called "reduced" or "deflected" (or "projected," as in Chapter 9) gradient techniques.

11.1.1 Zoutendijk's Method of Feasible Directions

Finding a Feasible Direction   Consider a maximization problem with a nonlinear objective function and linear constraints: Maximize $f(\mathbf{X})$ subject to $\mathbf{AX} \leq \mathbf{B}$ and $\mathbf{X} \geq \mathbf{0}$. One procedure for finding *feasible directions* of movement, developed by Zoutendijk (1960), has the following general character. Given a current feasible point, $\mathbf{X}^k$, the direction of movement from $\mathbf{X}^k$ should be $\nabla f(\mathbf{X}^k)$—which we will denote simply as ∇^k—if that is possible (as in gradient approaches to the unconstrained case). However, if $\mathbf{X}^k$ is located on one or more of the boundaries, it is quite possible that the gradient there will point outward from the feasible region. If that is the case, Zoutendijk suggests that we find a direction, $\mathbf{D}^k$, that deviates from ∇^k as little as possible (which means that $\mathbf{D}^k$ should make as small an angle as possible with ∇^k), while at the same time assuring that *some* movement along this direction is feasible.

We saw in Section 6.2 that maximizing the inner product of two vectors was tantamount to maximizing the cosine of the angle included between them, which in turn was the same as *minimizing* the angle between them. Hence, in Zoutendijk's approach we want to find elements d_j^k of the vector $\mathbf{D}^k$ such that $\mathbf{D}^k \circ \nabla^k$ is maximized. Since the elements of ∇^k are constants, $\mathbf{D}^k \circ \nabla^k$ is linear, and we want to carry out this maximization subject to (1) some kind of normalization on the sizes of the elements d_j^k (since *direction* is all that we are interested in at the moment) and (2) the requirements imposed by the constraints themselves (feasibility). We consider these in turn:

1. A simple normalization is $-1 \leq d_j^k \leq 1$, for all j. Note that this is a set of very simple *linear* constraints. Note also that this kind of constraint would rule out a nonnegativity requirement on each unknown (each d_j^k). Or one could require that $\|\mathbf{D}^k\| \leq 1$, a *nonlinear* constraint. There are many other possibilities, which it is unnecessary to explore here. We are interested in the fundamental logic of the method, and the simple linear constraint on each d_j^k is entirely adequate for illustration.

2. For feasibility, we need to require that if a constraint was met as an equality at $\mathbf{X}^k$, the new direction, $\mathbf{D}^k$, is not outward from it. In terms of the *gradient of the constraint* (for linear functions, this is the *normal* to the constraint), this means that the new direction must make an angle of *at least* 90° with that gradient. (Modifications are necessary in the case of nonlinear constraints.) Again, representation of the cosine of the included angle between two vectors as their inner product divided by the product of their lengths is relevant. In the range from 0° to 90° to 180° (the possible range of an *included* angle), the cosine goes from 1 to 0 to −1; hence requiring the cosine to be *nonpositive* would assure that the angle between the vectors is at least 90°. Since the product of the two vector lengths is necessarily positive, the cosine will be nonpositive if, and only if, the inner product of the vectors is nonpositive.

---

[1]In problems with three unknowns, this could mean, for example, the line defined by the intersection of the two planes associated with binding constraints at $\mathbf{X}^k$.

Let $_iA$ denote the ith *row* of A in $AX \le B$. (We use this notation because it is standard to use A_i to denote the ith *column* of A.) The column vector that is the normal (gradient) to the ith constraint is then just $(_iA)'$. Therefore what must be required to maintain feasibility is, for all i for which $_iAX^k = b_i$, that $(_iA)' \circ D^k \le 0$ (or, not using inner product notation, $_iAD^k \le 0$). This assures that the elements of the direction vector will send us at least along an edge or face of the feasible region ($_iAD^k = 0$) and possibly back into the interior ($_iAD^k < 0$). Similarly, to maintain nonnegativity of the x's, we require $d_j^k \ge 0$, for those j for which $x_j^k = 0$. Notice that both kinds of feasibility constraints are linear in the unknowns, d_j^k.[2]

We can therefore find a feasible direction at each step by solving an associated *linear program* (by using a linear normalization constraint). In so doing, we deviate as little as possible from the gradient direction that would be optimal in an unconstrained maximization problem. (When the normalization constraints are nonlinear, the associated problem at each step is nonlinear in the constraints, although smaller than the original problem, and with a linear objective.)

In either case, the step length still must be determined, as with all gradient methods. In inequality constrained problems, the step length (once a feasible direction has been established) is generally determined by the nearer of the following two points: (a) where the direction vector leaves or nears the boundaries of the feasible region or (b) where $f(X)$ reaches a maximum on the direction vector. This is a one-dimensional (line search) maximization problem of the sort examined in Chapter 6.

Example 1

$$\text{Maximize}\quad f(X) = -x_1^2 - x_2^2 - x_3^2 + 4x_1 + 6x_2 + 10x_3$$
$$\text{subject to}\quad x_j \le 4 \quad (j = 1,2,3) \tag{11.1}$$
$$\text{and}\quad X \ge 0$$

The feasible region, in three-dimensional solution space, is a cube with a corner at the origin and each of whose edges is four units long. You can easily establish that, without any

constraints, $f(X)$ has a unique global maximum at $X^u = \begin{bmatrix} 2 \\ 3 \\ 5 \end{bmatrix}$, where $f(X^u) = 38$.

If we start at $X^0 = \begin{bmatrix} 4 \\ 4 \\ 4 \end{bmatrix}$, where $f(X^0) = 32$ and $\nabla^0 = \begin{bmatrix} -4 \\ -2 \\ 2 \end{bmatrix}$, the first Zoutendijk linear

program is

$$\text{Maximize}\quad -4d_1^1 - 2d_2^1 + 2d_3^1$$
$$\text{subject to}\quad \text{(a)}\ -1 \le d_j^1 \le 1 \quad (j = 1,2,3)$$
$$\text{and}\quad \text{(b)}\ \begin{bmatrix} 1 & 0 & 0 \\ 0 & 1 & 0 \\ 0 & 0 & 1 \end{bmatrix} \begin{bmatrix} d_1^1 \\ d_2^1 \\ d_3^1 \end{bmatrix} \le \begin{bmatrix} 0 \\ 0 \\ 0 \end{bmatrix}$$

---

[2]We can incorporate the nonnegativity requirement on the x's into the set of requirements given by $_iAD^k \le 0$. Assume that the nonnegativity constraints for the maximization problem have been written as $-X \le 0$. Then, for each x_j for which $x_j^k = 0$, the associated normal to the nonnegativity constraint is $_jA = [0, 0, \ldots, -1, \ldots, 0] = (-I_j)'$, and $_jAD^k \le 0$ ensures that $-d_j^k \le 0$ or $d_j^k \ge 0$, as required for nonnegativity of x_j^{k+1}.

The first three constraints, in (a), are simple normalizations for the three elements in the direction vector, $\mathbf{D}^1$. The last three constraints, in (b), reflect the fact that all three x's are currently at their upper limits; they assure that the new direction makes an angle of at least $90°$ with each of the currently binding constraints. Looking at the objective function, we see that we want d_1^1 and d_2^1 as negative as possible, with $d_3^1 = 0$ (rather than negative), so the optimal solution is easily seen to be

$$\begin{bmatrix} d_1^1 \\ d_2^1 \\ d_3^1 \end{bmatrix} = \begin{bmatrix} -1 \\ -1 \\ 0 \end{bmatrix}$$

and $\mathbf{X}^1$ is to be found as

$$\mathbf{X}^1 = \mathbf{X}^0 + \alpha_1 \mathbf{D}^1 = \begin{bmatrix} 4 \\ 4 \\ 4 \end{bmatrix} + \alpha_1 \begin{bmatrix} -1 \\ -1 \\ 0 \end{bmatrix}$$

If we make α_1 as large as possible, we will have $\alpha_1 = 4$, because then $\mathbf{X}^1 = \begin{bmatrix} 0 \\ 0 \\ 4 \end{bmatrix}$, and both x_1^1 and x_2^1 have reached their lower limits; at that point, $f(\mathbf{X}^1) = 24$, which is not an improvement over $f(\mathbf{X}^0) = 32$. The maximum value of $f(\mathbf{X})$ in the feasible region and in the direction defined by $\alpha_1 \begin{bmatrix} -1 \\ -1 \\ 0 \end{bmatrix}$ away from $\begin{bmatrix} 4 \\ 4 \\ 4 \end{bmatrix}$ turns out to be when $\alpha_1 = 1.5$, so that

$$\mathbf{X}^1 = \begin{bmatrix} 4 \\ 4 \\ 4 \end{bmatrix} + (1.5) \begin{bmatrix} -1 \\ -1 \\ 0 \end{bmatrix} = \begin{bmatrix} 2.5 \\ 2.5 \\ 4 \end{bmatrix}$$

where $f(\mathbf{X}^1) = 36.5$. (This value for α_1 can be found by a simple equal-interval search procedure, as in Chapter 6.)

At this point, $\nabla^1 = \begin{bmatrix} -1 \\ 1 \\ 2 \end{bmatrix}$ and the next Zoutendijk linear program is

$$\text{Maximize} \quad -d_1^2 + d_2^2 + 2d_3^2$$

$$\text{subject to} \quad -1 \le d_j^2 \le 1 \quad (j = 1,2,3)$$

$$\text{and} \quad d_3^2 \le 0$$

(The last constraint reflects the fact that x_3 is at its upper limit.) Again, the optimal solution can be found by inspection. Here

$$\begin{bmatrix} d_1^2 \\ d_2^2 \\ d_3^2 \end{bmatrix} = \begin{bmatrix} -1 \\ 1 \\ 0 \end{bmatrix}$$

so that

$$\mathbf{X}^2 = \begin{bmatrix} 2.5 \\ 2.5 \\ 4 \end{bmatrix} + \alpha_2 \begin{bmatrix} -1 \\ 1 \\ 0 \end{bmatrix}$$

At $\alpha_2 = 1.5$, x_2 hits its upper limit, giving $\mathbf{X}^2 = \begin{bmatrix} 1 \\ 4 \\ 4 \end{bmatrix}$ and $f(\mathbf{X}^2) = 35$, which is worse than

$f(\mathbf{X}^1)$. However, a line search reveals that for $\alpha_2 = 0.5$, $\mathbf{X}^2 = \begin{bmatrix} 2 \\ 3 \\ 4 \end{bmatrix}$ and $f(\mathbf{X}^2) = 37$. Now ∇^2

$= \begin{bmatrix} 0 \\ 0 \\ 2 \end{bmatrix}$, and so we stop, since there is no incentive to change x_1 or x_2 and it is impossible

to change x_3 in the current gradient direction. (You can establish this formally by writing the next Zoutendijk linear program; you will find that $d_3^3 = 0$ in the optimal solution to that program.)

11.1.2   Rosen's Gradient Projection Method

The approach suggested by Rosen (1960, 1961) to constrained nonlinear optimization problems is also designed to modify the gradient direction when it points out of a feasible region. For simplicity, we again consider a problem with a nonlinear objective function and linear constraints. If there are three unknowns, the boundaries of the feasible region are made up of faces (planes, which are two-dimensional regions), edges (lines, which are one-dimensional regions), and vertices (points, which are zero-dimensional regions).

Rosen suggests that when the current point, $\mathbf{X}^k$, is on a boundary and the gradient points outward, one should use as a direction of feasible movement the *projection* of the gradient onto the boundary of smallest dimension (but no less than 1) on which $\mathbf{X}^k$ lies. This means that if $\mathbf{X}^k$ is on a face but not at an edge, ∇^k should be projected onto the face; if $\mathbf{X}^k$ is on an edge, project onto the edge; and if $\mathbf{X}^k$ is at a vertex, also project onto an edge.[3] No optimization is involved; one simply finds an acceptable (feasible) direction along which the objective function changes in the correct manner. This is exactly the same kind of gradient projection logic that is employed in some interior point methods (such as affine scaling) for linear programs (Chapter 9).

To contrast the Zoutendijk and Rosen approaches, consider Figure 11.1, in which the gradient from the current $\mathbf{X}^k$, on an edge, points upward and outward from the feasible region (the interior and faces of the distorted cube). Zoutendijk's direction, $\mathbf{Z}$, will be on the upper face; Rosen's, $\mathbf{R}$, will be along the edge. (When $\mathbf{X}^k$ is on a face, the directions will be the same.)

---

[3]When the "surface" defined by the binding constraints is a point (e.g., when three linear constraints are binding in a three-variable problem), then it is necessary to search for the appropriate one-dimensional surface (a line, when there are three variables) onto which to project. If none can be found, then the search is among two-dimensional surfaces (planes in this example). We will meet this situation in the numerical example.

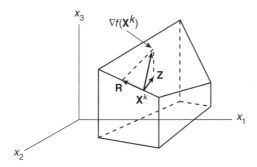

FIGURE 11.1   Zoutendijk's and Rosen's modification of the gradient.

Vector Projections   We saw in Chapter 9 that the *projection matrix*, $\mathbf{M}$, that drops an outward pointing three-element gradient, ∇, onto the plane defined by the single linear equation $\mathbf{AX} = b$, was [this is (9.8) in Chapter 9]

$$\mathbf{M} = [\mathbf{I} - \mathbf{A}'(\mathbf{AA}')^{-1}\mathbf{A}] \tag{11.2}$$

so that the projected gradient, $\mathbf{G}$, is

$$\mathbf{G} = \mathbf{M}\nabla = [\mathbf{I} - \mathbf{A}'(\mathbf{AA}')^{-1}\mathbf{A}]\nabla \tag{11.3}$$

If the gradient is to be projected onto a surface defined by more than one linear equation, the logic is the same as in (11.2) and (11.3), except that now $\mathbf{A}$ is not a row vector but a matrix with two or more rows.

Rosen's Projections   Here is how Rosen's approach works on the problem in (11.1). Starting again at $\mathbf{X}^0 = \begin{bmatrix} 4 \\ 4 \\ 4 \end{bmatrix}$, with $\nabla^0 = \begin{bmatrix} -4 \\ -2 \\ 2 \end{bmatrix}$ and $\mathbf{A}^0 = \begin{bmatrix} 1 & 0 & 0 \\ 0 & 1 & 0 \\ 0 & 0 & 1 \end{bmatrix}$, we find, as in (11.2), that

$$\mathbf{M}^0 = \left\{ \begin{bmatrix} 1 & 0 & 0 \\ 0 & 1 & 0 \\ 0 & 0 & 1 \end{bmatrix} - \begin{bmatrix} 1 & 0 & 0 \\ 0 & 1 & 0 \\ 0 & 0 & 1 \end{bmatrix} \right\} = \mathbf{0} \tag{11.4}$$

and therefore the projected gradient, $\mathbf{G}^0$ in (11.3), is a null vector.[4] This is simply a reflection of the fact that *three* binding constraints in *three*-dimensional solution space define only one point. Therefore, we must investigate the possibilities of projecting the gradient onto *pairs* of the binding constraints (edges of the feasible region, not a corner). There are three possibilities for choosing two of the three rows in $\mathbf{A}^0$ as the binding constraints onto which to project ∇^0.

---

[4]You can easily conclude that $\mathbf{M}$ in (11.2) will *always* be a null matrix whenever $\mathbf{A}$ is square and nonsingular, making use of the matrix algebra fact that $(\mathbf{EF})^{-1} = \mathbf{F}^{-1}\mathbf{E}^{-1}$ whenever $\mathbf{E}$ and $\mathbf{F}$ are square, conformable for multiplication, and nonsingular.

1. If we choose rows 1 and 2, $_1\mathbf{A}_0$ and $_2\mathbf{A}^0$, we find

$$\mathbf{M}^0 = \left\{ \begin{bmatrix} 1 & 0 & 0 \\ 0 & 1 & 0 \\ 0 & 0 & 1 \end{bmatrix} - \begin{bmatrix} 1 & 0 & 0 \\ 0 & 1 & 0 \\ 0 & 0 & 0 \end{bmatrix} \right\} = \begin{bmatrix} 0 & 0 & 0 \\ 0 & 0 & 0 \\ 0 & 0 & 1 \end{bmatrix} \tag{11.5}$$

and $\mathbf{G}^0 = \begin{bmatrix} 0 & 0 & 0 \\ 0 & 0 & 0 \\ 0 & 0 & 1 \end{bmatrix} \begin{bmatrix} -4 \\ -2 \\ 2 \end{bmatrix} = \begin{bmatrix} 0 \\ 0 \\ 2 \end{bmatrix}$. Movement in this direction is impossible since x_3 is already at its upper limit.

2. Using rows 1 and 3, we find $\mathbf{M}^0 = \begin{bmatrix} 0 & 0 & 0 \\ 0 & 1 & 0 \\ 0 & 0 & 0 \end{bmatrix}$ and $\mathbf{G}^0 = \begin{bmatrix} 0 \\ -2 \\ 0 \end{bmatrix}$,

so

$$\mathbf{X}^1 = \begin{bmatrix} 4 \\ 4 \\ 4 \end{bmatrix} + \alpha_0 \begin{bmatrix} 0 \\ -2 \\ 0 \end{bmatrix}$$

Finding the best value for α_0, using line search methods (again, as in Chapter 6), leads to $\alpha_0 = 0.5$ and so $\mathbf{X}^1 = \begin{bmatrix} 4 \\ 3 \\ 4 \end{bmatrix}$ with $f(\mathbf{X}^1) = 33$.

3. Using rows 2 and 3, $\mathbf{M}^0 = \begin{bmatrix} 1 & 0 & 0 \\ 0 & 0 & 0 \\ 0 & 0 & 0 \end{bmatrix}$ and $\mathbf{G}^0 = \begin{bmatrix} -4 \\ 0 \\ 0 \end{bmatrix}$, so

$$\mathbf{X}^1 = \begin{bmatrix} 4 \\ 4 \\ 4 \end{bmatrix} + \alpha_0 \begin{bmatrix} -4 \\ 0 \\ 0 \end{bmatrix}$$

and again it turns out that $\alpha_0 = 0.5$ is the best choice, giving $\mathbf{X}^1 = \begin{bmatrix} 2 \\ 4 \\ 4 \end{bmatrix}$ with $f(\mathbf{X}^1) = 36$.

(You can work through the mathematical details of these two possibilities.) So this is the best choice for $\mathbf{X}^1$.

At $\mathbf{X}^1 = \begin{bmatrix} 2 \\ 4 \\ 4 \end{bmatrix}$ we have $\nabla^1 = \begin{bmatrix} 0 \\ -2 \\ 2 \end{bmatrix}$ and $\mathbf{A}^1 = \begin{bmatrix} 0 & 1 & 0 \\ 0 & 0 & 1 \end{bmatrix}$, so now $\mathbf{M}^1 = \begin{bmatrix} 1 & 0 & 0 \\ 0 & 0 & 0 \\ 0 & 0 & 0 \end{bmatrix}$

leading to $\mathbf{G}^1 = \mathbf{M}^1 \nabla^1 = \begin{bmatrix} 0 \\ 0 \\ 0 \end{bmatrix}$. This is a signal that there is no way to increase $f(\mathbf{X})$ by trying to project onto an edge; it does not rule out the possibility that a better solution will be found by projecting onto a face—where only *one* constraint is binding.

There are now two options:

1. Choose row 1 from $\mathbf{A}^1$. This generates

$$\mathbf{M}^1 = \left\{ \begin{bmatrix} 1 & 0 & 0 \\ 0 & 1 & 0 \\ 0 & 0 & 1 \end{bmatrix} - \begin{bmatrix} 0 \\ 1 \\ 0 \end{bmatrix} \left\{ \begin{bmatrix} 0 & 1 & 0 \end{bmatrix} \begin{bmatrix} 0 \\ 1 \\ 0 \end{bmatrix} \right\}^{-1} \begin{bmatrix} 0 & 1 & 0 \end{bmatrix} \right\} = \begin{bmatrix} 1 & 0 & 0 \\ 0 & 0 & 0 \\ 0 & 0 & 1 \end{bmatrix} \qquad (11.6)$$

so that $\mathbf{G}^1 = \mathbf{M}^1 \boldsymbol{\nabla}^1 = \begin{bmatrix} 0 \\ 0 \\ 2 \end{bmatrix}$. This is impossible, since x_3 is already at its upper limit.

2. Using row 2 from $\mathbf{A}^1$, $\mathbf{M}^1 = \begin{bmatrix} 1 & 0 & 0 \\ 0 & 1 & 0 \\ 0 & 0 & 0 \end{bmatrix}$ and $\mathbf{G}^1 = \begin{bmatrix} 0 \\ -2 \\ 0 \end{bmatrix}$. Therefore, $\mathbf{X}^2 = \mathbf{X}^1 +$

$\alpha_1 \mathbf{G}^1 = \begin{bmatrix} 2 \\ 4 \\ 4 \end{bmatrix} + \alpha_1 \begin{bmatrix} 0 \\ -2 \\ 0 \end{bmatrix}$ and a line search for the optimal α_1 leads to $\alpha_1 = 0.5$, so $\mathbf{X}^2 =$

$\begin{bmatrix} 2 \\ 3 \\ 4 \end{bmatrix}$ and $f(\mathbf{X}^2) = 37$. Also, $\boldsymbol{\nabla}^2 = \begin{bmatrix} 0 \\ 0 \\ 2 \end{bmatrix}$, and there is no incentive to move from this (optimal) point.

11.1.3 Alternative Interior Point Methods for Nonlinear Programs

In Chapter 9 we explored a pair of methods for linear programming problems that moved through the feasible region, instead of around the edges, as in the simplex method. One of these methods (affine scaling) relies on projecting gradients at each step in an iterative procedure. There is currently a great deal of work in developing variants of that approach, as well as other path-following techniques, to use for nonlinear programming problems.

Either Zoutendijk's or Rosen's method could be used on linear programming problems. What generally happens is that one is almost immediately sent to the boundary of the feasible region, and from then on the directions are such that you find yourself crawling around the edges, just as in the simplex method. [Problem 11.3(b) asks you to apply Zoutendijk's directions and Rosen's projections to a linear programming problem to see how the edges assert themselves immediately.]

11.2 DUALITY IN NONLINEAR PROGRAMMING

In this section we take a quick look at some results that have been worked out for a kind of duality theory for nonlinear programming problems. These are quite parallel to those in the linear programming case (Section 7.4). Proofs of most of the statements are more complicated than in the linear case, and we omit them. There are two principal reasons for interest in nonlinear duality at this point: (1) for any feasible solutions to a pair of primal and dual nonlinear programming problems, the dual objective function value provides a limit on the value of the primal objective function (as with linear programs) and (2) for a pair of optimal solutions, the values of the dual variables may have the same kind of "shadow price" interpretation that we associate with the linear programming case—giving a positive marginal valuation to resources that are used up in the optimal solution and a value of zero to those resources that are in excess supply at an optimal solution.

The general nonlinear programming maximization problem [(10.2′) in the previous chapter] is

$$\text{Maximize} \quad \Pi(\mathbf{X}) = f(\mathbf{X})$$
$$\text{subject to} \quad \mathbf{h}(\mathbf{X}) \leq \mathbf{0} \tag{11.7}$$
$$\text{and} \quad \mathbf{X} \geq \mathbf{0}$$

where, as usual $\mathbf{h}(\mathbf{X}) = [h^1(\mathbf{X}), \ldots, h^m(\mathbf{X})]'$. The associated Lagrangian function, used primarily in development of Kuhn–Tucker conditions and in the discussion of saddle points, is

$$L(\mathbf{X},\mathbf{Y}) = f(\mathbf{X}) - \mathbf{Y}'[\mathbf{h}(\mathbf{X})] \tag{11.8}$$

where $\mathbf{Y} = [y_1, \ldots, y_m]'$. [This was (10.31) in Chapter 10.] The gradient of $L(\mathbf{X},\mathbf{Y})$ with respect to the y's is just $\nabla L_\mathbf{Y} = -\mathbf{h}(\mathbf{X})$, and so $f(\mathbf{X})$ in (11.7) could be expressed as

$$f(\mathbf{X}) = L(\mathbf{X},\mathbf{Y}) + \mathbf{Y}'[\mathbf{h}(\mathbf{X})] = L(\mathbf{X},\mathbf{Y}) - \mathbf{Y}'\nabla L_\mathbf{Y}$$

In addition, the requirements in $\mathbf{h}(\mathbf{X}) \leq \mathbf{0}$ can be expressed as $\nabla L_\mathbf{Y} \geq \mathbf{0}$, so the general nonlinear programming maximization problem in (11.7) could also be written

$$\text{Maximize} \quad L(\mathbf{X},\mathbf{Y}) - \mathbf{Y}'\nabla LY$$
$$\text{subject to} \quad -\nabla L_\mathbf{Y} \leq \mathbf{0} \tag{11.9}$$
$$\text{and} \quad \mathbf{X} \geq \mathbf{0}$$

This appears only to add complexity to the expression in (11.7), but it does suggest a "symmetric" problem:[5]

$$\text{Minimize} \quad L(\mathbf{X},\mathbf{Y}) - \mathbf{X}'\nabla L_\mathbf{X}$$
$$\text{subject to} \quad -\nabla L_\mathbf{X} \geq \mathbf{0} \tag{11.10}$$
$$\text{and} \quad \mathbf{Y} \geq \mathbf{0}$$

From $L(\mathbf{X},\mathbf{Y})$ in (11.8), and using the Jacobian matrix (Chapter 4)

$$\mathbf{J} = \begin{bmatrix} [\nabla h^1]' \\ \vdots \\ \vdots \\ \vdots \\ [\nabla h^m]' \end{bmatrix}$$

---

[5]There are other possibilities. Some more general developments in nonlinear duality have been accomplished via conjugate function theory, which goes beyond the level of this book.

we have

$$\nabla L_{\mathbf{X}} = \nabla f - \mathbf{J}'\mathbf{Y}$$

and so (11.10) in more detail is

$$\text{Minimize}\quad \Delta(\mathbf{X},\mathbf{Y}) = f(\mathbf{X}) - \mathbf{Y}'[\mathbf{h}(\mathbf{X})] - \mathbf{X}'[\nabla f - \mathbf{J}'\mathbf{Y}]$$
$$\text{subject to}\quad \mathbf{J}'\mathbf{Y} - \nabla f \geq \mathbf{0} \tag{11.11}$$
$$\text{and}\quad \mathbf{Y} \geq \mathbf{0}$$

Written out in more detail, this is

$$\text{Minimize}\quad f(\mathbf{X}) - \sum_{i=1}^{m} y_i h^i(\mathbf{X}) - \sum_{j=1}^{n} x_j \left(f_j - \sum_{i=1}^{m} y_i h_j^i\right)$$
$$\text{subject to}\quad \sum_{i=1}^{m} y_i h_j^i - f_j \geq 0 \qquad (j = 1, \ldots, n) \tag{11.11'}$$
$$\text{and}\quad y_i \geq 0 \qquad (i = 1, \ldots, m)$$

When $f(\mathbf{X})$ in the maximization problem is concave and each constraint is convex or quasiconvex, so that the conditions of Theorem SP4 in Chapter 10 constitute both necessary and sufficient conditions for a maximum to the nonlinear programming problem, then (11.11) [or (11.10)] is taken to be the dual to (11.7) [or (11.9)].[6] This is primarily because (1) a pair of dual *linear* programs corresponds precisely to (11.9) and (11.10), and (2) a set of theorems parallel to those in duality theory for linear programs can be derived.

Regarding the linear programming connection, the general linear programming maximization problem (Chapter 7) was

$$\text{Maximize}\quad \mathbf{P}'\mathbf{X}$$
$$\text{subject to}\quad \mathbf{AX} \leq \mathbf{B}$$
$$\text{and}\quad \mathbf{X} \geq \mathbf{0}$$

The associated Lagrangian function, as in (11.8), would be

$$L(\mathbf{X},\mathbf{Y}) = \mathbf{P}'\mathbf{X} - \mathbf{Y}'(\mathbf{AX} - \mathbf{B}) \tag{11.12}$$

and so $\nabla L_{\mathbf{Y}} = -[\mathbf{AX} - \mathbf{B}]$ and $\nabla L_{\mathbf{X}} = [\mathbf{P} - \mathbf{A}'\mathbf{Y}]$. (The latter gradient is defined as a column vector to conform to the convention of expressing gradients as columns.) The primal linear program can now be written in the style of (11.9) as

$$\text{Maximize}\quad \mathbf{P}'\mathbf{X} - \mathbf{Y}'[\mathbf{AX} - \mathbf{B}] + \mathbf{Y}'[\mathbf{AX} - \mathbf{B}]$$
$$\text{subject to}\quad \mathbf{AX} - \mathbf{B} \leq \mathbf{0} \quad\text{or}\quad \mathbf{AX} \leq \mathbf{B} \tag{11.13}$$
$$\text{and}\quad \mathbf{X} \geq \mathbf{0}$$

---

[6]As usual, we also assume that the problem satisfies the Kuhn–Tucker constraint qualification. These concavity–convexity requirements play a role in the proofs of the following theorems, which establish a "reasonable" primal–dual connection.

Corresponding to (11.10) we have, for this linear program,

$$\text{Minimize } \mathbf{P}'\mathbf{X} - \mathbf{Y}'[\mathbf{AX} - \mathbf{B}] - \mathbf{X}'[\mathbf{P} - \mathbf{A}'\mathbf{Y}]$$

or, since $\mathbf{P}'\mathbf{X} = \mathbf{X}'\mathbf{P}$, $\mathbf{Y}'\mathbf{B} = \mathbf{B}'\mathbf{Y}$ and $\mathbf{Y}'\mathbf{AX} = \mathbf{X}'\mathbf{A}'\mathbf{Y}$[7]

$$\text{Minimize } \mathbf{B}'\mathbf{Y}$$

subject to $\quad -\mathbf{P} + \mathbf{A}'\mathbf{Y} \geq \mathbf{0} \quad$ or $\quad \mathbf{A}'\mathbf{Y} \geq \mathbf{P} \qquad (11.14)$

and $\quad \mathbf{Y} \geq \mathbf{0}$

Clearly, (11.13) and (11.14) are exactly a pair of dual linear programs.

We now explore a set of primal–dual theorems for the nonlinear programming problems in (11.7) and (11.11). We use a "prime" for these in the nonlinear case; there are obvious parallels to the results for linear programs in Chapter 7.

Theorem PD1′   For any feasible solutions to the primal ($\mathbf{X}^p$) and dual ($\mathbf{X}^d$, $\mathbf{Y}^d$) problems in (11.7) and (11.11), $\Delta(\mathbf{X}^d, \mathbf{Y}^d) \geq \Pi(\mathbf{X}^p)$.

The proof is similar to that for Theorem PD1 in Chapter 7. It depends on concavity of $f(\mathbf{X})$ and convexity of each constraint in the primal problem. The theorem establishes for any feasible solutions an upper bound to the value of the objective function in a maximization problem and a lower bound to the objective function in a minimization problem. We will see in the next section that this provides useful information for establishing a stopping criterion for an approximation procedure for solving nonlinear programming problems. It also plays a role in some of the stopping criteria used by interior point methods for linear programs (Chapter 9).

Theorem PD2′   Feasible solutions to the primal and dual problems are optimal if and only if $\Pi(\mathbf{X}^p) = \Delta(\mathbf{X}^d, \mathbf{Y}^d)$.

Clearly, *if* $\Pi = \Delta$, then both objective functions have reached their limits (Theorem PD1′) and so the solutions are optimal. The *only if* part of the theorem relies in part on properties given in Theorem SP4 in Chapter 10. An important outcome of the proof is that, given an optimal $\mathbf{X}^*$ for the primal, a vector $\mathbf{Y}^*$ can be found such that $(\mathbf{X}^*, \mathbf{Y}^*)$ is an optimal solution to the dual.

Theorem PD3′   A pair of feasible solutions has $\Pi(\mathbf{X}^*) = \Delta(\mathbf{X}^*, \mathbf{Y}^*)$ if and only if (1) $(\mathbf{Y}^*)'[-\mathbf{h}(\mathbf{X}^*)] = 0$ and (2) $(\mathbf{X}^*)'[\nabla f^* - (\mathbf{J}^*)'\mathbf{Y}^*] = 0$.

Note that $[-\mathbf{h}(\mathbf{X}^*)]$ in (1) is the vector of slack variables for the constraints in (11.7), and $[\nabla f^* - (\mathbf{J}^*)'\mathbf{Y}^*]$ in (2) is likewise the vector of slack variables for the constraints in (11.11). So this theorem describes a kind of "complementary slackness" for optimal solutions that is parallel to that in the linear programming case. This property of the optimal solutions to the pair of dual nonlinear programs shows the equivalence of the dual vari-

---

[7]The first is just the sum (over all j) of all products $p_j x_j$ (a scalar), the second is a similar sum (over all i) of $b_i y_i$ (another scalar), and the third is the sum (over all i and j) of all products $y_i a_{ij} x_j$ (also a scalar).

ables (**Y**) and the Lagrange multipliers used in Chapter 10 in the development of Kuhn–Tucker conditions and the saddle point problem connection.

Theorem PD4′   Under certain conditions, $\partial \Pi(\mathbf{X}^*)/\partial b_i = y_i^*$.

This marginal valuation property of the dual variables at optimum held for linear programs, subject to the qualification that the appropriate derivatives existed. The same sort of requirement is needed here. Under those conditions, y_i^* is a measure of the impact on the optimum value of the primal objective function of a marginal change in b_i.

Because the pair of linear programs in (11.13) and (11.14) conform to the structures in (11.7) and (11.11), and because of the similarity of Theorems PD1′–PD4′ to their counterparts in the linear programming case, the nonlinear minimization problem in (11.11) has been viewed as the dual to the nonlinear maximization problem in (11.7). Then the connection with Kuhn–Tucker theory and saddle point problems is established, since a pair of *feasible* solutions to (11.7) and (11.11) satisfies parts (a)(i) and (b)(i) of Theorem SP4, and a pair of *optimal* solutions also satisfies (a)(ii) and (b)(ii). Thus the economic interpretation of the conditions of Theorem SP4 transfers to the constraints and variables in a pair of dual nonlinear programming problems and parallels the interpretation for the linear case in Chapter 7. Interpretation of the dual objective function is less clear.

11.3   ADDITIONAL COMPUTATIONAL METHODS FOR NONLINEAR PROGRAMS

In this section we explore the fundamental logic of three *approximation* procedures for solving nonlinear programming problems. The first two, separable programming and cutting-plane methods, use linear approximations for the nonlinearities in the problem. The final approach, the sequential unconstrained technique, replaces the original constrained nonlinear problem with an unconstrained but still nonlinear one.

11.3.1   Separable Programming

Creating the Linear Approximations   One approach to solving nonlinear programming problems is to approximate the nonlinear functions involved—objective function and/or constraints—with straight-line segments and then use methods that are appropriate for linear problems. In a completely general nonlinear programming problem—with an objective function like $3x_1^3 - 2x_1^2 x_2 + x_3^4$ or a constraint such as $e^{x_1^2} + 5 \log x_1 x_2 + x_3^2 \leq 4$—linear approximations are not easy to specify algebraically, even though the basic idea can still be appreciated. However, when the functions are *separable*, the mathematical specification is more easily done. It is then possible to use the simplex method (with slight modification) on a linear program that approximates the original nonlinear one. While the simplex method is exact, the solution only approximates that of the original problem; the accuracy of the solution depends on how well the approximating linear program represents the original nonlinear problem.

If a function $f(\mathbf{X})$ can be represented as $\sum_{j=1}^n f_j(x_j)$, it is said to be separable, meaning that it can be represented as the sum of several functions (generally nonlinear) of a single variable each. [In this section, the f_j in $f_j(x_j)$ will denote a particular function of a single variable (the *j*th) only and not $\partial f/\partial x_j$, as in previous chapters. This is done to keep the notation relatively simple, especially for problems with several nonlinear constraints.] Here is a separable function:

$$f(\mathbf{X}) = 3x_1^2 + 5x_2^2 + x_1 - 6x_2 \qquad (11.15)$$

Letting $f_1(x_1) = 3x_1^2 + x_1$ and $f_2(x_2) = 5x_2^2 - 6x_2$, we have $f(\mathbf{X}) = \Sigma_{j=1}^2 f_j(x_j)$.

Consider a programming problem in which the objective function, $f(\mathbf{X})$, and each constraint function, $g^i(\mathbf{X})$, is separable. [It turns out to be useful in separable programming problems to think of the constraints $h^i(\mathbf{X}) \le 0$ in their $g^i(\mathbf{X}) \le b_i$ form. For the ith constraint, we use $g_j^i(x_j)$ for that part of $g^i(\mathbf{X})$ that involves x_j only.] For example, the constraints in such a problem might be

$$g^1(\mathbf{X}) = x_1^2 + 6x_2 \le 10$$
$$g^2(\mathbf{X}) = x_1 + 4x_2^2 \le 30 \qquad (11.16)$$

for which $g_1^1 = x_1^2$, $g_2^1 = 6x_2$, $g_1^2 = x_1$, and $g_2^2 = 4x_2^2$, so that $g^1(\mathbf{X}) = \Sigma_{j=1}^2 g_j^1(x_j) \le 10$ and $g^2(\mathbf{X}) = \Sigma_{j=1}^2 g_j^2(x_j) \le 30$. So the general nonlinear separable programming maximization problem is

$$\text{Maximize} \quad \sum_{j=1}^n f_j(x_j)$$

$$\text{subject to} \quad \sum_{j=1}^n g_j^i(x_j) \le b_i \qquad (i = 1, \ldots, m) \qquad (11.17)$$

$$\text{and} \quad \mathbf{X} \ge \mathbf{0}$$

The approach to such a problem is designed to find an approximate solution using approximating linear programs for which the simplex method can be used.

The first issue is how to specify a piecewise linear approximation to any one of the separable functions—for example, to any of the $f_j(x_j)$ in the objective function. Figure 11.2 illustrates the problem for a given interval $0 \le x_j \le t_j$. Select a set of $(m + 1)$ points, x_{kj}, in sequence in the interval; let $x_{0j} = 0 < x_{1j} < \cdots < x_{kj} < \cdots < x_{mj} = t_j$. For each point, calculate $f_j(x_{kj})$ $(k = 0, \ldots, m)$; these are the vertical heights in Figure 11.2. Connect each pair of adjacent points on the function, $[x_{kj}, f_j(x_{kj})]$ and $[x_{k+1,j}, f_j(x_{k+1,j})]$, by a straight line—the

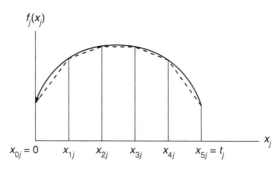

FIGURE 11.2   Approximating $f_j(x_j)$ over $0 \le x_j \le t_j$.

dashed lines in the figure—and denote this piecewise linear approximation to $f_j(x_j)$ by $\hat{f}_j(x_j)$. The smaller the distances between adjacent x_{kj}'s, the better this approximation will be.

Suppose that such approximations have been found for *each* $f_j(x_j)$ ($j = 1, \ldots, n$) and for *each* $g_j^i(x_j)$ ($i = 1, \ldots, m; j = 1, \ldots, n$) in each of the constraints by selecting *mesh points* x_{kj} for each variable x_j. Then the *approximation* to the problem in (11.17) is

$$\text{Maximize} \quad \hat{f}(\mathbf{X}) = \sum_{j=1}^{n} \hat{f}_j(x_j)$$

$$\text{subject to} \quad \sum_{j=1}^{n} \hat{g}_j^i(x_j) \leq b_i \qquad (i = 1, \ldots, m) \tag{11.18}$$

$$\text{and} \quad \mathbf{X} \geq \mathbf{0}$$

The next task is to find an optimal solution to this approximating problem. To do so, we need to be precise about how to write the approximating functions $\hat{f}_j$ and $\hat{g}_j^i$. First, any x_j in the interval between adjacent mesh points, $x_{kj} \leq x_j \leq x_{k+1,j}$, can be expressed as a convex combination of the endpoints of the interval:

$$x_j = \alpha x_{kj} + (1 - \alpha)x_{k+1,j} \tag{11.19}$$

for $0 \leq \alpha \leq 1$. And similarly, $\hat{f}_j(x_j)$, which is the point directly above x_j and on the line between $f_j(x_{kj})$ and $f_j(x_{k+1,j})$, is just

$$\hat{f}_j(x_j) = \alpha f_j(x_{kj}) + (1 - \alpha)f_j(x_{k+1,j}) \tag{11.20}$$

The same convex combination that defines x_j relative to its endpoints, x_{kj} and $x_{k+1,j}$, defines the value of the linear approximating function at that x_j, relative to its endpoints, which are the values of the original $f(\mathbf{X})$ at those two points.[8] This holds for any x_j that lies between x_{kj} and $x_{k+1,j}$, so we can denote the α in (11.19) and (11.20) by α_{kj} and let $(1 - \alpha)$ be $\alpha_{k+1,j}$. Then, for x_j in that interval, we have

$$x_j = \alpha_{kj}x_{kj} + \alpha_{k+1,j}x_{k+1,j} \tag{11.21}$$

and

$$\hat{f}_j(x_j) = \alpha_{kj}f_j(x_{kj}) + \alpha_{k+1,j}f_j(x_{k+1,j}) \tag{11.22}$$

Therefore, (11.21) and (11.22), with the additional requirements that $\alpha_{kj} + \alpha_{k+1,j} = 1$, $\alpha_{kj} \geq 0$, and $\alpha_{k+1,j} \geq 0$, serve to describe the approximating function $\hat{f}_j(x_j)$ anywhere in the interval $x_{kj} \leq x_j \leq x_{k+1,j}$ in terms of *known* endpoint values $f_j(x_{kj})$ and $f_j(x_{k+1,j})$. All that need to be specified are α_{kj} and $\alpha_{k+1,j}$; these identify the exact location of x_j in the interval.

Even more generally, for any x_j in the entire interval $x_{0j} = 0 \leq x_j \leq x_{mj} = t_j$ (Fig. 11.2), we can write, parallel to (11.21)

$$x_j = \sum_{k=0}^{m} \alpha_{kj}x_{kj} \tag{11.23}$$

[8]We used this fact in exploring the geometry of the definitions of concavity and convexity of $f(x)$ in Chapter 3, Section 3.2. It is based on a straightforward application of properties of similar triangles.

and, in place of (11.22)

$$\hat{f}_j(x_j) = \sum_{k=0}^{m} \alpha_{kj} f_j(x_{kj}) \tag{11.24}$$

provided that

$$\sum_{k=0}^{m} \alpha_{kj} = 1 \qquad \alpha_{kj} \geq 0 \qquad (k = 0, \ldots, m) \tag{11.25}$$

and that

(1) no more than two α_{kj} are positive *and* (2) *if* two α_{kj} are positive, they must be *adjacent*. $\qquad$ (11.26)

The approximating problem in (11.18) was composed of n such linear approximating functions, $\hat{f}_j(x_j)$, in the objective function and nm linear approximating functions, $\hat{g}_j^i(x_j)$, in the constraints. Each of these could be expressed in the same way as $\hat{f}_j(x_j)$ in (11.24).

Suppose that we can find a maximum value, t_j, for each x_j in the problem. (This is usually possible in the context of the problem.) For example, from the two constraints in (11.16), knowing that $\mathbf{X} \geq \mathbf{0}$ and setting first $x_2 = 0$ and then $x_1 = 0$, we find that

$$t_1 = \min [\sqrt{10}, 30] = \min [3.16, 30] = 3.16$$

$$t_2 = \min [10/6, \sqrt{7.5}] = \min [1.66, 2.74] = 1.66$$

Then subdivide the range for *each* x_j from 0 to t_j by mesh points x_{kj} ($k = 0, \ldots, m_j$). This simply makes a division similar to that in Figure 11.2 for each variable. Use the same mesh points (the same intervals) for f_j and for all constraints g_j^i for any particular x_j.[9] Then, following (11.23)–(11.25), we have

$$x_j = \sum_{k=0}^{m_j} \alpha_{kj} x_{kj} \tag{11.23'}$$

$$\hat{f}_j(x_j) = \sum_{k=0}^{m_j} \alpha_{kj} f_j(x_{kj})$$
$$\tag{11.24'}$$
$$\hat{g}_j^i(x_j) = \sum_{k=0}^{m_j} \alpha_{kj} g_j^i(x_{kj})$$

$$\sum_{k=0}^{m_j} \alpha_{kj} = 1, \qquad \alpha_{kj} \geq 0 \qquad (j = 1, \ldots, n; \quad k = 0, \ldots, m_j) \tag{11.25'}$$

and the requirements in (11.26).

---

[9]For example, for x_1, $t_1 = 3.16$, so if $m_1 = 5$, this indicates that the interval $0 \leq x_1 \leq 3.16$ is divided into five subintervals with mesh points $x_{01} = 0 < x_{11} \cdots < x_{41} < x_{51} = 3.16$. For x_2, $t_2 = 1.66$, and if $m_2 = 5$ also, the smaller interval $0 \leq x_2 \leq 1.66$ would be divided into five (different) subintervals $x_{02} = 0 < x_{12} \cdots < x_{42} < x_{52} = 1.66$.

This means that the approximating problem in (11.18) becomes

$$\text{Maximize} \quad \hat{f}(\mathbf{X}) = \sum_{j=1}^{n} \sum_{k=0}^{m_j} f_j(x_{kj}) \alpha_{kj}$$

$$\text{subject to} \quad \sum_{j=1}^{n} \sum_{k=0}^{m_j} g_j^i(x_{kj}) \alpha_{kj} \le b_i \qquad (i = 1, \ldots, m)$$

$$\sum_{k=0}^{m_j} \alpha_{kj} = 1 \qquad (j = 1, \ldots, n) \tag{11.27}$$

$$\alpha_{kj} \ge 0 \qquad (k = 0, \ldots, m_j; j = 1, \ldots, n)$$

and (1) no more than two α_{kj} are positive *and* (2) if two α_{kj} are positive, they must be *adjacent*.

Except for these last conditions, this is a straightforward *linear* program in the variables α_{kj}. This means that the simplex method or any of the more modern interior point methods for linear programs can be used.[10] Once the optimal α_{kj} are found, the associated x_j in the original problem, (11.17), are found from (11.23′). Note that the size of the problem has increased substantially. Whereas (11.17) had m constraints (excluding nonnegativities) and n variables (excluding slacks), the associated linear program in (11.27) has $(m + n)$ constraints (excluding nonnegativities) and $\Sigma_{j=1}^{n}(m_j + 1) = \Sigma_{j=1}^{n} m_j + n$ variables (excluding slacks). A separable nonlinear programming problem with, say, 10 variables and 5 constraints (plus 10 nonnegativities), in which the range of each variable was divided into nine intervals (hence 10 x_{kj} for each x_j),[11] would be converted to an approximating linear program with 100 variables and 15 constraints (plus 100 nonnegativities).

Example 2 The objective function in this illustration is linear, but the constraint is nonlinear but separable:

$$\text{Maximize} \quad f(\mathbf{X}) = 3x_1 + 2x_2$$

$$\text{subject to} \quad g^1(\mathbf{X}) = 4x_1^2 + x_2^2 \le 16 \tag{11.28}$$

$$\text{and} \quad \mathbf{X} \ge \mathbf{0}$$

It is easy to tell from the constraint that $x_1 \le 2$ and $x_2 \le 4$. Now let $t_1 = t_2 = 4$ and divide the closed interval $[0, 4]$ into four equally sized subintervals for both x_1 and x_2. This is not necessary; it just simplifies presentation of the results. Here

$$f_1(x_1) = 3x_1, \quad f_2(x_2) = 2x_2, \quad g_1^1(x_1) = 4x_1^2 \quad \text{and} \quad g_2^1(x_2) = x_2^2$$

so the problem, as in (11.17), is

---

[10]The fact that (11.27) contains both equality and inequality constraints need not be troublesome, as we saw in Chapter 8.
[11]There is no need for the same number of intervals for each variable; it only serves to simplify the illustration.

TABLE 11.1   Data for Separable Programming Approximation to (11.29)

	$x_j\,(j=1,2)$	$f_1(x_1)$	$g_1^1(x_1)$	$f_2(x_2)$	$g_2^1(x_2)$
x_{0j}	0	0	0	0	0
x_{1j}	1	3	4	2	1
x_{2j}	2	6	16	4	4
x_{3j}	3	9	36	6	9
x_{4j}	4	12	64	8	16

$$\text{Maximize}\quad f(\mathbf{X}) = \sum_{j=1}^{2} f_j(x_j)$$

$$\text{subject to}\quad \sum_{j=1}^{2} g_j^1(x_j) \le 16 \tag{11.29}$$

$$\text{and}\quad \mathbf{X} \ge \mathbf{0}$$

To operate the approximating linear program, in (11.27), we need the values of the f_j and g_j^1 for the mesh points. These are shown in Table 11.1. Therefore, written out completely, the approximating linear program is

$$\text{Maximize}\quad \hat{f}(\mathbf{X}) = \Pi(\alpha) = 0\alpha_{01} + 3\alpha_{11} + 6\alpha_{21} + 9\alpha_{31} + 12\alpha_{41}$$
$$+ 0\alpha_{02} + 2\alpha_{12} + 4\alpha_{22} + 6\alpha_{32} + 8\alpha_{42}$$

$$\text{subject to}\quad (1)\ \ 0\alpha_{01} + 4\alpha_{11} + 16\alpha_{21} + 36\alpha_{31} + 64\alpha_{41}$$
$$+ 0\alpha_{02} + \alpha_{12} + 4\alpha_{22} + 9\alpha_{32} + 16\alpha_{42} \le 16 \tag{11.30}$$

$$(2)\ \left\{ \begin{array}{l} \alpha_{01} + \alpha_{11} + \alpha_{21} + \alpha_{31} + \alpha_{41} = 1 \\ \alpha_{02} + \alpha_{12} + \alpha_{22} + \alpha_{32} + \alpha_{42} = 1 \end{array} \right.$$

$$\text{and}\quad \alpha_{kj} \ge 0 \qquad (j = 1, 2;\ k = 0, \ldots, 4)$$

The equations among the constraints, in (2) here, could be reexpressed as two inequalities each, and then some sort of artificial variable technique could be used (Section 8.2); here, instead, we use a two-phase procedure on the equations directly (Section 8.3).[12] In this example, α_{01} and α_{02} can play the roles of slack variables for the constraints in (2): they are absent from constraint (1) and from the objective function (coefficients of zero) and each enters its own constraint with a coefficient of one. Using s_1 for the slack variable in constraint (1), the initial simplex table is as shown in Table 11.2. The simplex method applies, but with "restricted basis entry" to accommodate the requirements of (11.26).

The initial basic solution is therefore $s_1 = 16$, $\alpha_{01} = \alpha_{02} = 1$ and $\Pi = 0$. The most attractive nonbasic variable is α_{41} (largest positive coefficient in the top row). However, α_{41} can be brought into the next basis only if it replaces α_{01}; otherwise, two nonadjacent α_{k1} would be positive, violating (11.26). But simplex rules would force s_1 out of the basis as α_{41} enters. The next most attractive nonbasic variable is α_{31} for which the same kind of

---

[12]In fact, we essentially combine the two phases by pivoting in order to maximize the objective function in (11.30)—which is formally the phase II objective—and noting that in the process $\alpha_{01} + \alpha_{02} = 0$—which is really the phase I objective.

TABLE 11.2   Initial Simplex Table for the Problem in (11.30)

		α_{11}	α_{21}	α_{31}	α_{41}	α_{12}	α_{22}	α_{32}	α_{42}
Π	0	3	6	9	12	2	4	6	8
s_1	16	-4	-16	-36	-64	-1	-4	-9	-16
α_{01}	1	-1	-1	-1	-1	0	0	0	0
α_{02}	1	0	0	0	0	-1	-1	-1	-1

problem exists. Next, then, consider α_{42}, with the third-largest top-row element. Now α_{02} can be removed when α_{42} is introduced into the basis, and the new simplex table is shown in Table 11.3, where $s_1 = 0$, $\alpha_{01} = \alpha_{42} = 1$ and $\Pi = 8$.

Continuing in this way, using the simplex method but paying attention to the requirements in (11.26), we will eventually find an optimal solution in which $\alpha^*_{11} = 1$, $\alpha^*_{32} = 0.57$, and $\alpha^*_{42} = 0.43$. Therefore, from (11.23) and the values in Table 11.1, we obtain

$$x_1^* = \sum_{k=0}^{4} \alpha^*_{k1} x_{k1} = (1)(1) = 1$$

and

$$x_2^* = \sum_{k=0}^{4} \alpha^*_{k2} x_{k2} = (3)(0.57) + (4)(0.43) = 3.43$$

and so, in (11.28), $f(\mathbf{X}^*) = 3(1) + 2(3.43) = 9.86$.

The exact solution to this problem is easily found, via the Kuhn–Tucker approach; it is $\mathbf{X}^{**} = \begin{bmatrix} 1.2 \\ 3.2 \end{bmatrix}$, with $f(\mathbf{X}^{**}) = 10$. The approximating problem has underestimated x_1^{**} and overestimated x_2^{**}, leading to an underestimate of $f(\mathbf{X}^{**})$. Finer division of the ranges for x_1 and x_2 (more mesh points) would increase the accuracy of the approximation—but also the number of variables in the linear programming problem. For example, if we let $t_1 = 2$, $t_2 = 4$ but divide both intervals into half-unit subintervals (0, 0.5, 1, 1.5, . . .), the solution to that approximating problem would be $\mathbf{X}^* = \begin{bmatrix} 1 \\ 3.46 \end{bmatrix}$ with $f(\mathbf{X}^*) = 9.92$. Finer interval divisions would get us closer to the exact optimum, but of course the point is that this technique will be used in cases when the exact optimum is not known, and so the tradeoff is

TABLE 11.3   Second Simplex Table for the Problem in (11.30)

		α_{11}	α_{21}	α_{31}	α_{41}	α_{12}	α_{22}	α_{32}	α_{02}
Π	8	3	6	9	12	-6	-4	-2	-8
s_1	0	-4	-16	-36	-64	15	12	7	16
α_{01}	1	-1	-1	-1	-1	0	0	0	0
α_{42}	1	0	0	0	0	-1	-1	-1	-1

between accuracy of the linear approximations (more mesh points are better) and size of the approximating problem (fewer mesh points are better).

11.3.2 Cutting-Plane Methods

Separable programming techniques rely on linear approximations to the nonlinearities in a programming problem, in either the objective function, the constraints, or both. Cutting-plane methods also use linear approximations; here the idea is to represent a feasible region defined by nonlinear constraints as the intersection of a number of *linear* inequalities that contain it. Figure 11.3 illustrates a two-variable maximization problem with a linear objective function (dashed-line contour) and one nonlinear constraint (along with nonnegativity). The lines c_1, c_2, and c_3 represent three artificially constructed boundaries; the area defined below them completely encloses the true feasible region (shaded). The point N represents the true optimum; L is the optimum to the linear program with the same objective and the three linear constraints c_1 through c_3.

The object is to create a sequence of linear constraints ("cutting planes," or lines in two dimensions), one at a time, so that L approaches N. Note that the solutions to the linear approximating problems will be infeasible in terms of the original nonlinear program; note also that in general they will always be only approximations. Hence a rule for stopping is usually formulated in terms of the amount by which the nonlinear constraints (in Fig. 11.3 there is only one) are violated. When L lies outside the original feasible region by no more than some prespecified small number in a particular solution, that solution is accepted as the approximation to the optimum of the original problem. We make this procedure specific in what follows.

Creating a Simple Linear Objective Function We mentioned only nonlinear *constraints*. This is because there is a clever and simple method for conversion of any programming problem with a nonlinear objective function *and* nonlinear constraints to a problem with a *linear* objective function and nonlinear constraints. Let the original problem be [as in (11.7)]

$$\text{Maximize} \quad f(\mathbf{X})$$
$$\text{subject to} \quad h^i(\mathbf{X}) \leq 0 \qquad (i = 1, \ldots, m) \qquad (11.31)$$
$$\text{and} \quad \mathbf{X} \geq 0$$

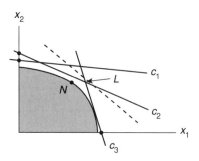

FIGURE 11.3 Linear inequalities containing a nonlinear feasible region.

Define a new variable, x_{n+1}, through the inequality

$$x_{n+1} \leq f(\mathbf{X}) \qquad \text{so that} \qquad -f(\mathbf{X}) + x_{n+1} \leq 0$$

Add this as the $(m+1)$st *constraint* in (11.31), and replace the objective function with x_{n+1}, giving

$$\begin{aligned}
&\text{Maximize} \quad x_{n+1} \\
&\text{subject to} \quad h^i(\mathbf{X}) \leq 0 \qquad (i = 1, \ldots, m) \\
&\qquad\qquad\quad -f(\mathbf{X}) + x_{n+1} \leq 0 \\
&\text{and} \quad \mathbf{X} \geq \mathbf{0}
\end{aligned} \tag{11.32}$$

This poses the same problem as (11.31) with a (very simple) linear objective function and one more nonlinear constraint. Thus it is adequate for a computational method to be concerned exclusively with nonlinearities in the constraints, as is the case for the cutting plane approach. If the objective function was originally linear, then the conversion from (11.31) to (11.32) is unnecessary; if it was not, we may assume that (11.31) represents the converted form. That is, we may think of $f(\mathbf{X})$ in (11.31) as either x_{n+1} or else as a linear function of the x_j's. If it is x_{n+1}, then we may assume that $-f(\mathbf{X}) + x_{n+1} \leq 0$ is included in the set of constraints in (11.31). (A similar transformation is possible for the general nonlinear programming *minimization* problem.)

To be convinced that the problems in (11.31) and (11.32) indeed lead to the same optimal solution, consider a specific example (this is from Section 4.5):

$$\begin{aligned}
&\text{Maximize} \quad -x_1^2 - x_2^2 + 4x_1 + 6x_2 \\
&\text{subject to} \quad x_1 + x_2 \leq 2
\end{aligned} \tag{11.33}$$

The Kuhn–Tucker conditions for this problem are:

$$\begin{aligned}
-2x_1 + 4 - \lambda &= 0 \\
-2x_2 + 6 - \lambda &= 0
\end{aligned} \tag{11.33$'$a}$$

$$\lambda(x_1 + x_2 - 2) = 0 \tag{11.33$'$b}$$

$$x_1 + x_2 \leq 2 \tag{11.33$'$c}$$

$$\lambda \geq 0 \tag{11.33$'$d}$$

It is easily established that only $\mathbf{X}^* = \begin{bmatrix} 0.5 \\ 1.5 \end{bmatrix}$ satisfies all these necessary conditions. Sufficiency is taken care of by the observations that the objective function is concave (its Hessian is $\begin{bmatrix} -2 & 0 \\ 0 & -2 \end{bmatrix}$) and the constraint, being linear, is convex.

Transforming this problem into the style of (11.32), we have

$$\text{Maximize} \quad x_3$$

$$\text{subject to} \quad x_1 + x_2 \leq 2 \tag{11.34}$$

$$\text{and} \quad x_1^2 + x_2^2 - 4x_1 - 6x_2 + x_3 \leq 0$$

The Kuhn–Tucker conditions for this problem are easily seen to be

$$0 - \lambda_1 - \lambda_2(2x_1 - 4) = 0$$

$$0 - \lambda_1 - \lambda_2(2x_2 - 6) = 0 \tag{11.34'a}$$

$$1 - 0 - \lambda_2 = 0$$

$$\lambda_1(x_1 + x_2 - 2) = 0$$

$$\lambda_2(x_1^2 + x_2^2 - 4x_1 - 6x_2 + x_3) = 0 \tag{11.34'b}$$

$$x_1 + x_2 \leq 2$$

$$x_1^2 + x_2^2 - 4x_1 - 6x_2 + x_3 \leq 0 \tag{11.34'c}$$

$$\lambda_1, \lambda_2 \geq 0 \tag{11.34'd}$$

From the third equation in (11.34'a), $\lambda_2 = 1$. Then, from the second condition in (11.34'b), it must be true that

$$-x_1^2 - x_2^2 + 4x_1 + 6x_2 = x_3$$

That is, x_3 will be exactly equal to $f(\mathbf{X})$ in the original problem, (11.33); the objective functions in the two problems are identical. Note that the second conditions in (11.34'b)–(11.34'd) are automatically satisfied, once it is established that $\lambda_2 = 1$.

The first two equations in (11.34'a) then become

$$-2x_1 + 4 - \lambda_1 = 0$$

$$-2x_2 + 6 - \lambda_1 = 0$$

and these are exactly the same as the Kuhn–Tucker requirements in (11.33'a) for the original problem, with the present $\lambda_1 = \lambda$ from before. And the remaining requirements in (11.34'b)–(11.34'd) are also exactly the same as those in (11.33') for the original problem, again with $\lambda_1 = \lambda$.

Therefore, the two problems have the same objective function and the same Kuhn–Tucker conditions on the original unknowns, $x_1, \ldots, x_n$. Points that satisfy (11.33') will be exactly the same as those that satisfy (11.34'). Sufficiency requirements for (11.34) are the same as those for (11.33), since the objective function in (11.34) is linear and hence concave and the added constraint will be convex as long as $-f(\mathbf{X})$ is convex, which means as long as $f(\mathbf{X})$ is concave. Therefore the two problems have identical optimal solutions. (You are asked to demonstrate this equivalence in several of the problems at the end of this chapter. The general nonlinear programming minimization problem can be similarly transformed.)

Creating the Cutting Planes   Returning to the cutting-plane approach, the essential issue is how to construct the cutting planes. It turns out that they are merely *linearizations* of the constraints expressed by using a first-order (linear) Taylor's series approximation at the current point (see Appendix 3.2).

As a concrete illustration, consider only one constraint, $h(\mathbf{X}) \leq 0$, that depends on two variables, as in Figure 11.3, so that the *function* $h(\mathbf{X})$ is representable in three-dimensional $x_1, x_2, h(\mathbf{X})$ space. Suppose, consistent with the maximization problem suggested in Figure 11.3, that $h(\mathbf{X})$ has the shape of a bowl, the bottom of which is directly below the origin [so the minimum value of $h(\mathbf{X})$ is negative]. Then, for values of x_1 and x_2 that satisfy the constraint (the shaded area in the nonnegative quadrant in Fig. 11.3), $h(\mathbf{X})$ will, indeed, be negative (the bowl drops below the x_1, x_2 plane).

Consider a feasible point, $\mathbf{X}^k$, where both $x_1 > 0$ and $x_2 > 0$ and where $h(\mathbf{X}^k) < 0$. Drop that point down to the surface of the bowl, below the x_1, x_2 plane; then construct a plane that is tangent to the bowl at that point, and observe the intersection of that plane with the x_1, x_2 plane. This tangent plane is exactly the first-order Taylor's series linearization of $h(\mathbf{X})$ at the point $\mathbf{X}^k$ (Chapter 3, Section 3.5). The intersection will be a straight line, since it is formed by one plane cutting another. In particular, the constraint function, $h(\mathbf{X}^k)$, would be approximated for $\mathbf{X}$ in the neighborhood of $\mathbf{X}^k$ by $h(\mathbf{X}) = h(\mathbf{X}^k) + [\nabla h(\mathbf{X}^k)]'(\mathbf{X} - \mathbf{X}^k)$ and so the linearization of the inequality constraint around $\mathbf{X}^k$ is

$$h(\mathbf{X}^k) + [\nabla h(\mathbf{X}^k)]'(\mathbf{X} - \mathbf{X}^k) \leq 0$$

The cutting-plane approach, for a problem with m nonlinear inequality constraints, proceeds as follows:

1. Find some initial point, $\mathbf{X}^0$; it need not even be feasible but suppose that it is.

2. Solve the following *linear* program, in which each constraint is replaced by its first-order Taylor's series approximation around $\mathbf{X}^0$:

$$\text{Maximize} \quad f(\mathbf{X})$$

$$\text{subject to} \quad h^i(\mathbf{X}^0) + [\nabla h^i(\mathbf{X}^0)]'(\mathbf{X} - \mathbf{X}^0) \leq 0 \quad (i = 1, \ldots, m)$$

$$\text{and} \quad \mathbf{X} \geq \mathbf{0}$$

Denote the optimal solution to this problem by $\mathbf{X}^1$.

3. Find the constraint that is most violated by $\mathbf{X}^1$—the $h^i(\mathbf{X}^1)$ that is max $[h^1(\mathbf{X}^1), \ldots, h^m(\mathbf{X}^1)]$. Linearize that constraint around $\mathbf{X}^1$ and add it to the set of m linear constraints (and also the nonnegativity requirements) already used in step 2. Solve the resulting linear program, and denote the optimal solution $\mathbf{X}^2$.

4. Suppose that a sequence of solutions, $\mathbf{X}^0, \mathbf{X}^1, \mathbf{X}^2, \ldots, \mathbf{X}^k$, has been found. If $h^i(\mathbf{X}^k) \leq \varepsilon$ [for all $i = 1, \ldots, m+(k-1)$], where ε is some predetermined (small) positive number, then the procedure can stop. If not, find among the constraints the one that is most violated at $\mathbf{X}^k$, linearize it around that point, add that converted constraint to the linear program

whose solution gave $\mathbf{X}^k$, and find the optimum, $\mathbf{X}^{k+1}$, to this new linear program with $m + k$ linear constraints and n nonnegativity requirements.

5. Continue until the solution values of the variables are "sufficiently close" to the feasible region of the original problem, as measured by a criterion such as ε in step 4.

This procedure is fine only when, for the original maximum problem (11.31), $f(\mathbf{X})$ is concave and all $h^i(\mathbf{X})$ are convex functions of $\mathbf{X}$. Recall that these are exactly the conditions under which Kuhn–Tucker necessary conditions are also sufficient for a maximum. Only then are we assured that the Taylor's series approximations of the constraints never remove any of the feasible region.

Example 3   We use the problem in (11.28) once again to illustrate the procedure:

$$\text{Maximize}\quad 3x_1 + 2x_2$$
$$\text{subject to}\quad 4x_1^2 + x_2^2 \le 16 \tag{11.28}$$
$$\text{and}\quad \mathbf{X} \ge \mathbf{0}$$

The convexity test for this single constraint is easily carried out; it is convex. Figure 11.4 illustrates the programming problem.

Steps in the solution procedure are:

1. Choose an $\mathbf{X}^0$; here, let $\mathbf{X}^0 = \begin{bmatrix} 1 \\ 1 \end{bmatrix}$. This is a feasible solution to (11.28), although feasibility is not a requirement for the initial value of $\mathbf{X}$.

2. For the first linear program, we need $h(\mathbf{X}^0)$ and $\nabla h(\mathbf{X}^0)$. Here $\nabla h(\mathbf{X}) = [8x_1 \quad 2x_2]'$, so $\nabla h(\mathbf{X}^0) = [8 \quad 2]'$ and $h(\mathbf{X}^0) = -11$. Therefore we want to

$$\text{Maximize}\quad 3x_1 + 2x_2$$
$$\text{subject to}\quad -11 + [8 \quad 2]\begin{bmatrix} x_1 - 1 \\ x_2 - 1 \end{bmatrix} \le 0$$
$$\text{which works out to be}\quad 8x_1 + 2x_2 \le 21$$
$$\text{and}\quad \mathbf{X} \ge \mathbf{0}$$

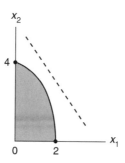

FIGURE 11.4   Feasible region and objective function contour for maximization problem (11.28).

In Figure 11.5, we add this cutting line to the original problem as it appeared in Figure 11.4. It is labeled c_1, the first of several added constraints. Figure 11.5 makes clear that the optimal solution will be at the point where c_1 cuts the x_2 axis, at $x_1 = 0$, $x_2 = 10.5$. (The simplex method would produce this answer in two iterations, since the larger coefficient on x_1 in the objective function would lead one to introduce it first; it is not needed here since we have the geometry in front of us.) Thus $\mathbf{X}^1 = \begin{bmatrix} 0 \\ 10.5 \end{bmatrix}$.

3. Let $\varepsilon = 0.0005$. Since $h(\mathbf{X}^1) = 94.25$, we appear to have a long way to go toward meeting our stopping criterion, and so we add a second constraint. We find that $\nabla h(\mathbf{X}^0) = \begin{bmatrix} 0 \\ 21 \end{bmatrix}$, and so the constraint to be added next is the linearization of $h(\mathbf{X})$ at the current point, $\mathbf{X}^1$, which is

$$94.25 + [0 \quad 21]\begin{bmatrix} x_1 - 0 \\ x_2 - 10.5 \end{bmatrix} \le 0$$

or $21x_2 \le 126.25$. The next linear program to be solved is therefore

$$\text{Maximize} \quad 3x_1 + 2x_2$$

$$\text{subject to} \quad \begin{cases} (c_1)\ 8x_1 + 2x_2 \le 21 \\ (c_2)\ x_2 \le 6.0119 \end{cases}$$

$$\text{and} \quad \mathbf{X} \ge \mathbf{0}$$

The new constraint, c_2, is added to the previous problem in Figure 11.6. In the solutions that follow we will show four places to the right of the decimal. Again, the solution

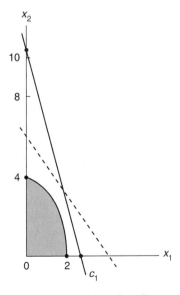

FIGURE 11.5   First cutting-plane linear program.

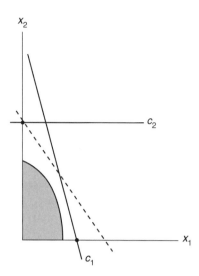

FIGURE 11.6   Second cutting-plane linear program.

is clear from the picture; it will now be at the corner defined by the intersection of c_1 and c_2, where $x_2 = 6.0119$ and $x_1 = 1.1220$. So $\mathbf{X}^2 = \begin{bmatrix} 1.1220 \\ 6.0119 \end{bmatrix}$ and, since $h(\mathbf{X}^2) = 25.1784$ ($> \varepsilon$), we continue.

The information developed during the sequence of calculations is shown in Table 11.4. Each new linear program contains all previous constraints plus one new one. The first seven of these cutting line constraints are shown in Figure 11.7; remember that they have been added one at a time to generate a *series* of linear programs of increasing size.

The calculations were stopped at $k = 12$, where $\mathbf{X}^{12} = \begin{bmatrix} 1.2013 \\ 3.1981 \end{bmatrix}$, because $h(\mathbf{X}^{12}) = 0.000329$. This is the first $\mathbf{X}^k$ for which $h(\mathbf{X}^k) < \varepsilon$, which we arbitrarily set at 0.0005. At this point, $f(\mathbf{X}^{12}) = 10.0001$. From earlier observations, we know that the exact solution to this problem is at $\mathbf{X}^* = \begin{bmatrix} 1.2 \\ 3.2 \end{bmatrix}$, where $f(\mathbf{X}^*) = 10$.

Note in Table 11.4 (and Fig. 11.7) that each new constraint becomes increasingly similar to its predecessor as k increases, and the right-hand side appears to approach 32. As the optimal solution to the linear programming approximation problem approaches the solution to the nonlinear programming problem, the cutting lines naturally become closer and closer to the linearization that would be obtained at the exact solution. For this problem, with $\mathbf{X}^* = \begin{bmatrix} 1.2 \\ 3.2 \end{bmatrix}$, the first-order Taylor's series would give

$$9.6x_1 + 6.4x_2 \leq 32$$

(If you want to add a few more cutting lines to the results in Table 11.4, you will see how they approach this constraint.)

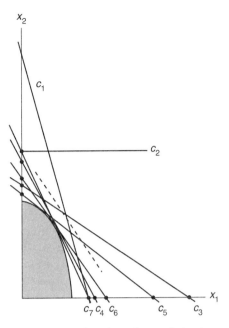

FIGURE 11.7   First seven cutting planes for maximization problem (11.28).

11.3.3   Sequential Unconstrained Techniques

Both cutting planes and separable programming approaches employ one or more approximating *linear* programming problems whose solutions can then be found by the simplex method. Sequential unconstrained techniques are based on an entirely different kind of approximation; rather than formulating linear programs, they use a series of *unconstrained nonlinear* maximization or minimization problems. In parallel with the original references, we explore a nonlinear minimization problem. The full title of this technique is *sequential unconstrained minimization technique or, as it is often denoted, SUMT.*[13] This is exactly the approach that is used in many of the path-following interior point methods for linear programs (Chapter 9). In fact, this method, with its use of *penalty* or *barrier functions*, provided the inspiration, albeit with a time lag, for much of the recent work on path-following techniques for both linear and nonlinear programs.

The major *computational* focus, then, is on techniques for unconstrained optimization, and the calculus-based approaches of Chapters 3 and 4, as well as the iterative methods of Chapter 6, are all potentially useful. Our initial interest is in how the unconstrained problem is formulated.

Creating the Unconstrained Problem   Consider the general nonlinear programming minimization problem in which the nonnegativity requirements on the variables have been incorporated in the constraint set:

---

[13]Sequential unconstrained *maximization* techniques could equally well be known as "SUMT." The work of Fiacco and McCormick, who are chiefly responsible for development of the method, concentrated on minimization problems [see Fiacco and McCormick (1968) for a full presentation of this work].

TABLE 11.4 Iterations in the Cutting-Plane Approach to Problem (11.28)

k	X^k	$\nabla h(X^k)$	$h(X^k)$	Final Form of Added Constraint $\{h(X^k) + [\nabla h(X^k)]'(X - X^k) \le 0\}$	Pair of Constraints Met as Equalities in Linear Program Solution	Linear Program Optimal Solution ($= X^{k+1}$)	Value of Objective Function at Current Solution [$f(X^k)$]
0	$\begin{bmatrix}1\\1\end{bmatrix}$	$\begin{bmatrix}8\\2\end{bmatrix}$	-11	$8x_1 + 2x_2 \le 21$	c_1 and x_2 axis	$\begin{bmatrix}0\\10.5\end{bmatrix}$	21
1	$\begin{bmatrix}0\\10.5\end{bmatrix}$	$\begin{bmatrix}0\\21\end{bmatrix}$	94.25	$x_2 \le 6.0119$	c_2 and c_1	$\begin{bmatrix}1.1220\\6.0119\end{bmatrix}$	15.3899
2	$\begin{bmatrix}1.1220\\6.0119\end{bmatrix}$	$\begin{bmatrix}8.9760\\12.0238\end{bmatrix}$	25.1784	$8.9760x_1 + 12.0238x_2 \le 57.1785$	c_3 and c_1	$\begin{bmatrix}1.7657\\3.4373\end{bmatrix}$	12.1716
3	$\begin{bmatrix}1.7657\\3.4373\end{bmatrix}$	$\begin{bmatrix}14.1256\\6.8746\end{bmatrix}$	8.2858	$14.1256x_1 + 6.8746x_2 \le 40.2853$	c_4 and c_3	$\begin{bmatrix}0.8442\\4.1252\end{bmatrix}$	10.7829
4	$\begin{bmatrix}0.8442\\4.1252\end{bmatrix}$	$\begin{bmatrix}6.7536\\8.2504\end{bmatrix}$	3.8680	$6.7536x_1 + 8.2504x_2 \le 35.8680$	c_5 and c_4	$\begin{bmatrix}1.2234\\3.3460\end{bmatrix}$	10.3622
5	$\begin{bmatrix}1.2234\\3.3460\end{bmatrix}$	$\begin{bmatrix}9.7872\\6.6920\end{bmatrix}$	1.1825	$9.7872x_1 + 6.6920x_2 \le 33.1826$	c_6 and c_4	$\begin{bmatrix}1.5218\\2.7330\end{bmatrix}$	10.0312
6	$\begin{bmatrix}1.5218\\2.7330\end{bmatrix}$	$\begin{bmatrix}12.1744\\5.4660\end{bmatrix}$	0.7328	$12.1744x_1 + 5.4660x_2 \le 32.7328$	c_7 and c_6	$\begin{bmatrix}1.3468\\2.9888\end{bmatrix}$	10.0181
7	$\begin{bmatrix}1.3468\\2.9888\end{bmatrix}$	$\begin{bmatrix}10.7744\\5.9776\end{bmatrix}$	0.1884	$10.7744x_1 + 5.9776x_2 \le 32.1884$	c_8 and c_6	$\begin{bmatrix}1.2536\\3.1251\end{bmatrix}$	10.0111
8	$\begin{bmatrix}1.2536\\3.1251\end{bmatrix}$	$\begin{bmatrix}10.0288\\6.2502\end{bmatrix}$	0.0523	$10.0288x_1 + 6.2502x_2 \le 32.0523$	c_9 and c_6	$\begin{bmatrix}1.1949\\3.2109\end{bmatrix}$	10.0067
9	$\begin{bmatrix}1.1949\\3.2109\end{bmatrix}$	$\begin{bmatrix}9.5592\\6.4218\end{bmatrix}$	0.0210	$9.5592x_1 + 6.4218x_2 \le 32.0210$	c_{10} and c_9	$\begin{bmatrix}1.2239\\3.1645\end{bmatrix}$	10.0006
10	$\begin{bmatrix}1.2239\\3.1645\end{bmatrix}$	$\begin{bmatrix}9.7912\\6.3290\end{bmatrix}$	0.0058	$9.7912x_1 + 6.3290x_2 \le 32.0058$	c_{11} and c_{10}	$\begin{bmatrix}1.2091\\3.1865\end{bmatrix}$	10.0003
11	$\begin{bmatrix}1.2091\\3.1865\end{bmatrix}$	$\begin{bmatrix}9.6728\\6.3730\end{bmatrix}$	0.0015	$9.6728x_1 + 6.3730x_2 \le 32.0015$	c_{12} and c_{10}	$\begin{bmatrix}1.2013\\3.1981\end{bmatrix}$	10.0001
12	$\begin{bmatrix}1.2013\\3.1981\end{bmatrix}$	$\begin{bmatrix}9.6104\\6.3962\end{bmatrix}$	0.0003	Stop, since $h(X^k) < \varepsilon = 0.0005$			

Minimize $f(\mathbf{X})$

subject to $h^i(\mathbf{X}) \geq 0$ $(i = 1, \ldots, m; m+1, \ldots, m+n)$ (11.35)

From Section 4.5, we know that the Kuhn–Tucker conditions necessary for a minimum of $f(\mathbf{X})$ subject to the constraints in (11.35) are that an $\mathbf{X}^* = [x_j^*]$ ($j = 1, \ldots, n$) and a $\mathbf{\Lambda}^* = [\lambda_i^*]$ ($i = 1, \ldots, m+n$) be found to satisfy

$$\nabla f - \sum_i \lambda_i \nabla h^i = \mathbf{0} \tag{11.36a}$$

$$\lambda_i[h^i(\mathbf{X})] = 0 \tag{11.36b}$$

$$h^i(\mathbf{X}) \geq 0 \tag{11.36c}$$

$$\lambda_i \geq 0 \tag{11.36d}$$

Suppose that there is a point near $\mathbf{X}^*, \mathbf{\Lambda}^*$—call it $[\mathbf{X}(\mu), \mathbf{\Lambda}(\mu)]$, since it will depend on a parameter μ—for which the following *perturbation* of (11.36) holds:

No change   (11.36′a)

$^{-}\lambda_i[h^i(\mathbf{X})] = \mu > 0$ (11.36′b)

$h^i(\mathbf{X}) > 0$ (11.36′c)

No change   (11.36′d)

From (11.36′c) we recognize that the point $[\mathbf{X}(\mu), \mathbf{\Lambda}(\mu)]$ is *strictly inside* the feasible region for the problem in (11.35).

In addition, imposition of the new requirement (11.36′b) in means that, in (11.36′d), $\lambda_i > 0$, since $\lambda_i = \mu/h^i(\mathbf{X})$, where both numerator and denominator are strictly positive. Putting this result for λ_i into (11.36a), we obtain

$$\nabla f[\mathbf{X}(\mu)] - \sum_i \left\{ \frac{\mu}{h^i[\mathbf{X}(\mu)]} \right\} \nabla h^i[\mathbf{X}(\mu)] = \mathbf{0} \tag{11.37}$$

Now the important observation is that (11.37) is precisely the *gradient* of the function $L(\mathbf{X};\mu)$, where

$$L(\mathbf{X};\mu) = f(\mathbf{X}) - \mu \sum_i \ln h^i(\mathbf{X}) \tag{11.38}$$

Thus, at $\mathbf{X}(\mu)$, $\nabla L(\mathbf{X};\mu) = \mathbf{0}$, by (11.37), and $\mathbf{X}(\mu)$ satisfies the first-order conditions for a local (unconstrained) minimum of $L(\mathbf{X};\mu)$; we denote it $\mathbf{X}^*(\mu)$ to indicate that it is a point satisfying these conditions.

Different perturbations of the Kuhn–Tucker conditions in (11.36) will generate other unconstrained minimization problems. For example, if we define $\ell_i^2 = \lambda_i$, then (11.36d) is automatically satisfied. Then (11.36b) is equivalent to $\ell_i[h^i(\mathbf{X})] = 0$, and if we now perturb (11.36b) and (11.36c) as before, we have

$$\text{No change} \tag{11.36''a}$$

$$\ell_i[h^i(\mathbf{X})] = \mu > 0 \tag{11.36''b}$$

$$h^i(\mathbf{X}) > 0 \tag{1.36''c}$$

From these conditions, $\ell_i = \mu/h^i(\mathbf{X})$, and again both numerator and denominator are strictly positive. Then $\lambda_i = \mu^2/[h^i(\mathbf{X})]^2$, and this, substituted into (11.36a), gives

$$\nabla f[\mathbf{X}(\mu)] - \sum_i \left[\frac{\mu^2}{\{h^i[\mathbf{X}(\mu)]\}^2} \right] \nabla h^i[\mathbf{X}(\mu)] = \mathbf{0} \tag{11.39}$$

which is just the gradient of $\mathscr{L}(\mathbf{X};\mu)$, where

$$\mathscr{L}(\mathbf{X};\mu) = f(\mathbf{X}) + \mu^2 \sum_i \left[\frac{1}{h^i(\mathbf{X})} \right] \tag{11.40}$$

And, as above, since $\nabla L(\mathbf{X};\mu) = \mathbf{0}$ at $\mathbf{X}(\mu)$, by (11.39), $\mathbf{X}(\mu)$ satisfies the first-order conditions for an unconstrained local minimum of $\mathscr{L}(\mathbf{X};\mu)$. Again, then, the notation $\mathbf{X}^*(\mu)$ is appropriate.[14]

The functions $L(\mathbf{X};\mu)$ or $\mathscr{L}(\mathbf{X};\mu)$ provide the foundations for an *unconstrained* approach to the general nonlinear programming minimization problem. The basic idea is to define a *sequence* of functions, $L(\mathbf{X};\mu)$ [or $\mathscr{L}(\mathbf{X};\mu)$], through systematic variation of the parameter μ and to find the unconstrained minima of each of these functions. The functions are known as "penalty" or "barrier" functions (as in Chapter 9), and hence the approach is also known as the *penalty* or *barrier function approach*. The first, $L(\mathbf{X};\mu)$, is an example of a logarithmic penalty function; the second, $\mathscr{L}(\mathbf{X};\mu)$, is sometimes referred to as an inverse penalty function. There are other types of penalty functions also, such as quadratic. (Variations of this type of approach sometimes take a more positive attitude and construct what are known as "merit functions.")

For $L(\mathbf{X};\mu)$ in (11.38), as one or more of the $h^i(\mathbf{X}) \to 0$ (as $\mathbf{X}$ approaches the boundary set by one or more of the constraints), $\ln h^i(\mathbf{X}) \to -\infty$ and therefore, for $\mu > 0$, $-\mu \sum_i \ln h^i(\mathbf{X}) \to +\infty$. So an $\mathbf{X}$ that is on a boundary defined by one or more of the constraints could not possibly be a minimizer of $L(\mathbf{X};\mu)$, since the penalty for being on the boundary is to add an enormous amount to the value of the function that is being minimized. Similarly, for $\mathscr{L}(\mathbf{X};\mu)$ in (11.40), $\mu^2 \sum_i [1/h^i(\mathbf{X})] \to +\infty$ as one or more of the $h^i(\mathbf{X}) \to 0$, and this again prevents an $\mathbf{X}$ on the boundary from being a local minimizer for $\mathscr{L}(\mathbf{X};\mu)$.

The amount of the penalty that is added to the value of the penalty function is determined by μ. We know from (11.36'c) or (11.36''c) that all the constraints of the original nonlinear program are *strictly* satisfied at this minimum point. When μ is very small, the weight given to avoiding the boundary is also very small, and the local unconstrained

---

[14]Questions regarding the existence of an $\mathbf{X}(\mu)$ satisfying the perturbed conditions (11.36') or (11.36''), as well as conditions for positive definiteness of the Hessians of $L(\mathbf{X};\mu)$ and $\mathscr{L}(\mathbf{X};\mu)$, to ensure that the second total differential is positive (for a minimum point), are covered in detail in Fiacco and McCormick (1968).

minimum will be close to the constrained minimum of (11.35), since the perturbation in (11.36′) or (11.36″) will be small.

The steps in this procedure are:

1. Choose an initial value for μ (= $\mu_1 > 0$)[15] and find $\mathbf{X}^*(\mu_1)$, the local minimum of $L(\mathbf{X};\mu_1)$ or $\mathscr{L}(\mathbf{X};\mu_1)$.
2. Move from $\mathbf{X}^*(\mu_1)$ to $\mathbf{X}^*(\mu_2)$, the local minimum of $L(\mathbf{X};\mu_2)$ [or $\mathscr{L}(\mathbf{X};\mu_2)$], where $\mu_1 > \mu_2 > 0$.
3. Continue moving from $\mathbf{X}^*(\mu_{k-1})$ to $\mathbf{X}^*(\mu_k)$, the local minimum of $L(\mathbf{X};\mu_k)$ [or $\mathscr{L}(\mathbf{X};\mu_k)$], where $\mu_1 > \mu_2 > \cdots > \mu_{k-1} > \mu_k > \cdots > 0$. Stop when some termination criterion is satisfied.

As usual, stopping criteria could be based on differences between successive values of $\mathbf{X}^*(\mu_{k-1})$ and $\mathbf{X}^*(\mu_k)$ or of $L(\mathbf{X}^*;\mu_{k-1})$ and $L(\mathbf{X}^*;\mu_k)$ [or of $\mathscr{L}(\mathbf{X}^*;\mu_{k-1})$ and $\mathscr{L}(\mathbf{X}^*;\mu_k)$]. Also, differences in L (or $\mathscr{L}$) could be normalized to take account of the current size of the function; for example

$$\frac{\Delta L_k}{0.5[L(\mathbf{X}^*;\ \mu_{k-1}) + L(\mathbf{X}^*;\ \mu_k)]}$$

where $\Delta L_k = L(\mathbf{X}^*;\mu_k) - L(\mathbf{X}^*;\mu_{k-1})$. This means that if $\Delta L_k = 10$, you would be more inclined to stop if the average of the current and previous values were 250,000 than if it were 25. These kinds of rules are the same as those that are used in general in iterative methods (Chapters 5, 6, and 9).

Another possible rule might be to stop when the value of the objective function, $f[\mathbf{X}^*(\mu_k)]$, deviates by less than some preselected small ε from the theoretical true minimum of $f(\mathbf{X})$. The duality theory of Section 11.2 provides guidance for a criterion of this sort. Since (11.35) is a minimization problem, its objective function was denoted by Δ in Section 11.2. Let the optimal value of Δ be given by $f(\mathbf{X}^\circ)$. The dual to (11.35) will be a maximization problem in which the constraints appear as *equalities,* since nonnegativity is not specified explicitly in (11.35):

$$\text{Maximize} \quad \Pi(\mathbf{X},\mathbf{Y}) = f(\mathbf{X}) - \mathbf{Y}'[\mathbf{h}(\mathbf{X})]$$

$$\text{subject to} \quad \mathbf{J}'\mathbf{Y} - \nabla f = \mathbf{0} \tag{11.41}$$

$$\text{and} \quad \mathbf{Y} \geq \mathbf{0}$$

The final term from the dual objective function (11.11) is absent here because the dual constraints assure that it is zero. We know that $f(\mathbf{X}) \geq \Pi(\mathbf{X},\mathbf{Y})$ for any feasible solutions $\mathbf{X}$ and $\mathbf{Y}$ to the two problems and, at optimum, $f(\mathbf{X}^\circ) = \Pi(\mathbf{X}^\circ, \mathbf{Y}^\circ)$. Thus $f(\mathbf{X}) \geq f(\mathbf{X}^\circ) \geq \Pi(\mathbf{X},\mathbf{Y})$; for the particular feasible solution involving $\mathbf{X}(\mu_k)$ and corresponding additional dual variables $\mathbf{Y}(\mu_k)$:

$$f[\mathbf{X}(\mu_k)] \geq f(\mathbf{X}^\circ) \geq \Pi[\mathbf{X}(\mu_k), \mathbf{Y}(\mu_k)] \tag{11.42}$$

---

[15]There are subtle ways of choosing an efficient initial value, μ_1, for the penalty weight, but these details need not concern us.

We want to stop calculations when

$$f[\mathbf{X}(\mu_k)] - f(\mathbf{X}^\circ) < \varepsilon \tag{11.43}$$

But from (11.42), if

$$f[\mathbf{X}(\mu_k)] - \Pi[\mathbf{X}(\mu_k), \mathbf{Y}(\mu_k)] < \varepsilon \tag{11.44}$$

then (11.43) will hold. The left-hand side is just the difference between primal and dual objective function values for "current" solutions $\mathbf{X}(\mu_k)$ and $[\mathbf{X}(\mu_k), \mathbf{Y}(\mu_k)]$; from the dual objective function this is just $[\mathbf{Y}(\mu_k)]'\mathbf{h}[\mathbf{X}(\mu_k)]$, or, explicitly

$$\sum_i y_i(\mu_k) h^i[\mathbf{X}(\mu_k)] \tag{11.45}$$

All that remains is to establish the feasibility of $\mathbf{Y}(\mu_k)$. We know that $\mathbf{X}(\mu_k)$ is feasible for the primal because of the way in which it was obtained. Feasibility in the dual requires that the constraints of (11.41) be met.

Consider first the form of penalty function illustrated in (11.40). In (11.39), which is always true for *feasible* $\mathbf{X}(\mu_k)$, if

$$y_i(\mu_k) = \frac{\mu_k^2}{\{h^i[\mathbf{X}(\mu_k)]\}^2} \tag{11.46}$$

we are assured of the feasibility of the $y_i(\mu_k)$ in the dual programming problem (11.41). (Nonnegativity is also assured, since both numerator and denominator are squared.) So the stopping criterion can be made explicit. We will be satisfied with the solution to the kth unconstrained minimization problem as soon as we have

$$\mu_k^2 \sum_i \frac{1}{h^i[\mathbf{X}(\mu_k)]} < \varepsilon \tag{11.47}$$

A similar argument, using the logarithmic form of the penalty function in (11.38), gives a stopping criterion that is independent of $h^i[\mathbf{X}(\mu)]$—namely, that one should stop when $\mu_k < \varepsilon$. (You might work through the argument to see why this is the case. There is a parallel in the path-following methods in Chapter 9.)

The procedures described above, for logarithmic or inverse penalty functions, are "interior point" methods, since the solutions $\mathbf{X}^*(\mu_1), \ldots, \mathbf{X}^*(\mu_n)$ are always inside the feasible region. The constraints in the original problem are $h^i(\mathbf{X}) \geq 0$; along with $\lambda_i > 0$, the perturbations $\lambda_i h^i[\mathbf{X}(\mu_k)] = \mu_k \neq 0$ ensure that $h^i[\mathbf{X}(\mu_k)] > 0$ and $\mu_k > 0$.

Consider a nonlinear programming *maximization* problem with only one constraint and with nonnegativity ignored, for simplicity:

$$\text{Maximize} \quad f(\mathbf{X})$$

$$\text{subject to} \quad h(\mathbf{X}) \leq 0$$

The parallel to the approach in (11.36′), now for a maximization problem, would use the following perturbation of standard Kuhn–Tucker conditions:

$\quad$ (a) $\nabla f - \lambda[\nabla h] = \mathbf{0}$

$\quad$ (b) $\lambda[h(\mathbf{X})] = \mu \neq 0$

$\quad$ (c) $h(\mathbf{X}) < 0$

$\quad$ (d) $\lambda > 0$

These differ from (11.36′) in that the sign of μ is not initially specified in (b) and the direction of the constraint in (c) reflects a maximization problem. The logarithmic penalty function for this problem would have the structure of (11.38); for only one constraint, this is

$$L(\mathbf{X};\mu) = f(\mathbf{X}) - \mu \ln h(\mathbf{X})$$

Since $\lambda > 0$ [from (d)] and $h(\mathbf{X}) < 0$ [from (c)], $\mu < 0$; however, the natural logarithm of a negative number does not exist (more correctly, it requires imaginary numbers) and so the $[-\mu \ln h(\mathbf{X})]$ term would cause mathematical problems.

$\quad$ One approach would simply convert the maximization problem to one of minimization through multiplication of the objective function and constraint(s) by -1. Then the results in (11.36′), for a logarithmic penalty function approach to a minimization problem, would apply directly to the converted (minimization) problem. An alternative would be to try to impose the conditions in (11.36′) for a minimization problem directly on the maximization problem. Since there is only one constraint, we have

$\quad$ (a) $\nabla f - \lambda[\nabla h] = \mathbf{0}$

$\quad$ (b) $\lambda[h(\mathbf{X})] = \mu > 0$

$\quad$ (c) $h(\mathbf{X}) > 0$

$\quad$ (d) $\lambda > 0$

This creates what is known as an "exterior point" method, since imposition of $h(\mathbf{X}) > 0$ ensures that the original constraint, $h(\mathbf{X}) \leq 0$, is violated. The logarithmic penalty function still has the form shown in (11.38), and the problem of logarithms of negative numbers is removed. This approach is also illustrated below.

Example 4 $\quad$ We turn again to the same example problem in (11.28). We ignore the non-negativity requirements, only to simplify the algebra and calculations for the illustration. To apply the interior point method, we multiply $f(\mathbf{X})$ and the constraint by -1 and hence work with the equivalent minimization problem:

$$\text{Minimize} \quad -f(\mathbf{X}) = -3x_1 - 2x_2$$

$$\text{subject to} \quad -4x_1^2 - x_2^2 + 16 \geq 0$$

$\hfill$ (11.48)

Using the logarithmic penalty function in (11.38), we have

$$L(\mathbf{X};\mu) = -3x_1 - 2x_2 - \mu \ln (-4x_1^2 - x_2^2 + 16) \hfill (11.49)$$

Results for the unconstrained minimization of this function, for a sequence of values of μ, are given in the upper panel of Table 11.5. The results in the lower panel of the table were derived using an exterior point approach, in which (11.36'b) and (11.36'c) are imposed *directly* on the maximization problem in (11.28), and again ignoring nonnegativity constraints, for simplicity. The last column in the table contains the value of the stopping criterion that was given in (11.47). Results, as usual, are shown to four decimal places; a higher degree of accuracy would delay the emergence of the problems created by $\ln(0)$ and division by 0.

For either of these versions of the problem, the direct calculus techniques of Chapter 3 could be applied. Taking first partial derivatives of the minimization version of the problem,

$$\frac{\partial L}{\partial x_1} = -3 + \frac{8\mu x_1}{-4x_1^2 - x_2^2 + 16}$$

$$\frac{\partial L}{\partial x_2} = -2 + \frac{2\mu x_2}{-4x_1^2 - x_2^2 + 16}$$

TABLE 11.5   Illustrations of SUMT Interior Point and Exterior Point Methods

Interior Point[a]

μ	x_1	x_2	L	$h(\mathbf{X})$	$f(\mathbf{X})$	$\mu^2/h(\mathbf{X})$
1.0	1.0860	2.8960	−10.1131	2.8956	−9.0500	0.3454
0.75	1.1134	2.9690	−9.8785	2.2264	−9.2782	0.2526
0.50	1.1415	3.0440	−9.7225	1.5220	−9.5125	0.1643
0.25	1.1704	3.1210	−9.6911	0.7800	−9.7532	0.0801
0.10	1.1881	3.1682	−9.7574	0.3162	−9.9007	0.0316
0.005	1.1994	3.1984	−9.9743	0.0160	−9.9950	0.0016
0.001	1.1999	3.1997	−9.9932	0.0029	−9.9991	0.0003
0.0005	1.1999	3.1998	−9.9962	0.0022	−9.9993	0.0001
0.0001	1.2000	3.2000	—[b]	0	−10.0000	—[c]

Exterior Point[d]

μ	x_1	x_2	L	$h(\mathbf{X})$	$f(\mathbf{X})$	$\mu^2/h(\mathbf{X})$
1.0	1.3260	3.5360	9.7869	3.5364	11.0500	0.2828
0.75	1.2934	3.4490	10.0652	2.5871	10.7782	0.2174
0.50	1.2615	3.3640	10.2525	1.6820	10.5125	0.1483
0.25	1.2304	3.2810	10.3026	0.8205	10.2532	0.0762
0.10	1.2121	3.2322	10.2134	0.3239	10.1007	0.0309
0.005	1.2006	3.2016	10.0256	0.0160	10.0050	0.0016
0.001	1.2001	3.2003	10.0067	0.0029	10.0009	0.0003
0.0005	1.2001	3.2002	10.0037	0.0022	10.0007	0.0001
0.0001	1.2000	3.2000	—[b]	0	10.0000	—[c]

[a]Minimize $f(\mathbf{X}) = -3x_1 - 2x_2$, subject to $h(\mathbf{X}) = -4x_1^2 - x_2^2 + 16 \geq 0$ [$L(\mathbf{X}; \mu) = -3x_1 - 2x_2 - \mu \ln (-4x_1^2 - x_2^2 + 16)$].
[b]$\ln (0)$ is required
[c]Division by 0 is required.
[d]Maximize $f(\mathbf{X}) = 3x_1 + 2x_2$, subject to $h(\mathbf{X}) = 4x_1^2 + x_2^2 - 16 \leq 0$ [$L(\mathbf{X}; \mu) = 3x_1 + 2x_2 - \mu \ln (4x_1^2 + x_2^2 - 16)$].

Setting these equal to zero and solving simultaneously gives

$$x_1 = \frac{-6\mu \pm \sqrt{36\mu^2 + 3600}}{50}$$

and

$$x_2 = \frac{-16\mu \pm \sqrt{256\mu^2 + 25600}}{50}$$

(11.50)

It is possible to show (using computer software makes the task much easier) the conditions under which the Hessian matrix for (11.49) is positive definite (assuring a minimum) as long as x_1 and x_2 have the same sign (and $\mu > 0$). We could thus compensate for the omission of the nonnegativity constraints by using the positive square roots in (11.50). As $\mu \to 0$, and using the positive roots, it is clear that $x_1 \to 1.2$ and $x_2 \to 3.2$. (Explicit inclusion of the two nonnegativity requirements creates a problem whose algebra is much more complicated. This is left as a exercise for anyone who is interested.)

11.4 QUADRATIC PROGRAMMING

For "quadratic" nonlinear programming problems, it turns out that an extension of the simplex method can be used to arrive at an exact solution. This is an advantage that quadratic programs have over general nonlinear problems. It is sometimes useful, in cases where nonlinearities are clearly present and the restrictions of a linear programming framework would be too severe, to consider approximating an inherently nonlinear program with a quadratic one.

11.4.1 Structure

In a quadratic program, all constraints are linear and the objective function is quadratic. This means that variables in $f(\mathbf{X})$ may be squared and pairs of variables may appear multiplied together (a quadratic form); there may also be a linear part and a constant term. The general quadratic programming maximization problem is

$$\text{Maximize} \quad f(\mathbf{X}) = \sum_{j=1}^{n} \sum_{k=1}^{n} p_{jk} x_j x_k + \sum_{j=1}^{n} p_j x_j + c$$

$$\text{subject to} \quad \sum_{j=1}^{n} a_{ij} x_j \le b_i \quad (i = 1, \ldots, m)$$

$$\text{and} \quad x_j \ge 0 \quad (j = 1, \ldots, n)$$

(11.51)

where c represents any constant. The general matrix representation of a quadratic function was given in Section 6.2 of Chapter 6 in the discussion of conjugate direction methods for maximization and minimization. Using $\mathbf{Q} = [q_{jk}]$ and $\mathbf{P} = [p_1, \ldots, p_n]'$, the problem in (11.51) is

$$\text{Maximize} \quad f(\mathbf{X}) = \tfrac{1}{2}\mathbf{X}'\mathbf{QX} + \mathbf{P}'\mathbf{X} + c$$

$$\text{subject to} \quad \mathbf{AX} \le \mathbf{B} \quad \text{or} \quad \mathbf{h}(\mathbf{X}) = \mathbf{AX} - \mathbf{B} \le \mathbf{0}$$

$$\text{and} \quad \mathbf{X} \ge \mathbf{0}$$

(11.52)

where $\mathbf{Q}$ is a symmetric matrix. This is always true for quadratic functions but involves some rewriting of the original coefficients, so we distinguish the elements q_{jk} from the original p_{jk} in (11.51). As an example, $\Sigma_{k=1}^n \Sigma_{j=1}^n p_{jk}x_jx_k = x_1^2 + 6x_1x_2 + 4x_2^2$ can be expressed in many alternative ways, including

$$\mathbf{X}'\begin{bmatrix} 1 & 6 \\ 0 & 4 \end{bmatrix}\mathbf{X} = \mathbf{X}'\begin{bmatrix} 1 & 0 \\ 6 & 4 \end{bmatrix}\mathbf{X} = \mathbf{X}'\begin{bmatrix} 1 & 3 \\ 3 & 4 \end{bmatrix}\mathbf{X} = \tfrac{1}{2}\mathbf{X}'\begin{bmatrix} 2 & 6 \\ 6 & 8 \end{bmatrix}\mathbf{X}$$

and in this illustration $\mathbf{Q} = \begin{bmatrix} 2 & 6 \\ 6 & 8 \end{bmatrix}$. The $\tfrac{1}{2}\mathbf{X}'\mathbf{Q}\mathbf{X}$ form provides a useful mathematical convenience; if $f(\mathbf{X}) = \tfrac{1}{2}\mathbf{X}'\mathbf{Q}\mathbf{X} + \mathbf{P}'\mathbf{X} + c$, then the Hessian matrix of $f(\mathbf{X})$ is just $\mathbf{Q}$.

Since the linear inequalities of the problem define a convex feasible region, we need to require only concavity or quasiconcavity of the objective function to be assured that a local maximum is also the global one. The linear part, $\mathbf{P}'\mathbf{X}$, does not alter the basic functional shape; hence $f(\mathbf{X})$ is (quasi)concave if its quadratic part is (quasi)concave. Geometrically, for the case of two variables, we are looking for the highest point on an inverted bowl or a bell above a convex feasible region whose boundaries are straight lines. A negative semidefinite (negative definite) quadratic form $\mathbf{X}'\mathbf{Q}\mathbf{X}$ is a concave (strictly concave) function of $\mathbf{X}$, so if $\mathbf{Q}$ is negative definite or semidefinite, $f(\mathbf{X})$ will be strictly concave or concave.

In examining the Kuhn–Tucker necessary conditions for a problem with this structure, it turns out to be useful to use (10.6′a)–(10.6′c) and (10.7) in Chapter 10 (even though they include the λ_{m+j}'s associated with nonnegativity constraints), because the conditions in (10.6′a) take the form of *equations* rather than inequalities [as they are in (10.8) after removal of those λ's]. In matrix form, denote the vector of these λ_{m+j}'s as

$$\tilde{\mathbf{\Lambda}} = [\lambda_{m+1}, \ldots, \lambda_{m+n}]'$$

These are to be distinguished from the λ's associated with the original constraints, $h^1(\mathbf{X}), \ldots, h^m(\mathbf{X})$:

$$\mathbf{\Lambda} = [\lambda_1, \ldots, \lambda_m]'$$

Also, recall that the Jacobian matrix of the constraints is represented as

$$\mathbf{J} = \begin{bmatrix} [\nabla h^1]' \\ \vdots \\ [\nabla h^m]' \end{bmatrix}$$

The Kuhn–Tucker conditions in (10.6′) and (10.7) are then

$$\nabla f - \mathbf{J}'\mathbf{\Lambda} + \tilde{\mathbf{\Lambda}} = \mathbf{0} \tag{11.53a}$$

$$\begin{cases} \mathbf{\Lambda} \circ \mathbf{h}(\mathbf{X}) = 0 \\ -\tilde{\mathbf{\Lambda}} \circ \mathbf{X} = 0 \end{cases} \tag{11.53b}$$

$$\begin{cases} \mathbf{h}(\mathbf{X}) \leq \mathbf{0} \\ \mathbf{X}, \mathbf{\Lambda}, \tilde{\mathbf{\Lambda}} \geq \mathbf{0} \end{cases} \tag{11.53c}$$

[The nonnegativity on $\mathbf{\Lambda}$ and $\tilde{\mathbf{\Lambda}}$ comes from (10.7).] Convert each $h^i(\mathbf{X}) \le 0$ [in (11.53c)] into an equation, exactly as in the linear programming case, by adding a nonnegative slack variable, s_i, to the left-hand side of $h^i(\mathbf{X})$. Letting $\mathbf{S} = [s_1, \dots, s_m]'$, we have, from (11.52), $h(\mathbf{X}) + \mathbf{S} = \mathbf{AX} - \mathbf{B} + \mathbf{S} = \mathbf{0}$, so $h(\mathbf{X}) = -\mathbf{S}$. The constraints $h(\mathbf{X}) \le \mathbf{0}$ are fully captured by nonnegativity requirements on the slack variables, $\mathbf{S} \ge \mathbf{0}$, and the conditions in (11.53) can now be expressed as

$$\nabla f - \mathbf{J}'\mathbf{\Lambda} + \tilde{\mathbf{\Lambda}} = \mathbf{0} \tag{11.53$'$a}$$

$$-\mathbf{\Lambda} \circ \mathbf{S} = 0$$
$$-\tilde{\mathbf{\Lambda}} \circ \mathbf{X} = 0 \tag{11.53$'$b}$$

$$\mathbf{S} \ge \mathbf{0}$$
$$\mathbf{X}, \mathbf{\Lambda}, \tilde{\mathbf{\Lambda}} \ge \mathbf{0} \tag{11.53$'$c}$$

$$\mathbf{S} = \mathbf{B} - \mathbf{AX} \tag{11.53$'$d}$$

where (11.53$'$d) defines the new slack variables.

When the $h^i(\mathbf{X})$ are all linear functions, $\mathbf{J} = \mathbf{A}$, and when $f(\mathbf{X})$ is a quadratic function, $\nabla f = \mathbf{QX} + \mathbf{P}$, so conditions (11.53$'$) become

$$\mathbf{QX} + \mathbf{P} - \mathbf{A}'\mathbf{\Lambda} + \tilde{\mathbf{\Lambda}} = \mathbf{0} \tag{11.53$''$a}$$

$$\mathbf{\Lambda} \circ \mathbf{S} = 0$$
$$\tilde{\mathbf{\Lambda}} \circ \mathbf{X} = 0 \tag{11.53$''$b}$$

$$\mathbf{S} \ge \mathbf{0}$$
$$\mathbf{X}, \mathbf{\Lambda}, \tilde{\mathbf{\Lambda}} \ge \mathbf{0} \tag{11.53$''$c}$$

$$\mathbf{S} = \mathbf{B} - \mathbf{AX} \tag{11.53$''$d}$$

[The minuses have been removed from (11.53$''$b) since they play no role.]

The important observation is that (11.53$''$a) and (11.53$''$d) are $n + m$ *linear* equations with $2n + 2m$ unknowns—$\mathbf{X}$, $\mathbf{S}$, $\mathbf{\Lambda}$, and $\tilde{\mathbf{\Lambda}}$. These equations are (including all unknowns in all equations by using null matrices and appropriately sized identity matrices where needed)

$$\mathbf{QX} - \mathbf{A}'\mathbf{\Lambda} + \mathbf{I}\tilde{\mathbf{\Lambda}} + \mathbf{0S} = -\mathbf{P} \tag{11.54a}$$

$$\mathbf{AX} + \mathbf{0\Lambda} + \mathbf{0}\tilde{\mathbf{\Lambda}} + \mathbf{IS} = \mathbf{B} \tag{11.54b}$$

Using partitioned matrices, let

$$\mathbf{M} = \left[\begin{array}{c|c|c|c} \mathbf{Q} & -\mathbf{A}' & \mathbf{I} & \mathbf{0} \\ \hline \mathbf{A} & \mathbf{0} & \mathbf{0} & \mathbf{I} \end{array}\right], \qquad \mathbf{V} = \begin{bmatrix} \mathbf{X} \\ \mathbf{\Lambda} \\ \tilde{\mathbf{\Lambda}} \\ \mathbf{S} \end{bmatrix}, \qquad \text{and} \qquad \mathbf{N} = \begin{bmatrix} -\mathbf{P} \\ \mathbf{B} \end{bmatrix}$$

where $\mathbf{M}$ has dimensions $(m+n) \times (2m+2n)$, $\mathbf{V}$ is a $(2m+2n)$-element vector, and $\mathbf{N}$ is an $(m+n)$-element vector, so the linear equation system in (11.54) is simply

$$\mathbf{MV} = \mathbf{N} \qquad (11.55)$$

In any solution to these equations, all the variables must be nonnegative, from (11.53″c), and at least $n + m$ must be zero, from (11.53″b). This boils down to the fact that to satisfy all the requirements in (11.53″), we want *nonnegative basic* solutions to the set of linear equations in (11.54) or (11.55). This suggests that a kind of simplex method approach could be appropriate, and indeed it is. Notice that the requirements in (11.53″b) stipulate more than just that *any* $n + m$ of the variables must be zero. Specifically, if $x_1 \neq 0$, then from among the λ_{m+j}'s it is precisely λ_{m+1} that must be zero; if $\lambda_2 \neq 0$, then of all the s_i's, it is s_2 that must be zero; and so on.

11.4.2   Computational Methods

A number of computer software packages solve quadratic programming problems by taking advantage of this structure of the Kuhn–Tucker conditions: fewer linear equations than unknowns requiring nonnegative basic solutions. They use variations on the simplex method for the purpose of generating the appropriate set of basic solutions.

Finding nonnegative basic solutions to a set of linear equations can be an *exact* mathematical procedure, at least in those cases where an *exact* inverse is found. We find a set of basic solutions to (11.55) using square $(m+n) \times (m+n)$ submatrices from $\mathbf{M}$, call them $\mathbf{M}^{(b)}$, formed by eliminating some set of $m + n$ columns of $\mathbf{M}$. Then these basic solutions are

$$\mathbf{V}^{(1)} = [\mathbf{M}^{(1)}]^{-1}\mathbf{N}, \ldots, \mathbf{V}^{(k)} = [\mathbf{M}^{(k)}]^{-1}\mathbf{N}$$

Of course, for large $m + n$, each of these basic solutions might be found from $[\mathbf{M}^{(j)}]\mathbf{V}^{(j)} = \mathbf{N}$ not with an inverse but through the use of iterative methods (Chapter 5), and then this approach could be considered to be a kind of approximation technique.

Recall (Chapters 2 and 7) that there will be a maximum of $C_{(m+n)}^{(2m+2n)}$ possible basic solutions to (11.55), so $k \leq C_{(m+n)}^{(2m+2n)}$. (This is not necessarily good news—for a problem with 20 unknowns and 10 constraints, $C_{30}^{60} = 118$ quadrillion, an impossibly large number.) Some of the $\mathbf{M}^{(b)}$ matrices may be singular, so an associated basic solution cannot be found. For some of the nonsingular $\mathbf{M}^{(b)}$ matrices, the resulting solution will contain one or more negative values. Those solutions are unacceptable, since they violate (11.53″c). Of the remaining, which will be nonnegative basic solutions, there will generally be several for which conditions (11.53″b) are not satisfied. Even though they are basic solutions with $n + m$ variables at zero value, the zero-valued variables are not the right ones. (We will encounter this situation in the numerical illustrations, below.[16]) If $f(\mathbf{X})$ is strictly (quasi)concave, so that the necessary conditions are also sufficient, then we realize that there will be only one nonnegative basic solution to (11.54) [or (11.55)] that also satisfies the requirements in (11.53″b).

---

[16]The heart of a computer software package for quadratic programs is an efficient procedure for sorting through all possible nonnegative basic solutions to (11.54), identifying and eliminating those that fail to satisfy the "one of each pair must be zero" requirements of (11.53″b). However, identifying potential solutions from among trillions or quadrillions of candidates can still be an overwhelming task.

Other solution approaches could also be used, including the following:

(a) Gradient-based or SUMT-like procedures could be applied. In fact, any of a variety of interior point methods (Chapter 9) are also possible for the quadratic programming problem.

(b) If Q is a diagonal matrix [meaning that there are no $x_j x_k$ terms in $f(\mathbf{X})$], then separable programming techniques could be used.

(c) Cutting-plane methods would have the advantage that the original feasible region already has linear sides. Addition of the constraint formed from the rewritten objective function, $-f(\mathbf{X}) + x_{n+1} \leq 0$ [as in (11.34)], adds a nonlinearity, however.

11.4.3   Duality in Quadratic Programming Problems

As long as $\mathbf{Q}$ in $f(\mathbf{X})$ in (11.52) is negative (semi)definite, the concavity/convexity requirements for duality (Section 11.2) will be met, and it is easily established that the dual to the quadratic programming maximization problem in (11.52) would be[17]

$$\text{Minimize} \quad \Delta(\mathbf{X},\mathbf{Y}) = \mathbf{B}'\mathbf{Y} - \tfrac{1}{2}\mathbf{X}'\mathbf{Q}\mathbf{X} + c$$
$$\text{subject to} \quad \mathbf{A}'\mathbf{Y} - \mathbf{Q}\mathbf{X} \geq \mathbf{P} \tag{11.56}$$
$$\text{and} \quad \mathbf{Y} \geq \mathbf{0}$$

We know that if $\mathbf{X}^*$ is an optimal solution to (11.52), then that $\mathbf{X}^*$ and some $\mathbf{Y}^*$ will be an optimal solution to (11.56). And when $\mathbf{X}^*$ is known, (11.56) is just a *linear programming* problem in the unknowns $\mathbf{Y}$.

If we introduce a vector of nonnegative surplus variables, $\mathbf{T} = [t_1, \ldots, t_n]'$ into the constraints (they must be subtracted, since these are "$\geq$" constraints), we have $\mathbf{A}'\mathbf{Y} - \mathbf{Q}\mathbf{X} - \mathbf{T} = \mathbf{P}$, or $\mathbf{T} = \mathbf{A}'\mathbf{Y} - \mathbf{Q}\mathbf{X} - \mathbf{P}$. In conjunction with the results in Theorem PD3′, we see that $\mathbf{T}$ plays exactly the role of the $\tilde{\mathbf{\Lambda}}$ vector and $\mathbf{Y}$ is the $\mathbf{\Lambda}$ vector from (11.53″). Rewriting, we obtain

$$\mathbf{T} = \mathbf{A}'\mathbf{Y} - \mathbf{Q}\mathbf{X} - \mathbf{P} \tag{11.53'''a}$$

$$\mathbf{Y} \circ \mathbf{S} = 0$$
$$\mathbf{T} \circ \mathbf{X} = 0 \tag{11.53'''b}$$

$$\mathbf{X},\mathbf{Y},\mathbf{S},\mathbf{T} \geq \mathbf{0} \tag{11.53'''c}$$

$$\mathbf{S} = \mathbf{B} - \mathbf{A}\mathbf{X} \tag{11.53'''d}$$

The requirements in (11.53′′′b) are exactly the complementary slackness results for optimal solutions to a pair of primal and dual *linear* programs, (11.53′′′a) and (11.53′′′d) simply define the surplus and slack variables, and (11.53′′′c) requires nonnegativity of everything. (At an optimal solution, all these variables will be starred.)

---

[17]It is also clear that the Kuhn–Tucker constraint qualification is satisfied, since all the constraints are linear.

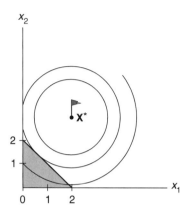

FIGURE 11.8   Feasible region and objective function contours for problem in Example 5.

11.4.4   Numerical Examples

Example 5

$$\text{Maximize} \quad f(\mathbf{X}) = -x_1^2 - x_2^2 + 4x_1 + 6x_2$$

$$\text{subject to} \quad x_1 + x_2 \leq 2 \tag{11.57}$$

$$\text{and} \quad \mathbf{X} \geq \mathbf{0}$$

We saw this problem in Chapter 4, without the nonnegativity requirements. The geometry is shown in Figure 11.8. Letting

$$\mathbf{Q} = \begin{bmatrix} -2 & 0 \\ 0 & -2 \end{bmatrix}, \qquad \mathbf{P} = \begin{bmatrix} 4 \\ 6 \end{bmatrix}, \qquad c = 0$$

then

$$f(\mathbf{X}) = \tfrac{1}{2}\mathbf{X}'\mathbf{Q}\mathbf{X} + \mathbf{P}'\mathbf{X} + c$$

The appropriate pieces for the equations in (11.55) are

$$\mathbf{M} = \left[\begin{array}{cc:c:ccc} -2 & 0 & -1 & 1 & 0 & 0 \\ \hdashline 0 & -2 & -1 & 0 & 1 & 0 \\ \hdashline 1 & 1 & 0 & 0 & 0 & 1 \end{array} \right], \qquad \mathbf{V} = \begin{bmatrix} x_1 \\ x_2 \\ \lambda_1 \\ \lambda_2 \\ \lambda_3 \\ s \end{bmatrix}, \qquad \mathbf{N} = \begin{bmatrix} -4 \\ -6 \\ 2 \end{bmatrix}$$

In this case—three linear equations with six variables—we know that the maximum number of possible basic solutions is $C_3^6 = 20$.[18] It is not very difficult to examine all 20 of

---

[18]Note that in the original problem, in (11.57), we have $n = 2$ and $m = 1$, and the maximum number of basic solutions to the related linear equation system in (11.55) is, indeed, $C_{m+n}^{2m+2n} = C_3^6$.

the $\mathbf{M}^{(b)}$ matrices using linear algebra software on a computer. Three of them turn out to be singular, so there are no associated basic solutions. Of the remaining 17 that are non-singular, only five satisfy the nonnegativity conditions. But four of these five cases fail to satisfy one of the conditions in (11.53''b). Written out explicitly, these conditions for this problem are

    (i) $\lambda_1 s_1 = 0$

    (ii) $\lambda_2 x_1 = 0$

    (iii) $\lambda_3 x_2 = 0$

(Note here that $\lambda_{m+1} = \lambda_2$ and $\lambda_{m+2} = \lambda_3$.) We are left, then, with only one solution to all of the conditions in (11.53'') for the quadratic program in (11.57):

$$\mathbf{X^*} = \begin{bmatrix} 0.5 \\ 1.5 \end{bmatrix} \quad \text{and} \quad f(\mathbf{X^*}) = 8.5$$

and we know, from earlier explorations of this problem (and from the geometry in Fig. 11.8), that this is the correct constrained maximum point.[19] The optimal values of the remaining variables are $\lambda_1^* = 3$; $\lambda_2^* = 0$, $\lambda_3^* = 0$ and $s^* = 0$.

Following (11.56), the dual to (11.57) is

$$\text{Minimize} \quad 2y + \tfrac{5}{2}$$

$$\text{subject to} \quad \begin{bmatrix} 1 \\ 1 \end{bmatrix} y \geq \begin{bmatrix} 3 \\ 3 \end{bmatrix}$$

$$\text{which means simply} \quad y \geq 3$$

$$\text{and} \quad y \geq 0$$

which is redundant. (The constant term in the objective function is irrelevant.) It is easily established that $y^* = \lambda_1^* = 3$, $\mathbf{T^*} = [t_1^*, t_2^*]' = [\lambda_2^*, \lambda_3^*]' = [0 \quad 0]$ and that all conditions in (11.53''') are met.[20]

This suggests that the λ's of the Lagrange approach to quadratic programming problems have exactly the interpretation of dual variables in linear programs. In both linear and quadratic programming problems, if the ith constraint is not binding in an optimal solution ($s_i^* > 0$), then additional units of resource i (increases in the right-hand side of that constraint) are valueless; this is reflected in $y_i^* = 0$ in a (dual) linear program and in $\lambda_i^* = 0$ in a (primal) quadratic program. Conversely, in both linear and quadratic programming problems, if the jth resource is given a positive "shadow price" ($y_j^* > 0$ or $\lambda_j^* > 0$, respectively) at the optimal solution, it must be the case that resource j is used up completely by that solution ($s_j^* = 0$).

In this example, with only one constraint, $\lambda_1^* = 3$. The unconstrained maximum for this problem is at $\mathbf{X}^u = \begin{bmatrix} 2 \\ 3 \end{bmatrix}$, outside the boundary set by the constraint. You can easily check

---

[19]Since $\mathbf{Q}$ is negative definite, $f(\mathbf{X})$ is strictly concave, and so we expect to find a single solution to the conditions in (11.53'').

[20]Note that there are two elements in $\mathbf{T}$ because the constraint(s) are $\begin{bmatrix} 1 \\ 1 \end{bmatrix} y \geq \begin{bmatrix} 3 \\ 3 \end{bmatrix}$.

that if the right-hand side of the constraint were 1, $\mathbf{X^{**}} = \begin{bmatrix} 0 \\ 1 \end{bmatrix}$ and $f(\mathbf{X^{**}}) = 5$; if the right-hand side were 3, $\mathbf{X^{***}} = \begin{bmatrix} 1 \\ 2 \end{bmatrix}$ and $f(\mathbf{X^{***}}) = 11$. So a decrease of the right-hand side of the constraint from 2 to 1 produces a decrease in $f(\mathbf{X^*})$ of 3.5; a unit increase of the right-hand side brings with it an increase in $f(\mathbf{X^*})$ of 2.5. The *average* of these changes in $f(\mathbf{X^*})$ for a unit change in the right-hand side of the constraint is 3, which is just λ_1^*.[21]

Example 6 Here is an example with two constraints, to illustrate how the size of the linear equations problem increases with an additional constraint:

$$\text{Maximize} \quad f(\mathbf{X}) = 2x_1 + x_2 - x_1^2$$
$$\text{subject to} \quad 2x_1 + 3x_2 \leq 6$$
$$2x_1 + x_2 \leq 4 \tag{11.58}$$
$$\text{and} \quad \mathbf{X} \geq \mathbf{0}$$

The geometry is indicated in Figure 11.9. In this case, the appropriate pieces corresponding to the equations in (11.55) are

$$\mathbf{M} = \begin{bmatrix} -2 & 0 & -2 & -2 & 1 & 0 & 0 & 0 \\ 0 & 0 & 3 & -1 & 0 & 1 & 0 & 0 \\ 2 & 3 & 0 & 0 & 0 & 0 & 1 & 0 \\ 2 & 1 & 0 & 0 & 0 & 0 & 0 & 1 \end{bmatrix}, \quad \mathbf{V} = \begin{bmatrix} x_1 \\ x_2 \\ \lambda_1 \\ \lambda_2 \\ \lambda_3 \\ \lambda_4 \\ s_1 \\ s_2 \end{bmatrix}, \quad \mathbf{N} = \begin{bmatrix} -2 \\ -1 \\ 6 \\ 4 \end{bmatrix}$$

Note that $\mathbf{M}$ is now a 4×8 matrix; adding a second constraint has increased the dimensions of $\mathbf{M}$ by one row and two columns. More to the point, the number of square 4×4 $\mathbf{M}^{(b)}$ submatrices in $\mathbf{M}$ has increased more than threefold over the previous example, to $C_4^8 = 70$. Again, with appropriate linear algebra computer software (and patience) you can establish that

(i) Of the 70, 26 are singular, and so no basic solution can be associated with them.

(ii) Of the remaining 44, $\mathbf{M}^{(b)}$ matrices for which basic solutions can be found, 28 fail the nonnegativity requirements.

(iii) Of the remaining 16 nonnegative basic solutions, 15 fail one of the four complementary slackness requirements.

---

[21]We saw earlier, in the nonlinear problems in Chapters 4 and 10, that Lagrange multiplier information at an optimal solution is most reasonably treated as an indicator of the *direction* of the effect on f^* of a change in the right-hand side of a constraint. In the all-linear case (linear programming) it is also an exact indicator of the *magnitude* of the effect on f^* for unit changes in the right-hand side, up to a point where other constraints become binding.

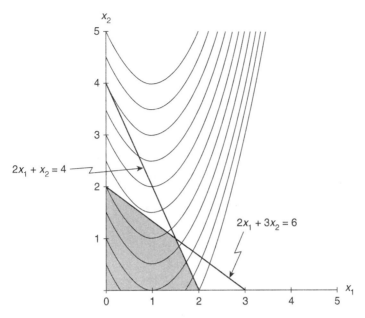

FIGURE 11.9 Feasible region and objective function contours for maximiation problem in Example 6.

The single remaining basic solution is the optimal solution, $\mathbf{X}^* = \begin{bmatrix} 0.6667 \\ 1.5556 \end{bmatrix}$, where λ_1^* = 0.3333, $\lambda_2^* = 0$, $\lambda_3^* = 0$, $\lambda_4^* = 0$, $s_1^* = 0$, $s_2^* = 1.1111$, and $f(\mathbf{X}^*) = 2.4444$ (Fig. 11.9).

The dual to the quadratic maximization problem in (11.58) is

$$\text{Minimize} \quad \Delta(\mathbf{X},\mathbf{Y}) = [6 \quad 4]\begin{bmatrix} y_1 \\ y_2 \end{bmatrix} + \tfrac{4}{9}$$

$$\text{subject to} \quad \begin{bmatrix} 2 & 2 \\ 3 & 1 \end{bmatrix}\begin{bmatrix} y_1 \\ y_2 \end{bmatrix} \geq \begin{bmatrix} \tfrac{2}{3} \\ 1 \end{bmatrix} \tag{11.59}$$

$$\text{and} \quad \mathbf{Y} \geq \mathbf{0}$$

The straightforward geometry of this linear program is indicated in Figure 11.10. The (degenerate) optimal solution is easily found to be $\mathbf{Y}^* = \begin{bmatrix} 0.3333 \\ 0 \end{bmatrix}$, with $\Delta(\mathbf{X}^*,\mathbf{Y}^*) = 2.4444$, the same as $\Pi(\mathbf{X}^*)$, as we would expect from Theorem PD2′, and all the complementary slackness conditions hold.

It is clear from Examples 5 and 6 that the simplexlike approach to a quadratic programming problem is very much a brute-force method. With two unknowns and two constraints we needed to examine 70 potential solutions.[22] [Recall that for a *linear* programming problem with two unknowns and two constraints, the simplex method would only need to examine (a nonnegative subset of) $C_m^{m+n} = C_2^4 = 6$ possible basic solutions.] If we

---

[22]For 20 unknowns and 10 constraints, the number is astronomical.

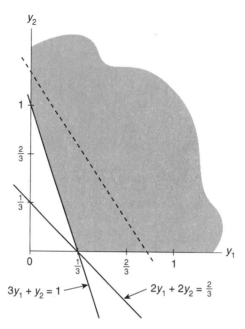

FIGURE 11.10 Feasible region and objective function contour for minimization problem in Example 6.

were to employ the straightforward approach of Section 4.5, we would have only four potential solutions to examine, associated with the cases in which none, one (of the two), or both constraints were imposed as equalities, and only the case in which the first constraint was an equation would generate the optimal solution.

The simplexlike attack on the problem exploits the linearity of the quadratic programming problem constraints and the fact that the gradient of a quadratic function is linear in the unknowns. When coupled with the complementary slackness requirements in (11.53''b) and the requirements of nonnegativity of all variables, the simplex algebra of finding nonnegative basic solutions to linear equations can be used. There may be many solutions to sort through, but this can often be done very quickly.

11.5 SUMMARY

In this chapter we have explored a number of alternative approaches to solving nonlinear programming problems, both in the general case and for quadratic programs in particular. We have also looked at the notion of a dual to a nonlinear programming problem.

Initially we examined the possibility of using gradientlike methods for constrained nonlinear maximization and minimization problems. Instead of always moving in the unconstrained "best" (gradient) direction, these approaches move in a constrained "best" (possibly altered gradient) direction, where the criterion is that the next point reached remains feasible. If modification is necessary because of boundaries, there are many possibilities. We briefly studied the operation of two of these, due to Zoutendijk and Rosen, illustrated in a simplified problem in which there were only linear constraints.

Many variants of "boundary rules" to avoid leaving the feasible region are possible. Nonlinear constraints further complicate a gradient-based approach, although some methods use appropriately constructed linear approximations to the constraints. We have avoided a detailed examination of methods for trying to determine the optimal feasible "step length," once the direction of movement has been established. These are line maximization or minimization problems, but with constraints, and techniques of the kind described in Section 6.1 can be useful.

Next we examined the logic of duality in the nonlinear case. As with linear programming, the dual program itself may not always be relevant in a particular problem, yet because of the parallel kinds of primal–dual relations (Theorems PD1′–PD4′), the optimal values of the dual variables themselves may be useful. Not all relationships carry over from the linear case; for example, the dual of the dual in (11.11) is not the primal problem in (11.7). In the process, we saw that the saddle point conditions of Chapter 10 developed exactly the characteristics that we want to final in a pair of mutually dual nonlinear programs.

Next, we examined a collection of solution approaches for nonlinear programming problems that rely on various kinds of *approximating* problems. The separable programming technique approximates both the objective function and the constraints by a series of straight-line segments and then employs a variant of the simplex method to solve the (linearized) problem. Accuracy (and also the size of the linear problem) is increased with an increase in the number of approximating line segments. Cutting-plane methods also employ linear approximations to the nonlinear constraints of the original problem, using a series of first-order Taylor series approximations, and again variations on the simplex method can be used to solve the approximating problems. By contrast, SUMT creates a series of approximations in the form of unconstrained but nonlinear maximization or minimization problems. Unconstrained optimization techniques can then be applied to these approximations.

Finally, we examined quadratic programming problems and their duals. The simplex method is saved once again, this time thanks to the structure of the Kuhn–Tucker conditions for nonlinear programming problems and to the fact that first partial derivatives of a quadratic function are themselves linear functions. The optimal solution is found as one of the nonnegative basic solutions to a set of linear equations, so the simplexlike approach to examining such basic solutions assures an exact result, unlike the approximating techniques for general nonlinear programming problems. Also, unlike the gradient-based (Section 11.1) and linear approximating techniques (Section 11.3), an optimal solution to a quadratic programming problem derived via simplexlike techniques provides optimal values to the Lagrange multipliers associated with the constraints in the problem. These have the usual interpretations of dual variables in linear programs, namely, as marginal valuations on the right-hand sides of the constraints.

REFERENCES

Fiacco, Anthony V., and Garth P. McCormick, *Nonlinear Programming: Sequential Unconstrained Minimization Techniques*, Wiley, New York, 1968.

Rosen, J. B., "The Gradient Projection Method for Nonlinear Programming: Part I, Linear Constraints," *SIAM J.* **8**, 181–217 (1960).

———, "The Gradient Projection Method for Nonlinear Programming: Part II, Nonlinear Constraints," *SIAM J.* **9**, 514–532 (1961).

Zoutendijk, Guus, *Methods of Feasible Directions; a Study in Linear and Nonlinear Programming*, Elsevier, New York, 1960.

PROBLEMS

11.1   Construct the dual to each of the following nonlinear programming primal problems:
(a) Problem 10.1 in Chapter 10.
(b) Problem 10.2 in Chapter 10.
(c) Problem 10.3 in Chapter 10.

11.2   Solve the dual problems that you constructed in Problems 11.1(a) and (b), using any approach that you wish. Check Theorems PD3′ and PD4′, using the optimal solutions from the primals (Problems 10.1 and 10.2) in conjunction with those for the duals, found now. [Suggestion: do Problem 11.1(b) first, since it is the smaller of the two.]

11.3   Use both Zoutendijk's method and Rosen's method on
(a) Problems 11.1(a)–(b).
(b) The linear program in (7.6) in Chapter 7. The object is to trace out the way in which Zoutendijk's and Rosen's approaches work on all-linear problems.

11.4   (a) Investigate the steps in Zoutendijk's feasible directions method for the problem in (11.1), starting now at $\mathbf{X}^0 = \begin{bmatrix} 0 \\ 0 \\ 0 \end{bmatrix}$. (In the illustration in the text, $\mathbf{X}^0 = \begin{bmatrix} 4 \\ 4 \\ 4 \end{bmatrix}$.)

(b) Do the same for Rosen's gradient projection method, from the new $\mathbf{X}^0$.

11.5   Find the maximum of $f(\mathbf{X}) = 16 - 2(x_1 - 3)^2 - (x_2 - 7)^2$ subject to $x_1^2 + x_2 \le 16$ and $x_1, x_2 \ge 0$:
(a) Using the separable programming approach, with mesh points of 0, 1, 2, 3, and 4 for x_1 and 0, 2, 4, 6, and 8 for x_2. [*Hint:* Define $f_1(x_1)$ as $8 - 2(x_1 - 3)^2$ and $f_2(x_2)$ as $8 - (x_2 - 7)^2$.]
(b) Directly, ignoring the nonnegativity requirements [i.e., try each of the only two options for the constraint, which are that it is ineffective (ignore it) or that it is binding (impose it as an equation)].

11.6   Solve the following linear programming problem:

$$\text{Maximize} \quad 2x_1 + 4x_2$$
$$\text{subject to} \quad x_1 + x_2 \le 10$$
$$\text{and} \quad \mathbf{X} \ge \mathbf{0}$$

Next, solve the following associated linear programming problem in which a third variable has been defined as *no larger than* the objective function above, namely, $x_3 \le 2x_1 + 4x_2$:

Maximize x_3

subject to $x_1 + x_2 \leq 0$

$-2x_1 - 4x_2 + x_3 \leq 0$ (the rewritten definition of x_3)

and $\mathbf{X} \geq \mathbf{0}$

The optimal values of x_1 and x_2, and of the objective function, should be the same in both problems. This *illustrates*, in the context of a *linear* programming problem, the idea that a multivariable objective function

$$\text{Maximize}\quad f(x_1, \ldots, x_n)$$

can be replaced by the objective

$$\text{Maximize}\quad x_{n+1}$$

in a maximization problem if the restriction $x_{n+1} \leq f(\mathbf{X})$ is added as a constraint. In a minimization problem, the added restriction is $x_{n+1} \geq f(\mathbf{X})$, and the new objective is simply

$$\text{Minimize}\quad x_{n+1}$$

11.7   Use Kuhn–Tucker conditions to solve the problem

$$\text{Minimize}\quad 3x_1^2 + 2x_2^2$$

subject to $x_1 + x_2 \geq 10$

and $\mathbf{X} \geq \mathbf{0}$

Next, use Kuhn–Tucker conditions to solve the associated problem in which x is required to satisfy $x_3 \geq 3x_1^2 + 2x_2^2$:

Minimize x_3

subject to $x_1 + x_2 \geq 10$

$-3x_1^2 - 2x_2^2 + x_3 \geq 0$ (rewritten definition of x_3)

and $\mathbf{X} \geq \mathbf{0}$

Confirm that the optimal values of x_1 and x_2 are 4 and 6, respectively, in both problems, leading to a minimized objective function value, in both problems, of 120. This *illustrates* the same principle as in Problem 11.6, only now for a *nonlinear* programming problem.

11.8   Write out the complete set of Kuhn–Tucker conditions for the transformation of the general nonlinear programming problem

$$\text{Maximize}\quad f(\mathbf{X})$$

subject to $h^i(\mathbf{X}) \leq 0$

and $\mathbf{X} \geq \mathbf{0}$

as in Problem 11.7—that is, where $f(\mathbf{X})$ is replaced by x_{n+1} and $-f(\mathbf{X}) + x_{n+1} \leq 0$ is added to the constraints. Consider nonnegativity requirements explicitly, as in (10.8) in Chapter 10, but note that x_{n+1} is unrestricted in sign. (Why?) Show that these requirements for the transformed problem are exactly the same as (10.8) applied to the original problem.

11.9   Solve Problem 11.8 for the general nonlinear programming *minimization* problem with nonnegativities considered explicitly.

11.10   Consider the problem

$$\text{Maximize} \quad 6x_1 + 8x_2$$
$$\text{subject to} \quad x_1^2 + x_2^2 \leq 16$$
$$\text{and} \quad x_1, x_2 \geq 0$$

(a) Use the cutting-plane method, with $\varepsilon = 1.0$ (i.e., stop when the constraint is exceeded by less than one unit) and let $\mathbf{X}^0 = \begin{bmatrix} 2 \\ 2 \end{bmatrix}$. (*Hint:* Draw a fairly accurate solution-space picture so that it will be clear which constraints will be met as equalities in each of the series of linear programming problems.)

(b) Find the exact solution (without the nonnegativity requirements) by (i) ignoring the constraint and then (ii) imposing the constraint as an equation.

11.11   Find an approximate solution to the following problem, using the cutting-plane approach, starting at $\mathbf{X}^0 = [1 \quad 2]'$ and using a stopping criterion of $\varepsilon = 0.025$:

$$\text{Maximize} \quad x_1 + x_2$$
$$\text{subject to} \quad 3x_1^2 + 2x_2^2 \leq 18$$
$$\text{and} \quad \mathbf{X} \geq \mathbf{0}$$

Compare your approximate optimal solution for $\mathbf{X}$ and $f(\mathbf{X})$ with the exact answer, which is easily found using standard Kuhn–Tucker conditions.

11.12   Approximate a solution to the nonlinear program in Problem 11.10, using the sequential unconstrained minimization technique on $-f(\mathbf{X})$ and a logarithmic penalty function:

(a) Evaluate at $\mu = 1.0, 0.1$, and 0.01 only.

(b) Find the exact answer. [If you did Problem 11.10(b), you have already done this.] Observe that as $\mu \to 0$, the solution values for x_1 and x_2 approach the exact answer. (Of course, SUMT and, indeed, all nonlinear programming approximation techniques are designed for problems for which an exact solution cannot be found.)

11.13   Find an approximate solution to the nonlinear program in Problem 11.11 using the *interior* point SUMT approach and ignoring the nonnegativity requirements. Examine the increase in accuracy as you go from $\mu = 1.0$ to $\mu = 0.0001$ by selecting several intermediate values of μ.

11.14 Find an approximate solution to the nonlinear programming *minimization* problem that is the same as the maximization problem in Problem 11.11, using an *exterior* point SUMT approach and ignoring the nonnegativity conditions. As in Problem 11.13, examine the evolution of your approximate solutions as μ decreases from μ = 1.0 to μ = 0.0001 by choosing a number of intermediate values of μ.

11.15 Maximize $f(\mathbf{X})$ from Problem 11.4, this time subject to $x_1 + x_2 \leq 10$ and $\mathbf{X} \geq \mathbf{0}$.

(a) Since this is a quadratic program, use the modified simplex method approach for quadratic programming problems.

(b) Since the objective function and the constraint are also separable, find the solution via separable programming. (Note that this is very close to the work involved in Problem 11.4.)

(c) Solve exactly, ignoring nonnegativity, as a check on (a). [Again, this is very close to (b) in Problem 11.4.]

11.16 Write down and solve the dual to the quadratic program in Problem 11.15. Check that the optimal primal and dual values satisfy the properties of Theorems PD3$'$ and PD4$'$.

ANSWERS TO SELECTED PROBLEMS

CHAPTER 1

1.2. **(a)** $\mathbf{B}$ must be a 3×2 matrix. **(b)** $\mathbf{B} = \begin{bmatrix} 10 & -2 \\ 10 & -2 \\ 10 & -2 \end{bmatrix}$. **(c)** $\mathbf{A} + \mathbf{B} = \begin{bmatrix} 21 & -3 \\ 10 & -2 \\ 7 & 3 \end{bmatrix}$.

1.3. $(\mathbf{A} + \mathbf{B}) - \mathbf{C} = \begin{bmatrix} 24 & -5 \\ 7 & 4 \\ 2 & 7 \end{bmatrix}$.

1.5. Since $\mathbf{A}$ has dimensions 2×3, $\mathbf{B}$ must be $3 \times n$ in order to postmultiply $\mathbf{A}$ and $\mathbf{B}$ must be $m \times 2$ to premultiply $\mathbf{A}$. **(a)** $\mathbf{BA}$ is possible, $\mathbf{AB}$ is not. **(b)** Both $\mathbf{AB}$ and $\mathbf{BA}$ are possible. **(c)** Neither is possible. **(d)** As in **(b)**, both are possible. **(e)** Neither is possible.

1.6. $\mathbf{AB} = \begin{bmatrix} 17 & -16 & 31 \\ 45 & -32 & 83 \end{bmatrix}$, $\mathbf{AC} = \begin{bmatrix} 12 & 9 & 12 \\ 28 & 37 & 44 \end{bmatrix}$, $\mathbf{AB} + \mathbf{AC} = \begin{bmatrix} 29 & -7 & 43 \\ 73 & 5 & 127 \end{bmatrix}$;

$\mathbf{B} + \mathbf{C} = \begin{bmatrix} 2 & 8 & 10 \\ 9 & -5 & 11 \end{bmatrix}$, $\mathbf{A}(\mathbf{B} + \mathbf{C}) = \begin{bmatrix} 29 & -7 & 43 \\ 73 & 5 & 127 \end{bmatrix}$.

1.7. $\mathbf{ABC} = -700$.

1.9. **(a)** -2. **(b)** 1. **(c)** -4.

1.10. **(a)** Columns (and rows) equal. **(b)** Column 3 = 3 × column 1. **(c)** Row of zeros. **(d)** Column 2 = 5 × column 3.

1.11. **(a)** $|\mathbf{A}| = -1 \neq 0$, so $\mathbf{A}^{-1}$ exists; there is a unique solution. **(b)** Same as **(a)**. **(c)** $|\mathbf{A}| = 0$; equation 2 is four times equation 1, so this is one equation in two unknowns. **(d)** $|\mathbf{A}| = 0$; $\mathbf{A}$ is the same as in **(c)**, but now $b_2 = 25$, which is not four times b_1. These are two parallel lines in solution space.

1.12. $\mathbf{X} = \begin{bmatrix} 4 \\ 14 \end{bmatrix}$.

1.13. **(a)** $\mathbf{A}^{-1} = \begin{bmatrix} 4/11 & 1/11 \\ -3/11 & 1/22 \end{bmatrix}$. **(b)** $|\mathbf{B}| = 0$ so there is no inverse. **(c)** $\mathbf{C}^{-1} = \begin{bmatrix} 1 & 0 & 0 \\ 0 & 1 & 0 \\ 0 & 0 & 1/2 \end{bmatrix}$.

This illustrates the general rule for inverses of diagonal matrices that the inverse contains reciprocals of the original elements. **(d)** $\mathbf{D}^{-1} = \begin{bmatrix} 0.5 & 0.25 & -1 \\ 0 & -0.5 & 1 \\ -0.5 & -0.25 & 2 \end{bmatrix}$.

1.14. **(a)** $\rho(\mathbf{A}) = 1$. **(b)** $\rho(\mathbf{B}) = 2$. **(c)** $\rho(\mathbf{C}) = 1$. **(d)** $\rho(\mathbf{D}) = 1$. **(e)** $\rho(\mathbf{E}) = 2$. **(f)** $\rho(\mathbf{F}) = 3$. **(g)** $\rho(\mathbf{G}) = 2$.

1.22. $\mathbf{A} = \begin{bmatrix} 1 & 4 & 3 \\ 2 & 3 & 4 \\ 3 & 4 & 5 \end{bmatrix}$; $\mathbf{A}^{-1} = \begin{bmatrix} -0.25 & -2 & 1.75 \\ 0.5 & -1 & 0.5 \\ -0.25 & 2 & -1.25 \end{bmatrix}$. Let $\alpha = 2$; $(2\mathbf{A}) = \begin{bmatrix} 2 & 8 & 6 \\ 4 & 6 & 8 \\ 6 & 8 & 10 \end{bmatrix}$

and $(2\mathbf{A})^{-1} = \begin{bmatrix} -0.125 & -1 & 0.875 \\ 0.25 & -0.5 & 0.25 \\ -0.125 & 1 & -0.625 \end{bmatrix}$, which is clearly $(1/2)\mathbf{A}^{-1}$. For the general case, $(\alpha\mathbf{A})^{-1} = \mathbf{A}^{-1}\alpha^{-1} = \alpha^{-1}\mathbf{A}^{-1}$ (since α is a scalar) $= (1/\alpha)\mathbf{A}^{-1}$.

1.23. **(a)** Yes, since $|\mathbf{A}| \neq 0$. **(c)** No; the 4×4 matrix made up of these four columns is singular. **(d)** No; there are too few vectors. Three linearly independent vectors are needed for three dimensional vector space.

CHAPTER 2

2.1. **(a)** $\rho(\mathbf{A}) = 1$, $\rho(\mathbf{A} \vdots \mathbf{B}) = 2$. Inconsistent system; the equations are parallel lines in solution space. **(b)** $\rho(\mathbf{A}) = 2$, $\rho(\mathbf{A} \vdots \mathbf{B}) = 2$. Homogeneous equations; the system is consistent. Since $\rho(\mathbf{A}) < 3$, there are multiple solutions. The two planes intersect in a line in solution space. **(c)** $\rho(\mathbf{A}) = 2$, $\rho(\mathbf{A} \vdots \mathbf{B}) = 2$. Consistent system. Any one equation is redundant; the remaining set of two equations in three unknowns has multiple solutions. **(d)** $\rho(\mathbf{A}) = 2$, $\rho(\mathbf{A} \vdots \mathbf{B}) = 3$. Inconsistent system; no solutions.

2.2. **(a)** There will be a maximum of $C_2^3 = 3$ basic solutions. (i) Set $x_1 = 0$; then $x_2 = -10$ and $x_3 = 5$. (ii) Set $x_2 = 0$; then the coefficient matrix of the remaining 2×2 system is singular. (iii) Set $x_3 = 0$; then $x_1 = 20$, $x_2 = -10$.

2.3. **(a)** $\rho(\mathbf{A}) = 2 \, (= n)$, $\rho(\mathbf{A} \vdots \mathbf{B}) = 2$. Consistent system with a unique solution; but equation $3 = 3 \times$ equation 1, so either can be dropped. **(b)** $\rho(\mathbf{A}) = 2 \, (= n)$, $\rho(\mathbf{A} \vdots \mathbf{B}) = 3$. Inconsistent system; if any one equation is ignored, the remaining 2×2 system has a unique solution. **(d)** $\rho(\mathbf{A}) = 2 \, (= n)$, $\rho(\mathbf{A} \vdots \mathbf{B}) = 2$. Consistent system; if *any* equation is dropped, the remaining system can be solved for a unique solution. [If equation 3 is dropped, this is just the system in part **(a)**, above.]

2.4. **(a)** Since $|\mathbf{A}| = 0$, the answer is no. **(b)** $|\mathbf{A}| \neq 0$, so there is a unique solution. **(c)** Same as **(b)**; unique solution.

2.6. (a) For a homogeneous system, $c = 0$. Since $|\mathbf{A}| = ab - 10$, multiple solutions occur when $ab = 10$, so $|\mathbf{A}| = 0$. (b) For a nonhomogeneous system, $c \neq 0$. Using Cramer's rule, $x_1 = -c/(ab - 10)$ and $x_2 = ac/(ab - 10)$, so $x_2 = 0$ if and only if $a = 0$.

2.10. (a) Since $\rho(\mathbf{A}) = 3$ ($< n = 4$), the possible kinds of solutions depend on $\rho(\mathbf{A} \vdots \mathbf{B})$. If $\rho(\mathbf{A} \vdots \mathbf{B}) = 4$, the system is inconsistent. If $\rho(\mathbf{A} \vdots \mathbf{B}) = 3$, there are multiple solutions. (b) Here $\rho(\mathbf{A}) = 4$, so there will be a unique solution. (c) As in (a), $\rho(\mathbf{A}) = 3$; now $\rho(\mathbf{A} \vdots \mathbf{B}) = 3$, so there are multiple solutions. (d) Again, $\rho(\mathbf{A}) = 3$ but now $\rho(\mathbf{A} \vdots \mathbf{B}) = 4$, so the system is inconsistent.

2.12. Here $\rho(\mathbf{A}) = 2$ and $\rho(\mathbf{A} \vdots \mathbf{B}) = 3$; inconsistent system with partial solutions. For equations 1 and 3, $\mathbf{X} = \begin{bmatrix} 0.4 \\ 0.4 \end{bmatrix}$; for equations 1 and 4, $\mathbf{X} = \begin{bmatrix} 0.5 \\ 1.0 \end{bmatrix}$; for equations 3 and 4, $\mathbf{X} = \begin{bmatrix} 0.5 \\ 0.5 \end{bmatrix}$.

2.14. (a) $\mathbf{X} = [24 \quad 48 \quad 48]'$ (in millions of dollars). (b) $\mathbf{X} = [0.2\beta \quad 0.4\beta \quad 0.4\beta]'$ (in millions of dollars).

2.17. (a) $\mathbf{X} = \begin{bmatrix} 102 \\ -23 \end{bmatrix}$. (b) Now $\rho(\mathbf{A}) = 1$, $\rho(\mathbf{A} \vdots \mathbf{B}) = 2$ and the system is inconsistent. (c) As in part (b), $\rho(\mathbf{A}) = 1$, since $\mathbf{A}$ is the same. Now $b_1 = 200$, and the factor of proportionality between rows 1 and 2 of $\mathbf{A}$ is 200, so b_2 must be 40,000.

2.18. Use x_1 and x_2 for *changes in* the city's holdings of poorly developed and undeveloped land, respectively; $\mathbf{X} = \begin{bmatrix} -260 \\ 150 \end{bmatrix}$.

2.19. Let x_1, x_2 and x_3 denote the three kinds of projects planned for the coming year; then $\mathbf{X} = [54 \quad -19 \quad -17]'$.

2.21. Here $(\mathbf{A} \vdots \mathbf{B}) = \begin{bmatrix} 1 & 2 & 3 & 3 \\ 2 & 4 & 6 & 3 \\ 1 & 1 & 1 & 3 \end{bmatrix}$; after two elementary row operations to generate zeros in positions a_{21} and a_{31}, we have $\begin{bmatrix} 1 & 2 & 3 & 3 \\ 0 & 0 & 0 & -3 \\ 0 & -1 & -2 & 0 \end{bmatrix}$. Dividing row 3 by -1 and interchanging rows 2 and 3 leads to $\begin{bmatrix} 1 & 2 & 3 & 3 \\ 0 & 1 & 2 & 0 \\ 0 & 0 & 0 & -3 \end{bmatrix}$. The inconsistency in this equation system is clear from row 3, where $0x_1 + 0x_2 + 0x_3 = -3$.

2.23. Forcing a_{31} to zero (subtracting row 1 from row 3) generates $\begin{bmatrix} 1 & 0 & 2 \\ 0 & 1 & 2 \\ 0 & 1 & 2 \end{bmatrix}$. This shows that one equation, either the second or the third, is unnecessary (redundant); these rows now contain exactly the same information. The solution is $\mathbf{X} = \begin{bmatrix} 2 \\ 2 \end{bmatrix}$.

2.25. (a) $(A \vdots B) = \begin{bmatrix} 3 & 2 & 7 \\ 2 & 5 & 12 \end{bmatrix}$ which can be reduced to $\begin{bmatrix} 1 & 0.667 & 2.333 \\ 0 & 1 & 2 \end{bmatrix}$ (echelon

form). (b) The reduced matrix in (a) further reduces to $\begin{bmatrix} 1 & 0 & 1 \\ 0 & 1 & 2 \end{bmatrix}$ (reduced eche-

lon form). (c) $L = \begin{bmatrix} 1 & 0 \\ 3 & 1 \end{bmatrix}$ and $U = \begin{bmatrix} -1 & 1 \\ 0 & -1 \end{bmatrix}$ and indeed $X = U^{-1}L^{-1}B = \begin{bmatrix} 4 \\ 14 \end{bmatrix}$.

CHAPTER 3

3.1. (a) $f(x_0) = 4x_0^2 + 2x_0 + 1$; $f(x_0 + \Delta x) = 4(x_0 + \Delta x)^2 + 2(x_0 + \Delta x) + 1$. Therefore $\Delta y = f(x_0 + \Delta x) - f(x_0) = 4x_0^2 + 8x_0\Delta x + 4(\Delta x)^2 + 2x_0 + 2\Delta x + 1 - 4x_0^2 - 2x_0 - 1$ and so $\Delta y/\Delta x = 8x_0 + 4\Delta x + 2$. Taking the limit, as $\Delta x \to 0$, and dropping the subscript, since this is valid for any particular x_0, $dy/dx = 8x + 2$.

3.2. (a) $f'(x) = 6x + 12 = 0$, so $x^* = -2$. Since $f'(-3) = -6$ and $f'(1) = 6$, x^* is a minimum point.

3.3. (a) $y - 430 = 72(x - 10)$, or $72x - y = 290$.

3.4. (a) $f' = 6x^2 + 6x - 12 = 0$ leads to $x^* = -2$ or $x^{**} = 1$. $f'' = 12x + 6$, so $f''(x^*) < 0$ (x^* is a relative maximum) and $f''(x^{**}) > 0$, (x^{**} is a relative minimum). These extrema are relative, not absolute; the function continues upward toward $+\infty$ to the right of x^{**} and downward toward $-\infty$ to the left of x^*. (b) Setting $f'(x) = 3x^2(4x - 1) = 0$ yields $x^* = 0$ and $x^{**} = 0.25$. $f''(x^*) = 0$; taking higher-order derivatives, $f'''(x^*) = -6$ so x^* is a point of inflection. $f''(x^{**}) = 0$, so x^{**} is a minimum; $f(x)$ has no finite maximum but it has an absolute minimum at x^{**}.

3.6. (a) $x^* = 0.5$. (b) $x_1^* = x_2^* = 20$.

3.7. (a) $X^* = \begin{bmatrix} -0.3333 \\ 0.6667 \end{bmatrix}$ is a minimum point. (b) $X^* = \begin{bmatrix} 3 \\ 4 \end{bmatrix}$ is a minimum point. (c) The

unique solution to $\nabla f = 0$ is $X^* = \begin{bmatrix} 1 \\ -2 \end{bmatrix}$, and $H^* = \begin{bmatrix} 18 & 0 \\ 0 & -32 \end{bmatrix}$, so X^* is neither a maximum nor a minimum.

3.9. $x^* = 150$; maximized rent is \$1125.

3.13. Here $f'(x) = -4x + 90 = 0$ leads to $x^* = 22.5$—but $f''(x^*) = -4$, so x^* is a maximum point; then, examine $f(x)$ at the two endpoints. Since $f(0) = 600$ and $f(50) = 100$, we prefer $x = 50$. If $0 \le x \le 40$, since $f(40) = 1000$, the answer is now $x = 0$.

3.16. The problem is to find a value of t to maximize $P = e^{(-0.08)t}(6(2.5)^{0.5})$, for which

$$P' = e^{0.9163(t)0.5 - (2t/25)}[(2.7489/t^{0.5}) - (12/25)]$$

Setting $P' = 0$ and using computer software leads to $t^* \cong 32.7963$. Evaluation of $P''(t^*)$ is tedious algebra, even with a computer, but it establishes that $P''(t^*) < 0$, so t^* represents a maximum.

CHAPTER 4

4.1. Using Lagrange multipliers, $L = 6x_1^2 + 5x_2^2 - \lambda(x_1 + 5x_2 - 3)$. Solving $\nabla L = 0$, $\mathbf{X}^* =$

$\begin{bmatrix} 3/31 \\ 18/31 \end{bmatrix}$, but for this problem $\overline{\mathbf{V}}^* = \begin{bmatrix} 0 & 1 & 5 \\ 1 & 12 & 0 \\ 5 & 0 & 10 \end{bmatrix}$, and since $|\overline{\mathbf{V}}_3^*| = |\overline{\mathbf{V}}^*| = -310$,

$\mathbf{X}^*$ represents a minimum, not a maximum. In fact, the function has no finite maximum. Increasingly large positive values of x_1 and appropriately correspondingly large negative values of x_2 (to satisfy the constraint) lead to increasingly large values of $f(\mathbf{X})$.

4.3. $\mathbf{X}^* = [20/14 \quad 30/14 \quad 10/14]'$; $|\mathbf{V}_3^*| = 26$, $|\mathbf{V}_4^*| = -56$ so $\mathbf{X}^*$ represents a maximum to the function, subject to the constraint.

4.5. $\mathbf{X}^* = \begin{bmatrix} 11/3 \\ 10/3 \\ 8 \end{bmatrix}$ is a minimum point.

4.8. (a) $\mathbf{X}^* = \begin{bmatrix} 44 \\ 2 \end{bmatrix}$. (b) From (a), you can find that $\lambda^* = 100$, so it would be correct to treat this constraint as binding in an inequality constrained maximization problem; a local maximum occurs on the boundary at $\mathbf{X}^*$ from part (a). It is clear that $f(\mathbf{X})$ has no finite maximum, whether unconstrained or subject to the inequality constraint; letting $x_1 \rightarrow -\infty$ and $x_2 = 0$ satisfies the constraint and leads to larger and larger values of $f(\mathbf{X})$.

4.12. $\mathbf{X}^* = \begin{bmatrix} 25 \\ 25 \end{bmatrix}$.

4.13. (a) $\mathbf{X}^* = [10 \quad 10 \quad 10]'$.

4.15. If the amount of light is a function of the window area, then the problem is to maximize $2rh + (0.5)\pi r^2$, subject to $2h + 2r + \pi r = 2h + (2 + \pi)r = c$, for which $h^* = r^*$ $= \lambda^* = c/(4 + \pi)$. This is clearly a maximum, since letting $h = c/2$ and $r = 0$ would satisfy the constraint and give a window area of zero.

CHAPTER 5

5.2. (a) Newton–Raphson from $x^0 = 2$ finds $x^1 = 1.25$, $x^2 = 1.025$, $x^3 = 1.00030$ and $x^4 = 1.00000$ before stopping. (b) The secant method, from $x^0 = 2$ and $x^1 = 1.25$, finds $x^2 = 1.0769$, $x^3 = 1.0083$, $x^4 = 1.0003$ and $x^5 = 1.0000$. (c) Modified Newton takes 15 iterations. (d) Bisection requires a pair of initial points that bracket a root. Using $x^0 = 2$ and $x^1 = 0.5$, $x = 1.0000$ is reached in 18 iterations; from $x^0 = 2$ and $x^1 = -3$, $x = -1.0000$ takes 21.

5.3. (a) Newton–Raphson and modified Newton fail because of the division by zero problem. Secant and bisection methods are unaffected. (b) For Newton–Raphson, you end up at $x = -1$ in about 10 iterations. The secant method leads you to -1 or $+1$ depending on what you use for x^1, in conjunction with $x^0 = -0.01$. Modified Newton doesn't work; it explodes toward $x = +\infty$. From $x^0 = -0.01$ and $x^1 = 2$, bisection reaches $x = 1$ in about 17 iterations.

5.13. Substitution of $f'(x^k) \cong [f(x^k + \alpha) - f(x^k)]/\alpha$ and $f'(x^k + \beta) \cong [f(x^k + \beta + \alpha) - f(x^k + \beta)]/\alpha$ into $f''(x^k) \cong [f'(x^k + \beta) - f'(x^k)]/\beta$ yields the result. When $\alpha = 0.1$, $\beta = 0.2$ and $x^k = 2$, the estimate is $(3.29 - 2.84 - 2.41 + 2)/0.02 = 2$, which is exactly right; in this example $f''(x) = 2$ everywhere.

CHAPTER 6

6.1. $I^0 = (0, x_n)$, with midpoint x_1. $I^1 = (0, x_1 + (\delta/2)) = (0.5)I^0 + (\delta/2)$. $I^2 = (0.5)I^1 + (\delta/2)$, and substituting I^1 gives $I^2 = (0.5)^2 I^0 + (0.5)(\delta/2) + (\delta/2)$. The results for I^3, $I^4, \ldots, I^k$ follow directly.

6.4. Setting $\nabla f(X) = 0$ leads immediately to $X^* = \begin{bmatrix} 3 \\ 7 \end{bmatrix}$. Since $H = \begin{bmatrix} -4 & 0 \\ 0 & -2 \end{bmatrix}$, X^* represents a maximum. A simple gradient approach—$X^{k+1} = X^k + (0.25)\nabla^k$—from $X^0 = 0$, leads to the solution in 21 iterations, using a stopping criterion that $\nabla^k \leq 10^{-5}$.

6.5. (a) Here $\nabla f(X) = 0$ leads to $X^* = \begin{bmatrix} -0.3333 \\ 0.6667 \end{bmatrix}$ which is easily verified to be a minimum. Using the gradient approach, as in (6.1), with $\alpha = 0.5$, leads to this solution in 17 iterations, from $X^0 = 0$ and with a stopping criterion that $\nabla^k \leq 10^{-5}$.

CHAPTER 7

7.1. $X^* = [0 \quad 0 \quad 3.6]'$, $S^* = [8.4 \quad 0]'$ and $f(X^*) = 54$.

7.2. $X^* = \begin{bmatrix} 0 \\ 1 \, 5/7 \end{bmatrix}$, $S^* = \begin{bmatrix} 0 \\ 5/7 \end{bmatrix}$ and $f(X^*) = 13 \, 5/7$.

7.3. $X^* = \begin{bmatrix} 0 \\ 1.5 \end{bmatrix}$, $S^* = \begin{bmatrix} 4 \\ 0 \end{bmatrix}$ and $f(X^*) = 13.5$.

7.4. $X^* = [0 \quad 5 \quad 10 \quad 0]'$, $S^* = 0$ and $f(X^*) = 140$.

7.5. Unbounded feasible region; there is no finite maximum.

7.7. (a) Minimize $12y_1 + 18y_2$ subject to $y_1 + 2y_2 \geq 3$, $y_1 + y_2 \geq 2$, $y_1 + 5y_2 \geq 15$ and $Y \geq 0$. $Y^* = \begin{bmatrix} 0 \\ 3 \end{bmatrix}$, $T^* = [3 \quad 1 \quad 0]'$ and $\Delta^* = 54$. Recalling that $X^* = [0 \quad 0 \quad 3.6]'$ and $S^* = [8.4 \quad 0]'$, it is obvious by inspection that $Y^* \circ S^* = 0$ and $X^* \circ T^* = 0$.

7.11. (a) The solution-space geometry for the primal will be unchanged so the optimal values of the primal variables would not be affected. (b) In the solution-space picture for the dual, the right-hand sides of all constraints would be (1/10)th as large so optimal values of dual variables would also be (1/10)th as large.

7.11. Consider a small general problem.

$$\text{Maximize} \quad p_1 x_1 + p_2 x_2$$
$$\text{subject to} \quad a_{11}x_1 + a_{12}x_2 \leq b_1$$
$$a_{21}x_1 + a_{22}x_2 \leq b_2$$

and $X \geq 0$. Converting to augmented form

$$\text{Maximize} \quad p_1 x_1 + p_2 x_2 + 0s_1 + 0s_2$$

$$a_{11}x_1 + a_{12}x_2 + s_1 = b_1$$

$$a_{21}x_1 + a_{22}x_2 + s_2 = b_2$$

The dual constraints associated with the coefficient columns for the slack variables are

$$1y_1 + 0y_2 \geq 0; \text{ that is, } y_1 \geq 0$$

$$0y_1 + 1y_2 \geq 0; \text{ that is, } y_2 \geq 0$$

7.13. 140 Clippers, 150 Cruisers, $\Pi^* = 430{,}000$.

7.14. 100 Clippers, 200 Cruisers, $\Pi^* = 300{,}000$.

7.15. (a) $X^* = [x_A^* \quad x_B^*]' = \begin{bmatrix} 40 \\ 18 \end{bmatrix}$ and $\Pi^* = 24{,}600$. (b) $y_1^* = 7$; the range over which this marginal figure is valid is $1200 \leq b_1 \leq 3200$. An additional 40 person-hours would be worth \$280. (c) $y_2^* = 4.5$ and $y_3^* = 0$; it should be made available in region A.

7.16. (a) No. An optimal solution is a basic feasible solution; at most two of the three unknowns will be nonzero. (b) $X^* = [x_A \quad x_B \quad x_C]' = [30\ 15\ 0]'$, $\Pi^* = 24{,}000$. (c) There is no conflict. The optimal solution in (b) uses 810 pounds of recycled plastic. (d) Decreased by 555.56. (e) $y_2^* = 33.33$, with a range $180 \leq b_2 \leq 450$. So the strike would cost 333.33 $[= (360 - 350)y_2^*]$. If b_2 went down to 150, that is below the range of validity for y_2^*. (f) The range over which changes in the value of p_3 will not alter the current optimal solution has an upper limit of 755.56, so increases to 500 or to 750 would not change X^*.

CHAPTER 8

8.1. The optimal *basic* solutions are $X^* = \begin{bmatrix} 2 \\ 6 \end{bmatrix}$ and $X^{**} = \begin{bmatrix} 8 \\ 2 \end{bmatrix}$; in both cases, $\Pi^* = 66$. All *nonbasic* solutions on the line connecting X^* and X^{**} are also equally optimal. (The objective function in this problem is parallel to the third constraint.)

8.3. $X^* = \begin{bmatrix} 6.872 \\ 7.436 \end{bmatrix}$ and $\Pi^* = 21.179$.

8.4. $X^* = \begin{bmatrix} 5 \\ 2 \end{bmatrix}$ and $\Pi^* = 64$.

8.5. $X^* = [0 \quad 3 \quad 3]'$ and $\Pi^* = 51$.

8.8. $x_{11}^* = 15, x_{12}^* = 0, x_{13}^* = 15, x_{21}^* = 5, x_{22}^* = 45, x_{23}^* = 0$; minimal costs are 855.

8.9. There are multiple optima. One solution is $x_{12}^* = 4, x_{14}^* = 10, x_{16}^* = 4, x_{21}^* = 8, x_{22}^* = 2, x_{23}^* = 6, x_{25}^* = 7, x_{36}^* = 6$, and all other variables are zero; minimal costs are 54.

8.10. Truck 1 to D, Truck 2 to C, Truck 3 to B and Truck 4 to A; minimized costs are 4300.

8.11. W_1 to L_5, W_2 to L_3, W_3 gets no assignment, W_4 to L_4, W_5 to L_1, W_r to L_2; total cost $= 210$.

8.13. $x^*_{12} = 6$, $x^*_{14} = 8$, $x^*_{16} = 4$, $x^*_{21} = 8$, $x^*_{23} = 6$, $x^*_{24} = 2$, $x^*_{25} = 7$, $x^*_{36} = 6$; minimized cost is 60.

8.14. Using OW^*_{ij} for the optimal shipment from origin i to warehouse j and WD^*_{jk} for the optimal shipment from warehouse j to destination k, $x^*_{12} = 6$, $x^*_{14} = 10$, $x^*_{15} = 2$, $x^*_{23} = 6$, $x^*_{36} = 6$, $OW^*_{22} = 17$, $WD^*_{21} = 8$, $WD^*_{25} = 5$, $WD^*_{26} = 4$; total cost is now 36.75.

8.16. **(a)** Plants 1 and 3 are both cheapest at location B; locations A and C are both cheapest for plant 2. **(b)** The optimal assignment is: 1 to B, 2 to C and 3 to A; total cost $= 12$. **(c)** In this case, 3 should also go to B; total cost $= 10$. **(d)** Now the optimal assignment is 1 to D, 2 to C, 3 to B; total cost $= 9$.

CHAPTER 9

9.1. After seven pivots, $\mathbf{X}^* = [10{,}000 \quad 0 \quad 0]'$, $\Pi^* = 10{,}000$.

9.4. $a^* = b^* = c^* = (2/3)\bar{s}$. Second-order conditions would be very complicated, but you can be convinced that this is a maximum and not a minimum; letting $a = b = \bar{s}$ and $c = 0$ satisfies the constraint and gives the objective function a value of zero.

9.6. **(a)** $\mathbf{A_R} = (\mathbf{A}')(\mathbf{AA}')^{-1} = \begin{bmatrix} 1 \\ 2 \\ 1 \end{bmatrix} \left\{ \begin{bmatrix} 1 & 2 & 1 \end{bmatrix} \begin{bmatrix} 1 \\ 2 \\ 1 \end{bmatrix} \right\}^{-1} = \begin{bmatrix} 1/6 \\ 2/6 \\ 1/6 \end{bmatrix}$, so $\tilde{\mathbf{X}}^0 = (\mathbf{A}')(\mathbf{AA}')^{-1}$

$\mathbf{B} = \begin{bmatrix} 10/6 \\ 20/6 \\ 10/6 \end{bmatrix}$. The distance from this $\tilde{\mathbf{X}}^0$ to the origin is $[(10/6)^2 + (20/6)^2 +$

$(10/6)^2]^{0.5} = 4.08$. From each of the endpoints of the constraint, $\tilde{\mathbf{X}}^0 = \begin{bmatrix} 10 \\ 0 \\ 0 \end{bmatrix}$,

$\begin{bmatrix} 0 \\ 5 \\ 0 \end{bmatrix}$ or $\begin{bmatrix} 0 \\ 0 \\ 10 \end{bmatrix}$, the distances to the origin are simply 10, 5 and 10, respectively.

from a randomly chosen alternative point on the constraint, $\tilde{\mathbf{X}}^0 = \begin{bmatrix} 3 \\ 1 \\ 5 \end{bmatrix}$, the distance

to the origin is 5.92. Alternative points on the constraint will all be more than 4.08 units from the origin.

CHAPTER 10

10.1. The Kuhn–Tucker conditions for this problem are:

$$
\text{(a)} \begin{cases} -2x_1 + 4 - \lambda_1 - 2\lambda_2 \le 0 \\ -2x_2 + 6 - \lambda_1 - 3\lambda_2 \le 0 \\ -2x_3 \le 0 \end{cases}
$$

$$
\text{(b)} \begin{cases} -x_1(-2x_1 + 4 - \lambda_1 - 2\lambda_2) = 0 \\ x_2(-2x_2 + 6 - \lambda_1 - 3\lambda_2) = 0 \\ x_3(-2x_3) = 0 \end{cases}
$$

$$
\text{(c)} \begin{cases} x_1 + x_2 - 2 \le 0 \\ 2x_1 + 3x_2 - 12 \le 0 \end{cases}
$$

$$
\text{(d)} \begin{cases} \lambda_1(x_1 + x_2 - 2) = 0 \\ \lambda_2(2x_1 + 3x_2 - 12) = 0 \end{cases}
$$

$$
\text{(e)} \; x_1 \ge 0, x_2 \ge 0, x_3 \ge 0; \lambda_1 \ge 0, \lambda_2 \ge 0
$$

It is obvious, from (**b**), that $x_3 = 0$. If you start with the implications of (**d**), there are four possible cases.

Case 1: $\lambda_1 \ne 0$, $\lambda_2 \ne 0$, so $x_1 + x_2 = 2$ and $2x_1 + 3x_2 = 12$, from which $x_1 = -6$ and $x_2 = 8$. A negative x_1 violates (**e**).

Case 2: $\lambda_1 = 0$, $\lambda_2 = 0$. From (**a**), $x_1 = 2$, $x_2 = 3$, but this violates both inequalities in (**c**).

Case 3: $\lambda_1 = 0$, $\lambda_2 \ne 0$. From (**d**), $2x_1 + 3x_2 = 12$. With $\lambda_1 = 0$, the first two equations in (**a**) lead to $-6x_1 + 4x_2 = 0$. These two linear equations have a unique solution $x_1 = 1.846$, $x_2 = 2.769$. This violates the first inequality in (**c**).

Case 4: $\lambda_1 \ne 0$, $\lambda_2 = 0$. The procedure is the same as in Case 3. From (**d**), $x_1 + x_2 = 2$. With $\lambda_2 = 0$, from the first two equations in (**a**) we find $-x_1 + x_2 = 1$. These two linear equations have a unique solution $x_1 = 0.5$, $x_2 = 1.5$. All other requirements in (**a**) through (**e**) are met.

Since the constraints are linear and therefore convex, these conditions will be sufficient as well if $f(\mathbf{X})$ is concave. Using the principal minors test on the Hessian matrix, it is easily established that $f(\mathbf{X})$ is indeed concave, so $\mathbf{X}^* = [0.5 \quad 1.5 \quad 0]'$ satisfies all necessary and sufficient conditions for a maximum of $f(\mathbf{X})$ subject to the two linear constraints.

10.4. (**a**) $L(\mathbf{X}, \mathbf{Y}) = -x_1^2 - x_2^2 - x_3^2 - y_1(x_1 + x_2 - 2) - y_2(2x_1 + 3x_2 - 12)$. From Problem 10.1, we know that $\mathbf{X}^* = [0.5 \quad 1.5 \quad 0]'$; also from those results, using y's instead of λ's, $\mathbf{Y}^* = [3 \quad 0]'$.

10.5. (**a**) Problem 7.1. $\mathbf{X}^* = [0 \quad 0 \quad 3.6]'$, $\mathbf{S}^* = [8.4 \quad 0]'$, $\mathbf{Y}^* = [0 \quad 3]'$ and $\mathbf{T}^* = [-3 \quad -1 \quad 0]'$. For all of these problems with linear objective functions, $\nabla f(\mathbf{X}^*) = \nabla f(\mathbf{X})$. The constraints that are binding at the optimal solution are $h^2(\mathbf{X})$ along with the lower limits on x_1 and x_2, so $\mathbf{M^*} = \begin{bmatrix} 2 & 1 & 0 \\ 1 & 0 & 1 \\ 5 & 0 & 0 \end{bmatrix}$, and $\nabla f(\mathbf{X}) = \begin{bmatrix} 3 \\ 2 \\ 15 \end{bmatrix} = (y_2^*)$

$$\nabla h^2(\mathbf{X^*}) + (t_1^*)\mathbf{I}_1 + (t_2^*)\mathbf{I}_2 = (3)\begin{bmatrix} 2 \\ 1 \\ 5 \end{bmatrix} + (-3)\begin{bmatrix} 1 \\ 0 \\ 0 \end{bmatrix} + (-1)\begin{bmatrix} 0 \\ 1 \\ 0 \end{bmatrix}.$$

10.6. (a) Example 1; $\mathbf{M^*} = \begin{bmatrix} -1 \\ 0 \end{bmatrix}$ and $\rho(\mathbf{M^*}) = 1$. **(c)** Example 3; $\mathbf{M^*} = \begin{bmatrix} 1 & 1 \\ 2 & 10 \end{bmatrix}$ and

$\rho(\mathbf{M^*}) = 2$. **(e)** Example 5; $\mathbf{M^*} = \begin{bmatrix} -1 & 0 \\ 0 & 0 \end{bmatrix}$ and $\rho(\mathbf{M^*}) = 1 \neq 2$. **(f)** Example 6; $\mathbf{M^*}$

$= \begin{bmatrix} 0 & 0 \\ 1 & -1 \end{bmatrix}$ and $\rho(\mathbf{M^*}) = 1 \neq 2$.

10.7. (a) $\mathbf{X^*} = \begin{bmatrix} 20 \\ 0 \end{bmatrix}$. **(b)** You might guess that, with a new constraint of $x_1 \leq 12$, the optimal solution will be on this boundary, since the minimum without the constraint is outside the boundary. Forming the Lagrangian and checking the sign of λ^* will show this to be correct; $\mathbf{X^*} = \begin{bmatrix} 12 \\ 0 \end{bmatrix}$ and $\lambda^* = -32 < 0$, indicating a minimum on the boundary. **(c)** With the new estimate of the pollution function, $\nabla f(\mathbf{X}) = \mathbf{0}$ leads to $\mathbf{X^*} = \begin{bmatrix} 80/3 \\ 20/3 \end{bmatrix}$; since these are hours per day, the value for x_1 is impossible. Using the full set of Kuhn–Tucker conditions, including the constraints $x_1 \leq 24$ and $x_2 \leq 24$, only $\mathbf{X} = \begin{bmatrix} 24 \\ 6 \end{bmatrix}$ satisfies all conditions. The constraints are linear; $\mathbf{H} = \begin{bmatrix} 4 & -4 \\ -4 & 16 \end{bmatrix}$ so $f(\mathbf{X})$ is convex, and these conditions are both necessary and sufficient for a minimum to $f(\mathbf{X})$ subject to the constraints.

CHAPTER 11

11.1. (b) Here and in part **(c)**, you can rewrite the nonlinear primal problem as one of maximization of $-f(\mathbf{X})$ and, in part **(b)**, reverse the direction of the primal inequality by multiplying through by -1. Then the dual will be a minimization problem, formulated as in the text. In that case, $\mathbf{H} = \begin{bmatrix} -12 & 0 \\ 0 & -10 \end{bmatrix}$, so we have a concave objective for the nonlinear primal maximization problem, with a linear constraint, and the structure of (11.11) indicates the dual. Alternatively, you could work out the nonlinear *maximization* problem that is dual to a nonlinear primal *minimization* problem. Using the former approach (there is only one y since there is only one constraint in the primal problem):

Minimize $\Delta(\mathbf{X}, y) = -6x_1^2 - 5x_2^2 - y(-x_1 - 5x_2 + 3) - x_1(-12x_1 + y) - x_2(-10x_2 + 5y)$, subject to $-y + 12x_1 \geq 0$, $-5y + 10x_2 \geq 0$ and $y \geq 0$.

11.4. (a) Zoutendijk. From $\mathbf{X}^0 = \mathbf{0}$, where $\nabla^0 = [4 \quad 6 \quad 10]'$, the initial Zoutendijk linear program is

$$\text{Maximize} \quad 4d_1^0 + 6d_2^0 + 10d_3^0$$

$$\text{subject to} \quad -1 \le d_1^0 \le 1$$

$$-1 \le d_2^0 \le 1$$

$$-1 \le d_3^0 \le 1$$

$$\begin{bmatrix} -1 & 0 & 0 \\ 0 & -1 & 0 \\ 0 & 0 & -1 \end{bmatrix} \mathbf{D}^0 \le \mathbf{0}$$

The last three constraints reflect the fact that currently all x's are at their lower limits. From inspection, it is clear that $\mathbf{D}^0 = [1 \quad 1 \quad 1]'$, so $\mathbf{X}^1 = \begin{bmatrix} 0 \\ 0 \\ 0 \end{bmatrix} + \alpha_0 \begin{bmatrix} 1 \\ 1 \\ 1 \end{bmatrix}$. A line search indicates $f(\mathbf{X})$ is maximized when $\alpha_0 = 3.3$, so $\mathbf{X}^1 = \begin{bmatrix} 3.3 \\ 3.3 \\ 3.3 \end{bmatrix}$; $\nabla^1 = \begin{bmatrix} -2.6 \\ -0.6 \\ 3.4 \end{bmatrix}$,

and the next linear program is

$$\text{Maximize} \quad -2.6d_1^1 - 0.6d_2^1 + 3.4d_3^1$$

$$\text{subject to} \quad -1 \le d_1^1 \le 1$$

$$-1 \le d_2^1 \le 1$$

$$-1 \le d_3^1 \le 1$$

for which $\mathbf{D}^1 = \begin{bmatrix} -1 \\ -1 \\ 1 \end{bmatrix}$, so $\mathbf{X}^2 = \begin{bmatrix} 3.3 \\ 3.3 \\ 3.3 \end{bmatrix} + \alpha_1 \begin{bmatrix} -1 \\ -1 \\ 1 \end{bmatrix}$. A line search for a maximum along this line takes us beyond $x_3 = 4$, so we stop when that constraint is hit ($\alpha_1 = 0.7$), giving $\mathbf{X}^2 = \begin{bmatrix} 2.6 \\ 2.6 \\ 4 \end{bmatrix}$.

Now $\nabla^2 = \begin{bmatrix} -1.2 \\ 0.8 \\ 2 \end{bmatrix}$ and the new Zoutendijk linear program, including the new constraint $d_3^2 \le 0$ (because we are at the upper limit of x_3), has an optimal solution $\mathbf{D}^2 = \begin{bmatrix} -1 \\ 1 \\ 0 \end{bmatrix}$, so $\mathbf{X}^3 = \begin{bmatrix} 2.6 \\ 2.6 \\ 4 \end{bmatrix} + \alpha_2 \begin{bmatrix} -1 \\ 1 \\ 0 \end{bmatrix}$. The maximum for $f(\mathbf{X})$ along this direction is when $\alpha_2 = 0.5$, giving $\mathbf{X}^3 = \begin{bmatrix} 2.1 \\ 3.1 \\ 4 \end{bmatrix}$. Here $\nabla^3 = \begin{bmatrix} -0.2 \\ -0.2 \\ 2 \end{bmatrix}$ and the solution of the next Zoutendijk linear program is $\mathbf{D}^3 = \begin{bmatrix} -1 \\ -1 \\ 0 \end{bmatrix}$ and, with an optimal $\alpha_3 = 0.1$, we find $X^4 = \begin{bmatrix} 2.1 \\ 3.1 \\ 4 \end{bmatrix} + (0.1) \begin{bmatrix} -1 \\ -1 \\ 0 \end{bmatrix} = \begin{bmatrix} 2 \\ 3 \\ 4 \end{bmatrix}$. At this point, $\nabla^4 = \begin{bmatrix} 0 \\ 0 \\ 2 \end{bmatrix}$, indicating that this is the optimal solution. You don't want to move away from the current x_1 or x_2, and you would like to increase x_3 but you can't, since it is already at its upper limit.

(b) Rosen. The route to $X^1 = \begin{bmatrix} 3.3 \\ 3.3 \\ 3.3 \end{bmatrix}$ and then to $X^2 = \begin{bmatrix} 2.6 \\ 2.6 \\ 4 \end{bmatrix}$ will be exactly the same as in Zoutendijk's approach. At X^2, $\nabla^2 = \begin{bmatrix} -1.2 \\ 0.8 \\ 2 \end{bmatrix}$, and x_3 is at its upper limit,

so $A^2 = [0 \quad 0 \quad 1]$, $M^2 = \begin{bmatrix} 1 & 0 & 0 \\ 0 & 1 & 0 \\ 0 & 0 & 0 \end{bmatrix}$ and $G^2 = \begin{bmatrix} -1.2 \\ 0.8 \\ 0 \end{bmatrix}$. $X^3 = \begin{bmatrix} 2.6 \\ 2.6 \\ 4 \end{bmatrix} + \alpha_2 \begin{bmatrix} -1.2 \\ 0.8 \\ 0 \end{bmatrix}$,

and a line search will turn up a maximum when $\alpha_2 = 0.5$, so that $X^3 = \begin{bmatrix} 2 \\ 3 \\ 4 \end{bmatrix}$. At this

point, $\nabla^3 = \begin{bmatrix} 0 \\ 0 \\ 2 \end{bmatrix}$, and, as with Zoutendijk, we are finished.

11.6.   For the original problem:

		x_1	x_2
Π	0	2	4
s_1	10	-1	-1^*

		x_1	s_1
Π	40	-2	-4
x_2	10	-1	-1

The optimal solution is $X^* = \begin{bmatrix} 0 \\ 10 \end{bmatrix}$ and $\Pi^* = 40$.

For the rewritten version of the problem, two pivots are required to reach X^* $= \begin{bmatrix} 0 \\ 10 \\ 40 \end{bmatrix}$ and $\Pi^* = 40$. Note that $x_3^* = \Pi^*$, as we would expect.

		x_1	x_2	x_3
Π	0	0	0	1
s_1	10	-1	-1	0
s_2	0	2	4	-1^*

		x_1	x_2	s_2
Π	0	2	4	-1
s_1	10	-1	-1^*	0
x_3	0	2	4	-1

		x_1	s_1	s_2
Π	40	-2	-4	-1
x_2	10	-1	-1	0
x_3	40	-2	-4	-1

11.11. Here is the series of cutting lines to create a series of linear programming problems, with $f(\mathbf{X}) = x_1 + x_2$, along with other relevant information at each step:

k	Constraint	$\mathbf{X}^k$	$h(\mathbf{X}^k)$	$\nabla h(\mathbf{X}^k)$
0		$\begin{bmatrix} 1 \\ 2 \end{bmatrix}$	-7	$\begin{bmatrix} 6 \\ 8 \end{bmatrix}$
1	$c_1\ 6x_1 + 8x_2 \leq 29$	$\begin{bmatrix} 4.83 \\ 0 \end{bmatrix}$	52.08	$\begin{bmatrix} 29 \\ 0 \end{bmatrix}$
2	$c_2\ x_1 \leq 3.03$	$\begin{bmatrix} 3.03 \\ 1.35 \end{bmatrix}$	13.19	$\begin{bmatrix} 18.18 \\ 5.4 \end{bmatrix}$
3	$c_3\ 18.18x_1 + 5.40x_2 \leq 48.64$	$\begin{bmatrix} 2.06 \\ 2.08 \end{bmatrix}$	3.38	$\begin{bmatrix} 12.36 \\ 8.32 \end{bmatrix}$
4	$c_4\ 12.36x_1 + 8.32x_2 \leq 39.39$	$\begin{bmatrix} 1.51 \\ 2.49 \end{bmatrix}$	1.24	$\begin{bmatrix} 9.06 \\ 9.96 \end{bmatrix}$
5	$c_5\ 9.06x_1 + 9.96x_2 \leq 37.24$	$\begin{bmatrix} 1.74 \\ 2.16 \end{bmatrix}$	0.41	$\begin{bmatrix} 10.44 \\ 8.64 \end{bmatrix}$
6	$c_6\ 10.44x_1 + 8.64x_2 \leq 36.42$	$\begin{bmatrix} 1.59 \\ 2.29 \end{bmatrix}$	0.07	$\begin{bmatrix} 9.54 \\ 9.16 \end{bmatrix}$
7	$c_7\ 9.54x_1 + 9.16x_2 \leq 36.07$	$\begin{bmatrix} 1.508 \\ 2.367 \end{bmatrix}$	0.028	$\begin{bmatrix} 9.06 \\ 9.48 \end{bmatrix}$
8	$c_8\ 9.06x_1 + 9.48x_2 \leq 36.074$	$\begin{bmatrix} 1.5445 \\ 2.3292 \end{bmatrix}$	0.007	so stop

The optimal solution to this problem using exact methods (not approximations) is $\mathbf{X}^* = \begin{bmatrix} 1.5492 \\ 2.3238 \end{bmatrix}$.

11.12. (a) Here are results for

$$L(\mathbf{X}; \mu) = -6x_1 - 8x_2 - \mu \ln (-x_1^2 - x_2^2 + 16)$$

(ignoring explicit consideration of $\mathbf{X} \geq \mathbf{0}$).

μ	x_1	x_2	$f(\mathbf{X})$
1.0	2.3408	3.1210	39.0128
0.1	2.3940	3.1920	39.9000
0.01	2.3973	3.2008	39.9902

(b) The exact answer is easily found to be $\mathbf{X}^* = \begin{bmatrix} 2.4 \\ 3.2 \end{bmatrix}$.

11.15. (a) For this quadratic program we have

$$\mathbf{M} = \begin{bmatrix} -4 & 0 & -1 & 1 & 0 & 0 \\ 0 & -2 & -1 & 0 & 1 & 0 \\ 1 & 1 & 0 & 0 & 0 & 1 \end{bmatrix} \text{ and } \mathbf{N} = \begin{bmatrix} -12 \\ -14 \\ 5 \end{bmatrix}$$

and $\mathbf{V} = [x_1 \; x_2 \; \lambda_1 \; \lambda_2 \; \lambda_3 \; s]'$. There are up to $C_3^6 = 20$ basic solutions to the system $\mathbf{MV} = \mathbf{N}$, as in (11.55). If you examine these, three have singular coefficient matrices and of the remaining 17, 12 generate solutions that fail nonnegativity requirements. Of the remaining five, four fail one of the conditions in (11.53''b). The one that is left is $\mathbf{X}^* = \begin{bmatrix} 1.333 \\ 3.667 \end{bmatrix}$, which is the correct solution.

INDEX